Quantum Theory and Symmetries

Proceedings of the Second International Symposium on

Quantum Theory and Symmetries

Kraków, Poland July 18–21, 2001

Editors

Edward Kapuścik

Andrzej Horzela

H. Niewodniczański Institute of Nuclear Physics
Kraków, Poland

World Scientific
New Jersey • London • Singapore • Hong Kong

Published by

World Scientific Publishing Co. Pte. Ltd.

P O Box 128, Farrer Road, Singapore 912805

USA office: Suite 1B, 1060 Main Street, River Edge, NJ 07661

UK office: 57 Shelton Street, Covent Garden, London WC2H 9HE

British Library Cataloguing-in-Publication Data
A catalogue record for this book is available from the British Library.

QUANTUM THEORY AND SYMMETRIES
Proceedings of the Second International Symposium

ISBN 981-02-4887-3

Printed in Singapore by Mainland Press

PREFACE

It is our great pleasure to present to the Physics community the *Proceedings of the Second International Symposium* ***Quantum Theory and Symmetries*** held at the Henryk Niewodniczański Institute of Nuclear Physics in Kraków from 18 till 21 July 2001. The Kraków Symposium was the second one in a series of bi-annual symposia presided over by the **International Board** consisting of: H.-D. Doebner (Chairman), S.-T. Ali, L. L. Boyle, V. S. Dobrev, E. Kapuścik, V. I. Man'ko, F. Mansouri, G. Marmo, M. A. del Olmo, G. S. Pogosyan, M. Rasetti, T. H. Seligman and A. I. Solomon. The first symposium of the series took place in Goslar (Germany) in 1999 and its proceedings were published by World Scientific in 2000.

The members of the Advisory Committee of the Kraków symposium were: A. Bohm, C. Burdik, Z. Haba, G. C. Hegerfeldt, A. Jamiołkowski, R. Kerner, Y. S. Kim, A. U. Klimyk, A. Kossakowski, M. A. Mańko, B. Mielnik, G. Marmo, Y. Ne'eman, A. Odzijewicz, L. O'Raifeartaigh, J. Rembieliński, G. Rudolph, A. S. Shumovsky, A. N. Sissakian and I. Todorov. While preparing the conference, in November 2000, we got sad news that Professor Lochlain O'Raifeartaigh, a great man and scientist and our long-time friend, had passed away. Professors H.-D. Doebner and V. I. Man'ko paid a Memorial Tribute to him just after the Symposium opening, while a Memorial Session dedicated to him, arranged on the second day of the symposium, reminded the participants of this good and modest man whom we owe so much to and whom we badly miss.

The participants of the Symposium presented over 100 talks: 20 plenary lectures and more than 80 contributions given during parallel sessions, and covering aspects belonging to the foundations of quantum theory, quantum optics, quantum computing and control, the Lie algebraic methods, quantum groups and noncommutative geometry, gauge, field and string theories, integrable systems, symmetries in molecular physics, as well as discrete periodic and aperiodic systems. When selecting the material for the present volume, we decided to publish all contributions in alphabetical order, distingushing only between plenary lectures and contributions. In this way, any confusing overlaps between different sessions have been avoided. However, we would like to mention two associated workshops held as parallel sessions of the Symposium. These were: the 4th Workshop on Lie Theory and its Applications in Physics, organized by Professors H-D. Doebner and V. S. Dobrev, and the 3rd Workshop on Extensions of Quantum Theories, organized by one of us (A. H.). Unfortunately, some of the participants were not able to send their contributions in time, and thus these could not be included in this volume.

We would like to thank our sponsoring institutions whose help and financial support made the Symposium possible, namely the Polish State Commitee for Scientific Research, the Henryk Niewodniczański Institute of Nuclear Physics, the Cracow Pedagogical Academy, the College of Computer Science in Łódź, and the Joint Institute of Nuclear Research in Dubna. We are personally indepted to Professor Andrzej Budzanowski, Director General, and to Dr. Maria Pollak - Stachura, Deputy Director, both of the Henryk Niewodniczański Institute of Nuclear Physics, Professor Michał Śliwa, Rector of the Cracow Pedagogical Academy, Ms. Aniela Bednarek, Director of the College of Computer Science, and to Professor Alexey N. Sissakian, Deputy Director General of the Joint Institute of Nuclear Research. We also would like to thank all the members of the Local Organizing Committee: K. Bajan, A. Baran, A. Bieniek, P. Błasiak, W. Kantor, A. Korzeniowska, P. Łach, E. Pagaczewska, W. Polak, D. Wcisło and D. Żenczykowska, whose enthusiastic involvement contributed greatly to the success of the Symposium.

We have no doubt that all readers will find this comprehensive volume interesting, providing new insights into various problems of quantum physics and its most powerful tool, the theory of symmetries. A more extended discussion of the various related problems can be found in the Special Issue of Journal of Physics A entitled *Foundations of Quantum Theory*, with Professor Vladimir I. Man'ko and both of us acting as Guest Editors. To this respect let us express our gratitude to the Editorial Board of the Journal of Physics A (London, UK).

We are convinced that the idea of bi-annual meetings devoted specifically to the subject in question will be successfully continued and that all the participants of the Kraków conference will meet again as planned during the Third International Symposium on **Quantum Theory and Symmetries** to take place in Cincinatti (USA) in 2003.

Kraków, January 2002

Andrzej Horzela and Edward Kapuścik

Symposium Organizers

H.Niewodniczański Institute of Nuclear Physics

Kraków, Poland

http://qts2.ifj.edu.pl

CONTENTS

Contributions

THE INTERACTING FOCK SPACE OF HALDANE'S EXCLUSION STATISTICS

L. ACCARDI AND M. NHANI

Centro Vito Volterra, Università degli Studi di Roma "Tor Vergata" Roma, 00133, Italia

E–mail: accardi@volterra.mat.uniroma2.it

http://volterra.mat.uniroma2.it/

We prove that Haldane exclusion statistics (ES) can be included in the more general class of standard interacting Fock spaces (IFS). This result provides a standard rule to assign in a consistent way the parameters of the most general exclusion statistics and, through the results of [4], it shows that there is a canonical one–to–one correspondence between these parameters and the families of orthogonal polynomials associated to an arbitrary measure with moments of all orders. Such a correspondence has been discovered in a multiplicity of particular cases (see for example [5], [9]). Thus our result gives a rule to identify immediately the family of orthogonal polynomials associated to a given ES. Moreover it allows to extend the ES in a natural way to infinitely many species and in particular it suggests a standard way to take the thermodynamic limit of these statistics.

1 Introduction

Let us begin by a brief exposition of the main idea on Haldane exclusion statistics (ES) [6] as described in [2]. In Quantum Statistical Mechanics (QSM) any statistics is described by a *partition function*. Consider a system of N identical particles, each of which can occupy K independent states corresponding to energies $\varepsilon_1 \dots \varepsilon_K$ (alternatively we may think of the index $j = 1, \dots, K$ as labeling *sites* or *species*, ...). For such a system the partition function has the following structure:

$$Q_N(x_1 \dots x_K) = \sum_{\substack{n_i \in \mathbb{N} \\ n_1 + \dots + n_K = N}} \alpha_N(n_1 \dots n_K) x_1^{n_1} \dots x_K^{n_K} \,(I) \tag{1}$$

where $x_i \equiv \exp(-\varepsilon_i/\varepsilon_T)$, $i = 1 \dots K$. A K–uple $(n_1, \dots, n_K)$ of natural integers, i.e.a sequence of occupation numbers for the different states, will be called a *configuration*. The function $\alpha(n_1 \dots n_K)$ which weights all possibles configuration has the following properties (see [1]):

(a) $\alpha(n_1 \dots n_K) \geq 0$

(b) $\alpha(n_1 \dots n_K)$ is a symmetric function [ensuring that no particular ε_i is preferred].

A function $\alpha : \mathbb{N}^K \to \mathbb{R}$ satisfying these properties defines the statistics which the identical particles obey. In the case of Haldane's ES the function $\alpha(n_1 \dots n_K)$ must satisfy two additional conditions:

(c) $\alpha(n_1 \dots n_i + 1 \dots n_K) \leq \alpha(n_1 \dots n_i \dots n_K) \; \forall \, i = 1 \dots K$

(d) there exists a number $p \in \mathbb{N}$ such that $\alpha(p+1, , n_2 \dots n_K) = 0$ and $\alpha(n_1 \dots n_K) > 0$ if $\max\{n_i \; : \; i \leq K\} \leq p$.

Ilinski, Gunn and Ilinskaia (see [2]) have constructed an Hilbert space $\mathcal{H}$, called *fractional–dimensional Fock space* (FDFS), with the property that (1) is a *weighted partition function*

$$Z = Q_N(x_1 \dots x_K) = Tr(f(N)e^{-\beta H})$$

where the trace is meant in $\mathcal{H}$,

$$H = \sum_j \varepsilon_j N_j$$

is the usual Hamiltonian of an ideal gas, N_j ($j = 1, \dots, K$) is the number operator counting the number of particles in the j–th species) and $f(N)$ (denoted Id in [2]) is a self–adjoint operator (a function of the occupation numbers of each species) characterizing the ES. The trace of the operator Id (always finite because of property (d) above) has been called in [2] the *fractional dimension* of the Hilbert space of these configurations.

We show in this paper that the *fractional–dimensional Fock space* is in fact a particular case of a standard interacting Fock space.

2 One mode interacting Fock spaces (IFS)

In this section we recall some properties of the one mode interacting Fock spaces and we refer to the survey paper [3] for a discussion of the general case. The one mode interacting Fock spaces $\Gamma(\mathbb{C})$ is defined as follows: as a vector space $\Gamma(\mathbb{C})$ is the direct sum of 1–dimensional sub–spaces

$$\Gamma(\mathbb{C}) = \bigoplus_{n=0}^{\infty} \mathbb{C}|n> \tag{1}$$

The space $\mathbb{C}|n>$ is called the *n–particle sub–space.* On $\Gamma(\mathbb{C})$ one defines a Hilbert space structure by prescribing that: (i) the sum in (1) is orthogonal, i.e. different n–particle sub–spaces are orthogonal; (ii) the norm of the vector $|n\rangle$ is given by

$$\langle n|n\rangle = \lambda_n \tag{2}$$

(iii) the creation operator is well defined by the prescription

$$a^+|n>=|n+1> \tag{3}$$

Notice that, because of (2), condition (iii) implies that, if λ_n is zero, then all λ_k with $k>n$ must also be zero Therefore the adjoint of the creation operator (annihilator) is uniquely defined by the condition:

$$a|n+1>=\frac{\lambda_{n+1}}{\lambda_n}|n>$$

with the convenction that $0/0=0$. Moreover one has

$$\langle n,n\rangle=\langle a^+(n-1),n\rangle=\langle (n-1),an\rangle=\frac{\lambda_n}{\lambda_{n-1}}\langle n-1,n-1\rangle=\dots$$

$$\|\,|n>\|^2=\frac{\lambda_n}{\lambda_{n-1}}\cdot\frac{\lambda_{n-1}}{\lambda_{n-2}}\cdots\frac{\lambda_1}{\lambda_0}=\frac{\lambda_n}{\lambda_0}$$

which, together with (2) gives $\lambda_0=1$.

3 Inclusion of the one mode FDFS into the one mode IFS

In order to establish the connection between IFS, orthogonal polynomyals (see [4]) and the ES it is useful to introduce the numbers

$$\omega_j=\frac{\lambda_j}{\lambda_{j-1}}\ ;\quad \lambda_0=1$$

so that

$$\lambda_n=\prod_{j=0}^{n}\omega_j$$

It is clear that the sequence (ω_j) uniquely determines the IFS. Moreover

$$a|n+1>=\omega_{n+1}|n>$$

$$a^+a|n>=\omega_n|n>$$

In [2] the additional condition

$$N|n>=a_0^+a_0|n>=n|n> \tag{1}$$

is required. In order to satisfy also this condition we introduce a new creation operator a_0^+, defined by the condition:

$$a_0^+|n>:=\alpha_n|n+1>=\alpha_n a^+|n>$$

thus, in the notation (1):

$$a_0^+ = a^+ \alpha_N$$

since N, hence α_N, is self–adjoint also for the interacting scalar product, the adjoint of a_j^+ is

$$a_0 = \alpha_N a$$

Therefore

$$a_0^+ a_0 |n >= a_0^+ \alpha_N a |n >= \omega_n a_0^+ \alpha_N |n-1 >= \omega_n \alpha_{n-1} a^+ \alpha_N |n-1 >=$$
$$\omega_n \alpha_{n-1}^2 a^+ |n-1 >= \omega_n \alpha_{n-1}^2 |n >$$

If we want

$$a_0^+ a_0 = N$$

we must then have

$$\omega_n \alpha_{n-1}^2 = n$$

or

$$\alpha_{n-1} = \left(\frac{n}{\omega_n}\right)^{1/2}$$

Equivalently

$$\alpha_n = \left(\frac{n+1}{\omega_{n+1}}\right)^{1/2}$$

which, given (2) below, coincides with the result of [2]. Summing up:

LEMMA (1) The FDFS of [2] (hence Haldane ES in the case of a single species) is isomorphic to the IFS uniquely defined by the prescriptions:

$$\omega_{n+1} = 1 - gn \tag{2}$$

and $\omega_{n+1} = 0$ as soon as the right hand side of (2) becomes negative.

PROOF From the above discussion.

REMARK The right hand side of (2) shows why in the Haldane statistics we must restrict to finite dimensional spaces. In fact if $n > 1/g$ the right hand side of (2) becomes negative and this is incompatible with the positivity of the scalar product which requires

$$\omega_n \geq 0 \; ; \qquad \forall n$$

The general connection between the ω_n's and the Jacobi coefficients of orthogonal polynomials is described in [4].

4 IFS for many species

We fix a number $K < +\infty$, representing the number of species, (the case $K \leq +\infty$ can be treated in a similar way) and, for each $j = 1, \ldots, K$ let be given a one mode IFS $\{\Gamma(\mathbb{C}), (\lambda_n^{(j)})_n\}$. On the algebraic tensor product of vector spaces

$$\mathcal{H} = \otimes^K \Gamma(\mathbb{C}) = \Gamma(\mathbb{C}) \otimes \ldots \otimes \Gamma(\mathbb{C})$$

we consider the Hilbert space structure induced by the tensor product of the scalar products defined by the $(\lambda_n^{(j)})_n$ as explained in section (**??**). Then we define the global N–particle structure as follows: we refer each IFS

$$\{\Gamma(\mathbb{C}), (\lambda_n^{(j)})_n\}$$

to an orthonormal basis

$$|n_j> \quad ; \qquad n_j \in \mathbb{N} \quad ; \qquad j = 1, \ldots, K$$

and we define the global N–particle space by

$$\mathcal{H}_N = \bigoplus_{n_1 + \ldots + n_K = N} \mathbb{C}|n_1 > \otimes \ldots \otimes |n_K >$$

an orthonormal basis of $\mathcal{H}_N$ is

$$|n_1, \ldots, n_K >= |n_1 > \otimes \ldots \otimes |n_K > ; \qquad n_1 + \ldots + n_k = N$$

Introducing the multiple number operator

$$N := N_1 \otimes \ldots \otimes N_K$$

where N_j $(j = 1, \ldots, K)$ is the number operator of the j–th species in $\mathcal{H}$:

$$N_j|n_1, \ldots, n_K >= n_j|n_1, \ldots, n_K >$$

and the operator

$$W_N := \sum_{n_1 + \ldots + n_K} \left[\prod_{i=1}^{K} \lambda_{n_i}^{(i)} \right] |n_1, \ldots, n_K\rangle\langle n_1, \ldots, n_K| \tag{1}$$

we see that the scalar product on $\mathcal{H}_N$ is defined by

$$\langle \xi, \eta \rangle_{\mathcal{H}_N} = \langle \xi, W_N \eta \rangle_{\otimes^N \Gamma(\mathbb{C})} = \langle \xi, W_N \eta \rangle_{\otimes^N \Gamma(\mathbb{C})}$$

where $\langle \xi, \ \cdot \ \eta \rangle_{\otimes^N \Gamma(\mathbb{C})}$ denotes the usual scalar product on $\otimes^N \Gamma(\mathbb{C})$ (i.e. the one coming from the tensor product of the usual scalar products). With these notations it is easy to see that

$$\mathcal{H} := \bigoplus_{N \geq 0} \mathcal{H}_N$$

is a standard IFS with weights W_N given by (1). The fact that the operator W_N does not factorize: this fact corresponds to an interaction, among the different species, of a purely statistical nature, i.e. not encoded into the dynamics, and justifies the name *interacting Fock space.*

REMARK A more general interacting Fock space can be obtained by replacing, in (1), the operator W_N by

$$W_N := \sum_{n_1+\ldots+n_K=N} \alpha_N(n_1,\ldots,n_K)|n_1,\ldots,n_K><n_K,\ldots,n_1|$$

where α is a function of $(n_1,\ldots,n_K)$. In the notations of [2], the operator Id is the direct sum of all the operators W_N. The case (1) corresponds to the factorization

$$\alpha_N(n_1,\ldots,n_K) = \prod_{i=1}^{K} \alpha(n_i)$$

5 The IFS of Haldane's ES: the case of many one mode species

Consider now a system of identical particles obeying Haldane's ES with parameter g. We suppose first that $g = \frac{1}{p}$, $p \geq 1$. Let $K \in N$ be the dimension of the states Hilbert space H of a single-particle. As we said in the introduction we can as well say that we have K species of particles $j = 1,...,K$ each of which has a state space of dimension 1: $H_j = Ce_j$. All the species having the same statistical parameter g. To each species $j = 1,...,K$ we associate the IFS $\{\Gamma(\mathbb{C}), (\lambda_n^{(j)})_n\}$ as in section (4). We define the Hilbert space of the system with K species to be the space introduced in section (4), i.e.:

$$\mathcal{H} := \otimes\Gamma(C) = \Gamma_1 \otimes \Gamma_2 \otimes ... \otimes \Gamma_K$$

Introducing the global N–particle spaces as in section (4), we define the scalar products:

$$\forall N \in \mathbb{N}\,, \quad \forall u,v \in H_N \quad (u,v)_N = \langle W_N(u), v\rangle_N$$

where $\langle\cdot\rangle_N$ is the natural scalar product on H_N and

$$W_N|n_1,\ldots,n_K\rangle = \lambda_{n_1}^{(1)}\lambda_{n_2}^{(2)}\ldots\lambda_{n_k}^{(k)}|n_1,\ldots,n_K\rangle$$

REMARKS:

(i) The function $\alpha : \mathbb{N}^K \to \mathbb{R}$ which weights the different occupation numbers in the partition function (1.1) is determined by the sequences $(\lambda_{n_j}^{(j)}\ ,\ n_j = 1\ldots p)$, $j = 1,\ldots,K$, by:

$$\alpha(n_1\ldots n_K) = \lambda_{n_1}^{(1)}\lambda_{n_2}^{(2)}\ldots\lambda_{n_K}^{(K)}\ , \qquad \forall\,(n_1\ldots n_K) \in \mathbb{N}^K$$

(ii) In the Ilinski, Gunn, Ilinskaia's interpretation of Haldane's ES the $\lambda_{n_j}^{(j)}$ are given by:

$$\lambda_{n_j}^{(j)} = (1-g)(1-2g)\dots(1-(n_j-1)g) \ , \qquad j = 1\dots K$$

with the convection that $\omega_n = 0$ if the right hand side of the above identity is negative or equivalently

$$\lambda_n^{(j)} = 0$$

if $n > 1/g$. Any way both the function α and the λ_{n_j} $j = 1, \dots, K$ depend essentially on the parameter g.

(iii) One easily verifies that $W_N : H_N \to H_N$ is a positive self–adjoint operator. Moreover, $\forall\, u \in H$ and $\forall\, F \in H_N$

$$\langle F, F\rangle_N = 0 \Rightarrow \langle u \otimes F, W_{N+1}(u \otimes F)\rangle_{N+1} = 0$$

This property ensures that $\forall\, u \in H$. The *creation operator* $a^+(u) : H_N \to H_{N+1}$ is well defined. For all $u \in H$, the *annihilation operator* is, by definition, $a(u) : H_{N+1} \to H_N$ the adjoint of $a^+(u)$ for the scalar product $(\cdot)_N$. As usual we note $a_i^+ = a^+(e_i)$, $a_i = a(e_i)$ $i = 1\dots K$, where $\{e_1 \dots e_K\}$ is a fixed orthonormal basis of H. We shall denote $\Gamma_I(H, (W_N)_{N\geq 1}, g)$ the standard interacting Fock space constructed in this way.

We will now prove that the Haldane exclusion g–statistics coincides with the vacuum statistics of the operators

$$e^{-\beta H_N} = e^{-\beta N_1} \otimes \dots \otimes e^{-\beta N_K}$$

THEOREM: A system of identical particles obeying Haldane's exclusion statistics with statistical parameter g is the standard IFS $\Gamma_I(H; (W_N)_{N\geq i}; g)$ over the Hilbert space of the single–particle state.

The corresponding creation and annihilation operators verify the following relations:

$$a_i|n_1 \dots n_i \dots n_K\rangle = \frac{\lambda_{n_i}}{\lambda_{(n_i-1)}} |n_1 \dots\dots n_i - 1, \dots, n_K > \tag{1}$$

$$[a_i, a_j^+] = 0 \qquad \text{if} \qquad i \neq j \tag{2}$$

We have also

$$[a_i, a_j] = [a_i^+, a_j^+] = 0 \qquad \text{if} \qquad i \neq j \tag{3}$$

$$a_i^+ a_i = \frac{\lambda_{N_i}}{\lambda_{(N_i-1)}} \quad , \qquad a_i a_i^+ = \frac{\lambda_{(N_i+1)}}{\lambda_{N_i}} \tag{4}$$

$$a_i^{p+1} = a_i^{+(p+1)} = 0 \quad , \qquad \forall i,j = 1 \dots K \tag{5}$$

$$a_i|0\rangle = 0 \quad , \qquad \forall i = 1 \dots K \tag{6}$$

Proof. To proof that Haldane's ES is realized in $\Gamma_I(H, (W_N)_{N\geq 1}, g)$ we just have to show that the partition function corresponding to this standard IFS is the same as the one of Haldane's ES. Generally for an ideal gas the partition function of N particles is given by:

$$Q_N = Tr(e^{-\beta \mathcal{H}})$$

where $\mathcal{H} = \sum_{i=1}^K \varepsilon_i N_i$ is the Hamiltonian, N_i is the number operator of the i–th species, $N = \sum_{i=1}^K N_i$ and the trace T_r is expressed in any orthonormal basis of the Hilbert space of the states of the N particles (see K. Huang: Statistical Mechanics, p. 196, p. 213). In our case this gives:

$$\begin{aligned} Q_N &= \sum_{n_1+\dots n_K=N} \langle n_1 \dots n_K | e^{-\beta \mathcal{H}} | n_1 \dots n_K \rangle_N \\ &= \sum_{n_1+\dots n_K=N} \alpha(n_1 \dots n_K) e^{-\beta \varepsilon n_1} \dots e^{-\beta \varepsilon_K n_K} \end{aligned}$$

Given our choice of $\alpha(n_1 \dots n_K)$ Q_N has been shown in [2] to be exactly the partition function of Haldane's ES.

The relations (2)–(6) are easily verified.

6 The IFS of Haldane's ES: the case of many species, many modes

In the case $g = \frac{m}{p}$, $m > 1$ and with m and p coprime, we can suppose that we have K species of particles each of which with m independent states, i.e. $H_j = \mathbb{C}^m$. We can also say that we have p blocks each of which has w species of particles with Hilbert space of dimension 1 for each species. Generalizing the above procedure we fix an ortho–normal basis $(e_k^{(j)})$ of $H_j = \mathbb{C}^m$, we denote $a_{j,k}^+ := a^+(e_k^{(j)})$ and we say that this creator generates the sub–species (j,k) of the j–th species. We assume for simplicity that each species (and sub–species) has the same statistical parameter $g' = 1/p$, but with essentially no changes in the construction below we could as well deal with the case in which the the (j,k) sub–species has parameter $g'_{j,k} = 1/p_{j,k}$. This would include the most general statistics considered by Haldane. With these notations we get:

1) For each species $j = 1 \dots K$ we have a standard IFS

$$\Gamma_j = \Gamma_j \left(K, (W_{N_j})_{N_j \geq 1}, \frac{1}{p} \right) = \oplus_{N_j} K_{N_j}$$

of the type considered in section (5). More explicitly:

$$K_{N_j} = \text{span} \left\{ a_{j,1}^{+n_{j1}} \dots a_{j,m}^{+n_{jm}} |0\rangle = |n_{j,1}^1, \dots, n_{j,m}^m\rangle \; : \; \sum_{i=1}^m n_{ji} = N_j \right\}$$

2) The standard IFS for the whole population of K species is:

$$\begin{aligned} \Gamma = \bigotimes_{j=1}^K \Gamma_j &= \bigotimes_{j=1}^K \left(\bigoplus_{N_j} K_{N_j} \right) \\ &= \bigoplus_{N \in \mathbb{N}} \left[\bigoplus_{N_1 + \dots + N_K = N} K_{N_1} \otimes K_{N_2} \otimes \dots \otimes K_{N_K} \right] \\ &= \bigoplus_{N \in \mathbb{N}} \mathcal{H}_N \end{aligned}$$

where $\forall N$, the weight operator $W_N : H_N \to H_N$, defining the scalar product on $\mathcal{H}_N$ through the identity:

$$(u, v)_N = \langle W_N(u), v \rangle$$

($\langle \cdot, \cdot \rangle$ being the usual scalar product) is given by

$$W_N = \sum_{N_1 + \dots + N_K = N}^{\oplus} W_{N_1} \otimes \dots \otimes W_{N_K}$$

and the operators W_{N_j} are defined as in section (5).

REMARK. The above construction allows to realize any partition function of a system of identical particles, whose structure is given by the equation (**??**) by means of a standard IFS.

References

1. S. Chaturvedi and V. Srinivasan V., *Phys. Rev. Lett.* **78**, 4316 (1997).
2. K.N. Ilinski, J.M.F. Gunn, A.V. Ilinskaia, Phys. Rev. B **53**, 2615 (1996).
3. L. Accardi, Y.G. Lu, I. Volovich, *The QED Hilbert module and Interacting Fock spaces*, Publications of IIAS (Kyoto) (1997).
4. L. Accardi, M. Bożejko, Interacting Fock spaces and Gaussianization of probability measures, *Infinite dimensional analysis, quantum probability and related topics*, 2, N. 4 (1998), Volterra preprint N. 321, (1997).
5. Z.N.C. Ha, *Phys. Rev. Lett.* **73**, 1574 (1994).
6. F.D.M. Haldane, *Phys. Rev. Lett.* **67**, 937 (1991).
7. Yong-Shi Wu, *Phys. Rev. Lett.* **73**, 922 (1994).
8. K.N. Ilinski, J.M.F. Gunn, *Phys. Lett. A* **210**, 168 (1996).
9. Z.N.C. Ha, *Nucl. Phys. B* **435**, 604 (1995).

COMPLEX HAMILTONIANS HAVING REAL SPECTRA

CARL M. BENDER
Physics Department, Washington University,
St. Louis, MO 63130, USA
E-mail: cmb@howdy.wustl.edu

A fundamental discrete symmetry of quantum field theory, namely $\mathcal{PCT}$ symmetry, is derived from the assumption of Lorentz invariance and positivity of the spectrum of the Hamiltonian. What happens if we assume only Lorentz invariance and $\mathcal{PCT}$ symmetry? Hamiltonians having this property need not be Hermitian, but, except when $\mathcal{PCT}$ is spontaneously broken, the energy levels of such Hamiltonians are all real and positive! In this talk I examine quantum mechanical and quantum field theoretic systems whose Hamiltonians are non-Hermitian but obey PCT symmetry. These systems have weird and remarkable properties, and the classical theories underlying these quantum systems also have strange behaviors. Simple examples of such Hamiltonians are $H = p^2 + ix^3$ and $H = p^2 - x^4$. Hamiltonians such as these may be thought of as *complex deformations* of conventional Hermitian Hamiltonians. Thus, in this talk I will be studying the analytic continuation of conventional classical mechanics and quantum mechanics into the complex plane.

1 Introduction

In a recent series of papers we showed that when properly defined, the ix^3 and $-gx^4$ potentials in quantum mechanics possess positive definite spectra. The positivity of the spectrum is apparently due to the $\mathcal{PT}$ symmetry of the Hamiltonian.[1,2,3,4,5,6,7,8,9,10,11,12,13,14,15,16,17,18,19,20,21,22,23,24] In the previous Paris conference we spoke on non-Hermitian quantum mechanical Hamiltonians.[9] In this talk we extend these ideas to quantum field theory. We argue that in a $-x^4$ theory the expectation value $\langle x \rangle$ is *not* zero. The corresponding result for a $-g\phi^4$ quantum field theory in D-dimensional Euclidean space is that the one-point Green's function $G_1 = \langle \phi \rangle$ is also nonzero. This finding may allow us to construct new models for the Higgs boson. We also examine bound states in a $-g\phi^4$ quantum field theory.

In 1952 Dyson argued heuristically that Rayleigh-Schrödinger perturbation theory for quantum electrodynamics is divergent.[25] Applied to the quantum anharmonic oscillator, whose Hamiltonian is

$$H = p^2/2 + m^2x^2/2 + gx^4/4 \qquad (g > 0), \tag{1}$$

the argument goes as follows: If the coupling constant g is replaced by $-g$, then the potential is no longer bounded below, so the resulting theory has no ground state. Thus, the ground-state energy $E_0(g)$ has an abrupt transition at $g = 0$. If we represent $E_0(g)$ as a series in powers of g, this series must have a zero radius of convergence because $E_0(g)$ has a singularity at $g = 0$ in the complex-coupling-constant plane. Hence, the perturbation series must diverge for all $g \neq 0$. While the perturbation series does indeed diverge, this heuristic argument is flawed because the spectrum of the Hamiltonian

$$H = p^2/2 + m^2x^2/2 - gx^4/4 \qquad (g > 0) \tag{2}$$

is ambiguous. This ambiguity is due to the absence of well-specified boundary conditions that the wave functions must satisfy. The spectrum depends crucially on how this Hamiltonian with a negative coupling constant is obtained.

There are many ways to obtain the Hamiltonian (2). For example, one can substitute $g = |g|e^{i\theta}$ into the Hamiltonian (1) and rotate from $\theta = 0$ to $\theta = \pi$. Under this rotation, the ground-state energy $E_0(g)$ becomes complex. Evidently, $E_0(g)$ is real and positive when $g > 0$ and complex when $g < 0$.[a] One can also obtain (2) as a limit of the Hamiltonian

$$H = p^2/2 + m^2x^2/2 + gx^2(ix)^\alpha/4 \qquad (g > 0) \tag{3}$$

as $\alpha : 0 \rightarrow 2$. Having studied Hamiltonians like that in (3) in great detail, we and others have shown that for $\alpha \geq 0$ the spectra of such Hamiltonians are real, positive, and discrete. The spectrum of the limiting Hamiltonian (2) obtained in this manner is similar to that of the Hamiltonian in (1); it is entirely real, positive, and discrete. Very recently, the reality and positivity of the spectra have been established rigorously.[26,27]

In Fig. 1 we plot the eigenvalues of the massless version of the Hamiltonian in (3), namely

$$H = p^2 - (ix)^N$$

, as a function of the parameter N. Note that when $N \geq 2$, the entire spectrum is real and positive. The special case $N = 2$, corresponds to the harmonic oscillator, whose energy levels are $E_n = 2n + 1$. When $1 < N < 2$, there are a finite number of positive real eigenvalues and infinitely many complex conjugate pairs of eigenvalues. In this region the $\mathcal{PT}$ symmetry of the Hamiltonian is spontaneously broken. As N decreases from 2 to 1, the number of real eigenvalues decreases; when $N \leq 1.42207$, the only real eigenvalue is the ground-state energy.

How can the Hamiltonian (2) possess two such astonishingly different spectra? The answer lies in the boundary conditions satisfied by the wave functions

[a] Rotating from $\theta = 0$ to $\theta = -\pi$, we obtain the same Hamiltonian as in (2) but the spectrum is the complex conjugate of the spectrum obtained when we rotate from $\theta = 0$ to $\theta = \pi$.

$\psi_n(x)$. In the first case, in which $\theta = \arg g$ is rotated in the complex-g plane from 0 to π, $\psi_n(x)$ vanishes in the complex-x plane as $|x| \to \infty$ inside the wedges $-\pi/3 < \arg x < 0$ and $-4\pi/3 < \arg x < -\pi$. In the second case, in which the exponent α ranges from 0 to 2, $\psi_n(x)$ vanishes in the complex-x plane as $|x| \to \infty$ inside the wedges $-\pi/3 < \arg x < 0$ and $-\pi < \arg x < -2\pi/3$. In this second case the boundary conditions hold in wedges that are symmetric with respect to the imaginary axis; these boundary conditions enforce the $\mathcal{PT}$ symmetry of H and are responsible for the reality of the energy spectrum.

There is another striking difference between the two theories corresponding to H in (2). We define the one-point Green's function $G_1(g)$ [the expectation value of the operator x in the ground-state wave function $\psi_0(x)$] by

$$G_1(g) = \langle 0|x|0\rangle / \langle 0|0\rangle \equiv \int_C dx\, x\psi_0^2(x) \Big/ \int_C dx\, \psi_0^2(x), \tag{4}$$

where C is a contour that lies in the asymptotic wedges described above. The value of $G_1(g)$ for H in (2) depends on the limiting process by which we obtain H. If we substitute $g = g_0 e^{i\theta}$ into the Hamiltonian (1) and rotate from $\theta = 0$ to $\theta = \pi$, we find by an elementary symmetry argument that $G_1(g) = 0$ for all g

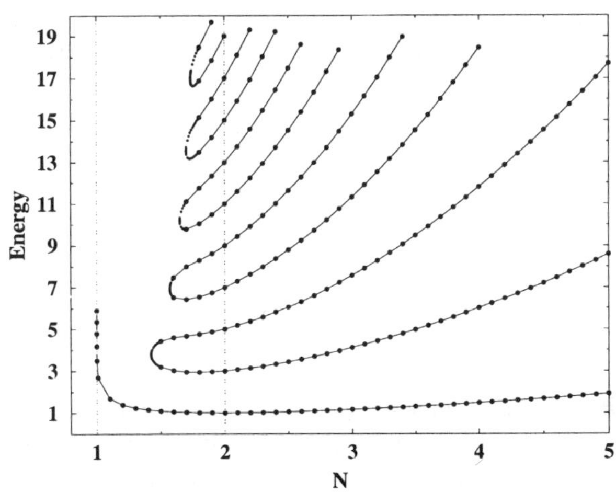

Figure 1: Real energy levels of the Hamiltonian $H = p^2 - (ix)^N$ as a function of the parameter N. When $N \geq 2$, the entire spectrum is real and positive. The lower bound of this region, $N = 2$, corresponds to the harmonic oscillator, whose energy levels are $E_n = 2n + 1$. When $1 < N < 2$, there are a finite number of positive real eigenvalues and infinitely many complex conjugate pairs of eigenvalues. As N decreases from 2 to 1, the number of real eigenvalues decreases; when $N \leq 1.42207$, the only real eigenvalue is the ground-state energy. As $N \to 1^+$, the ground-state energy diverges. For $N \leq 1$ there are no real eigenvalues.

on the semicircle in the complex-g plane. Thus, this rotation in the complex-g plane preserves parity symmetry ($x \to -x$). However, if we define H in (2) by using the Hamiltonian in (3) and by allowing α to range from 0 to 2, we find that $G_1(g) \neq 0$. Indeed, $G_1(g) \neq 0$ for *all* values of $\alpha > 0$. Thus, in this theory $\mathcal{PT}$ symmetry (reflection about the imaginary axis, $x \to -x^*$) is preserved, but parity symmetry is permanently broken.

These two different results for $G_1(g)$ emphasize the ambiguity in Dyson's argument and show that the boundary conditions in the integrals in (4) are crucial for determining the one-point Green's function. We are concerned in this talk with the theory that preserves $\mathcal{PT}$ symmetry. In this theory the energy spectrum is real and positive and $G_1(g)$ is nonzero.

We have extended these quantum-mechanical arguments to the quantum field theory whose D-dimensional Euclidean space Lagrangian density is

$$\mathcal{L} = (\nabla\phi)^2/2 + m^2\phi^2/2 - g\phi^4/4. \tag{5}$$

What is remarkable about this "wrong-sign" field theory is that, when it is obtained using the $\mathcal{PT}$-symmetric limit, the energy spectrum is real and positive, and the one-point Green's function is nonzero. Furthermore, the field theory is renormalizable, and in four dimensions is asymptotically free (and thus nontrivial). Based on these features of the theory, we believe that the theory may provide a useful setting to describe the Higgs particle.

2 Weak-Coupling Calculation of G_1

We have calculated G_1 in both the weak- and the strong-coupling limit. [17] The results for the weak-coupling limit are as follows. The one-point Green's function G_1 is defined as a *complex* functional integral in Euclidean space:

$$G_1 = \int_C \mathcal{D}\phi\, \phi(0) e^{-L[\phi]} \Big/ \int_C \mathcal{D}\phi\, e^{-L[\phi]}. \tag{6}$$

Here, $L[\phi] = \int d^D x\, \mathcal{L}$ with $\mathcal{L}$ given in (5) and C is a contour in the complex-ϕ plane defined as follows: The functional integrals in (6) are infinite products of ordinary integrals, one integral for each lattice point in Euclidean space. For these ordinary integrals the contour of integration must lie within 45° wedges that lie in the lower-half plane and are centered about −45° and −135°. These wedges are symmetrically placed about the imaginary axis.

In D-dimensional space we use $\epsilon = gm^{D-4}/4$ to represent the dimensionless coupling constant. The small-ϵ asymptotic behavior of G_1 is determined by a *soliton* (not an instanton). In general, G_1 has a *negative imaginary* value.

Table 1: Numerical values of $\Lambda[D]$ in (8) for various values of the Euclidean space-time dimension D. Note that $\Lambda[D]$ rises to a maximum value between $D = 3.5$ and $D = 4$ and then starts to decrease.

D	$\Lambda[D]$	D	$\Lambda[D]$	D	$\Lambda[D]$	D	$\Lambda[D]$
0	1/4	1.5	2.8755	3.0	18.9023	3.9	29.3452
0.5	0.5853	2.0	5.8505	3.5	27.6012	3.95	28.3897
1.0	4/3	2.5	11.0599	3.8	30.0387	4.0	26.3209

For $D = 0$ and $D = 1$ we have

$$D=0: \; G_1 \sim -\frac{i}{m} 2^{-1/2} \epsilon^{-1/2} e^{-1/\epsilon} \quad (\epsilon \to 0^+);$$

$$D=1: \; G_1 \sim -\frac{i}{\sqrt{m}} 16\sqrt{\pi e}(2/\epsilon)^{2/3} e^{-16/(3\epsilon)} 3^{-1/6}/\Gamma^2(1/3) \quad (\epsilon \to 0^+). \tag{7}$$

In dimensions other than $D = 0$ and $D = 1$ the controlling factor of the asymptotic behavior of G_1 for small ϵ has the form

$$G_1 \sim e^{-4\Lambda[D]/\epsilon} \quad (\epsilon \to 0), \tag{8}$$

where $\Lambda[D]$ is determined by a spherically symmetric soliton. Numerical values of $\Lambda[D]$ for various values of D are given in Table 1.

3 Bound States

Another significant difference between the spectra of the Hermitian Hamiltonian (1) and the $\mathcal{PT}$-symmetric Hamiltonian (2) is that when g is sufficiently small, the latter Hamiltonian possesses bound states while the former does not. The bound states persist in the corresponding non-Hermitian $\mathcal{PT}$-symmetric $-g\phi^4$ quantum field theory (5) for all dimensions $0 \leq D < 3$ but are not present in the conventional Hermitian $g\phi^4$ field theory.

We calculate the bound states perturbatively. For the conventional quantum anharmonic oscillator Hamiltonian (1), the small-g Rayleigh-Schrödinger perturbation series for the kth energy level E_k is

$$E_k \sim m[k + 1/2 + 3(2k^2 + 2k + 1)\epsilon/4 + \mathrm{O}(\epsilon^2)] \qquad (\epsilon \to 0^+), \tag{9}$$

where $\epsilon = g/4m^3$. The *renormalized mass* M is defined as the first excitation above the ground state: $M \equiv E_1 - E_0 \sim m[1 + 3\epsilon + \mathrm{O}(\epsilon^2)]$ as $\epsilon \to 0^+$.

To determine if the two-particle state is bound, we examine the second excitation above the ground state using (9). We define

$$B_2 \equiv E_2 - E_0 \sim m\left[2 + 9\epsilon + \mathrm{O}(\epsilon^2)\right] \qquad (\epsilon \to 0^+). \tag{10}$$

If $B_2 < 2M$, then a two-particle bound state exists and the (negative) binding energy is $B_2 - 2M$. If $B_2 > 2M$, then the second excitation above the vacuum is interpreted as an unbound two-particle state. We see from (10) that in the small-coupling region, where perturbation theory is valid, the conventional anharmonic oscillator does not possess a bound state. Indeed, using WKB, variational methods, or numerical calculations one can show that there is no two-particle bound state for any value of $g > 0$. Because there is no bound state the gx^4 interaction may be considered to represent a repulsive force.[b]

We obtain the perturbation series for the Hamiltonian H in (2) from the perturbation series for the conventional anharmonic oscillator by replacing ϵ with $-\epsilon$. Thus, while the conventional anharmonic oscillator does not possess a two-particle bound state, the $\mathcal{PT}$-symmetric oscillator does indeed possess such a state. We measure the binding energy of this state in units of the renormalized mass M and we define the *dimensionless* binding energy Δ_2 by

$$\Delta_2 \equiv (B_2 - 2M)/M \sim -3\epsilon + O(\epsilon^2) \qquad (\epsilon \to 0^+). \tag{11}$$

This bound state disappears when ϵ increases beyond $\epsilon = 0.0465\ldots$. As ϵ continues to increase, Δ_2 reaches a maximum value of 0.427 at $\epsilon = 0.13$ and then approaches the limiting value 0.28 as $\epsilon \to \infty$.

In the $\mathcal{PT}$-symmetric anharmonic oscillator, there are not only two-particle bound states for small coupling constant but also k-particle bound states for all $k \geq 2$. The dimensionless binding energies are

$$\Delta_k \equiv (B_k - kM)/M \sim -3k(k-1)\epsilon/2 + O(\epsilon^2) \qquad (\epsilon \to 0+). \tag{12}$$

The key feature of this equation is that the coefficient of ϵ is negative. Since the dimensionless binding energy becomes negative as ϵ increases from 0, there is a k-particle bound state. The dimensionless binding energies for the first five k-particle bound states are shown in Fig. 2. Note that the higher multiparticle bound states cease to be bound for smaller values of ϵ; starting with the three-particle bound state, the binding energy of these states becomes positive as ϵ increases past 0.039, 0.034, 0.030, and 0.027.

Figure 2 shows that for any value of ϵ there are always a finite number of bound states and an infinite number of unbound states. The number of bound states decreases with increasing ϵ until there are no bound states at all.

[b]In general, a repulsive force in a quantum field theory is represented by an energy dependence in which the energy of a two-particle state decreases with separation. The conventional anharmonic oscillator Hamiltonian corresponds to a field theory in one space-time dimension, where there cannot be any spatial dependence. In this case the repulsive nature of the force is understood to mean that the energy B_2 needed to create two particles at a given time is more than twice the energy M needed to create one particle.

Observe that there is a range of ϵ for which there are only two- and three-particle bound states. This situation is analogous to the physical world in which one observes only states of two and three bound quarks. In this range of ϵ if one has an initial state containing a number of particles (renormalized masses), these particles will clump together into bound states, releasing energy in the process. Depending on the value of ϵ, the final state will consist either of two- or of three-particle bound states, whichever is energetically favored. Note also that there is a special value of ϵ for which two- and three-particle bound states can exist in thermodynamic equilibrium.

These results generalize from quantum mechanics to the D-dimensional $\mathcal{PT}$-symmetric quantum field theory in (5). To show that there is a bound state we demonstrate that *to leading order in the dimensionless coupling constant* ϵ the binding energy becomes negative as ϵ increases from 0. We calculate the energy of the bound state by summing all "sausage-link" graphs and identifying the bound-state pole in the geometric series. The dimensionless binding energy to leading order ϵ is

$$\Delta_2 \sim -(4\pi)^{(D-1)/(D-3)}[3\Gamma(3/2-D/2)]^{2/(3-D)}\epsilon^{2/(3-D)}, \tag{13}$$

which reduces to (11) when $D = 1$. Equation (13) is valid for $0 \leq D < 3$ because we have performed a mass renormalization to leading order, but not a wave function or coupling-constant renormalization.

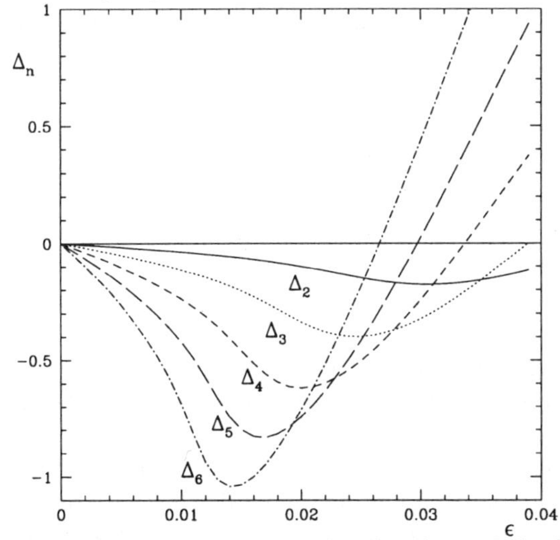

Figure 2: Dimensionless binding energies Δ_2, Δ_3, Δ_4, Δ_5, and Δ_6 for the two-, three-, four-, five-, and six-particle bound states plotted as functions of the dimensionless coupling constant ϵ. Note that the multiparticle bound states cease to be bound as ϵ increases past 0.0465, 0.039, 0.034, 0.030, and 0.027.

We conclude by comparing a $g\phi^3$ theory with a $g\phi^4$ theory: A $g\phi^3$ theory represents an attractive force. The bound states that arise as a consequence of this force can be found by using the Bethe-Salpeter equation. However, the $g\phi^3$ field theory is unacceptable because the spectrum is not bounded below. If we replace g by ig, the spectrum becomes real and positive, but the force becomes repulsive and there are no bound states. The same is true for a two-scalar theory with interaction of the form $ig\phi^2\chi$. This latter theory is an acceptable model of scalar electrodynamics, but has no analog of positronium.

Acknowledgments

I thank my coauthors M. Berry, S. Boettcher, F. Cooper, G. Dunne, H. Jones, P. Meisinger, K. Milton, V. Savage, M. Şimşek, Q. Wang, E. J. Weniger, and H. Yang for their contributions to this research. This work was supported by the U.S. Department of Energy.

References

1. C. M. Bender and K. A. Milton, *Phys. Rev. D* **55**, R3255 (1997).
2. C. M. Bender and S. Boettcher, *J. Phys. A: Math. Gen.* **31**, L273 (1998).
3. C. M. Bender and K. A. Milton, *Phys. Rev. D* **57**, 3595 (1998).
4. C. M. Bender and S. Boettcher, *Phys. Rev. Lett.* **80**, 5243 (1998).
5. C. M. Bender, Invited talk published in the proceedings of the conference Particles, Fields & Gravitation '98, held at the University of Lodz, Poland, April 1998, AIP Conference Proceedings 453, Ed. by J. Rembielinski, American Institute of Physics, New York (1998), p. 167.
6. C. M. Bender, G. V. Dunne, and P. N. Meisinger, *Phys. Lett. A* **252**, 272 (1999).
7. C. M. Bender, S. Boettcher, and P. N. Meisinger, *J. Math. Phys.* **40**, 2201 (1999).
8. C. M. Bender, Invited talk at the Memorial Symposium in Honor of Richard Slansky, Los Alamos National Laboratory, May 1998, *Phys. Rep.* **315**, 27 (1999).
9. C. M. Bender, Invited talk published in the proceedings of the Fourth Workshop on Quantum Chromodynamics held at the American University of Paris, June 1998, Ed. by H. M. Fried and B. Müller, World Scientific, Singapore (1999), p. 435.
10. C. M. Bender and K. A. Milton, *J. Phys. A: Math. Gen.* **32**, L87 (1999).
11. C. M. Bender and G. V. Dunne, *J. Math. Phys.* **40**, 4616 (1999).

12. C. M. Bender, F. Cooper, P. N. Meisinger, and V. M. Savage, *Phys. Lett. A* **259**, 224 (1999).
13. C. M. Bender, S. Boettcher, H. F. Jones, and V. M. Savage, *J. Phys. A: Math. Gen.* **32**, 6771 (1999).
14. C. M. Bender and H. F. Jones, *Found. Phys.* **30**, 393 (2000).
15. C. M. Bender, S. Boettcher, and V. M. Savage, *J. Math. Phys.* **41**, 6381 (2000).
16. C. M. Bender, K. A. Milton, and V. M. Savage, *Phys. Rev. D* **62**, 85001 (2000).
17. C. M. Bender, P. Meisinger, and H. Yang, *Phys. Rev. D* **63**, 45001 (2001).
18. C. M. Bender and Q. Wang, *J. Phys. A: Math. Gen.* **34**, 3325 (2001).
19. C. M. Bender, S. Boettcher, and P. N. Meisinger, to be published in *Fluctuating Paths and Fields*, World Scientific, Singapore.
20. C. M. Bender, S. Boettcher, H. F. Jones, and P. N. Meisinger, *J. Math. Phys.* **42**, 1960 (2001).
21. C. M. Bender and E. J. Weniger, *J. Math. Phys.* **42**, 2167 (2001).
22. C. M. Bender, M. Berry, P. N. Meisinger, V. M. Savage, and M. Şimşek, *J. Phys. A: Math. Gen.* **34**, L31 (2001).
23. C. M. Bender, G. V. Dunne, P. N. Meisinger, and M. Şimşek, *Phys. Lett. A* **281**, 311 (2001).
24. C. M. Bender, S. Boettcher, H. F. Jones, P. N. Meisinger, and M. Şimşek, *Phys. Lett. A*, in press.
25. F. J. Dyson, *Phys. Rev.* **85**, 631 (1952).
26. P. Dorey, C. Dunning, and R. Tatao, *J. Phys. A: Math. Gen.* **34**, L391 (2001) and **34**, 5679 (2001).
27. P. Dorey, C. Dunning, and R. Tatao, *J. Phys. A: Math. Gen.* (to appear).

FROM RESONANCES TO POINCARÉ SEMIGROUPS

A. R. BOHM, B. A. TAY,
Center for Particle Physics,
The University of Texas at Austin, Austin, Texas 78712-1081, USA
Email: bohm@ph.utexas.edu

P. KIELANOWSKI*
Departamento de Física,
Centro de Investigación y de Estudios Avanzados del IPN, México City, México

Resonances and decaying particles are characterized by two numbers, either by the mass m (or resonance energy E_R) and the Breit-Wigner width Γ or by the mass m and the decay rate (inverse lifetime) $1/\tau$. In non-relativistic quantum mechanics, it is always assumed that $\Gamma = 1/\tau$, though this is only based on an approximation (Weisskopf-Wigner). For relativistic resonances, opinions are more divided and some believe that the lineshape has a more complicated form than the Breit-Wigner, or at least that there is an energy dependent width $\Gamma(\mathsf{s})$. This has led to disputes about the correct definition of mass and width for a relativistic resonance, in particular for the Z-boson. An answer to this problem is provided if, in analogy to Wigner's definition of stable relativistic particles, one defines relativistic resonances by irreducible representations of the Poincaré group. Since the decay of resonances is an irreversible process, one needs to generalize the unitary representations (m, j) of Wigner to semigroup representations characterized by spin j and a complex number for the invariant mass squared $\mathsf{s}_R = (m - i\Gamma/2)^2$. These semigroup representations and their transformation properties under causal Poincaré transformations are presented. They provide a unique definition of mass, width and lineshape for a relativistic resonance if one accepts that the relation $\Gamma = 1/\tau$ holds also for relativistic resonances.

1 Introduction - Non-relativistic resonance

Resonances and quasistable particles are characterized by two sets of real numbers (E_R, Γ) or $(E_R, R = \frac{1}{\tau})$, respectively. Lifetime τ and its inverse $R \equiv \frac{1}{\tau}$, the initial decay rate, are measured by fits of the counting rate of the decay products to the exponential law

$$\frac{1}{N}\frac{\Delta N_\eta(t)}{\Delta t} \approx R_\eta(t) = R_\eta e^{-Rt}\,, \tag{1}$$

*also at Institute of Theoretical Physics, University of Białystok, Poland whereas the width Γ is measured by fits of the cross section to the Lorentzian (Breit-Wigner) energy distribution

$$\sigma_j^{\mathrm{BW}}(E) \sim |a_j^{\mathrm{BW}}(E)|^2 = \left| \frac{r_\eta}{E-(E_R - i\frac{\Gamma}{2})} \right|^2 \sim \frac{1}{(E-E_R)^2 - (\frac{\Gamma}{2})^2}, \quad \text{with} \quad 0 \leq E < \infty . \tag{2}$$

(plus usually some background term $B(E)$). Decay rate $R = \sum_\eta R_\eta(t)$ and resonance width Γ are thus different quantities, $R = \frac{1}{\tau}$ is connected with the exponential time evolution (1), and Γ is connected with the Lorentzian energy distribution. But usually one does not even distinguish between R and the width Γ; one identifies $\frac{\hbar}{\tau} \equiv R$ with Γ

$$\Gamma = \frac{\hbar}{\tau} \equiv R \tag{3}$$

and uses the words rate and width for either of them.

In non-relativistic physics the relation (3) was justified by the Wigner-Weisskopf approximation [1], which is really not an approximation of standard quantum mechanics (based on the Hilbert space axiom), because in Hilbert space $\mathcal{H}$ there is no state vector $\psi^D(t) \in \mathcal{H}$ whose rate for the Born probability $P_\eta(t)$

$$R_\eta(t) = \frac{d}{dt} P_\eta(t) \equiv \frac{d}{dt}(\mathrm{Tr}(\Lambda_\eta |\psi^D(t)\rangle\langle\psi^D(t)|)) \tag{4}$$

obeys the exponential law (1)[2].

Recently a non-relativistic theory which fulfills

$$\frac{\hbar}{\Gamma} = \tau . \tag{5}$$

has been constructed using a slight modification of the standard Hilbert space axiom [3,4,5].

The mathematical axioms of standard quantum mechanics in the Hilbert space are not completely in agreement with the calculational tools (like for instance Dirac kets) that physicists use in their practical calculations. As a consequence of von Neumann's mathematical assumptions, standard quantum mechanics is time symmetric, because the Hilbert Space does not allow time asymmetric boundary conditions for the time symmetric dynamical equations, and the time evolution and all other symmetry transformations of non-relativistic and relativistic space time are described by unitary group transformations. In practice, however, one uses properties of time asymmetry, as is e.g. expressed by the $\pm i\epsilon$ of the Dirac kets in the Lippmann-Schwinger equation. Though orthodox quantum theory would require that the set of all in-states $\{\phi^+\}$ and the set of all out-states $\{\psi^-\}$ is the same Hilbert space $\mathcal{H}$ (or the same dense subspace thereof, asymptotic completeness), the Lippmann-Schwinger equation suggests a different hypothesis. If one modifies the Hilbert Space axiom slightly and replaces it by the new hypothesis that the set of in-states defined by the preparation apparatus (accelerator) and the set of out-states or observables defined by the registration apparatus (detector) form different dense subspaces of the Hilbert Space, $\mathbf{\Phi}_-$ and $\mathbf{\Phi}_+$, respectively, one obtains asymmetric boundary conditions. Guided by the Lippmann-Schwinger equations we require that the space of states $\mathbf{\Phi}_- = \{\phi^+\}$ is a Hardy class space of the lower complex energy half-plane, and the space of observables $\mathbf{\Phi}_+ = \{\psi^-\}$ is a Hardy space of the upper complex energy half-plane. This assumption allows a mathematically rigorous definition of the Dirac-Lippmann-Schwinger kets as continuous antilinear functionals, $|p, \alpha^+\rangle \in \mathbf{\Phi}_-^\times$ and $|p, \alpha^-\rangle \in \mathbf{\Phi}_+^\times$, where $\Phi_-^\times$ and $\Phi_+^\times$ are the spaces of continuous antilinear functionals over Φ_- and Φ_+ respectively. The Hardy class boundary condition thus makes the singular Lippmann-Schwinger equation mathematically rigorous. But it does more, because in addition to the Dirac-Lippmann-Schwinger kets for real energy E, $|E^-\rangle \in \mathbf{\Phi}_+^\times$, the space $\mathbf{\Phi}_+^\times$ contains also Gamow kets $\psi^G = |E_R - i\frac{\Gamma}{2}^-\rangle$, which are generalized eigenvectors of the self-adjoint Hamiltonian operator $H = H_0 + V$ with complex eigenvalue $z_R = E_R - i\Gamma/2$. If one uses these Gamow states ψ^G in place of ψ^D in (4) then $R_\eta(t)$ fulfills the exponential law (1) and the Breit-Wigner energy distribution (2), where R of (1) is related to Γ of (2) by (3).

This is the theoretical foundation of the opinion favored by almost everyone in non-relativistic physics, that resonances for which one usually measures Γ by (2), and decaying states for which one usually measures $\tau = \frac{1}{R}$ by (1), are just two different appearance of one and the same micro-physical state. The resulting exact theoretical identity (3) has recently also been confirmed experimentally in atomic physics [6].

2 Relativistic Resonances

In relativistic physics the situation is quite different. Based on the perturbation theoretical definition by the self-energy of the propagators, resonances and decaying states are considered as complicated objects that cannot be described as an exponentially decaying state or as a state characterized by two numbers like (E, Γ).

The jth partial scattering amplitude in a relativistic resonance formation process $a_j(\mathsf{s})$ is a function of invariant mass square $\mathsf{s} = (p_1^\mu + p_2^\mu)^2 = (E_1^{cm} + E_2^{cm})^2$, where p_1^μ, p_2^μ are the momenta of the two incoming (or outgoing) particles. One writes the amplitude of a resonance scattering process as

$$a_j(\mathsf{s}) = a_j^{\mathrm{res}}(\mathsf{s}) + B_j(\mathsf{s}) \tag{6}$$

where $B_j(\mathsf{s})$ is the non-resonant (constant) background and a_j^{res} is the contribution of the resonance for which one uses something like the Breit-Wigner of (2). However in contrast to the non-relativistic case, where one defines resonance energy and resonance width by (2), in the relativistic case one does not have an universally agreed upon definition of the resonance amplitude and of mass and width.

The Review of Particle Properties [7] gives two definitions of the mass and width of the Z-boson and lists two different values which are obtained from the fit of two different formulas for the lineshape to the same experimental data. The value M_Z is obtained from the fit to the "relativistic Breit-Wigner with energy dependent width" of the on-shell renormalization scheme

$$\begin{aligned} a_j^{\mathrm{om}}(\mathsf{s}) = \frac{-\sqrt{\mathsf{s}}\sqrt{\Gamma_e(\mathsf{s})\Gamma_f(\mathsf{s})}}{\mathsf{s} - M_Z^2 + i\sqrt{\mathsf{s}}\Gamma_Z(\mathsf{s})} \approx \frac{-M_Z B_{ef}\Gamma_Z}{\mathsf{s} - M_Z^2 + i\frac{\mathsf{s}}{M_Z}\Gamma_Z} = \\ \frac{R_Z}{\mathsf{s} - M_Z^2 + i\frac{\mathsf{s}}{M_Z}\Gamma_Z}, \quad m_0^2 \leq \mathsf{s} < \infty \,. \end{aligned} \tag{7}$$

The value $\bar{M}_Z$ is obtained from the relativistic Breit-Wigner of the S-matrix pole

$$a_j^{\mathrm{BW}}(\mathsf{s}) = \frac{R_Z}{\mathsf{s} - \mathsf{s}_R} = \frac{R_Z}{\mathsf{s} - \bar{M}_Z^2 + i\bar{M}_Z\bar{\Gamma}_Z} = \frac{R_Z}{\mathsf{s} - (M_R - i\frac{\Gamma_R}{2})^2}, \quad m_0^2 \leq \mathsf{s} < \infty \,. \tag{8}$$

Both lineshape formulas (7) and (8) fit the experimental data equally well [8,9,10]. But they lead to values of the mass parameters M_Z and $\bar{M}_Z$ which differ from each other by about 10 times the experimental error. From $a_j^{\mathrm{om}}(\mathsf{s})$ the fit gives the values

$$M_Z = (91.1871 \pm 0.0021)\ \mathrm{GeV}\,, \quad \Gamma_Z = (2.4945 \pm 0.0024)\ \mathrm{GeV}\,. \tag{9}$$

From $a_j^{\mathrm{BW}}(\mathsf{s})$ one obtains the values $\bar{M}_Z, \bar{\Gamma}_Z$ or the values

$$M_R = (91.1626 \pm 0.0031)\ \mathrm{GeV}\,, \quad \Gamma_R = (2.4934 \pm 0.0024)\ \mathrm{GeV}\,, \tag{10}$$

from a fit to the same data. Numerically these parameters differ from each other by:

$$M_R = M_Z - 0.026\ \mathrm{GeV} = M_Z - 10 \times \Delta m_{exp}\,, \quad \Gamma_R = \Gamma_Z - 1.2\ \mathrm{MeV}\,, \tag{11}$$

or

$$\bar{M}_Z = M_Z - 34 \text{ MeV}\,. \tag{12}$$

The question thus is: What is the right definition of the Z-boson mass and width and therefore the right numerical value of the mass of the Z-boson?

Even if one discards the on-shell definition (7) and chooses the S-matrix definition (8) because it is gauge invariant [11,12,13], the complex parameter s_R in Eq. (8) can be expressed in terms of the real parameters mass and width in many different ways leading to many arbitrary definitions of the Z-boson mass. Some of these mentioned in the literature are

1. $(\bar{M}_Z, \bar{\Gamma}_Z)$ (also called (m_2, Γ_2) [11])

$$\mathsf{s}_R = \bar{M}_Z^2 - i\bar{M}_Z\bar{\Gamma}_Z\,. \tag{13}$$

2. (M_R, Γ_R) [13] which is often used but not dictated by the analytic S-matrix theory ([14], in particular p. 248)

$$\mathsf{s}_R = \left(M_R - i\frac{\Gamma_R}{2}\right)^2\,. \tag{14}$$

This is related to $(\bar{M}_Z, \bar{\Gamma}_Z)$ by the algebraic identity

$$\bar{M}_Z = M_R\left(1 - \frac{1}{4}\left(\frac{\Gamma_R}{M_R}\right)^2\right)^{\frac{1}{2}} \quad \text{and} \quad \bar{\Gamma}_Z = \Gamma_R\left(1 - \frac{1}{4}\left(\frac{\Gamma_R}{M_R}\right)^2\right)^{-\frac{1}{2}}\,. \tag{15}$$

3. (m_1, Γ_1) [11] which is numerically very close to the conventional Standard Model values (M_Z, Γ_Z). It can be defined in terms of $\bar{M}_Z$ and $\bar{\Gamma}_Z$ by:

$$\bar{M}_Z^2 = m_1^2\left(1 + \left(\frac{\Gamma_1}{m_1}\right)^2\right)^{-1} \quad \text{and} \quad \bar{\Gamma}_Z^2 = \Gamma_1^2\left(1 + \left(\frac{\Gamma_1}{m_1}\right)^2\right)^{-1}, \tag{16}$$

such that m_1 is numerically the same as M_Z of (9), if one identifies the maximum of $|a_j^{\text{om}}(\mathsf{s})|^2$, which is $M_Z\left(1 + (\Gamma_Z/M_Z)^2\right)^{-1}$, with the maximum of $|a_j^{\text{BW}}(\mathsf{s})|^2$, which is $\bar{M}_Z^2$,

$$m_1 \approx M_Z\,, \quad \Gamma_1 \approx \Gamma_Z\,. \tag{17}$$

The extraordinary accuracy of the Z-lineshape data, obtained with great expense of time and effort, allowed for the first time to discuss the problem of the definition of mass and width for an unstable relativistic particle and go beyond the level of precision given by the Weisskopf-Wigner approximation and one-loop effects. This has opened up a new tier of inquiries, as witnessed by [11,12,13], and led to the question: what is a relativistic unstable particle?

Our answer to this question is given, in analogy to the case of stable relativistic particles, in terms of representations of transformations of relativistic space-time (Poincaré transformations). The states of stable elementary particles are vectors of an irreducible representation space $[j, m^2]$ of the Poincaré group $\mathcal{P}$ [15] from which one can define the fields [16]. This should not be restricted to interaction free, asymptotic states, but apply also to the exact states and to Poincaré transformations generated by the (interaction-incorporating) "exact generators" $P_0 = H = H_0 + V$, P^i, $J^{\mu\nu}$ [16a].

It has been claimed [16] that the exact in-states $|\alpha^+\rangle$ and the out-states $|\alpha^-\rangle$ that fulfill the Lippmann-Schwinger equation transform under the Poincaré group in the same way as the free states. This is not quite correct but under a precisely defined mathematical hypothesis, which replaces the Hilbert space axiom of standard quantum mechanics, one can show that the in- and out-states $|\alpha^+\rangle$ and $|\alpha^-\rangle$ of the Lippmann-Schwinger equation span an irreducible representation space of Poincaré-*semigroup* transformations into the backward and forward light cone, respectively [19]. The *semigroup* representations into the forward light cone are characterized by the angular momentum j (of the partial wave) and by the center-of-mass energy square s (which we will just call energy in the following) where $m_0^2 \leq \mathsf{s} < \infty$. Thus we denote $|\alpha^-\rangle = |[j, \mathsf{s}], b^-\rangle$ where b are some degeneracy parameters (e.g. j_3 and momentum $\mathbf{p}$; as in Wigner's basis vectors for unitary representations $[j, m^2]$ of the Poincaré group.) However, one can also take the 4-velocities $\hat{\mathbf{p}} = \frac{\mathbf{p}}{\sqrt{\mathsf{s}}} = \gamma \mathbf{v}$; $\gamma = \hat{p}^0 = \sqrt{1 + \hat{v}^2}$. The kets $|[j, \mathsf{s}], b = \hat{\mathbf{p}}^-\rangle$, and therewith the semigroup representations $[j, \mathsf{s}]$, will be analytically continued into the lower half plane, second Riemann sheet of the partial S-matrix $S_j(\mathsf{s})$ to the resonance pole position $\mathsf{s} = \mathsf{s}_R$. The resulting semigroup representations $[j, \mathsf{s}_R]$, obtained by integrating around the resonance pole, are the resonance analogue of Wigner's stable particle representations $[j, m^2]$, and we use the representation $[j, \mathsf{s}_R]$ as definition of the relativistic unstable particle in the same way as the unitary representation $[j, m^2]$ serves as definition of the relativistic stable particle.

This does not yet fix the definition of resonance mass and the resonance width, because the complex number s_R can be written in many ways in terms of the two real parameters M and Γ, of which (14), (15) and (16) are three examples. From the transformation property of the relativistic Gamow vectors

---

[a] For the conventional quantities we follow here fairly closely the notation of Ref. [16], but omit the letters Ψ and use Dirac's notation instead; thus $|\alpha^\pm\rangle = \Psi_\alpha^\pm$ of [16].

$|[j, \mathsf{s}_R], b^-\rangle$ under Poincaré transformation we will see that only Γ_R can fulfill the lifetime-width relation (3) $\Gamma_R = \frac{1}{\tau}$. Therefore (M_R, Γ_R) and the relativistic Breit-Wigner amplitude (8) are the preferred choice for the definition of the mass and width and of the lineshape for a relativistic resonance.

3 From the S-matrix pole to the Gamow vector

In order to obtain the representation of the Poincaré transformation associated with the resonance pole we start with the S-matrix element between a prepared in-state ϕ^+ and a detected out-observable ψ^-.

$$
\begin{aligned}
(\psi^{\mathrm{out}}, \phi^{\mathrm{out}}) &= (\psi^{\mathrm{out}}, S\phi^{\mathrm{in}}) = (\Omega^- \psi^{\mathrm{out}}, \Omega^+ \phi^{\mathrm{in}}) = (\psi^-, \phi^+) \\
&= \sum_{j, j_3, n} \int \frac{d^3\hat{\mathbf{p}}}{2\hat{E}} d\mathsf{s} \sum_{j', j_3', n'} \frac{d^3\hat{\mathbf{p}}'}{2\hat{E}'} d\mathsf{s}' \, \langle \psi^- | [j, \mathsf{s}], n, j_3, \hat{\mathbf{p}}^- \rangle \\
&\quad \times \langle \hat{\mathbf{p}}, j_3, [j, \mathsf{s}], n | S | [j', \mathsf{s}'], j_3', \hat{\mathbf{p}}', n' \rangle \langle {}^+ j_3', \hat{\mathbf{p}}', [j', \mathsf{s}'], n' | \phi^+ \rangle
\end{aligned} \tag{18}
$$

In this S-matrix element, ϕ^{in} describes the asymptotically free in-state e.g. the e^+e^- colliding beam that is prepared by the accelerator outside the interaction region. This ϕ^{in} becomes the ϕ^+ in the interaction region (of the two beams in $e^+e^- \to Z \to \bar{f}f$), the energy distribution in the beams is described by the wave function $\phi^{\mathrm{in}}(\mathsf{s}) = \phi^+(\mathsf{s})$. The out-state vector ψ^{out} describes the detected out-particles (e.g., a particular $\bar{f}f$) when they are asymptotically free. It comes from the ψ^- in the interaction region and its wave function $\psi^{\mathrm{out}}(\mathsf{s}) = \psi^-(\mathsf{s})$ describes the energy resolution of the detectors. ψ^- is defined by the registration apparatus (detector)—for which reason $|\psi^-\rangle\langle\psi^-|$ should be called observable rather than out-state. The kets $|[j, \mathsf{s}], b^\mp\rangle$ are the eigenvectors of the exact energy operator $P_\mu P^\mu = (P_{e^+} + P_{e^-})_\mu (P_{e^+} + P_{e^-})^\mu = (P_f + P_{\bar{f}})_\mu (P_f + P_{\bar{f}})^\mu$ with eigenvalue s and total angular momentum j. They span a representation of the Poincaré transformation for the $\bar{f}f$ and e^+e^- system, respectively and are eigenvectors of P_μ with eigenvalue $\hat{p}_\mu \sqrt{\mathsf{s}}$:

$$
P_\mu |[j, \mathsf{s}] \hat{\mathbf{p}} j_3, n^\pm\rangle = \sqrt{\mathsf{s}} \hat{p}_\mu |[j, \mathsf{s}] \hat{\mathbf{p}} j_3, n^\pm\rangle \tag{19}
$$

If one assumes the Hilbert space axiom of standard quantum mechanics or *asymptotic completeness* of scattering theory then

$$
\{\phi^+\} = \{\psi^-\} = \mathcal{H} \qquad \text{or} \qquad \{\phi^+\} = \{\psi^-\} = \mathcal{D} \subset \mathcal{H} \,. \tag{20}
$$

We assume the asymmetric boundary conditions $\phi^+ \in \mathbf{\Phi}_-$ and $\psi^- \in \mathbf{\Phi}_+$ and use in place of (20) a new assumption:

The space of prepared states $|\phi^+\rangle\langle\phi^+|$ defined by the preparation apparatus is $\{\phi^+\} = \mathbf{\Phi}_- \subset \mathcal{H}$,

The space of registered observables $|\psi^-\rangle\langle\psi^-|$ defined by the registration apparatus is $\{\psi^-\} = \mathbf{\Phi}_+ \subset \mathcal{H}$, (21)

where $\Phi_-(\Phi_+)$ is the (dense) Hardy subspace of analytic functions of the lower(upper) complex energy half-plane. The eigenkets in (19) are continuous antilinear functionals, $|[j,\mathsf{s}]b^\mp\rangle \in \Phi_\pm^\times$, on $\Phi_\pm$ respectively.

This new hypothesis replaces the Hilbert space hypothesis or the asymptotic completeness assumption (20). The new hypothesis allows the continuation of the Lippmann-Schwinger kets $|[j,\mathsf{s}]b^-\rangle$ ($|[j,\mathsf{s}]b^+\rangle$) into the lower (upper) half complex energy plane for which we choose the second sheet of the S-matrix. It also allows the Gamow kets to be obtained from the resonance poles of the S-matrix. The Hardy class axiom (21) means that the energy wave functions

$$\langle^-b,[j,\mathsf{s}]|\psi^-\rangle \equiv \psi^-(\mathsf{s})\,, \quad \text{and} \quad \langle^+b,[j,\mathsf{s}]|\phi^+\rangle \equiv \phi^+(\mathsf{s}) \tag{22}$$

have, in addition to the property that $|\psi^-(\mathsf{s})|^2$ and $|\phi^+(\mathsf{s})|^2$ be smooth functions of s (since they describe apparatus resolutions), also certain analyticity properties (Hardy class), namely $\psi^-(\mathsf{s})(\phi^+(\mathsf{s}))$ are analytic functions in the upper(lower) complex energy plane. Precisely,

$$\psi^-(\mathsf{s}) \in (\mathcal{S}\cap\mathcal{H}_+^2)||_{\mathbb{R}_+} \qquad \phi^+(\mathsf{s}) \in \mathcal{S}\cap\mathcal{H}_-^2||_{\mathbb{R}_-} \tag{23}$$

where $\mathcal{S}$ is the Schwartz space and the $\mathcal{H}_\pm^2$ are the Hardy class functions in the upper and lower complex plane, respectively [17].

The distinction between prepared in-states $\mathbf{\Phi}_-$ and registered out-observables $\mathbf{\Phi}_+$, by the use of two different dense (complete in a different topology than the $\mathcal{H}$ topology) subspaces of the *same* $\mathcal{H}$, is the only modification in the foundation of the theory; the dynamical equations and the algebra of observables, like the commutation relations of the Poincaré transformations, remain the same. All other novel conclusions, like the exact exponential decay law, the precise definition of width and mass of an S-matrix pole resonance and the quantum mechanical time asymmetry are mathematical consequences of this new hypothesis (21).

From the invariance of the S-operator with respect to Poincaré transformations one can show that the S-matrix element in (18) can be written as

$$\langle\hat{\mathbf{p}},j_3,[j,\mathsf{s}],n|S|\hat{\mathbf{p}}',j_3',[j',\mathsf{s}'],n'\rangle = 2\hat{E}(\hat{p})\delta^3(\hat{p}-\hat{p}')\delta(\mathsf{s}-\mathsf{s}')\delta_{j_3j_3'}\delta_{jj'}\langle n||S_j(\mathsf{s})||n'\rangle\,, \tag{24}$$

where $\langle n\|S_j(\mathsf{s})\|n'\rangle$ is the reduced S-matrix element which depends upon j (which labels the partial wave; it is the total orbital angular momentum for the case without spins, e.g., the $\pi^+\pi^-$ system) and the particle species and channel quantum numbers n, n'. For a fixed initial state n' it is written as

$$\langle n\|S_j(\mathsf{s})\|n'\rangle = S_j(\mathsf{s}) = \begin{cases} 2ia_j(\mathsf{s})+1 & \text{for elastic scattering } n=n' \\ 2ia_j^{(n)}(\mathsf{s}) & \text{for reaction from } n' \\ & \text{into the channel } n\,, \end{cases} \tag{25}$$

where $a_j(\mathsf{s})$ and $a_j^{(n)}(\mathsf{s})$ are the partial wave amplitudes of (7) and (8) (for $e^+e^- \to Z \to e^+e^-$ and $e^+e^- \to Z \to \mu\bar{\mu}$, etc.,). We insert (24) and (25) into (18) and obtain for the S-matrix element (omitting the additional quantum numbers n):

$$(\psi^-,\phi^+) = \sum_{j,j_3}\int_{m_0^2}^{\infty} d\mathsf{s} \int \frac{d^3\hat{\mathbf{p}}}{2\hat{E}} \langle \psi^-|[j,\mathsf{s}],j_3,\hat{\mathbf{p}}^-\rangle S_j(\mathsf{s})\langle^+ j_3,\hat{\mathbf{p}},[j,\mathsf{s}]|\phi^+\rangle\,. \tag{26}$$

After we have made use of the Poincaré invariance of the S-matrix (24) we ignore the degeneracy quantum numbers $b = \hat{\mathbf{p}}, j_3, \ldots$. Further we consider only the j-th partial S-matrix element (where j is the spin-parity of the resonance) in the sum (26). This matrix element we write as

$$(\psi^-,\phi^+)_j = \int_{m_0^2}^{\infty} d\mathsf{s}\,\langle\psi^-|\mathsf{s}^-\rangle S_j(\mathsf{s})\langle^+\mathsf{s}|\phi^+\rangle\,, \tag{27}$$

where the wave functions are

$$\psi^-(\mathsf{s}) = \langle^-\mathsf{s}|\psi^-\rangle = \langle^- b,[j,\mathsf{s}]|\psi^-\rangle\,, \tag{28}$$

$$\phi^+(\mathsf{s}) = \langle^+\mathsf{s}|\phi^+\rangle = \langle^+ b,[j,\mathsf{s}]|\phi^+\rangle\,. \tag{29}$$

They have, according to our new hypothesis, the Hardy property (23). ϕ^+ describes the prepared in-state (e^+e^-) and $|\phi^+(\mathsf{s})|^2 = |\phi^{\text{in}}(\mathsf{s})|^2$ describes the energy distribution of the beam. ψ^- describes the observed out-observable (e^+e^-, $\mu\bar{\mu}$, $\tau\bar{\tau}$,...) which is registered by the detector, and $|\psi^-(\mathsf{s})|^2 = |\psi^{\text{out}}(\mathsf{s})|^2$ describes the detector efficiency.

In order to be specific we shall consider the case that there are $N = 2$ resonances in the j-th partial wave, each described by a first order pole at the position $\mathsf{s} = \mathsf{s}_{R_1}$ and $\mathsf{s} = \mathsf{s}_{R_2}$ in the second sheet. The integrand in (27) is thus analytic in the lower half second sheet except for the two poles at $\mathsf{s} = \mathsf{s}_{R_i}$, and we can deform the contour of integration in (27) from the positive real line through the cut into the lower half plane of the second sheet. The r.h.s.

of (27) becomes (dropping the j notation)

$$(\psi^-, \phi^+) = \int_0^{-\infty_{II}} d\mathsf{s}\, \langle\psi^-|\mathsf{s}^-\rangle S_{II}(\mathsf{s})\langle^+\mathsf{s}|\phi^+\rangle$$
$$+ \oint_{C_1} d\mathsf{s}\, \langle\psi^-|\mathsf{s}^-\rangle S_{II}(\mathsf{s})\langle^+\mathsf{s}|\phi^+\rangle$$
$$+ \oint_{C_2} d\mathsf{s}\, \langle\psi^-|\mathsf{s}^-\rangle S_{II}(\mathsf{s})\langle^+\mathsf{s}|\phi^+\rangle\,, \tag{30}$$

where C_i is the circle around the pole at s_{R_i}, and the first integral extends along the negative real axis in the second sheet (indicated by $-\infty_{II}$); the integral along the infinite semicircle is zero due to the assumed property of the integrand. The first term has nothing to do with any of the resonances, it is the non-resonant background term corresponding to the $B_j(\mathsf{s})$ of (6),

$$\int_{m_0^2}^{-\infty_{II}} d\mathsf{s}\, \langle\psi^-|\mathsf{s}^-\rangle S_{II}(\mathsf{s})\langle^+\mathsf{s}|\phi^+\rangle \equiv \langle\psi^-|\phi^{\mathrm{bg}}\rangle\,, \tag{31}$$

which we express as the matrix element of ψ^- with a generalized vector ϕ^{bg} that is defined by it.

We now use the expansion around the pole s_{R_i}

$$S(\mathsf{s}) = \frac{R^{(i)}}{\mathsf{s} - \mathsf{s}_{R_i}} + R_0 + R_1(\mathsf{s} - \mathsf{s}_{R_i}) + \cdots \tag{32}$$

for each of the two (or N) integrals *separately*. The integrals around each of the poles, the pole terms, are calculated in the following way:

$$(\psi^-, \phi^+)_{\text{pole term}} \equiv \oint_{\hookleftarrow C_i} d\mathsf{s}\, \langle\psi^-|\mathsf{s}^-\rangle S(\mathsf{s})\langle^+\mathsf{s}|\phi^+\rangle \tag{33}$$
$$= \oint_{\hookleftarrow C_i} d\mathsf{s}\, \langle\psi^-|\mathsf{s}^-\rangle \frac{R^{(i)}}{\mathsf{s} - \mathsf{s}_{R_i}} \langle^+\mathsf{s}|\phi^+\rangle \tag{34}$$
$$= -2\pi i R^{(i)} \langle\psi^-|\mathsf{s}^-_{R_i}\rangle\langle^+\mathsf{s}_{R_i}|\phi^+\rangle \tag{35}$$
$$= \int_{-\infty_{II}}^{\infty} d\mathsf{s}\, \langle\psi^-|\mathsf{s}^-\rangle\langle^+\mathsf{s}|\phi^+\rangle \frac{R^{(i)}}{\mathsf{s} - \mathsf{s}_{R_i}}\,. \tag{36}$$

To get from (34) to (35), the Cauchy theorem has been applied; to get from (34) to (36), the contour C_i of each integral separately has been deformed into the integral along the real axis from $-\infty_{II} < \mathsf{s} < +\infty$ and an integral along the infinite semicircle, which vanishes, because of the Hardy class property. The equality (35) and (36) is the Titchmarsh theorem for Hardy class functions, [17] or Appendix in [3].

The integral (36) extends from $\mathsf{s} = -\infty_{II}$ in the second sheet along the real axis to $\mathsf{s} = 0$ and then from $\mathsf{s} = 0$ to $\mathsf{s} = +\infty$ in either sheet. (It does not matter whether we take the second part of the integral over the physical values of s, $m_0^2 \leq \mathsf{s} < \infty$, immediately below the real axis in the second sheet or in the first sheet immediately above the real axis). The major contribution to the integral comes from the physical values $m_0^2 \leq \mathsf{s} < \infty$, if s_{R_i} is not too far from the real axis. The integral in (36) contains the Breit-Wigner amplitude or Lorentzian

$$a_j^{\mathrm{BW}_i} = \frac{R^{(i)}}{\mathsf{s} - \mathsf{s}_{R_i}}, \quad \text{but with} - \infty_{II} < \mathsf{s} < +\infty \,. \tag{37}$$

Unlike the conventional Breit-Wigner of (8), for which s is taken (if one worries about these mathematical details) over $m_0^2 \leq \mathsf{s} < +\infty$, the Breit-Wigner (37) is an idealized Breit-Wigner whose domain extends to $-\infty_{II}$ in the second (unphysical) sheet. This idealized Breit-Wigner describes the resonance per se (without background).

By (36) we have associated each resonance at s_{R_i} to an idealized Breit-Wigner (37). By (35) we have associated each resonance at s_{R_i} to a vector $|\mathsf{s}_{R_i}^-\rangle = |[j, \mathsf{s}_{R_i}], b^-\rangle$ which we call Gamow ket.

We obtain a representation of the Gamow ket by using the equality (35)=(36) and omitting the arbitrary $\psi^- \in \mathbf{\Phi}_+$ (which represents the decay products defined by the detector). For this fundamental defining relation of the relativistic Gamow kets we shall restore the notation that includes the degeneracy quantum numbers b:

$$\begin{aligned} |[j, \mathsf{s}_{R_i}], b^-\rangle &= \frac{i}{2\pi} \int_{-\infty}^{\infty} d\mathsf{s}\, |[j, \mathsf{s}], b^-\rangle \frac{1}{\mathsf{s} - \mathsf{s}_{R_i}} \frac{\langle^+\mathsf{s}|\phi^+\rangle}{\langle^+\mathsf{s}_{R_i}|\phi^+\rangle} \\ &= \frac{i}{2\pi} \int_{-\infty}^{\infty} d\mathsf{s}\, |[j, \mathsf{s}], b^-\rangle \frac{1}{\mathsf{s} - \mathsf{s}_{R_i}} \,. \end{aligned} \tag{38}$$

The Gamow kets (38) are a superposition of the exact—not asymptotically free [18]—"out states" $|[j, \mathsf{s}], b^-\rangle$. The degeneracy quantum numbers b of the Gamow kets $|[j, \mathsf{s}_{R_i}], b^-\rangle$ are the same as the ones chosen for the Dirac-Lippmann-Schwinger kets $|[j, \mathsf{s}], b^-\rangle$. However, whereas for the Dirac-Lippmann-Schwinger kets one can choose for $b = b_1, \ldots, b_n$ the eigenvectors of any complete set of observables, one does not have the same freedom for the b in the Gamow kets, since in the contour deformations that one uses to get from (27) to (30) and ultimately to (33)-(36) one makes an analytic continuation in the variable s to complex values. If one chooses for b quantum numbers that also change when s is analytically continued, b could not be kept at one and the same value during this analytic continuation and the Gamow vector on the l.h.s. of (38) would be a complicated (continuous) superposition (integral) over different values of b and not just a superposition over different values of s. For this reason, the momentum $\mathbf{p}$ is not a good choice for the quantum numbers b in $|[j,\mathsf{s}],b^-\rangle$, because the momentum will also become complex if the energy in the center of mass rest frame becomes complex. This is the reason why we choose the space components of the 4-velocity $\hat{\mathbf{p}} = \mathbf{p}/\sqrt{\mathsf{s}}$ rather than the momentum $\mathbf{p}$ as the additional quantum numbers b in (19) and (18). Then we can impose the condition that $\mathbf{p}$ will become complex in the analytic continuation in such a way that $\hat{p}^\mu = p^\mu/\sqrt{\mathsf{s}}$ will always remains real. This condition restricts the arbitrariness of the analytic continuation, it makes the momentum only "minimally complex" and keeps the representations of the Lorentz subgroup of the Poincaré group $\mathcal{P}$ unitary. Only representations of the space-time translations turn into semigroup representations. The homogeneous Lorentz transformations $U(\Lambda)$ are the same as in Wigner's representations. We will call this subclass of semigroup representations of $\mathcal{P}$ minimally complex [19,20].

With (37) and (38) we have obtained for each resonance defined by the pole of the j-th partial S-matrix at $\mathsf{s} = \mathsf{s}_R$ an "exact" Breit-Wigner (37) and associated to it a set of "exact" Gamow kets (38). These Gamow kets (38) span, the space of an irreducible representation $[j, \mathsf{s}_R]$ of Poincaré transformations, but the representation space spanned by the kets of (38) is not the representation space of a unitary group. Thus we have the correspondence

Exact Breit-Wigner	$\Longleftrightarrow$	Exact Gamow vectors	
$a_j^{\mathrm{BW}_i}(\mathsf{s}) = \frac{R_i}{\mathsf{s}-\mathsf{s}_R}$	$\Longleftrightarrow$	$\|[j,\mathsf{s}_R],f^-\rangle = \int d\mu(b)\,\|[j,\mathsf{s}_R],b^-\rangle f(b)$	(39)
for $-\infty_{II} < \mathsf{s} < +\infty$		for all $f(b) \in \mathcal{S}(\mathbb{R}^3), -j \leq j_3 \leq j\,.$	

The Gamow vectors $|[j,\mathsf{s}_R],f^-\rangle$ have, according to (38), the representation

$$|[j,\mathsf{s}_R],f^-\rangle = \frac{i}{2\pi}\int_{-\infty_{II}}^{+\infty} d\mathsf{s}\,|[j,\mathsf{s}],f^-\rangle \frac{1}{\mathsf{s}-\mathsf{s}_R}\,. \tag{40}$$

They are functionals on the Hardy space $\mathbf{\Phi}_+$, i.e., $|[j,\mathsf{s}_R],f^-\rangle \in \mathbf{\Phi}_+^\times$.

The "wave function" $\psi^G(\mathsf{s}) \equiv \frac{i}{2\pi}\frac{1}{\mathsf{s}-\mathsf{s}_R}$ in (40) is not a very well behaved, Hardy class wave function. Also in the exact Breit-Wigner "wave function" $\psi^G(\mathsf{s})$ in (40), the variable s extends over $-\infty_{II} < \mathsf{s} < +\infty$. Thus the continuous linear superpositions (40), which define the relativistic Gamow vectors, are entirely different mathematical entities than the $\psi^- \in \Phi_+$. The Gamow vectors $|[j,\mathsf{s}_R],f^-\rangle$ and also the Gamow kets $|[j,\mathsf{s}_R],b^-\rangle$ of (38) are functionals over the space $\mathbf{\Phi}_+$ (the $|[j,\mathsf{s}_R],b^-\rangle$ are in addition functionals over the Schwartz space, i.e., $\langle\psi^-|[j,\mathsf{s}_R],b^-\rangle = f(b,j_3) \in \mathcal{S}(\mathbb{R}^3)$ (for fixed value s)). The equation (40) and (38) are functional equations over the space $\mathbf{\Phi}_+$, and (38) can be stated in terms of the smooth Hardy class functions $\overline{\psi}^-(\mathsf{s}) \equiv \langle\psi^-|[j,\mathsf{s}],b^-\rangle \in \mathcal{S}\cap\mathcal{H}^2_-$ as

$$\langle\psi^-|[j,\mathsf{s}_R],b^-\rangle = \frac{i}{2\pi}\int_{-\infty_{II}}^{+\infty} d\mathsf{s}\,\langle\psi^-|[j,\mathsf{s}],b^-\rangle\,\frac{1}{\mathsf{s}-\mathsf{s}_R}\,, \tag{41}$$

for all $\psi^- \in \mathbf{\Phi}_+$, and similarly for the vectors (40).

The equality (41) is the Titchmarsh theorem for the Hardy class function $\overline{\psi}^-(\mathsf{s}) = \langle\psi^-|[j,\mathsf{s}]b^-\rangle$ in the lower half plane of the second sheet. The integration path extends as in (36) along the real axis in the second sheet.

From these Gamow vectors we can now calculate consequences without any further mathematical assumption. These predictions are that (40) and/or (38) are generalized eigenvectors of the operators

$$P_\mu P^\mu\,,\ P^0 = H = H_0 + V\,,\ e^{iHt}(e^{-iH^\times t})\,. \tag{42}$$

We shall now show this for one example $P_\mu P^\mu$. We consider the vector $\psi'^- \equiv P_\mu P^\mu\psi^-$; this makes sense because $\psi'^- \in \mathbf{\Phi}_+$ for any $\psi^- \in \mathbf{\Phi}_+$, because $\mathbf{\Phi}_+$ is constructed such that all observables are continuous operators with respect to the topology in $\mathbf{\Phi}_+$. We now use (41) for $\psi'^- = P_\mu P^\mu\psi^-$ (cf. Appendix for the definition of the conjugate operator $(P_\mu P^\mu)^\times$):

$$\begin{aligned}
\langle P_\mu P^\mu\psi^-|[j,\mathsf{s}_R],b^-\rangle &= \langle\psi^-|(P_\mu P^\mu)^\times[j,\mathsf{s}_R],b^-\rangle \\
&= \frac{i}{2\pi}\int_{-\infty_{II}}^{+\infty} d\mathsf{s}\,\langle P_\mu P^\mu\psi^-|[j,\mathsf{s}],b^-\rangle\frac{1}{\mathsf{s}-\mathsf{s}_R} \\
&= \frac{i}{2\pi}\int_{-\infty_{II}}^{+\infty} d\mathsf{s}\,\langle\psi^-|(P_\mu P^\mu)^\times[j,\mathsf{s}],b^-\rangle\,\frac{1}{\mathsf{s}-\mathsf{s}_R} \\
&= \frac{i}{2\pi}\int_{-\infty_{II}}^{+\infty} d\mathsf{s}\,\langle\psi^-|[j,\mathsf{s}],b^-\rangle\ \mathsf{s}\ \frac{1}{\mathsf{s}-\mathsf{s}_R} \\
&= \mathsf{s}_R\langle\psi^-|[j,\mathsf{s}_R],b^-\rangle\,.
\end{aligned} \tag{43}$$

The next to the last step makes use of (19) and the last step is again the Titchmarsh theorem as in (41) but this time for the function $\overline{\psi}'^-(\mathsf{s}) \equiv \overline{\psi}^-(\mathsf{s})\,\mathsf{s}$, which is also a Hardy class function, $\overline{\psi}^-(\mathsf{s})\,\mathsf{s} \in \mathcal{S}\cap\mathcal{H}^2_-$ if $\overline{\psi}^-(\mathsf{s}) \in \mathcal{S}\cap\mathcal{H}^2_-$. We write (43) as a functional equation over the space $\mathbf{\Phi}_+$ omitting the arbitrary $\psi^- \in \mathbf{\Phi}_+$,

$$(P^\mu P_\mu)^\times|[j,\mathsf{s}_R],b,j_3^-\rangle = \mathsf{s}_R\,|[j,\mathsf{s}_R],b,j_3^-\rangle \tag{44}$$

Similarly, we find from (19) for the full Hamiltonian $P^0 = H = H_0 + V$ at rest:

$$H^{\times}|[j, \mathsf{s}_R], b_{\text{rest}}, j_3^-\rangle = \sqrt{\mathsf{s}_R}\,|[j, \mathsf{s}_R], b_{\text{rest}}, j_3^-\rangle \,. \tag{45}$$

If we use the parameterizations (13)-(17), then the eigenvalue of H can be written as

$$\text{eigenv}(H) = \sqrt{\mathsf{s}_R} = (M_R - i\frac{\Gamma_R}{2}) = \sqrt{\bar{M}_Z^2 - i\bar{M}_Z^2\bar{\Gamma}_Z} = \sqrt{\frac{(m_1^2 - im_1\Gamma_1)}{(1 + \frac{\Gamma_1}{m_1})}}\,. \tag{46}$$

The results (43)-(45) do not yet distinguish any of the parameterizations of s_R. In order to address the question of whether there is any preferred physical significance to one of the parameterizations in (46), we shall consider the Poincaré transformations (a, Λ), where a is the 4-vector of space time translations and Λ is a 4×4 Lorentz matrix. We shall give here only the action of time translations $(a = (t, 0, 0, 0), \Lambda = \mathbf{1})$ on the basis vectors at rest $|[j, \mathsf{s}_R], b^-_{\text{rest}}\rangle$. For the general transformation property of the Poincaré semigroup and a detailed proof of the semigroup property we refer to the forthcoming [20].

We know that in the stable particle representations $[j, m^2]$ the time translation is represented by the unitary group operator $U((t, \mathbf{0}), \mathbf{1}) \equiv U(t) = e^{iHt}$ in the Hilbert space $\mathcal{H}(j, m^2)$ and therefore also in the space $\mathcal{H}$ of (21). The time translation in the dense subspace $\mathbf{\Phi}_+$ will therefore be the restriction $U_+(t) = U(t)_{\mathbf{\Phi}_+} = e^{iHt}|_{\mathbf{\Phi}_+}$ to this subspace. The time translation in the space $\mathbf{\Phi}_+^\times$ will be the extension of the unitary group operator $U^\dagger(t) = e^{-iHt} \subset U_+^\times(t)$, whenever it can be defined. In order that $(e^{iHt})^\times = U_+^\times(t)$ can be defined the operator $U_+(t)$ needs to be a continuous operator with respect to the topology in $\mathbf{\Phi}_+$ (cf. Appendix). Therefore the question is: for which value of the parameter t is $U_+(t)$ (and therewith also $U_+^\times(t)$) a continuous operator in Φ_+,

$$\mathbf{\Phi}_+ \ni \psi^- \to U_+(t)\psi^- \equiv \psi'^- \in \mathbf{\Phi}_+ \,. \tag{47}$$

It can be shown [20,21] that (47) holds only for $t \geq 0$ so that also $U_+^\times(t)$ can only be defined for $t \geq 0$. For those values of t for which $U_+(t)$ and $U_+^\times(t)$ is defined, we have for all $m_0^2 \leq \mathsf{s} < \infty$ and then also for all real s [b]:

$$\begin{aligned}\langle \psi'^{-}|[j,\mathsf{s}], b^{-}_{\text{rest}}\rangle &= \langle U_{+}(t)\psi^{-}|[j,\mathsf{s}], b^{-}_{\text{rest}}\rangle \\ &= \langle \psi^{-}|U^{\times}_{+}(t)|[j,\mathsf{s}], b^{-}_{\text{rest}}\rangle \\ &= \langle e^{iHt}\psi^{-}|[j,\mathsf{s}], b^{-}_{\text{rest}}\rangle \\ &= e^{-i\sqrt{\mathsf{s}}\,t}\langle \psi^{-}|[j,\mathsf{s}], b^{-}_{\text{rest}}\rangle \text{ for all } \psi^{-} \in \mathbf{\Phi}_{+} \end{aligned} \tag{48}$$

Omitting the arbitrary $\psi^{-} \in \mathbf{\Phi}_{+}$ we write this as a functional equation:

$$e^{-iH^{\times}_{+}t}|[j,\mathsf{s}], b^{-}_{\text{rest}}\rangle = e^{-i\sqrt{\mathsf{s}}\,t}|[j,\mathsf{s}], b^{-}_{\text{rest}}\rangle\,, \quad \text{for } t \geq 0 \text{ only}, \tag{49}$$

where

$$e^{-iH^{\times}_{+}t} \equiv U^{\times}_{+}(t) = (e^{iHt})^{\times}\,, \quad t \geq 0\,. \tag{50}$$

The first term in (50) is so far only a definition, but one can show that $H^{\times}_{+} \supset H^{\dagger}$, the conjugate operator of $H|_{\mathbf{\Phi}_{+}} \equiv H_{+}$, and extension of the operator $H^{\dagger} = H$ to $\mathbf{\Phi}^{\times}_{+}$ [cf Appendix] is indeed the generator of the semigroup $U^{\times}_{+}(t)$ [21].

The solutions of the other Lippmann-Schwinger equation with $+i\epsilon$, $|[j,\mathsf{s}], b^{+}\rangle \in \mathbf{\Phi}^{\times}_{-}$, have a time evolution given by the other semigroup $U^{\times}_{-}(t)$ with $t \leq 0$,

$$e^{-iH^{\times}_{-}t}|[j,\mathsf{s}], b^{+}_{\text{rest}}\rangle = e^{-i\sqrt{\mathsf{s}}t}|[j,\mathsf{s}], b^{+}_{\text{rest}}\rangle \;\text{ for } t \leq 0 \text{ only.} \tag{51}$$

In here $U^{\times}_{-}(t) = e^{-iH^{\times}_{-}t}$ (and $H^{\times}_{-}$) are defined as the extension of the $U^{\dagger}(t)$ (and of $H^{\dagger}$) to the space $\mathbf{\Phi}^{\times}_{-}$.

We obtain now the action of $e^{-iH^{\times}t}$ on the Gamow ket at rest using (41) for the vector $e^{iHt}\psi^{-} \equiv \psi'^{-}(t)$, where t is the time in the rest frame of the resonance:

$$\begin{aligned}\langle \psi^{-}|e^{-iH^{\times}t}|[j,\mathsf{s}_R], b^{-}_{\text{rest}}\rangle &= \langle e^{iHt}\psi^{-}|[j,\mathsf{s}_R], b^{-}_{\text{rest}}\rangle \\ &= \frac{i}{2\pi}\int_{-\infty_{II}}^{+\infty} d\mathsf{s}\,\langle e^{iHt}\psi^{-}|[j,\mathsf{s}], b^{-}_{\text{rest}}\rangle \frac{1}{\mathsf{s}-\mathsf{s}_R} \\ &= \frac{i}{2\pi}\int_{-\infty_{II}}^{+\infty} d\mathsf{s}\,\langle \psi^{-}|e^{-iH^{\times}t}|[j,\mathsf{s}], b^{-}_{\text{rest}}\rangle\,\frac{1}{\mathsf{s}-\mathsf{s}_R} \\ &= \frac{i}{2\pi}\int_{-\infty_{II}}^{+\infty} d\mathsf{s}\,\langle \psi^{-}|[j,\mathsf{s}], b^{-}_{\text{rest}}\rangle e^{-i\sqrt{\mathsf{s}}t}\,\frac{1}{\mathsf{s}-\mathsf{s}_R} \\ &= e^{-i\sqrt{\mathsf{s}_R}t}\langle \psi^{-}|[j,\mathsf{s}_R], b^{-}_{\text{rest}}\rangle\,, \quad \text{for } t \geq 0\,. \end{aligned} \tag{52}$$

[b] Using the van Winter theorem for Hardy class functions Appendix A2 of [3].

The next to the last step makes use of (49) and the last step is again the Titchmarsh theorem as in (41) or (43) but this time for the function $\overline{\psi}'^{-}(\mathsf{s}) \equiv \overline{\psi}^{-}(\mathsf{s})e^{-i\sqrt{\mathsf{s}}t} \in \mathcal{S} \cap \mathcal{H}^2_-$, which is also a Hardy class function if $\overline{\psi}^{-}(\mathsf{s})$ is; however $\psi'^{-}(\mathsf{s})$ is Hardy class only if $t \geq 0$. For $t < 0$ the time evolution operator $e^{-iH^{\times}t}$ on $|[j,\mathsf{s}],b^{-}\rangle$ and on $|[j,\mathsf{s}_R],b^{-}\rangle$ does not exist.

We write (52) as a functional equation over the space $\mathbf{\Phi}_+$ omitting the arbitrary $\psi^- \in \mathbf{\Phi}_+$ and using (14),

$$|\psi^G_{j,\mathsf{s}_R}(t)^-\rangle \equiv e^{-iH^{\times}t}|[j,\mathsf{s}_R],b^-_{\text{rest}}\rangle = e^{-iM_Rt}e^{-\frac{\Gamma_Rt}{2}}|[j,\mathsf{s}_R],b^-_{\text{rest}}\rangle\,, \quad t \geq 0\,. \tag{53}$$

The result (53) is a mathematical consequence of the new hypothesis of (21). It has two physical consequences:

1. The time evolution of the Gamow vectors has a preferred direction of time, i.e., it is not reversible.
2. The time evolution of the relativistic Gamow vectors is exponential with the decay constant (inverse lifetime) Γ_R.

The time asymmetry, $t \geq 0$, has been discussed in detail elsewhere [22] and also in [23,24] and is not the main subject of our discussion here. Though the irreversible character of quantum decay has been mentioned in the past [25], it often has been attributed to other principles (effect of the environment, collapse of the wave function, decoherence), probably because of the misconception that time evolution in quantum mechanics must be given by the reversible unitary group as is dictated by the mathematics of the Hilbert space. The irreversibility on the microphysical level expressed by the time asymmetry (53) was the first unintended and the most surprising result of time asymmetric quantum theory based on the hypothesis (21) [26]. It is a consequence of the time asymmetric boundary conditions (21) and has nothing to do with violation of time reversal invariance (which is a statement about the T-transformation property of the Hamiltonian H) and it is not known to us whether it can be connected to entropy increase [27]. In retrospect this time asymmetry is no longer shocking, since Maxwell's theory and general relativity theory also use time asymmetric boundary conditions for time symmetric dynamical equations. Time asymmetric (purely outgoing) boundary conditions in place of the Hilbert space assumption have also been suggested in the past for quantum mechanics [28]. The germ of this time asymmetry is already inherent in the Lippmann-Schwinger equations [29] and in the Feynman rules, by the infinitesimal $i\epsilon$, though this has apparently not yet been related to a semigroup evolution (49) and (51). This semigroup evolution of the in-

and out-planewave states is a manifestation of a fundamental time asymmetry in quantum scattering – independently of whether resonance formulation is involved or not.

It can be shown that the time evolution semigroup (53) is the rest frame version of a causal semigroup $P_+ = (a_+, \Lambda)$ of Poincaré transformations where Λ is a proper orthochronous Lorentz transformation and $a_+ = (a_0, \mathbf{a})$ is a 4-vector that fulfills $b \cdot a = \sqrt{1+\mathbf{b}^2}\, a_0 - \mathbf{b} \cdot \mathbf{a} \geq 0$ for any $\mathbf{b} \in \mathbb{R}^3$. It means that the relativistic Gamow vectors can only undergo Poincaré transformations into the forward light cone [20].

The second consequence of (53), the exponential time evolution of the Gamow vectors, is not a surprise, because it was the purpose for which the Gamow vectors of the non-relativistic theory were originally constructed [26]. However that the exponential decay constant is Γ_R, and not any of the other Γ's of (46), is a new prediction of the *relativistic* Gamow vector. It follows from (53) in the following way:
The Born probability $P_\eta(t) = |\langle \psi_\eta^- | \psi_{js_R}^G(t) \rangle|^2$ is the probability to observe the decay products described by the observable $|\psi_\eta^-\rangle\langle\psi_\eta^-|$, $\psi_\eta^- \in \Phi_+$, in the Gamow state $\psi_{js_R}^G(t) \in \Phi_+^\times$ at a time t in the rest frame of the resonance state. The result (52) (53) states that this transition probability obeys the exponential law

$$|\langle \psi_\eta^- | \psi_{js_R}^G(t) \rangle|^2 = e^{-\Gamma_R t} |\langle \psi_\eta^- | \psi_{js_R}^G(0) \rangle|^2 \quad \text{for} \quad t > 0 \tag{54}$$

where $t = 0$ is the time at which the decaying resonance state $\psi_{js_R}^G$ was created (e.g. at a target in a scattering experiment). From (54) follows the exponential law for the partial decay rate

$$R_\eta(t) = \frac{d}{dt} P_\eta(t) = e^{-\Gamma_R t} R_\eta(0) \tag{55}$$

where Γ_R is the parameter defined in (14) and not any of the other possible width parameters. The decay rate is measured as the counting rate (1) of the decay products η, $\frac{\Delta N_\eta(t)}{\Delta t}$. Therefore we conclude from our results (52) (53) that the lifetime $\tau \equiv \frac{1}{R}$ (average lifetime in the rest frame) of the relativistic resonance is given by

$$\tau = \frac{1}{\Gamma_R} \tag{56}$$

where Γ_R is the width defined by the lineshape formula (8) of the relativistic resonance R with the parameterization (14).

The lifetime τ of a relativistic decaying state is measured by the counting rate (1) as a function of time t in the rest frame or (practically always) as a function of the distance $d = \hat{\mathbf{p}}t = \gamma \mathbf{v}t$ which the quasistable particle with space components of the 4-velocity $\hat{\mathbf{p}} = \gamma\mathbf{v} = (1-v^2)^{-1/2}\mathbf{v} = \gamma\beta c$ travels in the lab frame:

$$\frac{1}{\dot{N}}\frac{\Delta N_\eta(t)}{\Delta t} = e^{-t/\tau} = e^{-d\frac{1}{\hat{p}}\frac{1}{\tau}} = e^{-d\frac{1}{\gamma|\mathbf{v}|}\frac{1}{\tau}} = e^{-\frac{d}{\gamma\beta c}\Gamma_R}\,. \tag{57}$$

For the Z-boson (and for hadron resonances) $\hbar/\Gamma_R$ is too small to be measured in this way and one can only measure Γ_R from the lineshape (8). For other weakly decaying particles [30], like the K^0, one measures the lifetime τ using (57) but one cannot resolve the lineshape (7) because $\hbar/\tau$ is too small. Thus there may be no way to test the relation (56) in the near future, as has been done in atomic physics for the non-relativistic case of (3) in [6].

The width-lifetime relation (56) is a result for the relativistic Gamow vectors. Without Gamow vectors neither (54) nor (56) can be proven as an exact relation. The difference between the relativistic and non-relativistic case is that in the latter there was never any doubt regarding the meaning of the width, it is the full width at half-maximum of the non-relativistic Breit-Wigner (2). For the relativistic resonance one was not sure which part of the j-th partial wave amplitude (25) one should assign to the resonance per se, i.e., whether (7) or (8) or any other part of $a_j(\mathsf{s})$ describes the resonance [11,12,13]. Neither was one sure, that even when (8) was chosen (using the S-matrix pole definition), which of the parameterizations (13), (14), (16) should be used to define the "width" and the real mass of the resonance. The result (53) for the relativistic Gamow vector fixes this: the mass is $M_R = \mathrm{Re}(\sqrt{\mathsf{s}_R})$ (the coefficient of the phase for the time evolution (53)) and the width is $\Gamma_R = -2\mathrm{Im}(\sqrt{\mathsf{s}_R})$ (the coefficient of the exponential decay). Then (3) holds universally. This means that if for every resonance and every decaying state one wants (3) to hold independently of whether both are measurable, then the relativistic Gamow vectors (38), (40) predict that M_R is the mass and Γ_R is the "width" of the relativistic quasistable particle. For the Z-boson, this means that neither $(\bar{M}_Z, \bar{\Gamma}_Z)$ nor $(m_1 = M_Z, \Gamma_1 = \Gamma_Z)$, as quoted in the Particle Data Table [7], is the Z-boson mass and width, but (M_R, Γ_R) of (14).

4 Conclusion

The Gamow kets, spanning the irreducible representation space $[j, \mathsf{s}_R]$ of the Poincaré semigroup transformation into the forward light-cone, can be obtained from the pole position $\mathsf{s}_R = (M_R - i\Gamma_R/2)^2$ of the j-th partial S-matrix $S_j(\mathsf{s})$ (or scattering amplitude $a_j(\mathsf{s})$) and correspond to the Breit-Wigner part $a_j^{\mathrm{BW}}(\mathsf{s})$ of the scattering amplitude.

From the Poincaré semigroup transformation of the Gamow kets, or more specifically from the time evolution of the Gamow kets in the rest frame, one obtains the exponential decay law $e^{-\Gamma_R t}$ for the Gamow states $\psi^G_{js_R} = |s^-_R\rangle$. This means that the parameter of exponential time evolution, the inverse lifetime, is predicted to be $\hbar/\tau = \Gamma_R = -2\mathrm{Im}\sqrt{s_R}$. The Gamow vector unites the notion of a resonance described by the Breit-Wigner amplitude (2) and the notion of decaying particle described by the exponential time evolution (1) into one and the same entity, the quasistable (relativistic) particle, with $\Gamma_R = 1/\tau$. From this we obtain the definition of "mass" and "width" for a relativistic resonance as $M_R = \mathrm{Re}\sqrt{s_R}$ and $\Gamma_R = -2\mathrm{Im}(\sqrt{s_R})$. If one requires the exact and universal validity of the lifetime-width relation (3) then any other definition of "mass" and "width", e.g. those used for the Z-boson in the particle data table [7] are excluded.

In summary, a relativistic time asymmetric quantum theory that is consistent with the purpose of the the Lippmann-Schwinger equations leads to a definition of resonance mass M_R and resonance width Γ_R for which $1/\Gamma_R$ is the lifetime and for which the Z-boson mass is given by the value $M_R = (91.1626 \pm 0.0031)$ GeV, which is different from either of the two values quoted in the Particle Data Table [7] for the Z-boson mass.

Acknowledgments

This work was supported by the Welch Foundation, the Humboldt Foundation and CoNaCyT (Mexico). Part of this work was done while one of the authors A.B. was visiting the Institute for Quantum Physics Ulm University and we are grateful to Professor W. Schleich for the hospitality extended to him. We also acknowledge valuable discussions on the subject of this paper with W. Blum, W. Drechsler, N. Harshman, G. Höhler, T. Riemann, H. Saller and S. Wickramasekara.

Appendix

Rigged Hilbert Spaces

Let $\mathbf{\Phi}$ be a linear space with scalar product $(\cdot,\cdot)$ (this is the space which most physicists call Hilbert space, but which mathematicians call a pre-Hilbert space). Then the triplet of spaces that form a Rigged Hilbert Space [31] are obtained by the completion of $\mathbf{\Phi}$ with respect to three different topologies (meanings of convergence):

$$\phi \in \mathbf{\Phi} \subset \mathcal{H} \subset \mathbf{\Phi}^\times \ni |F\rangle.$$

The space $\mathcal{H}$ is the mathematician's Hilbert space and it is complete with respect to the norm convergence $\tau_{\mathcal{H}}$. The space $\mathbf{\Phi}$ is complete with respect to a stronger convergence $\tau_{\mathbf{\Phi}}$ (test function space). The space $\mathbf{\Phi}^\times$ is the space of $\tau_{\mathbf{\Phi}}$-continuous, antilinear functionals on the space $\mathbf{\Phi}$: $|F\rangle \in \mathbf{\Phi}^\times$ maps $\phi \in \mathbf{\Phi}$ into $\mathbb{C}$, i.e. $F(\phi) = \langle\phi|F\rangle \in \mathbb{C}$. The space $\mathcal{H}^\times$ of $\tau_{\mathcal{H}}$-continuous functionals $f : h \in \mathcal{H} \to f(h) \in \mathbb{C}$ can be identified with $\mathcal{H}$ by $f(h) = (h, f)$ and then $\langle\cdot|\cdot\rangle$ is the extension of $(\cdot,\cdot)$ to those F which are not elements of $\mathcal{H}$.

For a $\tau_{\mathbf{\Phi}}$-*continuous* operator A there is also a triplet:

$$A^\dagger|_{\mathbf{\Phi}} \subset A^\dagger \subset A^\times.$$

Here $A^\dagger$ is the Hilbert space adjoint operator and $A^\times$ is the conjugate operator. $A^\times$ is defined by:

$$\langle A\phi|F\rangle = \langle\phi|A^\times|F\rangle, \ \forall\phi \in \mathbf{\Phi} \text{ and } \forall F \in \mathbf{\Phi}^\times.$$

A vector $|F\rangle \in \mathbf{\Phi}^\times$ is a generalized eigenvector of the $\tau_{\mathbf{\Phi}}$-continuous operator A if for some complex number ω

$$\langle A\phi|F\rangle = \langle\phi|A^\times|F\rangle = \omega\langle\phi|F\rangle, \ \forall\phi \in \mathbf{\Phi}.$$

This is also written as $A^\times|F\rangle = \omega|F\rangle$, or $A|F\rangle = \omega|F\rangle$ if $A^\dagger$ is self-adjoint. An example is the Dirac ket

$$H^\times|E\rangle = E|E\rangle, \quad E \geq 0.$$

In the Rigged Hilbert Space one can prove the Nuclear Spectral Theorem (referring to the nuclearity of the topological spaces $\mathbf{\Phi}$) which is the most important property for quantum theory and another Golden Rule of Dirac: For every self-adjoint operator $H^\dagger$ there exists a complete set of generalized eigenvectors $|E\rangle \in \mathbf{\Phi}^\times$ such that every vector $\phi \in \mathbf{\Phi}$ can be written as

$$\phi = \int \rho(E)dE\,|E\rangle\langle E|\phi\rangle$$

where the (wave) functions $\langle E|\phi\rangle \equiv \phi(E)$ are elements of a space of well-behaved functions (e.g. the Schwartz space of infinitely differentiable rapidly decreasing functions). $\rho(E)$ is an integrable function and determines the normalization of the eigenvectors

$$\langle E'|E\rangle = \rho^{-1}(E)\,\delta(E' - E).$$

The integration can – in all known cases important for physics – be chosen to be Riemann integration, and the set of wave functions $\phi(E)$ define the space of vectors $\phi \in \mathbf{\Phi}$ and vice-versa. This theorem generalizes the basis vector expansion in 3- or N-dimensional spaces

$$\vec{x} = \sum_{i=1}^{N} \vec{e}_i(\vec{e}_i \cdot \vec{x}) \quad , \quad \vec{e}_i \cdot \vec{x} = x^i$$

to continuously infinite dimensions and is therefore also called Dirac's basis vector expansion. Note that it represents $\phi \in \mathbf{\Phi}$ as a continuous linear superposition of basis vectors $|E\rangle \in \mathbf{\Phi}^\times$ which are not elements of $\mathbf{\Phi}$. This may appear counter intuitive but has its justification in the choice of a set of very special functions $\phi(E)$ as possible "coordinates" for a vector $\phi \in \mathbf{\Phi}$.

Depending on the choice of the space $\mathbf{\Phi}$, one can also have generalized eigenvectors with complex eigenvalues for an operator H that is essentially self-adjoint (and bounded from below, which we always assume). The Gamow vectors are an example:

$$H^\times|E_R - i\Gamma_R/2^-\rangle = (E_R - i\Gamma_R/2)|E_R - i\Gamma_R/2^-\rangle.$$

References

1. V. Weisskopf, E. P. Wigner, *Z. f. Physik* **63**, 54 (1930); **65**, 18 (1930). See also M. L. Goldberger, K. M. Watson, *Collision Theory*, Wiley, New York (1964), chap. 8.
2. In the conventional Hilbert space theory one proves the mathematical theorem that the survival probability and thus the partial decay rate $\dot{P}_\eta(t)$ (which is measured as the counting rate $\dot{N}_\eta(t)$) must have some deviations from the exponential law. Cf. L. A. Khalfin, *JETP Lett.* **15**, 388 (1972). However, in practical calculations one uses many concepts, like the solutions of the Lippmann-Schwinger equations, that lie outside the Hilbert Space.
3. A. Bohm, S. Maxson, M. Loewe, M. Gadella, *Physica A* **236**, 485 (1997), and references therein.
4. A. Bohm, N. L. Harshman, in *Irreversibility and Causality*, A. Bohm, H. D. Doebner, P. Kielanowski, Eds., Springer, Berlin (1998), p. 181 and references therein.
5. A. Bohm, *Quantum Mechanics—Foundations and Applications*, 3rd edition, Springer, New York (1994), Chapter XXI.
6. U. Volz *et al.*, *Phys. Rev. Lett.* **76**, 2262 (1996); C. W. Oates *et. al.*, *Phys. Rev. Lett.* **76**, 2866 (1996).
7. C. Caso *et al.*, em The European Physical Journal C **15**, 1-878 (2000).

8. T. Riemann, in *Irreversibility and Causality*, A. Bohm, H.-D. Doebner, P. Kielanowski, Eds., Springer, Berlin (1998), p. 157, and references thereof.
9. Martin W. Grünwald, *Phys. Rep.* **322**, 125 (1999).
10. W. Hollik and G. Duckeck, *Electroweak Precision Tests at LEP1*, Springer Tracts in Modern Physics 162, Springer, New York (2000).
11. A. Sirlin, *Phys. Rev. Lett.* **67**, 2127 (1991); *Phys. Lett. B* **267**, 240 (1991).
12. R. G. Stuart, *Phys. Lett. B* **262**, 113 (1991); **272**, 353 (1991).
13. S. Willenbrock, G. Valencia, *Phys. Lett. B* **259**, 373 (1991).
14. R. J. Eden, P. V. Landshoff, P. J. Olive and J. C. Polikinghorne, *The Analytic S-Matrix*, Cambridge University Press, Cambridge (1996).
15. E. P. Wigner, *Ann. Math. (2)* **40**, 149 (1939).
16. S. Weinberg, *The Quantum Theory of Fields*, Cambridge Univ. Press, Cambridge (1995), vol. I, chap. 3.
17. P. L. Duren, $\mathcal{H}^p$ *spaces*, Academic Press, New York (1970). For the relativistic case these Hardy class spaces are not precisely the $\mathcal{S} \cap \mathcal{H}^2_\pm$ but they are of the same class; for details see [21]
18. That it is important not to use asymptotically free states for resonances has also been emphasized by A. Sirlin, *Proceedings of the Fourth International Symposium on Radiative Corrections (RADCOR 98)*, Barcelona, Spain, 8-12 September 1998, J. Sola Ed.; M. Passera, A. Sirlin, *Phys. Rev. D* **58**, 113010 (1998).
19. A. Bohm, H. Kaldass, S. Wickramasekara, P. Kielanowski, *Phys. Lett. A* **264**, 425 (2000).
20. A. Bohm, S. Wickramasekara and H. Kaldass, Relativistic Gamow Vectors I: Derivation from Poles of the S-Matrix, preprint, The University of Texas at Austin (1999); S. Wickramasekara and A. Bohm, Relativistic Gamow Vectors II: The Rigged Hilbert Space, preprint, The University of Texas at Austin; A. Bohm, S. Wickramasekara and H. Kaldass, Relativistic Gamow Vectors II: Transformation Under the Causal Poincaré Semigroup, preprint, The University of Texas at Austin.
21. S. Wickramasekara, *Differential Representations of Finite Dimensional Lie Groups in Rigged Hilbert Spaces*, Dissertation, The University of Texas at Austin (1999).
22. A. Bohm, *Phys. Rev. A* **60**, 861 (1999); A.Bohm, I. Antoniou, P. Kielanowski, *Phys. Lett. A* **189**, 442 (1994).
23. M. Gell-Mann, J. B. Hartle, in *Physical Origins of Time Asymmetry*, J. J. Halliwell, *et al.*, Eds., Cambridge University Press (1994), p. 311.
24. R. Haag, *Comm. Math. Phys.* **132**, 245 (1990), and Lectures at the Max Born Symposium on "Quantum Future", Przieka (1997).

25. N. van Kampen, *Proceedings of the XXI Solvay Conference*, K. Kitahara *et al.*, Eds., *Adv. Chem. Phys.*, (to appear). C. Cohen-Tannoudji, *et al.*, *Quantum Mechanics*, Vol. II, p. 1345, 1353-54, Wiley, New York (1977); T. D. Lee, *Particle Physics and Introduction to Field Theory*, Chap. 13, Harwood Academic, New York (1981).
26. A. Bohm, *J. Math. Phys.* **22**, 2813 (1981); *Lett. Math. Phys.* **3**, 455 (1978).
27. For a contrary statement see M. Gell-Mann, J. B. Hartle; *Proceedings 4th Drexel Symposium on Quantum Non-Integrability*, preprint UCSB TH-95-28, gr-qc/9509054.
28. A. F. J. Siegert, *Phys. Rev.* **56**, 750 (1939); R. Peierls, *Proceedings of the 1954 Glasgow Conference on Nuclear and Meson Physics*, E. M. Belamy *et al.*, Eds., Pergamon Press, New York (1955); A. Mondragón, E. Hernández, J. M. Velázquez Arcos, *Ann. Phys.* **7**, 48, 503 (1991).
29. B. A. Lippmann and J. Schwinger, *Phys. Rev.* **79**, 469 (1950); M. Gell-Mann and H. L. Goldberger, *Phys. Rev.* **91**, 398 (1953); W. Brenig and R. Haag, *Fortschr. Physik* **7**, 183 (1959).
30. For the example of $K^0_{S,L} \to \pi^+\pi^-$ see: K.Kleinknecht, in *CP violation*, C. Jarlskog, Ed., World Scientific (1989), p. 41 and references therein; K. L. Gibbons, et al., *Phys. Rev. D* **55**, 6625 (1997). Or concerning the exponential law in general: V. L. Fitch et al., *Phys. Rev. B* **140**, 1088 (1965); N. N. Nikolaev, *Sov. Phys. Usp.* **11**, 522 (1968), and references therein; E. B. Norman, *Phys. Rev. Lett.* **60**, 2246 (1988).
31. A.Bohm and M. Gadella, *Dirac Kets, Gamow Vectors and Gel'fand Triplets*, Lecture Notes in Physics, Volume 348, Springer-Verlag, Berlin (1989).

QUANTUM FIELD THEORY AS DYNAMICAL SYSTEM

H.J. BORCHERS

Institut für Theoretische Physik Universität Göttingen Bunsenstrasse 9, D 37073 Göttingen

A review on the representations of symmetries in quantum field theory of local observables will be given. One part is concerned with the interplay of the locality condition in configuration space (Einstein causality) and the spectrum condition in momentum space (positive energy). Another part uses Tomita's theory of modular Hilbert algebras (Tomita-Takesaki theory) to induce symmetries for the quantum field theory.

1 Introduction

There exists an extended literature on the so called dynamical systems. Many of these articles treat C^*-dynamical systems. They consist of triples

$$(\mathcal{A}, G, \alpha),$$

where $\mathcal{A}$ is a C^*-algebra, G a topological group (discrete topology not excluded) and α a group homomorphism of G into the automorphisms of $\mathcal{A}$,

$$\alpha: \quad G \quad \longrightarrow \quad \mathrm{A}ut(\mathcal{A}). \tag{1}$$

A covariant representation (π, U) of $(\mathcal{A}, G, \alpha)$ consists of a non-degenerate representation π of $\mathcal{A}$ on a Hilbert space $\mathcal{H}$ and a continuous unitary representation $U(g)$ implementing the automorphisms α_g on the same Hilbert space,

$$U(g)\pi(A)U^*(g) = \pi(\alpha_g A); \qquad A \in \mathcal{A}, \quad g \in G. \tag{2}$$

If one is dealing with a covariant representation, then the automorphisms α_g extend to the von Neumann algebra $\pi(\mathcal{A})$", and one becomes a W^*-dynamical system. This is a triple

$$(\mathcal{M}, G, \alpha)$$

where one has replaced the C^*-algebra $\mathcal{A}$ by a von Neumann algebra $\mathcal{M}$.

Dealing with a C^*- or W^*-dynamical system there are three families of questions:

- Find conditions implying that the symmetry group can be implemented.
- Derive a detailed structure of the group representation.

- Does the detailed sub-structure of the algebra with some requirement on its representation imply the existence of additional symmetry groups?

2 On cinematical problems

In the special situation that:

1) The group G is locally compact.

2) The group of automorphisms, α_g, $g \in G$, act strongly continuous, i.e. for every $A \in \mathcal{A}$ the function

$$g \longrightarrow \alpha_g(A)$$

is continuous in the norm topology of $\mathcal{A}$.

Then one can form the crossed product of the group with the algebra

$$G \underset{\alpha}{\times} \mathcal{A}.$$

Sometimes this crossed product is called the covariance algebra. Doplicher, Kastler and Robinson [1] have shown the following result:

Theorem 2.1 *The covariant representations of the system $(\mathcal{A}, G, \alpha)$ are in one-to-one correspondence with the representations of $G \underset{\alpha}{\times} \mathcal{A}$.*

For an extensive discussion of crossed products see G.K. Pedersen [2]. This theorem does not answer all questions, e.g.:

(i) The requirement that the group acts strongly continuous can not be fulfilled in many cases. Such situation appears if the algebra contains projections.

(ii) Not every group appearing in physics is locally compact. E.g. gauge groups in a field theory containing zero mass particles, diffeomorphisms in general relativity.

(iii) When can a given representation of the C^*-algebra of a dynamical system be extended to a covariant representation?

If one wants to extend a given representation π on $\mathcal{H}$ to a covariant representation then there are two obstructions:

(α) The multiplicity problem.

An example shall demonstrate the situation. One can represent the algebra $\mathcal{L}^\infty(\mathbb{R})$ on $\mathcal{L}^2(\mathbb{R}^+) \otimes \mathbb{C}^2 \oplus \mathcal{L}^2(\mathbb{R}^-)$, but the change of the multiplicity is an obstruction for the representation of the translations.

(β) It can happen that the automorphisms are represented by a projective representation. Quantum mechanics is such a case.

Both obstructions can be cured by passing from a given representation to a quasi-equivalent one.

Definition 2.2 (1) *Two representations π_1, π_2 of $\mathcal{A}$ are called quasi-equivalent if the von Neumann algebras $\pi_1(\mathcal{A})$" and $\pi_2(\mathcal{A})$" are isomorphic. This is equivalent to the requirement that $\pi_1(\mathcal{A})$ and $\pi_2(\mathcal{A})$ have the same set of normal states. This family is called the* folium $F(\pi)$ *of π.*
(2) *Let π be a representation of the algebra of the triple $(\mathcal{A}, G, \alpha)$, then π is called quasi-covariant if there exists a covariant representation (π_1, U) such that π and π_1 are quasi-equivalent.*

With this notation one obtains:

Theorem 2.3 *Let $(\mathcal{A}, G, \alpha)$ be a C^*-dynamical system and π be a representation of $\mathcal{A}$. Then π is quasi-covariant iff*
1) *The folium $F(\pi)$ is invariant under the transposed action α'_g.*
2) *For every $\omega \in F(\pi)$ the function*

$$g \quad \longrightarrow \quad \alpha'_g \omega$$

is continuous in the norm-topology of $\mathcal{A}^*$. ($\mathcal{A}^*$ is the topological dual of $\mathcal{A}$).

In the case of a W^*-dynamical system $(\mathcal{M}, G, \alpha)$ one has to replace the dual space by the pre-dual space $\mathcal{M}_*$. This result was shown [3] for locally compact groups. For arbitrary groups it is a consequence of the Tomita-Takesaki-theory [4].

For obtaining a continuous unitary representation it was necessary and sufficient that the group acts strongly continuous on the folium of the representation. Therefore, the following space is of interest:

Definition 2.4 *By $\mathcal{A}^*_c$ we denote the set of $\phi \in \mathcal{A}^*$, such that for every $\epsilon > 0$ exists a neighbourhood $\mathcal{U}$ of the identity of G such that*

$$||\phi \circ \alpha_g - \phi|| \leq \epsilon$$

holds for $g \in \mathcal{U}$.

Some properties of this set are described in the following

Proposition 2.5 *Let $(\mathcal{A}, G, \alpha)$ be a C^*-dynamical system and assume $G(\tau)$ is a topological group, then the space $\mathcal{A}^*_c$ has the following properties:*

(i) *$\mathcal{A}_c^*$ is a linear norm–closed space.*

(ii) *$\mathcal{A}_c^*$ is invariant under the action of the group i.e. $\phi \in \mathcal{A}_c^*$ implies $\phi \circ \alpha_g \in \mathcal{A}_c^*$ for every $g \in G$.*

(iii) *With $\phi \in \mathcal{A}_c^*$ one finds also that ϕ^* and $|\phi|$ belong to $\mathcal{A}_c^*$. $\mathcal{A}_c^*$ is generated by its positive elements.*

Using Connes' characterization of von Neumann algebras on $\mathcal{H}$ by self-dual, oriented cones in a Hilbert space [5] one can show [6]

Theorem 2.6 *Let $(\mathcal{A}, G, \alpha)$ be a C^*–dynamical system, with G a topological group. Then $\mathcal{A}_c^*$ is the predual of a von Neumann algebra $\mathcal{M}_c \subset \mathcal{A}^{**}$. This von Neumann algebra coincides with $\mathcal{M}_m^{(red)}$ where*

$$\mathcal{M}_m = \{A \in \mathcal{A}^{**}; A\omega \in \mathcal{A}_c^*,\ \omega A \in \mathcal{A}_c^*,\ \forall \omega \in \mathcal{A}_c^*\}.$$

The suffix "red" means division by the two-sided annihilator of $\mathcal{A}_c^*$.

Of great interest in physics is the case where G is the translation group of $\mathbb{R}^d$ with the requirement that the spectrum of the translations $T(a)$ is contained in a cone $C \in \mathbb{R}^d$. This is the positive energy condition. The first result in this direction is due to Kadison [7] and Sakai [8]. They showed: If a one-parametric symmetry group α_t acts norm-continuous on $\pi(\mathcal{A})$ then there exists a bounded self-adjoint operator $H \in \pi(\mathcal{A})$" with

$$\mathrm{e}^{\mathrm{i}Ht}\pi(A)\mathrm{e}^{-\mathrm{i}Ht} = \pi(\alpha_t A). \tag{3}$$

This result has been generalized [9] to one-parametric groups with semi-bounded spectrum. It took a long time to extend this to the higher dimensional situation with an arbitrary convex cone [10].

Theorem 2.7 *Let $(\mathcal{A}, \mathbb{R}^d, \alpha)$ be a C^*-dynamical system and C be a proper, convex, closed cone in $\mathbb{R}^d$ with interior points. Let π be a representation of $\mathcal{A}$ and assume on $\mathcal{H}_\pi$ exists a continuous unitary representation $V(a)$ of the translation group satisfying:*

(i) The spectrum of $V(a)$ is contained in the cone C.

(ii) $V(a)$ implements the automorphisms α_a.

Then there exists a continuous unitary representation $T(a)$ satisfying

(α) $T(a) \in \pi(\mathcal{A})$".

(β) $T(a)$ fulfils the conditions (i) and (ii).

$V(a)$ and $T(a)$ can be analytically extended into the tube $\tau(\hat{C}) = \{z;\ \Im m\, z \in \hat{C}\}$ where $\hat{C}$ is the (open) dual cone of C. This has several consequences:

If one adds to the definition of $\mathcal{A}_c^*$ also the requirement of the spectrum condition, then exists a projection $P \in \mathcal{Z}(\mathcal{A}^{**})$ such that

$$\mathcal{M}_m^{red} = P\mathcal{A}^{**}$$

holds. Moreover, this set $\mathcal{A}_c^*$ can be characterized by analyticity properties of a dense set of elements in $\mathcal{A}_c^*$. For details see [10].

3 Quantum field theory and dynamical consequences

Relativistic quantum field theory exists in two settings: Wightman field theory (see e.g. Streater and Wightman [11]) which uses unbounded operators, and the theory of local observables of Araki, Haag and Kastler (see Haag [12]), where bounded operators are used. It is widely believed that both theories coincide in their consequences for physics. Here the field theory of local rings will be used. Quantum field theory of local observables is concerned with C^*-algebras $\mathcal{A}(O)$, associated with bounded open regions $O \subset \mathbb{R}^d$, which have a common unit. These algebras shall fulfil isotony, i.e. $O_1 \subset O_2$ implies $\mathcal{A}(O_1) \subset \mathcal{A}(O_2)$, and locality, i.e. if O_1 and O_2 are space-like separated, then the algebras $\mathcal{A}(O_1)$ and $\mathcal{A}(O_2)$ commute element-wise. In order that this statement makes sense the space $\mathbb{R}^d$ must be furnished with the Minkowski metric. The global algebra $\mathcal{A}$ is defined as the C^*-inductive limit of $\{\mathcal{A}(O)\}$. We also assume that the net $\{\mathcal{A}(O)\}$ is covariant under a representation of the translation group of $\mathbb{R}^d$ by automorphisms, i.e. to $a \in \mathbb{R}^d$ exists an automorphism α_a with

$$\alpha_a \mathcal{A}(O) = \mathcal{A}(O + a).$$

A covariant representation π of $\mathcal{A}$ in the representation space $\mathcal{H}$ is such that there exists a continuous unitary representation $T(a)$ of the translation group of $\mathbb{R}^d$ implementing α_a

$$T(a)\pi(A)T(-a) = \pi(\alpha_a(A)), \qquad A \in \mathcal{A}.$$

In addition $T(a)$ shall fulfil the spectrum condition, i.e. spectrum $T(\mathbb{R}^d)$ is contained in the closed forward light-cone. Sometimes (not here) also the covariance under the Poincaré group $\mathcal{P}_+^\uparrow$ is required. From Thm. 2.7 it follows, that $T(a)$ can be chosen in $\pi(\mathcal{A})$". The multiplication of $T(a)$ by a phase-factor $e^{i(a,p_0)}$ corresponds to a shift of the spectrum of $T(a)$ by the value p_0. (p_0 might be $\pi(\mathcal{A})$"-center valued.) If this vector belongs to the forward light-cone than $e^{i(a,p_0)}T(a)$ fulfils again the spectrum condition. Now one can ask for the converse problem: Does there exist a representation such that the spectrum of $T(a)$ is minimal? The existence of a minimal representation in one dimension is obvious, by pushing the spectrum as far down as possible without violating the spectrum condition. This has been used by D. Olesen and G.K. Pedersen [13].

If the spectrum is in a cone, then no general result is possible since there is more then one direction in which one can push to the tip of the cone. One has to use the detailed structure of the local algebras. Let $T(a), T_1(a)$ both be representations of the translations implementing α_a belonging to $\pi(\mathcal{A})''$. In this situation exists an element P_0 affiliated with the center of $\pi(\mathcal{A})''$ with

$$T_1(a) = \mathrm{e}^{\mathrm{i}(a,P_0)}T(a).$$

Assume the spectrum of $T(a)$ is contained in C, then $T(a)$ is called the *unique minimal representation* if for every $T_1(a)$ which also has its spectrum in C it follows that the spectrum of $\mathrm{e}^{\mathrm{i}(a,P_0)}$ is contained in C. In the case of quantum field theory of local observables the existence of a unique minimal representation can be shown. An indirect argument is given in [14] and a direct proof can be found in [15].

Theorem 3.1 *Let $(\{\mathcal{A}(O)\}, \mathcal{A}, \mathbb{R}^d, \alpha)$ be a theory of local observables and $(\pi, T(a))$ be a covariant representation with $T(a)$ fulfilling the spectrum condition and $T(a) \in \pi(\mathcal{A})''$. Then exists a unique minimal representation.*

If one is working with a vacuum representation, and such representations which are generated from the vacuum by charged fields, then the translations one gets are automatically the minimal one.

In general theories one cannot expect that the representations obtained from a vacuum sector are all representations fulfilling the spectrum condition. Knowing the existence of a unique minimal representation has the consequence, that in arbitrary covariant representations with spectrum condition one can define energy and momentum unambiguously. Therefore, every result about the spectrum of the translations (the minimal representation) has a universal meaning. The following result can be found in [14] and [16]:

Theorem 3.2 *Let $(\{\mathcal{A}(O)\}, \mathcal{A}, \mathbb{R}^d, \alpha)$ be a theory of local observables and $(\pi, T(a))$ be a covariant representation such that $T(a)$ fulfils the spectrum condition and $T(a) \in \pi(\mathcal{A})''$. Moreover, let $T(a)$ be the minimal representation. Then the support of the spectrum is invariant under Lorentz transformations.*

It is remarkable that Lorentz covariance of the theory has not been used. Necessary for this result is the locality property in configuration space, which often is called Einstein causality, together with the spectrum condition. This makes it understandable why also for one electron the formula $p^2 = m^2$ is correct although the Lorentz covariance is broken in the presence of the electromagnetic field. See Fröhlich, Morchio and Strocchi [17], or Buchholz [18].

Up to now we have looked at general properties of the spectrum of the translations. There are several results about the details of the structure of the spectrum, the first result is due to Wightman [19].

Theorem 3.3 *Let $(\pi_0, T(a))$ be a vacuum representation of a quantum field theory of local observables $(\{\mathcal{A}(O)\}, \mathcal{A}, \mathbb{R}^d, \alpha)$, this means $T(a)$ does not only fulfil the spectrum condition, but $\pi_0(\mathcal{A})$ is generated by a translational invariant vector Ω. Assume that, up to a scale-factor, Ω is the only translational invariant vector. Then the support of the spectrum of $T(a)$ is additive. If S denotes the support of the spectrum then there holds*

$$S + S \subset S. \tag{4}$$

This result is based on the cluster property, which is a consequence of the locality and the uniqueness of the vacuum vector. It also works if the operators anti-commute. Therefore, Eq. (4) can be generalized to the presence of charged fields. In field theory factor representations are characterized by a charge-quantum-number Q. If the representations are generated from the vacuum by charged fields, and if the support of the spectrum of the translations in that representation space is S_Q, then Eq. (4) can be generalized to

$$S_{Q_1} + S_{Q_2} \subset S_{Q_1+Q_2}, \tag{5}$$

which holds for additive charges. See [20] and [21], [22].

For factor representations which are not generated from the vacuum a result like Eq. (5) does not exist. The following relation has been conjectured in [23]

$$S + S + S \subset S. \tag{6}$$

Unfortunately this has not been proved. But there exist two results showing into the direction of the conjecture. If the spectrum starts at m_1 and if (3.3) is correct then the continuous spectrum has to start at least at $3m_1$. This is the content of the first result [23].

Theorem 3.4 *Let π be a translation covariant factor representation of the theory of local observables fulfilling the spectrum condition, and let $T(a)$ be the unique minimal representation of the translations. Assume the spectrum of $T(a)$ consists of two parts:*

(a) *Isolated hyperbolas with the masses $m_1 < m_2 ... < m_i < ...$*

(*b*) *The rest starting at* $m_c > m_i$ *where* m_c *denotes the beginning of the continuous spectrum or the first accumulation point of the* m_i*'s.*

Then

$$3m_1 > m_c.$$

One would like to show that above $3m_1$ there is no gap in the spectrum. Unfortunately this has not been done. Since one knows from examples that there can be a gap between m_1 and $3m_1$ one needs a special technique for showing that result. Such method of proof must be able to distinguish between the regions below $3m_1$ and above $3m_1$. Up to now such technique does not exist. Therefore, only the following result has been shown [24].

Theorem 3.5 *Let* π *be a factor representation of the theory of local observables covariant under translation fulfilling the spectrum condition, and let* $T(a)$ *be the unique minimal representation of the translations. Let the spectrum start at* m_1*, then there is no gap with width larger than* $2m_1$*.*

From this theorem the following result holds:

Corollary 3.6 *With the assumption of the last theorem assume that the spectrum starts at* $m_1 = 0$*. Then there is no gap in the spectrum.*

4 Symmetries as consequences of the dynamics

In Thm. 3.2 it was shown that the spectrum of the minimal representation of the translations is located on a Lorentz invariant set. This is a consequence of the dynamical requirements. Therefore, one might hope to derive the symmetry itself, if one imposes slight improved dynamical requirements. This is indeed the case if the representations are in the vacuum sector. All these results are based on the Tomita-Takesaki theory, which we assume to be known.

These results of Tomita and Takesaki apply in particular to the theory of local observables in a vacuum representation. Together with some weak additivity property the Reeh-Schlieder theorem [25] implies that the vacuum vector is cyclic and separating for every algebra $\mathcal{M}(G)$, where G is any domain which has a space-like complement with interior points.

A domain of special importance is the *wedge*. Such a domain can be characterized as follows:
Let t, s be two perpendicular vectors in $\mathbb{R}^d$. i.e. $(t, s) = 0$, such that $t^2 = 1$ and t belongs to the forward light cone and $s^2 = -1$ is space-like. In this situation one defines

$$W(t,s) := \{a \in \mathbb{R}^d; |(a,t)| < -(a,s)\}. \tag{7}$$

If, for instance, t is the time direction and s is the 1-direction then this becomes $W_R = \{a; |a_0| < a_1\}$.
The two-plane spanned by $\{t, s\}$ is called the characteristic two-plane of the wedge.

If π is a positive energy representation and $\psi \in \mathcal{H}_\pi$ a vector analytic for the energy, then this is cyclic and separating for every $\pi(\mathcal{A}(W))$ (without weak additivity) because every bounded open region can be shifted into W.

The first explicit determination of a modular group is due to Bisognano and Wichmann [26], [27]. They assumed that the local algebras are generated by Lorentz covariant Wightman fields over the four-dimensional Minkowski space, and showed that in this situation the modular group of any wedge coincides with the corresponing group of Lorentz boosts.

Passing from Wightman fields to local rings there appears a problem of the size of the commutant. In order to avoid these problems we will work with *maximal* local von Neumann algebras. They are defined as $\mathcal{M}(D) = \bigcap_{D\subset W} \mathcal{M}(W)$. In the future the maximal algebras will be used without further mentioning. Another concept, which often will be used, is the *wedge duality*. One says a representation π of the theory of local observables fulfils wedge duality if for every wedge the commutant of $\mathcal{M}(W)$ coincides with $\mathcal{M}(W')$, where W' is the (open) space-like complement of W. If $\{0\}$ is contained in the edge of W then $W' = -W$ holds.

The result of Bisognano and Wichmann suggests the following notion:

Definition 4.1 *A representation of a quantum field theory of local observables fulfils the Bisognano-Wichmann property, if:*
(i) There exists a vector $\Omega \in \mathcal{H}$ which is cyclic and separating for every wedge algebra $\mathcal{M}(W)$.
(ii) The modular group of any wedge acts local like the associated group of Lorentz boosts. This means the modular group acts on the underlying Minkowski space like the Lorentz boosts.

Since in a general quantum field theory of local observables exists no condition implying that the theory is generated from a finite number of Wightman fields, it took several years to find corresponding conditions. The first result can be found in [29]:

Theorem 4.2 *Assume we are dealing with a vacuum representation of a quantum field theory of local observables over the* two-dimensional *Minkowski space. This theory shall only be covariant under translations. In addition assume that wedge duality holds, i.e.*

$$\pi(\mathcal{A}(W))' = \pi(\mathcal{A}(W'))''.$$

Then the theory is covariant under the Poincaré group $\mathcal{P}_+$.

In this situation the modular conjugation acts as total reflection. This contains the time-reversal, and hence one obtains two components of the Poincaré group.

In the last result only translational covariance, spectrum condition and wedge duality has been used. Since double cones are defined by intersections of wedges, also covariance of the maximal double cone algebras is implied. A generalization of Thm. 4.2 to higher dimensions is still missing. In order to obtain Poincaré covariance one has to assume that the modular groups of all wedges act local. The following result has been shown by Brunetti, Guido and Longo [30] and by Guido and Longo [31]. In the first quoted paper the constructed group could have been the covering of the Poincaré group $\mathcal{P}_+^\uparrow$. In the second paper it is shown, that one deals with a true representation. In [32] a direct proof of the following result was given.

Theorem 4.3 *Assume the Bisognano-Wichmann property, i.e. the modular group of every wedge algebra $\mathcal{M}(W[t,s,a])$ act on every double cone algebra like the associated group of Lorentz boosts of the wedge $W[t,s,a]$. Then the modular groups $\{\Delta^{it}[t,s]\}$ define a representation of the Poincaré group.*

$W[t,s,a]$ denotes the shifted wedge $W[t,s]+a$. It should be remarked that the conditions of Thm. 4.3 imply that wedge duality holds. An extension of Thm. 4.2 to higher dimensions has not been achieved. There is a result due to Gaier and Yngvason [33] showing that such generalization might be possible. They used generalized free fields.

A different approach to the construction of the Poincaré symmetry is due to Buchholz, Summers and co-workers [34], [35]. One knows that the Poincaré group $\mathcal{P}_+$ is generated by reflections at planes of co-dimension 2. In the situation discussed before such reflections are due to the modular conjugations of wedge algebras. For the program of Buchholz, Summers and co-workers we refer to the literature.

One important symmetry for physics is the PCT-symmetry. The demonstration of this symmetry has a long history which shall not be explained here. For a review on this subject see Yngvason and the author [36].

Acknowledgments

Parts of this review has been prepared at the Erwin Schr"odinger Institute. I am grateful for their kind hospitality.

References

1. S. Doplicher, D. Kastler and D.W. Robinson, *Commun. Math. Phys.* **3**, 1 (1966).
2. G.K. Pedersen, *C^*-Algebras and their Automorphism Groups*, Academic Press, London, New York, San Francisco (1979).
3. H.J. Borchers,*Commun. Math. Phys.* **14**, 305 (1969).
4. M. Takesaki, *Tomita's Theory of Modular Hilbert Algebras and its Applications*, Lecture Notes in Mathematics, Vol. **128** Springer-Verlag Berlin, Heidelberg, New York (1970).
5. A. Connes, *Ann. Inst. Fourier* **26**, 121 (1974).
6. H.-J. Borchers,*Commun. Math. Phys.* **88**, 95 (1983).
7. R.V. Kadison, *Ann. of Math.* **83**, 280 (1966).
8. S. Sakai, *Ann. of Math.* **83**, 273 (1966).
9. H.J. Borchers, *Commun. Math. Phys.* **2**, 49-54 (1966).
10. H.-J. Borchers, *Translation Group and Particle Representations in Quantum Field Theory*, Lecture Notes in Physics **m40** Springer, Heidelberg (1996).
11. R.F. Streater and A.S. Wightman, *PCT, Spin and Statistics, and All That*, Benjamin, New York (1964).
12. R. Haag, *Local Quantum Physics*, Springer Verlag, Berlin-Heidelberg-New York (1992).
13. D. Olesen and G.K. Pedersen, *Pacific J. Math.* **53**, 563 (1974).
14. H.J. Borchers and D. Buchholz, *Commun. Math. Phys.* **97**, 169 (1985).
15. H.J. Borchers, *Contemporary Mathematics* **62**, 143 (1987).
16. H.J. Borchers, *Fizika* **17**, 289 (1985).
17. J. Fröhlich, G. Morchio, and F. Strocchi, *Phys.Lett.B* **89**, 61 (1979).
18. D. Buchholz, *Phys.Lett.B* **174**, 331 (1986).
19. A.S. Wightman, *Ann.Inst.Henri Poincaré, A* **I**, 403 (1964).
20. H.J. Borchers, *Commun. Math. Phys.* **1**, 281 (1965).
21. S. Doplicher, R. Haag. J.E. Roberts, *Commun. Math. Phys.* **13**, 1 (1969).
22. S. Doplicher, R. Haag. J.E. Roberts, *Commun. Math. Phys.* **15**, 173 (1969).
23. H.J. Borchers, *A Remark on Antiparticles*, in Lecture Notes in Physics Vol **257**, (1986), p.268.
24. H.J. Borchers, *On the Gap in the Spectrum of the Translations*, Preprint (2001); http://www.lqp.Uni-Goettingen.DE
25. H. Reeh and S. Schlieder, *Nuovo Cimento* **22**, 1051 (1961).
26. J. Bisognano and E.H. Wichmann, *J. Math. Phys.* **16**, 985 (1975).

27. J. Bisognano and E.H. Wichmann, *J. Math. Phys.* **17**, 303 (1976).
28. R. Jost, *Helv. Phys. Acta* **30**, 409 (1957).
29. H.-J. Borchers, *Commun. Math. Phys.* **143**, 315 (1992).
30. R. Brunetti, D. Guido and R. Longo, *Rev. Math. Phys.* **7**, 57 (1994).
31. D. Guido and R. Longo, *Commun. Math. Phys.* **172**, 517 (1995).
32. H.-J. Borchers, *Commun. Math. Phys.* **179**, 703 (1996).
33. J. Gaier and J. Yngvason, *J. Math. Phys.* **41**, 5910 (2000).
34. D. Buchholz, O. Dreyer, M. Florig and S. Summers, *Rev. Math. Phys.* **4**, 475 (2000).
35. D. Buchholz, M. Florig and S. Summers,*Lett.Math.Phys.* **49**, 337 (1999).
36. H.J. Borchers and J. Yngvason: On the PCT–Theorem in the Theory of Local Observables, preprint (2001); math-ph/0012020.

BETA-LATTICES FOR APERIODIC ORDER

JEAN PIERRE GAZEAU
Laboratoire de Physique Théorique de la Matière Condensée, Boite 7020,
Université Paris 7 Denis Diderot,
F-75251 Paris Cedex 05, France
E-mail: gazeau@ccr.jussieu.fr

We give a survey of recent results on counting systems, so-called beta-integers, based on Pisot-Vijayaraghavan algebraic integers $\beta > 1$ and leading to the notion of beta-lattices in $\mathbb{R}^d$. These beta-lattices may be equipped with an abelian group structure and reveal themselves as ideal labelling frames for discrete self-similar quasiperiodic structures with scaling invariant β.

1 Introduction

Everyone is familiar with periodic structures in which *repetitivity* is synonymous of translational symmetry. Geometrical models underlying such objects are (possibly decorated) $\mathbb{Z}$-modules or lattices: an elementary *cell* or finite pattern is infinitely repeated along precise directions in space. The variety of those directional arrangements compatible with translational symmetry have given rise to famous crystallographic classifications, well known from physicists and chemists. Needless to say that the role of integers is crucial here!

Actually, those crystalline properties like repetitivity and long-range order are not specific to periodic objects. Many aperiodic structures encountered in Nature or in Art also have these properties, one of the most popular recent examples being afforded by quasicrystals. In many cases, these properties are explained by the existence of an inflationary or scaling symmetry in which the scaling factor is a certain irrational algebraic integer $\beta > 1$, and it can occur that deep algebraic features of this β are the mark of (local) n-fold rotational symmetries in the structure.

The aim of this contribution is to give a survey of recent results that pertain to algebraic number theory and algorithmics of numeration systems and which lead to the notion of beta-lattices. The latter are relevant to the description of certain aperiodic structures showing inflationary symmetry possibly combined with n-fold symmetry. These structures can be quasicrystals or quasicrystalline diffraction patterns, of course. Nevertheless, they can be encountered in various other fields of Physics, like dynamical systems or hydrodynamics. We refer to the contribution of El Kharrat *et al* [1] in the present volume in which more details with figures are given on hydrodynamical examples (see also [2] for other examples like Penrose tilings and their diffraction patterns).

We shall restrict our presentation to one- and two-dimensional structures. In Section 2 will be given some basic definitions concerning discrete point sets of interest for our purpose and which are nowadays in use. Section 3 will be devoted to cyclotomic structures in the plane which are natural support for n-fold aperiodic point sets and from which the notion of *cyclotomic Pisot-Vijayaraghavan* number emerges as the fundamental object at the root of the construction of the latter. In Section 4, we explain the important notions of *beta-expansions* and *beta-integers*, the latter sets taking naturally the place of the integers in the definition of *beta-lattices* in $\mathbb{R}^d$. In Section 5, we give a set of results with illustrative examples concerning the arithmetics and the algebra of these *beta-integers*. In Section 6 are shown some applications of beta-integers to the construction and the description of one-dimensional *model sets* (or *cut-and-project sets* in a more widely accepted terminology). We shall finally end up by presenting and discussing the important concept of beta-lattice derived from the beta-integers and we give an outline of possible applications in the next future.

2 Discrete point sets: basic definitions

Let us start out with the notion of *Delone set* which is the obvious model of any large grouping of atoms defining a material structure. By Delone set $\Lambda \subset \mathbb{R}^d$ we mean that Λ is *uniformly discrete* (there exists $r > 0$ such that each ball of radius r contains at most one point of Λ) and *relatively dense* (there exists $R > 0$ such that each ball of radius R contains at least one point of Λ). A Delone set Λ is a *Meyer set* if $\Lambda - \Lambda$ is a Delone set or equivalently if there exists a finite set F such that $\Lambda - \Lambda \subset \Lambda + F$.

The most popular geometrical models for quasicrystals are the so-called *cut-and-project* sets. These sets actually were previously introduced by Meyer [3,4] in the context of Harmonic Analysis and Number Theory and christened by him *model sets*. First, a *cut and projection scheme* is the following

$$\begin{array}{ccccc} \mathbb{R}^d & \xleftarrow{\pi_1} & \mathbb{R}^d \times G & \xrightarrow{\pi_2} & G \\ & & \cup & & \\ & & D & & \end{array}$$

where G is a locally compact abelian group, called the *internal space*, $\mathbb{R}^d$ is called the *physical space*, D is a lattice, *i.e.* a discrete subgroup of $\mathbb{R}^d \times G$ such that $(\mathbb{R}^d \times G)/D$ is compact. The projection $\pi_1|_D$ is 1-to-1, and $\pi_2(D)$ is dense in G.

Let $M = \pi_1(D)$ and set $* = \pi_2 \circ (\pi_1|_D)^{-1} : M \longrightarrow G$. The set $\Lambda \subset \mathbb{R}^d$ is a *model set* if there exist a cut and projection scheme and a relatively compact set $\Omega \subset G$ of non-empty interior such that $\Lambda = \{x \in M \mid x^* \in \Omega\}$. The set Ω is called a *window*.

The following was proved by Meyer: *A model set is a Meyer set. Conversely if Λ is a Meyer set, there exist a finite set F and a model set Λ_0 such that $\Lambda \subset \Lambda_0 + F$.*

We say that a Delone set Λ is self-similar of scaling factor $\theta > 1$ (*inflation factor*) if $\theta\Lambda \subset \Lambda$.

We now recall some definitions on numbers. A *Pisot-Vijayaraghavan number* or PV number is an algebraic integer > 1, solution to $X^m + c_{m-1}X^{m-1} + \cdots + c_1X + c_0 = 0$ such that all the other roots or *Galois conjugates* have modulus strictly less than one. A *Salem number* is an algebraic intege > 1 such that every conjugate has modulus smaller than or equal to 1, and at least one of them has modulus 1.

Another result from Meyer makes the connection between Meyer sets and those algebraic integers: *If $\Lambda \subset \mathbb{R}^d$ is a Meyer set and if $\theta > 1$ is a real number such that $\theta\Lambda \subset \Lambda$ then θ is a Pisot or a Salem number. Conversely for each d and for each Pisot or Salem number θ, there exists a Meyer set $\Lambda \subset \mathbb{R}^d$ such that $\theta\Lambda \subset \Lambda$.*

Let us illustrate this construction with one type of Fibonacci chain, well known as an one-dimensional toy model for quasicrystals [5]. Consider the following cut and projection scheme:

$$\begin{array}{c} \mathbb{R} \xleftarrow{\pi_1} \mathbb{R} \times \mathbb{R} \xrightarrow{\pi_2} \mathbb{R} \\ \cup \\ \mathbb{Z}^2 \end{array} \tag{1}$$

Here $\pi_1(\mathbb{Z}^2) \sim \mathbb{Z}[\tau] = \{a + b\tau \mid a, b \in \mathbb{Z}\}$, with $\tau = \frac{1+\sqrt{5}}{2}$.

(One version of) the Fibonacci chain $\mathcal{F}$ is

$$\begin{aligned} \mathcal{F} &= \{x = a + b\tau \mid x' = a - \frac{b}{\tau} \in \Omega = [0, 1)\} \\ &= \{\ldots, -\tau^3, -\tau, 0, \tau^2, \tau^3 + 1, \tau^4, \ldots\} \end{aligned} \tag{2}$$

It is the set of left endpoints of a quasiperiodic tiling of $\mathbb{R}$ with 2 tiles L et S, of respective length τ^2 and τ, generated by the substitution rules $L \mapsto LLS$, $S \mapsto LS$. Starting from SL in both directions we get a biinfinite word $\cdots LLSLS \mid LLSLLSLS \cdots$ Note that $\tau^2\mathcal{F} \subset \mathcal{F}$ and that this tiling is, by construction, *stone inflation*, which means that all the tiles when scaled by the factor τ^2 can be packed face-to-face from the original ones.

3 Cyclotomic point sets and PV numbers

It is well known that the condition $2\cos\frac{2\pi}{n}\in\mathbb{Z}$ characterizes n-fold Bravais lattices in $\mathbb{R}^2$ (and in $\mathbb{R}^3$). Let us put $\zeta = e^{\frac{2\pi i}{n}}$, $\zeta^n = 1$. If we consider the $\mathbb{Z}$-module in the plane :

$$\mathbb{Z}[\zeta] \;=\; \mathbb{Z}+\mathbb{Z}\zeta+\mathbb{Z}\zeta^2+\cdots+\mathbb{Z}\zeta^{n-1} = \mathbb{Z}[2\cos\frac{2\pi}{n}]+\mathbb{Z}[2\cos\frac{2\pi}{n}]\zeta, \qquad (3)$$

we get the cyclotomic ring of order n. This n-fold structure is generically dense in $\mathbb{C}$, except precisely for the crystallographic cases. We indeed check that $\mathbb{Z}[\zeta] = \mathbb{Z}$ for $n = 1$ or 2, $\mathbb{Z}[\zeta] = \mathbb{Z}+\mathbb{Z}i$ for $n = 4$ (square lattice), and $\mathbb{Z}[\zeta] = \mathbb{Z}+\mathbb{Z}e^{\frac{i\pi}{3}}$ for the triangular and hexagonal cases $n = 3$ and $n = 6$. If n is not crystallographic, $2\cos\frac{2\pi}{n}$ is an algebraic integer of degree $m = \varphi(n)/2 \le \lfloor (n-1)/2\rfloor$ where φ is the Euler function and $\lfloor y\rfloor$ denotes the integer part of a real number y. A *cyclotomic PV number* with symmetry of order n is a PV number β such that

$$\mathbb{Z}[2\cos\frac{2\pi}{n}] = \mathbb{Z}[\beta]. \qquad (4)$$

Then $\mathbb{Z}[\zeta] = \mathbb{Z}[\beta]+\mathbb{Z}[\beta]\zeta$ is a ring invariant under rotation of order n (see [6]). This ring is the natural framework for two-dimensional n-fold structures having β as scaling factor.

Let us give some examples of those numbers related to noncrystallographic cyclotomic structures in the plane.

$$\begin{array}{lll} n=5 & \beta=\tau=\dfrac{1+\sqrt{5}}{2}=1+2\cos\dfrac{2\pi}{5} & \text{(pentagonal case)}\\ n=10 & \beta=\tau=\dfrac{1+\sqrt{5}}{2}=2\cos\dfrac{\pi}{5} & \text{(decagonal case)}\\ n=8 & \beta=1+\sqrt{2}=1+2\cos\dfrac{\pi}{4} & \text{(octogonal case)}\\ n=12 & \beta=2+\sqrt{3}=2+2\cos\dfrac{\pi}{6} & \text{(dodecagonal case).} \end{array} \qquad (5)$$

These numbers are all quadratic integers, more precisely they are *quadratic PV units*, for obeying an equation of the type

$$X^2 - aX \pm 1 = 0, \quad a \ge 1\,\text{if}\,-, \quad a \ge 3\,\text{if}\,+. \qquad (6)$$

In the case $n = 7$, we have $\beta = 1+2\cos\frac{2\pi}{7}$ which is solution to the cubic equation $X^3 - 2X^2 - X + 1 = 0$.

At this point, we should be aware that finding a PV number such that the cyclotomic condition (4) is fulfilled for $n \ge 16$ is an open problem!

4 Beta-expansions and beta-integers

When a number $\beta > 1$ appears as a kind of fundamental invariant in a given structure, it is tempting to introduce into the procedure of understanding the latter a *counting system* based precisely on this β. Let us explain here what we mean by counting system.

Among all *beta-representations* of a real number $x \geq 0$, *i.e.* infinite sequences $(x_i)_{i\leq k}$, such that $x = \sum_{i\leq k} x_i\beta^i$ for a certain integer k, there exists a particular one, called the *beta-expansion*, which is obtained through the "greedy algorithm" (see [7] and [8]). Denote by $\{y\}$ the fractional part of a real number y. There exists $k \in \mathbb{Z}$ such that $\beta^k \leq x < \beta^{k+1}$. Let $x_k = \lfloor x/\beta^k \rfloor$ and $r_k = \{x/\beta^k\}$. Then for $i < k$, put $x_i = \lfloor \beta r_{i+1} \rfloor$, and $r_i = \{\beta r_{i+1}\}$. We get an expansion $x = x_k\beta^k + x_{k-1}\beta^{k-1} + \cdots$.

For instance, if $\beta = \tau \approx 1\cdot 618\cdots$, then $x_i \in \{0,1\}$. The tau-expansion of, say, 4 is $4 = \tau^2 + 1 + 1/\tau^2$, *i.e.* $\langle 4\rangle_\tau = 101\cdot 01$.

If $x < 1$ then $k < 0$, and we put $x_0 = x_{-1} = \cdots = x_{k+1} = 0$. The beta-expansion of x is denoted by

$$\langle x\rangle_\beta = x_k x_{k-1}\ldots x_1 x_0 \cdot x_{-1}x_{-2}\cdots \tag{7}$$

The digits x_i obtained by this algorithm are integers from the set $A = \{0,\ldots,\lceil\beta\rceil - 1\}$, called the *canonical alphabet*, where $\lceil\beta\rceil$ denotes the closest integer to β from above. If an expansion ends in infinitely many zeros, it is said to be *finite*, and the ending zeros are omitted.

There is a representation which plays an important role in the theory. The *beta-expansion* of 1, denoted by $d_\beta(1)$, is computed by the following process [7]. Let the *beta-transform* be defined on $[0,1]$ by $T_\beta(x) = \beta x \bmod 1$. Then $d_\beta(1) = (t_i)_{i\geq 1}$, where $t_i = \lfloor \beta T_\beta^{i-1}(1)\rfloor$. Bertrand has proved that if β is a PV number, then $d_\beta(1)$ is eventually periodic [9]. For instance, $d_\tau(1) = 11$, $d_{2+\sqrt{3}}(1) = 322\cdots \equiv 32^\omega$. A number β such that $d_\beta(1)$ is eventually periodic is traditionally called a *beta-number*. Since these numbers were introduced by Parry [8], we propose to call them *Parry numbers*. When $d_\beta(1)$ is finite, β is said to be a *simple* Parry number.

We now come to the important notion of *beta-integer*. The set of beta-integers is the set of real numbers such that their beta-expansion is polynomial,

$$\mathbb{Z}_\beta = \{x \in \mathbb{R} \mid \langle |x|\rangle_\beta = x_k \cdots x_0\} \tag{8}$$
$$= \mathbb{Z}_\beta^+ \cup (-\mathbb{Z}_\beta^+) \tag{9}$$

where $\mathbb{Z}_\beta^+$ is the set of non-negative beta-integers. For instance, the first positive tau-integers are given by

$$\begin{array}{ccccccccccc} & \overset{1}{\frown} & & \overset{\frac{1}{\tau}}{\frown} & & \overset{1}{\frown} & & \overset{1}{\frown} & & \overset{\frac{1}{\tau}}{\frown} & \\ 0 & & 1 & & \tau & & \tau^2 & & \tau^2+1 & & \tau^3 \\ (0) & & (1) & & (10) & & (100) & & (101) & & (1000). \end{array} \tag{10}$$

In the dodecagonal case, $\beta = 2+\sqrt{3} \equiv \theta$, the first positive beta-integers read

$$\begin{array}{ccccccccccc} & \overset{1}{\frown} & & \overset{1}{\frown} & & \overset{1}{\frown} & & \overset{\theta-3}{\frown} & & \overset{1}{\frown} & \\ 0 & & 1 & & 2 & & 3 & & \theta & & \theta+1 \\ (0) & & (1) & & (2) & & (3) & & (10) & & (11). \end{array} \tag{11}$$

The set $\mathbb{Z}_\beta$ is self-similar and symmetrical with respect to the origin

$$\beta\mathbb{Z}_\beta \subset \mathbb{Z}_\beta\,, \; \mathbb{Z}_\beta = -\mathbb{Z}_\beta.$$

5 Arithmetics and algebra of beta-integers

It has been shown in [10] that if β is a PV number then $\mathbb{Z}_\beta$ is a Meyer set, which means that there exists a finite set F such that $\mathbb{Z}_\beta - \mathbb{Z}_\beta = \mathbb{Z}_\beta + \mathbb{Z}_\beta \subset \mathbb{Z}_\beta + F$. Hence, one can see that the set $\mathbb{Z}_\beta$ is "quasiadditive" to a certain extent. The question to be next addressed is to characterize completely this beta-dependent F in order to see to what extent beta-integers differ from ordinary integers concerning additive (and multiplicative) group structures. This problem is solved in [10,11,12] for all quadratic PV units and for a few higher-degree cases (see also [13]). Let us give here an outline of these results, referring to the quoted papers for proofs and technical details. The demonstrations mainly rest upon the algebraic properties of $\mathbb{Z}_\beta$, themselves mostly consequences of the bounded nature of the Galois conjugates of β: $\beta', \beta'', \ldots \in (-1,1)$. As a matter of fact, $\mathbb{Z}_\beta$ is by construction a subset of the extension ring $\mathbb{Z}[\beta] = \sum_{k\geq 0} \mathbb{Z}\beta^k$ and has all its Galois conjugate sets $\mathbb{Z}'_\beta$, $\mathbb{Z}''_\beta, \ldots$ bounded in their respective extension rings $\mathbb{Z}[\beta']$, $\mathbb{Z}[\beta''], \ldots$.

We now restrict the presentation to quadratic Pisot units.

5.1 *Preliminary algebra*

Let us introduce the *algebraic model set*

$$\Sigma^\Omega = \{x = m + n\beta \in \mathbb{Z}[\beta] \mid x' = m + n\beta \in \Omega\}, \tag{12}$$

where the closure $\overline{\Omega}$ is compact and the interior $\overset{\circ}{\Omega}$ is not empty. Then,

- for quadratic Pisot unit β solution to $X^2 - aX - 1 = 0, a \geq 1$ (the Galois conjugate is $\beta' = -1/\beta$), for example $\beta = \tau$, we have

$$\mathbb{Z}_\beta^+ = \Sigma^{(-1,\beta)} \cap \mathbb{R}^+, \ \mathbb{Z}_\beta^- = -\mathbb{Z}_\beta^+ = \Sigma^{(-\beta,1)} \cap \mathbb{R}^+, \tag{13}$$

and this implies

$$\mathbb{Z}'_\beta \subset (-\beta, \beta) \ \text{ and } \ \mathbb{Z}'_\beta \cap (-1,1) = \mathbb{Z}[\beta] \cap (-1,1). \tag{14}$$

- for quadratic Pisot unit β solution to $X^2 - aX + 1 = 0, a \geq 3$ (the Galois conjugate is $\beta' = 1/\beta$), for example $\beta = \theta$,

$$\mathbb{Z}_\beta^+ = \Sigma^{[0,\beta)} \cap \mathbb{R}^+ \ \text{ and } \ \mathbb{Z}_\beta^- = \Sigma^{(-\beta,0]} \cap \mathbb{R}^-, \tag{15}$$

$$\mathbb{Z}_\beta \subset \Sigma^{(-\beta,\beta)} \subset \mathbb{Z}_\beta + \{0, \pm\frac{1}{\beta}\}. \tag{16}$$

5.2 *And now, some arithmetics*

The above results lead to the following characterizations for addition and multiplication of beta-integers.

- For quadratic Pisot unit β solution to $X^2 - aX - 1 = 0$, $a \geq 1$, we have

$$\mathbb{Z}_\beta + \mathbb{Z}_\beta \subset \mathbb{Z}_\beta + \{0, \pm(1 - \frac{1}{\beta})\} \subset \mathbb{Z}_\beta/\beta^2, \tag{17}$$

$$\mathbb{Z}_\beta \times \mathbb{Z}_\beta \subset \mathbb{Z}_\beta + \{0, \pm\frac{1}{\beta}, \ldots, \pm\frac{a}{\beta}\} \subset \mathbb{Z}_\beta/\beta^2. \tag{18}$$

  For instance, for $\beta = \tau$, $1 + 1 = 2 = \tau + \frac{1}{\tau^2} = \tau + (1 - \frac{1}{\tau})$, and $(\tau^2 + 1)(\tau^2 + 1) = \tau^5 + \tau + \frac{1}{\tau^2} = \tau^5 + \tau^2 - \frac{1}{\tau}$.

- For quadratic Pisot unit β solution to $X^2 - aX + 1 = 0$, $a \geq 3$, we have

$$\mathbb{Z}_\beta + \mathbb{Z}_\beta \subset \mathbb{Z}_\beta + \{0, \pm\frac{1}{\beta}\}, \ \mathbb{Z}_\beta^+ + \mathbb{Z}_\beta^+ \subset \mathbb{Z}_\beta^+/\beta, \ \mathbb{Z}_\beta \times \mathbb{Z}_\beta \subset \mathbb{Z}_\beta/\beta. \tag{19}$$

  For instance, for $\beta = \theta$, $2 + 2 = \theta + \frac{1}{\theta} = 2 \times 2$.

5.3 *Beta-integers as an additive group*

One can equip the set of beta-integers with an abelian group structure [11,12]. Actually, we can achieve this for any countable strictly increasing sequence $U = \{u_n\}_{n \in \mathbb{Z}}$ of numbers, $\mathbb{Z} \ni n \to u_n \in \mathbb{R}$, $u_0 = 0$, by simple isomorphic transport of the additive group structure of the integers: the additive law of U is defined by

$$u_m \oplus u_n \stackrel{def}{=} u_{m+n}. \tag{20}$$

However, there is no reason that the law $\oplus$ be consistent in general with the ordinary addition $+$ in the real numbers. Consistency means that *for any* $(m,n) \in \mathbb{Z}^2$ *such that there exists* $k \in \mathbb{Z}$ *for which* $u_m + u_n = u_k$, *then* $k = m + n$, *or equivalently* $u_m + u_n = u_m \oplus u_n$. It is remarkable that this consistency holds true for the sequence of beta-integers $\mathbb{Z}_\beta$ if β is PV quadratic unit. Due to the above listed properties of these numbers, we can also assert that there is between both laws *minimal distortion* in the following sense: for all $(b_m, b_n) \in \mathbb{Z}_\beta$ with β quadratic PV number,

$$b_m + b_n - (b_m \oplus b_n) \in \begin{cases} \{0, \pm(1 - \frac{1}{\beta})\} & \text{if } \beta^2 - a\beta - 1 = 0 \; a \geq 1, \\ \{0, \pm\frac{1}{\beta}\} & \text{if } \beta^2 - a\beta + 1 = 0 \; a \geq 3. \end{cases} \tag{21}$$

For instance, if $\beta = \tau$, then $1 \oplus 1 = \tau$ and $2 - \tau = \frac{1}{\tau^2}$, and if $\beta = \theta$, then $2 \oplus 2 = \theta$ and $4 - \theta = \frac{1}{\theta}$.

5.4 Internal multiplicative law for beta-integers

We could attempt to play the same game with multiplication by defining

$$b_m \text{“}\times\text{”} b_n \stackrel{def}{=} b_{mn},$$

for all $(b_m, b_n) \in \mathbb{Z}_\beta^2$. However, this definition is quitely not appropriate for being, unlike (20), not compatible with the ordinary multiplication. For instance, for $\beta = \tau$, $b_2 \times b_2 = \tau \times \tau = \tau^2 = b_3 \neq b_4$. A consistent definition is instead given by [12]:

$$b_m \otimes b_n = \begin{cases} b_{(mn - a\rho_S(m)\rho_S(n))} & \text{if } \beta^2 - a\beta - 1 = 0, \; a \geq 1, \\ b_{(mn - a\rho_S(m)\rho_S(n))} & \text{if } \beta^2 - a\beta + 1 = 0, \; a \geq 3, \end{cases} \tag{22}$$

where $\rho_S(n)$ is the function counting the number of small tiles appearing in the string defined by $|b_n|$. For instance, refering to (10), $\rho_S(5) = 2$ while for (11) $\rho_S(5) = 1$. Such a definition (22) is not only consistent with ordinary multiplication of real numbers, but it has also minimal distortion with respect to the latter:

$$b_m b_n - (b_m \otimes b_n) \in \begin{cases} \{(0, \pm 1, \ldots, \pm a)(1 - \frac{1}{\beta})\} & \text{if } \beta^2 - a\beta - 1 = 0 \; a \geq 1, \\ \{(0, \; 1, \ldots, a-1)\frac{\text{sgn}(b_m b_n)}{\beta}\} & \text{if } \beta^2 - a\beta + 1 = 0 \; a \geq 3. \end{cases} \tag{23}$$

However, it should be noticed that the internal law (22) does not define a group for being not associative. It is also not distributive with respect to the law $\oplus$. So it seems to be hopeless to get in this way a ring structure like we have with integers.

6 One-dimensional model sets as subsets of beta-integers

We now give some hints about what can be the relationship between sets of beta-integers and more generic model sets. For sake of simplicity, let us consider one-dimensional model sets like the Fibonacci chain presented in (2). The latter $\mathcal{F}$ reads as $\Sigma^{[0,1)}$ in the notations of (12). More generally, we shall denote by $\mathcal{F}_\beta$ the model set associated to a PV quadratic unit β and defined algebraically by

$$\mathcal{F}_\beta = \{x \in \mathbb{Z}[\beta] \mid x' \in [0,1)\}. \tag{24}$$

The tiling associated with it has 2 tiles, and is not symmetrical. No point is favoured. In this sense we can say that this chain is an homogeneous set. The set $\mathcal{F}_\beta \cup (-\mathcal{F}_\beta)$ is also a model set, $\mathcal{F}_\beta \cup (-\mathcal{F}_\beta) = \{x \in \mathbb{Z}[\beta] \mid x' \in (-1,1)\} = \Sigma^{(-1,1)}$. In constrast, this set has a specific point, the origin! However, this symmetrical tiling is more complex, in the sense that it has three tiles.

On the other hand, the set $\mathbb{Z}_\beta$ of β-integers is symmetrical. The tiling associated with it has 2 tiles. This set is not a model set, but it is a Meyer set. Let us now show that the β-integers are above all perfect labelling frames for model sets like (24).

Suppose that β is a quadratic PV number of the first category, *i.e.* $\beta^2 - a\beta - 1 = 0$, $a \geq 1$. We easily derive from (13) and (14)

$$\mathcal{F}_\beta = \{x \in \mathbb{Z}_\beta \mid x' \in [0,1)\},\ \mathcal{F}_\beta \cup (-\mathcal{F}_\beta) = \{x \in \mathbb{Z}_\beta \mid x' \in (-1,1)\} \tag{25}$$

which means that the points of $\mathbb{Z}_\beta$ enumerate the model sets $\mathcal{F}_\beta$ and $\mathcal{F}_\beta \cup (-\mathcal{F}_\beta) = \mathcal{F}_\beta - \mathcal{F}_\beta$. Note that the scaling invariance reads here as $\beta^2\mathcal{F}_\beta \subset \mathcal{F}_\beta$. This sieving procedure from $\mathbb{Z}_\beta$ allowing to get the nodes of $\mathcal{F}_\beta$ in an easy manner is illustrated by the following example in which $\beta = \tau$ and involving a few numbers in $\mathbb{Z}_\tau$. The numbers belonging to $\mathcal{F}_\beta = \mathcal{F}$ are underlined.
$\ldots,\ -\tau^4-\tau^2-1,\ -\tau^4-\tau^2,\ \underline{-\tau^4-\tau},\ -\tau^4-1,\ -\tau^4,\ \underline{-\tau^3-\tau},\ -\tau^3-1,\ \underline{-\tau^3},$ $-\tau^2-1,\ -\tau^2,\ \underline{-\tau},\ -1,\ \underline{0},\ 1,\ \tau,\ \underline{\tau^2},\ \tau^2+1,\ \tau^3,\ \underline{\tau^3+1},\ \tau^3+\tau,\ \underline{\tau^4},\ \tau^4+1,$ $\tau^4+\tau,\ \underline{\tau^4+\tau^2},\ \tau^4+\tau^2+1,\ldots$

From (25) follows that, for each x_0 in $\mathcal{F}_\beta$, $\mathcal{F}_\beta \subset \mathbb{Z}_\beta + \{x_0\}$. Thus, if we choose a point in $\mathcal{F}_\beta$ as origin point of $\mathbb{Z}_\beta$, then all the points of $\mathcal{F}_\beta$ are in $\mathbb{Z}_\beta$, and therefore the β-integers form a labelling of the chain $\mathcal{F}_\beta$.

Let us now turn our attention to quadratic PV numbers β in the second category, *i.e.* $\beta^2 - a\beta + 1 = 0$, $a \geq 3$. Because of (15), it is necessary to introduce a *decorated* or *completed* version of $\mathbb{Z}_\beta$, namely

$$\tilde{\mathbb{Z}_\beta} = \mathbb{Z}_\beta + \{0, \pm\frac{1}{\beta}\}, \tag{26}$$

in order to carry out the same sieving technique as in the above case. Then we can derive from (15) and (16)

$$\mathcal{F}_\beta = \{x \in \tilde{\mathbb{Z}_\beta} \mid x' \in [0,1)\},\ \mathcal{F}_\beta \cup (-\mathcal{F}_\beta) = \{x \in \tilde{\mathbb{Z}_\beta} \mid x' \in (-1,1)\} \tag{27}$$

This noticeably simplifies if the window is suitably chosen. For example, (15) leads straightforwardly to

$$\Sigma^{[0,\beta)} = \mathbb{Z}_\beta^+ \cup ((-\mathbb{Z}_\beta^+)/0 + \{\frac{1}{\beta}\}). \tag{28}$$

Applied to the number τ^2 which is in the second category with $a = 3$, this yields a simpler representation of the *contracted* Fibonacci chain: $\mathcal{F}/\tau^2 = \mathbb{Z}_{\tau^2}^+ \cup ((-\mathbb{Z}_{\tau^2}^+) \setminus \{0\} + \{\frac{1}{\tau^2}\})$.

More generally, let Ω be a bounded window in $\mathbb{R}$. It is always possible to find a λ in $\mathbb{Z}[\beta]$ and an integer j in $\mathbb{Z}$ such that $\Omega \subset (-\beta^j, \beta^j) + \lambda$. Let $\Delta = \beta^{-j}(\Omega - \lambda) \subset (-1,1)$. Then, one can easily prove that

$$\Sigma^\Omega = (-\,(\text{resp.}\ +1))^j \beta^{-j} \{x \in \mathbb{Z}_\beta\,(\text{resp.}\,\tilde{\mathbb{Z}_\beta}) \mid x' \in \Delta\} + \lambda'.$$

Therefore $\mathbb{Z}_\beta$ (resp. $\tilde{\mathbb{Z}_\beta}$) enumerates the model set Σ^Ω.

7 Toward the operational concept of beta-lattice

Given a PV or more generally a Parry number β, and d linearly independent vectors $\boldsymbol{e}_i$, we define a *beta-lattice* in $\mathbb{R}^d$ as a Delone point set of the type

$$\Lambda_j = \sum_{i=1}^{d} \mathbb{Z}_\beta/\beta^j\, \boldsymbol{e}_i \text{ and more generally } \tilde{\Lambda}_j = \sum_{i=1}^{d} \tilde{\mathbb{Z}_\beta}/\beta^j\, \boldsymbol{e}_i, \tag{29}$$

with $j \in \mathbb{Z}$ and where $\tilde{\mathbb{Z}}_\beta = \mathbb{Z}_\beta + F$ is some decorated version of the set of beta-integers in the sense that F contains the origin and is a finite symmetrical subset of $\mathbb{Z}_\beta/\beta^l$ for a certain $l \in \mathbb{N}$. It is clear that Λ_j inheritates an abelian group structure from $\mathbb{Z}_\beta$ if the latter is equipped with such a structure, as it is the case for quadratic PV numbers. From the self-similarity of $\mathbb{Z}_\beta$, $\beta\mathbb{Z}_\beta \subset \mathbb{Z}_\beta$, we get infinite increasing sequences of discretizations of $\mathbb{R}^d$:

$$\cdots \subset \tilde{\Lambda}_{j-1} \subset \tilde{\Lambda}_j \subset \tilde{\Lambda}_{j+1} \subset \cdots,\ \mathbb{R}^d = \overline{\bigcup_{j\in\mathbb{Z}} \tilde{\Lambda}_j},\ \{0\} = \bigcap_{j\in\mathbb{Z}} \tilde{\Lambda}_j. \tag{30}$$

In various circumstances, these sequences efficiently play the role of natural discrete frames for structures with aperiodic order based on the invariant β, like model sets of Section 6, like diffraction patterns of real quasicrystals or of Penrose tilings in [10,2], like Faraday instability patterns in [1]. Monte-Carlo simulations for quasicrystalline ground states have been carried out in [14] by using these beta-lattices in place of usual lattices or random explorations. Finally, discretizations (30) are at the basis of multiresolution analysis for discrete wavelet families organized according to their own long-range aperiodic order [15].

References

1. A.-D. El Kharrat, J.-P. Gazeau, M. Guidy, and S. Métens, this volume
2. J.-P. Gazeau and R. Krejcar, in *From Quasicrystals to More Complex Systems*, F. Axel, F. Denoyer, and J.-P. Gazeau eds., *Les éditions de physique*, Springer-Verlag (2000).
3. Y. Meyer, in *Beyond Quasicrystals*, F. Axel and D. Gratias eds., *Les éditions de physique*, Springer-Verlag (1995).
4. R. Moody, Model Sets: A Survey, in *From Quasicrystals to More Complex Systems*, F. Axel, F. Denoyer and J.-P. Gazeau eds., *Les éditions de physique*, Springer-Verlag (2000).
5. C. Janot, *Quasicrystals, A Primer*, Oxford U. P. (1996).
6. D. Barache, B. Champagne and J.-P. Gazeau, in *Quasicrystals and Discrete Geometry*, J. Patera ed., *Fields Institute Monograph Series*, Volume 10, Amer. Math. Soc. (1998).
7. A. Rényi, *Acta Math. Acad. Sci. Hungar.* **8**, 477 (1957).
8. W. Parry, *Acta Math. Acad. Sci. Hungar.* **11**, 401 (1960).
9. A. Bertrand, *C.R. Acad. Sc., Paris* **285**, 419 (1977).
10. Č. Burdík, Ch. Frougny, J.-P. Gazeau and R. Krejcar, *J. of Physics A: Math. Gen.* **31**, 6449 (1998).
11. Č. Burdík, Ch. Frougny, J.-P. Gazeau and R. Krejcar, in *Dynamical Systems: From Crystal to Chaos*, World Scientific (2000), p. 125.
12. Ch. Frougny, J.-P. Gazeau and R. Krejcar, *Theor. Comp. Sci.*, to appear (2002).
13. J.-L. Verger-Gaugry and J.-P. Gazeau, *Institut J. Fourier preprint* (2001).
14. J.-P. Gazeau, R. Krejcar and J. Miękisz, *Material Science and Engineering A*, **294-296**, 425 (2002).
15. M. Andrle, this volume.

GENERALIZED SYMMETRIES AND TIME

Michael HELLER[a]
Vatican Observatory, V-00120 Vatican City State

A model unifying general relativity and quantum mechanics is proposed the fundamental symmetry of which is not that of a group but rather that of a groupoid. With this groupoid there is associated a noncommutative C^*-algebra $\mathcal{A}$. It defines a certain noncommutative space. It is shown that both general relativity and quantum mechanics can be recovered from $\mathcal{A}$. Noncommutative space defined by $\mathcal{A}$ is nonlocal. The usual concepts of space and time, as composed of points and instants, are meaningless in it. In spite of this, a generalized dynamics can be defined, but it is intrinsically probabilistic. It is demonstrated how, in the process of transition from the noncommutative regime to the commutative regime, the usual dynamics and time emerge.

1 Introduction

A dogma is functioning in contemporary physics that at the very basis of physical laws are hidden symmetries. If we are looking for the "final unification of physics", we almost automatically think of a "primordial supersymmetry", the subsequent spontaneous breaking of which have given us the variety of forms and forces shaping the present Universe. This belief is well founded in great successes of modern physics which have also thought us that the real progress in this branch of science is usually connected with generalization of concepts. In the present paper, I explore a certain generalization of the concept of symmetry.

The idea of symmetry is mathematically implemented in the group structure and this is why the group theory belongs to the core of contemporary mathematical physics. The most obvious — but by no means trivial — generalization of the group concept is that of a groupoid, and it is exactly this concept which, in the present essay, is in the center of my interest. With every groupoid there is associated a C^*-algebra (which is, in general, noncommutative), and I shall demonstrate that, only by exploring the fact that in two special cases it must reproduce general relativity and quantum mechanics, it is possible to learn a handful of interesting things about the mathematical structure which coherently unifies in itself these two physical theories.

The organization of my material is the following. In section 2, I recollect, mainly in a descriptive way, some facts about groupoids. In section 3, I define

[a]Correspondence address: ul. Powstańców Warszawy 13/94, 33-110 Tarnów, Poland; e-mail: mheller@wsd.tarnow.pl

a noncommutative algebra $\mathcal{A}$ on a groupoid Γ, and show how to obtain from it the standard space-time geometry. The von Neumann algebra generated by $\mathcal{A}$ unifies in itself the noncommutative dynamics and the generalized probability measure. This is discussed in section 4. In section 5, I demonstrate how the usual quantum mechanics can be obtained from the algebra $\mathcal{A}$. In this way, the noncommutative geometry based on $\mathcal{A}$ can be regarded as unifying in itself general relativity and quantum mechanics. Section 7 contains some comments referring to atemporal aspects of the fundamental level as it is suggested by the proposed model.

2 Groupoids and Their Symmetries

Groupoid is a small category every morphism of which is an isomorphism – this is the most elegant, but perhaps the less informative, groupoid definition. An intuitive description says that a groupoid is "a set with partially defined multiplication for which the usual properties of a group hold whenever they make sense". [1] The strict definition, but nevertheless "giving some insight", can be found in the literature. [1,2,3]

Groupoids can be regarded as generalizing (and unifying) the properties of groups and equivalence relations. If one thinks of a group as describing symmetries defined on the whole set, then a groupoid (as a "group with many units") should be regarded as describing partial symmetries defined on this set. A comment of Weinstein: "In fact, though groups are indeed sufficient to characterize homogeneous structures, there are plenty of objects which exhibit what we clearly recognize as symmetry, but which admit few or no nontrivial automorphisms". [3] Such symmetries are met, for instance, in quasicrystals. [1] In connection with this, the necessity has been recognized to reformulate Klein's Erlangen program in terms of groupoids. [4] This would correspond to the transition from classical geometries to "quasicrystal-like geometries".

Another important difference between groups and groupoids is that geometry underlying groups is the one of sets with distinguished base point, whereas the geometry that underlies groupoids is that of directed graphs. A comment by Brown: "An argument usually made for groups is that they give the mathematics of reversible processes, and hence have a strong connection with symmetry. This argument applies even more strongly for groupoids. For groups, the processes all start and return to the same position. For groupoids, the processes may start at one point and finish at another". [5]

This directedness of groupoids becomes evident if we focus on the fact that they are also generalizations of equivalence relations (think on their transitivity). Again Weinstein: "From this point of view, a groupoid over B tells us not only *which* elements of B are equivalent to one another (or "isomorphic"), but it also parametrizes the *different ways* ("isomorphisms") in which two elements can be equivalent".[3] In fact, each equivalence relation is trivially a groupoid.

With every groupoid there can be associated an algebra, in fact a C^*-algebra (in general, noncommutative). The main idea of the present essay is to treat this algebra as defining a certain (noncommutative) space in the spirit of Connes' noncommutative geometry.[6] If we define groupoid as a small category in which every morphism is an isomorphism then, with the standard understanding of morphisms, the association of a C^*-algebra with the groupoid is not functorial. However, one can equivalently define a groupoid in terms of relations rather than mappings[7] and, accordingly, morphisms as relations rather than mappings (this definition of morphisms is not equivalent to the standard one). In this category of groupoids the association of C^*-algebras with groupoids is functorial.[8]

In what follows, the important role is played by groupoids of a special kind, called *transformation groupoids* (or *action groupoids*).[9] If E is a differential manifold and G a Lie group acting on E, $E \times G \rightarrow E$, then $\Gamma = E \times G$ can be naturally given the groupoid structure. We shall consider the case when E is the total space of a principal fibre bundle over space-time M. In this paper, we assume that G is a compact group.

Let $\mathcal{A}$ be the C^*-algebra of the groupoid Γ (it will be explicitly constructed below). The algebra $\mathcal{A}$ is, in general, noncommutative. I shall demonstrate that it has the following remarkable properties:

1. The groupoid $\Gamma = E \times G$ (understood as a Cartesian product) has two canonical projections: $\pi_M : \Gamma \rightarrow M$ and $\pi_G : \Gamma \rightarrow G$. With the projection π_M there is associated the algebra $\mathcal{A}_{proj} \simeq C^\infty(M)$ of smooth functions on space-time $M = E/G$. This allows us to recover general relativity, essentially along Geroch's program.[10]

2. With the projection π_G there is associated a C^*-algebra $\mathcal{A}_G$ which allows us to recover quantum mechanics, essentially in the quantization program developed by Haag and Kastler.[11]

The nature of these two associations will be elucidated below. We shall treat $\mathcal{A}$ as an algebra of "smooth" functions defining a certain noncommutative space (à la Connes[6]). The geometry of this space unifies in itself general relativity and quantum mechanics.

3 Noncommutative Geometry of the Groupoid Γ

Let then $\Gamma = E \times G$ be the transformation groupoid with E – the total space of a principal fibre bundle such that the orbits of the action of the group G on E form a differential manifold M, interpreted as space-time. Let also Γ_p be the fiber over $p \in E$ (in the Cartesian product $E \times G$). Two fibres Γ_p and Γ_q, $p, q \in E$ are called *equivalent* if there is $g \in G$ such that $q = pg$. Let $\mathcal{A} = C_c^\infty(\Gamma, \mathbf{C})$ be the algebra of compactly supported complex valued functions on the groupoid Γ. Multiplication in $\mathcal{A}$ is defined by

$$(a \cdot b)(\gamma) = \int_{\Gamma_{d(\gamma)}} a(\gamma\gamma_1^{-1})b(\gamma_1)d\mu(\gamma_1), \tag{1}$$

$a, b \in \mathcal{A}, \gamma \in \Gamma$, where $d\mu$ is the Haar measure, and the mapping $d : \Gamma \to E$ is defined by $d(\gamma) = \gamma^{-1}\gamma$. Involution in $\mathcal{A}$ is defined by $a^*(\gamma) = \overline{a(\gamma^{-1})}$.

Let further $\mathcal{A}_{proj}$ be the subalgebra of $\mathcal{A}$ consisting of (bounded) functions (with the usual pointwise addition and multiplication) which are constant on the equivalence classes of fibres (as defined above). The functions belonging to $\mathcal{A}_{proj}$ are called *projectible* functions. Since the algebra $\mathcal{A}_{proj}$ is isomorphic to the algebra $C^\infty(M)$, it is clear that by restricting the algebra $\mathcal{A}$ to $\mathcal{A}_{proj}$ we recover the usual space-time geometry, and by implementing Geroch's program[10] we obtain the standard general theory of relativity.

The quantum sector of the model is obtained by representing the algebra $\mathcal{A}$ in the Hilbert space $\mathcal{H}_q = L^2(\Gamma_{d(\gamma)}) = L^2(\Gamma_q)$, where $\gamma = (q, g)$, by

$$\pi_q(a)(\xi) = \int_{\Gamma_q} a(\gamma\gamma_1^{-1})\xi(\gamma_1)d\gamma_1. \tag{2}$$

We shall also consider the "integrated representation" of $\mathcal{A}$, $\pi = \bigoplus_{q \in E} \pi_q$, which is, in fact, a one-parameter family of representations.

We define the norm in the algebra $\mathcal{A}$

$$||a|| = \sup_{q \in E} || \pi_q(a) || \, . \tag{3}$$

The completion of $\mathcal{A}$ with respect to this norm is the C^*-algebra. Now, we can quantize the system with the help of the standard algebraic method.[11] Let S denote the space of states on the algebra $\mathcal{A}$. We assume that elements of S represent states of the system and pure states of S represent pure states of the system. If a is a Hermitian element of $\mathcal{A}$, and φ a state on $\mathcal{A}$ then $\varphi(a)$ is the expectation value of the observable a when the system is in the state φ. As we shall see below, if we suitably restrict the quantum sector of the model, we obtain the standard quantum mechanics.

4 Noncommutative Probability and Dynamics

In this section, we focus on some unifying aspects of the proposed model. It does not only unify general relativity with quantum mechanics (as we have seen above), but it also unifies some important theoretical concepts which so far were considered to be essentially independent of each other. The point is that the algebra $\pi_p(\mathcal{A})$ generates the von Neumann algebra $\mathcal{N} = (\pi(\mathcal{A}))''$ (which is the double commutant of $\pi(\mathcal{A})$) and, as we shall see, this von Neumann algebra is both the noncommutative probability measure and a "dynamic object" representing the generalized dynamics of the model.

Let us notice that for each $a \in \mathcal{A}$ there is a function ρ_a on E with values in the operator space given by

$$\rho_a(p) = \pi_p(a) \tag{4}$$

for $p \in E$. If $\rho_a = \rho_b$ then $a = b$ almost everywhere. It can be demonstrated that the operator ρ_a is a *random operator*, where by the latter we understand a one parameter family $r = (r_p)_{p \in E}$ of operators on a Hilbert space with some natural measurability conditions. [6] The equivalence classes of random operators, modulo everywhere, equipped with the obvious operations, form a von Neumann algebra. In our case, the random operators ρ_a, $a \in \mathcal{A}$, form the von Neumann algebra of the foliation groupoid (Γ is foliated by $\Gamma^p, p \in E$).[12] In this way, the noncommutative space defined by $\mathcal{A}$ is intrinsically a probabilistic object, and every element of $\mathcal{A}$ can be interpreted as a random operator. A randomicity should be understood here in a generalized sense, with the von Neumann algebra playing the role of a generalized probability measure. Indeed, in the commutative case von Neumann algebras are naturally isomorphic to the algebra of bounded measurable functions on a suitable Hilbert space.

Let $\rho_a = (\pi_p)_{p \in E}$ be a random operator, an element of the von Neumann algebra $\mathcal{N}$ of the foliation of Γ. On the strength of the Tomita-Takesaki theorem [13], with every state φ on $\mathcal{N}$ there is associated the one-parameter group

$$\alpha_t^{\varphi}(\rho_a) = e^{-it \ln \Delta} \rho_a e^{it \ln \Delta} \tag{5}$$

of automorphisms of $\mathcal{N}$, where $\Delta = S^* S$, and the operator $S : \mathcal{N} \to \mathcal{N}$ is given by $S(\rho_a)(\xi_\varphi) = \rho_a^*(\xi_\varphi)$. This is the dynamic equation of the model. It describes the "evolution" of random operators starting from $\alpha_0^{\varphi}(\rho_a)$. The following circumstance is worth noticing. In our model there is no time in the usual sense since the noncommutative space, as defined by the algebra $\mathcal{A}$, is essentially nonlocal. In this sense, such concepts as that of point in space or that of instant in time have no meaning. In spite of this, there is a generalized dynamics which, however, depends on the state φ.

After differentiating Eq. (5) and multiplying it by $i\hbar$, we obtain the dynamical equation of the more familiar form

$$i\hbar\frac{d}{dt}|_{t=0}\alpha_t^{\varphi}(\rho_a) = [\rho_a, -\hbar\ln\Delta]. \tag{6}$$

Let us notice that the "Hamiltonian" $-\hbar\ln\Delta$ depends on φ (through the dependence on the operator S).

5 Emergence of Quantum Mechanics

In order to obtain the usual time evolution we must get rid of the dependence of one-parameter groups on φ. How to do this? The set

$$U_G = \{u \in \mathcal{A} : uu^* = u^*u = \mathbf{I}\} \tag{7}$$

forms the unitary group of the algebra $\mathcal{A}$, and we have $\pi(U_G) \subset \pi(\mathcal{A})$. Two automorphisms α' and α'' of the von Neumann algebra $\mathcal{N}$ are said to be *inner equivalent* if there is an element $u \in \pi(U)$ such that $u\alpha''(a) = \alpha'(a)u$ for every $a \in \mathcal{N}$. The set of equivalence classes of this relation is called the *group of outer automorphisms* and is denoted by Out$\mathcal{N}$. It can be shown that if a one-parameter group α_t^{φ} is canonically projected onto a one-parameter group α_t in Out$\mathcal{N}$, it is independent of φ.[14] In this way, the von Neumann algebra $\mathcal{N}$ becomes more coarse (some of its elements are glued together with the help of inner equivalence) producing the usual time evolution. In this process, Eq. (6) looses its dependence on φ and becomes even more similar to the standard Schrödinger equation. But it is still valid only for random operators. However, if we restrict the algebra $\mathcal{A}$ to $\mathcal{A}_G = \{f \circ \pi_G : f \in C^\infty(G, \mathbf{C})\}$, where $\pi_G : \Gamma \to G$ is the obvious projection, then the operators of the family $(\pi_p(a))_{p\in E}$, $a \in \mathcal{A}_G$, forming the random operator ρ_a, are a constant family of operators (because of the natural isomorphisms of leaves of the groupoid foliation) which are projectible to the "typical leaf" of Γ, i.e., to the group G. The operator to which ρ_a projects will be denoted by a_Γ. It acts on the Hilbert space $L^2(G)$ and is no longer a random operator. If we now "project" Eq. (6) to $L^2(G)$, it assumes the familiar form of the Schrödinger equation (in the Heisenberg picture of quantum mechanics)

$$i\hbar\frac{d}{dt}|_{t=0}\alpha_t(a_\Gamma) = [a_\Gamma, H] \tag{8}$$

where $H = -\hbar[\ln\Delta]$. We thus obtain quantum mechanics on the group G. Its local version, in the neighborhood of unity, gives the standard version of quantum mechanics on $\mathbf{R}^n$. This reconstruction of quantum mechanics can be elegantly done by changing from the Lie groupoid Γ to its local version – the corresponding Lie algebroid.

Let us also mention that the proposed model is not "empty". Some simple examples and "toy solutions" to its equations have been found. [12,15,16]

6 Physics without Time

A remarkable property of the proposed model is that it presents the fundamental level of physics (below the Planck threshold) as existing with no space and no time in their usual meaning. The algebra $\mathcal{A}$ has no maximal ideals, and consequently in the noncommutative space corresponding to $\mathcal{A}$ the concept of point is meaningless. It turns out, however, that even in such "exotic" circumstances a generalized dynamics is possible.

Let us consider the triple $(\mathcal{A}, G, \alpha)$ where $\mathcal{A}$ is a C^*-algebra, G is a locally compact group, and $\alpha : G \to \mathrm{Aut}(\mathcal{A})$, given by $t \mapsto \alpha_t(a)$, $t \in G$, is a continuous mapping from the group G into $\mathrm{Aut}(\mathcal{A})$, for every $a \in \mathcal{A}$. In other words, $\alpha_t(a)$ is a family of automorphisms of $\mathcal{A}$, parametrized by elements of the group G. Such a triple $(\mathcal{A}, G, \alpha)$ is called a *generalized dynamical system.*[9] If $G = \mathbf{R}$ we obtain the dynamical system in the usual sense.

It is straightforward to identify in our model all elements of a generalized dynamical system. Such a system does not evolve in time. Elements of the group G, which parametrize the generalized dynamics, do not give any temporal order. We can improve the situation by restricting the dynamics to one-parameter subgroups of G but, in such a case, we obtain many "temporal regimes" (many one-parameter subgroups) unrelated with each other. The situation improves if we change to the von Neumann algebra (generated by $\mathcal{A}$) of operators on a suitable Hilbert space as it was discussed in section 4. Then, on the strength of the Tomita-Takesaki theorem, different temporal regimes correspond to states on the von Neumann algebra.

Another remarkable property of our model is that the von Neumann algebra \ is both a "dynamic object" and a "probabilistic object". It is a noncommutative counterpart of the probability measure and its elements are random operators. In this model, every dynamics is probabilistic, and every probabilistic measure defines a certain generalized dynamics. In the standard quantum mechanics we find a trace of this property. If we make suitable identifications of elements of $\mathcal{A}$ (with the help of the unitary group of $\mathcal{A}$), and if we change to the local version of the groupoid Γ, we obtain the usual quantum mechanical unitary evolution in which probabilities evolve rather than "quantum objects" themselves. But only if we restrict the algebra $\mathcal{A}$ to $\mathcal{A}_{proj}$, we obtain the usual spatio-temporal world of our everyday experience.

References

1. A.L.T. Paterson, *Groupoids, Inverse Semigroups, and Their Operator Algebras*, Birkhäuser, Boston – Basel – Berlin (1999).
2. J. Renault, *Groupoid Approach to C^*-Algebras* Springer, Berlin (1980).
3. A. Weinstein, *Not. Amer. Math. Soc.* **43**, 745 (1996).
4. M. Lawson, *Inverse Semigroups: The Theory of Partial Symmetries*, World Scientific, Singapore (1998).
5. R. Brown, *Higher dimensional group theory*, (Preprint: www.bangor.ac.uk).
6. A. Connes, *Noncommutative Geometry*, Academic Press, New York – London (1994).
7. S. Zakrzewski, *Comm. Math. Phys.* I: **134**, 347 (1990); II: **134**, 371 (1990).
8. P. Stachura, *Banach Center Publications*, **51**, 263 (2000).
9. L.P. Landsman, *Mathematical Topics Between Classical and Quantum Mechanics*, Springer, New York – Berlin – Heidelberg, (1998).
10. R.P. Geroch, *Comm. Math. Phys.* **26**, 271 (1972).
11. R. Haag and D. Kastler, *J. Math. Phys.* **5**, 848 (1964).
12. M. Heller *et al.*, *J. Math. Phys.* **41**, 5168 (2000).
13. V.S. Sunder, *An Invitation to von Neumann Algebras*, Springer, New York – Berlin – Heidelberg (1987).
14. A. Connes and C. Rovelli, *Classical and Quantum Gravity* **11**, 2899 (1994).
15. M. Heller *et al.*, *J. Math. Phys.*, **38**, 5840 (1997).
16. M. Heller and W. Sasin, *Int. J. Theor. Phys.*, **38**, 1619 (1999).

INTEGRABLE HIERARCHIES AND THE WDVV-EQUATIONS

G.F. HELMINCK

Department of Mathematics, Universiteit Twente, Enschede, The Netherlands
Email: helminck@math.utwente.nl

In this paper an overview is given of connections between the Witten-Dijkgraaf-Verlinde-Verlinde equations and its generalization and various integrable systems. In particular its link with the n-component KP-hierarchy is treated in a fairly detailed way.

1 Introduction

There are many instances where soliton equations of KdV-type play a role in quantum field theory, one of the most prominent ones being the Witten conjectures that relate partition functions of topological field theories to τ-functions of integrable hierarchies. Here we give an overview of connections between the Witten- Dijkgraaf-Verlinde-Verlinde-equations from 2-dimensional topological $N = 2$ superconformal field theory and various integrable hierarchies, more in particular with the n-component KP-hierarchy.

The contents of the remaining sections is as follows. The second section gives a quick introduction to the WDVV-equations and their generalization that shows up in Seiberg-Witten theory. The first system can be given a coordinate-free description as a Frobenius manifold, of which we recall the essentials in the first part of the third section. Next we limit our attention to a subclass, the semisimple Frobenius manifolds. This leads naturally to the Darboux-Egoroff system and its linearization. In the fourth section we point out a number of links that have been laid between integrable hierarchies and the WDVV-equations and vice-versa. Since we intend to elaborate here on the connection with the n-component KP-hierarchy, the next section is devoted to a crash course on the essentials of this integrable system. It starts with a description of the equations and their linearization. Next it is shown that in order to obtain solutions of the hierarchy it suffices to construct so-called wavefunctions. A useful characterization of these wavefunctions is by means of the bilinear relations and a pure algebraic setting for them is presented. The section concludes with the Grassmannian construction of wavefunctions. The last section shows how one can obtain within the Sato Grassmannian the different relations in the Darboux-Egoroff system and how to construct from the wavefunction a solution of the linearization of the Darboux-Egoroff system.

2 The WDVV-equations and its generalization

To give a direct idea of the form of the equations in question, we present here its local form that allows already interesting generalizations. The whole question concerns a holomorphic function $F(x_1,\ldots,x_n)$ in n complex variables. For each $i, 1 \le i \le n$, we denote by $F(i)$ the Hessian matrix of $\frac{\partial F}{\partial x_i}$, i.e.

$$F(i)_{jk} = \frac{\partial^3 F}{\partial x_i \partial x_j \partial x_k}.$$

Now we have the following

Definition 2.1 *We call F a solution of the WDVV-equations or Associativity equations if the following relations hold for F:*

1. *The matrix $F(1)$ is constant and invertible.*

2. *For all $i, 1 \le i \le n$, and $j, 1 \le j \le n$, $F(i)F(1)^{-1}F(j) = F(j)F(1)^{-1}F(i)$.*

Remark 2.2 *Witten, Dijkgraaf, Verlinde and Verlinde obtained matrices satisfying these equations as the three point correlators of the primary fields of a perturbed two dimensional topological field theory. The function F corresponds then to the primary free energy of the model.*

To explain the second name in the definition, let $C(i) := F(1)^{-1}F(i), 1 \le i \le n$. We can use these matrices to put a product structure on the tangent space in a point by putting

$$\frac{\partial}{\partial x_i} \circ \frac{\partial}{\partial x_k} := \sum_{j=1}^{n} C(i)_{jk} \frac{\partial}{\partial x_j}. \tag{1}$$

One verifies directly that this product structure is commutative and associative thanks to the second property, which can also be written as $C(i)C(j) = C(j)C(i)$ for all i and j. Clearly, differentiation w.r.t. x_1 is the unit element for this product structure.

Remark 2.3 *For the solutions of the WDVV-equations that relate to topological field theory, one requires still invariance w.r.t. a number of scaling transformations and this results in the condition: the function F is quasi-homogeneous, i.e.*

$$F(\lambda^{d_1}x_1,\ldots,\lambda^{d_r}x_n) = \lambda^{d_F}F(x_1,\ldots,x_n),$$

for certain numbers $d_F, d_1, \ldots, d_n$. In this paper we restrict our attention to the 2 conditions in the definition 2.1 and leave this third condition out of consideration.

If there is not a privileged direction like x_1, then the system of equations in 2.1 can be generalized as follows

Definition 2.4 *We call F a solution of the generalized WDVV-equations if the following relation hold for F: there is an invertible matrix $G = \sum_{i=1}^{n} a_i F(i)$ such that for all $i, 1 \le i \le n$, and $j, 1 \le j \le n$,*

$$F(i)G^{-1}F(j) = F(j)G^{-1}F(i).$$

Also this system of equations corresponds to an associativity condition: simply replace $F(1)$ by G in the definition of $C(i)$ above. Remarkably, it possesses also solutions coming from models in physics. It was shown by Marshakov, Mironov and Morozov that the full prepotential for $N = 2$ Super-Yang-Mills theory for the simple classical Lie algebras satisfies this generalized WDVV-system, see [18]. By a different method Ito and Yang, see [13], showed this to be true also for the groups of type E and the last nontrivial remaining case, that of F_4, was solved in [12].

3 Geometric Description

3.1 Frobenius manifolds

To globalize the structure occurring in the first section, one considers a complex manifold M with its structure sheaf $\mathcal{O}_M$. The tangent sheaf on M we denote by $\mathcal{T}_M$ and the cotangent sheaf by $\mathcal{T}_M^*$. The first piece of structure we need is

Definition 3.2 *An affine flat structure on M is a subsheaf $\mathcal{T}_M^f$ of linear spaces of pairwise commuting vectorfields such that $\mathcal{T}_M = \mathcal{O}_M \otimes_{\mathbb{C}} \mathcal{T}_M^f$. The sections of $\mathcal{T}_M^f$ are called flat vector fields.*

Locally, one can find coordinates $x_1, \ldots, x_n$ on M, the so-called *flat coordinates*, such that the $\{\tilde{\partial}_i := \frac{\partial}{\partial x_i}\}$ form a basis of $\mathcal{T}_M^f$.

The second ingredient is a metric on M, i.e. a symmetric pairing $g : S^2(\mathcal{T}_M) \to \mathcal{O}_M$. This induces an isomorphism $\tilde{g} : \mathcal{T}_M \to \mathcal{T}_M^*$. We want the metric g to be compatible with the affine structure $\mathcal{T}_M^f$, i.e. $g(X,Y)$ is constant for flat vector fields X and Y. The constant matrix $\{g_{ab} := g(\tilde{\partial}_a, \tilde{\partial}_b)\}$ corresponds then with the matrix $F(1)$ in the privileged direction.

To get to the multiplication on the tangent spaces, one needs a symmetric tensor $A : S^3(\mathcal{T}_M) \to \mathcal{O}_M$. This leads to the following implicit definition of the multiplication $\circ$ on $\mathcal{T}_M$:

$$g(X \circ Y, Z) := A(X, Y, Z), \tag{2}$$

for all vector fields X, Y and Z. Note that this definition implies that the metric is invariant w.r.t. this multiplication, i.e.

$$g(X \circ Y, Z) = g(X, Y \circ Z).$$

If we write $A_{abc} := A(\tilde{\partial}_a, \tilde{\partial}_b, \tilde{\partial}_c)$ and (g^{ab}) for the inverse of the matrix (g_{ab}), then this multiplication is locally given by

$$\tilde{\partial}_a \circ \tilde{\partial}_b = \sum_{c=1}^{n} \{ \sum_{e=1}^{n} A_{abe} g^{ec} \} \tilde{\partial}_c. \tag{3}$$

Since A is symmetric this multiplication $\circ$ is commutative, but the associativity does not have to hold. Also there does not have to be, locally, a function F such that

$$A_{abc} = \frac{\partial^3 F}{\partial x_a \partial x_b \partial x_c}.$$

By adding these two properties to the structures just introduced, one arrives at the global object corresponding to the WDVV-equations

Definition 3.3 *Let M be a complex manifold. We say that the triple $(\mathcal{T}_M^f, g, A)$ makes M into a Frobenius manifold, if the multiplication "$\circ$" is associative and the tensor A is locally potential, i.e. for all flat vector fields X, Y and Z*

$$A(X, Y, Z) = (XYZ)F,$$

for some holomorphic function F.

Remark 3.4 *The structure of Frobenius manifolds was treated extensively in the lecture notes [6]. The present short presentation of Frobenius manifolds is based on [17]. There one can also find a description of the role they play in quantum cohomology and at Gromov-Witten invariants. The proper globalization of the generalized WDVV-equations seems to be open still.*

Dubrovin showed that Frobenius manifolds form the moduli space of 2-dimensional topological field theories. As such any link to these models means a connection with the WDVV-equations.

3.5 Semisimplicity

An important class of Frobenius manifolds are the following ones

Definition 3.6 *Let $(M, \mathcal{T}_M^f, g, A)$ be a Frobenius manifold. It is called semisimple, if there exists locally a basis $X_1, \ldots, X_r$ of $\mathcal{T}_M$ such that the multiplication on $\mathcal{T}_M$ is given by*

$$\sum_{i=1}^{n} f_i X_i \circ \sum_{i=1}^{n} g_i X_i = \sum_{i=1}^{n} f_i g_i X_i.$$

In other words the tangent spaces at every point of M are semi-simple algebras. Dubrovin gave in [6] a complete description of this class and the link with the Darboux-Egoroff system. We resume the results from him that we need in a

Theorem 3.7 *For a semi-simple Frobenius manifold* $(M, \mathcal{T}_M^f, g, A)$ *there holds*

1. *Around each point of M there are local coordinates $(u_1, \ldots, u_n)$ such that, if $\partial_i := \frac{\partial}{\partial u_i}$, then $\partial_i \circ \partial_j = \delta_{ij}\partial_i$.*

2. *If the (ν_i) are the one-forms dual to the (∂_i), then the metric g and the tensor A have locally the form*

$$g = \sum_{i=1}^{n} h_i(u)^2 (\nu_i)^2 \tag{4}$$

$$A = \sum_{i=1}^{n} h_i(u)^2 (\nu_i)^3. \tag{5}$$

3. *Let $\gamma_{ij} := h_j^{-1}\partial_j(h_i), i \neq j$, be the rotation coefficients of the metric g. Then the flatness of g is equivalent to the (γ_{ij}) being a solution of the Darboux-Egoroff system, i.e.*

$$\begin{aligned} \partial_k(\gamma_{ij}) &= \gamma_{ik}\gamma_{kj}, \quad k \neq i \neq j \neq k, \\ \sum_{k=1}^{n} \partial_k(\gamma_{ij}) &= 0, \\ \gamma_{ij} &= \gamma_{ji}. \end{aligned} \tag{6}$$

3.8 The linearization of the Darboux-Egoroff system

The system of equations (6) already occurred in the nineteenth century in the work of Darboux on the description of orthogonal curvilinear coordinate systems in $\mathbb{R}^n$. It was Egoroff, who gave a complete solution of this problem for symmetric rotation coefficients. According to [6], the linearization of the system (6) is

$$\begin{aligned} \partial_k(\psi_{ij}) &= \gamma_{ik}\psi_{kj}, \quad k \neq i, \\ \sum_{k=1}^{n} \partial_k(\psi_{ij}) &= 0, \end{aligned} \tag{7}$$

Moreover he showed in [6] that if one has a solution $\{\gamma_{ij}\}$ of the Darboux-Egoroff system and one $\{\psi_{jk}\}$ of the associated linear system, then one has locally a solution of the WDVV equations

Theorem 3.9 *If the $\{\gamma_{ij}\}$ satisfy the equations (6) and the $\{\psi_{jk}\}$ the equations (7), then there corresponds locally a solution of the WDVV-equations to these data. On the domain determined by the inequalities $u_i \neq u_j, i \neq j$, and $\psi_{11}\ldots\psi_{1n} \neq 0$, one has the following expressions for the metric g and the tensor A*

$$h_i(u) = \psi_{i1}(u) \tag{8}$$

$$g_{ab} = \sum_{k=1}^{n} \psi_{ka}(u)\psi_{kb}(u) \tag{9}$$

$$A_{abc} = \sum_{k=1}^{n} \frac{\psi_{ka}(u)\psi_{kb}(u)\psi_{kc}(u)}{\psi_{k1}(u)} \tag{10}$$

This is exactly the point where the link between the WDVV-equations and the n-component KP-hierarchy will be made. The first step will be to determine the candidate rotation coefficients inside this hierarchy. Next one enforces conditions that will imply the equations of the DE-system and one concludes with finding solutions of the linear system.

4 Integrable systems linked to the WDVV-equations

There are many connections between integrable systems and the WDVV-equations. To give you an impression of what was done in the last decennium, we discuss a number of links and elaborate on one that permits a short description. It all started out with Dubrovin who constructed in [5] for each solution of the WDVV-equation a hierarchy of integrable bihamiltonian equations of hydrodynamic type such that the τ-function of a particular solution of the hierarchy coincides with the genus zero approximation of the corresponding topological field theory model. In [7] the genus one correction to this integrable hierarchy was computed. The remaining connections that we will present go in the opposite direction.

Around the same time Kricever showed in [14] that certain exact solutions of the Khoklov-Zabolotskaya hierarchy, which is the hierarchy connected with the dispersionless KP-equation, coincide with the perturbed superpotential of the minimal A_n model. This connection was put in a broader perspective in [15], where it was shown that so-called algebraic orbits in the moduli space of the universal Whitham hierarchy can be identified with various topological matter models coupled with gravity.

In [16] and its sequel [20] an orbit of a twisted loop group was shown to yield local Frobenius manifolds. Since this connection can be shown in a comprehensive way without an overload of technical data, we will outline the intermediate steps in the next sections.

The dispersionless limit considered in [14] was carried out in [21] for the generalized Drinfeld-Sokolov hierarchies, associated to three data: a non twisted affine Lie algebra $\hat{\mathfrak{g}}$, a Heisenberg subalgebra of $\hat{\mathfrak{g}}$ corresponding to a regular primitive conjugacy class in the Weyl group of $\mathfrak{g}$ and a regular element from it. This leads again to algebraic solutions to the WDVV-equations.

A recent construction, see [2], of local Frobenius manifolds starts out with the so-called constrained KP-hierarchy. By using the geometric description of this hierarchy from [8] and the additional symmetries from [1], they obtain solutions that go beyond the dispersionless limit.

Also the generalized WDVV-equations are intimately linked with integrable systems. For an overview of the various relations and the conjectured ones, we refer to [19].

5 The n-component KP-hierarchy

5.1 The equations and their linearization

The n-component KP-hierarchy is a matrix version of the KP-hierarchy and the variables involved are x and the $\{t_{i\alpha} \| i \geq 1, 1 \leq \alpha \leq n\}$, all corresponding to a group of commuting flows inherent to the system. A convenient form in which the equations of the n-component KP-hierarchy can be formulated is the so-called Lax form. It consists of a tower of identities between pseudodifferential operators with matrix coefficients, whose coefficients belong to a ring R and we start with a description of this ring. Consider the ring $\mathbb{C}[[x,t]]$ of formal powerseries in x and $t = (t_{i\alpha})$. Let θ be an element in $\mathbb{C}[[x,t]]$ of the form

$$\theta(x,0) = x^N + \sum_{j>N} a_j x^j. \tag{11}$$

In the sequel the ring R will always be the localization of $\mathbb{C}[[x,t]]$ w.r.t. the multiplicative subset $S_\theta = \{\theta^m | m \geq 0\}$ for some θ in $\mathbb{C}[[x,t]]$ of the form (11). On R we have the derivations $\partial_{i\alpha} = \dfrac{\partial}{\partial t_{i\alpha}}$ and $\partial = \dfrac{\partial}{\partial x}$. In the ring $M_n(R)[\partial]$ of differential operators in ∂ with coefficients from $M_n(R)$, one can, in general, not take roots of monic operators. Thereto one passes to the extension $M_n(R)[\partial, \partial^{-1})$ of all pseudodifferential operators in ∂ with coefficients from

$M_n(R)$. It consists of all expressions

$$\sum_{i=-\infty}^{N} a_i \partial^i, a_i \in M_n(R) \text{ for all } i,$$

that are added coefficientwise and that are multiplied according to

$$\partial^j \cdot a\partial^i = \sum_{k=0}^{\infty} \binom{j}{k} \partial^k(a)\partial^{i+j-k}$$

Each operator $A = \sum a_j \partial^j$ decomposes as $A = A_+ + A_-$ with $A_+ = \sum_{j\geq 0} a_j \partial^j$ its differential operator part and $A_- = \sum_{j<0} a_j \partial^j$ its pure integral operator part. Let $E_{\alpha\beta}$ be the matrix in $M_n(R)$ with its $(\alpha\beta)$-entry equal to 1 and the other entries equal to zero. Inside $M_n(R)[\partial, \partial^{-1})$ we consider operators L and U_α, $1 \leq \alpha \leq n$, of the form

$$L = \partial + \sum_{j\geq 1} l_j \partial^{-j} \text{ and } U_\alpha = E_{\alpha\alpha} + \sum_{j\geq 1} u_{\alpha,j}\partial^{-j}. \tag{12}$$

We want them to satisfy the following system of nonlinear equations inside $M_n(R)[\partial, \partial^{-1})$

$$[L, U_\alpha] = [U_\alpha, U_\beta] = 0, \tag{13}$$

$$\partial_{i\alpha}(L) = [(L^i U_\alpha)_+, L], \tag{14}$$

$$\partial_{i\alpha}(U_\beta) = [(L^i U_\alpha)_+, U_\beta], \tag{15}$$

for all α and β in $\{1, \ldots, n\}$. These are the equations of the n-component KP-hierarchy. Given R, we call a set of operators L and U_α, $1 \leq \alpha \leq n$, in $M_n(R)[\partial, \partial^{-1})$ of the form (12) that satisfy these equations, a *solution* of the n-component KP-hierarchy. One can see a solution L and $\{U_\alpha\}$ of the n-component KP-hierarchy as a deformation of the trivial solution $L = \partial$ and $U_\alpha = E_{\alpha\alpha}, 1 \leq \alpha \leq n$.

As in the one-dimensional case, there exists a linearization of the n-component KP-hierarchy from which the Lax equations (14) and (15) follow as compatibility conditions. Consider namely

$$L\psi = z\psi, \qquad U_\alpha \psi = \psi E_{\alpha\alpha} \tag{16}$$

and

$$\partial_{i\alpha}(\psi) = (L^i U_\alpha)_+(\psi) =: P_{i\alpha}(\psi) \qquad \text{for all } i \geq 1, 1 \leq \alpha \leq n. \tag{17}$$

If one applies the operator $\partial_{i\alpha}$ to both sides of both equations in (16) and performs the following manipulations

$$\begin{aligned} \partial_{i\alpha}(L)\psi + L\partial_{i\alpha}(\psi) = \{\partial_{i\alpha}(L) + LP_{i\alpha}\}\psi = \\ z\partial_{i\alpha}(\psi) = P_{i\alpha}(z\psi) = \{P_{i\alpha}L\}\psi, \qquad (18) \\ \partial_{i\alpha}(U_\beta)\psi + U_\beta\partial_{i\alpha}(\psi) = \{\partial_{i\alpha}(U_\beta) + U_\beta P_{i\alpha}\}\psi = \\ \partial_{i\alpha}(\psi)E_{\beta\beta} = \{P_{i\alpha}U_\beta\}\psi, \qquad (19) \end{aligned}$$

then one ends up with the equations

$$\left(\partial_{i\alpha}(L) - [(L^iU_\alpha)_+, L]\right)\psi = 0 \text{ and } \left(\partial_{i\alpha}(U_\beta) - [(L^iU_\alpha)_+, U_\beta]\right)\psi = 0. \quad (20)$$

So, if we can scratch the function ψ in both expressions, we get the equations (14) and (15) for L and the $\{U_\alpha\}$. Once one has obtained with this procedure operators L and U_α that satisfy (14) and (15), one still has to verify the equations (13). This will be immediate from our construction. We proceed now with specifying somewhat more the functions ψ in the linearization. This is done in the next subsection.

5.2 Wavefunctions

From the equations (16) and (17), we see that the functions ψ must be susceptible for left actions of various pseudodifferential operators with coefficients from $M_n(R)$ and for right actions with elements from $M_n(\mathbb{C})$. Therefore it is not surprising that they will be matrix functions. To motivate the form of the matrix functions ψ that we will choose, we consider the trivial solution $L = \partial$ and $U_\alpha = E_{\alpha\alpha}$. Then the equations (16) reduce to

$$\partial(\psi) = z\psi \quad , \quad E_{\alpha\alpha}\psi = \psi E_{\alpha\alpha} \quad \text{and} \quad \partial_{i\alpha}(\psi) = z^i E_{\alpha\alpha}\psi. \qquad (21)$$

Hence, for the trivial solution of the n-component KP-hierarchy, the matrix function

$$\gamma(x,t) = \exp\left(xz + \sum_{\alpha=1}^{n}\sum_{i=1}^{\infty} t_{i\alpha}E_{\alpha\alpha}z^i\right) = \sum_{k=0}^{\infty} \underline{p}_k(x,t)z^k, \qquad (22)$$

satisfies the equation in (16). Next we introduce a space of matrix functions for which all the manipulations at the end of subsection 5.1 make sense and whose elements can be seen as perturbations of the solution (22). This space M of so-called oscillating matrix functions, is defined as

$$M = \left\{ \left(\sum_{j\leq N} a_j z^j\right) \exp\left(xz + \sum_{\alpha=1}^{n}\sum_{i=1}^{\infty} t_{i\alpha}E_{\alpha\alpha}z^i\right) \middle| a_j \in M_n(R) \right\}.$$

Note that we will treat in this paper the product in the elements of M of a factor meromorphic at infinity and the exponential factor corresponding to the trivial solution as a formal one. It would no longer be a formal matter, if one can make sense in $M_n(R)$ of all the

$$\sum_{k\in\mathbb{Z}} a_{l-k}\underline{p}_{\ k}, \qquad l\in\mathbb{Z}, \tag{23}$$

and this clearly requires convergence considerations. They have been given in [10]. Here we present a more general algebraic approach.

The space M can be made into a module for the ring $M_n(R)[\partial,\partial^{-1})$ of pseudodifferential operators in ∂ with coefficients from $M_n(R)$. If one puts the actions of the elements $b\in M_n(R)$ and the derivations ∂ and $\partial_{k\beta}$ as follows

$$b\left[\left(\sum_j a_j z^j\right)\exp\left(xz+\sum_{\alpha=1}^{n}\sum_{i\geq 1} t_{i\alpha}E_{\alpha\alpha}z^i\right)\right]=\left(\sum_j ba_j z^j\right)\gamma(x,t)$$

$$\partial\left\{\left(\sum_j a_j z^j\right)\exp\left(xz+\sum_{\alpha=1}^{n}\sum_{i\geq 1} t_{i\alpha}E_{\alpha\alpha}z^i\right)\right\}=$$
$$\left(\sum_j \partial(a_j)+\sum_j a_j z^{1+j}\right)\gamma(x,t),$$

$$\partial_{k\beta}\left\{\left(\sum_j a_j z^j\right)\exp\left(xz+\sum_{\alpha=1}^{n}\sum_{i\geq 1} t_{i\alpha}E_{\alpha\alpha}z^i\right)\right\}=$$
$$\left(\sum_j \partial_{k\beta}(a_j)+\sum_j a_j E_{\beta\beta}z^{k+j}\right)\gamma(x,t),$$

then one sees immediately that ∂ acts by an invertible transformation on M. Hence these actions extend in a natural way to one of $M_n(R)[\partial,\partial^{-1})$. Thus M becomes even a free left $M_n(R)[\partial,\partial^{-1})$-module, for

$$\left(\sum p_j\partial^j\right)\cdot\exp\left(xz+\sum_{\alpha,i} t_{i\alpha}E_{\alpha\alpha}z^i\right)=\left(\sum p_j z^j\right)\exp\left(xz+\sum_{\alpha,i} t_{i\alpha}E_{\alpha\alpha}z^i\right).$$

M is also in a natural way a right module for the diagonal matrices in $M_n(\mathbb{C})$.

For, if $A = \sum_{i=1}^{n} a_i E_{ii}, a_i \in \mathbb{C}$, then we put

$$\left\{\sum a_j z^j\right\} \exp\left(xz + \sum t_{i\alpha} z^i E_{\alpha\alpha}\right) \cdot A := \left\{\sum_j a_j A z^j\right\} \exp\left(xz + \sum t_{i\alpha} z^i E_{\alpha\alpha}\right).$$

Since all the operations in the formulae (16) and (17) now make sense in M, one can look for solutions of these equations inside M. Note that the equations (16) impose still another relation on the operators U_α. For, we have

$$(\sum_{\alpha=1}^{n} U_\alpha)\psi = \psi(\sum_{\alpha=1}^{n} E_{\alpha\alpha}) = \psi, \tag{24}$$

and, since M is a free $M_n(R)[\partial, \partial^{-1})$-module, this implies that

$$\sum_{\alpha=1}^{n} U_\alpha = Id. \tag{25}$$

Hence the solutions of the n-component KP-hierarchy that one obtains via the linearization satisfy still this additional condition.

There are two types of transformations that map solutions ψ in M of the equations (16) and (17) into new ones for the same operators L and U_α. First there is the right action of the diagonal matrices in $M_n(\mathbb{C})$. Secondly there is an action of a discrete group. For, if $\delta = (\delta_1, \ldots, \delta_n) \in \mathbb{Z}^n$ and we write z^δ for the element

$$\operatorname{diag}\left(z^{\delta_1}, \ldots, z^{\delta_n}\right) = \begin{pmatrix} z^{\delta_1} & & 0 \\ & \ddots & \\ 0 & & z^{\delta_n} \end{pmatrix}.$$

Then the element $z^\delta \exp(\sum t_{i\alpha} z^i E_{\alpha\alpha})$ belongs to M and we have

$$z^\delta \gamma(x,t) = \left(\sum_{\alpha=1}^{n} E_{\alpha\alpha} \partial_{1\alpha}^{\delta_\alpha}\right) \exp(xz + \sum t_{i\alpha} z^i E_{\alpha\alpha}) =: R(\delta) \cdot \gamma(x,t)$$

Clearly the element $R(\delta)$ in $M_n(R)[\partial, \partial^{-1})$ is invertible with inverse $R(\delta)^{-1} = R(-\delta)$. Note that, if $\psi = P \cdot \exp(xz + \sum t_{i\alpha} z^i E_{\alpha\alpha}) \in M$ satisfies the equations in (16) and (17), then also

$$\psi^{(\delta)} = P \cdot R(\delta) \cdot \exp(xz + \sum t_{i\alpha} z^i E_{\alpha\alpha}). \tag{26}$$

Any matrix function ψ in M can without any restriction be written in the form (26) with P of the form $P = \sum_{j \leq 0} p_j \partial^j$. We assume now that the leading coefficient of P is equal to the identity. This gauge carries a special name: an element ψ in M is called an oscillating matrix function *of type* z^δ if it has the form

$$\psi = P \cdot z^\delta \gamma(t) = P \cdot R(\delta) \cdot \exp(xz + \sum t_{i\alpha} z^i E_{\alpha\alpha}) \text{ with } P = \mathrm{Id} + \sum_{j<0} p_j \partial^j .$$

The connection between oscillating matrix functions of type z^δ satisfying (16) and (17) and solutions of the n-component KP-hierarchy is given by

Proposition 5.3 *Let* $\psi = P \cdot R(\delta) \cdot \exp(xz + \sum t_{i\alpha} z^i E_{\alpha\alpha})$ *in* M *be an oscillating matrix function of type* z^δ . *Assume, we have operators* L *and* $U_\alpha, 1 \leq \alpha \leq n$, *in* $M_n(R)[\partial, \partial^{-1})$ *of the form (12), such that the equations in (16) hold and such that for* $i \geq 1$ *and all* $\beta, 1 \leq \beta \leq n$, *there holds*

$$\partial_{i\beta}(\psi) = B_{i\beta}(\psi) \text{ for some } B_{i\beta} \in M_n(R)[\partial].$$

Then we have $L = P\partial P^{-1}$, $U_\alpha = PE_{\alpha\alpha}P^{-1}$ *and* $B_{i\beta} = P_{i\beta} = (L^i U_\beta)_+$ *for all* i *and* β. *Moreover the operators* L *and the* $\{U_\alpha\}$ *are a solution of the* n*-component* KP*-hierarchy.*

For a proof we refer the reader to [9]. A useful consequence of this proposition is

Corollary 5.4 *An oscillating matrix function* $\psi = PR(\delta)\exp(xz + \sum t_{i\alpha} z^i E_{\alpha\alpha})$ *of type* z^δ, *satisfies equation (17) if and only if* P *satisfies the so-called Sato-Wilson equations:*

$$\partial_{i\alpha}(P)P^{-1} = -(P\partial^i E_{\alpha\alpha} P^{-1})_-, \text{ for all } i \geq 1, 1 \leq \alpha \leq n. \tag{27}$$

Such a function ψ *is called a wavefunction of the* n*-component* KP*-hierarchy.*

5.5 The bilinear equations

At the description of the bilinear form of the n-component KP-hierarchy and at the construction of Darboux transformations of this system of equations it is convenient to have besides M also at one's disposal its adjoint space M^* consisting of all formal products

$$\{\sum_{j\leq N} a_j z^j\} \exp(-xz - \sum_{\alpha=1}^{n}\sum_{i=1}^{\infty} t_{i\alpha} z^i E_{\alpha\alpha}) = \{\sum_{j\leq N} a_j z^j\}\gamma(-x,-t),$$

where $a_j \in M_n(R)$ for all j. On the space M^* we define the following actions of $M_n(R)$, ∂ and the $\{\partial_{k\beta}\}$

$$b\{(\sum_j a_j z^j)\gamma(-x,-t)\} = (\sum_j b a_j z^j)\gamma(-x,-t)$$

$$\partial_{k\beta}\{(\sum_j a_j z^j)\gamma(-x,-t)\} = (\sum_j \partial_{k\beta}(a_j)z^j - \sum_j a_j E_{\beta\beta} z^{k+j})\gamma(-x,-t)$$

$$\partial\{(\sum_j a_j z^j)\gamma(-x,-t)\} = (\sum_j \partial_{k\beta}(a_j)z^j - \sum_j a_j z^{1+j})\gamma(-x,-t)$$

In particular we see that ∂ acts invertible on M^*, so that the natural extension of these actions leads to an $M_n(R)[\partial,\partial^{-1})$-module structure on M^*, which is again free, since we have

$$\sum_j P_j(-\partial)^j \cdot \exp(-xz - \sum t_{i\alpha} z^i E_{\alpha\alpha}) = \{\sum_j P_j z^j\}\exp(-xz - \sum t_{i\alpha} z^i E_{\alpha\alpha}).$$

Now we have a bilinear pairing $\mathcal{R} : M \times M^* \to M_n(R)$ defined as follows: if $\varphi = \varphi(x,t,z) = (\sum_j a_j(x,t)z^j)\exp(xz + \sum t_{i\alpha} z^i E_{\alpha\alpha})$ belongs to M and $\psi = \psi(x,t,z) = (\sum_k b_k(x,t)z^k)\exp(-xz - \sum t_{i\alpha} z^i E_{\alpha\alpha})$ is an element of M^*, then we put

$$\begin{aligned}\mathcal{R}(\varphi,\psi) :&= \mathrm{Res}_z((\sum_j a_j(x,t)z^j)(\sum b_k(x,t)^T z^k))\\ &= \sum_{k\in\mathbb{Z}} a_{-k-1}(x,t)b_k(x,t)^T,\end{aligned}$$

where a^T for each a in $M_n(R)$ denotes the transposed of a. Clearly the defining sum in $\mathcal{R}$ is finite, hence belongs to $M_n(R)$. On the algebra $M_n(R)[\partial,\partial^{-1})$ we have a $\mathbb{C}$-linear anti-algebra morphism called "taking the adjoint". The adjoint of $P = \sum P_j\partial^j$ is given by

$$\begin{aligned}P^* &= \sum_j(-\partial)^j P_j^T = \sum_j(-1)^j \sum_{k\geq 0}\binom{j}{k}\partial^k(P_j^T)\partial^{j-k}\\ &= \sum_l\left\{\sum_{k\geq 0}(-1)^{l+k}\binom{l+k}{k}\partial^k(P_{l+k}^T)\right\}\partial^l.\end{aligned}$$

The following theorem gives an important connection between the bilinear form $\mathcal{R}$ and taking the adjoint. Its proof can be found in [9]

Theorem 5.6 *Let* $\varphi(x,t,z) = P(x,t,\partial)\cdot\exp(xz+\sum t_{i\alpha}z^i E_{\alpha\alpha}) \in M$ *and* $\psi(t,z) = Q(x,t,\partial)\cdot\exp(-xz-\sum t_{i\alpha}z^i E_{\alpha\alpha}) \in M^*$, *then*

$$\begin{aligned}(P(x,t,\partial)Q(x,t,\partial)^*)_- &= -\sum_{n=0}^{\infty}\mathcal{R}(\varphi(x,t,z),\partial^n(\psi(x,t,z)))(-\partial)^{-n-1} \\ &=: \mathcal{R}(\varphi(x,t,z),\partial^{-1}\circ\psi(x,t,z))\end{aligned} \tag{28}$$

Like we have in M the notion of oscillating matrix function of type z^δ, we have in M^* that of dual oscillating matrix function of type $z^{-\delta}$. This is by definition an element ψ in M^* of the form

$$\begin{aligned}\psi(x,t,z) &= Q(x,t,z)\cdot R(-\delta)^*\cdot\exp(-xz-\sum t_{i\alpha}z^iE_{\alpha\alpha}) \\ &= \left\{(1+\sum_{j<0}g_jz^j)z^{-\delta}\right\}\exp(-xz-\sum t_{i\alpha}z^iE_{\alpha\alpha})\end{aligned}$$

If $\varphi = P(x,t,\partial)R(\delta)\cdot\exp(xz+\sum t_{i\alpha}z^iE_{\alpha\alpha})$ in M is an oscillating matrix function of type z^δ, then

$$\varphi^* = (P(t,\partial)^*)^{-1}R(-\delta)^*\cdot\exp(-xz-\sum t_{i\alpha}z^iE_{\alpha\alpha}) \tag{29}$$

is clearly a dual oscillating matrix function of type $z^{-\delta}$. It is called the adjoint of φ. Since

$$(P(x,t,\partial)((P(x,t,\partial)^*)^{-1})^*)_- = 0,$$

the foregoing theorem shows that for all $n\geq 0$

$$\mathcal{R}(\varphi,\partial^n(\varphi^*)) = 0. \tag{30}$$

This property even characterizes φ^* among the dual oscillating matrix functions of type $z^{-\delta}$. For, if $\psi = Q(x,t,\partial)R(-\delta)^*\exp(-xz-\sum t_{i\alpha}z^iE_{\alpha\alpha}) \in M^*$ is such a function satisfying (30) with φ^* replaced by ψ, then we have according to the theorem

$$PR(\delta)(R(-\delta)^*)^*Q^* = PQ^* = \partial^0 + (PQ^*)_- = 1.$$

In other words, $Q = (P^*)^{-1}$ and ψ is the adjoint of φ. We will use this criterion later on and therefore we resume it in a

Corollary 5.7 *Let* φ *be an oscillating matrix function of type* z^δ *and let* ψ *be a dual oscillating matrix function of type* $z^{-\delta}$. *Then* ψ *is the adjoint of* φ *if and only if it satisfies*

$$\mathcal{R}(\varphi(x,t,z),\partial^n(\psi(x,t,z))) = 0 \text{ for all } n\geq 0.$$

If φ is a wavefunction of the n-component KP-hierarchy, then its adjoint φ^* is called a dual wavefunction of the n-component KP-hierarchy. It satisfies a set of linear equations similar to (17), see [9].

The bilinear form $\mathcal{R}$ plays also an important role at the interpretation of the famous bilinear identities for an oscillating function ψ and and its dual, (see [3]):

$$\mathrm{Res}_z\, \psi(x,t,z)\psi^*(y,s,z)^T = 0. \tag{31}$$

First we have a look at the formal product $\psi(x,t,z)\phi(x,s,z)^T$, where $\psi(x,t,z) = \{\sum_k a_k(x,t)z^k\}\gamma(x,t) \in M$ and $\phi(y,t,z) = \{\sum_m b_m(y,s)z^m\}\gamma(-y,-t) \in M^*$. If we write $e^{xz+\sum t_{i\alpha}z^i E_{\alpha\alpha}} = \sum_{r=0}^{\infty} p_r(x,(t_{i\alpha}))z^r$, then this formal product becomes

$$(\sum_k a_k(x,t)z^k)(\sum_{r=0}^{\infty} p_r(x-y,(t_{i\alpha}-s_{i\alpha}))z^r)(\sum_m b_m(y,s)^T z^m) \tag{32}$$

Hence the equation (31) for $\phi = \psi^*$ boils formally down to

$$\sum_{r=0}^{\infty}\sum_k a_k(x,t)p_r(x-y,(t_{i\alpha}-s_{i\alpha}))b_{-r-1-k}(y,s)^T = 0,$$

for all relevant $t = (t_{i\alpha})$ and $s = (s_{i\alpha})$. In order to avoid convergence considerations, we look at a few differential consequences of these relations that always exist. If one differentiates the equation namely with respect to x or some of the parameters $(t_{i\alpha})$ or $(s_{i\alpha})$ and substitutes next $t_{i\alpha} = s_{i\alpha}$ for all $i \geq 1$, then one ends up with finite expressions in the coefficients of ψ and ψ^*. Thus we get for example for all $k \geq 0$, $i \geq 1$, $1 \leq \alpha \leq n$, and all $m \geq 0$ the relations

$$\mathcal{R}(\partial^k(\psi(x,t,z)), \partial^m(\psi^*(x,t,z))) = 0, \tag{33}$$

$$\mathcal{R}(\partial_{i\alpha}(\psi(x,t,z)), \partial^m(\psi^*(x,t,z))) = 0. \tag{34}$$

Note that this first equation can also be obtained by applying ∂ several times to the relations (30). Assume now that an oscillating matrix function ψ of type z^δ and its dual ψ^* satisfy the equations (33) and (34). Let $L = P\partial P^{-1}$ and $U_\alpha = PE_{\alpha\alpha}P^{-1}$ be the operators associated with ψ. From the form of the action of the $\partial_{i\alpha}$ on ψ one sees that there exists for each $i \geq 1$ and all $\alpha, 1 \leq \alpha \leq n$, a differential operator $Q_{i\alpha}$ in $M_n(R)[\partial]$ of degree i such that

$$\partial_{i\alpha}(\psi) - Q_{i\alpha}(\psi) = G_{i\alpha}\partial^l e^{xz+\sum_i t_i z^i}, \text{ with } G_{i\alpha} \in M_n(R)[\partial,\partial^{-1}] \text{ of degree} < 0.$$

Because of the relations in (33) and (34) we have then for all $m \geq 0$ that

$$\mathcal{R}(\partial_{i\alpha}(\psi(x,t,z)) - (Q_{i\alpha})(\psi(x,t,z)), \partial^m(\psi^*(x,t,z))) = 0. \tag{35}$$

According to theorem 5.6, we get then that $(G_{i\alpha}\partial^l P^{-1})_- = G_{i\alpha}\partial^l P^{-1} = 0$. Thus we have obtained that $\partial_{i\alpha}(\psi(x,t,z)) - (Q_{i\alpha})(\psi(x,t,z)) = 0$. From proposition 5.3 we know then that this implies for all $i \geq 1$ and all α that the operator $Q_{i\alpha}$ is equal to $(L^i U_\alpha)_+$. Hence ψ is a wavefunction of the n-component KP-hierarchy. Reversely, if ψ is a wavefunction of the the n-component KP-hierarchy, then the first equation in (33) holds and since $\partial_{i\alpha}$ acts on ψ as $(L^i U_\alpha)_+$ also the second relation holds. Thus we have found the following characterization of the wavefunctions in M

Proposition 5.8 *An oscillating matrix function ψ of type z^l is a wavefunction of the n-component KP-hierarchy if and only if ψ and its dual ψ^* satisfy the equations (33) and (34).*

A consequence of this proposition is that wavefunctions of the n-component KP-hierarchy are characterized by the following set of equations

$$\mathfrak{R}(\Delta_1(\psi(x,t,z)), \Delta_2(\psi^*(x,t,z))) = 0, \tag{36}$$

where Δ_1 and Δ_2 are arbitrary finite products of the $\{\partial_{i\alpha}\}$ and ∂. These equations are the algebraic version of the equations (31).

5.9 Grassmannian construction of solutions

In this subsection we describe the Sato Grassmannian, from which one can construct matrix wavefunctions of the n-component KP-hierarchy. It is a vector form of the line set out in [22]. Since the rows of these functions consist of infinite series in z and z^{-1} with coefficients from $\mathbb{C}^n$, it is not so strange that the basic manifold for the relevant subspaces consists of a class of series in z and z^{-1}. Consider the spaces

$$H_- = (z^{-1}\mathbb{C}[[z^{-1}]])^n = \{\sum_{j=1}^{\infty} a_j z^{-j} | a_j \in \mathbb{C}^n\} \quad \text{and}$$

$$H_+ = (\mathbb{C}[z])^n = \{\sum_{i=0}^{m} b_i z^i | b_i \in \mathbb{C}^n\}.$$

Hence $H = H_+ \oplus H_-$ is equal to $(\mathbb{C}((z^{-1})))^n$, where $\mathbb{C}((z^{-1}))$ is the quotient field of $\mathbb{C}[[z^{-1}]]$. First we fix some notations in the space H. For l, $0 \leq l \leq n-1$, let f_l be the vector in $\mathbb{C}^n$ with a one at the $l+1$-th place and elsewhere zeros. Then we denote for all $m \in \mathbb{Z}$ and all l, $0 \leq l \leq n-1$, the vector $f_l z^m$ by e_{l+mn}. Inside H we consider the subspaces $H_k, k \in \mathbb{Z}$ given by

$$H_k := \{\sum_{r \geq k} a_r e_r, \text{ with } a_r \in \mathbb{C} \text{ and } \sum_{r \geq k} a_r e_r \in H\}. \tag{37}$$

For each $k \in \mathbb{Z}$ let $p_k : H \to H_k$ be the projection

$$p_k(\sum a_j e_j) = \sum_{j \geq k} a_j e_k.$$

Then the Grassmann manifold of Sato consists of all subspaces of H that are of a size comparable to H_+. More precisely, it is given by

$$\mathrm{Gr}(H) = \left\{ W \,\middle|\, \begin{array}{l} W \subset H, p_0 : W \to H_+ \text{ has a finite} \\ \text{dimensional kernel and cokernel} \end{array} \right\}.$$

The space $\mathrm{Gr}(H)$ has a subdivision into different components according to the index of $p_0|_W$ that is defined by

$$\mathrm{ind}\,(p_0|_W) = \dim(\mathrm{Ker}(p_0|_W)) - \dim(\mathrm{Coker}(p_0|_W)).$$

We denote these components as follows

$$\mathrm{Gr}^{(k)}(H) = \{W | W \in \mathrm{Gr}(H), \mathrm{ind}(p_0|_W) = -k\}.$$

Clearly, the subspace H_k belongs to $\mathrm{Gr}^{(k)}(H)$ and one easily verifies that the component $\mathrm{Gr}^{(k)}(H)$ can also be described as

$$\mathrm{Gr}^{(k)}(H) = \{W | W \in \mathrm{Gr}(H), \mathrm{ind}(p_k|_W) = 0\}.$$

The "big cell" in $\mathrm{Gr}^{(k)}(H)$ consists of all subspaces in $\mathrm{Gr}(H)$ that project bijectively onto H_k, i.e. all W in

$$O^{(k)} = \left\{ W \,\middle|\, \begin{array}{l} W \in \mathrm{Gr}(H), w \mapsto p_k(w) \text{ is a} \\ \text{bijection: } W \to H_k \end{array} \right\}.$$

On the space H we have a bilinear form B that plays a role in the sequel. Namely, if $f = \sum_j a_j z^j$ and $g = \sum_j b_j z^j$ are in H, then we define

$$B(f, g) = \mathrm{Res}_z(f(z)g(z)) = \sum_j a_j b^T_{-j-1}, \tag{38}$$

where we have written the elements of $\mathbb{C}^n$ as rows and A^T denotes the transposed of a matrix A. For W in $\mathrm{Gr}(H)$, let $W^\perp$ be the orthocomplement of W in H w.r.t.this form B. With the above given description of spaces in $\mathrm{Gr}(H)$ one verifies that $W^\perp$ also belongs to $\mathrm{Gr}(H)$.

Our next step will be to show how the subspaces from $\mathrm{Gr}(H)$ occur as the span of the rows of the Laurent coefficients of certain oscillating functions.

Consider a matrix wavefunction ψ of the n-component KP-hierarchy and its dual ψ^* that have the form

$$\psi(x,t,z) = \{\sum_{j\le 0} a_j(x,t)z^j\}z^\delta \exp(xz + \sum t_{i\alpha}z^i E_{\alpha\alpha}) \tag{39}$$

$$\psi^*(x,s,z) = \{\sum_{m\le 0} b_m(x,t)z^m\}z^{-\delta} \exp(-xz - \sum t_{i\alpha}z^i E_{\alpha\alpha}). \tag{40}$$

The class of wavefunctions we will consider in this paper that satisfies the condition that there is an θ in $\mathbb{C}[[x,t]]$ of the form (11) such that for all $m \le 0$ and all $j \le 0$

$$\theta(x,t)a_j(x,t) \in M_n(\mathbb{C}[[x,t]]) \text{ and } \theta(x,t)b_m(x,t) \in M_n(\mathbb{C}[[x,t]]). \tag{41}$$

These matrix wavefunctions are called *regularizable*. For regularizable wavefunctions the Laurent series in x of ψ and ψ^* have the form

$$\psi(x,t,z) = \sum_{j\ge -N} w_j(t,z)x^j \exp(xz + \sum t_{i\alpha}z^i E_{\alpha\alpha}),$$

where $w_j(t,z) = \sum_{l=-\infty}^{N_1} v_l z^l$, with $v_l \in \mathbb{C}[[t]]^n$,

$$\psi^*(x,t,z) = \sum_{j\ge -N} w_j^*(t,z)x^j \exp(-xz - \sum t_{i\alpha}z^i E_{\alpha\alpha}),$$

where $w_j^*(t,z) = \sum_{l=-\infty}^{N_2} v_l^* z^l$, with $v_l^* \in \mathbb{C}[[t]]^n$. It is not hard to show that both spaces

$$W = \text{Span}\{\text{ rows of } w_j(0,z),\ j \ge -N\}$$

and

$$W^* = \text{Span}\{\text{ rows of } w_j^*(0,z),\ j \ge -N\}$$

belong to $\text{Gr}(H)$. Following ideas of Sato, one can show, see [3], that the space W even determines ψ, for there holds

Proposition 5.10 *The map B_δ that associates to a regularizable wavefunction ψ of type z^δ of the n-component KP-hierarchy the span of the rows of the coefficients in $t = 0$ of the Laurent series of ψ in x is an injection from the class of matrix wavefunctions of type z^δ into* $\text{Gr}(H)$. *All planes in* $\text{Gr}(H)$ *occur in the image of this map for a certain type z^δ. The same properties hold for the map B_δ^* that associates to a regularizable dual wavefunction ψ^* of type $z^{-\delta}$ of the n-component KP-hierarchy the span of the rows of the coefficients in $t = 0$ of the Laurent series of ψ^* in x.*

It might happen that there correspond to a certain plane, matrix wavefunctions of various types, see [11]. The wavefunctions that satisfy the conditions in (41) for $\theta = 1$ correspond to the union of the big cells. For each z^δ and each $W \in \mathrm{Gr}(H)$ in the image of B_δ, we write the matrix wavefunction of type z^δ corresponding to W and its dual matrix wavefunction as

$$\psi_W^{(\delta)} = \hat{\psi}_W^{(\delta)} z^\delta \exp(xz + \sum t_{i\alpha} z^i E_{\alpha\alpha}) \text{ resp.} \tag{42}$$

$$(\psi_W^{(\delta)})^* = (\hat{\psi}_W^{(\delta)})^* z^{-\delta} \exp(-xz - \sum t_{i\alpha} z^i E_{\alpha\alpha}). \tag{43}$$

Now that we have this link between matrix wavefunctions and the Grassmann manifold, we can also give a geometric description of the dual wavefunction of $\psi_W^{(\delta)}$. From the characterizing properties in (33), follows

Proposition 5.11 *Let W belong to the image of B_δ and let $\tilde{W}$ be a subspace in $\mathrm{Gr}(H)$. Then $\tilde{W}$ is the space W^* corresponding to the dual wavefunction $(\psi_W^{(\delta)})^*$, if and only if $\tilde{W} = W^\perp$ with $W^\perp$ the orthocomplement of W w.r.t. the bilinear form B on H.*

6 From n-component KP to WDVV

6.1 The relation $\partial_{\mathbf{k}}(\gamma_{\mathbf{ij}}) = \gamma_{\mathbf{ik}}\gamma_{\mathbf{kj}}$

Let $\psi = P \cdot z^\delta \gamma(t) = P \cdot R(\delta) \cdot \exp(xz + \sum t_{i\alpha} z^i E_{\alpha\alpha})$ with $P = \mathrm{Id} + \sum_{j<0} p_j \partial^j$ be a wavefunction of the n-component KP-hierarchy. Then we know from corollary 5.4 that P satisfies the Sato equations

$$\partial_{i\alpha}(P) = -(P\partial^i E_{\alpha\alpha} P^{-1})_- P \tag{44}$$

$$= -P\partial^i E_{\alpha\alpha} + (P\partial^i E_{\alpha\alpha} P^{-1})_+ P. \tag{45}$$

We want to show now that the coefficients of p_{-1} are the candidate rotation coefficients and that the $\{t_{1\alpha}\}$ will serve as the coordinates u_α from theorem 3.7. A direct computation shows that

$$(P\partial E_{\alpha\alpha} P^{-1})_+ = \partial E_{\alpha\alpha} + [p_{-1}, E_{\alpha\alpha}].$$

Substituting this in the Sato equation for $i = 1$ and looking at the coefficient of ∂^{-1} results in the matrix equality

$$\partial_{i\alpha}(p_{-1}) = [E_{\alpha\alpha}, p_{-2}] + E_{\alpha\alpha}\partial(p_{-1}) + [p_{-1}, E_{\alpha\alpha}]p_{-1}.$$

This relation implies for the off-diagonal entries of p_{-1} that for $\alpha \neq \beta \neq \gamma \neq \alpha$

$$\partial_{i\alpha}((p_{-1})_{\beta\gamma}) = 0 + 0 + (p_{-1})_{\beta\alpha}(p_{-1})_{\alpha\gamma} + 0,$$

which is exactly the relation we are looking for.

6.2 The dilatation invariance

Now that we have our candidate rotation coefficients, the $(p_{-1})_{\beta\alpha}, \beta \neq \alpha$, and know that $u_\alpha = t_{1\alpha}$, we have $\sum_k \partial_k = \sum_\alpha \partial_{i\alpha} = \partial$. The action of ∂ on ψ is

$$\partial(\psi) = \{\sum_{i<0} \partial(p_i) z^i\} z^\delta \gamma(x,t,z) + \{\sum_{i<0} p_i z^{i+1}\} z^\delta \gamma(x,t,z).$$

Hence we see that the condition $\partial(\psi) = z\psi$ is equivalent to $\partial(p_i) = 0$ for all $i < 0$. This is more than sufficient to get the dilatation invariance. Let $W = B_\delta(\psi)$. Then it is well-known that the relation $\partial(\psi) = z\psi$, amounts to the geometric property $zW \subset W$. Examples of such subspaces are the

$$W = g(z)H_+, \text{ where } g(z) \in L(S^1, Gl_n(\mathbb{C})), \tag{46}$$

the loopgroup of $Gl_n(\mathbb{C})$.

6.3 The symmetry of the rotation coefficients

To get the symmetry of the rotation coefficients we have to restrain the planes from the Grassmannian still more. Thereto we consider here besides the wavefunction ψ also its dual wavefunction

$$\psi^* = (P^*)^{-1}.z^{-\delta}\gamma(-x,-t,z) = \{Id - p_{-1}^T z^{-1} + \ldots\} z^{-\delta}\gamma(-x,-t,z)$$

Compare this function with

$$\psi(-x,t,-z) = \{\sum_i (-1)^i p_i z^i\}(-z)^\delta \exp\left(-xz + \sum_{\alpha=1}^{n}\sum_{i=1}^{\infty}(-1)^i t_{i\alpha} E_{\alpha\alpha} z^i\right).$$

Then we see that, if we choose $\delta = 0$, $\tilde{t} = (t_{1\alpha}, 0, t_{3\alpha}, 0, \ldots)$, the equality

$$\psi(-x,\tilde{t},-z) = \psi^*(x,\tilde{t},z)$$

implies that $p_{-1} = p_{-1}^T$ and this is the result we want. Recall that the dual wavefunction was characterized by the bilinear equations from 5.7. Thus one obtains the equivalence

$$\mathcal{R}(\psi(x,\tilde{t},z), \partial^n(\psi(-x,\tilde{t},-z))) = 0 \text{ for all } n \geq 0 \Leftrightarrow \psi(-x,\tilde{t},-z) = \psi^*(x,\tilde{t},z). \tag{47}$$

By using this characterization, one verifies that all the

$$W = g(z)H_+, \quad g(z) \in L(S^1, Gl_n(\mathbb{C})) \text{ and } g(z)^{-1} = g(-z)^T, \tag{48}$$

satisfy these conditions and hence supply a large collection of solutions of the Darboux-Egoroff equations.

6.4 *The linearization*

In view of the foregoing results we start with a plane $W = g(z)H_+,\ \ g(z) \in L(S^1, Gl_n(\mathbb{C}))$ and $g(z)^{-1} = g(-z)^T$. The corresponding wavefunction of type Id satisfies the equations

$$\sum_{\alpha=1}^{n} \partial_{1\alpha}\psi = z\psi \tag{49}$$

$$\partial_{1\alpha}(\psi) = \{zE_{\alpha\alpha} + [p_{-1}, E_{\alpha\alpha}]\}\psi. \tag{50}$$

Like in the foregoing section we put the even numbered flows equal to zero and introduce the function

$$\theta(x, \tilde{t}, z) := z^{-1}\psi(x, \tilde{t}, z)g(z).$$

It satisfies the same equations as ψ

$$\sum_{\alpha=1}^{n} \partial_{1\alpha}(\theta(x, \tilde{t}, z)) = z\theta(x, \tilde{t}, z) \tag{51}$$

$$\partial_{1\alpha}(\theta(x, \tilde{t}, z)) = \{zE_{\alpha\alpha} + [p_{-1}, E_{\alpha\alpha}]\}\theta(x, \tilde{t}, z).$$

From the bilinear relations one deduces , see [20],

Proposition 6.5 *For the function $\theta(x, \tilde{t}, z)$ hold the following matrix identities*

$$\mathrm{Res}_z(\theta(x, \tilde{t}, z)) \neq 0 \ \textit{and}\ \mathrm{Res}_z(z\theta(x, \tilde{t}, z)) = 0.$$

The first says that, if we define $\Psi(x, \tilde{t}) := \mathrm{Res}_z(\theta(x, \tilde{t}, z)) = (\Psi_{ij}(x, \tilde{t}))$, this is a nonzero matrix function. Substitution of the second property in the equations (51) gives that the $(\Psi_{ij}(\tilde{t}))$ satisfy

$$\sum_{\alpha=1}^{n} \partial_{1\alpha}(\Psi_{ij}(x, \tilde{t})) = 0 \tag{52}$$

$$\partial_{1\alpha}(\Psi_{ij}) = \sum_{s} [p_{-1}, E_{\alpha\alpha}]_{is}\Psi_{sj} = (p_{-1})_{i\alpha}\Psi_{\alpha j}, \alpha \neq i. \tag{53}$$

In other words they are a solution of the linearization of the Darboux-Egoroff equations for the rotation coefficients $(p_{-1})_{ij}$.

References

1. H. Aratyn, J. F. Gomes, E. Nissimov, S. Pacheva, A. H. Zimmerman, Symmetry flows, conservation laws and dressing approach to the integrable models, in *Integrable Hierachies and Modern Physical Theories*, Kluwer Academic Publishers, Dordrecht (2001).

2. H. Aratyn, J. van de Leur, Solutions of the WDVV Equations and Integrable Hierarchies of KP type, hep-th/0104092.
3. E. Date, M. Jimbo, M. Kashiwara, T. Miwa, *J. Phys. Soc. Japan* **50**, 3806 (1981).
4. B. Dubrovin, *Nucl. Phys. B* **379**, 627 (1992).
5. B. Dubrovin, Integrable systems and classification of 2-dimensional topological field theories, *Integrable systems (Luminy, 1991)*, Birkhäuser, Boston, MA, p.313.
6. B. Dubrovin, Geometry of 2D topological field theories, em Integrable systems and quantum groups (Montecatini Terme, 1993), Springer, Berlin (1996), p. 120.
7. B . Dubrovin, Y . Zhang, *Comm. Math. Phys.* **198**, 311 (1998).
8. G. F. Helminck and J. W. van de Leur, *Comm. Math. Phys.* **193**, 627 (1998).
9. G. F. Helminck, J. W. van de Leur, *A geometric construction of the Dual Wavefunction of the n-component KP-hierarchy*, in *Lie Theory and its Applications III*, World Scientific, Singapore (2000), p.187.
10. G. F. Helminck, G. F. Post, *Trans. Amer. Math. Soc.* **324**, 187 (1991).
11. G. F. Helminck, G. F. Post, *Indag. Mathem., N. S.*, **5**, 411 (1994).
12. L. K. Hoevenaars, P. H. M. Kersten, R. Martini, *Physics Letters B* **503**, 189 (2001).
13. K. Ito, S. -K. Yang, *Physics Letters B* **433**, 56 (1998).
14. I. M. Krichever, *Comm. Math. Phys.* **143**, 415 (1992).
15. I. M. Krichever, *Comm. Pure Appl. Math.* **47**, 437 (1992).
16. J. W. van de Leur, R. Martini, *Comm. Math. Phys.* **205**, 587 (1999).
17. Y. Manin, *Frobenius manifolds, quantum cohomology, and moduli spaces, American Mathematical Society Colloquium Publications* **47** American Mathematical Society, Providence, RI (1999).
18. A. Marshakov, A. Mironov, A. Morozov, *Phys Letters B* **389**, 43 (1996).
19. A. Mironov, Seiberg-Witten Theories, Integrable models and Perturbative Prepotentials, hep-th/0010078.
20. J. van de Leur, *Internat. Math. Res. Notices* **2001**, 551 (2001).
21. O. Pavlyk, math-ph/0003020.
22. T. Shiota, Prym varieties and soliton equations, in: *Infinite-dimensional Lie algebras and groups, Luminy-Marseille, 1988, Adv. Ser. Math. Phys.* **7**, World Sci. Publishing, Teaneck, NJ (1989), p.407.

GLOBAL GAUSS LAW FOR LATTICE QCD

J. KIJOWSKI

Center for Theoretical Physics, Polish Academy of Sciences, al. Lotnikow 32/46, 02-668 Warsaw, Poland
E-mail: kijowski@cft.edu.pl

G. RUDOLPH

Institute for Theoretical Physics, University of Leipzig, Augustusplatz 10, 04109 Leipzig, Germany
E-mail: rudolph@rz.uni-Leipzig.de

It is shown that the local Gauss law of quantum chromodynamics on a finite lattice implies a gauge invariant, additive law giving rise to a gauge invariant $\mathbf{Z}_3$-valued global charge in QCD. The total charge contained in a region of the lattice is equal to the flux through its boundary of a certain $\mathbf{Z}_3$-valued, additive quantity.

1 Introduction

In the present paper we analyze the gauge invariant content of the Gauss law and the notion of global charge in lattice quantum chromodynamics (QCD). For this purpose, we discuss the field algebra and the observable algebra of this theory in some detail. This contribution should be understood as a first step in the direction of investigating the nonperturbative structure of QCD, the final aim being an effective microscopic theory of interacting hadrons.

We stress that standard methods from algebraic quantum field theory for models which do not contain massless particles, see [2], do not apply here. Some progress towards an implementation of similar ideas for theories with massless particles has been made, for the case of quantum electrodynamics (QED) see [3] and [4] and further references therein. For some attempts to deal with the nonabelian case we refer to papers by Strocchi and Wightman, see [5] and [6], and for the study of a $\mathbf{Z}_2$-gauge theory see [7].

In QED, the notion of *global* (electric) charge is easy to understand. This is due to the fact, that in this theory we have a *local* Gauss law, which is built from gauge invariant operators and which is linear. Thus, one can "sum up" the local Gauss laws over all points of a given (spacelike) hyperplane in space time yielding the following gauge invariant conservation law: The global electric charge is equal to the electric flux through a 2-sphere at infinity. On the contrary, in QCD the local Gauss law is neither built from gauge invariant operators nor is it linear. The main point of the present paper is to show that it is possible to extract from the local Gauss equation of QCD a gauge invariant, additive law for operators with eigenvalues in $\mathbf{Z}_3$ (which is canonically identified with the dual of the center of $SU(3)$). This implies – as in QED – a gauge invariant conservation law: The global $\mathbf{Z}_3$-valued charge is equal to a $\mathbf{Z}_3$-valued gauge invariant quantity obtained from the color electric flux at infinity.

We stress that our main observation is universal in the following sense: Suppose somebody had constructed a nonperturbative version of QCD rigorously. Then our construction of the global charge would apply. Since, however, such a theory is not at our disposal, for a rigorous discussion we have to restrict ourselves to the case of QCD approximated on a finite lattice. For basic notions concerning lattice gauge theories (including fermions) we refer to [1] and references therein. Thus, we consider QCD on a finite, regular cubic lattice in the Hamiltonian approach. Our starting point is the notion of the algebra of field operators. By imposing the local Gauss law and gauge invariance we obtain the observable algebra $\mathcal{O}(\Lambda)$ for QCD on the lattice. For an analogous investigation of lattice QED we refer to [8] and [10], and for a (rather heuristic) discussion of continuum QCD within the functional integral approach (using only gauge invariant quantities) we refer to [9].

2 The Field Algebra for Lattice QCD

We consider QCD in the Hamiltonian framework on a finite, regular 3-dimensional lattice Λ. We denote the set of oriented, i-dimensional elements of Λ by Λ^i, $i = 0, 1, 2, 3$. Such elements are (in increasing order of i) called sites, links, plaquettes and cubes. The set of non-oriented i-dimensional elements will be denoted by $|\Lambda|^i$. If, for instance, $(x, y) \in \Lambda^1$ is an oriented link, then by $|(x, y)| \in |\Lambda|^1$ we mean the corresponding non-oriented link.

The basic fields of lattice QCD are quarks living at lattice sites and gluons living on links. The field algebra is thus, by definition, the tensor product of local fermionic and bosonic algebras:

$$\mathcal{F}(\Lambda) := \bigotimes_{x\in|\Lambda|^0} \mathcal{F}_x \bigotimes_{|(x,y)|\in|\Lambda|^1} \mathcal{F}_{|(x,y)|} \,. \tag{1}$$

We impose locality of the lattice quantum fields by postulating that the local algebras corresponding to different elements of Λ commute with each other.

The fermionic field algebra $\mathcal{F}_x$ associated with a lattice site x is generated from the algebra $\mathcal{L}_x$ of canonical anticommutation relations of quarks. In terms of coordinates, the quark field is given by

$$|\Lambda|^0 \ni x \to (\psi^{aA}(x)) \in \mathcal{L}_x \,, \tag{2}$$

where a stands for bispinorial and (possibly) flavour degrees of freedom and $A = 1, 2, 3$ is the colour index corresponding to the fundamental representation of $SU(3)$. The conjugate quark field is denoted by $\psi^{*a}{}_A(x)$, where we raise and lower indices by the help of the canonical hermitian metric tensor g_{AB} in $\mathbf{C}^3$. Finally, the (nontrivial) canonical anticommutation relations for elements of $\mathcal{L}_x$ read:

$$[\psi^{*a}{}_A(x), \psi^{bB}(x)]_+ = \delta^B{}_A \delta^{ab} . \tag{3}$$

Passing to self-adjoint generators (real and imaginary parts of ψ), we observe that $\mathcal{F}_x$ is a Clifford algebra.

The bosonic field algebra $\mathcal{F}_{|(x,y)|}$ associated with the non-oriented link $|(x, y)|$ will be constructed in terms of its equivalent copies $\mathcal{F}_{(x,y)}$ and $\mathcal{F}_{(y,x)}$, corresponding to the two orientations of the link (x, y). The bosonic field algebra $\mathcal{F}_{(x,y)}$ is, by definition, the tensor product

$$\mathcal{F}_{(x,y)} := \tilde{\mathcal{E}}_{(x,y)} \otimes \mathcal{A}_{(x,y)} , \tag{4}$$

where $\tilde{\mathcal{E}}_{(x,y)}$ is the enveloping algebra of the (real) Lie algebra $\mathcal{E}_{(x,y)} \cong su(3)$ and $\mathcal{A}_{(x,y)} \cong C^\infty(SU(3))$ is the commutative $*$-algebra of smooth functions on the Lie group $SU(3)$. We note that $\mathcal{F}_{(x,y)}$ can be identified with the algebra of differential operators on the group manifold. For our purposes, we endow $\mathcal{F}_{(x,y)}$ with a Lie algebra structure. Thus, we have to define the commutator between a generator $e \in \mathcal{E}_{(x,y)}$ of $\tilde{\mathcal{E}}_{(x,y)}$ and a function $f \in \mathcal{A}_{(x,y)}$,

$$[e, f] := e^R(f) , \tag{5}$$

where $e^R(f)$ denotes the derivative of f with respect to the right-invariant vector field e^R generated by e.

We give an explicit description of $\mathcal{F}_{(x,y)}$ in terms of generators and defining relations. The algebra $\mathcal{A}_{(x,y)}$ is generated by matrix elements of the gluonic gauge potential on the link (x, y),

$$\Lambda^1 \ni (x, y) \to U^A{}_B(x, y) \in \mathcal{A}_{(x,y)} , \tag{6}$$

where $A, B = 1, 2, 3$ are colour indices. Being functions on $SU(3)$, these generators fulfil the following relations:

$$(U^A{}_B(x, y))^* U_A{}^C(x, y) = \delta^C{}_B \mathbf{1} , \tag{7}$$

$$\epsilon_{ABC}\, U^A{}_D(x, y) U^B{}_E(x, y) U^C{}_F(x, y) = \epsilon_{DEF} \mathbf{1} . \tag{8}$$

The algebra $\tilde{\mathcal{E}}_{(x,y)}$ is generated by colour electric fields, spanning the Lie algebra $\mathcal{E}_{(x,y)}$. Choosing a basis $\{t_i\}$, $i = 1, \ldots, 8$, of $su(3)$ we denote by $\{E_i(x, y)\}$ the corresponding basis of $\mathcal{E}_{(x,y)}$,

$$\Lambda^1 \ni (x, y) \to E_i(x, y) := t_i \in \mathcal{E}_{(x,y)} . \tag{9}$$

These generators are self-adjoint (real), $E_i^* = E_i$. In the sequel we take as the basis the hermitean, traceless Gell-Mann matrices ${t_i}^A{}_B$, normalized as follows: ${t_i}^A{}_B\, {t_j}^B{}_A = \delta_{ij}$. We will also use the following traceless matrix built from these fields

$$E^A{}_B(x,y) = \sum_i E_i(x,y){t_i}^A{}_B \,. \tag{10}$$

Since the coefficients ${t_i}^A{}_B$ are complex, these fields are no longer self-adjoint:

$$(E^A{}_B(x,y))^* = E_B{}^A(x,y)\,. \tag{11}$$

The commutation relations between elements of $su(3)$ translated to the language of these fields read

$$[E^A{}_B(x,y), E^C{}_D(x,y)] = \delta^C{}_B E^A{}_D(x,y) - \delta^A{}_D E^C{}_B(x,y)\ , \tag{12}$$

whereas the commutation relations (5) between Lie algebra elements and functions, rewritten in terms of generators, take the following form:

$$[E^A{}_B(x,y), U^C{}_D(x,y)] = \delta^C{}_B U^A{}_D(x,y) - \frac{1}{3}\delta^A{}_B U^C{}_D(x,y)\,. \tag{13}$$

Now, observe that the (classical) $SU(3)$-valued parallel transporter $g(x,y)$ on (x,y) transforms to $g^{-1}(x,y)$ under the change of the link orientation. Thus, the change of orientation on (x,y) induces the transformation

$$SU(3) \ni g \to i(g) = g^{-1} \in SU(3) \tag{14}$$

on configuration space. This transformation induces an isomorphism

$$\mathcal{I}_{(x,y)} : \mathcal{F}_{(x,y)} \to \mathcal{F}_{(y,x)} \tag{15}$$

of field algebras given by

$$\mathcal{F}_{(x,y)} \ni (e,f) \to (-i_*(e), i^*(f)) \in \mathcal{F}_{(y,x)}. \tag{16}$$

We have $i^*(f)(g) = f(g^{-1})$ and $-i_*(e) = e^L$, where e^L is the left invariant vector field on $SU(3)$ generated by $-e$.

The bosonic field algebra $\mathcal{F}_{|(x,y)|}$, associated with the non-oriented link $|(x,y)|$, is defined by:

$$\mathcal{F}_{|(x,y)|} := \left\{ (l_{(x,y)}, k_{(y,x)}) \in \mathcal{F}_{(x,y)} \times \mathcal{F}_{(y,x)} \ : \ k_{(y,x)} = \mathcal{I}_{(x,y)}(l_{(x,y)}) \right\} . \tag{17}$$

Projection onto the first (resp. second) component gives us an isomorphism of $\mathcal{F}_{|(x,y)|}$ with $\mathcal{F}_{(x,y)}$ (resp. $\mathcal{F}_{(y,x)}$).

The above transformation law yields relations between generators of the two algebras. For functions on $SU(3)$, formula (16), together with (7) gives:

$$U^A{}_B(y,x) = (U_B{}^A(x,y))^*\ . \tag{18}$$

For Lie algebra elements, formula (16) gives

$$E^A{}_B(y,x) = -U^A{}_D(y,x)U^C{}_B(x,y)E^D{}_C(x,y)\ . \tag{19}$$

To summarize, the field algebra $\mathcal{F}(\Lambda)$ is the $*$-algebra, generated by elements

$$\{\psi^{aA}(x)\, ,\, U^A{}_B(x,y)\, ,\, E^A{}_B(x,y)\}\ ,$$

fulfilling relations (7), (8), (11), (18) and (19), together with canonical (anti-) commutation relations (3), (12) and (13).

3 The Observable Algebra

The observable algebra $\mathcal{O}(\Lambda)$ is defined by imposing the local Gauss law and gauge invariance.

It is easy to check that the group of local gauge transformations acting on $\mathcal{F}(\Lambda)$ by automorphisms is generated by the elements

$$\mathcal{G}^A{}_B(x) := \rho^A{}_B(x) - \sum_y E^A{}_B(x,y)\,, \tag{20}$$

where

$$\rho^A{}_B(x) = \sum_a \left(\psi^{*aA}(x)\psi^a{}_B(x) - \frac{1}{3}\delta^A_B\psi^{*aC}(x)\psi^a{}_C(x)\right) \tag{21}$$

is the local matter charge density. Observe that $\rho^A{}_A(x) = 0$.

To implement gauge invariance we have to take those elements of $\mathcal{F}(\Lambda)$, which commute with all generators $\mathcal{G}^A{}_B(x)$. Thus, the subalgebra of gauge invariant fields is, by definition, the commutant $\mathcal{A}(\mathcal{G})'$ of the algebra $\mathcal{A}(\mathcal{G})$, generated by the set $\{\mathcal{G}^A{}_B(x)\}$.

The local Gauss law at $x \in \Lambda^0$ has the form

$$\sum_y E^A{}_B(x,y) = \rho^A{}_B(x)\ , \tag{22}$$

with the sum taken over all points y adjacent to x. Imposing it on the subalgebra $\mathcal{A}(\mathcal{G})'$ of gauge invariant fields means factorizing the latter with respect to $\mathcal{I}(\Lambda)\cap\mathcal{A}(\mathcal{G})'$, where $\mathcal{I}(\Lambda)$ is the ideal generated by (22). Thus, the observable algebra is defined as follows:

$$\mathcal{O}(\Lambda) := \mathcal{A}(\mathcal{G})'/\{\mathcal{I}(\Lambda)\cap\mathcal{A}(\mathcal{G})'\}\,. \tag{23}$$

Obviously, every element of $\mathcal{O}(\Lambda)$ is represented by a gauge invariant element of $\mathcal{F}(\Lambda)$ and $\mathcal{O}(\Lambda)$ can be viewed as a $*$-subalgebra of $\mathcal{F}(\Lambda)$ generated by gauge invariant bosonic combinations of U and E and by gauge invariant combinations of ψ and ψ^* of mesonic and baryonic type, with the Gauss law inducing some identities between those generators.

It will be shown in the next section that there is a (gauge invariant), *additive* law, obtained by combining these two equations, which characterizes the local colour charge density carried by the lattice quantum fields. Additivity will allow us to obtain the global colour charge by adding up the local Gauss laws. In a subsequent paper we will show that the irreducible representations of $\mathcal{O}(\Lambda)$ are labeled by this global charge. This way, we get the physical Hilbert space $\mathcal{H}^{phys}$ as a direct sum of colour charge superselection sectors. The similar problem for QED on the lattice was solved in [8].

4 The Local Charge density

In this section we are going to analyze the local Gauss law (22). Suppose, for that purpose, that we are given a collection of operators $F^A{}_B$ in a Hilbert space $\mathcal{H}$, fulfilling $F^A{}_A = 0$ and $(F^A{}_B)^* = F_B{}^A$, realizing the canonical commutation relations for the Lie algebra $su(3)$:

$$[F^A{}_B, F^C{}_D] = \delta^C{}_B F^A{}_D - \delta^A{}_D F^C{}_B \,. \tag{24}$$

The field algebra $\mathcal{F}(\Lambda)$ provides us with two basic examples of this type: the electric field $E^A{}_B(x,y)$ on each lattice link, see (12), and the charge operator $\rho^A{}_B(x)$ at each lattice site, given by formula (21). Indeed, due to canonical anticommutation relations (3), the latter fulfills (24). Thus, the operators occurring on both sides of the local Gauss law fulfill (24).

Throughout this paper, we assume integrability of the Lie algebra representations under consideration. This means that for each F there exists a unitary representation of the group $SU(3)$

$$SU(3) \ni g \to \bar{F}(g) \in B(\mathcal{H}) \,, \tag{25}$$

associated with F.

It is easy to check that if F and G are two commuting representations of $su(3)$ then so is $F+G$. Indeed, if $\bar{F}(g)$ and $\bar{G}(g)$ are representations of $SU(3)$ corresponding to F and G, then $F+G$ may be obtained by differentiating the representation $SU(3) \ni g \to \bar{F}(g)\bar{G}(g) \in B(\mathcal{H})$, where $B(\mathcal{H})$ denotes the C^*-algebra of bounded operators on $\mathcal{H}$. Moreover, $-F^*$ is also a representation of $su(3)$, corresponding to the following representation of $SU(3)$: $SU(3) \ni g \to (\bar{F}(g^{-1}))^* \in B(\mathcal{H})$.

Such a collection of operators is an *operator domain* in the sense of Woronowicz (see [13]). We are going to construct an operator function on this domain, i. e. a mapping $F \to \varphi(F)$ which satisfies $\varphi(UFU^{-1}) = U\varphi(F)U^{-1}$

for an arbitrary isometry U. For this purpose, consider the two gauge-invariant, self-adjoint and commuting (Casimir) operators K_2 and K_3 of F:

$$K_2 = F^A{}_B F^B{}_A \tag{26}$$

$$K_3 = \frac{1}{2}\left(F^A{}_B F^B{}_C F^C{}_A + F^A{}_B F^C{}_A F^B{}_C\right). \tag{27}$$

The Hilbert space $\mathcal{H}$ splits into the direct sum of subspaces $\mathcal{H}_\alpha$ on which F acts irreducibly. Each of these subspaces is a common eigenspace of K_2 and K_3. Denoting the highest weight characterizing a given irreducible representation by (m,n), with m and n being nonnegative integers, the eigenvalues k_2 and k_3 of K_2 and K_3 are given by:

$$k_2 = \frac{2}{3}(m^2 + mn + n^2 + 3m + 3n), \tag{28}$$

$$k_3 = \frac{1}{9}(m-n)(3+2m+n)(3+m+2n). \tag{29}$$

It is easy to check that the above formulae may be uniquely solved with respect to m and n yielding:

$$m = M(k_2,k_3) := \sqrt{\tfrac{2}{3}(k_2+2)}\left(\cos\left(\frac{1}{3}\arccos\frac{\sqrt{6}k_3}{\sqrt{(k_2+2)^3}} + \frac{2}{3}\pi\right) + 2\cos\left(\frac{1}{3}\arccos\frac{\sqrt{6}k_3}{\sqrt{(k_2+2)^3}}\right)\right) - 1, \tag{30}$$

$$n = N(k_2,k_3) := -\sqrt{\tfrac{2}{3}(k_2+2)}\left(2\cos\left(\frac{1}{3}\arccos\frac{\sqrt{6}k_3}{\sqrt{(k_2+2)^3}} + \frac{2}{3}\pi\right) + \cos\left(\frac{1}{3}\arccos\frac{\sqrt{6}k_3}{\sqrt{(k_2+2)^3}}\right)\right) - 1. \tag{31}$$

Using these functions we may define a function with values in $\mathbf{Z}_3$, which will be identified with the dual to the center $\mathcal{C}$ of the gauge group $SU(3)$:

$$f(k_2,k_3) := (M(k_2,k_3) - N(k_2,k_3)) \,\mathbf{mod}\, 3. \tag{32}$$

For our purposes it is convenient to use the parametrization $\mathbf{Z}_3 = (-1,0,1)$. Obviously, we have

$$f(k_2,-k_3) = -f(k_2,k_3). \tag{33}$$

Since K_2 and K_3 are commuting and self-adjoint, there exists an operator-valued function:

$$\varphi(F) = f(K_2(F), K_3(F)). \tag{34}$$

This means that $\varphi(F)$ may take eigenvalues $-1, 0, 1$ and that every irreducible subspace $\mathcal{H}_\alpha$ is an eigenspace of $\varphi(F)$ with eigenvalue $m - n$ **mod** 3.

Theorem: The operator function φ defined above has the following properties:

$$\varphi(-F^*) = -\varphi(F), \tag{35}$$

$$\varphi(F+G) = \varphi(F) + \varphi(G), \tag{36}$$

for commuting F and G.

Proof: see [11]

Applying the operator function φ to the local Gauss law (22) and using additivity (36) of φ we obtain:

$$\sum_y \varphi(E(x,y)) = \varphi(\rho(x)) \ . \tag{37}$$

This is a gauge invariant equation for operators with eigenvalues in $\mathbf{Z}_3$, valid at every lattice site x. The quantity on the right hand side is the (gauge invariant) local colour charge density carried by the quark field.

5 The Global Charge and the Flux Law

Using the transformation law (19), one can show that

$$K_2(E(x,y)) = K_2(E(y,x)) \ , \qquad K_3(E(x,y)) = -K_3(E(y,x)) \ . \tag{38}$$

Hence, equation (33) implies:

$$\varphi(E(x,y)) + \varphi(E(y,x)) = 0 \, , \tag{39}$$

for every lattice bond (x, y).

Now we take the sum of equations (37) over all lattice sites $x \in \Lambda$. Due to the above identity, all terms on the left hand side cancel, except for contributions coming from the boundary. This way we obtain the total flux through the boundary $\partial\Lambda$ of Λ:

$$\Phi_{\partial\Lambda} := \sum_{x\in\partial\Lambda} \varphi(E(x,\infty)) \ , \tag{40}$$

where by $E(x, \infty)$ we denote the colour electric charge along the external link, connecting the point x on the boundary of Λ with the "rest of the world". On the right hand side we get the (gauge invariant) global colour charge, carried by the matter field

$$Q_\Lambda = \sum_{x\in\Lambda} \varphi(\rho(x)) \, . \tag{41}$$

Both quantities appearing in the global Gauss law

$$\Phi_{\partial\Lambda} = Q_\Lambda \,, \tag{42}$$

take values in the center $\mathbf{Z}_3$ of $SU(3)$. The "sum modulo three" is the composition law in $\mathbf{Z}_3$.

In the remainder we make some remarks on the nature of the boundary data $E(x, \infty)$.

1. If one treats Λ as a piece of a bigger lattice $\widetilde{\Lambda}$, the boundary flux operators $E(x, \infty)$ belong to $\mathcal{F}(\widetilde{\Lambda})$ and commute with $\mathcal{F}(\Lambda)$, (and also with $\mathcal{O}(\Lambda)$). Thus, the invariant operators $N(E(x, \infty))$ and $M(E(x, \infty))$, constructed from external fluxes should commute with all elements of the observable algebra $\mathcal{O}(\Lambda)$. This corresponds to treating external fluxes as purely *classical objects*, describing extra superselection rules, cf. [4], [5] and [6].

2. There is, however, a second option. In a subsequent paper we shall prove that all irreducible representations of the observable algebra $\mathcal{O}(\Lambda)$ of QCD on a finite lattice are classified by the value of the global colour charge Q_Λ, yielding three different superselection sectors labeled by elements of $\mathbf{Z}_3$. However, the local distribution of the (gauge invariant) gluon and anti-gluon fluxes $M(E(x, \infty))$ and $N(E(x, \infty))$ over the boundary $\partial\Lambda$ may be arbitrarily changed within one sector. The redistribution of fluxes along an oriented path from $\xi \in \partial\Lambda$ to $\eta \in \partial\Lambda$ can be described explicitly, see [11]. One can show that – under this redistribution – the value of $\varphi(E(\xi, \infty))$ changes by plus one and the value of $\varphi(E(\eta, \infty))$ changes by minus one. This suggests, that these objects could also be treated as *quantum* and not as classical quantities. Then, only their sum, the global flux $\Phi_{\partial\Lambda}$, would be a classical object proportional to the identity on every superselection sector.

3. Finally, we stress that it does not make sense to attribute any physical meaning to the external gluon or anti-gluon fluxes M and N themselves. It is only the quantity $(M - N)$ "modulo three" which makes sense.

Acknowledgments

The authors are very much indebted to J. Dittmann, C. Śliwa, M. Schmidt and I.P. Volobuev for helpful discussions and remarks. This work was supported in part by the Polish KBN Grant Nr. 2 P03A 047 15.

References

1. E. Seiler, *Gauge Theories as a Problem of Constructive Quantum Field Theory and Statistical Mechanics*, Lecture Notes in Phys., vol. 159, Springer (1982)
   E. Seiler, "Constructive Quantum Field Theory: Fermions", in *Gauge Theories: Fundamental Interactions and Rigorous Results*, eds. P. Dita, V. Georgescu, R. Purice.
2. S. Doplicher, R. Haag and J. Roberts, *Commun. Math. Phys.* **23**, 199 (1971).
3. J. Fröhlich, *Commun. Math. Phys.* **66**, 223 (1979).
4. D. Buchholz, *Commun. Math. Phys.* **85**, 49 (1982).
   D. Buchholz, *Phys. Lett. B* **174**, 331 (1986).
5. F. Strocchi and A. Wightman, *J. Math. Phys.* **15**, 2198 (1974).
6. F. Strocchi, *Phys. Rev. D* **17**, 2010 (1978).
7. K. Fredenhagen, M. Marcu, *Commun. Math. Phys.* **92**, 81 (1983).
8. J. Kijowski, G. Rudolph and A. Thielman, *Commun. Math. Phys.* **188**, 535 (1997).
9. J. Kijowski, G. Rudolph and M. Rudolph, *Ann. Inst. H. Poincaré* Vol. 68, n. 3, 285 (1998).
10. J. Kijowski, G. Rudolph and C. Sliwa, *Lett. Math. Phys.* **43**, 299 (1998).
11. J. Kijowski, G. Rudolph, *On the Gauss Law and Global Charge for QCD*, hep-th/0104052
12. A. Klimyk, K. Schmüdgen, *Quantum Groups and Their Representations*, Springer (1997).
13. S. L. Woronowicz, *Duality in the C^*-algebra theory, Proc. Intern. Congr. Mathematicians Warsaw 1983*, p. 1347, PWN Warszawa.
    S. L. Woronowicz, *Rev. Math. Phys.* **7**, 481 (1995).

QUANTUM ENTANGLEMENT AND SYMMETRIES

MAREK KUŚ
Center for Theoretical Physics, Polish Academy of Sciences, Warszawa Poland
E-mail: marek@cft.edu.pl

1 Introduction

The existence of entangled states, i.e., roughly speaking, the states of a composite system which exhibit quantum correlations among the subsystems, appeared recently to be extremely important in rapidly developing field of quantum communication. Non-classical properties of entangled states which came to prominence already in 1935 in the famous paper of Einstein, Podolsky, and Rosen[1] are also essential for various schemes of quantum computing[2], quantum cryptography[3] and quantum teleportation[4]. For this reasons characterizing and quantifying quantum mechanical entanglement became important problems of quantum information theory. Various symmetries play a significant role in the investigations - I shall concentrate on two symmetry aspects of entanglement 1) the equivalence classes of entangled states upon identification by local (i.e. restricted to subsystems) unitary transformations and 2) exchange symmetries of identical particles (fermions nad bosons).

2 Quantum states

In most applications which are interesting for the quantum information theory, quantum systems are characterized in terms of states in finite-dimensional Hilbert spaces - one should think some generalized spin systems as models of them. Hence in fact the corresponding Hilbert space is simply a many dimensional complex space ($\mathcal{H} = \mathbb{C}^N$, say), with the natural scalar product $\langle \cdot | \cdot \rangle$ structure. The complete characterization of a state of the system is given by a density matrix ρ i.e. a hermitian ($\rho^\dagger = \rho$), positive-definite ($\langle x|\rho|x\rangle \geq 0$ for each $|x\rangle \in \mathcal{H}$), trace-one ($\mathrm{Tr}\rho = 1$) operator acting on $\mathcal{H}$ ($\rho : \mathcal{H} \to \mathcal{H}$).

A particular situation occurs when the state ρ is a projection operator

$$\rho = \rho^2, \tag{1}$$

so the eigenvalues of ρ are equal 0 or 1. Since, from the definition, $\mathrm{Tr}\rho = 1$, only one eigenvalue equals 1, the operator ρ is a projection on a one-dimensional space

$$\rho = |\psi\rangle\langle\psi|. \tag{2}$$

In this case the state is called *pure* and is customary identified with the vector $|\psi\rangle \in \mathcal{H}$ (we assume $||\psi||^2 := \langle\psi|\psi\rangle = 1$ and identify vectors differing by a phase factor). All states which are not pure (i.e. do not fulfill (1) are called *mixed*.

3 States of composite quantum systems

Consider now two quantum systems A and B described by the states in, respectively, Hilbert spaces $\mathcal{H}_A = \mathbb{C}^M$ and $\mathcal{H}_B = \mathbb{C}^N$. The Hilbert space of the composite system $A + B$ is the tensor product of $\mathcal{H}_A$ and $\mathcal{H}_B$

$$\mathcal{H} := \mathcal{H}_{A+B} = \mathcal{H}_A \otimes \mathcal{H}_B, \tag{3}$$

and its states are operators acting in $\mathcal{H}$.

It is now clear that quantum mechanics admits more states for a composite system than the classical physics. Indeed, in the classical case if states of the subsystem A and B are described by sets of coordinates $\{x_\mu^{(A)}\}$ and $\{x_\nu^{(B)}\}$ (e.g. phase space canonical variables) then the state of the composite system is characterized by the set of coordinates $\{x_\mu^{(A)}, x_\nu^{(B)}\}$. It means that the (phase) space of the composite system has the structure of the Cartesian product of the spaces of the subsystems, rather than the tensor product of the quantum case. The, so called, product states of the form

$$|\Psi\rangle = |\psi\rangle \otimes |\phi\rangle, \quad |\psi\rangle \in \mathcal{H}_A, \quad |\phi\rangle \in \mathcal{H}_B, \tag{4}$$

have their counterparts on the classical level, whereas all other quantum states lack straightforward classical interpretation. States fulfilling (4) are also called *separable*, all other states, i.e. those which can not be written as simple tensors (tensor products of two states) but rather as nontrivial linear combinations of products like (4) go under the name of *entangled* states.

Pure states are only particular and rather special instances of states available for quantum systems. In particular if the system in question interacts or interacted with some other system(s) on which we perform measurements the final state of our system will be, in general, a mixed rather than a pure one. The generalization of the notion of entanglement to mixed states was proposed by Werner[6]. A mixed state ρ of a bipartite quantum system is separable if it is a convex combination of product states

$$\rho = \sum_i^k p_i \rho_i^A \otimes \rho_i^B, \quad p_i \geq 0, \quad \sum_i^k p_i = 1, \tag{5}$$

where ρ_i^A, ρ_i^B are legitimate (i.e. hermitian and positive definite) density matrices of the subsystems. States which are not separable are, by definition, entangled. If ρ is pure the above definitions of separability and entanglement coincide with the original ones for pure states. Moreover, according to this definition, separable states can be obtained "from scratch", i.e. from an initially uncorrelated state by local manipulations of in each subspace (i.e. admissible quantum operations, like quantum evolution, measurements etc. applied separately to each part), and, what is more important can not exhibit correlations which can not be reproduced by classical systems[6].

It is easy to check whether a *pure* state of a composite system is separable or entangled. Indeed, a general pure state of the composite system $\mathcal{H} = \mathcal{H}_A \otimes \mathcal{H}_B$ can be written as

$$|\psi\rangle = \sum_{ij} c_{ij} |e_i\rangle \otimes |f_j\rangle, \tag{6}$$

with $c_{ij} \in \mathbb{C}$. A change of the bases in both spaces in (6) by local unitary transformations:

$$|e_i\rangle = \sum_k^M U_{ik} |e'_k\rangle, \quad |f_j\rangle = \sum_l^N V_{jl} |f'_l\rangle, \tag{7}$$

amounts to

$$|\psi\rangle = \sum_{ij} (U^T C V)_{ij} |e'_i\rangle \otimes |f'_j\rangle, \tag{8}$$

where ^{T} means transposition and C is the matrix of the coefficients c_{ij}. We can choose the unitary matrices U^T and V to achieve the principal value decomposition of C i.e. $U^T C V = D$ where $D_{ij} = 0$ for $i \neq j$ and $\lambda_i := D_{ii} \geq 0$ for $i = 1, 2, \ldots, \min(M, N)$) arriving at the so called, Schmidt decomposition of the pure state $|\psi\rangle$.

$$|\psi\rangle = \sum_{i=1}^{\min(M,N)} \lambda_i |e'_i\rangle \otimes |f'_i\rangle. \tag{9}$$

The numbers λ_i are called Schmidt coefficients. It is now clear that $|\psi\rangle$ is separable if and only if there only one Schmidt coefficient does not vanish. Various degrees of entanglement of pure states can be characterized by the Schmidt rank of a pure state defined as the number of its non-vanishing Schmidt coefficients.

The situation complicates in the case of mixed states. A simple and practical necessary criterium of separability is known, but there are no known effective sufficient conditions for higher dimensional composite systems. To formulate Peres[7] necessary condition for separability let us make the following observations: first an arbitrary state ρ can be written in the form

$$\rho = \sum_{i}^{k} p_i \rho_i^A \otimes \rho_i^B, \tag{10}$$

by choosing some bases $\{\rho_l^A\}$ and $\{\rho_k^B\}$ in the spaces of linear Hermitian operators acting in $\mathcal{H}_A$ and $\mathcal{H}_B$. The coefficients p_i are real but otherwise restricted only by the (unimportant in the present argumentation) trace condition. Also neither ρ_i^A nor ρ_j^B need all to be positive definite (the positive definiteness is obeyed only by ρ itself). We define the *partial transposition* as

$$\rho^{T_B} := \sum_{i}^{k} p_i \rho_i^A \otimes (\rho_i^B)^T, \tag{11}$$

Now, as observed by Peres, if ρ is separable i.e. (5) is fulfilled, then ρ^{T_B} is positive definite, i.e. is also a legitimate density matrix for the composite system. For low dimensional ($M = N = 2$, $M = 2$, $N = 3$, and $M = 3$, $N = 2$) systems the above condition is also sufficient[8].

4 Entanglement measures and geometry of entangled states

As mentioned above the criterion based on positivity of partial transpose is, in general, only a necessary one. The quest for separability criteria and checks remains one of the major effort of quantum information theory (see eg. Lewenstein et al.[9] for a survey of recent results and developments). Another issue is how to quantify entanglement. It is, namely, reasonable to pose the question of the amount of entanglement in a particular state or, in other words to ask if a particular states is "more entangled" than the other one. Such a measure of entanglement should fulfill the following three conditions: it vanishes if and only if ρ is separable, is invariant under local unitary operations and its expectation value is non-increasing under general local operations[11]. Such a measure in the two qubits (i.e. $\mathbb{C}^2 \otimes \mathbb{C}^2$ case was constructed by Wootters[12] in the form of the *concurrence* defined as

$$c(\rho) = max\{0, c_1 - c_2 - c_3 - c_4\}, \tag{12}$$

where $c_1 \geq c_2 \geq c_3 \geq c_4$ are the square roots of the (real and positive) eigenvalues of the matrix

$$X := \Sigma\rho^*\Sigma\rho, \quad \Sigma = \sigma_2 \otimes \sigma_2, \quad \sigma_2 := \begin{bmatrix} 0 & -i \\ i & 0 \end{bmatrix}. \tag{13}$$

The problem of extension of the above construction to higher dimensional systems remains open.

An interesting approach to this problem was proposed by Lewenstein and Sanpera[10] where it was proven that for any state ρ there exists a unique decomposition

$$\rho = \lambda \rho_s + (1 - \lambda)\delta\rho, \tag{14}$$

where ρ_s is a separable state and λ is maximal (i.e. it is impossible to find such a decomposition with larger λ without violating positivity or separability of ρ_s). Such an optimal decomposition (14) can be interpreted as concentrating the whole entanglement in $(1 - \lambda)\delta\rho$ or, in other words, subtracting from ρ the maximal possible separable part $\lambda\rho_s$. The parameter $(1 - \lambda)$ could thus serve as a quantification of the entanglement contained in ρ. To make such a characterization fully functional one should find an effective, analytic method of finding the optimal decomposition (14). Until now it was possible only in the simplest case of two qubits[13], where as it was shown[10] the entangled part $\delta\rho$ in (14) is a pure state

$$\delta\rho = |\psi\rangle\langle\psi|. \tag{15}$$

Moreover, we were able to show that:

1. if the rank (the number of non-vanishing eigenvalues) of ρ_s is maximal (i.e. equal to 4) $\delta\rho$ is a maximally entangled pure state: i.e. a state which can be obtained from the singlet by local unitary transformations in each space

   $$|\psi\rangle = U_1 \otimes U_2 \Big(|0\rangle \otimes |1\rangle - |1\rangle \otimes |0\rangle \Big). \tag{16}$$

   The concurrence $c(\psi) := c\big(|\psi\rangle\langle\psi|\big)$ of such state equals 1 (the maximal allowed value of the concurrence) what explains the name (*maximally entangled*).

2. in the same case of maximal rank of ρ_s the quantity $1 - \lambda$ coincides with the above defined concurrence $c(\rho)$

3. in a general case $c(\rho)$ and $(1 - \lambda)c(\psi)$ are two inequivalent measures of entanglement connected, however, by the inequality

   $$c(\rho) \leq (1 - \lambda)c(\psi). \tag{17}$$

Similar results for larger dimensions are not known.

Two other concepts proved to be useful in characterizing entanglement[14,15]:

1. The Schmidt number s of a mixed state ρ is defined as the minimal Schmidt rank of pure states needed to decompose ρ in a convex sum of projectors $\rho = \sum_i p_i |\psi_i^{r_i}\rangle\langle\psi_i^{r_i}|$, where where r_i denotes the Schmidt rank of $|\psi_i^{r_i}\rangle$ i.e. $s = \min\{r_{\max}\}$, where $r_{\max}$ is the maximum Schmidt rank within a decomposition, and the minimum is taken over all decompositions. The states with Schmidt number s form a convex set S_s and, obviously, $S_s \subset S_{s+1}$.

2. The Schmidt number witness of class S_s is a hermitian operator W such that $\mathrm{Tr}(W\sigma) \geq 0$ for all $\sigma \in S_{s-1}$ and there exists at least one $\rho \in S_s$ such that $\mathrm{Tr}(W\sigma) < 0$. Such a witness testifies that ρ has the Schmidt number larger than $s-1$ and is thus "more entangled" than states from S_s.

5 Manifolds of equivalent states

By definition, entanglement measures do not discriminate between the states which can be connected by local unitary transformations. It is thus natural to ask "how many" states are equivalent to a given one with respect to the amount of entanglement. Since such states form a continuous family this is, in fact, a question about the dimensionality and geometry of the manifolds of states obtained from a given state by applying arbitrary local unitary transformations. We were able to fully characterize such manifolds in the simplest case of two qubits[16] and for the pure states of $N \times N$ system[17] (a generalization for $M \times N$, $M \neq N$ is straightforward). In the case of mixed states two qubits the manifold of states with the same amount of entanglement is, generally, six-dimensional, but for some states[16] its dimensionality can shrink to $5, 4, 3$ or 0. The manifolds of pure states of $N \times N$ system equivalent to a given one $|\psi\rangle$ are fully characterized by its Schmidt coefficients. More precisely let

$$|\psi\rangle = \sum_{i=1}^{N} \lambda_i |e_i\rangle \otimes |f_i\rangle \tag{18}$$

and let us order the Schmidt coefficients in the following way $\Lambda = (0 \leq \lambda_1 \leq \lambda_2 \leq \cdots \leq \lambda_N)$. In order to describe the character of the degeneracy we rename them into $\Lambda = (0, \cdots, 0, \nu_1, \cdots, \nu_1, \nu_2, \cdots, \nu_2, \ldots, \nu_K, \cdots, \nu_K)$ where each value ν_n occurs m_n times and m_0 is the number of vanishing Schmidt coefficients. Obviously $m_0 + \sum_{n=1}^{K} m_n = N$, and m_0 might be equal to zero.

Then the dimension of the manifold $\mathcal{M}_\psi$ of states equivalent to $|\psi\rangle$ equals

$$\dim(\mathcal{M}_\psi) = 2N^2 - 2m_0^2 - \sum_{n=1}^{K} m_n^2 - 1. \tag{19}$$

In fact the manifold $\mathcal{M}_\psi$ has a structure of a Cartesian product of two manifolds[18]

$$\mathcal{M}_\psi = \frac{U(N)}{G_1} \times \frac{U(N)}{G_2}, \tag{20}$$

Here $\frac{U(N)}{G}$ denotes the quotient manifold obtained by identifying unitary matrices (elements of the group $U(N)$) if they differ by multiplication by an unitary matrix belonging to the subgroup $G \subset U(N)$, and G_1, G_2 are subgroups of $U(N)$:

$$G_1 = \left\{ U : U = \begin{bmatrix} u_0 & & & \\ & u_1 & & \\ & & \ddots & \\ & & & u_K \end{bmatrix} \right\}, \quad G_2 = \left\{ U : U = \begin{bmatrix} v_0 & \\ & e^{i\phi} I \end{bmatrix} \right\}, \tag{21}$$

where u_0 and v_0 are arbitrary unitary matrices of dimension m_0, $u_1, \ldots, u_K$ arbitrary unitary matrices of dimensions $m_1, \ldots, m_K$, respectively, $e^{i\phi}$ - arbitrary phase factor (an one-dimensional unitary matrix), and I - the identity matrix of the dimension $N - m_0$. The generalization of the above results to arbitrary mixed states is not known.

6 Identical particles

While entanglement plays an essential role in quantum communication between parties separated by macroscopic distances, the characterization of quantum correlations at short distances is also an open problem, which has received much less attention so far. In this case the indistinguishable character of the particles involved (electrons, photons,...) has to be taken into account.

For indistinguishable particles a pure quantum state must be formulated in terms of Slater determinants or Slater permanents for fermions and bosons, respectively. Generically, a Slater determinant contains correlations due to the exchange statistics of the indistinguishable fermions. As the simplest possible example consider a wavefunction of two fermions,

$$\Psi(\vec{r}_1, \vec{r}_2) = \frac{1}{\sqrt{2}} [\phi(\vec{r}_1)\chi(\vec{r}_2) - \phi(\vec{r}_2)\chi(\vec{r}_1)] \tag{22}$$

with two orthormal single-particle wavefunctions $\phi(\vec{r})$, $\chi(\vec{r})$. Operator matrix elements between such single Slater determinants contain terms due to the antisymmetrization of coordinates. However, if $\phi(\vec{r})$, $\chi(\vec{r})$ have only vanishingly small overlap, these exchange correlations will also tend to zero for any physically meaningful operator. This situation is generically realized if the supports of the single-particle wavefunctions are essentially centered around locations being sufficiently apart from each other, or the particles are separated by a sufficiently large energy barrier. In this case the antisymmetrization present in Eq. (22) has no physical effect. Such observations clearly justify the treatment of indistinguishable particles separated by macroscopic distances as effectively distinguishable objects.

The situation is different if the particles are close together and possibly coupled in some computational process. This the case for all proposals of quantum information processing based on quantum dots technology [19,20]. Here qubits are realized by the spins of electrons living in a system of quantum dots. The electrons have the possibility of tunneling eventually from one dot to the other with a probability which can be modified by varying external parameters such as gate voltages and magnetic field. In such a situation the fermionic statistics of electrons is clearly essential.

Additional correlations in many-fermion-systems arise if more than one Slater determinant is involved, i.e. if there is no single-particle basis such that a given state of N indistinguishable fermions can be represented as an elementary Slater determinant (i.e. fully antisymmetric combination of N orthogonal single-particle states). These correlations are the analog of quantum entanglement in separated systems and are essential for quantum information processing in non-separated systems.

We consider two indistinguishable fermions each of which lives in the single-particle Hilbert space $\mathcal{C}^{2K}$. This situation is given, eg., in a system of two electrons in K neighboring quantum dots where only the orbital ground state of each dot is taken into account. Alternatively one may think of, say, two quantum dots with an appropriate number of orbital states available for the two fermions.

The states (density matrices) in such a system are positive self-adjoint operators acting on the antisymmetric space $\mathcal{A}(\mathcal{C}^{2K} \otimes \mathcal{C}^{2K})$. Let us first consider pure states, i.e. projectors on a vector $|\Psi\rangle \in \mathcal{A}(\mathbb{C}^{2K} \otimes \mathbb{C}^{2K})$. Let f_a, $f_a^\dagger$, $a = 1, \ldots, 2K$, denote the fermionic annihilation and creation operators of single-particle states forming an orthonormal basis in $\mathbb{C}^{2K}$, and $|\Omega\rangle$ denotes the vacuum state. Each vector in the two-electron space can be represented as $|\Psi\rangle = \sum_{a,b} w_{ab} f_a^\dagger f_b^\dagger |\Omega\rangle$, where $w_{ab} = -w_{ba}$ is an antisymmetric matrix. We have the following fermionic analog of the Schmidt decomposition[21]. Every vector in the antisymmetric space $\mathcal{A}(\mathbb{C}^{2K} \otimes \mathbb{C}^{2K})$ can be represented in an appropriately chosen basis in $\mathbb{C}^{2K}$ in a form of the Slater decomposition

$$|\Psi\rangle = \frac{1}{\sqrt{\sum_{i=1}^{K} |z_i|^2}} \sum_{i=1}^{K} z_i f^\dagger_{a_1(i)} f^\dagger_{a_2(i)} |\Omega\rangle, \tag{23}$$

where the states $f^\dagger_{a_1(i)}|\Omega\rangle$, $f^\dagger_{a_2(i)}|\Omega\rangle$, $i = 1, \ldots, K$, form an orthonormal basis in $\mathbb{C}^{2K}$, i.e. each of these single-particle states occurs only in one term in the summation (23). The number of non-vanishing coefficients z_i (i.e. the number of elementary Slater determinants required to construct $|\Psi\rangle$) is called the Slater rank.

Let us now generalize the concepts introduced above to the case of mixed states. To this end, we define the Slater number of a mixed state, in analogy to the Schmidt number for the case of distinguishable parties[21] :

Consider a density matrix ρ of a two fermion system, and all its possible convex decompositions in terms of pure states, i.e. $\rho = \sum_i p_i |\psi_i^{r_i}\rangle\langle\psi_i^{r_i}|$, where r_i denotes the Slater rank of $|\psi_i^{r_i}\rangle$; the Slater number of ρ, k, is defined as $k = \min\{r_{\max}\}$, where $r_{\max}$ is the maximum Slater rank within a decomposition, and the minimum is taken over all decompositions.

Many of the results concerning Schmidt numbers can be transferred directly to the Slater number. For instance, let us denote the whole space of density matrices in $\mathcal{A}(\mathbb{C}^{2K} \otimes \mathbb{C}^{2K})$ by Sl_K, and the set of density matrices that have Slater number k or less, by Sl_k. Sl_k is a convex compact subset of Sl_K; a state from Sl_k will be called a state of (Slater) class k. Sets of increasing Slater number are embedded into each other, i.e. $Sl_1 \subset Sl_2 \subset ...Sl_k... \subset Sl_K$. In particular, Sl_1 is the set of states that can be written as a convex combination of elementary Slater determinants; Sl_2 is the set of states of Slater number 2, i.e. those that require at least one pure state of Slater rank 2 for their formation, etc.

The determination of the Slater number of a given state is in general a very difficult task. However, as in the case of separability of mixed states of low dimensional composite systems of distinguishable particles (Peres-Horodecki criterion mentioned in Section 3, the situation is particularly simple in the case of small K. One should note, however, that in the considered case of fermionic states there exists no simple analogy of the partial transposition. Instead one observes that the Peres-Horodecki criterion is equivalent to Wootters' result[12], relating separability to the concurrence (see above, Section 4), and constructs a similar measure in the fermionic case[21], using the following

Theorem: Let the mixed state acting in $\mathcal{A}(\mathbb{C}^4 \otimes \mathbb{C}^4)$ have a spectral decomposition $\rho = \sum_{i=1}^{r} |\Psi_i\rangle\langle\Psi_i|$, where r is the rank of ρ, and the eigenvectors $|\Psi_i\rangle$ belonging to nonzero eigenvalues λ_i are normalized as $\langle\Psi_i|\Psi_j\rangle = \lambda_i\delta_{ij}$. Let $|\Psi_i\rangle = \sum_{a,b} w^i_{ab} f^\dagger_a f^\dagger_b |\Omega\rangle$ in some basis, and define the complex symmetric $r \times r$ matrix C by

$$C_{ij} = \sum_{abcd} \epsilon^{abcd} w^i_{ab} w^j_{cd}, \tag{24}$$

which can be represented using a unitary matrix as $C = UC_dU^T$, with $C_d = \text{diag}[c_1, c_2, \ldots, c_r]$ diagonal and $|c_1| \geq |c_2| \geq \cdots \geq |c_r|$. The state ρ has Slater number 1 iff

$$|c_1| \leq \sum_{i=2}^{r} |c_i|. \tag{25}$$

In analogy with the case of distinguishable particles case, one has also at disposal a tool in form of the $k-$Slater witness defined as a hermitian operator W such that $\text{Tr}(W\sigma) \geq 0$ for all $\sigma \in Sl_{k-1}$, and there exists at least one $\rho \in Sl_k$ such that $\text{Tr}(W\rho) < 0.$, which can be used to distinguish states of increasing Slater correlations[21].

7 Outlook

The geometry of entanglement, although seemingly simple, poses a lot of unsolved and challenging problems. The above presented and outlined results exhibit the role of various symmetries in search of effective characterization and quantification of entangled states. The continuation along this line of investigation is in progress[17].

References

1. A. Einstein, B. Podolsky, and N. Rosen, *Phys. Rev.* **47**, 777 (1935).
2. P. W. Shor, in *Proc. 35th Annual Symp. on Foundations of Computer Science*, IEEE Computer Society Press, Santa Fe (1994), p. 124.
3. A. Ekert, *Phys. Rev. Lett.* **67**, 661 (1991).
4. C. H. Bennet, G. Brassard, C. Crépeau, R. Jozsa, A. Peres and W. K. Wootters, *Phys. Rev. Lett.* **70**, 1895 (1993).
5. V. Vedral and M. Plenio, *Prog. Quant. Electron.* **22**, 1 (1998).
6. R. F. Werner, *Phys. Rev. A* **40**, 4277 (1989).
7. A. Peres, *Phys. Rev. Lett.* **77,** 1413 (1996).
8. M. Horodecki, P. Horodecki, and R. Horodecki, *Phys. Lett. A* **223**, 1 (1996).

9. M. Lewenstein, D. Bruß, J. i. Cirac, B. Kraus. M. Kuś, J. Samsonowicz, A. Sanpera, and R. Tarrach, *J. Mod. Opt.* **47**, 2481 (2000).
10. M. Lewenstein and A. Sanpera, *Phys. Rev. Lett.* **80**, 2261 (1998).
11. V. Vedral and M. B. Plenio, *Phys. Rev. A* **57**, 1619 (1998).
12. W. K. Wootters, *Phys. Rev. Lett.* **80**, 2245 (1998).
13. T. Wellens and M. Kuś, *Phys. Rev. A* **64**, 052302 (2001).
14. B. M. Terhal, P. Horodecki, *Phys. Rev. A* **61**, 040301 (2000).
15. A. Sanpera, D. Bruß, M. Lewenstein, *Phys. Rev. A* **63**, 050301 (2001).
16. M. Kuś and K. Życzkowski, *Phys. Rev. A* **63**, 032307 (2001).
17. M. M. Sinołęcka, K. Życzkowski, and M. Kuś, Manifolds of interconvertible pure states, to be published.
18. P. Heinzner, A. T. Huckleberry, M. Kuś, and K. Życzkowski, unpublished.
19. D. Loss, D. P. DiVincenzo, *Phys. Rev. A* **57**, 120 (1998).
20. J. Schliemann, D. Loss, A. H. MacDonald, *Phys. Rev. B* **63**, 085311 (2001).
21. J. Schliemann, J. I. Cirac, M. Kuś, M. Lewenstein, and D. Loss, *Phys. Rev. A* **64**, 022303 (2001).

FROM NONCOMMUTATIVE SPACE-TIME TO QUANTUM RELATIVISTIC SYMMETRIES WITH FUNDAMENTAL MASS PARAMETER

JERZY LUKIERSKI

Institute for Theoretical Physics, University of Wrocław,
pl. Maxa Borna 9, 50-204 Wrocław, Poland
E-mail: lukier@ift.uni.wroc.pl

We consider the simplest class of Lie-algebraic deformations of space-time algebra, with the selection of κ-deformations as providing quantum deformation of relativistic framework. We recall that the κ-deformation along any direction in Minkowski space can be obtained. Some problems of the formalism of κ-deformations will be considered. We shall comment on the conformal extension of light-like κ-deformation as well as on the applications to astrophysical problems.

1 Introduction

Recently the noncommutative modification of space-time geometry gained remarkable attention mainly due to two reasons:

i) It has been argued (see e.g [1,2,3,4]) that the quantum gravity corrections can be algebraically described by the noncommutative modification of Minkowski space-time geometry

$$[x_\mu, x_\nu] = 0 \quad \longrightarrow \quad [\hat{x}_\mu, \hat{x}_\nu] = i\frac{1}{\kappa}\theta_{\mu\nu}(\frac{\hat{x}}{\kappa}) \tag{1}$$

where κ is fundamental mass-like parameter and in the general case

$$\hat{\theta}_{\mu\nu}(\hat{x}) = \theta^{(0)}_{\mu\nu} + \theta^{(1)\lambda}_{\mu\nu}\hat{x}_\lambda + \theta^{(2)\lambda\tau}_{\mu\nu}\hat{x}_\lambda\hat{x}_\tau + \ldots \tag{2}$$

ii) The noncommutativity of space-time coordinates can be obtained also from the quantization of open string in $D = 10$ in the presence of non-vanishing 2-tensor background field $B_{\mu\nu}(x)$ (see e.g. [5,6,7]). If $B_{\mu\nu}$ is constant, it implies also constant value of $\theta_{\mu\nu}$. In general case the relation (1) should be treated as a part of deformed relativistic phase space algebra with generators $(\hat{x}_\mu, \hat{p}_\mu)$ and local function $\theta_{\mu\nu}(\hat{x})$ extended to the functional $\theta_{\mu\nu}[\hat{x}, \hat{p}]$ of eight phase space generators.

The most popular case of deformation (1) is provided by constant value of the commutator $(\theta_{\mu\nu}(\hat{x}) = \theta^{(0)}_{\mu\nu})$. Such assumption is also made in original Doplicher-Fredenhagen-Roberts (DFR) model [1]. In such a case the relativistic symmetries remain classical, however the presence of constant two-tensor $\theta_{\mu\nu}^{(0)}$ implies the breaking of classical Lorentz symmetry[a]. In local QFT models on deformed Minkowski space with constant value of $\theta_{\mu\nu}$ one obtains the standard relativistic action supplemented with Lorentz symmetry breaking terms, proportional to inverse powers of κ.

The next case is described by commutator (1) with linear term in $\hat{x}_\mu$.

$$[\hat{x}_\mu, \hat{x}_\nu] = i\frac{1}{\kappa}\theta_{\mu\nu}^{(1)\,\rho}\hat{x}_\rho\,. \tag{3}$$

This is the main subject of our considerations. In order to obtain the translational invariance of the theory

$$\hat{x}_\mu \to \hat{x}_\mu + \hat{v}_\mu \tag{4}$$

the basic algebra (3) imply that

$$[\hat{v}_\mu, \hat{v}_\nu] = i\frac{1}{\kappa}\theta_{\mu\nu}^{(1)\,\rho}\hat{v}_\rho \tag{5}$$

and

$$[\hat{x}_\mu, \hat{v}_\nu] = 0\,. \tag{6}$$

The relations (5)-(6) imply that the addition formula (4) can be described as

$$\Delta(\hat{x}_\mu) = \hat{x}_\mu \otimes 1 + 1 \otimes x_\mu \tag{7}$$

where $\hat{x}_\mu = \hat{x}_\mu \otimes 1$, $\hat{v}_\mu = 1 \otimes \hat{x}_\mu$. We see that the relations (3)-(7) describe the Hopf algebra of noncommutative translations, with primitive coproduct (7).

An interesting question is to ask whether the noncommutative relations (3) can be extended do $D = 4$ quantum Poincaré group, with generators $(\hat{x}_\mu, \hat{\Lambda}_\mu{}^\nu)$ endowed with classical coproduct

$$\begin{aligned}\Delta(\hat{\Lambda}_\mu{}^\rho) &= \hat{\Lambda}_\mu{}^\nu \dot{\otimes} \Lambda_\nu{}^\rho \\ \Delta(\hat{x}_\mu) &= \hat{x}_\rho \otimes \Lambda^\rho{}_\mu + 1 \otimes x_\mu\,.\end{aligned} \tag{8}$$

The general answer to this question is contained in the paper by Podleś and Woronowicz [8]. In the case of Lie-algebraic deformation (3) it appears [9,10,11] that the most general choice permitting to extend (3) to Poincaré quantum group depends on a four-vector a^μ in the following way:

$$\theta_{\mu\nu}^{(1)\,\rho} = a_\mu \delta_\nu{}^\rho - a_\nu \delta_\mu{}^\rho\,. \tag{9}$$

One can consider the following three cases:

[a] It should be mentioned that one can avoid Lorentz symmetry breaking in quantum theory, by assuming that $\theta_{\mu\nu}^{(0)}$ is a constant tensor operator transforming as relativistic two-tensor under unitary rotations of Lorentz group in extended Hilbert space.

i) If $a = (1,0,0,0)$ one obtains standard κ-deformation of Minkowski space, with quantum time coordinate $\hat{t}$ ($\hat{x}_0 = c\hat{t}$)

$$[\hat{x}_0, \hat{x}_i] = \frac{i}{\kappa}\hat{x}_i\,, \qquad [\hat{x}_i, \hat{x}_j] = 0\,. \tag{10}$$

ii) If $a = \frac{1}{\sqrt{2}}(1,1,0,0)$ (light-like four-vector; $a_\mu a^\mu = 0$) one gets ($\hat{x}_\pm = \frac{1}{\sqrt{2}}(\hat{x}_0 \pm \hat{x}_3)$; $r, s = 1, 2$)

$$\begin{aligned} [\hat{x}_+, \hat{x}_-] &= \frac{i}{\kappa}\hat{x}_-\,, \qquad & [\hat{x}_+, \hat{x}_r] &= \frac{i}{\kappa}\hat{x}_r\,. \\ [\hat{x}_-, \hat{x}_r] &= 0\,, & [\hat{x}_r, \hat{x}_s] &= 0\,. \end{aligned} \tag{11}$$

iii) If $a = (0,1,0,0)$ one obtains the tachyonic κ-deformation, with quantum third space coordinate

$$\begin{aligned} [\hat{x}_3, \hat{x}_0] &= \frac{i}{\kappa}\hat{x}_0\,, \qquad & [\hat{x}_3, \hat{x}_r] &= \frac{i}{\kappa}\hat{x}_r\,, \\ [\hat{x}_0, \hat{x}_r] &= 0\,, & [\hat{x}_r, \hat{x}_s] &= 0\,. \end{aligned} \tag{12}$$

The generalized κ-deformations, described by the choice (9) of noncommutative space-time was proposed firstly in [9] (see also [10,11,12]).

In Sect. 2 we shall describe some properties of standard κ-deformed symmetries. We shall consider as basic characterization of these quantum symmetries the κ-deformed Poincaré algebra which is dual as a Hopf algebra (see e.g. [13]) to the κ-deformed Poincaré group. We shall notice here some new results concerning finite κ-deformed Lorentz transformations [14].

In Sect. 3 we shall discuss the light-like κ-deformation and its extension to κ-deformation odd $= 4$ conformal symmetries. It appears that only the a_μ-dependent κ-deformations with $a_\mu a^\mu = 0$ are described by classical r-matrices satisfying the classical Yang-Baxter equation. It has been shown [15] that such r-matrix can be quantized by the extended Jordanian twist method [16,17,18].

In Sect. 4 we discuss some recent ideas related with the astrophysical applications of κ-deformed space-time.

2 Standard κ-deformation

Standard κ-deformation has been obtained firstly (see [19]) by contraction of q-deformed anti-de-Sitter algebra. In such a formulation of the κ-deformed Poincaré algebra the nonrelativistic $E(3)$ algebra ($O(3)$ generators M_i + space translation generators P_i) supplemented by P_0 remains classical. We call such a basis standard. Subsequently by suitable change of basis [20] there was obtained κ-deformed Poincaré algebra in bicrossproduct basis:

a) algebraic sector ($M_{\mu\nu} = (M_i = \frac{1}{2}\epsilon_{ijk}M_{jk},\ N_i = M_{i0}),\ P_\mu = (P_i, P_0)$)

$$
\begin{aligned}
[M_{\mu\nu}, M_{\rho\tau}] &= i(\eta_{\mu\tau}M_{\nu\rho} - \eta_{\mu\rho}M_{\nu\tau} + \eta_{\nu\rho}M_{\mu\tau} - \eta_{\nu\tau}M_{\mu\rho}) \\
[M_i, P_j] &= i\epsilon_{ijk}P_k\,, \qquad [M_i, P_0] = 0\,, \\
[N_i, P_j] &= i\delta_{ij}[\tfrac{\kappa}{2}(1 - e^{-\frac{2P_0}{\kappa}}) + \tfrac{1}{2\kappa}\vec{P}^2 + \tfrac{1}{\kappa}P_iP_j\,. \\
[N_i, P_0] &= iP_i\,, \\
[P_\mu, P_\nu] &= 0\,.
\end{aligned}
\tag{13}
$$

b) coalgebraic sector

$$
\begin{aligned}
\Delta M_i &= M_i \otimes 1 + 1 \otimes M_i\,, \\
\Delta N_i &= N_i \otimes 1 + e^{-\frac{P_0}{\kappa}} \otimes N_i + \tfrac{1}{\kappa}\epsilon_{ijk}P_j \otimes M_k\,, \\
\Delta P_i &= P_i \otimes 1 + e^{-\frac{P_0}{\kappa}} \otimes P_i\,, \\
\Delta P_0 &= P_0 \otimes 1 + 1 \otimes P_0\,,
\end{aligned}
\tag{14}
$$

c) antipodes

$$
\begin{aligned}
S(M_i) &= -M_i\,, \\
S(N_i) &= -e^{\frac{P_0}{\kappa}}N_i + \tfrac{1}{\kappa}\epsilon_{ijk}e^{\frac{P_0}{\kappa}}P_jM_k \\
S(P_i) &= -e^{\frac{P_0}{\kappa}}P_i\,, \\
S(P_0) &= -P_0\,.
\end{aligned}
\tag{15}
$$

One can show that the Hopf algebra (13)-(15) is described infinitesimally by the following classical r-matrix

$$
r_1 = \frac{1}{\kappa}(N_i \wedge P_i)
\tag{16}
$$

satisfying modified Yang-Baxter equation. In bicrossproduct basis we see that in comparison with classical Poincaré symmetries only the boost transformations of three-momenta are deformed. Further in bicrossproduct basis one can introduce consistently the finite Lorentz transformations of four-momentum generators by standard formulae:

$$
P_\mu(\vec{\beta}) = e^{i\vec{\beta}\vec{N}}P_\mu e^{-i\vec{\beta}\vec{N}} = \sum_{n=0}^{\infty}\frac{1^n}{n!}[\vec{\beta}\vec{N}, [\vec{\beta}\vec{N}, \ldots [\vec{\beta}\vec{N}, P_\mu]]\ldots]
\tag{17}
$$

The basis relation in bicrossproduct basis which justifies the formula (16) as consistent with Hopf algebra structure is the following equality for quantum adjoint operator (see [21], chapt 2d)

$$\mathrm{ad}_{N_i} f(P_\mu) = N_i^{(1)} f(P_\mu) S(N_i^{(2)}) = [N_i, f(P_\mu)]\,, \tag{18}$$

where $\Delta N_i = N_i^{(1)} \otimes N_i^{(2)}$ (no summation!) is given by (14). The calculation of the function $P_\mu(\vec{\beta})$ for the choice of boost $\vec{\beta} = (0,0,\beta)$ has been recently performed explicitly in [14].

The deformation of $D = 4$ Poincaré algebra can be removed if we introduce nonlinear realization of the four-momentum operators [21,22,23 b]

$$\begin{aligned} P_i^{(0)} &= e^{\frac{P_0}{2\kappa}} P_i\,, \\ P_0^{(0)} &= \kappa \sinh \tfrac{P_0}{\kappa} + \tfrac{1}{2\kappa} \vec{P}^2\,. \end{aligned} \tag{19}$$

The generators $(M_{\mu\nu}, P_\mu^{(0)})$ satisfy classical Poincaré algebra, with quite complicated coproduct [12]. Because the Lorentz transformations of the four-momenta $P_\mu^{(0)}$ are classical, it follows from the formulae (18)-(19) that the finite Lorentz transformations in bicrossproduct basis can be obtained from classical Lorentz transformations by deformation map, described by the inverse formulae to the relations (19):

$$\begin{aligned} P_i &= \kappa \left(P_0^{(0)} + \sqrt{P_\mu^{(0)} P^{(0)\mu} + \kappa^2} \right)^{-1} \cdot P_i^{(0)}\,, \\ P_0 &= \kappa \ln \left(\frac{P_0^{(0)} + \sqrt{P_\mu^{(0)} P^{(0)\mu} + \kappa^2}}{\kappa} \right). \end{aligned} \tag{20}$$

In the classical Poincaré algebra basis $(M_{\mu\nu},\ P_\mu^{(0)})$ the whole effect of deformation is contained in the coproduct. In particular we stress that the coproduct of the energy operator $P_0^{(0)}$ is nonprimitive (nonabelian). We would like to recall that in [12] an attempt was undertaken to interpret the noncocommutative coproduct for the energy operator $P_0^{(0)}$ as describing geometric two-particle interactions.

3 Light-like κ-deformation in standard basis

The κ-deformation in standard basis leading to "quantum" light-cone direction $x_+ = \frac{1}{\sqrt{2}}(x_0 + x_3)$ (see (11-12)) was firstly obtained in [24]. Further in [9] such a deformation was obtained in bicrossproduct basis as a special case

---

[b] We choose particular deformation map. The general case is discussed in [23].

of generalized κ-deformation, depending on arbitrary four-vector a_μ. The equivalence of two formulations has been shown in [25] (see also [26]).

Let us introduce the light-cone basis ($a_\mu = \frac{1}{\sqrt{2}}(1,1,0,0)$, $\widetilde{a}_\mu = \frac{1}{\sqrt{2}}(1,-1,0,0)$; $a_\mu a^\mu = \widetilde{a}_\mu \widetilde{a}^\mu = 0$, $a_\mu \widetilde{a}^\mu = 1$; $r = 1,2$)

$$
\begin{aligned}
&\widetilde{P}_0 = \widetilde{a}^\mu P_\mu = P_0 + P_3 \,, && \widetilde{P}_3 = a^\mu P_\mu = P_0 - P_3 \,, && \widetilde{P}_r = P_r \,, \\
&\widetilde{M}_{03} = \widetilde{a}^\mu a^\nu M_{\mu\nu} \,, && \widetilde{M}_{0r} = \widetilde{a}^\mu M_{\mu r} \,, && \\
&\widetilde{M}_{3r} = a^\mu M_{\mu r} \,, && \widetilde{M}_{12} = M_{12} \,. &&
\end{aligned}
\tag{21}
$$

The light-like κ-deformation of Poincaré algebra related by duality with non-commutative translation sector described by eq. (11) is given by the following Hopf algebra.

a) algebraic sector $\widetilde{M}_i = \frac{1}{2}\epsilon_{ijk}\widetilde{M}_{jk}$; $\widetilde{N}_i = \widetilde{M}_{i0}$ Besides the classical Lorentz algebra relations and classical commutativity of four-momenta we obtain

$$
\begin{aligned}
&[\widetilde{M}_i, \widetilde{P}_0] = 0 \,, \\
&[\widetilde{M}_i, \widetilde{P}_j] = i\kappa\epsilon_{ij}(1 - e^{-\frac{\widetilde{P}_0}{\kappa}}) + i\epsilon_{ijk}\widetilde{P}_k \,, \\
&[\widetilde{N}_i, \widetilde{P}_0] = i\kappa\delta_{i3}(1 - e^{-\frac{\widetilde{P}_0}{\kappa}}) + i(1-\delta_{i3})\widetilde{P}_i \,, \\
&[\widetilde{N}_i, \widetilde{P}_k] = i\delta_{ik}\widetilde{P}_3 e^{-\frac{\widetilde{P}_0}{\kappa}} + i\delta_{i3}\widetilde{P}_k(e^{-\frac{\widetilde{P}_0}{\kappa}} - 1) + \tfrac{i}{2\kappa}\delta_{ik}(\widetilde{P}_1^2 + \widetilde{P}_2^2) \\
&\qquad - \tfrac{i}{\kappa}(1-\delta_{i3})\widetilde{P}_i\widetilde{P}_k \,,
\end{aligned}
\tag{22}
$$

b) coalgebraic sector and antipodes are provided by the relations (14)-(15) with the replacement $M_i' \to \widetilde{M}_i$, $N_i \to \widetilde{N}_i$ and $P_\mu \to \widetilde{P}_\mu$.

The above Hopf algebra is described by the following classical r-matrix (see also [27])

$$
r_2 = \frac{1}{\kappa}(\widetilde{N}_i \wedge \widetilde{P}_i) = \frac{1}{\kappa\sqrt{2}}[N_3 \wedge (P_0 - P_3) + (N_2 - M_1) \wedge P_2 + (N_1 + M_2) \wedge P_1] \tag{23}
$$

which satisfies classical Yang-Baxter equation

$$
[[r_2, r_2]] \equiv [r_2^{(12)}, r_2^{(13)}] + [r_2^{(12)}, r_2^{(23)}] + [r_2^{(13)}, r_2^{(23)}] = 0 \tag{24}
$$

The classical r-matrix (23) can be promoted to the r-matrix for $D = 4$ conformal group, described by $SU(4) \simeq O(4,2)$ algebra. It appears that the generators ($\widetilde{N}_i$,$\widetilde{P}_i$) spanning classical r-matrix (23) can be described by the generators from the positive Borel subalgebra of $sl(4)$. With suitable reality conditions (see e.g. [27]) we obtain

$$r_2 = \frac{1}{\kappa} \left[\tfrac{1}{2}(h_1 + h_3) \wedge e_6 + e_1 \wedge e_5 - e_3 \wedge e_4 \right] , \tag{25}$$

where the root generators of $sl(4)$ enter into the fundamental 4×4 matrix representation as follows (see e.g. [28])

$$\begin{pmatrix} h_1 & e_1 & e_4 & e_6 \\ e_{-1} & h_2 & e_2 & e_5 \\ e_{-4} & e_{-2} & h_3 & e_3 \\ e_{-6} & e_{-5} & e_{-3} & h_4 \end{pmatrix} \tag{26}$$

Using results presented in [16,17,18] one can calculate the generalized twist function, providing the full Hopf algebra structure of κ-deformed $D = 4$ conformal algebra [15]. In fact, we have looked also on multiparameter solution of classical YB equation, spanned by the generators belonging to Weyl algebra (Poincaré generator + dilatations). We found for the classical r-matrices carrying mass dimension the corresponding quantum κ-deformed conformal algebras described by suitable extended twist functions.

4 Outlook

Recently κ-deformations as an example of noncommutative geometry supplemented with quantum covariance group were tried as describing quantum gravity effects (see e.g. [29,30]) and were applied to astrophysical phenomena (see e.g. [31,32,33]). In commutative four-momentum framework the following two ingredients of κ-deformed geometry are used:

i) The deformation of mass-shell condition[c]

$$\vec{P}^2 e^{\pm \frac{P_0}{\kappa}} - (2\kappa \sinh \tfrac{P_0}{\kappa})^2 = -M^2 \tag{27}$$

ii) The nonsymmetric, nonabelian composition law for the four-momenta, described by the coproduct of momentum generators (see (14)).

Both these elements of new theory lead at present to some conceptual problems. If we assume that the parameter κ is universal (besides $\hbar$ and c as the third fundamental constant) the straightforward application of formula (27) to macroscopic bodies leads to obvious contradiction with experimental results[d].

[c] In bicrossproduct basis one can consider two deformed mass-shell formulae related by the inversion of κ: $\kappa \to \kappa^{-1}$ (see [34], formula (2.34)).

[d] This difficulty was pointed out to the author firstly by I. Białynicki-Birula.

We see that the range of applicability of the formula (27) is still not understood what is related with some new notion of elementarity in microworld physics still to be discovered. The difficulty with consistent description by κ-physics of macroworld phenomena is also connected with the appearance of nonsymmetric, nonabelian addition law for the momenta. The problem how to incorporate modified momentum conservation law into physical models remains to be solved before we can apply κ-deformed kinematics e.g. to scattering processes and description of κ-deformed Feynman diagrams [35]. At present stage there were proposed some new ideas (see e.g. [36]) but according to our estimate they are quite preliminary.

Acknowledgment

Support by the KBN grant 5P03B05620 is acknowledged.

References

1. S. Doplicher, K. Fredenhagen, and J.E. Roberts, *Phys.Lett.B* **331**, 39 (1994); *Comm. Math. Phys.* **172**, 187 (1995).
2. D.V. Ahluwalia, *Phys.Lett.B* **33(**, 301 (1994).
3. G. Amelino-Camelia, *Mod. Phys. Lett.* A **9**, 3415 (1994); *Mod. Phys.Lett A* **11**, 111 (1996).
4. L.J. Garay, *Int. Journ. Mod. Phys.* **10A**, 145 (1995).
5. C.S. Chu and P.M. Ho, *Nucl. Phys. B* **550**, 151 (1999); hep-th/9812219.
6. N. Seiberg and E. Witten, *High Energy Phys.* **9909**, 032 (1991); hep-th/9908142.
7. R.J. Szabo, *Quantum Field Theory on Noncommutative spaces*; hep-th/0109162.
8. P. Podleś and S. Woronowicz, *Comm. Math. Phys.* **178**, 61 (1996).
9. P. Kosiński and P. Maślanka, in *From Field Theory to Quantum Groups*, eds. B. Jancewicz, J.Sobczyk, World Scientific (1996), p. 11.
10. J. Lukierski, in *Proc of Alushta Conference on Recent Problems in QFT*, May 1996, ed. D. Shirkov, D.I. Kazakov, and A.A. Vladimirov, Dubna (1996), p.82.
11. K. Przanowski, *Czech. J. Phys.* **47**, 107 (1997)
12. J. Lukierski, *κ-Deformations of relativistic symmetries: recent developments*, in *Proc. of Quantum Group Symposium*, July 1996, Goslar, ed. H.-D. Doebner and V.K. Dobrev, Heron Press, Sofia (1997), p. 173.
13. S. Majid, *Foundations of Quantum Groups*, Cambridge Univ. Press (1995).

14. N.R. Bruno, G. Amelino-Camelia, and J. Kowalski-Glikman, Deformed Boost Transformations that saturate at the Planck scale, `hep-th/0107039`
15. J. Lukierski, V. Lyakhovsky, and M. Mozrzymas, in preparation
16. P.P. Kulish, V.D. Lyakhovsky, and A.I. Mudrov, *J. Math. Phys.* **40**, 4569 (1999); math.QA/9806014.
17. P.P. Kulish, V.D. Lyakhovsky, and M.A. del Olmo, *J. Phys. A* **32**, 8671 (1999).
18. V.D. Lyakhovsky and M.A. del Olmo, math.QA/9903065.
19. J. Lukierski, A. Nowicki, H. Ruegg, and V.N. Tolstoy, *Phys.Lett.B* **264**, 331 (1991).
20. S. Majid and H. Ruegg, *Phys.Lett.B* **334**, 348 (1994).
21. S. Majid, H. Ruegg, and V.N. Tolstoy, in *Quantum Groups: Formalism and Applications*, Proceed. of XXX Karpacz Winter School, February 1994, Polish Scientific Publishers (1994), p. 359.
22. H. Ruegg and V.N. Tolstoy, *Lett. Math. Phys.* **32**, 85 (1995)
23. P. Kosiński, J. Lukierski, P. Maślanka, and J. Sobczyk, *Mod. Phys. Lett.* **10A**, 2599 (1995)
24. A. Ballesteros, E. Herranz, M.A. del Olmo, and M. Santander, *Phys.Lett.B* **351**, 137 (1995).
25. K. Przanowski, *Acta Phys. Polon.* B **28**, 1635 (1997).
26. O. Arratia, F.Herranz, and M.A. del Olmo, *J. Phys. A* **31**, L1 (1998); `q-alg/9707025`.
27. J. Lukierski, P. Minnaert, and M. Mozrzymas, *Phys.Lett.B* **371**, 215 (1996).
28. J. Lukierski, A. Nowicki, and J. Sobczyk, *J. Phys. A* **26**, 4047 (1993).
29. G. Amelino-Camelia, *Phys.Lett.B* **392**, 283 (1997).
30. G. Amelino-Camelia, J. Lukierski, and A. Nowicki, *Yad. Fizyka* **61**, 1925 (2000)
31. R. Alonso, P. Blasi, P.L. Ghia, and A.F. Grillo, *Phys. Rev.D* **62**, 053010 (2000).
32. J. Kowalski-Glikman, *Phys.Lett.B* **499**, 1 (2001)
33. G. Amelino-Camelia, J. Lukierski, and A. Nowicki, hep-th/0103227; *Czech. J. Phys.*, in press
34. J. Lukierski, H. Ruegg, and W. Zakrzewski, *Ann. Phys.* **234**, 90 (1995).
35. P. Kosiński, J. Lukierski, and P. Maślanka, *Phys. Rev. D* **62**, 025004 (2000).
36. G. Amelino-Camelia and M. Arzano, hep-th/0105120.

TOMOGRAPHIC MAP WITHIN THE FRAMEWORK OF STAR-PRODUCT QUANTIZATION

OLGA V. MAN'KO and VLADIMIR I. MAN'KO
P. N. Lebedev Physical Institute, Leninskii prospect 53, 119991 Moscow, Russia

GIUSEPPE MARMO
Dipartimento di Scienze Fisiche, Università "Federico II" di Napoli and Istituto Nazionale di Fisica Nucleare, Sezione di Napoli Complesso Universitario di Monte S'Angelo, via Cintia, 80126 Napoli, Italy

Tomograms introduced for the description of quantum states in terms of probability distributions are shown to be related to a standard star-product quantization with appropriate kernels. Examples of symplectic tomograms and spin tomograms are presented.

1 Introduction

Star-product quantization[1] provides us the a formulation of quantum mechanics where quantum observables are presented by functions on the phase space instead of operators acting on some Hilbert space. The product of functions is not the point-wise product. This induced product is nonlocal and it is defined by some kernel. Recently the tomographic description of quantum states has been introduced to associate states with standard probability distributions instead of density operators.[2,3] The tomographic description was introduced also for spin.[4,5] Properties of tomographic maps and their relation with Heisenberg–Weyl and SU(2) groups were studied.[6–8] The relation of tomographic representation to the star-product quantization was established.[8,9] The aim of this contribution is to review this approach[8,9] and to study properties of the star-product of symbols of quantum observables. We show that the tomographic map can be discussed within the framework of the standard star-product quantization procedure.

2 Star-product

In quantum mechanics, observables are described by operators acting on the Hilbert space of states. In order to consider observables as functions on a phase space, we review first a general construction and provide general relations and properties of a map from operators onto functions without a concrete realization of the map. Given a Hilbert space H and an operator $\hat{A}$ acting on this space, let us suppose that we have a set of operators $\hat{U}(\mathbf{x})$ and a n-

dimensional vector $\mathbf{x} = (x_1, x_2, \ldots, x_n)$. We construct the c-number function $f_{\hat{A}}(\mathbf{x})$ (we call it the symbol of operator $\hat{A}$)

$$f_{\hat{A}}(\mathbf{x}) = \mathrm{Tr}\left[\hat{A}\hat{U}(\mathbf{x})\right].$$

Let us suppose that the relation has an inverse, i.e., there exists a set of operators $\hat{D}(\mathbf{x})$ such that

$$\hat{A} = \int f_{\hat{A}}(\mathbf{x})\hat{D}(\mathbf{x})\, d\mathbf{x}.$$

In fact, we could consider maps of the form

$$\hat{A} \to f_{\hat{A}}(\mathbf{x}) \qquad \text{and} \qquad f_{\hat{A}}(\mathbf{x}) \to \hat{A},$$

and require them to be one the inverse of the other.

The most important property is the existence of associative product (star-product) of the symbols induced by the product of operators. We introduce the product (star-product) of two functions $f_{\hat{A}}(\mathbf{x})$ and $f_{\hat{B}}(\mathbf{x})$ corresponding to two operators $\hat{A}$ and $\hat{B}$ by the relations

$$f_{\hat{A}\hat{B}}(\mathbf{x}) = f_{\hat{A}}(\mathbf{x}) * f_{\hat{B}}(\mathbf{x}) := \mathrm{Tr}\left[\hat{A}\hat{B}\hat{U}(\mathbf{x})\right].$$

Since the standard product of operators on a Hilbert space is an associative product, i.e. $\hat{A}(\hat{B}\hat{C}) = (\hat{A}\hat{B})\hat{C}$, it is obvious that the formula defines an associative product for the functions $f_{\hat{A}}(\mathbf{x})$, namely,

$$f_{\hat{A}}(\mathbf{x}) * \left(f_{\hat{B}}(\mathbf{x}) * f_{\hat{C}}(\mathbf{x})\right) = \left(f_{\hat{A}}(\mathbf{x}) * f_{\hat{B}}(\mathbf{x})\right) * f_{\hat{C}}(\mathbf{x}).$$

The commutator of two operators

$$\hat{C} = [\hat{A}, \hat{B}] = \hat{A}\hat{B} - \hat{B}\hat{A}$$

is mapped onto the quantum Poisson bracket[10] $f_{\hat{C}}(\mathbf{x})$ of two symbols $f_{\hat{A}}(\mathbf{x})$ and $f_{\hat{B}}(\mathbf{x})$

$$f_{\hat{C}}(\mathbf{x}) = \left\{f_{\hat{A}}(\mathbf{x}), f_{\hat{B}}(\mathbf{x})\right\}_* = \mathrm{Tr}\left[[\hat{A}, \hat{B}]\hat{U}(\mathbf{x})\right].$$

Since the Jacobi identity is fulfilled for the commutator of the operators, the Jacobi identity is also fulfilled for the Poisson bracket of the functions $f_{\hat{A}}(\mathbf{x})$ and $f_{\hat{B}}(\mathbf{x})$. Since for operators one has the derivation property with respectto the associative product, the Poisson brackets reproduce this property

$$\left\{f_{\hat{A}}(\mathbf{x}), f_{\hat{B}}(\mathbf{x}) * f_{\hat{C}}(\mathbf{x})\right\}_* = \left\{f_{\hat{A}}(\mathbf{x}), f_{\hat{B}}(\mathbf{x})\right\}_* * f_{\hat{C}}(\mathbf{x}) + f_{\hat{B}}(\mathbf{x}) * \left\{f_{\hat{A}}(\mathbf{x}), f_{\hat{C}}(\mathbf{x})\right\}_*$$

this property qualifies it as a *quantum Poisson bracket.*

The evolution in the space of observables can be described by the Heisenberg equations of motion

$$\dot{\hat{A}} = i[\hat{H}, \hat{A}] \qquad (\hbar = 1),$$

where $\hat{H}$ is the Hamiltonian of the system and $\hat{A}$ is a generic observable. This equation can be rewritten in terms of the functions $f_{\hat{A}}(\mathbf{x})$ and $f_{\hat{H}}(\mathbf{x})$, where

$$f_{\hat{H}}(\mathbf{x}) = \operatorname{Tr}\left[\hat{H}\hat{U}(\mathbf{x})\right]$$

corresponds to the Hamiltonian, in the form

$$\dot{f}_{\hat{A}}(\mathbf{x}, t) = i\Big\{f_{\hat{H}}(\mathbf{x}, t), f_{\hat{A}}(\mathbf{x}, t)\Big\}_{*}$$

via the quantum Poisson bracket we have defined.

Let us suppose that there exist other maps, we choose two different ones. One map is described by a vector $\mathbf{x} = (x_1, x_2, \ldots, x_n)$ and operators $\hat{U}(\mathbf{x})$ and $\hat{D}(\mathbf{x})$. Another map is described by a vector $\mathbf{y} = (y_1, y_2, \cdots, y_m)$ and operators $\hat{U}_1(\mathbf{y})$ and $\hat{D}_1(\mathbf{y})$. For given operator $\hat{A}$, one has the function

$$\phi_{\hat{A}}(\mathbf{y}) = \operatorname{Tr}\left[\hat{A}\hat{U}_1(\mathbf{y})\right]$$

and the inverse relation

$$\hat{A} = \int \phi_A(\mathbf{y})\hat{D}_1(\mathbf{y})\, d\mathbf{y}.$$

One can obtain a relation of the function $f_{\hat{A}}(\mathbf{x})$ to the function $\phi_{\hat{A}}(\mathbf{y})$ in the form

$$\phi_{\hat{A}}(\mathbf{y}) = \int f_{\hat{A}}(\mathbf{x}) \operatorname{Tr}\left[\hat{D}(\mathbf{x})\hat{U}_1(\mathbf{y})\right] d\mathbf{x}$$

and the inverse relation

$$f_{\hat{A}}(\mathbf{x}) = \int \phi_{\hat{A}}(\mathbf{y}) \operatorname{Tr}\left[\hat{D}_1(\mathbf{y})\hat{U}(\mathbf{x})\right] d\mathbf{y}.$$

We see that functions $f_{\hat{A}}(\mathbf{x})$ and $\phi_{\hat{A}}(\mathbf{y})$ corresponding to different maps are connected by means of the invertible integral transform. These transforms are determined by means of intertwining kernels

$$K_1(\mathbf{x}, \mathbf{y}) = \operatorname{Tr}\left[\hat{D}(\mathbf{x})\hat{U}_1(\mathbf{y})\right] \qquad \text{and} \qquad K_2(\mathbf{x}, \mathbf{y}) = \operatorname{Tr}\left[\hat{D}_1(\mathbf{y})\hat{U}(\mathbf{x})\right].$$

One can write down a composition rule for two symbols $f_{\hat{A}}(\mathbf{x})$ and $f_{\hat{B}}(\mathbf{x})$, which determines the star-product of these symbols. The composition rule is described by the formula

$$f_{\hat{A}}(\mathbf{x}) * f_{\hat{B}}(\mathbf{x}) = \int f_{\hat{A}}(\mathbf{x}'') f_{\hat{B}}(\mathbf{x}') K\left(\mathbf{x}'', \mathbf{x}', \mathbf{x}\right) d\mathbf{x}'\, d\mathbf{x}''.$$

The kernel under the integral is determined by the trace of product of the basic operators, which we use to construct the map

$$K(\mathbf{x}'', \mathbf{x}', \mathbf{x}) = \mathrm{Tr}\left[\hat{D}(\mathbf{x}'')\hat{D}(\mathbf{x}')\hat{U}(\mathbf{x})\right].$$

3 Symplectic tomograms

According to the general scheme one can introduce for the operator $\hat{A}$ the function $f_{\hat{A}}(\mathbf{x})$, where $\mathbf{x} = (x_1, x_2, x_3) \equiv (X, \mu, \nu)$, which we denote here as $w_{\hat{A}}(X, \mu, \nu)$ depending on the position X and the parameters μ and ν of the reference frame

$$w_{\hat{A}}(X, \mu, \nu) = \mathrm{Tr}\left[\hat{A}\hat{U}(\mathbf{x})\right].$$

We call the function $w_{\hat{A}}(X, \mu, \nu)$ the tomographic symbol of the operator $\hat{A}$. The operator $\hat{U}(\mathbf{x})$ is given by

$$\hat{U}(\mathbf{x}) \equiv \hat{U}(X, \mu, \nu) = \delta\left(X - \mu\hat{q} - \nu\hat{p}\right),$$

where $\hat{q}$ and $\hat{p}$ are position and momentum operators.

In the case under consideration, the inverse transform will be of the form

$$\hat{A} = \int w_{\hat{A}}(X, \mu, \nu)\hat{D}(X, \mu, \nu)\, dX\, d\mu\, d\nu,$$

where[11]

$$\hat{D}(\mathbf{x}) \equiv \hat{D}(X, \mu, \nu) = \frac{1}{2\pi} \exp\left(iX - i\nu\hat{p} - i\mu\hat{q}\right).$$

If one takes two operators $\hat{A}_1$ and $\hat{A}_2$, which are expressed through the corresponding functions by the formulae

$$\hat{A}_1 = \int w_{\hat{A}_1}(X', \mu', \nu')\hat{D}(X', \mu', \nu')\, dX'\, d\mu'\, d\nu',$$

$$\hat{A}_2 = \int w_{\hat{A}_2}(X'', \mu'', \nu'')\hat{D}(X'', \mu'', \nu'') dX''\, d\mu''\, d\nu'',$$

and $\hat{A}$ denotes the product of $\hat{A}_1$ and $\hat{A}_2$, then the function $w_{\hat{A}}(X,\mu,\nu)$, which correspond s to $\hat{A}$, is the star-product of functions $w_{\hat{A}_1}(X,\mu,\nu)$ and $w_{\hat{A}_2}(X,\mu,\nu)$, i.e.,

$$w_{\hat{A}}(X,\mu,\nu) = w_{\hat{A}_1}(X,\mu,\nu) * w_{\hat{A}_2}(X,\mu,\nu)$$

reads

$$w_{\hat{A}}(X,\mu,\nu) = \int w_{\hat{A}_1}(\mathbf{x}'')w_{\hat{A}_2}(\mathbf{x}')K(\mathbf{x}'',\mathbf{x}',\mathbf{x})\,d\mathbf{x}''\,d\mathbf{x}',$$

with the kernel given by

$$K(\mathbf{x}'',\mathbf{x}',\mathbf{x}) = \mathrm{Tr}\left[\hat{D}(X'',\mu'',\nu'')\hat{D}(X',\mu',\nu')\hat{U}(X,\mu,\nu)\right].$$

The explicit form of the kernel reads

$$K(X_1,\mu_1,\nu_1,X_2,\mu_2,\nu_2,X\mu,\nu) = \frac{\delta\Big(\mu(\nu_1+\nu_2)-\nu(\mu_1+\mu_2)\Big)}{4\pi^2}$$
$$\times \exp\left[\frac{i}{2}\Big((\nu_1\mu_2-\nu_2\mu_1)+2X_1+2X_2-\frac{2\,(\nu_1+\nu_2)}{\nu}\,X\Big)\right].$$

By using the oscillator's ground state

$$w_0 = \left[\pi\left(\mu^2+\nu^2\right)\right]^{-1/2}\exp\left(-\frac{X^2}{\mu^2+\nu^2}\right),$$

one can check that this kernel satisfies the obvious relation

$$w_0 * w_0 = w_0.$$

4 Spin tomograms

In order to construct spin-tomogram according to the general considerations, let us introduce the operator U of finite rotation of the SU(2) group . Following [4,5] we will derive the expression for an arbitrary operator acting on spin states in terms of measurable mean values of the operator for the spin projection on a given direction, considered in a rotated reference frame. For arbitrary values of spin, let the operator $\widehat{A}^{(j)}$ be represented by the matrix

$$A^{(j)}_{mm'} = \langle jm \mid \widehat{A}^{(j)} \mid jm'\rangle\,, \qquad m=-j,-j+1,\ldots,j-1,j\,,$$

where

$$\hat{\jmath}_3 \mid jm\rangle = m \mid jm\rangle\,; \qquad \hat{\jmath}^2 \mid jm\rangle = j(j+1) \mid jm\rangle\,,$$

and

$$\widehat{A}^{(j)} = \sum_{m=-j}^{j} \sum_{m'=-j}^{j} A^{(j)}_{mm'} \mid jm\rangle\langle jm' \mid ; \qquad A^{(j)}_{mm} = w_0(m) .$$

Let us condsider the diagonal elements of the operator in another reference frame

$$A^{(j)}_{m_1\, m_1} = \left(UAU^{\dagger}\right)_{m_1\, m_1} .$$

The unitary rotation transform U depends on the Euler angles α, β, γ.

Below we introduce new notation for the diagonal matrix elements of the operator and rewrite the above equality in the form

$$\widetilde{w}\,(m_1, \alpha, \beta) = \sum_{m_1'=-j}^{j} \sum_{m_2'=-j}^{j} D^{(j)}_{m_1 m_1'}(\alpha, \beta, \gamma)\, A^{(j)}_{m_1' m_2'}\, D^{(j)*}_{m_1 m_2'}(\alpha, \beta, \gamma) .$$

The defined function is called the tomographic symbol of the operator. In view of the structure of the above formula, the tomogram depends only on two Euler angles. Here the matrix elements $D^{(j)}_{m_1\, m_1'}\,(\alpha, \beta, \gamma)$ are the matrix elements of the rotation transform U

$$D^{(j)}_{m'm}(\alpha, \beta, \gamma) = e^{im'\gamma}\, d^{(j)}_{m'm}(\beta)\, e^{im\alpha} ,$$

where

$$d^{(j)}_{m'm}(\beta) = \left[\frac{(j+m')!(j-m')!}{(j+m)!(j-m)!}\right]^{1/2} \left(\cos\frac{\beta}{2}\right)^{m'+m} \left(\sin\frac{\beta}{2}\right)^{m'-m} \times P^{(m'-m, m'+m)}_{j-m'}(\cos\beta)$$

and $P_n^{(a,b)}(x)$ is the Jacobi polynomial.

The inverse relation reads [8]

$$\sum_{j_3=0}^{2j} \sum_{m_3=-j_3}^{j_3} (2j_3+1)^2 \sum_{m_1=-j}^{j} \int (-1)^{m_1}\, \widetilde{w}\,(m_1, \alpha, \beta)\; D^{(j_3)}_{0\, m_3}\,(\alpha, \beta, \gamma)$$

$$\otimes \begin{pmatrix} j & j & j_3 \\ m_1 & -m_1 & 0 \end{pmatrix} \begin{pmatrix} j & j & j_3 \\ m_1' & -m_2' & m_3 \end{pmatrix} \frac{d\omega}{8\,\pi^2} = (-1)^{m_2'}\, A^{(j)}_{m_1' m_2'} ,$$

where $d\omega$ is the volume element in terms of Euler angles. One can obtain an invariant form of this relation. To do this, let us introduce the function on the sphere

$$\Phi^{(j_3)}_{jm_1' m_2'}\,(\alpha, \beta) = (-1)^{m_2'} \sum_{m_3=-j_3}^{j_3} D^{(j_3)}_{0\, m_3}\,(\alpha, \beta, \gamma) \begin{pmatrix} j & j & j_3 \\ m_1' & -m_2' & m_3 \end{pmatrix}$$

and the operator on the sphere

$$\widehat{A}_j^{(j_3)}(\alpha,\beta) = (2j_3+1)^2 \sum_{m_1'=-j}^{j} \sum_{m_2'=-j}^{j} |\, jm_1'\rangle\, \Phi_{jm_1'm_2'}^{(j_3)}(\alpha,\beta)\, \langle jm_2'\,| \,.$$

Then using another operator $\mathbf{x} = (m_1, \alpha, \beta)$

$$\widehat{B}^{(j)}(\mathbf{x}) \equiv \widehat{B}_{m_1}^{(j)}(\alpha,\beta) = (-1)^{m_1} \sum_{j_3=0}^{2j} \begin{pmatrix} j & j & j_3 \\ m_1 & -m_1 & 0 \end{pmatrix} \widehat{A}_j^{(j_3)}(\alpha,\beta)\,,$$

one can write for the operator under study the invariant expression: [5]

$$\widehat{A}^{(j)} = \sum_{m_1=-j}^{j} \int \frac{d\omega}{8\,\pi^2}\, \widetilde{w}\,(m_1,\alpha,\beta)\; \widehat{B}_{m_1}^{(j)}(\alpha,\beta)\,.$$

Let us consider the tomographic star-product of two functions $W(m_1,\alpha,\beta)$ and $W'(m_1,\alpha,\beta)$. One has, by definition ,

$$(W \star W')\,(m_1,\alpha,\beta) = \sum_{\mathbf{x}',\mathbf{x}''} W\,(m_1',\alpha',\beta')\, W'\,(m_2'',\alpha'',\beta'')\, K(\mathbf{x},\mathbf{x}',\mathbf{x}''),$$

where the kernel has the form

$$K(\mathbf{x},\mathbf{x}',\mathbf{x}'') = \frac{1}{(8\,\pi^2)^2} \mathrm{Tr}\left[|jm_1\rangle\langle jm_1| U\, \widehat{B}^{(j)}(\mathbf{x}')\, \widehat{B}^{(j)}(\mathbf{x}'')\, U^\dagger\right].$$

The kernel can be given in explicit form. The tomographic star-product defined by means of the above kernel defines an associative product. So we have an invertible map between the operator $\widehat{A}^{(j)}$ and its symbol $\widetilde{w}(\mathbf{x})$:

$$\widehat{A}^{(j)} = \sum_{\mathbf{x}} \widetilde{w}(\mathbf{x})\widehat{D}(\mathbf{x}), \qquad \widehat{D}(\mathbf{x}) = \frac{1}{8\pi^2}\widehat{B}_{m_1}^{(j)}(\mathbf{x}),$$

$$\widetilde{w}(\mathbf{x}) = \mathrm{Tr}[\widehat{A}^{(j)}\widehat{\widetilde{U}}(\mathbf{x})], \qquad \widehat{\widetilde{U}}(\mathbf{x}) = U^\dagger |jm_1\rangle\langle jm_1| U\,,$$

where sum over $\mathbf{x}$ means the integral over Euler angles and sum over angular momentum projection.

5 Conclusions

To conclude, we have embedded the tomographic map into the general scheme of symbols and operators. When the kernel of star-product is known, Heisenberg equations of motion for observables written in the form of equations for their symbols can be presented, in some cases, in the form of partial differential equations. For example, in the tomographic representation, the evolution equation for density matrix takes the form of Fokker–Planck-type equation.[2] One can also obtain such differential form of the evolution equation for spin tomograms.

Acknowledgments

Olga V. M. and Vladimir I. M. thank the organizers for kind hospitality, the Russian Foundation for Basic Research for partial support under Projects Nos. 99-02-17753, 00-02-16516, and 01-02-17745and the Ministry for Industry, Sciences and Technology of the Russian Federation for the support within the framework of the Programs "Optics. Laser Physics" and "Fundamental Nuclear Physics." Vladimir I. Man'ko is grateful to the Russian Foundation for Basic Research for Travel Grant No. 01-02-26874.

References

1. F. Bayen, M. Flato, C. Fronsdal, A. Lechnerovicz, and D. Sternheimer, *Lett. Math. Phys.*, **1**, 521(1975).
2. S. Mancini, V. I. Man'ko, and P. Tombesi, *Phys. Lett.* A, **213**, 1 (1996).
3. S. Mancini, V. I. Man'ko, and P. Tombesi, *Found. Phys.*, **27**, 801 (1997).
4. V. V. Dodonov and V. I. Man'ko, *Phys. Lett.*, **239**, 335 (1997).
5. V. I. Man'ko and O. V. Man'ko, *JETP*, **85**, 430 (1997).
6. V. I. Man'ko and G. Marmo, *Phys. Scr.*, **58**, 224 (1998).
7. V. I. Man'ko and G. Marmo, *Phys. Scr.*, **60**, 111 (1999).
8. O. V. Man'ko, V. I. Man'ko, and G. Marmo, *Phys. Scr.*, **62**, 446 (2000).
9. O. V. Man'ko, V. I. Man'ko, and G. Marmo, *J. Phys.* A, (submitted).
10. P. A. M. Dirac, *The Principles of Quantum Mechanics*, 4th ed., Pergamon, Oxford (1958).
11. G. M. D'Ariano, S. Mancini, V. I. Man'ko, and P. Tombesi, *J. Opt.* B, **8**, 1017 (1996).

ALGORITHMIC COOLING AND SCALABLE QUANTUM COMPUTERS: WAYS TO IMPROVE THE SPACE-TIME REQUIREMENTS OF THE ALGORITHM

TAL MOR

Department of Computer Science, Technion,
Technion City, Haifa, Israel
and Electrical Engineering Department,
College of Judea and Samaria, Ariel, Israel
E-mail: talmo@cs.technion.ac.il

YOSSI WEINSTEIN

Department of Physics, Technion,
Technion City, Haifa, Israel
E-mail: yossiv@physics.technion.ac.il

The "algorithmic cooling" of Boykin, Mor, Roychowdhury, Vatan and Vrijen (BMRVV) is a powerful cooling method for obtaining a large number of highly polarized spins in liquid nuclear-spin systems at finite temperature. The cooling is done by compressing the antropy of a string into a portion of it and then swaping that portion with rapidly thermal-relaxing bits, which transfers the antropy quickly to the environment. Algorithmic cooling could enable cooling spins to a very low temperature without cooling the environment and could resolve the scaling problem of NMR quantum computing devices. The BMRVV cooling algorithm is described in a simplified model in which all spins have identical bias (the minimum over all the different biases of the spins), and using blocks of m bits, hence the resulting cooling technique is non-optimal. We present here initial analysis of two improvements of the algorithm: working with variable-size blocks in the cooling process, and using better the fact that initial biases are not equal. As result, much less cooling steps are required in order to yield more spins which are highly polarized (with the desired bias).

1 Introduction

Ensemble computing is based on a model comprised of a macroscopic number of computers, all performing the same operations simultaneously and giving an output signal with a magnitude linear in the number of computers. Much work has been done in this field with applications to NMR quantum computers [1], and computation using 7-8 qubits has already been achieved with liquid state NMR.

It is a widespread belief however, that liquid state NMR quantum computers are not scalable [2,3,4]. Mainly the problem is the output's signal-to-noise ratio which is exponentially small in the number of qubits n involved in the computation. This problem is an artifact of the current approach (known as the Pseudo-Pure-state approach) to liquid state NMR quantum computers, as has already been explained [5]. We refer to this well-known problem as the *scaling problem.*

In it's initial state every qubit in the computer is at a highly mixed state, due to the finite temperature of the system. It only has a very small bias towards the zero state. A thermal equilibrium density matrix of the qubit i is presented:

$$\rho_{\varepsilon_0^i} \equiv \frac{1}{2}\begin{pmatrix} 1+\varepsilon_0^i & 0 \\ 0 & 1-\varepsilon_0^i \end{pmatrix} = \frac{1}{2\cosh\left(\frac{\hbar\omega^i}{2KT}\right)}\begin{pmatrix} e^{\frac{\hbar\omega^i}{2KT}} & 0 \\ 0 & e^{-\frac{\hbar\omega^i}{2KT}} \end{pmatrix}. \tag{1}$$

Using Taylor's theorem on the right hand side we can easily show, that for room temperature the initial bias $\varepsilon_0^i \cong \frac{\hbar\omega^i}{2KT}$ where ω^i is the resonance frequency of the nuclei which is proportional to the external magnetic field, T is the temperature and K is Bolzman's coefficient. It is determined by the chemical surrounding inside the molecule, but it is mainly determined by the external magnetic field and the temperature and of course properties of the nuclei which changes from one element to another, which is why the bias may be very different for various qubits in the same external field and temperature. For instance a proton will have about 4 times the initial bias of a carbon nuclei. But the bias difference between nuclei of the same species, such as two carbon nuclei with different chemical surroundings, is smaller in two orders of magnitude and more. In that case we may regard the initial bias of both qubits as identical.

In the liquid NMR systems in use today [1] $\varepsilon_0^i \sim 10^{-6}$. Some techniques have been developed in order to increase the bias such as optical pumping [6], ENDOR (Electron Nuclear DOuble Resonance), And INEPT (Insensitive Nuclei Enhanced by Polarization Transfer). The last is usually used to transfer the polarization of Hydrogen nuclei (protons) to nuclei with lower magnetic signal [7]. The initial state of of an n-qubit liquid NMR system can be represented as a tensor product of all the initial single-qubit density matrices, or as a weighted sum of all the n-bit binary strings:

$$\rho_{init}^{\{n\}} = \rho_{\varepsilon_0^1} \otimes \rho_{\varepsilon_0^2} \otimes \cdots \otimes \rho_{\varepsilon_0^n} \equiv \sum_j^{2^n-1} P_{jj}\,|j\rangle\langle j| \ , \tag{2}$$

where $|j\rangle$ are the basis vectors of the system corresponding to n-bit binary strings running from $|000\cdots 0\rangle$ to $|111\cdots 1\rangle$. In the case of a two qubit system:

$$\rho\ _{init}^{\{n\}} = \rho_{\varepsilon_0^1} \otimes \rho_{\varepsilon_0^2} \tag{3}$$

$$= \frac{1}{4} \begin{pmatrix} \left(1+\varepsilon_0^1\right)\left(1+\varepsilon_0^2\right) & & & \\ & \left(1+\varepsilon_0^1\right)\left(1-\varepsilon_0^2\right) & & \\ & & \left(1-\varepsilon_0^1\right)\left(1+\varepsilon_0^2\right) & \\ & & & \left(1-\varepsilon_0^1\right)\left(1-\varepsilon_0^2\right) \end{pmatrix}$$

so that $P_{00} = \left(1+\varepsilon_0^1\right)\left(1+\varepsilon_0^2\right)/4$ and so on. Note that the superscript is an index, not a power.

In order to be able to perform computation with such a highly mixed state, the technique that is used today in NMR quantum computers is preparing the system in a state known as a "pseudo-pure state" or PPS which has the density matrix

$$\rho_{PPS}^{\{n\}} = \frac{\mathcal{I}_{2^n}}{2^n}(1-p) + p\,|\psi\rangle\langle\psi|\,, \tag{4}$$

where $\mathcal{I}_{2^n}$ is an identity matrix of order 2^n. This is a mixture of a totally mixed state and pure state $|\psi\rangle$. This is a very useful form, since all quantum computational operations then affect only the pure part of that state. Unfortunately, the coefficient p is exponentially small with the number n, and this is the scaling problem. This problem can be avoided [5] if a bias close to one is obtained prior to using the PPS technique.

Schulman and Vazirani presented a reversible "molecular scale heat engine" [8] (based on data compression techniques) which provides scalable NMR quantum computers which do not suffer from the scaling problem. However, the Shannon's bound on reversible data compression creates the need for a huge number of bits in order to cool a small fraction of them via that technique.

By exploiting *irreversible* data compression, Boykin, Mor, Roychowdhury, Vatan, and Vrijen (BMRVV) invented algorithmic cooling [5] which bypasses the Shannon's bound on data compression and allows cooling a large fraction of the bits to very low temperatures. Surprisingly, interaction with the environment (usually a highly undesired process) is used to our benefit in algorithmic cooling: using data compression to shift the entropy into some bits which become much hotter, and then adding a SWAP operation between these hotter computational qubits and "environment" qubits which have a very short thermalization time, we get an irreversible operation in which the system loses entropy to the environment and is ready for another similar sequence. That would bring the spins to a very low temperature without the need for a large number of bits to start with.

The BMRVV algorithmic cooling had been introduced using a system in which all the spins have the same initial polarization bias, and using blocks of exactly m bits for any purification procedure [5]. This made the analysis possible (and relatively simple). However there are two simple ways to improve the algorithm. Although the analysis might become much more difficult, it is still possible to show that the resulting algorithm is more efficient in terms of space and time considerations.

- In the experimental reality spins with identical initial polarization bias are indistinguishable, since the initial bias, as we saw from Eq. (1), is a function of the spin's resonance frequency ω^i, which is the spin's ID. The results of BMRVV were, in some sense, non-optimal in terms of time complexity, since the cooling achieved there is precise only in the case where all spins have the same initial bias, and the cooling achieved there presents a bound (upper bound on the number of cooling levels, and the number of steps in the algorithm) in case of different biases when ϵ_0 is then the minimal bias of all the spins. We generalize here the analysis and the results of BMRVV [5] to deal with algorithmic cooling with different biases (via some approximation). As a result, we calculate the required levels of purification more tightly, and we show that less levels of purification are required, and therefore *much less* steps are required in order to obtain some desired bias (that is, some desired spin polarization).

- In the BMRVV algorithmic cooling all procedures are performed on m-bit blocks in order to simplify the analysis. However, the original algorithm due to Mor [9] for algorithmic cooling used a more efficient approach in terms of space, but was not easy to analyze hence was replaced by the algorithm analyzed in BMRVV [5]. We shall present here the algorithm of [9] and explain why it is more efficient in terms of space.

2 Algorithmic Cooling and its Modifications

The bias is given by $\varepsilon^i = \frac{\hbar\omega_i}{2KT} = -\frac{\hbar\gamma_i H_0}{2KT}$, where H_0 is the amplitude of a static external magnetic field. Cooling the sample by decreasing the temperature corresponds to increasing the bias in a thermal equailibrium. In non-equilibrium we can still use the formula of ε^i, but now it is a definition of the spin's temperature, so increasing the polarization (the bias) is equivalent to cooling the spins.

In the algorithmic cooling one first compresses the entropy from a few spins into the rest of the qubits adiabatically and then one conveys this entropy into the environment isothermally, just like the way your refrigerator works, only here one will use a logical algorithm to perform this process.

2.1 Basic Compression Subroutine

Basic Compression Subroutine or BCS is an algorithm that achieves classical data compression, so from now on we would regard the participating nuclei as bits and not qubits. It was shown [5,8] that using this subroutine on a number of bits we can transfer some of the entropy from a few of those bits to the rest leaving the first few bits cooler.

Let's see what happens to a pair of bits that undergo a CNOT gate. We begin from a density matrix of two molecular spins in a mixed state as shown in Eq. (4) and apply a CNOT gate:

$$CNOT\,|c,t\rangle \longrightarrow |c,t'\rangle \ , \tag{5}$$

where (c), is the control bit (t) is the target bit and $t' = \frac{1-(-1)^{c+t}}{2}$. Now say we can separate the pairs of spins in which the t'=0 from those of which t'=1 and call them *purified*, let's see what is the probability of having c=0 in a purified couple. Well, in order to get t'=0 the original pair of spins $|ct\rangle$ has to equal either $|00\rangle$ or $|11\rangle$, and in order to also get c=0 the input of CNOT needs to be $|00\rangle$. So the probability of getting c=0 in a purified couple would be:

$$\frac{1}{2}\left(1+\varepsilon_1^1\right) \equiv \frac{P_{00}}{P_{00}+P_{11}} = \frac{\frac{1}{4}\left(1+\varepsilon_0^1\right)\left(1+\varepsilon_0^2\right)}{\frac{1}{4}\left(1+\varepsilon_0^1\right)\left(1+\varepsilon_0^2\right)+\frac{1}{4}\left(1-\varepsilon_0^1\right)\left(1-\varepsilon_0^2\right)} = $$
$$= \frac{\left(1+\varepsilon_0^1+\varepsilon_0^2+\varepsilon_0^1\varepsilon_0^2\right)}{2\left(1+\varepsilon_0^1\varepsilon_0^2\right)} = \frac{1}{2}\left(1+\frac{\varepsilon_0^1+\varepsilon_0^2}{1+\varepsilon_0^1\varepsilon_0^2}\right) \tag{6}$$

So that

$$\varepsilon_1^1 = \frac{\varepsilon_0^1+\varepsilon_0^2}{1+\varepsilon_0^1\varepsilon_0^2} \ . \tag{7}$$

Now we can take two spins which were purified to bias of ε_1^1 and ε_1^2, and redo the CNOT if we again get a purification (i.e. t'=0) we will have a purified c-bit with a bias of $\varepsilon_2^1 = \frac{\varepsilon_1^1+\varepsilon_1^2}{1+\varepsilon_1^1\varepsilon_1^2}$. Generally if we take two spins with bias ε_j^1 and ε_j^2 and apply a CNOT, the purified c bit would end up with a bias of

$$\varepsilon_{j+1}^1 = \frac{\varepsilon_j^1+\varepsilon_j^2}{1+\varepsilon_j^1\varepsilon_j^2} \tag{8}$$

and since $0 \leq \varepsilon_0^i \leq 1$ then ε_j^i will always be smaller than 1 and so the denominator will always be smaller than 2. Therefore for all j, $\frac{\varepsilon_j^1+\varepsilon_j^2}{2} \leq \varepsilon_{j+1}^1 \leq \varepsilon_j^1+\varepsilon_j^2$.

To reach a desired bias ε_{des}, a particular j_{final} is chosen such that $\varepsilon_{j_{final}} \geq \varepsilon_{j_{des}}$. Note that for small values of j, $\varepsilon_j \ll 1$, but not for $j \lesssim j_{final}$. For $\varepsilon_j^i \ll 1$, we get

$$\varepsilon_{j+1}^1 \approx \varepsilon_j^1 + \varepsilon_j^2 \ . \tag{9}$$

When the difference between the two ε_j is large, say $\varepsilon_j^1 \approx 3\varepsilon_j^2$, the equation here yields $\varepsilon_{j+1}^1 \approx 4\varepsilon_j^2$, while the BMRVV algorithmic cooling which defines $\varepsilon_j = \min\{\varepsilon_j^1, \varepsilon_j^2\} = \varepsilon_j^2$ yields $\varepsilon_{j+1}^1 \approx 2\varepsilon_j^2$. In the extreme case where each pair is built of two particles with such a different bias (e.g. a chain of CHCH$\cdots$CH), the gain is so big that one less cooling procedure is required, meaning that j_{final} can be reduced by one. Since the number of overall steps in the algorithm is exponential in j_{final}, such analysis saves a huge number of steps. Alternatively, if the same number of steps shall be done, the final polarization shall be much better. Note that we can use the above strategy only when ε is very small, that is, only in the initial cooling procedures, the advantage is that big.

If we take a string of bit pairs ($ctctctct\cdots$) applying CNOT to every pair, some will be purified (c=P, t'=0), and some dirtier (c=D, t'=1) for example ($D1P0D1P0\cdots$). Beside the dirtier c bits, in each pair of bits their is the target bit which got dirtier. From now we will call c the *adjusted* bit, and t' the *supervisor* bit. We would like to separate the cooler (cleaner) bits from the rest of the bits. The second part of the BCS 'pushes' the cleaner bits to one side (for example the left) of the string, sends the supervisor bits to the right and leaves the dirty adjusted bits in the middle. In order to do so we will use a $\overline{\text{C}}$SWAP gate - a gate that swaps the values of two bits when a third bit (i.e. control bit) at their right has the value 0.

$$\overline{\text{C}}\text{SWAP} \begin{cases} |xy0\rangle \longrightarrow |yx0\rangle \\ |xy1\rangle \longrightarrow |xy1\rangle \end{cases} . \tag{10}$$

Using the supervisor bit as the control the $\overline{\text{C}}$SWAP will only swap the purified bits with their left neighbors since only the purified pairs have a t'=0, as can be seen from figure 1. After the $\overline{\text{C}}$SWAP we perform a SWAP gate between the supervisor bit and it's left neighbor then the supervisor will assume the position of a control bit in another $\overline{\text{C}}$SWAP time and again, so if the supervisor's value is 0 it will keep 'pushing' the cooler bit to the left, and if it is 1 it will only push itself to the left.

When the supervisor reaches the third place from the left end of the strings and performed a $\overline{\text{C}}$SWAP, it cannot be a control for $\overline{\text{C}}$SWAP anymore, since if we SWAP it once more with it's left neighbor, there will be only one bit to it's left and it needs two for $\overline{\text{C}}$SWAP. So now the supervisor will SWAP back to the right end of the string or until it reaches the supervisor that has

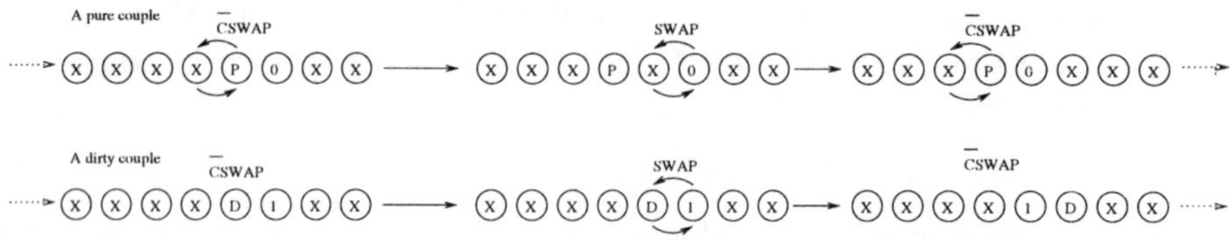

Figure 1. Basic Compression Subroutine or BCS. After the CNOT has been performed the control bit may become either purified ($t'= 0$) of dirty ($t'=1$) we would like to get the purified bits to the left side of the string so we will perform a sequence of $\overline{\text{C}}$SWAPs and SWAPs repeatedly. Where the t' will be used as the control to the $\overline{\text{C}}$SWAP. This way purified bits will be pushed to the left and the dirty ones would stay put.

done the described cycle previously, leaving the clean bit at the left end of the string if it has the value 0, or the dirty bit in it's previous place if it has the value 1.

Moving to the next pair of bits we need only to remember, that since the supervisor has moved from it's original place to the right and beyond all the pairs we will use next, all those pairs have shifted one place to the left, so we will need to consider that when we address them. After doing this process on all the pairs in the string we achieved a purified smaller string at the left end of the original string. For the analysis in the next section we need also to agree that the supervisor bits are at the right of the array (so the dirty adjusted bits are in the middle of the array).

2.2 Continuing to a higher bias

We will not be satisfied only with a bias of ε_1^i but would like to reach as high as possible, so it would seem like we should continue and take the purified bits that now make a substring and create a more purified sub-substring with a bias ε_2^i and so on. Lets see what is the length of a substring with a bias ε_{j+1}^i that was made using BCS on a n_j long string with a bias ε_j^i.

We begin from a string of length n_j so it contains $\frac{n_j}{2}$ c,t pairs each has a probability $(P_{00} + P_{11})$ to supply *one* purified bit. The expectation value of the length of a purified string created from one c,t pair will be

$$P_{00} + P_{11} = \frac{\left(1+\varepsilon_j^i\right)\left(1+\varepsilon_j^{i+1}\right)}{4} + \frac{\left(1-\varepsilon_j^i\right)\left(1-\varepsilon_j^{i+1}\right)}{4} = \frac{1+\varepsilon_j^i\varepsilon_j^{i+1}}{2} \tag{11}$$

were i always gets odd values (i.e. $i = 2k-1$ $k = 1, 2, ..\frac{n_j}{2}$) so the expectation value of the entire purified string's length would be:

$$\langle \mathcal{L}_{\mathrm{j+1}} \rangle = \sum_i \text{purification probablity for the i}^{th} \text{ pair} = \\ = \tfrac{1}{2} \sum_{k=1}^{n_j/2} \left(1 + \varepsilon_j^{2k-1} \varepsilon_j^{2k}\right) \tag{12}$$

Since $0 \leq \varepsilon_j^i \leq 1$ the value of the expression inside the parenthesis will be between 1 and 2 so

$$\frac{n_j}{4} \leq \langle \mathcal{L}_{\mathrm{j+1}} \rangle \leq \frac{n_j}{2} \tag{13}$$

2.3 Cooling Algorithm.

To go beyond the reversible data compression, BMRVV use another set of bits that are characterized by a very short thermalization time (T_1) in comparison to the thermalization time of the bits we used so far . These Rapidly Reaching Thermal Relaxation [5] (RRTR) bits will be SWAPed with the dirty bits thus receiving their antropy, and then quickly cooling down to thermal equilibrium with the environment getting ready to cool dirty bits again and again. In order to make this process technically simple we would want every computational bit to have an RRTR neighboring bit, the best candidates to be RRTR's in NMR would be the atoms electrons. Having learnt the BCS and the principle of SWAPing with RRTR bits we are ready to take one step further.

The goal now is to take a string of N bits and be able to cool them until we get m bits with a desired bias.

First define M_0 as an operation that takes an initial number of bits N_0 each having an RRTR neighbor and turning them to m bits with a bias ε_0^i. M_0 can simply be a SWAP between the m bits with the RRTR's, and that means that the number of initial bits we need is $N_0 = m$.

Note: the bias of the bits after M_0 is the initial bias of the RRTR's and not the computational bits. This is only for the best since short thermalization time usually corresponds with high resonance frequency and therefore high bias.

Now define a new operation M_1 that, similar to the previous operation, will be applied to N_1 bits and give an output of m bits with a bias ε_1^i to do that we shall begin with applying M_0 to $N_0 = m$ out of the N_1 bits resulting with m bits with a bias of ε_0^i, applying BCS on these m bits we will purify a string with length expectation value of

$$\langle \mathcal{L}_1 \rangle = \frac{1}{2} \sum_{k=1}^{m/2} \left(1 + \varepsilon_0^{2k-1} \varepsilon_0^{2k}\right) \tag{14}$$

We can see that $\langle \mathcal{L}_1 \rangle$ lies between $\frac{m}{4}$ and $\frac{m}{2}$. The rest of the bits ($\frac{m}{2}$ t'-bits) become dirtier.

Let a cooling step M_{j+1} take N_{j+1} bits with a bias ε_j^i and give out m bits at a bias ε_{j+1}^i. A new parameter ℓ, the *cooling depth* is required [5] for that operation. It is defined as the number of BCS procedures done during some cooling step M_{j+1}. Each such BCS operation is denoted as $\mathcal{B}_{\{\mathrm{j}\to\mathrm{j+1}\}}$. When we have made one $\mathcal{B}_{\{0\to1\}}$ on $N_0 = m$ bits, then ℓ=1. Now apply M_0 to N_0 bits starting from the $\frac{m}{2}$ place from the left ($i = \frac{m}{2} + 1 = 2k - 1 \Rightarrow k = \frac{m}{4} + 1$) and get another m bits with a bias ε_0^i. Applying $\mathcal{B}_{\{0\to1\}}$ to them we would end up with more adjusted bits and the the rest will be the dirty supervisor bits. Now the cooling depth will be ℓ=2, and average length of the purified string will be

$$\langle \mathcal{L}_1 \rangle = \frac{1}{2} \sum_{k=1}^{m/2} \left(1 + \varepsilon_0^{2k-1} \varepsilon_0^{2k}\right) + \frac{1}{2} \sum_{k=m/4+1}^{3m/4} \left(1 + \varepsilon_0^{2k-1} \varepsilon_0^{2k}\right) \geq \frac{m}{2} \tag{15}$$

If we keep on doing M_0 and $\mathcal{B}_{\{0\to1\}}$ ℓ times, we will be using $(\ell - 1)\frac{m}{2}$ bits, that will be adjusted before the last M_0, plus N_0 bits that will be used for the final M_0 and $\mathcal{B}_{\{0\to1\}}$ operations, which sums up to the total number of bits needed for M_1, $N_1 = (\ell - 1)\frac{m}{2} + N_0$.

Now a cooling depth $\ell \geq 4$ is chosen such that at least m purified bits are obtained. Now we *CUT* the first m bits from the rest with a good probability to have them all with a bias of ε_1^i. Choosing ℓ=5 will increase this probability significantly.

To present the recursive structure of this algorithm [5] we use the following notations. $[\mathcal{B}_{\{(k-1)\to k\}}]$—the BCS procedure purifying from ϵ_{k-1} to ϵ_k (followed by moving the purified bits to the relevant starting point). $[\mathcal{S}]$—SWAP m bits with the RRTR. $[\mathcal{C}_j]$—CUT, keep the first m bits from the starting point of the sub-array of the bits with a bias ϵ_j. Then, $M_0 \equiv \mathcal{S}$, and for $j \in \{1, \ldots, j_f\}$

$$M_j = \mathcal{C}_j \;\underbrace{\mathcal{B}_{\{(j-1)\to j\}} M_{j-1} \;\cdots\; \mathcal{B}_{\{(j-1)\to j\}} M_{j-1} \;\; \mathcal{B}_{\{(j-1)\to j\}} M_{j-1}}_{\ell \text{ times}} , \tag{16}$$

is the recursive formula describing the BMRVV algorithm.

The second modification we suggest here (following the first cooling algorithm [9]) is that instead of working on m-bit blocks, the processing shall be on all relevant bits. Therefore the above process shall contain a first application of M_0 on *all* the bits, a BCS on all these bits, followed by a CUT, then a second application of M_0 on all bits with ϵ_0, again followed by BCS and a

CUT. The full process is then described by

$$M_j = \mathcal{B}_{\{(j-1)\to j\}} M_{j-1} \underbrace{\mathcal{C}_{j-1}\ \mathcal{B}_{\{(j-1)\to j\}} M_{j-1} \cdots \mathcal{C}_{j-1}\ \mathcal{B}_{\{(j-1)\to j\}} M_{j-1}}_{\ell-1 \text{ times}}, \tag{17}$$

Via such a process, obtaining many bits for the next step shall be much faster (that is, a smaller cooling depth is required) and shall require less bits N.

3 Discussion

In this paper we suggested two modifications of algorithmic cooling, showing that better space/time results can be obtained. Further analysis is required.

References

1. N. A. Gershenfeld, I. L. Chuang, *Science* **275** 350 (1997). D.G. Cory, A. F. Fahmy T. F. Havel, *Proc. Natl. Acad. Sci.* **94**, 1634 (1977). J. A. Jones, M. Mosca, R. H. Hansen, *Nature***393**, 344 (1998). E. Knill, R. Laflamme, R. Martinez, C-H Tseng, *Nature*, **404**, 386 (2000).
2. W. S. Warren, *Science*, **277**, 1688 (1997)
3. D. DiVincenzo, *Nature*, **393**, 113 (1998)
4. S. Braunstein, C. Caves, R. Jozsa, N. Linden, S. Popescu & R. Schack, *Phys. Rev. Lett.* **83**, 1054 (1999).
5. P. O. Boykin, T. Mor, V. Roychowdhury, F. Vatan & R. Vrijen, LANL e-print quant-ph/0106093 (2001), submitted to *P*NAS; P. O. Boykin, T. Mor, V. Roychowdhury, F. Vatan, Algorithimic cooling and scalable ensemble quantum computers, submitted to *S*TOC 2002.
6. G. Navon Y.-Q. Song, T. Rõõm, S. Appelt, R. E. Taylor, and A. Pineset, *Science* **271**, 1848 (1996)
7. C. P. Slichter, *P*rinciples of magnetic resonance, Springer-Verlag USA (1990).
8. L. J. Schulman, U. Vazirany, in *P*roceedings of 31st ACM Symposium on Theory of Computing (1999) p. 322.
9. The first algorithm using interaction with the environment to produce a cooling mechanism is due to T. Mor, but it was never published, since that algorithm was too complicated to analyze. It was replaced by the m-bit blocks algorithm suggested by V. Roychowdhury, and presented and analyzed by BMRVV in [5].

QUANTUM THEORY ON THE TORUS WITH MAGNETIC FIELD

H. NARNHOFER

Institute for Theoretical Physics, University of Vienna, A 1090 Boltzmanngasse, Vienna, Austria
E-mail: narnh@ap.univie.ac.at

We discuss the reducible representation of a particle on a torus with magnetic field, where the gauge automorphisms are unitarily implementable. Identifying these gauge automorphisms for several particles demands a soldering of this representation. The soldering procedure is unique only for integer magnetic flux. For rational magnetic flux a kind of interaction between different particles results, that becomes negligible in the thermodynamic limit.

1 Introduction

In [1] the Quantum-Hall effect was studied from the point of view, that the system is given by a Weyl-algebra with appropriate commutation relations and that we have to look for representations of this algebra. This has the advantage that the Weyl-algebra depends only on the magnetic field whereas the Schrödinger-equation contains the vector potential which is not gauge invariant and in general on the torus not even globally defined. It turned out that there exist representations for every magnetic field, irreducible and reducible ones. But only if B is integer an irreducible representation can be realized on $L(\mathbf{T}^2)$. Otherwise the irreducible representations have to be constructed on a larger Hilbertspace, in this way loosing any continuity in B. But there is also a canonical reducible representation on $L(\mathbf{R}^2)$ that is continous in B.

The main problem arises if one tries to construct representations for several particles: starting with antisymmetric states in the tensor product of the irreducible representations produced the wrong filling factor, if the magnetic field is not integer. Also we were not able to find satisfying conditions on permitted states in tensor products of the canonical reducible representation.

The corresponding problem does not arise if one considers the simpler but nevertheless similar situation of a particle on $\mathbf{T}^1$ with several inequivalent irreducible representations: These representations can be combined in a reducible representation, where the gauge automorphism is unitarily represented. Passing to representations of several particles we have to solder i.e. identify the parameter that characterizes the gauge automorphism. Physically this means that soldering has the effect that the Bohm-Aharanov-phase is the same for all particles. The soldering is unique because the gauge automorphism group is abelian and the commutant in the reducible representation is commutative. The observable algebra for several particles can be described by creation and annihilation operators tensored with the center of the one particle representation. We will try to copy this soldering also for the representation of the Weyl-algebra with magnetic field on $L(\mathbf{R}^2)$. But here it is not evident how to solder. First we have to give up to some extent that the Weyl-algebra describes the system. We concentrate on the position operators and demand, that they keep their tensorproduct structure for several particles. Space translation and time evolution have to satisfy the appropriate commutation relations at least on the one particle level. Parameters in the representation additional to the position values have to be soldered. It turnes out that non trivial commutation relations between velocities of different particles cannot be avoided for general B and produce in this way a time evolution where the particles react on the existence of other particles. The amount of this reaction depends on B and on the strategy of soldering in the canonical reducible representation. There exists one soldering unique up to a kind of gauging where this reaction disappears for integer B and in an appropriate interpretation is minimal for arbitrary B.

2 Soldering on $\mathbf{T}^1$

Physics on $\mathbf{T}^1$ is described by the Weyl-algebra built by the operators $e^{2\pi inx}$ and $e^{i\alpha p}$, $n \in Z, \alpha \in R$. They satisfy the commutation relations

$$e^{i\alpha p} e^{2\pi inx} e^{-i\alpha p} e^{-2\pi inx} = e^{2\pi in\alpha}.$$

This algebra can be represented on $L^2(\mathbf{R}) \cong L^2(\mathbf{T}) \otimes L^2(\mathbf{T}) = L^2([0,1]) \otimes L^2([0,1])$ in the following way: we consider the functions $f(x) \in L^2(\mathbf{R})$ as functions on $< x > \in [0,1)$ and $[x] \in Z, x =< x > +[x]$ and then perform a Fourier transformation in $[x]$. On the so resulting functions in $L^2([0,1]) \otimes L^2([0,1])$ the Weyl-operators act as (compare [2])

$$e^{2\pi inx} f(\phi,\gamma) = e^{2\pi in\phi} f(\phi,\gamma),$$

$$e^{i\alpha p} f(\phi,\gamma) = e^{2\pi i\gamma[\phi+\alpha]} f(\langle\phi+\alpha\rangle,\gamma),$$

where with $n \in Z$ and $\alpha \in R$

$$\phi+\alpha = \langle\phi+\alpha\rangle + [\phi+\alpha],$$

$$0 \leq \langle\phi+\alpha\rangle < 1, [\phi+\alpha] \in Z,$$

with the possible extension for $\beta \in R$

$$e^{2\pi i\beta x} f(\phi,\gamma) = e^{2\pi i\beta\phi} f(\phi, \langle\gamma-\beta\rangle),$$

with $e^{2\pi i\beta x}$ belonging to $\mathcal{B}(\mathcal{H})$ but not to the weak closure of the Weyl algebra. Thus

$$e^{2\pi i\beta x} e^{i\alpha p} f(\phi,\gamma) = e^{2\pi i\beta x} e^{2\pi i\gamma[\phi+\alpha]} f(\langle\phi+\alpha\rangle,\gamma) =$$

$$= e^{2\pi i\beta\phi+2\pi i[\phi+\alpha](\gamma-\beta)} f(\langle\phi+\alpha\rangle, \langle\gamma-\beta\rangle),$$

whereas

$$e^{i\alpha p} e^{2\pi i\beta x} f(\phi,\gamma) = e^{i\alpha p} e^{2\pi i\beta\phi} f(\phi,\gamma-\beta) =$$

$$= e^{2\pi i\beta\langle\phi+\alpha\rangle+2\pi i\gamma[\phi+\alpha]} f(\phi+\alpha,\gamma-\beta),$$

so that, using $\langle\phi+\alpha\rangle = \phi+\alpha-[\phi+\alpha]$ and writing $f(\phi+\alpha)$ for $f(<\phi+\alpha>)$

$$e^{i\alpha p} e^{2\pi i\beta x} e^{-i\alpha p} e^{-2\pi i\beta x} = e^{2\pi i\beta\alpha}.$$

This representation is faithful and reducible with center $\mathcal{Z} = (e^{inp}) = (e^{in\gamma})$. Reduction with respect to the center gives irreducible representation where the momentum is given by the derivative with twisted boundary conditions

$$\pi_\gamma(p) = \frac{1}{i}\frac{\partial}{\partial\phi}|_\gamma,$$

$$\mathcal{D}(\pi_\gamma(p)) = (f(\phi); f(1) = e^{2\pi i\gamma} f(0)).$$

There exists also the possibility to represent the Weyl-operators by

$$e^{2\pi i\beta x} f(\phi,\gamma) = e^{2\pi i\phi[\gamma-\beta]} f(\phi,\gamma-\beta),$$

$$e^{i\alpha p} f(\phi,\gamma) = e^{i\alpha\gamma} f(\phi+\alpha,\gamma),$$

which is connected with the previous representation by the unitary transformation $e^{2\pi i\phi\gamma}$. The corresponding soldered representation for n particles is given on $L^2([0,1])^n \otimes L^2((0,1))$ by

$$e^{2\pi i\sum\beta_i x_i} f(\phi_1,...\phi_n,\gamma) = e^{2\pi i\sum\beta_i\phi_i} f(\phi_i,...\phi_n,\gamma-\sum\beta_i),$$

$$e^{i\sum\alpha_i p_i} f(\phi_1,...\phi_n,\gamma) = e^{2\pi i\gamma\sum[\phi_i+\alpha_i]} f(\phi_1+\alpha_1,...\phi_n+\alpha_n,\gamma).$$

The soldering is unique insofar as we can connect different possibilities by unitary transformations, e.g. by the unitary $e^{2\pi i\gamma\sum\phi_i}$.

3 The representation for the Quantum Hall effect inherited from $\mathbf{T}^1$

The Weyl algebra for the quantum Hall effect of one particle is built by the operators $(e^{2\pi inx}, e^{2\pi imy}, e^{i\alpha v_x}, e^{i\beta v_y})$, where they satisfy the commutation relations

$$[e^{2\pi inx}, e^{2\pi imy}] = 0,$$

$$e^{i\alpha v_x} e^{2\pi inx} e^{-i\alpha v_x} e^{-2\pi inx} = e^{2\pi in\alpha},$$

$$e^{i\beta v_y} e^{2\pi imy} e^{-i\beta v_y} e^{-2\pi imy} = e^{2\pi im\beta},$$

$$e^{i\alpha v_x} e^{i\beta v_y} e^{-i\alpha v_x} e^{-i\beta v_y} = e^{2\pi iB\alpha\beta}.$$

We can consider the algebra to be a subalgebra of $\mathcal{B}(L^2(\mathbf{R}))$ with

$$v_x = p_x - \pi By,$$

$$v_y = p_y + \pi Bx.$$

If we copy the representation in $L^2(\mathbf{R})$ used for the one dimensional torus this gives the following representation (f periodic)

$$e^{2\pi i\alpha x} f(\phi_1, \phi_2, \gamma_1, \gamma_2) = e^{2\pi i\alpha\phi_1} f(\phi_1, \phi_2, \gamma_1 - \alpha, \gamma_2),$$

$$e^{2\pi i\beta y} f(\phi_1, \phi_2, \gamma_1, \gamma_2) = e^{2\pi i\beta\phi_2} f(\phi_1, \phi_2, \gamma_1, \gamma_2 - \beta),$$

$$e^{2\pi i\alpha v_x} f(\phi_1, \phi_2, \gamma_1, \gamma_2) = e^{2\pi i\gamma_1[\phi_1+\alpha] - \pi iB\alpha\phi_2} f(\phi_1 + \alpha, \phi_2, \gamma_1, \gamma_2 + B\alpha/2),$$

$$e^{2\pi i\beta v_y} f(\phi_1, \phi_2, \gamma_1, \gamma_2) = e^{2\pi i\gamma_2[\phi_2+\beta] + \pi iB\beta\phi_1} f(\phi_1, \phi_2 + \beta, \gamma_1 - B\beta/2, \gamma_2).$$

This representation can be enlarged to a soldered representation of two particles on $L^2(\mathbf{T}^2)^2 \otimes L^2(\mathbf{T}^2)$ by defining

$$\begin{aligned}
&e^{2\pi i\alpha v_{1,x}} f(\phi_1, \phi_2, \psi_1, \psi_2, \gamma_1, \gamma_2) = \\
&\qquad e^{2\pi i\gamma_1[\phi_1+\alpha] - \pi iB\alpha\phi_2} f(\phi_1 + \alpha, \phi_2, \psi_1, \psi_2, \gamma_1, \gamma_2 + B\beta/2), \\
&e^{2\pi i\alpha v_{2,x}} f(\phi_1, \phi_2, \psi_1, \psi_2, \gamma_1, \gamma_2) = \\
&\qquad e^{2\pi i\gamma_1[\psi_1+\alpha] - \pi iB\alpha\psi_2} f(\phi_1, \phi_2, \psi_1 + \alpha, \psi_2, \gamma_1, \gamma_2 + B\alpha/2), \\
&e^{2\pi i\beta v_{1,y}} f(\phi_1, \phi_2, \psi_1, \psi_2, \gamma_1, \gamma_2) = \\
&\qquad e^{2\pi i\gamma_2[\phi_2+\beta] + \pi iB\beta\phi_1} f(\phi_1, \phi_2 + \beta, \psi_1, \psi_2, \gamma_1 - B\beta/2, \gamma_2), \\
&e^{2\pi i\beta v_{2,y}} f(\phi_1, \phi_2, \psi_1, \psi_2, \gamma_1, \gamma) = \\
&\qquad e^{2\pi i\gamma_2[\psi_2+\beta] + \pi iB\beta\psi_1} f(\phi_1, \phi_2, \psi_1, \psi_2 + \beta, \gamma_1 - B\beta/2, \gamma_2).
\end{aligned}$$

Evidently $[v_{1,x}, v_{2,x}] = 0$ and $[v_{1,y}, v_{2,y}] = 0$. Also

$$e^{i\alpha v_{2,j}} e^{2\pi i n x_1} e^{-i\alpha v_{2,j}} = e^{2\pi i n x_1}.$$

But

$$e^{2\pi i\alpha v_{1,x}} e^{2\pi i\beta v_{2,y}} f(\phi_1, \phi_2, \psi_1, \psi_2, \gamma_1, \gamma_2) = e^{2\pi i\gamma_1[\phi_1+\alpha]-\pi iB\alpha\phi_2+2\pi i(\gamma_2-B\alpha/2)[\psi_2+\beta]+\pi iB\beta\psi_1} \times f(\phi_1+\alpha, \phi_2, \psi_1, \psi_2+\beta, \gamma_1-B\beta/2, \gamma_2+B\alpha/2),$$

whereas

$$e^{2\pi i\beta v_{2,y}} e^{2\pi i\alpha v_{1,x}} f(\phi_1, \phi_2, \psi_1, \psi_2, \gamma_1, \gamma_2) = e^{2\pi i\gamma_2[\psi_2+\beta]+\pi iB\beta\psi_1+2\pi i(\gamma_1+B\beta/2)[\phi_1+\alpha]-\pi iB\alpha\phi_2} \times f(\phi_1+\alpha, \phi_2, \psi_1, \psi_2+\beta, \gamma_1-B\beta/2, \gamma_2+B\alpha/2).$$

This differs by a phase factor

$$e^{\pi iB\beta[\phi_1+\alpha]-\pi iB\alpha[\psi_2+\beta]}.$$

Therefore shifts of velocities commute only for steps of the size $\alpha = \frac{n}{B}$ and $\beta = \frac{m}{B}$. In the thermodynamic limit, when the size of the torus increases like [-L/2 ,L/2] and we consider the sequence of functions $f_L(\phi_1, \phi_2, \psi_1, \psi_2, \gamma_1, \gamma_2) = f(\phi_1/L, \phi_2/L, \psi_1/L, \psi_2/L)$ independent of γ_1 and γ_2 and localized around zero, we obtain strong resolvent convergence of $e^{i\alpha_L v_L}$ and therefore also of the time evolution of several particles in $\mathbf{R}^2$ with magnetic field B, since we have to scale $\alpha_L = \alpha/L$, $\phi_{i,L} = \phi_i/L$, $B_L = BL^2$. But in this frame work we cannot observe the special role of the integer magnetic flux, where we could start from an irreducible representation and pass to many particles by taking tensorproducts and demanding, that the wavefunctions have to be antisymmetric, so that the time evolution of the individual particle is not influenced by the existence of the other particles apart from the influence of the Pauli principle on the initial conditions.

Remark: We could also apply the soldering procedure to the functions $\bar{f} = e^{2\pi i\phi\gamma}$. Since the passage from f to $\bar{f}$ is induced by the unitary $e^{2\pi i\gamma\phi}$ which by soldering turns into $e^{2\pi i\sum(\phi_i)\gamma}$ we notice that this operator commutes with $e^{2\pi in\phi}$ and $e^{2\pi im\psi}$ and with the commutator of $e^{i\alpha v_1}$ and $e^{i\beta v_2}$. But the algebra of two (respectively of n particles) changes as explicit calculation shows. (Essentially ϕ and γ change their role in the definition of v_x, v_y ,x and y and therefore also in the commutant of the soldered velocities the additional phase becomes $e^{2\pi i(\beta[\gamma_1+\alpha B]+\alpha[\gamma_2-\beta B])}$. We can think of other unitaries U in $L^2(\mathbf{T}) \otimes L^2(\mathbf{T})$, that only have to satify that they commute with $e^{2\pi in\phi}$.

Different choices for these U lead to different identifications in the n-particle space and therefore to different algebras.

We have therefore to look for a preferable choice of such unitaries respectively for a preferable choice of an explicit representation, from which we start the soldering. A natural desire is that in such a representation we keep continuity in the magnetic field B but at the same time get independence of the time evolution of different particles if B is integer, i.e. if we could start the n-particle construction from an irreducible representation over $L^2(\mathbf{T}^2)$.

4 An alternative representation

If B is integer, there exists an irreducible representation on $L^2(\mathbf{T}^2)$ [1]. In this representation the effect of the magnetic field on the velocity is partly incorporated as a twist condition on the momentum operator. This twist condition is space dependent and only has the correct periodicity behaviour, if B is integer. Nevertheless this representation gives a hint how it can be generalized for arbitrary B: We give a representation on $L^2(\mathbf{T}^2) \otimes L^2(\mathbf{T}^2)$ by

$$e^{2\pi i n x_1} f(\phi_1, \phi_2, \gamma_1, \gamma_2) = e^{2\pi i n \phi_1} f(\phi_1, \phi_2, \gamma_1, \gamma_2),$$

$$e^{2\pi i n x_2} f(\phi_1, \phi_2, \gamma_1, \gamma_2) = e^{2\pi i n \phi_2} f(\phi_1, \phi_2, \gamma_1, \gamma_2),$$

$$e^{i\alpha v_1} f(\phi_1, \phi_2, \gamma_1, \gamma_2) = e^{-2\pi i \alpha \phi_2 + 2\pi i \gamma_2 [\phi_1 + \alpha]} f(\phi_1 + \alpha, \phi_2, \gamma_1 + B[\phi_1 + \alpha], \gamma_2),$$

$$e^{i\beta v_2} f(\phi_1, \phi_2, \gamma_1, \gamma_2) = e^{2\pi i (\phi_1 + \gamma_1 / B) B [\phi_2 + \beta]} f(\phi_1, \phi_2 + \beta, \gamma_1, \gamma_2).$$

The necessary commutation relations

$$[e^{2\pi i n x_1}, e^{2\pi i m x_2}] = [e^{2\pi i n x_1}, e^{i\beta v_2}] = [e^{2\pi i m x_2}, e^{i\alpha v_1}] = 0$$

are satisfied, because in the appropriate pairing all variables are either shifted or appear as a phase multiplication, but shift and phase multiplication are never combined. The only nontrivial commutation relation is obtained by comparing

$$\begin{aligned}
&e^{i\alpha v_1} e^{i\beta v_2} f(\phi_1, \phi_2, \gamma_1, \gamma_2) = \\
&e^{i\alpha v_1} e^{2\pi i \phi_1 B[\phi_2 + \beta] + 2\pi i \gamma_1 [\phi_2 + \beta]} f(\phi_1, \phi_2 + \beta, \gamma_1, \gamma_2) = \\
&\qquad e^{-2\pi i B \alpha \phi_2 + 2\pi i \gamma_2 [\phi_1 + \alpha] + 2\pi i \langle \phi_1 + \alpha \rangle B [\phi_2 + \beta] + 2\pi i \langle \gamma_1 + B[\phi_1 + \alpha] \rangle [\phi_2 + \beta]} \\
&\qquad \times f(\phi_1 + \alpha, \phi_2 + \beta, \gamma_1 + B[\phi_1 + \alpha], \gamma_2) = \\
&\qquad e^{-2\pi i B \alpha \phi_2 + 2\pi i \gamma_2 [\phi_1 + \alpha] + 2\pi i B \phi_1 [\phi_2 + \beta] + 2\pi i B \alpha [\phi_2 + \beta] - 2\pi i B [\phi_1 + \alpha][\phi_2 + \beta]} \\
&\qquad \times e^{2\pi i \gamma_1 [\phi_2 + \beta] + 2\pi i B [\phi_1 + \alpha][\phi_2 + \beta]} f(\phi_1 + \alpha, \phi_2 + \beta, \gamma_1 + B[\phi_1 + \alpha], \gamma_2).
\end{aligned}$$

with

$$e^{i\beta v_2} e^{i\alpha v_1} f(\phi_1, \phi_2, \gamma_1, \gamma_2) =$$

$$e^{i\beta v_2} e^{-2\pi i B\alpha\phi_2 + 2\pi i\gamma_2[\phi_1+\alpha]} f(\phi_1 + \alpha, \phi_2, \gamma_1 + B[\phi_1 + \alpha], \gamma_2) =$$

$$e^{2\pi i B\phi_1[\phi_2+\beta] + 2\pi i\gamma_1[\phi_2+\beta] - 2\pi i B\alpha\phi_2 - 2\pi i B\alpha\beta + 2\pi i B\alpha[\phi_2+\beta] + 2\pi i\gamma_2[\phi_1+\alpha]}$$

$$\times f(\phi_1 + \alpha, \phi_2 + \beta, \gamma_1 + B[\phi_1 + \alpha], \gamma_2).$$

Inspecting the phase factors this proves that we really have constructed a representation of the Weyl-algebra. It remains to argue that this representation is unitarily equivalent to the representation in $L^2(\mathbf{R}^2)$ respectively to the represention offered in the previous section. This can be done explicetely by first performing a gauge transformation to

$$v_x = p_x + 2\pi B y, v_y = p_y$$

and then applying the unitary operator $e^{2\pi i B\phi_1 p_{\gamma_1}}$ to the representation. It can also be seen by the following observation: in our present representation we can implement the (not inner) gauge automorphisms, that shift the velocities by the unitaries $e^{2\pi i\alpha x}$ and $e^{2\pi i\beta y}$ with $\alpha, \beta \in \mathbf{R}$ acting as

$$e^{2\pi i\alpha x} f(\phi_1, \phi_2, \gamma_1, \gamma_2) = e^{2\pi i\alpha\phi_1} f(\phi_1, \phi_2, \gamma_1, \gamma_2 - \alpha),$$

$$e^{2\pi i\beta y} f(\phi_1, \phi_2, \gamma_1, \gamma_2) = e^{2\pi i\beta\phi_2} f(\phi_1, \phi_2, \gamma_1 - \beta, \gamma_2)$$

simply as on the $\mathbf{T}^1 \otimes \mathbf{T}^1$ only that now γ_2 is connected with ϕ_1 respectively γ_1 with ϕ_2 in order to satisfy the correct commutation relations with the velocities. These unitaries donot belong to the Weyl algebra but can be used to construct the unitaries $e^{i(2\pi Bx - v_y)}$ and $e^{i(2\pi By + v_x)}$. As in $L^2(\mathbf{R})$ they form the commutant in the representation and act as

$$e^{i(2\pi Bx - v_y)} f(\phi_1, \phi_2, \gamma_1, \gamma_2) =$$

$$e^{2\pi i Bx} e^{-2\pi i B\phi_1 - 2\pi i\gamma_1} f(\phi_1, \phi_2, \gamma_1, \gamma_2) =$$

$$e^{-2\pi i\gamma_1} f(\phi_1, \phi_2, \gamma_1, \gamma_2 - B),$$

$$e^{2\pi i(By + v_x)} f(\phi_1, \phi_2, \gamma_1, \gamma_2) =$$

$$e^{2\pi i By} e^{-2\pi i B\phi_2 + 2\pi i\gamma_2} f(\phi_1, \phi_2, \gamma_1 + B, \gamma_2) =$$

$$e^{2\pi i\gamma_2} f(\phi_1, \phi_2, \gamma_1, \gamma_2).$$

Therefore the commutant acts only in the (γ_1, γ_2)space as the rotation-algebra $\mathcal{A}_\theta$with rotation parameter $\theta = B$ in its standard representation [3]. Together the operators of the algebra and of the commutator form a continuous Weyl-algebra over $\mathbf{R}^2$ that according to von Neumann has a unique representation. We can make the following observations: γ_2 enters only as a parameter and we could consider it to be fixed.Then it labels the irreducible representations if B is irrational. We only need it to be able to implement the corresponding gauge automorphism and to make the representation unitarily equivalent to the representation in section 3.

If B is integer also γ_1 is not shifted and therefore can be fixed. Then we regain the irreducible representations on $L^2(\mathbf{T}^2)$. If $B = g_1/g_2$ is rational then we get an irreducible representation of the Weyl-algebra on the Hilbert space $L^2(\mathbf{T}^2) \otimes H_{g_2}$ where H_{g_2} is g_2 dimensional and consists of the points $\gamma_{1,k} = \gamma_{1,0} + k/g_2$. This means that $e^{i\alpha v}$ does not implement a strongly continuous automorphism on the algebra $(e^{2\pi i(nx+my)})$ in $L^2(\mathbf{T}^2)$, we have to move to the larger space so that $e^{i\alpha v_1}$ rotates between (for rational B finitely many) irreducible representations. Only in the unification of these representation the time evolution is defined as automorphism over the Weyl-algebra, otherwise it is only defined as a map over the statespace of the algebra $(e^{2\pi i(nx+my)})$.

It remains to characterize the representation in such a way that it leads as best as possible to a unique soldering procedure. This can be done via the commutant. In our representation we have discribed the commutant as strong closure of the C^*-rotation- algebra [3] with rotation parameter B.If B is integer, this rotation algebra is abelian, if $B = g_1/g_2$ is rational, it is isomorphic to $M_{g_2} \otimes L^\infty$ and if B is irrational it is the type II_1 algebra in its standard representation.

In [1] we discussed the representation in $L^2(\mathbf{R}^2)$ in the following way: The Weyl-algebra $(e^{2\pi i(nx+my)}, e^{i\alpha v_x} e^{i\beta v_y})$ could be written as

$$(e^{i\alpha v_x} e^{i\beta v_y}) \otimes (e^{2\pi i(m\bar{x}+n\bar{y})}),$$

where $\bar{x}$ and $\bar{y}$ are the center of the Larmor precision. They form the rotation-algebra with rotationparameter $1/B$. They are represented in the form

$$\Pi(e^{i\alpha v_1} e^{i\beta v_2}) \otimes \Pi(\mathcal{A}_{1/B}) = \mathcal{B}(L^2(\mathbf{R})) \otimes II_1 \subset \mathcal{B}(L^2(\mathbf{R})) \otimes \mathcal{B}(L^2(\mathbf{R})).$$

In this representation $\Pi(\mathcal{A}_{1/B})' = \Pi(\mathcal{A}_B)$ so that the representation is not in its standard form, in which there exists a vector which is both cyclic and seperating for the algebra and its commutant, and therefore has a nontrivial coupling constant i.e. B. The passage from the representation in [1] to the present representation uses the fact that

$$I_\infty \otimes II_1 = II_\infty$$

and that the decomposition of the II_∞ factor into the tensor products is not unique. We have offered now another decomposition

$$I_\infty \otimes II_1 = II_\infty = \bar{I}_\infty \otimes \overline{II}_1$$

where now $\overline{II}_1$ is in its standard representation. According to [4] all decompositions of this type are spatially isomorphic and are determined by $(II_\infty)' = II_1$. These spatial isomorphisms are unique up to automorphisms of II_∞ that are all unitarily implementable in the standard representation and therefore preserve the splitting of the Hilbertspace into $H_{I_\infty} \otimes H_{II_1}$. The soldering corresponds to taking the tensor product

$$H = H_{I_\infty} \otimes H_{I_\infty} \otimes H_{II_1},$$

so that we are identifying H_{II_1} of the individual particles. Since they are naturally isomorphic the identification is independent of any automorphism of $\mathcal{A}'$. We have argued for irrational B, but the same considerations are true for rational $B = g_1/g_2$, where the $L^2(\mathbf{R}^2)$-representation can be interpreted as

$$B(L^2(\mathbf{R})) \otimes M_{g_1} \otimes \mathcal{Z} \bigvee M_{g_2} \otimes \mathcal{Z} = \mathcal{A} \bigvee \mathcal{A}',$$

where $\mathcal{A}$ and $\mathcal{A}'$ both include the common center $\mathcal{Z}$. Finally we use the spatial isomorphism

$$I_\infty \otimes M_{g_1} \cong I_\infty \otimes M_{g_2}.$$

Therefore we remain with the freedom of automorphisms of II_∞. Here we reduce the freedom by demanding that $(e^{2\pi i(nx+mx)}) \subset \mathcal{B}(H_{I_\infty})$ and that this algebra is maximally abelian. Therefore we are left with unitaries $U(\phi_1, \phi_2)$ in $\mathcal{B}(H_{II_1})$. On the position operators they act trivially:

$$U(\phi_1, \phi_2) e^{2\pi i(nx+my)} U(\phi_1, \phi_2)^* = e^{2\pi i(nx+my)}.$$

To discuss the effect of soldering on the velocity operators we first restrict the possible automorphisms to those implemented by unitaries in $1 \otimes \mathcal{A}_B$, i.e. on $e^{2\pi i \gamma_1}$ and $e^{2\pi i(\gamma_2 - Bp_{\gamma_1})}$; their effect on the velocities is

$$e^{2\pi i n \gamma_1} e^{i\alpha v_x} e^{-2\pi i n \gamma_1} = e^{-2\pi i n B[\phi_1 + \alpha]} e^{i\alpha v_x},$$

$$e^{2\pi i n \gamma_1} e^{i\beta v_y} e^{-2\pi i n \gamma_1} = e^{i\beta v_y},$$

$$e^{2\pi i m(\gamma_2 - Bp_{\gamma_1})} e^{i\alpha v_x} e^{-2\pi i m(\gamma_2 - Bp_{\gamma_1})} = e^{i\alpha v_x},$$

$$e^{2\pi im(\gamma_2-Bp_{\gamma_1})}e^{i\beta v_y}e^{-2\pi im(\gamma_2-Bp_{\gamma_1})}=e^{2\pi imB[\phi_2+\beta]}e^{i\beta v_y}.$$

For integer B these unitaries obviously commute with the velocities. These unitaries are also defined in $(H_{I_\infty})^n\otimes H_{II_1}$ and therefore do not change the commutation relations. The resulting N-particle algebras remain the same. If on the other hand we consider unitaries in $(H_{I_\infty})\otimes H_{II_1}$ of the type

$$U_1=\sum c_{nm}(\phi)e^{2\pi in\gamma_1}e^{2\pi im(\gamma_2-Bp_{\gamma_1})},$$

$$U_2=\sum c_{nm}(\psi)e^{2\pi in\gamma_1}e^{2\pi im(\gamma_2-Bp_{\gamma_1})}.$$

then this new soldering procedure is connected with the previous one by calculating

$$U_1e^{i\alpha v_1}U_1^*U_2e^{i\beta v_2}U_2^*=e^{i\alpha v_1}e^{if(\phi)}e^{i\beta v_2}e^{ig(\psi)},$$

where $f(\phi)$ resp. $g(\psi)$ act as multiplication in H_{I_∞} and more or less general operator in H_{II_1}. Since we cannot further reduce them they will change the commutation relations between the soldered velocities. Only if B is integer they reduce to a pure multiplication with functions in ϕ and ψ and therefore cancel in the commutation relations. Notice however, that in order that U_1 is unitary $c_{nm}(\phi)$ is either constant and therefore does not influence the commutation relations or is a highly non continuous function and thus destroys the simple structure of the representation of v_x and v_y. Therefore it seems advisable to concentrate on the explicetely given representation in which we will now evaluate the commutation relations between the soldered velocities.

Obviously

$$[e^{2\pi i(nx_1+my_1)},e^{i\alpha v_{2,k}}]=0,$$

also

$$[e^{i\alpha v_{1,x}},e^{i\beta v_{2,x}}]=0,$$

because shift and multiplication are not mixed. The only nontrivial communication relations result from

$$\begin{aligned}
&e^{i\alpha v_{1,x}}e^{i\beta v_{2,y}}f(\phi_1,\phi_2,\psi_1,\psi_2,\gamma_1,\gamma_2)=\\
&e^{i\alpha v_{1,x}}e^{2\pi i(\psi_1+\gamma_1/B)B[\psi_2+\beta]}f(\phi_1,\phi_2,\psi_1,\psi_2+\beta,\gamma_1,\gamma_2)=\\
&\qquad e^{-2\pi iB\phi_2+2\pi i\gamma_2[\phi_1+\alpha]+2\pi i\psi_1B[\psi_2+\beta]+2\pi i(\gamma_1+B[\phi_1+\alpha])[\psi_2+\beta]}\\
&\qquad\times\ f(\phi_1+\alpha,\phi_2,\psi_1,\psi_2+\beta,\gamma_1+B[\phi_1+\alpha],\gamma_2)=\\
&e^{i\beta v_{2,y}}e^{i\alpha v_{1,x}}f(\phi_1,\phi_2,\psi_1,\psi_2,\gamma_1,\gamma_2)=\\
&e^{i\beta v_{2,y}}e^{-2\pi iB\alpha\phi_2+2\pi i\gamma_2[\phi_1+\alpha]}f(\phi_1+\alpha,\phi_2,\psi_1,\psi_2,\gamma_1+B[\phi_1+\alpha],\gamma_2)=
\end{aligned}$$

$$e^{-2\pi iB\alpha\phi_2+2\pi i\gamma_2[\phi_1+\alpha]+2\pi iB\psi_1[\psi_2+\beta]+2\pi i\gamma_1[\psi_2+\beta]}$$
$$\times\ f(\phi_1+\alpha,\phi_2,\psi_1,\psi_2+\beta,\gamma_1+B[\phi_1+\alpha],\gamma_2).$$

Therefore

$$e^{-i\alpha v_{1,x}}e^{-i\beta v_{2,y}}e^{i\alpha v_{1,x}}e^{i\beta v_{2,y}}=e^{2\pi iB[\phi_1+\alpha][\psi_2+\beta]}.$$

We observe that different from the representation in section3 the velocities of different particles commute if B is integer. In this case the decomposition of the Hilbert space coincides with the central decomposition into irreducible representations so that the soldering only takes place in the commutative part of $\mathcal{A}$ and does not effect commutation relations.

5 Second Quantization

Already in [1] we formulated second quantization of the Weyl-algebra $\mathcal{W}$ over $L^2(\mathbf{T}^1)$: we started with the representation $\Pi(\mathcal{W})$ in $L^2(\mathbf{R})\cong L^2(\mathbf{T})\otimes L^2(\mathbf{T})$.The center of $\Pi(\mathcal{W})$ acts as maximally abelian multiplication algebra in the second factor. Therefore we can take as algebra of second quantization the algebra built by $(a^*(f)\otimes z_1, a(g)\otimes z_2)$, where

$$[a^*(f)\otimes z_1,a(g)\otimes z_2]_+=\langle g|f\rangle z_1z_2,$$

$$[a(f)\otimes z_1,a(g)\otimes z_2]_+=0,$$

$$z_1,z_2\in\mathcal{Z}.$$

In a similar way we can start with the soldering procedure offered in section 4 by fixing $L^2(\mathbf{R}^2)\cong L^2(\mathbf{T}^2)\otimes L^2(\mathbf{T}^2)$ and passing to the N-particle space $H^N_{I_\infty}\otimes H_{II_1}$ and then combine these N-particle spaces to a Fock-space over which the second quantized algebra acts, that permits the passage between different particle numbers . Thus we consider the algebra

$$\mathcal{A}(a^*(f)\otimes A,a(g)\otimes B),$$

where $A,B\in\Pi(\mathcal{A}_B)$, the II_1 algebra in its standard representation,$f\in L^2(\mathbf{T}^2)$ and

$$[a^*(f),a(g)]_+=\langle g|f\rangle,[a(f),a(g)]_+=0.$$

Therefore we can write e.g. the velocity operator in its second quantized form as an integral kernel

$$e^{i\alpha v_1}\Leftrightarrow\delta(\bar\phi_1-\phi_1-\alpha)\delta(\bar\phi_2-\phi_2)a^*_{\bar\phi}a_\phi\otimes e^{2\pi i\gamma_2[\phi_1+\alpha]+B[\phi_1+\alpha]p_{\gamma_1}}.$$

6 The thermodynamic limit

The present research is mainly motivated, that we want to be able to perform the thermodynamic limit for increasing tori in such a way, that in the limit essential features become independent of the precise choice of sequences of the length of the tori.Especially it should play no role, when the size of the torus tends to infinity, whether the ratio of the magnetic field to the length of the torus is integer. In [1] we failed insofar, as without soldering there was no natural restriction on permitted states on the tensorproducts of one particle representations, that gave the correct filling factor respectively the correct density dependence of the ground state energy.

For non interacting particles the degeneracy of the ground state in an irreducible representation determines the filling factor. But including the magnetic field this remains true only if the irreducible representation is given in $L^2(\mathbf{T}^2)$. Since by the soldering procedure we are introducing an effective interaction between particles, if the magnetic field is not integer, the degeneracy of the ground state of one particle becomes irrelevant. Instead we have to estimate the ground state energy for fixed particle numbers to get in the limit the correct density behaviour.. For all sequences of tori we expect the energy density to be linear in the density with varying linear coefficient, when the different energy levels for integer B are filled, i.e.

$$E(\rho) = \rho B/2 \text{ for } \rho \leq B,$$

$$E(\rho) = 3\rho B/2 - B^2 \text{ for } B \leq \rho \leq 2B.$$

Independence on periodic, Dirichlet or Neumann boundary conditions is a well established fact for systems without magnetic field. We will mimic the proof for the present situation. It will turn out that the special choice of the representation is again helpfull to control the effect of the boundary.

The time evolution is determined by the Hamiltonian

$$H = \sum (v_{i,x}^2 + v_{i,y}^2),$$

where the velocities $v_{i,j}$ depend on the magnetic field B. For the particle number $N = 1$ the groundstate is the same as in $L^2(\mathbf{R}^2)$ and therefore we can choose it continuous in B and especially we can choose the Gauss-function $e^{-(x^2+y^2)}$, but due to the degeneracy we can arrange a superposition so that for integer B it respects the tensor product structure $L^2(\mathbf{R}^2) \cong L^2(\mathbf{T}^2) \otimes L^2(\mathbf{T}^2)$, i.e. we take [1]

$$f_j(\phi_1, \phi_2, \gamma_1, \gamma_2) = \sum_{k \in Z} e^{2\pi i k(\phi_1 + j/B) - (k + B\phi_2)^2/2},$$

where $0 \leq j < B$, $j \in Z$. These functions are normalizable and explicit calculation shows that they are orthogonal for different j. With these eigenfunctions we can construct a ground state for the N-particle Hamiltonian for $N \leq B$ with

$$f(\phi_{1,x}, ...\phi_{N,y}) = \sum (-1)^{\Pi} \prod f_j(\phi_{\Pi(j)}).$$

For $N > B$ we have to move to higher Landau levels.

If B is not an integer then

$$\inf\langle\Psi|H_N|\Psi\rangle \geq \sum_i \inf\langle\Psi|H_i|\Psi\rangle = NB$$

as long as $N \leq B$.

To obtain an upper bound we have to notice that though $e^{i\alpha v_B}$ is strongly continuous in B and v_B for different B have a common core (remember that we can write $v_{B,x} = p_x + \pi B y$), f_j does not belong to this common core. Therefore we cannot apply Feynman-Hellmann- theorem for a first order estimate and cannot expect that the ground state is differentiable in B, only that it is continuous. To get an upper bound for general B we can map the ground state f built by the one particle eigenfunctions f_j of an integer B into a common domain, i.e. into a function that vanishes at the boundary for $\phi_2 = 0$.

In order to find a common domain for the velocities with varying B it is useful to express the velocities as derivatives with the appropriate boundary conditions: the representation can also be described by writing

$$v_x = 1/i \frac{\partial}{\partial \phi_1}|_{\gamma_2 + B p_{\gamma_1}} + B\phi_2,$$

$$v_y = 1/i \frac{\partial}{\partial \phi_2}|_{B\phi_1 + \gamma_1},$$

so that the domain of v_x coincides for integer B or for arbitrary B the intersection of the domain of v_x with the set of functions constant in γ_1 is independent of B and also of ϕ_2, whereas the boundary condition for v_y depends on ϕ_1. Therefore we first map the function $f(\phi_2)$ defined on the interval $[-1/2, 1/2]$ by a scaling transformation into a function f_ε with support in $[-1/2 + \varepsilon, 1/2 - \varepsilon]$. Now we define $\bar{f}$ by

$$\bar{f}(\phi) = f_\varepsilon \text{ for } -1/2 + 2\varepsilon < \phi < 1/2 - 2\varepsilon$$

$$\bar{f}(-1/2 + \delta) = g_1(\delta) f_\varepsilon(1/2 - 2\varepsilon + \delta) e^{-2\pi i(\gamma_1 + B\phi_1)},$$

$$\bar{f}(-1/2 + \varepsilon + \delta) = g_2(\delta) f_\varepsilon(-1/2 + \varepsilon + \delta),$$

$$\bar{f}(1/2-\varepsilon+\delta)=g_1(\delta)f_\varepsilon(-1/2+\varepsilon+\delta)e^{2\pi i(\gamma_1+B\phi_1)},$$

$$\bar{f}(1/2-2\varepsilon+\delta)=g_2(\delta)f_\varepsilon(1/2-2\varepsilon+\delta),$$

$$0<\delta<\varepsilon,$$

with g_1 differentiable,

$$g_1^2(\delta)+g_2^2(\delta)=1,$$

$$\lim_{\delta\to 0} g_1(\delta)=0 \text{ and } \lim_{\delta\to\varepsilon} g_1(\delta)=2^{-1/2},$$

so that by the continuity properties of f_ε the function $\bar{f}$ is everywhere differentiable and belongs for all B to the domain of v_x and v_y, provided we combine it with functions constant in γ_1. This isomorphic mapping can be applied to all ground state eigenfunctions for fixed B and N. Calculating the expectation value with this new function for $v_x^2+v_y^2$ we get, due to the functions g_1 and g_2 and the scaling an additional term of the order $O(1/\varepsilon^2)+O(\varepsilon^2)$. This term is not negligible for small tori. But if we consider the thermodynamic limit, in which the tori are $[-L/2,L/2]^{2N}$for fixed B or after a scaling transformation become again $[-1/2,1/2]^{2N}$, but now $B_L=BL^2$, then the additional contribution $\varepsilon^{-2}L^{-2}$ becomes negligible for fixed ε and increasing L. Every ground state energy for chosen B gives an upperbound plus this negligible contribution for all B in the neighborhood. Furthermore the points with$B_L=BL^2\in Z$ for which the energy is known become dense for $L\to\infty$ in B_L. Therefore in the thermodynamic limit the energy density of the groundstate becomes independent of the chosen sequence and the soldering procedure gives for the ground state energy density in the thermodynamic limit the correct answer.

7 The current algebra

It is a well established fact that in the thermodynamic limit the Quantum Hall effect can be explained by the existence of a current algebra that corresponds to a conformal quantum theory in 1+1 dimensions [5,6,7].In [8] this current algebra was obtained in a precisely defined limit for electrons confined to a local region by a harmonic oszillator whose strength was going to zero. Of course it is desirable that also in other settings of thermodynamic limits it is possible to construct this current. Especially our example of increasing tori mimics spacetranslation invariance and is therefore tailormade for the existence of a current. Nevertheless we will succeed in the construction only with some limitations, i.e. if with L the length of the torus BL^2 is an integer. If we consider only one particle the operators

$$e^{2\pi i n\bar{x}} = e^{2\pi i n(x - v_y/B)},$$

$$e^{2\pi i m\bar{y}} = e^{2\pi i m(y + v_x/B)},$$

commute with the Hamiltonian. They act as

$$e^{2\pi i n\bar{x}} f(\phi_1, \phi_2, \gamma_1, \gamma_2) = e^{2\pi i n\phi_1 + 2\pi i(\phi_1 + \gamma_1/B)B[\phi_2 - n/B]} f(\phi_1, \phi_2 - n/B, \gamma_1, \gamma_2),$$

$$e^{2\pi i m\bar{y}} f(\phi_1, \phi_2, \gamma_1, \gamma_2) = e^{2\pi i\gamma_2[\phi_1 + n/B]} f(\phi_1 + n/B, \phi_2, \gamma_1 + B[\phi_1 + n/B], \gamma_2).$$

Only for integer B their generalization to N particles commute with the N-particle Hamiltonian. Further in this case

$$e^{2\pi i B\bar{x}} f(\phi_1, \phi_2, \gamma_1, \gamma_2) = e^{-2\pi i\gamma_1} f(\phi_1, \phi_2, \gamma_1, \gamma_2),$$

$$e^{2\pi i B\bar{y}} f(\phi_1, \phi_2, \gamma_1, \gamma_2) = e^{2\pi i\gamma_2} f(\phi_1, \phi_2, \gamma_1, \gamma_2)$$

and these operators belong to the soldered center. Therefore we can write the algebra as

$$(e^{2\pi i n\bar{x}_i}, e^{2\pi i m\bar{y}_j}) \cong (M_{B\times B})^N_{as} \otimes \mathcal{Z}$$

and this algebra is pointwise invariant under the time evolution. If we switch on now a constant electric field the time evolution is implemented by

$$H_E = \sum_i (v_{x_i}^2 + v_{y_i}^2) + \sum_i Ex_i,$$

so that

$$e^{iH_E t} e^{2\pi i n\bar{x}_j} e^{-iH_E t} = e^{2\pi i n\bar{x}_j},$$

$$e^{iH_E t} e^{2\pi i n\bar{y}_j} e^{-iH_E t} = e^{2\pi i n(\bar{y}_j - Et/B)},$$

so that the perturbed timeevolution implements an automorphism on this subalgebra. For fixed N the perturbed Hamiltonian is unbounded from below, but if we define a groundstate as a state $f(\phi_{1,1}, ..\phi_{2,N})$ minimizing $\langle f\otimes g|H_E|f\otimes g\rangle$ for fixed $g(\gamma_1, \gamma_2)$, then this f exists and is independent of g, because $x = \phi - p_\gamma$. If E is sufficiently small then we choose $g = 1$ and the minimizing state will belong to the space built by the groundstate of the unperturbed Hamiltonian, i.e. there is an ordered basis in $M_{B\times B}$ that is filled up for increasing N. Therefore we are essentially in the same situation as in [8]: we have a basis f_n for $-N < n < B - N$ and associate to these f_n annihilationoperators a_n. After a Bogoliubov-transformation $a_n = b_n$ for $n \geq 0$ and $a_n^* = b_n$ for $n < 0$ [9] the groundstate reduces on the algebra built by these creation and annihilation operators to the quasifree state with $\langle b_m^* b_n \rangle = \delta_{nm}$. Now it is standard to take $a_x = \sum e^{2\pi i n x/L}$ and $\rho(x) =: a_x^* a_x :$ and in the thermodynamic limit $(L \mapsto \infty)$ we obtain the bosonic current algebra with

$$[\rho(x), \rho(x')] = \delta'(x - x'),$$

with time evolution

$$\tau_t \rho(x) = \rho(x + Et).$$

There remains the problem how far these considerations can be translated to rational and irrational sequences of tori. Evidently we can keep the algebra $(e^{2\pi i n \bar{x}_i}, e^{2\pi i m \bar{y}_j})$ as the relevant algebra that should reduce in the thermodynamic limit to the algebra corresponding to the current. But for finite torus size this algebra is not invariant under the unperturbed time evolution. We conjecture that in the thermodynamic limit this invariance is restored, but our present control on the groundstate energy is not sufficient for a proof of this conjecture.

Acknowledgements

It is a pleasure to thank Nevena Ilieva and Walter Thirring for their interest and many useful suggestions for improvements on the manuscript.

References

1. H. Narnhofer, W. Thirring, Representation of the quantum Hall system on $\mathbf{T}^2$ for several fermions, Vienna preprint (2001).
2. J. Zak, *Solid State Physics* **27**, 1 (1972).
3. M.A. Rieffel, *Pac. J. Math.* **93**, 415 (1981).
4. J.T. Schwarz, *W^*-algebras*, Nelson, Gordon and Breach (1967).
5. J. Fröhlich, T. Kerler, *Nucl. Phys. B* **354**, 369 (1991).
6. J. Fröhlich, A. Zee, *Nucl. Phys. B* **364**, 517 (1991).
7. X.G. Wen, *Phys. Rev. B* **43**, 11025 (1991).
8. N. Ilieva, W. Thirring, *Euro. Phys. J. C* **19**, 561 (2001).
9. D.C. Mattis, E.H. Lieb, *J. Math. Phys.* **6**, 304 (1965).

NONLOCAL REFLECTION BY PHOTONIC BARRIERS

GÜNTER NIMTZ AND ASTRID HAIBEL
Universität zu Köln, II. Physikalisches Institut,
Zülpicher Str. 77, D-50937 Köln, Germany

The time behaviour of partial reflection by opaque photonic barriers was measured with microwaves. It was observed that unlike the duration of partial reflection by dielectric sheets, the measured reflection duration of barriers is independent of their length. The experimental results point to a nonlocal behaviour of evanescent modes.

1 Introduction

We are used to measuring a reflection time determined by sheet thickness from partial reflection of light by a sheet of glass. The reflection is observed only after a time span corresponding to twice the layer thickness multiplied by the group velocity of light in glass. Three hundred years ago Newton conjectured that light was composed of corpuscles and argued in the case of partial reflection by two or more surfaces: Light striking the first surface sets off a kind of wave or field that travels along with the light and predisposes it to reflect or not reflect off the second surface. He called this process "fits of easy reflection or easy transmission" [1]. As theory and experiments have shown this is not true in the case of dielectric media with a real part of the refractive index.

Amazingly in the case of reflection by the surface of an opaque photonic barrier, where the refractive index is purely imaginary, Newton's conjecture seems to be close to reality: The partial reflection by barriers suffers a short and constant time delay independent of length. For the photonic barrier investigated here we found that the reflection duration equals the transmission time observed in photonic tunnelling experiments [2] (see Fig. 1).

We are going to explain the experimental set–up and the experiments and discuss the unexpected observation.

2 Experimental set-up

The experimental set–up is displayed in Fig. 2. Pulse–like microwave signals with a half width of 8.5 ns are transmitted from a parabolic antenna. The carrier frequency of the pulse is 9.15 GHz ($\lambda = 32.8$ mm) the frequency-band width is 80 MHz. The reflected signal is received with a second antenna

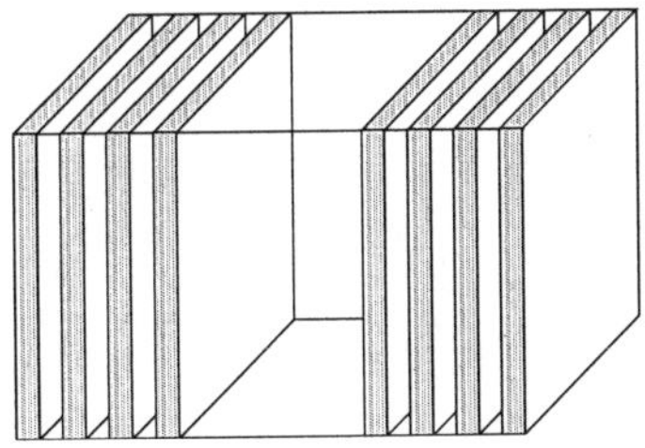

Figure 1. Two one–dimensional periodic quarter wavelength hetero–structures of perspex and air which are separated by a distance of 189 mm forming a resonant cavity (cross–section is 400×400 mm^2).

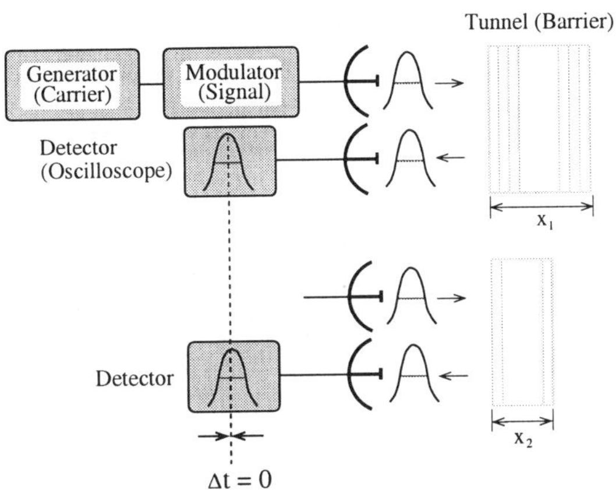

Figure 2. Experimental set-up for the periodic dielectric quarter wavelength heterostructure to measure the group velocity. The diagram shows two resonant photonic barriers with 2×4 and 2×2 perspex sheets separated by the same air gap.

and detected by an HP 54825 oscilloscope . The measurement is performed asymptotically, so that any coupling between generator, detector, and devices under test (a perspex sheet and photonic barriers) is avoided by the long optical distances ($\approx$ 3m) and by uniline devices in the microwave circuit. The time resolution of the set–up is $\pm$10 ps.

To check the experimental arrangement we measured the time response of partial reflection by the two surfaces of a perspex sheet of 0.8 m thickness. The second peak of the signal corresponding to the partial reflection by the back surface arrived $\approx$ 8.5 ns later than the first one related to the front surface reflection. For a clear demonstration of the partial reflection by the

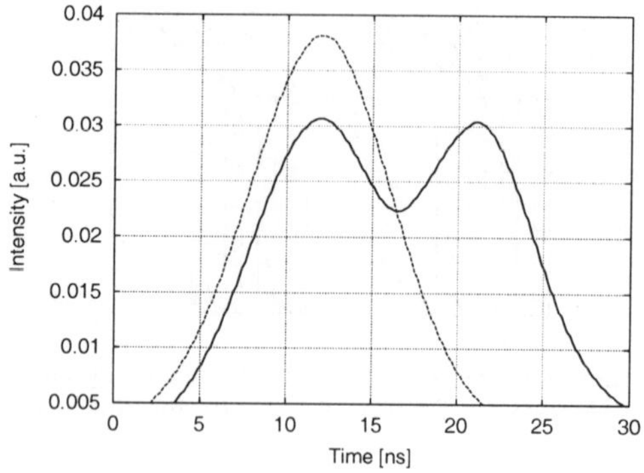

Figure 3. Partial reflection of a microwave pulse by a thick perspex sheet. The dashed pulse is the result of reflection by a metallic front surface only. The layer is 800 mm thick, its refractive index is 1.6. The double peak is due to the superposition of reflection of the signal pulse by front and back surfaces. The delay of the second peak is 8.5 ns in agreement with the propagation time in perspex. In order to enhance reflection the back surface was partially coated with a metal film.

two surfaces, the intensity reflected by the back surface was adjusted to that by the front side as shown in Fig. 3. This was obtained by a partial metallic coating of the perspex layer's back surface. The measured time delay of the two peaks is in agreement with the calculated propagation time 8.53 ns considering the refractive index n=1.6 of perspex.

3 Partial reflection by photonic barriers

The investigated photonic barrier device is sketched in Fig. 1. It consists of two photonic lattices each 45.5 mm long which are separated by an air gap of 189 mm resulting in a total length of 280 mm. Each lattice consists of four perspex slabs each 5 mm thick in an equidistant distance of 8.5 mm air.
For such a structure in the 'normal dielectric case' we would expect a broadened pulse composed of the partial reflections of all the slabs similar to the signal shown in Fig. 3. The last surface reflected signal should be seen ≈ 1.9 ns after the reflected one by the front surface. However, the partial reflection by photonic barriers revealed a strange behaviour. If the barrier is shortened to 4 or 2 sheets (see Figs. 2 and 4) the reflection duration keeps constant whereas the amplitude decreases as a result of the increasing transmission. The measured reflection duration is ≈ 100 ps. A back surface reflected signal from opaque barriers, however, has never been detected (see Fig. 4).

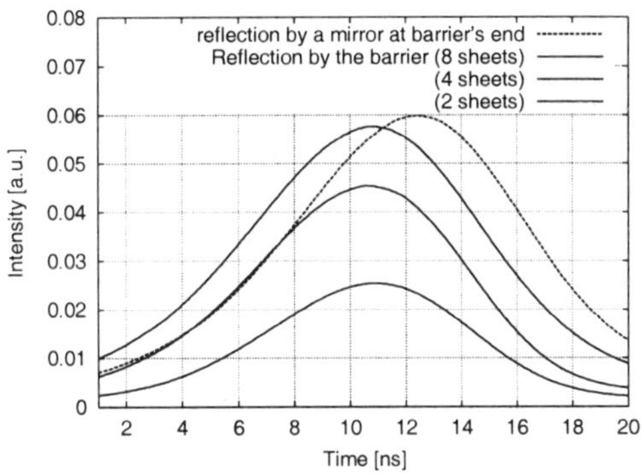

Figure 4. Signals reflected by barriers of different length. The largest one had a total barrier length of 280 mm, the two smaller one were recorded after the barrier was shortened to a total length of 226 mm and 199 mm, respectively. The procedure of shortening is illustrated in Fig. 2. For comparison the reflection time of a mirror (without barrier) at the back surface position of the longest photonic barrier is displayed. The expected travel time between front and end position of 1.87 ns has been in fact measured.

Transmission and reflection dispersion relations of the long barrier are displayed in Fig. 5. In Transmission there are five pronounced forbidden bands separated by resonance peaks in the frequency range displayed. The reflection behaviour in these forbidden bands was used for the measurements.

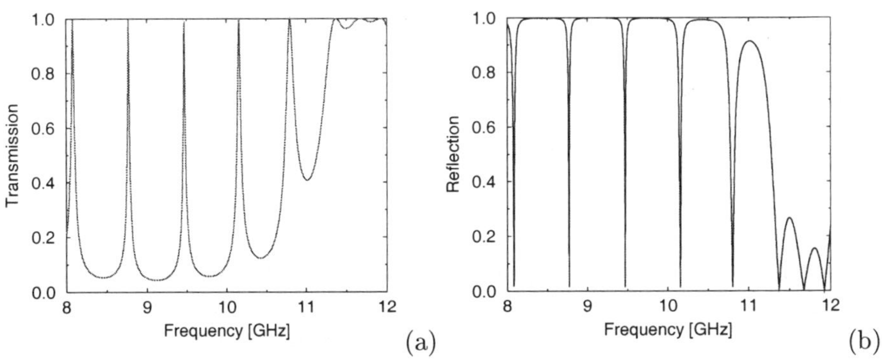

Figure 5. The graphs show the dispersion relations for the resonant heterostructure vs frequency of Fig. 1. (a) The transmission dispersion of the periodic heterostructure displays five forbidden gaps, which correspond to the photonic tunnelling regime, for details see Ref. [4]. (b) shows the reflection dispersion relation.

4 Conclusions

In measuring the reflection duration by photonic barriers we observed that the partial reflection by the back surface is an instantaneous effect on the amplitude, whereas the reflection duration is not changed. This strange behaviour is opposite to the measured partial reflection by a perspex layer (see Fig. 3). The behaviour may be explained by a nonlocality of evanescent modes. As a result of our experiments evanescent modes constituted by ensembles of photons behave like a quantum mechanical particle. Nonlocality and causality were investigated in Ref. [6] and quite recently with respect to superluminal photonic tunnelling nonlocality was discussed by Perel'man (Ref. [7]).
In our experiments the applied signal pulse had a carrier frequency of 9.15 GHz in the center of a forbidden band gap (see Fig. 5) and a narrow 1 % frequency-band width. Consequently all frequency components of the signal were evanescent. In this case there is no finite phase-time expected nor observed for a signal inside a barrier [8,9]. Such a behaviour seem to explain the experimental data of reflection by opaque barriers.
Obviously the information on photonic barrier length is available at the front surface already. This is a property which Newton suggested erroneously to explain partial reflection of corpuscles by dielectric layers [1]. Evanescent modes appear to be nonlocal at least within a range of some ten wavelengths as experiments have shown in this study [7]. The distance of observing nonlocality effects is limited by the exponential decay of the field intensity of evanescent modes, i.e. of the probability in the wave mechanical tunnelling analogy.

We gratefully acknowledge discussions with H. Aichmann, P. Mittelstaedt, A. Stahlhofen, and R.-M. Vetter. We thank M. E. Perel'man for giving us the paper on his investigation prior to publication.

References

1. R. P. Feynman, *QED, The strange Theory of Light and Matter*, Princeton University Press, Princeton NJ (1988), p.22.
2. A. Haibel and G. Nimtz, *Ann. Phys. (Leipzig)* **10**, number 8 (2001)
3. R. P. Feynman, R. B. Leighton, and M. Sands, *The Feyman Lectures on Physics*, Vol.II, Chapter 33, Addison–Wesley Publishing Company(1964).
4. G. Nimtz, A. Enders, and H. Spieker, *J. Phys. I., France* **4**, 565 (1994)
5. E. Merzbacher, *Quantum Mechanics*, John Wiley& Sons, NY (1970).
6. E. Recami, *Time's Arrows, Quantum Measurement and Superluminal Behavior*, ISBN 88 8080 024 8, pp. 17-35 (2000).
7. M. E. Perel'man, preprint (2001).
8. Th. Hartman, *J. Appl. Phys.* **33**, 3427 (1962).
9. G. Nimtz and W. Heitmann, *Prog. Quantum Electronics* **21**, 81 (1997).

LIGHTFRONT FORMALISM VERSUS HOLOGRAPHY&CHIRAL SCANNING *

BERT SCHROER

presently: *CBPF, Rua Dr.Xavier Sigaud, 22290-180 Rio de Janeiro, Brazil*
e-mail: schroer@cbpf.br
permanent address: *Institut für Theoretische Physik*
FU-Berlin, Arnimallee 14, 14195 Berlin, Germany

The limitations of the approach based on using fields restricted to the lightfront (Lightfront Quantization or p→ ∞ Frame Approach) which drive quantum fields towards canonical and ultimately free fields are well known. Here we propose a new concept which does not suffer from this limitation. It is based on a procedure which cannot be directly formulated in terms of pointlike fields but requires "holographic" manipulations of the algebras generated by those fields. We illustrate the new concepts in the setting of factorizable d=1+1 models and show that the known fact of absence of ultraviolet problems in those models (in the presence of higher than canonical dimensions) also passes to their holographic images. In higher spacetime dimensions d>1+1 the holographic image lacks the transversal localizability; however this can be remedied by doing holography on d-2 additional lightfronts which share one lightray (Scanning by d-1 chiral conformal theories).

1 A simple setting of the problem

Lightfront quantum field theory and the closely related p→ ∞ frame method have a long history which starts 1970 with papers trying to explain Bjoerken's scaling observation. The large number of articles on this subject since that time may be separated into two groups. On the one hand there are those papers whose aim is to show that such concepts constitute a potentially useful enrichment of standard local quantum physics [1] [2,3,4], but there are also innumerable attempts to use lightfront concepts as a starting point of more free-floating "effective" approximation ideas whose relations to causal and local quantum physics remained unclear. Here our interest is to extend the first mentioned results.

The main problem is to overcome the short distance limitations of the old lightfront quantization and the p→ ∞ frame method (which resulted in a severe restriction on the Kallen-Lehmann spectral function [2]) which unfortunately excludes all cases of genuine renormalizable interactions. This restriction is identical to that which canonical commutation relation (or the interpretation of the functional measure in the spirit of a quantum mechanical

*THIS WORK RECEIVED SUPPORT FROM THE CNPQ

Feynman-Kac euclidean representation) requires. But whereas the (perturbative) breakdown of canonicity causes no real harm (the canonical setting only serves as a mental starter, but plays no important role in the Feynman formalism) since the more general spacelike causality/locality structure is all one needs for dealing with renormalization, the restriction for the lightfront approach is a more serious matter. If the whole idea of lightfront-affiliated operators would be limited to free fields as indicated in various rigorous investigations about lightcone restrictions of Wightman fields[2], the subject would have only academic interest.

In a recent paper I have indicated how the field-coordinatization free approach of AQFT can overcome this undesired restriction [5]. Its intuitive basis is the radical different nature of the conformal structure which one encounters on the lightfront considered as a horizon of the wedge as compared to the massless conformal scaling limit; in fact the correct way to deal with lightfront localization is inexorably linked with the understanding of wedge localized operator algebras.. Whereas the scaling limit (which is closely linked to the renormalization group approach) admits a natural formulation in terms of pointlike fields, the present chiral conformal structure on a lightfront horizon of a wedge region associated with interacting fields with nonintegrable Kallen-Lehmann spectral function cannot be obtained simply as a restriction of such fields to the horizon. One rather needs to perform a reprocessing of the field content into the net of algebras and to apply a modular inclusion procedure to the wedge algebra. In fact this process supplies different spacetime indexed subalgebras; as global algebras the chiral horizon algebra is identical with the wedge algebra. Different from a restriction process, the notion of transversal localization is lost (i.e. the obtained chiral theory is a kind of kinematical coarse grained object) and has to be constructed by tilting the wedge as will be described in section 4.

To physicists familiar with the Unruh-Hawking effect the special role of the wedge algebra may not come totally unexpected, but the crucial role of operator algebras versus pointlike fields is somewhat surprising in view of the widespread opinion that their use is more a matter of mathematical conciseness and conceptual clarity rather than a matter of principle. Although concepts as Hawking temperature, horizons and holography originated with black hole physics, they have their natural analogs in Minkowski space[a]. Whereas

[a]Whereas in flat spacetime these local quantum physical concepts remain hidden and play no important role in particle physics, curved spacetime in special circumstances exposes them in a geometric quasiclassical veil.(in which they were discovered). The claims that holography is a characteristic manifestation of quantum gravity should be treated with cautious scepticism. In fact the present paper casts doubts on suggestions that such con-

for thermal aspects this had been known for some time, the presence of the other concepts is the main issue of this article.

Although our conceptual setting and the mathematical formalism goes beyond what the protagonist of "holographic projections" onto the horizon had in mind [6] and how this became subsequently related with lightfront quantization ideas [7], the intuitive physical content of our algebraic approach matches these ideas perfectly. For this reason (and also for the fact that we cannot think of any other way in which this idea could be furnished with a rigorous mathematical and conceptually tight local quantum physical meaning) the excellent name is maintained; in fact we think of our conceptual additions as the liberation of these ideas from their old short distance restrictions which did not leave any room for interactions.

The name holography is often used in a wider sense for an isomorphism of a local (massive) QFT with data which are associated with lower spacetime dimensions (In case of equal spacetime dimensions of the source and target theory the name transplantation has been used [8]). Not every aspect of such an isomorphism requires the full mathematical power of the field coordinatization free formulation of AQFT. For example the direction

$$AdS \to CQFT$$

in the Maldacena conjecture [9] can be seen by taking the limit of the spacial infinite remote Anti deSitter boundary in correlation functions of fields and is therefore susceptible to standard formulations of QFT. Only the existence of the inverse and the understanding of the detailed properties which one field theory inherits from the other requires a more ambitious mathematical theorem in the AQFT setting[b] [10]. In the present case of lightray holography the algebraic setting is crucial for both directions of the isomorphism. Holography in this sense (where localizations becomes scrambled up) is not useful for physical concepts which rely on spacetime localization between operators as e.g. time dependent scattering theory; rather their beneficial effect is expected to show up in the understanding of coarse-grain properties of a localized algebra as e.g. its relative localization entropy between different states where only the overall relative size of degrees of freedom matter, but not the sunlocalizations within that algebra. These matters will be discussed in a forthcoming paper.

The content is organized as follows. In the next section the properties of

---

cepts require new physical principles which go beyond those of local quantum physics; as the situation looks now it appears more and more like a lack of agreement between older conjectures versus more recent theorems.

[b]One rigorous result is that the Maldacena conjecture cannot hold between two Lagrangian (pointlike fields) quantum field theories [101211].

lightray restrictions in d=1+1 are recalled in the present setting. Section 3 illustrates the new construction based on modular wedge localization within the rich family of factorizing d=1+1 models. The additional problems posed by higher dimensional QFT including proposals for their solution are then addressed in section 4. The last section attempts to make connections to other fundamental problems of QFT for whose solution the new concepts are also expected to be important. In a mathematical appendix we collect some known mathematical results for the convenience of the reader.

2 Elementary review of d=1+1 lightray restriction

Let us consider the simplest case namely the lightray holography of a two-dimensional massive selfconjugate QFT. We first remind ourselves of the situation for a free massive scalar field

$$A(x) = \frac{1}{\sqrt{2\pi}} \int \left(e^{-ipx} a(\theta) + e^{ipx} a^*(\theta) \right) d\theta \tag{1}$$
$$p = m(ch\theta, sh\theta)$$

where for convenience we used the momentum space rapidity description. In order to approach the light ray $x_- = t - x = 0$ in such a way that $x_+ = t + x$ remains finite we approach the $x_+ > 0$ horizon of the right wedge $t^2 - x^2 < 0$, $x > 0$ by taking the $r \to 0$, $\chi = \hat{\chi} - lnr \to \hat{\chi} + \infty$ in the x-space rapidity parametrization

$$x = r(sh\chi, ch\chi), \;\; x \to (x_- = 0, x_+ \geq 0, \;\; finite) \tag{2}$$
$$A(x_+, x_- \to 0) \equiv A_+(x_+) = \frac{1}{\sqrt{2\pi}} \int \left(e^{-ip_- x_+} a(\theta) + e^{ip_- x_+} a^*(\theta) \right) d\theta$$
$$= \frac{1}{\sqrt{2\pi}} \int \left(e^{-ip_- x_+} a(p) + e^{ip_- x_+} a^*(p) \right) \frac{dp}{|p|}$$

where the last formula serves to make manifest that the limiting $A(x_+)$ field is a chiral conformal (gapless P_- spectrum) field; the mass in the exponent $p_- x_+ = mre^\theta e^{-\chi}$ is just a parameter which keeps track of the "engineering dimension" (the physical mass is the gap in the $P_- \cdot P_+$ spectrum). Since this limit only effects the numerical factors and not the Fock space operators $a^{\#}(\theta)$, we expect that there will be no problem with the horizontal restriction i.e. that the formal method (the last line in 2) agrees with the rigorous result. Up to a fine point which is related to the bad infrared behavior of a scalar $dimA = 0$, field this is indeed the case. Using the limiting χ-parametrization we see that for the smeared field with $supp\tilde{f} \in W$ one has the identity

$$\int A(x_+, x_-)\tilde{f}(x)d^2x = \int_C a(\theta)f(\theta) = \int A_+(x_+)\tilde{g}(x_+)dx_+ \tag{3}$$

$$\tilde{f}(x) = \int_C e^{ip(\theta)x} f(\theta)d\theta, \; \tilde{g}(x_+) = \int_C f(\theta)e^{ip_-(\theta)x_+}d\theta$$

These formulas warrant some explanation. The onshell character of free fields restricts the Fourier-transformed test function to their mass shell values $f(p)|_{p^2=m^2} = f(\theta)$ and the wedge support property is equivalent to the strip analyticity. The integration path C consists of the upper and lower rim of the strip and corresponds to the negative/positive frequency part of the Fourier transform. By introducing the test function $\tilde{g}(x_+)$ which is supported on the halfline $x_+ \geq 0$ it becomes manifest that the smeared field on the horizon rewritten in terms of the original Fourier transforms must vanish at $p = 0$ since

$$f(p)|_{p^2=m^2}\frac{dp}{\sqrt{p^2+m^2}} = f(\theta)d\theta \equiv g(\theta)d\theta = g(p)\frac{dp}{|p|} \tag{4}$$

$$\curvearrowright g(0) = 0$$

This infrared restriction is typical for $dimA = 0$ fields and would not occur for free fields with a nontrivial L-spin. The equality of the f-smeared $A(x)$ fields with the g-smeared $A_+(x_+)$ leads to (after clarification of some domain problems of these unbounded operators) the equality of the affiliated (weakly closed) operator algebras

$$\mathcal{A}(W) = \mathcal{A}(R_+) = \mathcal{A}(R_-) \tag{5}$$

Here the last equality expresses the fact we could have taken the lower horizon with the same result. This equality is the quantum version of the classical propagation property of characteristic data on the upper or lower lightfront of a wedge. With the exception of d=1+1 m=0 the amplitudes inside the causal shadow W of R_+ are uniquely determined by the lightfront data. Note that a finite interval on R_+ does not cast a 2-dimensional causal shadow; this is only the case for the full characteristic data. Related to this is fact that the opposite lightray translation

$$AdU_-(a)\mathcal{A}(R_+) \subset A(R_+) \tag{6}$$

$$U_-(a) = e^{-iP_-a}$$

is a totally fuzzy endomorphism of the $A(R_+)$ net whereas in the setting of the $\mathcal{A}(W)$ net spacetime indexing it is a geometric map.

It is very important to notice that even in the free case the horizontal limit is different from the scale invariant massless limit. The latter cannot be performed in the same Hilbert space since the $m \to 0$ limit needs a compensating $\ln m$ term in the momentum space rapidity θ of the operators $a^{\#}(\theta)$ whereas in the horizon limit it was only effecting the c-number factors. There is no problem of taking this massless limit in correlation functions if one uses smearing functions whose integral vanishes (or $f(p=0)=0$). The limiting correlation functions define via the GNS construction a new Hilbert space which contains two chiral copies of the conformal dimA=0 field corresponding to the right/left movers. In the case of interacting theories the appropriately defined horizontal holographic projection is different from the scaling limit by much more than just multiplicities as will be seen below.

For interacting fields a necessary condition for the above lightray restriction to work is the finiteness of the wave function renormalization constant which in terms of two-point function of the correctly normalized ($\langle p\,|A(0)|\,\Omega\rangle = \frac{1}{\sqrt{2\pi}}$) interpolating field is the convergence of the following integral over the Kallen-Lehmann spectral function

$$\int \rho(\kappa^2)d\kappa^2 < 0 \tag{7}$$

This condition is violated for all genuinely renormalizable (i.e. not super-renormalizable) theories. In this case the holographic reprocessing has to be done on the level of algebras according to the formal scheme

$$A(x) \to \mathcal{A}(W)\,net \stackrel{holography}{\longrightarrow} \mathcal{A}(R_+)\,net \to A_+(x_+)$$

Here $A(x)$ stands symbolically for a complete set of local fields which fulfill d=1+1 locality and $A_+(x_+)$ stand for possible R_+-local chiral fields.

3 Holography for factorizable models

The holography in the presence of renormalizable interactions which violate the finiteness condition (7) can be nicely illustrated for d=1+1 factorizable models. The most appropriate construction of these models starts from free fields with nonlocally modified commutations relations for the momentum space creation and annihilation operators [13]

$$A(x) = \frac{1}{\sqrt{2\pi}} \int \left(e^{-ipx} Z(\theta) + e^{ipx} Z^*(\theta)\right) d\theta, \;\; p = m(ch\theta, sh\theta)$$

$$A(\tilde{f}) = \frac{1}{\sqrt{2\pi}} \int_C Z(\theta) f(\theta) d\theta$$

where the Z's are defined on the incoming n-particle vectors by the following formula for the action of $Z^*(\theta)$ for the rapidity-ordering $\theta_i > \theta > \theta_{i+1}$, $\theta_1 > \theta_2 > ... > \theta_n$

$$Z^*(\theta)a^*(\theta_1)...a^*(\theta_i)...a^*(\theta_n)\Omega = \\ S(\theta-\theta_1)...S(\theta-\theta_i)a^*(\theta_1)...a^*(\theta_i)a^*(\theta)...a^*(\theta_n)\Omega \\ + contr.\, from\, bound\, states \tag{8}$$

In the absence of bound states this amounts to the commutation relations

$$Z^*(\theta)Z^*(\theta') = S(\theta-\theta')Z^*(\theta')Z^*(\theta),\ \theta<\theta' \\ Z(\theta)Z^*(\theta') = S(\theta'-\theta)Z^*(\theta')Z(\theta)+\delta(\theta-\theta') \tag{9}$$

A smeared field $A(\hat{f})$ applied to the vacuum creates a one-particle vector

$$A(\hat{f})\,|0\rangle = \int f(\theta-i\pi)\,|p(\theta)\rangle\,d\theta$$

Localized operators which applied to the vacuum create one-particle vectors without admixture of multiparticle vacuum polarization clouds are called polarization-free generators (PFG's). Their structure is restricted by the following theorems [13]

Theorem 1 *PFG's which are localized in regions whose causal completion is genuinely smaller than a wedge (e.g. spacelike cones, double cones) lead to interaction-free theories. On the other hand wedge-localized PFG's always exist even in the presence of interactions. They have Fourier transforms (tempered distributions) only in the case of d=1+1 with purely elastic scattering.*

According to this theorem we should ask the question whether the above PFG's are wedge-localized. The affirmative answer is contained in the following theorem

Theorem 2 *The PFG's with the above algebraic structure for the Z's are wedge-localized if and only if the structure coefficients S(θ) are meromorphic functions which fulfill crossing symmetry in the physical θ-strip i.e. the requirement of wedge localization converts the Z-algebra into a Zamolodchikov-Faddeev algebra.*

In this case the $A(\hat{f})$ are generators affiliated to the wedge algebra

$$\mathcal{A}(W) = alg\,\{A(f)|suppf \subset W\}$$

The most general operator A in $\mathcal{A}(W)$ is a LSZ-type power series in the Wick-ordered Z's

$$A = \sum \frac{1}{n!} \int_C \cdots \int_C a_n(\theta_1, ...\theta_n) : Z(\theta_1)...Z(\theta_n) d\theta_1 ... d\theta_n : \qquad (10)$$

with strip-analytic coefficient functions a_n which are related to the matrix elements of A between incoming ket and outgoing bra multiparticle state vectors. The integration path C consists of the real axis (associated with annihilation operators and the line $Im\theta = -i\pi$. Writing such power series without paying attention to domains of operators means that we are we are only dealing with bilinear forms whose operator status is still to be settled.. The bilinear forms which have their localization in double cones are characterized by their relative commutance (this formulation has to be changed for Fermions or more general objects) with shifted generators $A^{(a)}(f) \equiv U(a)A(f)U^*(a)$

$$\left[A, A^{(a)}(f)\right] = 0, \forall f \ suppf \subset W$$
$$A \subset \mathcal{A}_{bil.}(C_a)$$

where the subscript indicates that we are dealing with spaces of bilinear forms (formfactors of would-be operators localized in C_a) and not yet with unbounded operators and their affiliated von Neumann algebras. This relative commutant relation [13] on the level of bilinear forms is nothing but the famous "kinematical pole relations" which relate the even a_n to the residuum of a certain pole in the a_{n+2} meromorphic functions. The structure of these equations is the same as that for the formfactors of pointlike fields; but whereas the latter lead (after splitting off common factors [15] which are independent of the chosen field in the same superselection sector) to polynomial expressions with hard to control asymptotic behavior, the a_n of the double cone localized bilinear forms are solutions which have better asymptotic behavior which according to the Payley-Wiener-Schwartz theorem. We will not discuss here the problem of how this improvement can be used in order to convert the bilinear forms into genuine operators.

Suppose now that we consider a modular inclusion (see appendix) of wedges which in the above formalism just means that we take a as a lightlike vector $a \sim (1,1)$. In that case the relative commutant consists of bilinear forms which are interpreted to be localized in the interval $(0,1)$ on the lightray which according to general theorems are associated to a chiral conformal field theory. It is easily checked that the previous space of double cone localized space is reobtained (as expected) by applying a suitable opposite lightray translation to the interval localized space of bilinear forms. The total spaces of wedge-localized and lightray localized bilinear forms (10) which are the (weak closures of the) unions of the local spaces are the same (as expected from the classical picture that the causal shadow of the characteristic data on the halfline is the wedge i.e. $\mathcal{A}(R_+) = \mathcal{A}(W)$ whereas finite intervals cast no shadow).

According to the construction the net on the horizon is bosonic and since it is also chiral the spectrum of scale dimensions must be integer-valued. The massless limit of the pointlike field generators of the wedge theory on the other hand can have anomalous short distance behavior. This is yet another manifestation of the useful kinematical nature of the lightray algebra (the "scrambled up" wedge algebra). Important dynamical data have been transferred to the action of the opposite lightray translation which plays the role of a kind of hamiltonian and whose action destroys the conformal invariance and recreates the complications of the original massive theory. The fascinating aspects of the free field behavior of pointlike generators of the chiral $\mathcal{A}(R)$ net is the extreme nonlocality they must have with respect to the generators of the $\mathcal{A}(W)$ net together with the naturalness of their construction. Despite their simplicity their creation/annihilation operators must be infinite power series in the Z-operators. Factorizing models promise to play an important role in the better understanding of this horizont-localized chiral conformal field theory which is quite different from the conformal scaling limit theory (e.g. it has a richer supply of symmetries and maps). Automorphisms which act locally on the original set may become "fuzzy" on the horizon net (example: the opposite lightray translation) and vice versa (example the circular rotation on the conformal horizon).

4 Problems met in higher dimensional cases

When there are transversal spatial dimensions the restriction of fields to a lightfront continues to show the same problems in the presence of interactions. For a proof that (under very mild assumptions about operator domains) the fields must have the free canonical structure we refer to the second paper of Driessler. In the first paper the author also shows that the vacuum factorizes transversally[c] (a behavior otherwise met for spatially disjoint localized algebras in theories without vacuum polarization like second quantized QM) although it remains highly entangled with respect to longitudinal (along the light ray) disjointness where one needs the so-called split property in order to meet conditions which are similar to quantum mechanical (type I) algebras.

[c]This observation is closely related to the impossibility to associate an operator algebra on a lightray in the presence of transverse directions.

Before we turn to a field-free i.e. algebraic construction of a holographic projection we briefly present the lightfront formalism for free fields. The rapidity parametrization of a scalar free field with $x = r(sh\chi, ch\chi, x_\perp)$, $p = (m_{eff}ch\theta, m_{eff}sh\theta, p_\perp)$, $m_{eff}(p_\perp^2) = \sqrt{m^2 + p_\perp^2}$ leads to the following (upper) horizontal projection formula

$$A(x_+, x_\perp) = \frac{1}{(2\pi)^{\frac{d}{2}}} \int (e^{-ip_-x_+ + ip_\perp x_\perp} a(\theta, p_\perp) + e^{ip_-x_+ - ip_\perp x_\perp} a^*(\theta, p_\perp))d\theta$$
$$p_-x_+ = m_{eff}(p_\perp^2)e^\theta x_+ \tag{11}$$

Rewritten formally in terms of the momentum space measure

$$A(x_+, x_\perp) = \frac{1}{(2\pi)^{\frac{d}{2}}} \int (e^{-ip_-x_+ + ip_\perp x_\perp} a(p_-, p_\perp) + e^{ip_-x_+ - ip_\perp x_\perp} a^*(p_-, p_\perp)) \frac{dp_-}{|p_-|} \tag{12}$$

we encounter the typical infrared divergence which requires to use again the restricted test functions $f(x_+, x_\perp)$ (again only for s=0) with $\int f(x_+, x_\perp)dx_+ = 0$. The above rapidity representation reveals that the lightfront field almost looks like a continuous superposition of chiral fields at different scales set by the continuous values of the magnitude of the transverse momenta $|p_\perp|$.

As in the d=1+1 case in section 3, the algebraic holography starts from the modular inclusion of the standard wedge shifted along the x_+ lightray into itself (the reason we work with wedge algebras and not with the full algebra is that the modular prerequesites of absence of annihilation operators for the vacuum are violated for the latter). Unlike the lightfront restriction of free fields which together with the action of the opposite lightray translation allows to reconstruct the original d=1+1 theory, the modular inclusion in the present case yields a result which has no transverse localization, although the transverse translation $e^{iP_\perp x_\perp}$ acts on $A(R_+)$. Note that our notation is not meant to literally indicate a localization on the x_+ lightray but only indicates that the chiral holographic projection has no transversal localization concept (otherwise we could not describe it in terms of a chiral theory). This is the prize we have to pay for obtaining such a simple holographic image from a complicated higher dimensional QFT. Knowing the action of the translation into the opposite lightray direction does not help to resolve the lack of transversal localization. Whereas this may be enough in problems of degree of freedom counting (e.g. entropy discussions) for the reconstruction of the original net i.e. for the formulation of a holographic isomorphism one must get a control of transversal localization. This can be achieved by Lorentz tilting the standard wedge around its x_+ lightray. This is done by using the "translational" part of the Wigner little group of the light ray. In d dimensions there are d-2 such translations inside the homogenous Lorentz group and they act like a transversal Galilei transformation. Together with the longitudinal symmetries the geometric symmetry group of a lightfront is seven-dimensional [1]: the longitudinal dilation (alias L-boost) and translation and the two transversal translations as well as Galilei transformations and a transversal rotation (the wedge allows a natural association with an 8-parametric subgroup of the Poincaré group [14]).

In order to keep things geometrically simple let us choose in the following d=1+2. Then there is just one one-parametric group of tilting (Galileian) transformations such that the pair of the original wedge together with the tilted wedge forms a modular intersection in the sense of the appendix. This may be easily rephrased in terms of an additional localization structure on the upper lightfront horizon. The original transversal stripes which correspond to the longitudinal intervals are transformed into sloped stripes which intersect the original ones in parallelograms and form the localization regions of an algebraic net on the horizon. Again this is a fuzzy transformation of the chiral theory. Clearly the parallel translates of the tilted lightfront intersect the original lightfront in lines which are transversally shifted parallels of the original lightray. This can be used to generate a transversal localization on the original lightfront which generates a net structure on this lightfront. Together with the action of the opposite lightray translation this is sufficient to recuperate the original d=1+2 net. Instead one may also talk about reconstructing the original theory by scanning with two chiral theories, the second one resulting from the first by the tilting automorphism. The principle of generalization to d dimensions should be clear from our geometric interpretation.

5 Hopes based on modular localization

In these notes we have shown that (in the presence of interactions) modular localization and modular inclusions are essential tools in the formulation of ideas around "holography" as a significant extension of lightcone quantization. A closely related problem is that of modular symmetries beyond the geometric symmetries of Poincaré- or conformal invariance (which are also known to be of modular origin, in the case of the infinite chiral diffeomorphisms group this was shown in [16]). Fuzzy modular groups σ_t exist for each standard Reeh-Schlieder pair $(\mathcal{A}(\mathcal{O}),\Phi)$, they only depend on the state ϕ and not on its implementing vector Φ. They are consistent with causality because they keep the causal complement of $\mathcal{O}$ apart from $\mathcal{O}$ in analogy say to the wedge affiliated L-boost which does not mix $\mathcal{A}(W)$ observables with those of $\mathcal{A}(W')=\mathcal{A}(W)'$ and as globally defined automorphisms they maintain commutance if applied to originally commuting operators even if the original localization regions suffer a fuzzy t-dependent "diffusion". They do not exist in the corresponding classical theory (assuming that the correspondence principle creates such an object) and their concrete form depends much more on quantum dynamical aspects. The intrinsic understanding of interactions is related to the absence of better than wedge localized PFG's and therefore it seems to be a reasonable conjecture that the action of the modular groups on vectors created by the application of such PFG to the vacuum is related to the shape of their vacuum-polarization clouds. An intrinsic understanding of these interaction-caused clouds is of course an old but never fulfilled dreams of nonperturbative local quantum physics.

Another such dream (which was mentioned in the introduction) is the understanding of "entropy of localization" [17,13]; this is natural problem since "localization temperature" (the Unruh-Hawking effect) has already been successfully explained in terms of modular concepts [18]. In particular it would be of great interest to know whether localization entropy preempts the Bekenstein black hole behavior and is proportional to the area of the spatial boundary of the bounding causal horizon and if the analogs to the thermodynamical laws are in some sense quantum renormalized.

The formalism for factorizable models which we used for d=1+1 illustrations of holography in section 2 is of course interesting in its own right, since it offers a constructive approach without facing any ultraviolet problem just as it was expected way back in a pure S-matrix approach. Its pointlike formulation [19] has witnessed a quite impressive growth over many years and the present operator algebraic method adds to it a spacetime interpretation of the Smirnov axioms (in particular of the Zamolodchikov-Faddeev nonlocal creation and annihilation operators) and holds the promise of a construction of the operators behind the bilinear forms (formfactors). Operator constructions are easier for extended objects than for pointlike fields.

The insensitivity of modular-based concepts on ultraviolet behavior of pointlike fields calls for the reinvestigation of the perhaps most important question of the post Feynman era: does the standard renormalizabilty criterion really mark the short distance frontier of the underlying principles of local quantum physics or does it only limit the range of applicability of trying to understand interactions in terms of singular pointlike field coordinatizations? Whereas in the Lagrangian renormalization approach based on power counting massive free fields cannot have dimensions beyond one in d=1+3, the bootstrap-formfactor approach, wherever it works, does not know such bounds.

Unfortunately this onshell approach remains presently limited to factorizing models in d=1+1 where it constructs a tempered (polynomially bounded in momentum space) field theory [21] for each admissable (crossing symmetric, unitary) factorizing S-matrix; no a priory restriction on short distance behavior is necessary. This situation was what the S-matrix protagonists dreamed of in the 60ies [20]. Although they extracted most of their S-matrix properties (in particular the important crossing property) from the causality and spectral properties of local quantum physics they had the unfortunate idea that in order to be successful in their pursuit they had to liquidate QFT. One could of course adapt the pessimistic attitude that these factorizing models are too special since they all share the property that whenever they contain a variable coupling parameter (example: massive Thirring model) there is no genuine coupling constant renormalization and the Beta function vanishes.

There are however also encouraging indications from the perturbative aspects of higher spin theories whose couplings violate the orthodox power counting criterion. Consider for example interactions of massive d=1+3 vectormesons. Since the free field dimensions of the vector description is two instead of one (other associations of physical free fields with s=1 cannot lower the dimension), any Lorentz-invariant interaction involves has at least the powercounting dimension 5 and hence to the orthodox approach it is nonrenormalizable i.e. the number of counterterm structures increase in each order of perturbation theory which leads also to a nontempered behavior of correlation functions. On the other hand there is the following trick [22] (only available for massive vectorfields) which leads to a renormalizable solution i.e. one for which the operator dimension 2 of the vectormeson field (apart from the expected logarithmic corrections) is maintained after renormalization. The trick consists of representing the Wigner one particle space as a cohomology space in the spirit of a one particle version of the so-called BRST formalism and constitutes a generalization of the approach taken in [23]. The idea behind this trick is the same as in any application of BRST formalism: the combination of lowering the propagator dimensions in the cohomological extension together with the idea of stability of cohomological representations allows a safe return to physics after having renormalized the auxiliary correlation function of the extended correlation functions in the standard way (perturbative renormalization is a linear process which does not require positivity). The particular "one-particle version" of the BRST representation has two additional advantages. On the one hand it keeps the BRST "ghosts" away from self-interactions (their contribution to the action remains bilinear) and maintains the LSZ asymptotic structure throughout. On the other hand it requires the presence of additional interacting physical degrees of freedom which (unlike in the standard approach based on the mass generation by Higgs condensates) was not part of the input. The simplest (and probably only) realization of these additional degrees of freedom is via a standard scalar field i.e. one with vanishing vacuum expectations (i.e. not "Higgs" if this terminology refers to the condensate). There is of course no change of physical content as compared to the standard Higgs mechanism approach, but the perspective and physical interpretation is somewhat different. In particular one returns to the original physical question for renormalizable vectormesons in the pre-gauge spirit of Sakurai and Lewellyn-Smith. In this way the suspicion that there was only one renormalizable coupling involving massive vectormesons is confirmed and a gauge principle which selects between different possibilities[d] is not needed. From a particle physics point of view it makes more sense to emphasize the uniqueness of renormalizable vectormeson couplings than the differential geometric esthetical appeal of the gauge principle in the Higgs mechanism. One hopes that the present modular concepts may have will tell us eventually something about the true borders carved out by the underlying physical principles and perhaps show that the widening of renormalizability does not stop at the interaction of spin s=1 particles.

Another more mundane goal where one expects the new concepts to produce explicit answers is the local quantum physics of d=1+2 anyons and of the d=1+3 infinite spin zero mass Wigner representations. In both cases the multiparticle structure does not follow the usual tensor product structure and there exist no PFG's which have a better than wedge-like localization [24]. In fact the maintenance of the spin-statistics connection in the presence of braid group statistics requires even in the "free" case (conservation of real particles, even vanishing of the cross section) the presence of vacuum polarization clouds so that the nonrelativistic limit remains a QFT i.e. cannot be a QM. This connection is in particular not maintained in the QM of the Aharonov-Bohm like topological constructions.

We hope that we were able to convince the reader that there are plenty of deep and physically relevant local quantum physical problems in particle physics which fall into the range of the new method.

Acknowledgements: Some helpful informations by Detlev Buchholz concerning the older literature are greatfully acknowledges. The persuit of content of old papers also brought about an enjoyable electronic correspondence with Heinrich Leutwyler, a colleague and friend from the old "current algebra and lightcone days". He confirmed that the simplicity of lightfront algebras

[d]There are however several couplings involving (semi)classical vector fields. In that case the gauge principle selects the Maxwellian interactions. To the extend that the correspondence principle can be invoked, the classical gauge principle follows from renormalizability of the corresponding local quantum physics.

was the prime motive for their introduction, a motivation which the modern holography inherited.

6 Appendix: mathematical aspects of modular theory

For the benefit of the uninitiated reader we briefly collect and comment on some formulas from modular theory [25].

Let $(\mathcal{A}, \Omega)$ be a standard pair, i.e. an operator algebra in a Hilbert space H with a vector Ω on which the algebra acts cyclically ($\overline{\mathcal{A}\Omega} = H$) and on which there exist no annihilators ($A\Omega = 0$, $A \in \mathcal{A} \curvearrowright A = 0$, "$\Omega$ is separating")

Definition 3 *Tomita's involution:* $SA\Omega = A^*\Omega, \curvearrowright$ *S is closed, antilinear and involutive i.e.* $S^2 \subset \mathbf{1}$ *(involutive on the domain). The polar decomposition* $S = J\Delta^{\frac{1}{2}}$ *leads to an antiunitary J and a positive Δ which in turn gives rise to a one parametric unitary group*Δ^{it}*, the (Tomita) modular group*

Theorem 4 *(Tomita, Takesaki) The Ad-action of Δ^{it} defines an automorphism σ_t of the operator algebra $\mathcal{A}$ and the Ad -action of J maps $\mathcal{A}$ into its commutant algebra $\mathcal{A}'$*

$$\begin{aligned} \sigma_t(A) &:= Ad\Delta^{it}(A) \in \mathcal{A} \\ AdJ(\mathcal{A}) &= \mathcal{A}' \end{aligned} \tag{13}$$

The modular automorphism group σ_t depends only on the state $\omega(\cdot) = \langle \Omega |\cdot| \Omega \rangle$ and not its implementing vector Ω; they are related through the KMS property (strip-analyticity of the function F(z))

$$\begin{aligned} F(t) &= \omega(\sigma_t(A)B) \\ F(t+i) &= \omega(B\sigma_t(A)) \end{aligned}$$

The KMS property (which generalizes the Gibbs formula) characterizes the modular automorphism of $\mathcal{A}, \omega$

Theorem 5 *(Bisognano-Wichmann) If we take for $\mathcal{A}$ the wedge algebra $\mathcal{A} = \mathcal{A}(W)$ and for Ω the vacuum vector, the modular objects have the following physical interpretation*

$$\begin{aligned} \Delta^{it} &= U(\Lambda_W(-2\pi t)) \\ J &= TCP \cdot Rot_W(\pi) \end{aligned}$$

Here $\Lambda_W(\chi)$ is the wedge adapted L-boost, TCP the antiunitary TCP-transformation of local QFT and $Rot_W(\varphi)$ the rotation group around the spatial axis pointing into W.

Definition 6 *(Wiesbrock, Borchers* [25]*) An inclusion of operator algebras* $(\mathcal{A} \subset \mathcal{B}, \Omega)$ *is "modular" if* $(\mathcal{A},\Omega)$, $(\mathcal{B},\Omega)$ *are standard and* $\Delta_{\mathcal{B}}^{it}$ *acts for* $(t<0)$ *as a compression on* $\mathcal{A}$

$$Ad\Delta_{\mathcal{B}}^{it}\mathcal{A} \subset \mathcal{A} \tag{14}$$

A modular inclusion is standard if the relative commutant $(\mathcal{A}' \cap \mathcal{B}, \Omega)$ *is standard. In that case the conditional expectation E*

Remark 7 *In case the equality sign holds (i.e. the compression is an automorphism, a theorem of Takesaki states the equality* $\mathcal{B}=\mathcal{A}$*. Therefore the modular inclusion situation is a generalization of that studied by Takesaki.*

Theorem 8 *(Guido,Longo and Wiesbrock* [26]*) Standard modular inclusions are in correspondence with chiral AQFT*

An important physical application used in the main text is obtained by the following adaptation to a massive interacting AQFT

$$\begin{aligned} \mathcal{B} &= \mathcal{A}(W) \\ \mathcal{A} &= U(e_+)\mathcal{A}(W)U^*(e_+),\ e_+ = (1,1) \\ &\equiv \mathcal{A}(W_{e_+}) \end{aligned} \tag{15}$$

This is the inclusion of the algebra translated via a lightlike translation into itself so the geometrically the relative commutator

$$\mathcal{A}(W_{e_+})' \cap \mathcal{A}(W) \equiv \mathcal{A}(I(0,1)) \tag{16}$$

is by causality localized in the upper horizontal interval (0,1). The standardness of this inclusion then leads to a chiral conformal AQFT i.e. a net (more precisely a pre-cosheaf)

$$I \to \mathcal{A}(I),\ I \subset S^1 \tag{17}$$

$$\begin{aligned} \mathcal{A}(R_+) &= \overline{\cup_t Ad\Delta^{it}\mathcal{A}(I(0,1))} \\ \mathcal{A}(R) &= \mathcal{A}(R_+) \vee J\mathcal{A}(R_+)J \end{aligned} \tag{18}$$

on which the Moebius group which preserves the vacuum vector acts. With the help of an external (i.e. not in Moeb.) automorphism on $\mathcal{A}(R)$ implemented by the opposite lightray translation $U_-(a)$ we are able to return from the chiral net on the right upper horizon to the original 2-dim. net. We call this chiral theory supplemented by the opposite lightray automorphism the holographic image of the 2-dim. massive net. With this interpretation the holographic projection for massive d=1+1 theories is nothing else than the conceptually and mathematically tightened version of the old lightray/p$\to \infty$ frame quantization including the rules of how to reprocess back the holographic image into the original local d=1+1 theory. For the extension of holographic projection to higher dimensional theory one needs one more mathematical definition and theorem about "modular intersections"

Definition 9 *A* ($\pm$) *modular intersection is defined in terms of two standard pairs* $(\mathcal{N},\Omega),(\mathcal{M},\Omega)$ *whose intersection is also standard* $(\mathcal{N}\cap\mathcal{M},\Omega)$ *and which in addition fulfill*

$$((\mathcal{N}\cap\mathcal{M})\subset\mathcal{N},\otimes) \;\; and \;\; ((\mathcal{N}\cap\mathcal{M})\subset\mathcal{M},\otimes) \;\; is \;\; \pm \text{ mod} ular$$

$$J_{\mathcal{N}}(\lim_{t\to\mp\infty}\Delta_{\mathcal{N}}^{it}\Delta_{\mathcal{M}}^{-it})J_{\mathcal{N}}=\lim_{t\to\mp\infty}\Delta_{\mathcal{M}}^{it}\Delta_{\mathcal{N}}^{-it}=J_{\mathcal{M}}\lim_{t\to\mp\infty}\Delta_{\mathcal{N}}^{it}\Delta_{\mathcal{M}}^{-it})J_{\mathcal{M}}$$

All limits are in the modular setting are to be understood in the sense of strong convergence on Hilbert space vectors. In the geometric setting of local quantum physics the the modular intersection property is realized par excellence by the pair of intersecting wedge algebras $\mathcal{M}=\mathcal{A}(W),\mathcal{N}=AdU(\Lambda_{e_+}(a))\mathcal{M}$ together with the Ω =vacuum. Here $\Lambda_{e_+}(a)$ denotes a "translation" (transversal Galilei transformation) in the Wigner little group which fixes the lightray vector e_+, i.e. the Lorentz transformation which tilts $\mathcal{W}$ around this lightray. In fact the limit in the second line is geometrically nothing else but $\Lambda_{e_+}(a)$ and the commutation relation with $J_{\mathcal{N},\mathcal{M}}$ is easily checked as a geometric relation in the extended Lorentz group.

Modular intersections play an analogous role in the construction of 3- and higher- dimensional AQFT starting from a finite set of wedge algebras [27] and the related holographic isomorphisms as the modular inclusions used in section 3 for the 2-dimensional case.

References

1. H. Leutwyler, *Acta Phys. Austr., Suppl. V*; H. Leutwyler, in *Magic without Magic*, J. R. Klauder ed., Freeman (1972); F. Jegerlehner, *Helv. Phys. Acta* **46**, 824 (1974).
2. W. Driessler, *Acta Phys. Austr.* **46**, 63 (1977).
3. W. Driessler, *Acta Phys. Austr.* **46**, 163 (1977).
4. S. Schlieder and E. Seiler, *Comm. Math. Phys.* **25**, 62 (1972).
5. B. Schroer, Holography &Transplantation and All That (The tip of an iceberg for a paradigmatic change in QFT?), hep-th/06284.
6. G. 't Hooft, in *Salam-Festschrift*, A. Ali et al. eds., World Scientific (1993) p.284.
7. L. Susskind, *J. Math. Phys.* **36**, 6377 (1995).
8. D. Buchholz, J. Mund, S. J. Summers, Transplantation of Local Nets and Geometric Modular Action on Robertson-Walker Space-Times, hep-

th/0011237.
9. J. Maldacena, *Adv. Theor. Math. Phys.* **2**, 231 (1998); E. Witten, *Adv. Theor. Math. Phys.* **2**, 253 (1998); S. S. Gubser, I. R. Klebanov, A. M. Polyakov, *Phys.Lett. B* **428**, 105 (1998).
10. K.-H. Rehren, *Ann, Henri Poincaré* **1**, 607 (2000); K.-H. Rehren, Local Quantum Observables in the Anti de Sitter-Conformal QFT Correespondence, hep-th/0003120.
11. M. Arnsdorf and L. Smolin, The Maldacena Conjecture and Rehren Duality, hep-th/0106073.
12. B. Schroer, *Commun. Math. Phys.* **219**, 57 (2001).
13. B. Schroer, *J. Math. Phys.* **41**, 3801 (2000), and ealier papers of the author quoted therein.
14. B. Schroer and H-W Wiesbrock, *Rev. Math. Phys.* **12**, 139 (2000).
15. H. Babujian, A. Fring, M. Karowski and A. Zapletal, *Nucl.Phys. B* **538**, 535 (1999), and references quoted therein.
16. L. Fassarella and B. Schroer, The Fuzzy Analog of Chiral Diffeomorphisms in higher dimensional Quantum Field Theories, hep-th/0106064.
17. H. Narnhofer, in *The State of Matter*, ed. by M. Aizenman and H. Araki, World-Scientific, Singapore (1994).
18. G. L. Sewell, *Ann. Phys.* **141**, 201 (1982).
19. F. A. Smirnov, *Int.J. Mod. Phys. A* **4**, 156 (1989).
20. A. R.White, The Past and Future of S-Matrix Theory, hep-ph/0002303.
21. H.J. Borchers, D. Buchholz and B. Schroer, *Commun. Math. Phys.* **219**, 125 (2001), hep-th/0003243.
22. M. Duetsch and B. Schroer, *J. Phys. A. Math. Gen.* **33**, 4317 (2000), hep-th/9906089 and references therein.
23. G. Scharf, *Quantum Gauge Theory: a True Ghost Story*, John Wiley & Sons, Inc. (2001).
24. J. Mund, *Lett.Math.Phys.* **43**, 319 (1998).
25. H. J. Borchers, *J. Math. Phys.* **41**, 3604 (2000).
26. D. Guido, R. Longo and H.-W. Wiesbrock, *Commun. Math. Phys.* **192**, 217 (1998).
27. R. Kaehler, *J. Math. Phys.* **42**, 74 (2000).

BROKEN SYMMETRIES

W. THIRRING
Institute for Theoretical Physics, University of Vienna, A 1090 Boltzmanngasse, Vienna, Austria
E-mail: Walter.Thirring@univie.ac.at

Definitions

(i) A symmetry α is an automorphism of a C*-algebra $\mathcal{A}$.

(ii) A dynamical symmetry α of a dynamical system $(\mathcal{A}, \tau_t)$, τ_t the time evolution, is a symmetry with $\alpha \circ \tau = \tau \circ \alpha$.

(iii) The dynamical symmetry α is spontaneously broken by a state ω over $\mathcal{A}$, if $\omega = \omega \circ \tau \neq \omega \circ \alpha$.

(iv) The symmetry α is spontaneously destroyed by a representation Π if α is an automorphism of $\Pi(\mathcal{A})$ but cannot be extended to the weak closure $\Pi(\mathcal{A})''$.

(v) The symmetry α is completely destroyed by a representation Π, if it cannot be restricted to $\Pi(\mathcal{A})$.

Explanations

ad (iii) Since ω is time invariant τ is unitarily implemetable whereas α may or may not. For finite dimensional matrix algebras it certainly is and already then (iii) can be realized: Let $\vec{\sigma}_1, \vec{\sigma}_2$ be two sets of Pauli matrices, $\mathcal{A} = \{\mathbf{1}, \vec{\sigma}_1, \vec{\sigma}_2\}$. An example is provided by

$$\tau_t = \operatorname{ad} \exp[i\vec{\sigma}_1 \vec{\sigma}_2 t], \qquad \alpha = \operatorname{ad} \exp[i\gamma(\vec{\sigma}_{1z} + \vec{\sigma}_{2z})],$$

$$\omega(a) = \operatorname{Tr} \frac{1 + \sigma_{1x}}{2} \cdot \frac{1 + \sigma_{2x}}{2}\, a.$$

ad (iv) Here we have the anti-Wignerian situation where α cannot be unitarily implemented because otherwise it could be extended to $\Pi(\mathcal{A})''$. This does not happen for matrix algebras where the weak closure does not go beyond norm closure but can happen in quantum field theory. More generally, if α is unitarily implementable (iv) cannot happen but not all symmetries can be unitarily implemented. For instance, the algebra of the diagonal matrices

$$A(a,b) = \begin{pmatrix} a & 0 & 0 \\ 0 & a & 0 \\ 0 & 0 & b \end{pmatrix}$$

has the symmetry $A(a,b) \to A(b,a)$ but this is not unitarily implementable.

ad (v) This happens if $\mathcal{A}$ has a center on which α acts nontrivially and Π is not faithful. It occurs for for τ in the Weyl algebra over T^1 but also already in finite dimensions: Consider $\mathcal{A} = (\mathbf{1}, \vec{\sigma}, \tau_3)$, $\tau_3^2 = \mathbf{1}$. $\alpha(\mathbf{1}, \vec{\sigma}, \tau_3) = (\mathbf{1}, \vec{\sigma}, -\tau_3)$ cannot be realized in $\Pi(\mathbf{1}, \vec{\sigma}, \tau_3) = (\mathbf{1}, \vec{\sigma}, \mathbf{1})$.

Finitedimensional Algebras

$$\mathcal{A} = \bigoplus \sum_i M_{m_i} \otimes \mathbf{1}_{m_i}$$

acts in $C^{\sum n_i m_i}$, (n_i, m_i) lexicographicaaly ordered,

$$\overline{\mathcal{A}} = \bigoplus \sum_i M_{\overline{n}_i} \otimes \mathbf{1}_{\overline{m}_i}.$$

Proposition

(i) $\exists$ algebraic isomorphism $\mathcal{A} \leftrightarrow \overline{\mathcal{A}} \Leftrightarrow n_i = \overline{n}_i \; \forall\, i$

(ii) $\exists$ unitary $U : \overline{\mathcal{A}} = U\mathcal{A}U^{-1} \Leftrightarrow n_i = \overline{n}_i,\; m_i = \overline{m}_i \; \forall\, i$

(iii) $\exists\, \psi \in C^{\sum_i n_i m_i}$ cyclic and separating for $\mathcal{A} \Leftrightarrow m_i = n_i \; \forall\, i$

Corollary

(i) If $\mathcal{A}$ has a cyclic and separating vector all its isomorphism are unitarily implementable. (Also valid in the infinitedimensional case.)

(ii) Iff $n_i = n_j$ but $m_i \neq m_j$ $\mathcal{A}$ has an automorphism

$$M_{n_i} \otimes \mathbf{1}_{m_i} \leftrightarrow M_{n_j} \otimes \mathbf{1}_{m_j}$$

which is not unitarily implementable. If $m_i = m_j$ it is unitary but not inner. Hence it does not have the center elementwise invariant and it is completely destroyed in an irreducible representation.

Proposition

(i) If $\tau = \text{ad}\, e^{iHt}$, H has a pure point spectrum and ω is represented by a density matrix ρ then there is a spontaneously broken symmetry iff ρ is not the unit operator in all degeneracy spaces of H.

(ii) If $\mathcal{A}$ is a von Neumann algebra its weakly closed two-sided ideals are generated by central projections. Then there is no completely destroyed symmetry if the center $\mathcal{Z} = c\,\mathbf{1}$.

Subalgebras of the Weyl Algebra

The Weyl algebra $\mathcal{A}$ is characterized by the relations

$$W(\alpha,\beta) = e^{i\alpha x}e^{i\beta p} = e^{-\frac{i}{2}\alpha\beta}e^{i(\alpha x+\beta p)}, \qquad \alpha,\beta \in R.$$

It is a simple C*-algebra whose irreducible representations are all (quasi) equivalent to the (faithful) Π:

$$\Pi(e^{i\alpha x})\psi(x) = e^{i\alpha x}\,\psi(x), \quad \Pi(e^{i\beta p})\psi(x) = \psi(x+\beta), \quad \psi \in L^2(R).$$

The torus algebra $\mathcal{A}_T$ is

$$\{W(\alpha,\beta), \alpha = 2\pi n, n \in Z\} \subset \mathcal{A}$$

and the quantum cat (rotation) algebra $\mathcal{A}_\Theta$ is

$$\{W(\alpha,\beta), \alpha = 2\pi n, \beta = \Theta m, m \in Z\} \subset \mathcal{A}_T.$$

$\mathcal{A}_T$ is not simple, its center is $\mathcal{Z} = \{e^{imp}, m \in Z\}$. $\mathcal{A}_\Theta$ is simple for irrational Θ, if $\Theta = g_1/g_2$ it has the central elements with $m \in g_2 Z$. Since $\mathcal{Z}$ is in Π not represented by pure numbers it is reducible over $\mathcal{A}_T$ or $\mathcal{A}_\Theta$, if Θ is rational. The reduction is accomplished by dividing $R \ni x$ in its integer part $[x]$ and a rest $\langle x\rangle \in [0,1) : x = [x] + \langle x\rangle$ and thus factorizing $L^2(R) = \ell^2 \otimes L^2([0,1)) \simeq L^2(0,1) \otimes L^2(0,1)$ where $\simeq$ indicates the Fourier transform in the first factor

$$f(\langle x\rangle,\gamma) = \sum_{k\in Z} e^{2\pi ik\gamma}\,\psi(k+\langle x\rangle)$$

$$\psi(x) = \int_0^1 d\gamma\, e^{-2\pi i\gamma[x]}\, f(\langle x\rangle,\gamma).$$

Now x and p act as

$$(e^{2\pi inx}\, f)(\langle x\rangle,\gamma) = e^{2\pi in\langle x\rangle}\, f(\langle x\rangle,\gamma)$$

$$(e^{ip\beta}\, f)(\langle x\rangle,\gamma) = e^{-2\pi i\gamma[\beta+\langle x\rangle]}\, f(\langle\langle x\rangle+\beta\rangle,\gamma).$$

Since for $\beta = n \in Z$ we have $[\beta + \langle x \rangle] = \beta$, $\langle \beta + \langle x \rangle \rangle = \langle x \rangle$ the elements of the center are multiplication by $e^{-2\pi i n \gamma}$. For fixed γ the first factor of the tensor product furnishes an irreducible representation Π_γ where x is multiplication by an angle and the angular momentum p has the eigenvalues $2\pi Z + \gamma$. Thus γ can be identified with the Bohm–Aharonov phase and the change of γ : $p \to p + \gamma_0$ which is an automorphism of $\mathcal{A}_T$ is no longer an automorphism of $\Pi_\gamma(\mathcal{A}_T)$. The algebraic relation $\Pi_\gamma(e^{inp}) = e^{-2\pi i n \gamma} \cdot \mathbf{1}$ is not invariant and this symmetry is completely destroyed.

A spontaneous symmetry destruction occurs if Π_γ is further restricted to $\mathcal{A}_\Theta$, $\Theta =$ irrational. It has no center and is faithfully represented in Π_γ. Thus the shift $e^{im\Theta p} \to e^{im\Theta(p+\gamma_0)}$ is now an automorphism of $\Pi_\gamma(\mathcal{A}_\Theta)$, however it is not an automorphism, even of $\Pi_\gamma(\mathcal{A}_\Theta)''$ as the point spectrum of $\Pi_\gamma(e^{im\Theta p})$ is changed.

GAUGE THEORIES ON NON-COMMUTATIVE SPACES

JULIUS WESS

Sektion Physik der Ludwig-Maximilians-Universität
Theresienstr. 37, D-80333 München, Germany
and
Max-Planck-Institut für Physik
(Werner-Heisenberg-Institut)
Föhringer Ring 6, D-80805 München, Germany

1 The Algebraic Concept

Gauge theories are usually formulated on differentiable manifolds. This allows a very precise formulation but also makes a very strong assumption on the structure of space at very short distances. Such an assumption is plausible but needs an experimental verification if gauge theories are applied to physics. At present we know that a lattice structure at very short distances much smaller than the confinement range is compatible with our experimental knowledge. We have to probe these very small distances experimentally to see to what extent the notion of causal, continuous space is actually valid.

A lattice structure might very well manifest itself by an algebraic, non-commutative structure of the space coordinates. Non-commutativity generally leads to discrete spectra. Thus one would try to formulate a gauge theory on non-commutative spaces and see if such a theory leads to observable effects without having to verify the lattice structure by measurement on the space coordinates themselves. The space coordinates might not be the most relevant observables for a system at very small distances and thus at very high energies.

Let us first consider the commutative space from an algebraic point of view. We start with N coordinates $x^1, \ldots, x^N$ and consider the algebra freely generated by these elements. The commutativity we formulate via the relations

$$[x^i, x^j] = 0\,, \quad i, j = 1, \ldots, N\,. \tag{1}$$

We construct the ideal generated by these relations and divide the freely generated algebra by this ideal to obtain an algebra:

$$\mathcal{A}_x = \frac{[[x^1 \ldots x^N]]}{I_{[x^i, x^j]=0}}\,. \tag{2}$$

This is an infinite-dimensional algebra but we can get some control on the growth of this algebra by considering polynomials of fixed degree; they all form finite-dimensional subspaces. In each of these subspaces we can define a basis, e.g. by demanding a certain ordering of the elements. A complete symmetrization would be an obvious ordering. In this case we denote the elements of the basis by

$$(x^{i_1}\dots x^{i_k})_s = \frac{1}{k!}\sum_{P(i_1\dots i_k)} x^{i_1}\cdots x^{i_k}\,, \quad i_1,\dots,i_k = 1,\dots,N\,. \tag{3}$$

An element of the algebra can be expanded in this basis

$$\begin{aligned} &f(x) \in \mathcal{A}_x \\ &f(x) = \sum_{k=0}^{\infty}\sum_{i_1\dots i_k} C_{i_1\dots i_k}(x^{i_1}\cdots x^{i_k})_s\,. \end{aligned} \tag{4}$$

Formal power series are accepted, this is indicated by the double bracket in definition (2).

This algebraic concept can be easily generalized to non-commutative spaces. We consider algebras freely generated by elements $\hat{x}^1,\dots\hat{x}^n$, again calling them coordinates. But now we change the relations to arrive at non-commutative spaces:

$$\mathcal{R}_{\hat{x},\hat{x}}: \quad [\hat{x}^i,\hat{x}^j] = i\theta^{ij}(\hat{x}). \tag{5}$$

Following L.Landau, non-commutativity carries a hat.

Now we deal with the algebra:

$$\begin{aligned} &\mathcal{A}_{\hat{x}} = \frac{\mathbb{C} << \hat{x^1},\dots,\hat{x^n} >>}{I_{\mathcal{R}_{\hat{x},\hat{x}}}}, \\ &\hat{f} \in \mathcal{A}_{\hat{x}}. \end{aligned} \tag{6}$$

In the following we impose one more condition on the algebra: the dimension of the subspace of homogeneous polynomials should be the same as for commuting coordinates. This is the socalled Poincare-Birkhof-Witt property (PBW). Only algebras with this property will be considered, among them are the algebras where θ^{ij} is a constant:

Canonical structure, ref. [2]:

$$[\hat{x}^i,\hat{x}^j] = i\theta^{ij}, \tag{7}$$

where θ^{ij} is linear in $\hat{x}$:
<u>Lie structure</u>, ref. [3]:

$$[\hat{x}^i, \hat{x}^j] = i\theta^{ij}_k \hat{x}^k, \tag{8}$$

where θ^{ij} is quadratic in $\hat{x}$:
<u>Quantum space structure</u>, ref.[4]:

$$[\hat{x}^i, \hat{x}^j] = i\theta^{ij}_{kl} \hat{x}^k \hat{x}^l, \tag{9}$$

The constants θ^{ij}_k and θ^{ij}_{kl} are subject to conditions to guarantee PBW. For Lie structures this will be the Jacobi identity, for the quantum space structure the Yang-Baxter equation.

There is a natural vector space isomorphism between $\mathcal{A}_x$ and $\mathcal{A}_{\hat{x}}$. It is based on the isomorphism of the vector spaces of homogeneous polynomials that have the same degree due to the PBW property.

In order to establish the isomorphism we choose a particular basis in the vector space of homogeneous polynomials in the non-commuting variables $\hat{x}$ and characterize the elements of $\mathcal{A}_{\hat{x}}$ by the coefficient functions in this basis. The corresponding element in the algebra $\mathcal{A}_x$ of commuting variables is supposed to have the same coefficient function. The particular form of this isomorphism depends on the basis chosen.

The vector space isomorphism can be extended to an algebra isomorphism. To establish it we compute the coefficient function of the product of two elements in $\mathcal{A}_{\hat{x}}$ and map it to $\mathcal{A}_x$. This defines a product in $\mathcal{A}_{\S}$ that we denote as diamond product ($\diamond$ product). The algebra with this $\diamond$ product we call ${}^\diamond\mathcal{A}_x$. There is a natural isomorphism:

$$\mathcal{A}_{\hat{x}} \longleftrightarrow {}^\diamond\mathcal{A}_x. \tag{10}$$

The three structures that we have mentioned above have an even stronger property than PBW. It turns out that monomials in any well-defined ordering of the coordinates form a basis. Among them is an ordering as we have used it (5) or the completely symmetrized ordering of nomonials as well.

For such structures we shall denote the $\diamond$ product as $*$ product (star product), ref. [5].

For the <u>canonical structure</u> we obtain the Moyal-Weyl $*$ product, ref. [6], if we start from the basis of completely symmetrized monomials:

$$\begin{aligned}(f * g)(x) &= e^{\frac{i}{2}\frac{\partial}{\partial x^i}\theta^{ij}\frac{\partial}{\partial y^j}} f(x)g(y)\Big|_{y\Rightarrow x} \\ &= \int d^n y\, \delta^n(x-y) e^{\frac{i}{2}\frac{\partial}{\partial x^i}\theta^{ij}\frac{\partial}{\partial y^j}} f(x)g(y).\end{aligned} \tag{11}$$

For the <u>Lie structure</u> we can use the Baker-Campbell-Hausdorf formula:

$$e^{ik\cdot\hat{x}}e^{ip\cdot\hat{x}} = e^{i(k+p+\frac{1}{2}g(k,p))\cdot\hat{x}}. \tag{12}$$

This defines $g(k,p)$.

$$(f * g)(x) = e^{\frac{i}{2}x\cdot g(i\frac{\partial}{\partial y}, i\frac{\partial}{\partial z})} f(y)g(z)\Big|_{\substack{y\to x\\ z\to x}}. \tag{13}$$

For the <u>quantum plane</u> we consider the example of the Manin plane

$$\hat{x}\hat{y} = q\hat{y}\hat{x}, \tag{14}$$
$$(f * g)(x) = q^{-x'\frac{\partial}{\partial x'}y\frac{\partial}{\partial y}} f(x,y)g(x',y')\Big|_{\substack{x'\to x\\ y'\to y}}.$$

It is natural to use the elements of $^\diamond\mathcal{A}_x$ as objects in physics. Fields of a field theory will be such objects.

$$\phi(x) \in {}^\diamond\mathcal{A}_x. \tag{15}$$

The product of fields will always be the * product.

To formulate field equations we introduce derivatives. On the algebra $\mathcal{A}_{\hat{x}}$ this can be done on purely algebraic grounds. We have to extend the algebra $\mathcal{A}_{\hat{x}}$ by algebraic elements $\hat{\partial}_i$, ref. [7].

A generalized Leibniz rule will play the role of algebraic relations. Leibniz rule:

$$(\hat{\partial}_i\hat{f}\hat{g}) = (\hat{\partial}_i\hat{f})\hat{g} + O_i^l(\hat{f})\hat{\partial}_l\hat{g} \quad : \mathcal{R}_{\hat{x},\hat{\partial}}. \tag{16}$$

From the law of assotiativity in $\mathcal{A}_{\hat{x}}$ follows that the operation O has to be an algebra homomorphism:

$$O_j^i(\hat{f}\hat{g}) = O_l^i(\hat{f})O_j^l(\hat{g}). \tag{17}$$

But we shall restrict the Leibniz rule by an even stronger requirement. The ideal generated by the $\mathcal{R}_{\hat{x},\hat{x}}$ relations has to remain a two-sided ideal in the larger algebra generated by $\hat{x}$ and $\hat{\partial}$. This leads to socalled consistency relations.

Finally $\mathcal{R}_{\hat{\partial},\hat{\partial}}$ relations have to be defined. As conditions we consider the $\hat{\partial}$ subalgebra, demand PBW and derive consistency relations from $\mathcal{R}_{\hat{\partial},\hat{\partial}}$ and the Leibniz rule as before.

Derivatives defined that way induce a map from $\mathcal{A}_{\hat{x}}$ to $\mathcal{A}_{\hat{x}}$:

$$\hat{f} \in \mathcal{A}_{\hat{x}}\ , \quad (\hat{\partial}_i\hat{f}) \in \mathcal{A}_{\hat{x}}, \tag{18}$$
$$(\hat{\partial}_i\hat{f}) = \hat{\partial}_i\hat{f} - O_i^l(\hat{f})\hat{\partial}_l.$$

This algebraic concept of derivatives has been explained in ref[] and applied to quantum planes. Following the same strategy derivatives can be defined for the canonical structure as well.

For the rest of this talk we will restrict ourselves to the canonical case only.

The Leibniz rule for the canonical case is the usual one:

$$\hat{\partial}_i \hat{x}^j = \delta_i^j + \hat{x}^j \hat{\partial}_i. \tag{19}$$

It satisfies all the consistency relations of ref [].

As explained above, the derivatives induce a map on the algebra $\mathcal{A}_{\hat{x}}$:

$$\hat{f} \in \mathcal{A}_{\hat{x}} : \hat{f} \to [\hat{\partial}_i, \hat{f}] \in \mathcal{A}_{\hat{x}}. \tag{20}$$

This is the relation that we shall use to define derivatives on fields. For this purpose we map $\hat{\partial}$ to ${}^\diamond\mathcal{A}_x$. From (20) follows that it becomes the usual derivative in ${}^\diamond\mathcal{A}_x$:

$$f(x) \to \partial_i f(x). \tag{21}$$

From the definition of the * product follows:

$$\partial_i (f * g) = \partial_i f * g + f * \partial_i g. \tag{22}$$

This is the Leibniz rule (20) when mapped to the ${}^\diamond\mathcal{A}_x$ algebra. As a consequence of (20) and (7) we find that

$$\hat{x}^i - i\theta^{ij} \hat{\partial}_j \tag{23}$$

commutes with all coordinates. For invertible θ^{ij} this can be used to define the action of the derivative entirely in $\mathcal{A}_{\hat{x}}$

$$\hat{\partial}_i = -i\theta_{ij}^{-1} \hat{x}^j. \tag{24}$$

Tanslated to the ${}^\diamond\mathcal{A}_x$ algebra this implies:

$$\partial_i f(x) = -i\theta_{ij}^{-1} [x^j \overset{*}{,} f]. \tag{25}$$

As a consequence we derive

$$\hat{\partial}_j \hat{\partial}_k - \hat{\partial}_k \hat{\partial}_j = -i\theta_{jk}^{-1} \quad : \mathcal{R}_{\hat{\partial},\hat{\partial}}. \tag{26}$$

This $\mathcal{R}_{\hat{\partial},\hat{\partial}}$ relation satisfies all the requirements of (7).

To formulate a Lagrangian field theory we have to learn how to integrate. Whereas it was easier to formulate derivatives on objects of $\mathcal{A}_{\hat{x}}$ it is easier to formulate integration on objects of ${}^\diamond\mathcal{A}_x$. For the canonical structure we define:

$$\int \hat{f} = \int d^n x \, f(x), \qquad \hat{f} \in \mathcal{A}_{\hat{x}}, f \in {}^\diamond\mathcal{A}_x. \tag{27}$$

This is a linear map of the algebra $\mathcal{A}_{\hat{x}}$ into $\mathbb{C}$

$$S : \mathcal{A}_{\hat{x}} \to \mathbb{C}, \tag{28}$$
$$S(c_1\hat{f} + c_2\hat{g}) = c_1 \int \hat{f} + c_2 \int \hat{g},$$

and it has the trace property:

$$\int \hat{f}\hat{g} = \int \hat{g}\hat{f}. \tag{29}$$

This can be verified explicitely using the definition of the * product:

$$\int f * g = \int g * f = \int d^n x \ f(x)g(x). \tag{30}$$

For the quantum space structure the definition (27) for the integral does not have the trace property. There is, however, a measure for the integration that leads to an integral with the trace property.

$$\int \hat{f} \equiv \int d^n x \ \mu(x)f(x) \tag{31}$$

For the Manin plane we can verify explicitely that the measure

$$\mu(x, y) = \frac{1}{xy} \tag{32}$$

has this property.

In general we can construct Hilbert space representations of the algebra and define the integral as the trace. This will lead to infinite sums that can be interpreted as Riemannian sums for an integral and lead to the respective measure for the integration.

2 Gauge theories

Our aim is to formulate gauge theories. They will be based on a Lie algebra:

$$[T^a, T^b] = i f_c^{ab} T^c. \tag{33}$$

In a usual gauge theorie on $\mathbb{R}^n$ the fields will span a representation of the Lie algebra and transform under an infinitesimal gauge transformation:

$$\delta_{\alpha^0}\psi(x) = i\alpha^0(x)\psi(x). \tag{34}$$

The transformation parameters are Lie algebra valued:

$$\alpha^0(x) = \alpha_a^0 T^a \tag{35}$$

and consequently:

$$\begin{aligned}(\delta_{\alpha^0}\delta_{\beta^0} - \delta_{\beta^0}\delta_{\alpha^0})\psi &= -(\beta^0\alpha^0 - \alpha^0\beta^0)\psi \\ &= i(\alpha^0 \times \beta^0)\psi = \delta_{\alpha^0\times\beta^0}\psi, \\ \alpha^0 \times \beta^0 &\equiv \alpha^0_a\beta^0_b f^{ab}_c T^c.\end{aligned} \tag{36}$$

covariant derivatives are defined with the help of a Lie algebra valued gauge field a:

$$\begin{aligned}\mathcal{D}_i\psi &= (\partial_i - ia_i)\psi, \\ a_i &= a^a_i T_a.\end{aligned} \tag{37}$$

To obtain:

$$\delta_{\alpha^0}\mathcal{D}_i\psi = i\alpha^0\mathcal{D}_i\psi \tag{38}$$

we have to demand:

$$\begin{aligned}\delta a_i &= \partial_i\alpha^0 + i[\alpha^0, a_i], \\ \delta a_{i,a} &= \partial_i\alpha^0_a - \alpha^0_b f^{bc}_a a_{i,c}.\end{aligned} \tag{39}$$

To formulate a gauge theory on a non-commutative space we start with fields $\psi(x)$ that are elements of $^\diamond\mathcal{A}_{\hat{x}}$ and again span a representation of the Lie algebra (33). We demand the transformation law:

$$\delta_\alpha\psi(x) = i\alpha(x) * \psi(x) \tag{40}$$

in analogy to (34). But now we cannot demand α to be Lie algebra valued, we shall assume it to be enveloping algebra valued:

$$\alpha(x) = \alpha^0_a(x)T^a + \alpha^1_{ab}(x) :T^aT^b: + \ldots + \alpha^{n-1}_{a_1\ldots a_n}(x) :T^{a_1}\cdot\ldots\cdot T^{a_n}: + \cdots \tag{41}$$

This is in analogy to (35). We have adopted the :: notation for the basis elements of the enveloping algebra. We shall use the symmetrized polynomials as a basis:

$$\begin{aligned}:T^a: &= T^a, \\ :T^aT^b: &= \frac{1}{2}(T^aT^b + T^bT^a) \text{ etc.}\end{aligned} \tag{42}$$

In analogy to (36) we find

$$(\delta_\alpha\delta_\beta - \delta_\beta\delta_\alpha)\psi = [\alpha \stackrel{*}{,} \beta] * \psi. \tag{43}$$

Naturally, $[\alpha \stackrel{*}{,} \beta]$ will be an enveloping algebra valued element of $^\diamond\mathcal{A}_x$.

The unpleasant fact of the definition (41) of an enveloping algebra valued transformation parameter is that it depends on an infinite set of parameter fields $\alpha^n(x)$. In physics we would have to deal with an infinite set of fields when defining a covariant derivative, something we try to avoid.

However, a gauge transformation can be realized by transformation parameters that depend on x via the parameter field $\alpha^0(x)$, the gauge field $a_{i,a}(x)$ and their derivatives only. In the notation of eqn (41) we have

$$\alpha^n_{a_1\dots a_{n+1}}(x) = \alpha^n_{a_1\dots a_{n+1}}(\alpha^0_a(x), a^0_{i,a}(x), \partial_i \alpha^0_a(x), \dots). \tag{44}$$

Transformation parameters that are restricted that way we shall denote $\Lambda_{\alpha^0}(x)$. These parameters can be constructed in such a way that eqn (36) holds:

$$\begin{aligned} \delta_{\alpha^0}\psi(x) &= i\Lambda_{\alpha^0(x)}(x) * \psi(x), \\ (\delta_{\alpha^0}\delta_{\beta^0} - \delta_{\beta^0}\delta_{\alpha^0})\psi &= \delta_{\alpha^0\times\beta^0}\psi, \\ (\alpha^0 \times \beta^0)_a &= \alpha^0_b \beta^0_c f^{bc}_a. \end{aligned} \tag{45}$$

This together with the * product is the defining equations for the gauge transformations. That such parameters $\Lambda_{\alpha^0}(x)$ can be found is not obvious, it's rather a miracle in our present understanding of such gauge theories. Their existence is a consequence of the Seiberg-Witten map.

In the second variation of ψ we also have to account for the variation of Λ_{α^0} as it depends on $a_{i,a}$:

$$\begin{aligned} (\delta_{\alpha^0}\delta_{\beta^0} - \delta_{\beta^0}\delta_{\alpha^0})\psi &= i(\delta_{\alpha^0}\Lambda_{\beta^0} - \delta_{\beta^0}\Lambda_{\alpha^0}) * \psi + [\Lambda_{\alpha^0} \stackrel{*}{,} \Lambda_{\beta_0}] * \psi, \\ &= \delta_{\alpha^0\times\beta^0}\psi = i\Lambda_{\alpha^0\times\beta^0} * \psi \end{aligned} \tag{46}$$

We shall construct Λ_{α^0} in a power series expansion in θ. To illustrate the method we expand Λ_{α^0} to first order in θ

$$\Lambda_{\alpha^0} = \alpha^0_a T^a + \theta^{ij}\Lambda^1_{\alpha^0,ij} + \dots, \tag{47}$$

To be consistent we expand the * product in (46) also to first order in θ and compare powers of θ. the θ-independent term defines $\alpha^0 \times \beta^0$ as we have used it in (45). This had to be expected, this order is exactly the commutative case. To first order we obtain the equation:

$$\begin{aligned} \theta^{ij}\big((\delta_{\alpha^0}\Lambda^1_{\beta^0,ij} - \delta_{\beta^0}\Lambda^1_{\alpha^0,ij}) - i([\alpha^0, \Lambda^1_{\beta^0,ij}] - \\ - [\beta^0, \Lambda^1_{\alpha^0,ij}])\big) + \frac{1}{2}\partial_i\alpha^0_a\partial_j\beta^0_b : T^aT^b := \theta^{ij}\Lambda^1_{\alpha^0\times\beta^0,ij}. \end{aligned} \tag{48}$$

This equation has the solution:

$$\theta^{ij}\Lambda^1_{\alpha^0,ij} = \frac{1}{2}\theta^{ij}(\partial_i\alpha^0_a)a_{j,b} : T^aT^b : . \tag{49}$$

We see that Λ^1 is of second order in the generators T of the Lie algebra. The structure of eqn (46) allows a solution where Λ^n, the term in (47) of order $n-1$ in θ, is a polynomial of order n in T.

$$\Lambda_{\alpha^0} = \alpha_a^0 T^a + \frac{1}{2}\theta^{ij}(\partial_i \alpha_a^0) a_{j,b} : T^a T^b : + \dots \tag{50}$$

In a next step in the formulation of a gauge theory we introduce covariant derivatives. Eqn (24) shows that we can relate this problem to the construction of covariant coordinates. We try to define such coordinates with the help of a gauge field, in the same way as we did it for derivatives in eqn (37):

$$\begin{aligned} X^i &= x^i + A^i(x), \\ \delta_{\alpha^0} X^i * \psi &= i\Lambda_{\alpha^0} * X^i * \psi. \end{aligned} \tag{51}$$

This leads to a transformation law for the gauge field $A^i(x)$:

$$\delta A^i = -i[x^i \stackrel{*}{,} \Lambda_{\alpha^0}] + i[\Lambda_{\alpha^0} \stackrel{*}{,} A^i]. \tag{52}$$

We have to assume that A^i is enveloping algebra valued but we try to make an ansatz where all the coefficient functions only depend on $a_{i,a}$ and its derivatives:

$$\begin{aligned} A^i(x) &= A_a^{i,0}(x)T^a + A_{ab}^{i,1}(x) : T^a T^b : + \dots + A_{a_1\dots a_n}^{i,n-1}(x) : T^{a_1} \cdot \dots \cdot T^{a_n} : + \\ A^{i,n} &= A^{i,n}(a_{i,a}, \partial a_{i,a}, \dots). \end{aligned} \tag{53}$$

Now we expand (52) in θ, demand $A^{i,n}$ to be a polynomial of order n in θ and solve eqn (52),

$$\begin{aligned} A^i(x) &= \theta^{ij} V_j, \\ V_j(x) &= a_{j,a}T^a - \frac{1}{2}\theta^{ln} a_{l,a}(\partial_n a_{j,b} + F_{nj,b} : T^a T^b : + \dots, \\ F_{nj,b} &= \partial_n a_{j,b} - \partial_j a_{n,b} + f_b^{cd} a_{n,c} a_{j,d}. \end{aligned} \tag{54}$$

This together with (41) is known as Seiberg-Witten map for an abelian gauge group. We have constructed it for an arbitrary non-abelian gauge group as well.

Covariant derivatives follow from (37)

$$\begin{aligned} \mathcal{D}_i * \psi &= (\partial_i - iV_i) * \psi, \\ \delta_{\alpha^0} \mathcal{D}_i * \psi &= i\Lambda_{\alpha^0} * \mathcal{D}_i * \psi. \end{aligned} \tag{55}$$

We now proceed with the definition of tensors as in a usual gauge theory, keeping in mind (27)

$$\tilde{F}_{ij} = \mathcal{D}_i * \mathcal{D}_j - \mathcal{D}_j * \mathcal{D}_i - i\theta_{ij}^{-1}. \tag{56}$$

The transformation law of the tensor is

$$\delta_{\alpha^0}\tilde{F}_{ij} = i[\Lambda_{\alpha^0} \overset{*}{,} \tilde{F}_{ij}]. \tag{57}$$

This can be verified from (52) and the definition of $\tilde{F}$.

To first order in θ we find:

$$\tilde{F}_{ij} = F_{ij,a}T^a + \theta^{ln}(F_{il,a}F_{jn,l} - \frac{1}{2}a_{l,a}(2\partial_n F_{ij,b} + a_{n,c}F_{ij,d}f_e^{cd})) : T^aT^b : + \ldots. \tag{58}$$

We see that new "contact" terms appear in the field strength $\tilde{F}$.

A good candidate for a Lagrangian is

$$L = \frac{1}{4}\mathrm{Tr}F_{ij} * F^{ij}. \tag{59}$$

The trace is taken in the representation space of the generators T. The Lagrangian (59) is not invariant because the * product is not commutative:

$$\delta L = \frac{1}{4}\mathrm{Tr}\, i[\Lambda_{\alpha^0} \overset{*}{,} L]. \tag{60}$$

We know, however, that the integral has the trace property (31). This allows us to define the invariant action:

$$\begin{aligned} W &= \frac{1}{4}\int \mathrm{Tr}F_{ij} * F^{ij} \\ &= \frac{1}{4}\int \mathrm{Tr}F_{ij}F^{ij}. \end{aligned} \tag{61}$$

This action depends on the gauge field $a_{i,a}$ and its derivatives only. It can be considered as a gauge-invariant object if $a_{i,a}$ transforms according to (39). this implies that W satisfies the Ward identities.

$$\begin{aligned} \delta_{\alpha^0(x)}W &= 0, \\ \delta_{\alpha^0(x)} &= -\alpha_a^0(x)\left(\frac{\partial}{\partial x^i}\delta_d^a + a_{i,b}(x)f_d^{ab}\right)\frac{\delta}{\delta a_{i,d}(x)}. \end{aligned} \tag{62}$$

The Lagrangian expanded to all orders in θ, is a non-local object. It remains to be seen if it is acceptable for a quantum field theory or if it has to be viewed as an effective Lagrangian, ref. [8].

References

1. B. Jurčo, S. Schraml, P. Schupp and J. Wess, *Eur. Phys. J. C* **17**, 521 (2000), hep-th/0006246; J. Madore, S. Schraml, P. Schupp and J. Wess, *Eur. Phys. J. C* **16**, 161 (2000), hep-th/0001203; B. Jurčo, P. Schupp,

Eur. Phys. J. C **14**, 367 (2000), hep-th/0001032; B. Jurčo, P. Schupp and J. Wess, *Nucl. Phys. B* **584**, 784 (2000), hep-th/0005005; B. Jurčo, P. Schupp and J. Wess, Nonabelian noncommutative gauge theory and Seiberg-Witten map, in preparation.

2. N. Seiberg and E. Witten, *JHEP* **9909**, 032 (1999), hep-th/9908142.
3. A. Dimakis, J. Madore, *J.Math.Phys.* **37**, 4647 (1996); M. Dubois-Violette, R. Kerner, J. Madore, *Gauge bosons in a noncommutative geometry, Phys.Lett.B* **217**, 485 (1989); J. Hoppe, *Int.J.Mod.Phys. A* **4**, 5235 (1989); J. Madore, *An Introduction to Noncommutative Differential Geometry and it Physical Applications*, 2nd Edition, Cambridge University Press (1999);
   B. deWit, J. Hoppe, H. Nicolai, *Nucl.Phys.B* **305** [FS 23], 545 (1988); D. Kabat, W. Taylor IV, *Adv.Theor.Phys.* **2**, 181 (1998), hep-th 9711078.
4. J. Wess, q-deformed Heisenberg Algebras, in *Proceedings of the 38. Internationale Universitätswochen für Kern- und Teilchenphysik, Schladming, January 1999*, H. Gausterer, H. Grosse and L. Pittner, Eds., Lect. Notes in Phys. no. 543, Springer-Verlag (2000), math-ph/9910013.
5. F. Bayen, M. Flato, C. Fronsdal, A. Lichnerowicz, D. Sternheimer, ;, Ann. Physics **111**, 61 (1978)M. Kontsevitch, Deformation quantization of Poisson manifolds, I, q-alg/9709040; D. Sternheimer, Deformation Quantization: Twenty Years After, math/9809056.
6. H. Weyl, *Z. Physik* **46**, 1 (1927);*The theory of groups and quantum mechanics*, Dover, New-York (1931), translated from *Gruppentheorie und Quantenmechanik*, Hirzel Verlag, Leipzig (1928); J. E. Moyal, *Proc. Cambridge Phil. Soc.***45**, 99 (1949).
7. J. Wess and B. Zumino, *Nucl. Phys. Proc. Suppl. B* **18**, 302 (1991).
8. L. Bonora, M. Schnabl, M. M. Sheikh-Jabbari and A. Tomasiello, Noncommutative SO(n) and Sp(n) gauge theories, hep-th/0006091; I. Chepelev, R. Roiban, Convergence Theorem for Non-commutative Feynman Graphs and Renormalization, hep-th/0008090; A. Bichl, J.M. Grimstrup, V. Putz, M. Schweda, Perturbative Chern-Simons Theory on noncommutative $\mathbb{R}^3$, hep-th/0004071; A. Bichl, J.M. Grimstrup, H. Grosse, L. Popp, M. Schweda, R. Wulkenhaar, The Superfield Formalism Applied to the Non-commutative Wess-Zumino Model, hep-th/0007050.

ALGEBRA OF DIFFERENTIAL FORMS WITH EXTERIOR DIFFERENTIAL $d^3 = 0$ IN DIMENSION ONE

V. ABRAMOV, N. BAZUNOVA

Institute of Pure Mathematics, University of Tartu, Vanemuise 46, Tartu 51 014, Estonia,

E-mail:abramov@ut.ee, nadegda@ut.ee

In this work, we construct the algebra of differential forms with exterior differential d satisfying $d^3 = 0$ on one-dimensional space. We prove that this algebra is a graded q-differential algebra where q is a cubic root of unity. Since $d^2 \neq 0$ the algebra of differential forms is generated not only by the first order differential dx but also by the second order differential d^2x of a coordinate x. We study the bimodule generated by this second order differential, and show that its structure is similar to the structure of bimodule generated by the first order differential dx in the case of the anyonic line.

1 Introduction

It is well known that one of possible ways to generalize a classical Grassmann algebra is to increase the power of nilpotency of its generators. This means that a possible generalization can be defined as an associative unital algebra generated by $\theta_1, \theta_2, \ldots, \theta_n$ satisfying $\theta_i^N = 0, N > 2, i = 1, 2, \ldots, n$. In order to obtain an analogue of classical Grassmann algebra, one should add commutation relations between generators to the above mentioned definition, and this leads to different generalizations of classical Grassmann algebra in the case $n > 1$. If one imposes the commutation relations $\theta_i\theta_j = q\,\theta_j\theta_i\,(i < j)$, where q is an N-th primitive root of unity, then the corresponding structure is called a *generalized Grassmann algebra*[1,3]. Another approach based on ternary commutation relations and the representation of group of cyclic permutations Z_3 by cubic roots of unity was developed in[2,3] and the corresponding structure is called a *ternary Grassmann algebra.* It should be mentioned that both generalizations of classical Grassmann algebra mentioned above coincide in the trivial case of one generator θ satisfying $\theta^N = 0$, and this algebra known as anyonic line[4] is closely related to fractional supersymmetry[5].

A classical Grassmann algebra underlies an exterior calculus on a smooth manifold with exterior differential d satisfying $d^2 = 0$. Therefore the above mentioned generalizations of Grassmann algebra raise a natural question of possible generalizations of classical exterior calculus to one with exterior differential satisfying $d^N = 0,\ N > 2$. From an algebraic point of view, an adequate algebraic structure underlying an exterior calculus is the notion of graded differential algebra. Hence one can generalize a classical exterior calculus with the help of an appropriate generalization of graded differential algebra. This generalization called *graded q-differential algebra* was proposed and studied by M. Dubois-Violette in the series of papers [6], where the author constructed several realizations of graded q-differential algebra.

According to the definition given by M. Dubois-Violette, a graded q-differential algebra is an associative unital $\mathbf{N}$-graded algebra endowed with linear endomorphism d of degree 1 satisfying $d^N = 0$ and the graded q-Leibniz rule

$$d(\alpha\beta) = d(\alpha)\,\beta + q^a\;\alpha\,d(\beta), \tag{1}$$

where a is the grading of an element α, and q is a primitive N-th root of unity.

From a point of view of differential geometry, the above definition can be used to generalize the de Rham complex on a finite-dimensional smooth manifold. This question was studied in the series of papers [7], [8], where the authors used the notion of ternary Grassmann algebra (considering the first non-trivial generalization of classical exterior calculus corresponding to $N = 3$) to construct the algebra of differential forms. If one assumes $d^2 \neq 0$ replacing it by $d^3 = 0$, then in order to construct a self-consistent theory of differential forms, it is necessary to add to the first order differentials of local coordinates $dx^1, dx^2, \ldots, dx^m$ a set of *second order differentials* $d^2x^1, d^2x^2, \ldots, d^2x^m$ (and higher order differentials in the case of $N > 3$). Appearance of higher order differentials, which are missing in classical exterior calculus, is a peculiar property of a proposed generalization of differential forms. This has as a consequence certain problems. It is well known that in classical exterior calculus functions commute with differentials, i.e.

$$f\,dx^i = dx^i\,f, \qquad \forall i = 1, 2, \ldots, m, \tag{2}$$

where f is a smooth function on a manifold, or, from an algebraic point of view, the space of 1-forms is a free finite bimodule over the algebra of smooth functions generated by the first order differentials, and (2) shows how its left and right structures are related to each other. Now, assuming that d is no more classical exterior differential, i.e. $d^2 \neq 0$, and differentiating (2) with regard to graded q-Leibniz rule (1), one immediately obtains

$$d^2x^i\,f = f\,d^2x^i + [df, dx^i]_q, \tag{3}$$

where $[df, dx^i]_q = df\,dx^i - q\;dx^i\,df$, and we assign grade 1 to first order differentials dx^i. The above relations (3) are *not homogeneous* in the sense that the commutation relations between functions and second order differentials include first order differentials as well.

In this paper, we show that commutation relations between functions and second order differentials can be made homogeneous, i.e. they will not include first order differentials, if we take a non-commutative geometry point of view. In order to make our construction more transparent, we begin with the simplest case of one-dimensional space. We construct the differential forms on this space with exterior differential satisfying $d^3 = 0$ and show that they form a graded q-differential algebra. In our construction, we use a *coordinate calculi* developed in [9]. Then we study the commutation relations between functions and second order differentials and show that the requirement of homogeneity (i.e. vanishing of the second term in the right-hand side of (3)) implies that a coordinate calculi we are considering is the differential calculus on the anyonic line, which means that the commutation relations between functions and second order differentials are homogeneous, i.e they take on the form $d^2x\, f = f\, d^2x$, only in the case of the anyonic line (in dimension one).

2 Graded q-differential algebra on one-dimensional space

In this section, we construct a graded q-differential algebra of differential forms with exterior differential d satisfying $d^3 = 0$ in dimension one. We study the structure of a bimodule of second order differentials and show that it is homogeneous in the case of the anyonic line. In this section, q is a primitive cube root of unity, i.e. $q^3 = 1$.

Let $\mathcal{A}$ be a free unital associative $\mathbf{C}$-algebra generated by a variable x. If $\xi : \mathcal{A} \to \mathcal{A}$ is a homomorphism of this algebra and $\partial : \mathcal{A} \to \mathcal{A}$ is a linear map such that

$$\partial(x) = 1, \qquad \partial(f\, g) = \partial(f)\, g + \xi(f)\, \partial(g), \quad \forall f, g \in \mathcal{A}, \tag{4}$$

then according to coordinate calculi the map

$$d : f \to \partial(f)\, dx, \tag{5}$$

where dx is the first order differential of a variable x, is a coordinate differential, i.e. d is a linear map $d : \mathcal{A} \to D_\xi(\mathcal{A})$ satisfying the Leibniz rule

$$d(f\, g) = d(f)\, g + f\, d(g), \tag{6}$$

and $D_\xi(\mathcal{A})$ is a free left module over $\mathcal{A}$ generated by dx with the right module structure defined by the commutation rule

$$dx\, f = \xi(f)\, dx. \tag{7}$$

If $f = \sum_m \alpha_m x^m$ is an element of $\mathcal{A}$, then the derivative $\partial : \mathcal{A} \to \mathcal{A}$ can be written in explicit form

$$\partial(f) = \sum_{m\geq 1} \sum_{k=0}^{m-1} \alpha_m \, \xi^k(x) \, x^{m-k-1}. \tag{8}$$

In order to construct a generalization of exterior calculus with exterior differential d satisfying $d^2 \neq 0$, let us introduce the second order differential d^2x. Let $(dx)^k \, (d^2x)^m$ be monomials composed from first and second order differentials, where k, m nonnegative integers. As usual, we assume that $(dx)^0 = (d^2x)^0 = 1$ where 1 is the unit element of $\mathcal{A}$. Let $\Omega_\xi(\mathcal{A})$ be a free left module over the algebra $\mathcal{A}$ generated by the above introduced monomials. It is easy to see that $\mathcal{A} \subset \Omega_\xi(\mathcal{A})$, $D_\xi(\mathcal{A}) \subset \Omega_\xi(\mathcal{A})$. The module $\Omega_\xi(\mathcal{A})$ becomes an unital associative algebra if we define a multiplication law on $\Omega_\xi(\mathcal{A})$ by the relations

$$dx \, x = \xi(x) \, dx, \qquad d^2x \, f = \xi(f) \, d^2x + [\partial, \xi]_q \, (f) \, (dx)^2, \tag{9}$$

$$(dx)^3 = 0, \qquad d^2x \, dx = q^2 \, dx \, d^2x, \tag{10}$$

where $[\partial, \xi]_q(f) = \partial\,(\xi(f)) - q\,\xi\,(\partial(f))$, and $f \in \mathcal{A}$.

Analyzing the defining commutation relations (9,10) of the algebra $\Omega_\xi(\mathcal{A})$, one can note that the bimodule $D_\xi(\mathcal{A})$ of a coordinate calculi is a submodule of $\Omega_\xi(\mathcal{A})$, and the relation (7) between its left and right structures follows from the commutation relation between the first order differential dx and a variable x. The second remark concerns the structure of the algebra $\Omega_\xi(\mathcal{A})$ with respect to the second order differential. The relations (9, 10) show that any power of the second order differential does not vanish. Hence the algebra $\Omega_\xi(\mathcal{A})$ is an infinite-dimensional vector space, and an arbitrary element ω of this algebra can be written in the form

$$\omega = \sum_{m\geq 0} \sum_{k=0}^{2} f_{km} \, (dx)^k \, (d^2x)^m, \qquad f_{km} \in \mathcal{A}. \tag{11}$$

We shall call elements of the algebra $\Omega_\xi(\mathcal{A})$ differential forms on one-dimensional space generated by a variable x. The algebra of differential forms $\Omega_\xi(\mathcal{A})$ becomes a $\mathbf{N}$-graded algebra if we assign grading zero to each element of the algebra $\mathcal{A}$ and grading $k + 2m$ to monomial $(dx)^k (d^2x)^m$, i.e. we assume that a variable x has grading zero and the gradings of the differentials dx, d^2x are respectively 1, 2. Then the algebra of differential forms splits into the direct sum of its subspaces

$$\Omega_\xi(\mathcal{A}) = \bigoplus_{m=0}^{\infty} \Omega_\xi^m(\mathcal{A})$$

where

$$\begin{aligned}
\Omega^0_\xi(\mathcal{A}) &= \mathcal{A}, \\
\Omega^{2k}_\xi(\mathcal{A}) &= \{f\,(d^2x)^k + h\,(dx)^2\,(d^2x)^{k-1} : \; f,h \in \mathcal{A}\}, \; k=1,2,\ldots \quad (12)\\
\Omega^{2k+1}_\xi(\mathcal{A}) &= \{f\,dx\,(d^2x)^k : f \in \mathcal{A}\}, \quad k=0,1,\ldots. \quad (13)
\end{aligned}$$

We now extend the differential (5) of the coordinate calculi to the whole algebra $\Omega_\xi(\mathcal{A})$ as follows:

$$d(\omega) = (\partial\, f - h)\,dx\,(d^2x)^k, \qquad \omega \in \Omega^{2k}_\xi(\mathcal{A})$$

$$d(\omega) = f\,(d^2x)^{k+1} + \partial f\,(dx)^2\,(d^2x)^k, \quad \omega \in \Omega^{2k+1}_\xi(\mathcal{A}).$$

We shall call the above defined differential an exterior differential on the algebra of differential forms $\Omega_\xi(\mathcal{A})$. It follows from the definition that exterior differential is an endomorphism of degree 1 of the algebra $\Omega_\xi(\mathcal{A})$, i.e. $d : \Omega^m_\xi(\mathcal{A}) \to \Omega^{m+1}_\xi$.

Proposition 1 *The algebra of differential forms $\Omega_\xi(\mathcal{A})$ is a graded q-differential algebra with respect to exterior differential d, i.e. for any two differential forms ω, θ the exterior differential d satisfies*

$$d^3(\omega) = 0, \tag{14}$$

$$d(\omega\theta) = d(\omega)\,\theta + q^{|\omega|}\,\omega\, d(\theta), \tag{15}$$

where $|\omega|$ is the grading of a form ω.

Proof. It follows from (12,13) that any differential form ω can be decomposed into the sum of two forms ω_o, ω_e respectively of odd and even grading, where

$$\omega_e = \sum_{k\geq 1} [f_k\,(d^2x)^k + h_{k-1}\,(dx)^2\,(d^2x)^{k-1}], \tag{16}$$

$$\omega_o = \sum_{k\geq 0} g_k\,dx\,(d^2x)^k.$$

From the definition of the exterior differential, it also follows that a differential form of even grading(16) is d-closed ($d\omega_e = 0$) if it satisfies the following condition

$$\partial f_k = h_{k-1}. \tag{17}$$

Now it is easy to show that any form of odd grading is d^2-closed, i.e. $d^2\omega_o = 0$. Indeed, applying the exterior differential d to form ω_o, one obtains the form of even grading

$$d\omega_o = \sum_{k\geq 0} [\, g_k \, (d^2x)^{k+1} + \partial g_k \, (dx)^2 \, (d^2x)^k],$$

which is d-closed according to (17). Differentiating form (16) of even grading twice, one obtains the form

$$d^2\omega_e = \sum_{k\geq 1} [\, (\partial f_k - h_{k-1}) \, (d^2x)^{k+2} + (\partial^2 f_k - \partial h_{k-1}) \, (dx)^2 \, (d^2x)^{k+1}], \quad (18)$$

which is d-closed. Thus the cube nilpotency (14) of the exterior differential is proved. The q-Leibniz rule (15) can be verified by a direct calculation.

A homomorphism ξ plays a role of a parameter in the structure of the algebra of differential forms, and, choosing particular homomorphism, we can specify the structure of $\Omega_\xi(\mathcal{A})$. We remind that, according to the definition, the algebra of differential forms $\Omega_\xi(\mathcal{A})$ is a free left module over the algebra $\mathcal{A}$ generated by the monomials $(dx)^k(d^2x)^m$ with associative multiplication law determined by the relations (9, 10). Actually this algebra is generated by three generators x, dx, d^2x. Hence its structure will be more transparent if we define it by means of *commutation relations* imposed on the generators x, dx, d^2x. The only relation in (9, 10), which is not a commutation relation between generators, is the second relation in (9) containing an arbitrary element f of the algebra $\mathcal{A}$. The reason why it contains an arbitrary element f is that in contrast to the commutation relation $dx\, x = \xi(x)\, dx$ the right-hand side of this relation is not a homomorphism of the algebra $\mathcal{A}$ because of the non-homogeneous term $[\partial, \xi]_q(f)$. Obviously imposing the condition

$$[\partial, \xi]_q(f) = 0, \quad (19)$$

we can replace the second relation in (9) by the commutation relation

$$d^2x\, x = \xi(x)\, d^2x, \quad (20)$$

which has exactly the same form as the first one involving the first order differential.

Actually the choice for a homomorphism ξ is not very wide. Indeed every homomorphism ξ of the algebra $\mathcal{A}$ is determined by an element $h_\xi \in \mathcal{A}$ such that $\xi(x) = h_\xi$. From (4) it follows that if the derivative ∂ determined by a homomorphism ξ should satisfy $\partial(x^m) \sim x^{m-1}$, then $\xi(x) = \alpha_\xi\, x$ where α_ξ is a complex number. The condition (19) can be solved with regard to a homomorphism ξ, and the following proposition describes a structure, which is induced on one-dimensional space in this case.

Proposition 2 *The condition (19) is satisfied if and only if* $\xi(x) = q\,x$. *This solution leads to the q-differential calculus on anyonic line with derivative*

$$\partial(f) = \sum_{k\geq 1} \alpha_k \frac{x^{k-1}}{[k-1]_q!}, \quad f = \sum_{k\geq 0} \alpha_k \frac{x^k}{[k]_q!} \tag{21}$$

This means that one can consistently add the relation $x^3 = 0$ *to the relations (9, 10).*

Proof. Using the formula (8), one can find

$$\partial(\xi(f)) = \alpha_\xi \sum_m \alpha_m \sum_{k=0}^{m-1} \xi^k(h)\, h^{m-k-1},$$

$$q\,\xi(\partial(f)) = q \sum_m \alpha_m \sum_{k=0}^{m-1} \xi^k(h)\, h^{m-k-1}.$$

Thus

$$[\partial, \xi]_q(f) = (\alpha_\xi - q) \sum_m \alpha_m \sum_{k=0}^{m-1} \xi^k(h)\, h^{m-k-1}. \tag{22}$$

From the above formula it immediately follows that $\xi(x) = q\,x$ or $\xi(f(x)) = f(qx)$. Putting $\xi(x) = q\,x$ in (8), one obtains (21). It was explained in [5] that q-differential calculus determined by the derivative (21) is correctly defined at cube root of unity on one-dimensional space generated by a variable x only in the case when $x^3 = 0$. It is easy to show that the relation $x^3 = 0$ can be consistently added to the relations (9, 10).

Now the algebra of differential forms $\Omega_\xi(\mathcal{A})$ on anyonic line can be defined as an unital associative algebra generated by three generators x, dx, d^2x satisfying the following commutation relations:

$$x^3 = 0,$$

$$dx\,x = q\,x\,dx, \qquad d^2x\,x = q\,x\,d^2x \tag{23}$$

$$(dx)^3 = 0, \qquad d^2x\,dx = q^2\,dx\,d^2x. \tag{24}$$

Acknowledgments

The authors would like to acknowledge the financial support of Estonian Science Foundation under the grants No.4515 and No.1134.

References

1. N. Fleury et M. Rausch de Traubenberg, *J. Math. Phys.* **33**, 3356 (1992); *Adv. Clif. Alg.* **4**, 123, (1994).
2. R. Kerner, *C.R. Acad. Sci. Paris* **312, Sér. II**, 191 (1991); *J. Math. Phys.* **33**, 403, (1992); in *Proceedings of the Third Max Born Symposium*, 349 (1993).
3. V. Abramov, R. Kerner, B. Le Roy, *J. Math. Phys.* **38**, 1650 (1997).
4. S. Majid, in *Proceedings of 2nd Max Born Symposium*, 327, (1992).
5. J.A. de Azcárraga and A.J. Macfarlane, *J. Math. Phys.* **37**, 1115 (1996).
6. M. Dubois-Violette, *Czech. J. Phys.* **46**, 1227 (1997); LPTHE-ORSAY 97/53; LPTHE-ORSAY 97/59.
7. R. Kerner, *Lett. Math. Phys.* **36**, 441 (1996); *Class. Quant. Gravity* **14** A203, (1997).
8. R. Kerner, V. Abramov, *Rep. Math. Phys.* **43**, 179 (1999).
9. A. Borowiec, V.K. Kharchenko, *Sib. Adv. Math.* **5**, 10 (1995).

QUANTUM PROBABILITY AND THE NON–LOCALITY ISSUE IN QUANTUM THEORY

L. ACCARDI AND M. REGOLI

Centro Vito Volterra, Università degli Studi di Roma "Tor Vergata"

Three computers, with local independent choices, genereate the EPR correlations hence violating Bell's inequality.

1 Bell's inequality and its consequences according to Bell

Bell's inequality: Let A, B, C be random variables defined on the same probability space $(\Omega, \mathcal{F}, P)$ and with values in the set $\{-1, 1\}$. Denote

$$\langle AB \rangle := \int_{\Omega} A(\omega)B(\omega)P(d\omega)$$

the correlations (mean zero can be assumed without loss of generality). Then the following inequality holds:

$$|\langle AB \rangle - \langle CB \rangle| \leq 1 - \langle AC \rangle \tag{1}$$

Proof. Since the expectation value is linear

$$|\langle AB \rangle - \langle CB \rangle| = |\langle AB - CB \rangle| \tag{2}$$

Since $|\langle X \rangle| \leq \langle |X| \rangle$ and A, B, C are ± 1–valued (2) is

$$\leq \langle |AB - CB| \rangle = \langle |AB| \cdot |1 - AC| \rangle = \langle |1 - AC| \rangle = \langle 1 - AC \rangle = 1 - \langle AC \rangle$$

Corollary (Clauser-Horne-Shimony-Holt (CHSH) inequality): Let A, B, A', B' be random variables defined on the same probability space $(\Omega, \mathcal{F}, P)$ and with values in the set $\{-1, 1\}$. Then the following inequality holds:

$$|\langle AB \rangle - \langle A'B \rangle + \langle AB' \rangle + \langle A'B' \rangle| \leq 2 \tag{3}$$

Proof. Replace C with A' in (1). Then add to this what we obtain by replacing again A' with $-A'$ and B with B'.

Theorem 1: There cannot exist a stochastic process $S_a^{(1)}$, $S_b^{(2)} (a, b \in [0, 2\pi])$ defined on a probability space $(\Omega, \mathcal{F}, P)$ and with values in the set $\{\pm 1\}$, whose correlations are given by:

$$\langle S_a^{(1)} S_b^{(2)} \rangle = -\cos(a - b) \quad ; \quad a, b \in [0, 2\pi] \tag{4}$$

Proof. (4) implies that, for any $a \in [0, 2\pi]$, $S_a^{(1)} = -S_a^{(2)}$. Because of (2), if such a process exists, it should satisfy

$$|\langle S_a^{(1)} S_b^{(2)} \rangle - \langle S_c^{(1)} S_b^{(2)} \rangle| - \langle S_a^{(1)} S_c^{(2)} \rangle \leq 1 \tag{5}$$

But, because of (4), with $a = 0$, $b = \pi/2$, $c = \pi/4$, the left hand side of (5) is equal to $\sqrt{2}$.

Physical meaning (Bell [1]): "... the statistical predictions of quantum mechanics are incompatible with local predetermination ..."

2 Criticism to Bell's analysis

Observations:

(i) The contradiction, pointed out by Bell, arises only from his implicit postulate that 3 statistical correlations, coming from 3 mutually incompatible experiments, can be described within a single classical probabilistic model.

(ii) This implicit postulate is by no means a consequence of locality and reality.

Weak point of Bell's proof. The first step (2) of Bell's proof:
$\langle AB \rangle - \langle CB \rangle = \langle AB - CB \rangle$ reads as

$$\int_\Omega A(\omega)B(\omega)P_{a,b}(d\omega) - \int_\Omega C(\omega)B(\omega)P_{c,b}(d\omega) = \int_\Omega \left(A(\omega)B(\omega) - C(\omega)B(\omega)\right) P_{a,c,b}(d\omega) \ .$$

Thus, Bell's proof relies on the existence of the triple joint probability $P_{a,c,b}$. While the pair joint probabilities $P_{a,b}, P_{c,b}, \ldots$ are experimentally observable, there is no reason to postulate that the, experimentally unobservable, triple joint probabilities $P_{a,c,b}$ always exist. Indeed, it is well known from classical probability that there are constraints, i.e. compatibility conditions, which relate the pair with the triple joint probabilities and which are necessary conditions for the existence of the latter ones. This fact was first pointed out in [2].

It has been claimed that the existence of the triple joint probabilities is consequence of the "realism" assumption. This is true for "ballot box" (or Einstein) realism, not for "chemeleon realism".

3 "Chameleon (or adaptive) Reality"

Starting point: von Neumann's measurement theory

The joint evolution of system and apparatus should be taken into account to analyze measurement processes.

von Neumann's theory was extended in [3] (cf. also [4] [5]) to include in it the two basic conditions of locality and causality. To this goal the notion of "Chameleon reality" was introduced to denote those systems whose dynamics depends on the observables we want to measure (or the environment in general). For such systems what you measure is a *response to an interaction*. This is opposed to ballot box (or Einstein) reality in which you measure what was there (independently of the environment).

The law of classical statistics and probability were abstracted having in mind the "ballot box" reality model. However one can prove that some simple induction rules (e.g. the "counterfactual argument" cf. [4]) which are constantly used in classical statistics and probability, lead to wrong conclusions when applied to "chameleon reality". Therefore the statistics of adaptive systems must be based on different induction rules, hence on a different probability theory. Quantum probability is one of these non–Kolmogorovian models. Others are mathematically possible and we expect that they will find applications to other types of adaptive systems arising, for example, in medicine, in economics...

Remark 1: Such a dependence, of the dynamical evolution of a system on the observables we want to measure, is called "chameleon effect".

Remark 2: The chameleon effect can also be considered within the framework of classical mechanics.

Remark 3: A classical system realizing the chameleon effect can violate Bell's inequality while satisfying locality condition.

4 A local classical system violating Bell's inequality

(i) The classical system σ is composed of two sub-systems: σ_1 and σ_2

(ii) The states of σ_1 and σ_2 are numbers λ_1, λ_2 in $[0, 2\pi]$; the observables are ± 1–valued functions $\tilde{S}_a^{(1)}(\lambda_1)$, $\tilde{S}_b^{(2)}(\lambda_2)$ of the states (realistic description). Also a, b take values in $[0, 2\pi]$.

(iii) At time 0 the states (λ_1, λ_2), of (σ_1, σ_2), are randomly distributed with law

$$\psi_0(d\lambda_1, d\lambda_2) = (2\pi)^{-1}\delta(\lambda_1 - \lambda_2)d\lambda_1 d\lambda_2 \tag{6}$$

(iv) The subsystems σ_1 and σ_2 evolve independently (locality).

(v) Time is discretized and 1 is the instant of simultaneous measurement of both systems.

(vi) At time 1 two observables, $\tilde{S}_a^{(1)}$ of σ_1 and $\tilde{S}_b^{(2)}$ of σ_2 are measured and the result of the measurement is uniquely determined by the state (λ_1, λ_2) of the system at time 0 (pre-determination).

(vii) The (Heisenberg) time evolution of the system realizes the chameleon effect (i.e. depends on the observables we want to measure) and it is assumed to have the form

$$P(f \otimes g)(\lambda_1, \lambda_2) = T'_{1,a}(\lambda_1) f\left(T_{1,a}(\lambda_1)\right)\ T'_{2,b}(\lambda_2) g\left(T_{2,b}(\lambda_2)\right)$$

where

$$T'_{1,a}(\lambda_1) = \sqrt{2\pi} \qquad , \qquad T'_{2,b}(\lambda_2) = \frac{\sqrt{2\pi}}{4}\Big|\cos(\lambda_2 - b)\Big|$$

Notice that the dynamics is deterministic and entirely local: the evolution of particle 1 (resp. 2) depends only on a (resp. b). It is not an automorphism because it describes the reduced evolution after averaging over the reservoir degrees of freedom of the apparatus (cf. Section 3). Notice in addition that the dual evolution P^* $(P^*\psi(F) := \psi(P(F)))$ applied to the initial probability measure (6) still gives a probability measure. Thus P is a generalized Markovian operator (in the sense that $P^*\psi$ is a probability measure not for all probability measures ψ, but only for a convex subset $\mathcal{S}_o$ of them, called the set of admissible initial states). Within this class the interpretation of P as a reduced evolution is justified. Recently [AcImRe01] it has been proved that this evolution is indeed obtained by taking partial expectation of an appropriate automorphic evolution.

Under these conditions, if the observables are defined by the conditions

$$\tilde{S}_a^{(1)}\left(T_{1,a}(\lambda_1)\right) = \operatorname{sgn}\left(\cos(\lambda_1 - a)\right) \ , \quad \tilde{S}_b^{(2)}\left(T_{2,b}(\lambda_2)\right) = -\operatorname{sgn}\left(\cos(\lambda_2 - b)\right)$$

then their correlations

$$\langle S_a^{(1)} S_b^{(2)} \rangle := \int\int P\left(\tilde{S}_a^{(1)} \otimes \tilde{S}_b^{(2)}\right)(\lambda_1, \lambda_2)\ \psi_0(d\lambda_1, d\lambda_2) = \tag{7}$$

$$= \int_0^{2\pi} \frac{d\lambda}{2\pi}\, \tilde{S}_a^{(1)}\left(T_{1,a}(\lambda_1)\right) T'_{1,a}(\lambda) \tilde{S}_b^{(2)}\left(T_{2,b}(\lambda_2)\right) T'_{2,b}(\lambda)$$

satisfy condition (4) hence they violate Bell's inequality. The Schrödinger represenatation of the correlations (7) is

$$\langle S_a^{(1)} S_b^{(2)} \rangle := \int\int \tilde{S}_a^{(1)} \otimes \tilde{S}_b^{(2)}(\lambda_1, \lambda_2)\; P^*\left(\psi_0\right)(d\lambda_1, d\lambda_2) =$$

$$= \int\int \tilde{S}_a^{(1)}(\mu_1) \tilde{S}_b^{(2)}(\mu_2) \delta(T_{1,a}^{-1}\mu_1 - T_{2,b}^{-1}\mu_2) \frac{d\mu_1 d\mu_2}{2\pi} \tag{8}$$

which, since $P^*(\psi_0)$ is easily seen to be a probability measure, shows that our model indeed produces standard correlations of ± 1–valued observables.

Notice that, in terms of $S_x^{(j)}(\lambda) := \tilde{S}_x^{(j)}\left(T_{j,x}(\lambda)\right)$ $(j = 1, 2,\ x = a, b)$ and $P_{a,b}(d\lambda) := T'_{1,a}(\lambda) T'_{2,b}(\lambda) d\lambda / 2\pi$, (7) reads as an average over an observable-dependent probability measure:

$$\langle S_a^{(1)} S_b^{(2)} \rangle = \int_0^{2\pi} S_a^{(1)}(\lambda) S_b^{(2)}(\lambda) P_{a,b}(d\lambda)$$

The measure $P_{a,b}(d\lambda)$, which arises from integration of the δ–function, is a global object, however the factorization of its density $T'_{1,a}(\lambda) T'_{2,b}(\lambda)$ reflects the local action of the environment and allows local simulation.

In summary: we have proved that the correlations (6), of our dynamical system, are equal to (5) and we use (5) for the local simulation.

Further details on this model can be found in [5], [4].

Description of the experiment:

The experiment uses three computers, A, B and C, which can be situated in different towns and exchange data via internet. Due to the identity between the right hand side of (7) and the right hand side of (6), we use the form (6) of the integral, which is simpler to implement in the computer. It consists of the following steps:

(1) Computer C generates a random number $\lambda \in [0, 2\pi]$ and it is sent to computers A and B.

(2) Computer A choses a value $a \in [0, 2\pi]$ and evaluates $T'_{1,a}(\lambda) \tilde{S}_a^{(1)}\left(T_{1,a}(\lambda)\right)$ from the λ-value generated by Computer C. The value is sent back to Computer C.

(3) Computer B choses a value $b \in [0, 2\pi]$ and evaluates $T'_{2,b}(\lambda) \tilde{S}_b^{(2)}\left(T_{2,b}(\lambda)\right)$ from the λ-value generated by Computer C. The value is sent back to Computer C.

(4) Based on the data sent back from Computers A and B, the correlation $\langle S_a^{(1)} S_b^{(2)} \rangle$ is calculated following (5) and the violation of Bell's inequality is checked.

References

1. J.S. Bell, *Physics* **1**, 195 (1964).
2. L. Accardi, *Phys. Rep.* **77**, 169 (1981).
3. L. Accardi, Einstein-Bohr: one all, in: *The interpretation of quantum theory: where do we stand?* Acta Enciclopedica, Instituto dell'Enciclopedia Italiana (1994) p.95, Volterra preprint N. 174 (1993).
4. L. Accardi, M. Regoli, *Locality and Bell's inequality*, Volterra preprint N. 427 (2000); quant-ph/0007005; an extended version of this paper, including the description of the present experiment will appear in the proceedings of the conference "Foundations of Probability and Physics", Vaxjo University, Sweden, November 27 - December 1 (2000), A. Khrennikov (ed.), World Scientific (2001).
5. L. Accardi, M. Regoli, *The EPR correlations and the chameleon effect*, discussed during the "Japan-Italy Joint Workshop on Quantum Open Systems and Quantum Measurement", Waseda University, 27–29 September 2001, quant-ph/0110086.

ON THE ELECTROMAGNETIC RECIPROCITY-LIKE THEOREMS

G.N. AFANASIEV
Bogoliubov Laboratory of Theoretical Physics,
Joint Institute for Nuclear Research,
Dubna, Moscow District, 141980, Russia

We analyse the Lorentz and Feld-Tai lemmas (or reciprocity-like theorems) having numerous applications in electrodynamics, optics, radiophysics, electronics, etc. It is demonstrated that these lemmas are valid for more general time-dependences of the electromagnetic field sources than it was suggested up to now. It is shown also that the validity of reciprocity-like theorems is intimately related to the equality of electromagnetic action and reaction: both of them are fulfilled or violated under the same conditions. Conditions are stated under which reciprocity-like theorems can be violated. The concrete example of their violation is presented.

1 The Lorentz and Feld-Tai lemmas

The Lorentz [1] and Feld-Tai [2,3] lemmas state that $\mathcal{E}_{12} = \mathcal{E}_{21}$ and $\mathcal{H}_{12} = \mathcal{H}_{21}$, resp. where $\mathcal{E}_{12} = \int \vec{j}_1 \vec{E}_2 dV$, $\mathcal{E}_{21} = \int \vec{j}_2 \vec{E}_1 dV$, $\mathcal{H}_{12} = \int \vec{j}_1 \vec{H}_2 dV$, $\mathcal{H}_{21} = \int \vec{j}_2 \vec{H}_1 dV$. Here $\vec{j}_1$ and $\vec{j}_2$ are arbitrary current densities, $\vec{E}_1$, $\vec{E}_2$, $\vec{H}_1$ and $\vec{H}_2$ are electric and magnetic fields generated by them. In Ref. [4], applying these lemmas to particular electromagnetic field sources, we proved their violation. No we clarify this situation.

1.1 Lorentz lemma

Substituting the retarded electric fields into both sides of the Lorentz lemma one gets

$$\mathcal{E}_{12} = -\int \dot{\rho}_1(\vec{r}_1, t)\rho_2(\vec{r}_2, \tau)\delta(\tau - t + R_{12}/c)\frac{1}{R_{12}} dV_1 dV_2 d\tau +$$

$$+\frac{1}{c^2}\int \vec{j}_1(\vec{r}_1, t)\vec{j}_2(\vec{r}_2, \tau)\dot{\delta}(\tau - t + R_{12}/c)\frac{1}{R_{12}} dV_1 dV_2 d\tau,$$

$$\mathcal{E}_{21} = -\int \dot{\rho}_2(\vec{r}_2, t)\rho_1(\vec{r}_1, \tau)\delta(\tau - t + R_{12}/c)\frac{1}{R_{12}} dV_1 dV_2 d\tau +$$

$$+\frac{1}{c^2}\int \vec{j}_2(\vec{r}_2,t)\vec{j}_1(\vec{r}_1,\tau)\dot{\delta}(\tau-t+R_{12}/c)\frac{1}{R_{12}}dV_1dV_2d\tau,$$

where the dot above ρ means derivative w.r.t. t and the dot above the δ function means derivative w.r.t. its argument. We see that, in general, $\mathcal{E}_{12}(t) \neq \mathcal{E}_{21}(t)$. Let now time dependences in ρ and $\vec{j}$ be separated: $\rho_1(\vec{r}_1,t_1) = \rho_1(t_1)\rho_1(\vec{r}_1)$, $\rho_2(\vec{r}_2,t_2) = \rho_2(t_2)\rho_2(\vec{r}_2)$, $\vec{j}_1(\vec{r}_1,t_1) = j_1(t_1)\vec{j}_1(\vec{r}_1)$, $\vec{j}_2(\vec{r}_2,t_2) = j_2(t_2)\vec{j}_2(\vec{r}_2)$. Then,

$$\mathcal{E}_{12} = -\int \dot{\rho}_1(t)\rho_1(\vec{r}_1)\rho_2(\vec{r}_2)\rho_2(\tau)\frac{1}{R_{12}}\delta(\tau-t+R_{12}/c)dV_1dV_2d\tau+$$

$$+\frac{1}{c^2}\int j_1(t)\vec{j}_1(\vec{r}_1)\vec{j}_2(\vec{r}_2)j_2(\tau)\frac{1}{R_{12}}\dot{\delta}(\tau-t+R_{12}/c)dV_1dV_2d\tau,$$

$$\mathcal{E}_{21} = -\int \dot{\rho}_2(t)\rho_2(\vec{r}_2)\rho_1(\vec{r}_1)\rho_1(\tau)\frac{1}{R_{12}}\delta(\tau-t+R_{12}/c)dV_1dV_2d\tau+$$

$$+\frac{1}{c^2}\int j_2(t)\vec{j}_2(\vec{r}_2)\vec{j}_1(\vec{r}_1)j_1(\tau)]\frac{1}{R_{12}}\dot{\delta}(\tau-t+R_{12}/c)dV_1dV_2d\tau.$$

It follows from this that $\mathcal{E}_{12} = \mathcal{E}_{21}$ if the time dependences of sources 1 and 2 coincide, i.e., when $\rho_1(t) = \rho_2(t)$ and $j_1(t) = j_2(t)$, that if the time-dependences of sources 1 and 2 are separated and coincide.

1.2 Feld-Tai lemma

Direct evaluation of integrals entering into the Feld-Tai lemma gives

$$\mathcal{H}_{12} = \epsilon_{ijk}\int j_{1i}(\vec{r}_1,t)j_{2k}(\vec{r}_2,\tau)\frac{\partial}{\partial x_{1j}}\frac{1}{R_{12}}\delta(\tau-t+\frac{R_{12}}{c})dV_1dV_2d\tau,$$

$$\mathcal{H}_{21} = \epsilon_{ijk}\int j_{2i}(\vec{r}_2,t)j_{1k}(\vec{r}_1,\tau)\frac{\partial}{\partial x_{2j}}\frac{1}{R_{12}}\delta(\tau-t+\frac{R_{12}}{c})dV_1dV_2d\tau.$$

Here ϵ_{ijk} is the unit antisymmetrical tensor of the third rank. When the space-time dependences in current densities are separated ($\vec{j}(r,t) = j(t)\vec{j}(r)$), these equations are reduced to

$$\mathcal{H}_{12} = \epsilon_{ijk} j_1(t) \int j_2(\tau) j_{1i}(\vec{r}_1) j_{2k}(\vec{r}_2) \frac{\partial}{\partial x_{1j}} \frac{1}{R_{12}} \delta(\tau - t + R_{12}/c) dV_1 dV_2 d\tau,$$

$$\mathcal{H}_{21} = \epsilon_{ijk} j_2(t) \int j_1(\tau) j_{2i}(\vec{r}_2) j_{1k}(\vec{r}_1) \frac{\partial}{\partial x_{2j}} \frac{1}{R_{12}} \delta(\tau - t + R_{12}/c) dV_1 dV_2 d\tau.$$

Obviously, $\mathcal{H}_{12} = \mathcal{H}_{21}$ when the arbitrary time dependences of sources 1 and 2 coincide ($j_1(t) = j_2(t)$).

1.3 The physical meaning of the Lorentz and Feld-Tai lemma for the interacting current sources

We conclude: the Lorentz and Feld-Tai lemmas are fulfilled when the following two conditions are satisfied: i) Time dependences are separated from space variables in the charge-current densities. This means that the time dependence should be the same for all space points of a particular source; ii) The separated time dependence is the same for sources 1 and 2. The physical meaning of the Lorentz lemma is as follows [5,6]. The time- dependent magnetic flux penetrating a particular turn of winding, creates an electric field directed along this turn. Being summed, they give potential difference between ends of the winding if it is not closed and induce the current in the winding if it is closed. This voltage (or current) can be measured. To obtain voltage, we omit in $\mathcal{E}_{12}$ the time dependent current force I_1 (not the current density $\vec{j}_1$). Thus obtained $\mathcal{E}_{12}(t)$ gives time-dependent voltage induced in the winding 1 by the time-dependent current flowing in the winding 2. Similarly, if we omit in $\mathcal{E}_{21}$ the time dependent current force I_2, then $\mathcal{E}_{21}(t)$ gives time-dependent voltage induced in the winding 2 by the time-dependent current flowing in the winding 1. Thus obtained $\mathcal{E}_{12}$ and $\mathcal{E}_{21}$ coincide if $I_1 = I_2$. We observe that in the first case 1 is a receiver and 2 is a transmitter. In the second case, the situation is opposite. This means that an induced voltage is invariant under the replacement of the detector and transmitter. The physical meaning of the Feld-Tai lemma for interacting current sources is not clear to us. A time-dependent electric field penetrating a particular turn of winding, creates the magnetic field directed along this turn. If the free magnetic charges existed, then integrals entering into the Feld-Tai lemma (after omitting the corresponding factors as in the Lorentz lemma) would gave the magnetic voltage between the ends of the winding (if it is not closed). Their equality would gave the symmetry between the transmitter and receiver. Since the monopoles up to now were not found, this interpretation of the Feld-Tai lemma has no relation to reality. However, Lakhtakia [7] and Monzon [8], seem to have found numerous application of the Feld-Tai lemma.

1.4 Another viewpoint on the Lorentz and Feld-Tai lemmas

In the Fourier representation ($\vec{E}(t) = \int \vec{E}(\omega)\exp(i\omega t)d\omega$, etc.) the *curl* parts of Maxwell equations look like

$$curl\vec{E} = -ik\vec{H}, \quad curl\vec{H} = ik\vec{E} + \frac{4\pi}{c}\vec{j}, \quad k = \omega/c.$$

Then, the Lorentz and Feld-Tai lemmas are satisfied trivially. For example, the proof of the Lorentz lemma without appeal to the Maxwell equations takes three lines

$$\mathcal{E}_{12} = \int \vec{j}_1(\vec{r}_1)\vec{E}_{12}(\vec{r}_2)dV_1 = \int \vec{j}_1(\vec{r}_1)[-\vec{\nabla}\Phi_{12}(\vec{r}_1) - ik\vec{A}_{12}(\vec{r}_1)]dV_1 =$$

$$= -i\omega \int [\rho_1(\vec{r}_1)\Phi_{12}(\vec{r}_1) + \frac{1}{c}\vec{j}_1(\vec{r}_1)\vec{A}_{12}(\vec{r}_1)]dV_1 =$$

$$= -i\omega \int [\rho_1(\vec{r}_1)\rho_2(\vec{r}_2) + \frac{1}{c^2}\vec{j}_1(\vec{r}_1)\vec{j}_2(\vec{r}_2)]\frac{\exp(-ikR_{12})}{R_{12}}dV_1dV_2 = \mathcal{E}_{21}.$$

Therefore, the Lorentz and Feld-Tai lemmas may be viewed as the integral relations between the Fourier transforms of the current densities and field strengths. This, in its turn, may be used to derive new identities. For example, multiplying $\mathcal{E}_{12}$ by $\exp(i\omega t)$ and integrating over ω, one gets

$$\int \vec{j}_1(\vec{r}_1,\omega)\vec{E}_{12}(\vec{r}_1,\omega)\exp(i\omega t)dV_1d\omega =$$

$$= \frac{1}{4\pi^2}\int \vec{j}_1(\vec{r}_1,t')\vec{E}_{12}(\vec{r}_1,t'')\exp[i\omega(t-t'-t'')]dV_1d\omega dt'dt'' =$$

$$= \frac{1}{2\pi}\int \vec{j}_1(\vec{r}_1,t')\vec{E}_{12}(\vec{r}_1,t'')\delta(t-t'-t'')dV_1dt'dt'' = \frac{1}{2\pi}\int \vec{j}_1(\vec{r}_1,t-t')\vec{E}_{12}(\vec{r}_1,t')dV_1dt'. \tag{1.1}$$

Performing the same operation with $\mathcal{E}_{21}$ and equalizing the result to (1.1), one arrives at

$$\int \vec{j}_1(\vec{r}_1,t-t')\vec{E}_{12}(\vec{r}_1,t')dV_1dt' = \int \vec{j}_2(\vec{r}_2,t-t')\vec{E}_{21}(\vec{r}_2,t')dV_2dt'. \tag{1.2}$$

This equation was obtained by Feld ([9]). We make one step further, excluding electric strengths. Then, the LHS of (1.2) is reduced to

$$\frac{1}{2\pi}\frac{\partial}{\partial t}\int [\rho_1(\vec{r}_1,t-t')\rho_2(\vec{r}_2,t'-\frac{R_{12}}{c})+\frac{1}{c^2}\vec{j}_1(\vec{r}_1,t-t')\vec{j}_2(\vec{r}_2,t'-\frac{R_{12}}{c})]\frac{1}{R_{12}}dt'dV_1dV_2.$$

Therefore, the following equation should be satisfied:

$$\int [\rho_1(\vec{r}_1,t-t')\rho_2(\vec{r}_2,t'-\frac{R_{12}}{c})+\frac{1}{c^2}\vec{j}_1(\vec{r}_1,t-t')\vec{j}_2(\vec{r}_2,t'-\frac{R_{12}}{c})]\frac{1}{R_{12}}dt'dV_1dV_2 =$$

$$= \int [\rho_2(\vec{r}_2,t-t')\rho_1(\vec{r}_1,t'-\frac{R_{12}}{c})+\frac{1}{c^2}\vec{j}_2(\vec{r}_2,t-t')\vec{j}_1(\vec{r}_1,t'-\frac{R_{12}}{c})]\frac{1}{R_{12}}dt'dV_1dV_2. \tag{1.3}$$

For the integrals entering into the Feld-Tai lemma, one gets

$$\int \exp(i\omega t)\vec{j}_1(\vec{r}_1,\omega)\vec{H}_{12}(\vec{r}_1,\omega)d\omega dV_1 = \frac{1}{2\pi}\int \vec{j}_1(\vec{r}_1,t-t')\vec{H}_{12}(\vec{r}_1,t')dt'dV_1 =$$

$$= -\frac{1}{2\pi c}\int \frac{1}{R_{12}} curl\vec{j}_1(\vec{r}_1,t-t')\vec{j}_2(\vec{r}_2,t'-R_{12}/c)dV_1dV_2dt'.$$

Therefore, the following equalities should be fulfilled

$$\int \vec{j}_1(\vec{r}_1,t-t')\vec{H}_{12}(\vec{r}_1,t')dt'dV_1 = \int \vec{j}_2(\vec{r}_2,t-t')\vec{H}_{21}(\vec{r}_2,t')dt'dV_2,$$

$$\int \frac{1}{R_{12}} curl\vec{j}_1(\vec{r}_1,t-t')\vec{j}_2(\vec{r}_2,t'-R_{12}/c)dV_1dV_2dt' =$$

$$= \int \frac{1}{R_{12}} curl \vec{j}_2(\vec{r}_2, t - t')\vec{j}_1(\vec{r}_1, t' - R_{12}/c) dV_1 dV_2 dt'. \qquad (1.4)$$

It is important that Eqs.(1.2)-(1.4), contrary to the equations defining the Lorentz and Feld-Tai lemmas, are satisfied for any charge-current densities. No assumption on the separation of space and time dependences as well the equality of time dependences for two interacting sources is needed.

1.5 Violation of the Lorentz and Feld-Tai lemmas

Now we analyze assumption on the separability time and spatial variables in charge-current densities. Take at first the single circular current loop. Since there are no other turns, there is no resistive or capacity connections between them. Therefore, the current is the same along the whole wire (due to the continuity equation $div\vec{j} = 0$) and the time dependence is clearly separated from the space variables. On the other hand, consider the winding with many overlapping turns, e.g., the toroidal solenoid. If the turns are close to each other, there is a finite capacitance between them. For high frequencies, the leakage currents appear between particular turns and the current will be changed along the wire. This does not have any relation to the violation of the continuity equation $div\vec{j} = 0$. It will be fulfilled due to the presence of other $\vec{j}$ components, having the direction different from that of wire. Since the current density changes along the wire, the time dependence is not separated now. This should lead to the violation of reciprocity theorem. We conclude: the violation of the reciprocity is possible for high frequencies and large number of overlapping coils.

1.6 Concrete example of the reciprocity violation: interacting electric oscillator amd current loop

We demonstrate the violation of of the Lorentz and Feld-Tai lemmas using the interacting electric oscillator and infinitesimal current loop as an example. We consider the electric oscillator oriented along the z axis:

$$\rho_o = e\delta(x)\delta(y)\delta(z - a\cos\omega t), \quad j_z^o = -ea\omega\sin\omega t\delta(x)\delta(y)\delta(z - a\cos\omega t) \quad (1.5)$$

(o means oscillator). Let the origin of a current loop is represented by the vector $\vec{r}_L$ and let $\vec{n}_L$ be a vector normal to its plane. Then [10], $\vec{j}_L = f_L(t) curl(\vec{n}_L \delta^3(\vec{r} - \vec{r}_L))$. We evaluate at first the electromagnetic potentials of an electric dipole. They are obtained from general expressions

$$\Phi_o(\vec{r},t) = \int \frac{1}{R}\rho_o(\vec{r}',t')\delta(t'-t+\frac{R}{c})dV'dt',$$

$$A_z^o(\vec{r},t) = \int \frac{1}{R}j_o^z(\vec{r}',t')\delta(t'-t+\frac{R}{c})dV'dt', \quad R = |\vec{r}-\vec{r}'|$$

by substituting charge-current densities (1.5) into them and performing integration over space variables. This gives

$$\Phi_o(\vec{r},t) = \frac{e}{Q_1}, \quad A_z^o(\vec{r},t) = -\frac{e\beta_0 \sin\omega t_1}{Q_1},$$

where $Q_1 = R_1 + \beta_0(z - a\cos\omega t_1)\sin\omega t_1$, $R_1 = [x^2 + y^2 + (z - a\cos\omega t_1)^2]^{1/2}$, $\beta_0 = a\omega/c$,. The retarded time t_1 is found from the equation $c(t - t_1) = R_1$. Since the charge velocity $\beta = -\beta_0 \sin\omega t$ is less than 1, there is only one root of this equation.

Electromagnetic field strengths are given by

$$E_\rho = -\frac{\partial\Phi}{\partial\rho}, \quad E_z = -\frac{\partial\Phi}{\partial z} - \frac{\partial A_z}{c\partial t}, \quad H_\phi = -\frac{\partial A_z}{\partial\rho}.$$

When performing differentiation, one should take into account that t_1 depends on the space-time coordinates of the observation point. The corresponding derivatives are given by

$$\frac{dt_1}{dt} = \frac{R_1}{Q_1}, \quad \frac{dt_1}{d\rho} = -\frac{\rho}{cQ_1}, \quad \frac{dt_1}{dz} = -\frac{z - a\cos\omega t_1}{Q_1}.$$

Then,

$$E_\rho^o = \frac{e\rho}{Q_1^3}[1 - \beta_0^2\sin^2\omega t_1 - \beta_0^2\cos\omega t_1(\frac{z}{a} - \cos\omega t_1)],$$

$$E_z^o = \frac{ea}{Q_1^3}[(1 - \beta_0^2\sin^2\omega t_1)(\frac{z}{a} - \cos\omega t_1 - \beta_0\frac{R_1}{a}\sin\omega t_1) + \beta_0^2\frac{\rho^2}{a^2}\cos\omega t_1],$$

$$H_\phi^o = -\frac{e\beta_0}{Q_1^3}[\beta_0\frac{R_1}{a}\cos\omega t_1 + (1 - \beta_0^2\sin^2\omega t_1)\sin\omega t_1].$$

We need also EMF strengths of the current loop. They are given by [10]

$$\vec{E}_L = \frac{1}{c^3 r_L^2}\dot{D}_L((\vec{r}-\vec{r}_L)\times\vec{n}_L), \quad \vec{H}_L = \frac{1}{c^3 r_L}[\frac{((\vec{r}-\vec{r}_L)\vec{n}_L)}{r_L^2}(\vec{r}-\vec{r}_L)F_L - \vec{n}_L G_L],$$

where $r_L = |\vec{r} - \vec{r}_L|$ and

$$D_L = D(f_L) = \dot{f}_l + \frac{c}{r} f_L, \quad F_L = F(f_L) = \ddot{f}_l + \frac{3c}{r}\dot{f}_L + \frac{3c^2}{r^2} f_L,$$

$$G_L = G(f_L) = \ddot{f}_l + \frac{c}{r}\dot{f}_L + \frac{c^2}{r^2} f_L.$$

The arguments of f_L functions entering into D_L, F_L and G_L are $t_r = t - r_L/c$; dots above the f_L, D_L, F_L and G_L functions mean time derivatives. Lorentz lemma Direct evaluation of the integrals entering into the Lorentz lemma gives

$$\mathcal{E}(o, L) = \int \vec{j}_o \vec{E}_L dV = \frac{ea\omega \sin \omega t}{c^3 R_L^2} \dot{D}_L [x_L n_L^y - y_L n_L^x],$$

where x_L, y_L and z_L define the position of the current loop; $R_L = [x_L^2 + y_L^2 + (z_L - a\cos\omega t)^2]^{1/2}$; the argument of the D_L function is $t - R_L/c$. Further,

$$\mathcal{E}(L, o) = \int \vec{j}_L \vec{E}_o dV = e\beta_0 f_L(t)[x_L n_L^y - y_L n_L^x] \times$$

$$\frac{R_{1L}}{cQ_{1L}} \frac{d}{dt_1} \{ \frac{1}{Q_{1L}^3} [\beta_0 \frac{R_{1L}}{a} \cos\omega t_1 + (1 - \beta_0^2 \sin^2 \omega t_1) \sin \omega t_1]\}.$$

Here $R_{1L} = [x_L^2 + y_L^2 + (z_L - a\cos\omega t_1)^2]^{1/2}$, $Q_{1L} = R_{1L} + \beta_0(z_L - a\cos\omega t_1)\sin\omega t_1$ and t_1 is found from the equation: $t - t_1 = R_{1L}/c$. Now we choose the time dependence of the charge current loop to be the same as that of electric oscillator: $f_L(t) = f_0 \sin\omega t$. Then, equalizing $\mathcal{E}(L, o)$ and $\mathcal{E}(o, L)$, one gets

$$-\frac{k^2}{R_L^2}(\sin\omega t - \frac{1}{kR_L}\cos\omega t) = \frac{R_{1L}}{cQ_{1L}} \frac{d}{dt_1} \{ \frac{1}{Q_{1L}^3} [\beta_0 \frac{R_{1L}}{a} \cos\omega t_1 + (1 - \beta_0^2 \sin^2\omega t_1)\sin\omega t_1]\}.$$

Here $k = \omega/c$. This equality is not satisfied, and, therefore, the Lorentz lemma is not fulfilled for the interacting electric oscillator and current loop. Feld-Tai lemma For the integrals entering into Feld-Tai lemma one gets

$$\mathcal{H}(o, L) = -\frac{ea\omega \sin\omega t}{c^3 R_L} \{[(a\cos\omega t - z_L) n_L^z - x_L n_L^x - y_L n_L^y] \frac{a\cos\omega t - z_L}{R_L^2} F_L - n_L^z G_L\},$$

$$\mathcal{H}(L, o) = \frac{e f_L(t)}{c} \frac{R_{1L}}{Q_{1L}} \frac{d}{dt_1} \{(x_L n_L^x + y_L n_L^Y)[1 - \beta_0^2 \sin^2\omega t_1 - \beta_0^2 \cos\omega t_1 (\frac{z_0}{a} - \cos\omega t_1)] +$$

$$+an_L^z[(1-\beta_0^2\sin^2\omega t_1)(\frac{z_l}{a}-\cos\omega t_1-\beta_0\frac{R_{1L}}{a}\sin\omega t_1)+\beta_0^2\frac{\rho_L^2}{a^2}\cos\omega t_1]\}.$$

Putting $f_L(t) = f_0 \sin\omega t$ we find that Feld-Tai lemma cannot be satisfied for the interacting electric oscillator and current loop. It should be mentioned that more general reciprocity relations formulated above, are fulfilled for arbitrary time dependences and, in particular, for the interacting electric oscillator and current loop.

2 Discussion and Conclusion

Recently, we were aware on experiments with toroidal coils of higher orders. In them, the non-coincidence of voltages induced in them (i.e., the violation transmitter-receiver symmetry mentioned above) was observed for large frequencies. The reciprocity theorem seems to be so well established that the scientists performing this experiment did not dare to attribute this non-coincidence to its violation. However, the present consideration shows that this violation is, indeed, possible. It should be mentioned that violation of reciprocity will lead to dramatic consequences both in theoretical and experimental elecromagnetism. According to [7], "One of the basic and most important theorems of electromagnetic theory is the so-called reciprocity theorem. Its importance is evident from its wide range of applicability in all branches of electrical engineering".

We briefly enumerate the main results obtained:

1. It is shown that the proportionality of time derivatives of EMF strengths to the EMF strengths themselves is not a necessary condition for the fulfillment of the reciprocity theorem.

2. The alternative proof of the reciprocity theorem is given. It is shown that the reciprocity theorem works for more general time dependences than it was suggested up to now. The conditions for its validity are reduced to the two following ones: i) The time dependence should be separated from the spatial one in the charge-current densities of interacting sources; ii) The time dependences of these sources should be the same.

These conditions are essentially the same as ones needed for the equality of action and reaction between two interacting electromagnetic sources. The estimation of action-reaction violation for the interacting current loop and toroidal solenoid is given.

3. Conditions under which the reciprocity theorem can be violated are given. The concrete example is presented for which the reciprocity theorem is manifestly violated.

4. New reciprocity-like theorems valid for arbitrary space-time dependences (and, in particular, for those discussed in the previous item) of the interacting current densities are obtained. However, their physical meaning is not very clear.

Acknowledgments.

The author is deeply indebted to Prof. N.I. Zheludev from the Southampton University, England, and to Richard Wardle from the Southampton Chilworth Research Centre who, in fact, initiated this investigation.

References

1. H.A. Lorentz, *Amsterdammer Akademie der Wetenschappen* **176** (1895-1896), No 4.
2. Ya.N. Feld, *Sov.Phys.Doklady* **37** (1992), 235.
3. C.T. Tai, *IEEE Trans. Antennas Propagation* **40** (1992), 675.
4. G.N. Afanasiev, *J.Phys. D: Appl. Phys.* **34** (2001), 539.
5. V.H. Rumsey, *Phys. Rev.* **94** (1954), 1483.
6. S.A. Schelkunoff, *Electromagnetic Fields, Blaisdell*, New York (1965).
7. J.C. Monzon, *IEEE Trans. Microwave Theory Tech.* **44** (1996), 10.
8. A. Lakhtakia, *Beltrami fields in chiral media* World Scientific, Singapore (1994).
9. Ya.N. Feld, *Doklady Akad. Nauk SSSR* **318** (1991), 325.
10. G.N. Afanasiev, *Topological Effects in Quantum Mechanics*, Kluwer Acad. Publ., Dordrecht (1999).

INTERTWINING OPERATORS FOR SCHRÖDINGER ALGEBRAS AND HIERARCHY OF INVARIANT EQUATIONS

N. AIZAWA

Department of Applied Mathematics, Osaka Women's University, Sakai, Osaka 590-0035, Japan

V. K. DOBREV

School of Computing and Mathematics, University of Northumbria, Ellison Place, Newcastle upon Tyne, NE1 8ST, Newcastle upon Tyne, UK
and
Institute of Nuclear Research and Nuclear Energy, Bulgarian Academy of Sciences, 72 Tsarigradsko Chaussee, 1784 Sofia, Bulgaria (permanent address)

H.-D. DOEBNER

Arnold Sommerfeld Institute for Mathematical Physics, Technical University of Clausthal, Leibnizstr. 10, D38678 Clausthal-Zellerfeld, FRG

Hierarchies of equations invariant under the Schrödinger algebras in $(n+1)$-dimensional spacetime are constructed for $n = 1, 2, 3$. The heat/Schrödinger equation is found as a member of the hierarchies.

1 Introduction

Let G be a Lie group and $\mathcal{D}$ be a differential operator. Then a differential equation $\mathcal{D}\Psi = 0$ is said to be G-invariant, if the operator $\mathcal{D}$ is a intertwiner of some representations of G, namely, $\mathcal{D}$ satisfies the relation $T(g) \circ \mathcal{D} = \mathcal{D} \circ T'(g)$ for $\forall g \in G$, where T and T' are two representations of G. Let us recall that physically important equations have this invariant properties, *e.g.*, both the Maxwell equation and the massless Dirac equation are $SU(2,2)$-invariant. For the Lie algebra $\mathfrak{g}$ of G, we can define $\mathfrak{g}$-invariant equation in a similar way.

For semisimple Lie groups and algebras, there exists a canonical method to construct all intertwining differential operators[1]. This means we obtain all G-invariant differential equations by this method. Furthermore, this method was extended to quantum group setting[2,3]. In this note, we apply this method to the Schrödinger algebras in $(n+1)$-dimensional spacetime for $n = 1, 2, 3$ and we obtain a hierarchy of invariant equations. We choose the Schrödinger algebra because (i) it is one of the physically important algebras, (ii) it is non-semisimple. As we shall see, heat/Schrödinger equation is obtained as an invariant equation. This gives a new group theoretical viewpoint to the one of the most basic equations in physics. The success of application of the method implies it may be applicable to wider class of Lie algebras beyond semisimple ones.

2 Preliminaries

In this section, we outline the method[1] used in the subsequent sections. Let G be a complex semisimple Lie group and $\mathfrak{g}$ its Lie algebra. Let $\mathfrak{g} = \mathfrak{g}_+ \oplus \mathfrak{g}_0 \oplus \mathfrak{g}_-$ be the standard triangular decomposition. We consider representations of $\mathfrak{g}$ whose representation spaces $\mathcal{C}_\Lambda$ are $\mathbb{C}^\infty$ functions $\mathcal{F}$ on G with the property called *right covariance*

$$\mathcal{F}(gxg_-) = e^{\Lambda(H)}\mathcal{F}(g), \quad g \in G,\ x = e^H \in G_0,\ H \in \mathfrak{g}_0,\ g_- \in G_- \,, \tag{1}$$

where $\Lambda \in \mathfrak{g}_0^*$, $\Lambda(H) \in \mathbb{Z}$, $G_0 \equiv \exp \mathfrak{g}_0$, $G_\pm \equiv \exp \mathfrak{g}_\pm$. Thus the functions of $\mathcal{C}_\Lambda$ are actually functions on the coset G/B; $B \equiv G_0 G_-$ being a Borel subgroup, or equivalently[1], on G_+ which is dense in G/B. The restricted representation spaces of functions f on G_+, such that $f = \mathcal{F}|_{G_+}$, we denote by C_Λ. We introduce the left $\pi_L(X)$ and right $\pi_R(X)$ actions of $\mathfrak{g}$ on $\mathcal{C}_\Lambda$ by the standard formulae

$$\pi_L(X)\mathcal{F}(g) \equiv \frac{d}{d\tau}\mathcal{F}(e^{-\tau X}g)\Big|_{\tau=0}, \quad \pi_R(X)\mathcal{F}(g) \equiv \frac{d}{d\tau}\mathcal{F}(ge^{\tau X})\Big|_{\tau=0}, \tag{2}$$

where $X \in \mathfrak{g}$. The left action gives representations of $\mathfrak{g}$ by differential operators. It may be considered for arbitrary $\Lambda(H) \in \mathbb{C}$ and it may be restricted to C_Λ. On the other hand, the space $\mathcal{C}_\Lambda$ has a lowest weight structure with respect to the right action, since one can show, using the right covariance, that

$$\pi_R(H)\mathcal{F}(g) = \Lambda(H)\mathcal{F}(g), \qquad \pi_R(X)\mathcal{F}(g) = 0, \tag{3}$$

where $H \in \mathfrak{g}_0$ and $X \in \mathfrak{g}_-$. Thus we are prompted to employ properties of the Verma module V^Λ with the lowest weight Λ, such that $V^\Lambda \simeq U(\mathfrak{g}_+)v_0$, where v_0 is the lowest weight vector, $U(\mathfrak{g}_+)$ is the universal enveloping algebra of $\mathfrak{g}_+$. When a Verma module is reducible, it has (at least one) singular vector v_s such that: $Hv_s = \Lambda'(H)$, $\Lambda' \neq \Lambda$, $H \in \mathfrak{g}_0$, $Xv_s = 0$, $X \in \mathfrak{g}_-$. It has the general structure $v_s = P(\mathfrak{g}_+)v_0$, where $P(\mathfrak{g}_+)$ is a polynomial of the generators of $\mathfrak{g}_+$. Then it is shown that the same polynomial $P(\mathfrak{g}_+)$ gives rise to a $\mathfrak{g}$-invariant differential equation given explicitly by: $P(\pi_R(\mathfrak{g}_+))\psi = 0$, $\psi = \mathcal{F},\ f$.

3 Schrödinger algebra

The Schrödinger algebra $\mathfrak{s}(n)$ in $(n+1)$-dimensional spacetime is generated by time translation P_t, space translation P_k, Galilei boosts G_k, rotations $J_{k\ell} = -J_{\ell k}$ (which generate the subalgebra $so(n)$), dilatation D and conformal transformation K $(k, \ell = 1, \cdots n)$. The non-trivial commutation relations are[4]

$$\begin{array}{llll} [P_t, D] = 2P_t, & [P_t, G_k] = P_k, & [P_t, K] = D, & [P_k, D] = P_k, \\ [P_k, K] = G_k, & [D, G_k] = G_k, & [D, K] = 2K, & \\ [P_i, J_{k\ell}] = \delta_{i\ell} P_k - \delta_{ik} P_\ell, & & [G_i, J_{k\ell}] = \delta_{i\ell} G_k - \delta_{ik} G_\ell, & \\ [J_{ij}, J_{k\ell}] = \delta_{ik} J_{j\ell} + \delta_{j\ell} J_{ik} - \delta_{i\ell} J_{jk} - \delta_{jk} J_{i\ell}, & & & \end{array} \tag{4}$$

Actually, we shall work with the central extension $\hat{\mathfrak{s}}(n)$ of the Schrödinger algebra obtained by adding the central element M to $\mathfrak{s}(n)$ which enters the additional commutation relations: $[P_k, G_\ell] = \delta_{k\ell} M$. Physically, the central element M corresponds to mass. Some structural properties of the centrally extended Schrödinger algebra $\hat{\mathfrak{s}}(n)$ have already been discussed[5]. Amongst the properties, here we point out the two: (i) $\hat{\mathfrak{s}}(n)$ is not semisimple, (ii) it is possible to introduce the triangular decomposition to $\hat{\mathfrak{s}}(n)$ by regarding it as a complex Lie algebra and taking into account the triangular decomposition of $so(n)$: $\hat{\mathfrak{s}}(n) = \hat{\mathfrak{s}}(n)^+ \oplus \hat{\mathfrak{s}}(n)^0 \oplus \hat{\mathfrak{s}}(n)^-$ with

$$\begin{array}{ll} \hat{\mathfrak{s}}(n)^+ = \{G_k, K\} \oplus so(n),^+ & \hat{\mathfrak{s}}(n)^0 = \{D, M\} \oplus so(n)^0, \\ \hat{\mathfrak{s}}(n)^+ = \{P_k, P_t\} \oplus so(n),^- & \end{array} \tag{5}$$

where $\{\cdots\}$ denotes abelian subalgebras generated by the entries.

One can construct a finite dimensional representation of $\hat{\mathfrak{s}}(n)$ by embedding $\hat{\mathfrak{s}}(n)$ into $sl(2n+2)$. Let $E_{k,\ell}$ be the matrices with only one non-zero entry, namely, $(E_{k,\ell})_{ab} = \delta_{ka}\delta_{\ell b}$. Then the Schrödinger algebra $\hat{\mathfrak{s}}(n)$ has the representation given by

$$\begin{array}{ll} D = \sum_{k=1}^{n} (-E_{k,k} + E_{k+1,k+1}), & M = -2E_{1,2n+2}, \\ G_k = E_{1,2k} - E_{2n+3-2k,2n+2}, & P_k = E_{1,2n+3-2k} + E_{2k,2n+2}, \\ P_t = -\sum_{k=1}^{n} E_{2k,2n+3-2k}, & K = \sum_{k=1}^{n} E_{2k+1,2n+2-2k}, \\ J_{\ell k} = E_{2\ell,2k} - E_{2k,2\ell} - E_{2n+3-2k,2n+3-2\ell} + E_{2n+3-2\ell,2n+3-2k} & \end{array} \tag{6}$$

4 The case of (1+1)-dimensional spacetime

In this section, we investigate the case of (1+1)-dimensional spacetime. We repeat the procedure for semisimple Lie algebras which was recalled in §2 and we shall see that it also works for the Schrödinger algebra and group.

We first introduce the centrally extended Schrödinger group $\hat{S}(1)$ by exponentiating the matrix representation of generators of $\hat{\mathfrak{s}}(1)$: $g = g_+ \, g_0 \, g_- \in \hat{S}(1)$ where $g_+ = \exp(xG+tK)$, $g_0 = \exp(\delta D+mM)$ and $g_- = \exp(vP_x+\kappa P_t)$. The set of elements g that have such a decomposition is dense in $\hat{S}(1)$. The explicit matrix form of g is given by

$$g_+ = \begin{pmatrix} 1 & x & 0 & 0 \\ 0 & 1 & 0 & 0 \\ 0 & t & 1 & -x \\ 0 & 0 & 0 & 1 \end{pmatrix}, \quad g_0 = \begin{pmatrix} 1 & 0 & 0 & -2m \\ 0 & e^{-\delta} & 0 & 0 \\ 0 & 0 & e^{\delta} & 0 \\ 0 & 0 & 0 & 1 \end{pmatrix}, \quad g_- = \begin{pmatrix} 1 & 0 & v & 0 \\ 0 & 1 & -\kappa & v \\ 0 & 0 & 1 & 0 \\ 0 & 0 & 0 & 1 \end{pmatrix} \tag{7}$$

We next introduce the space $\mathcal{C}_\Lambda$ of $\mathbb{C}^\infty$ functions $\mathcal{F}$ on $\hat{S}(1)$ with the right covariance property (1). The left and right action of $\hat{\mathfrak{s}}(1)$ on the space $\mathcal{C}_\Lambda$ is defined in the same way as (2) and because of right covariance (3) also holds. We also introduce the space C_Λ of $\mathbb{C}^\infty$ functions f on $\hat{S}(1)^+$. The explicit form of the left action on C_Λ is given by

$$\begin{aligned} &\pi_L(G) = -\frac{\partial}{\partial x}, \quad \pi_L(K) = -\frac{\partial}{\partial t}, \quad \pi_L(M) = -\Lambda(M), \\ &\pi_L(D) = -\Lambda(D) - x\frac{\partial}{\partial x} - 2t\frac{\partial}{\partial t}, \quad \pi_L(P_x) = -x\Lambda(M) - t\frac{\partial}{\partial x}, \\ &\pi_L(P_t) = -t\Lambda(D) - \frac{x^2}{2}\Lambda(M) - tx\frac{\partial}{\partial x} - t^2\frac{\partial}{\partial t}, \end{aligned} \tag{8}$$

and the right action by

$$\pi_R(D) = \Lambda(D), \quad \pi_R(M) = \Lambda(M), \quad \pi_R(G) = \frac{\partial}{\partial x}, \quad \pi_R(K) = \frac{\partial}{\partial t}, \tag{9}$$

where $\Lambda(D), \Lambda(M) \in \mathbb{C}$.

Since the functions $\mathcal{F}$ behave like lowest weight vectors, we shall use as before Verma modules $V^\Lambda \simeq U(\hat{\mathfrak{s}}(1)^+)v_0$. To see the reducibility of this Verma module and to obtain the invariant equations, we need singular vectors. The singular vectors for $\hat{\mathfrak{s}}(1)$ are found in the literature[5]. The singular vectors of V^Λ over $\hat{\mathfrak{s}}(1)$ are given by (up to overall constant)

(i) for $\Lambda(M) \neq 0$, $\Lambda(D) = \frac{1}{2}(3-p)$ with p even

$$v_s^p = (G^2 - 2MK)^{p/2} \, v_0 \,, \tag{10}$$

(ii) for $\Lambda(M) = 0$ and arbitrary $\Lambda(D)$

$$v_s^p = G^p\, v_0\ , \tag{11}$$

where p is a natural number.

Using the explicit form of the right action (9), we obtain the following hierarchies of $\hat{\mathfrak{s}}(1)$-invariant equations

$$\pi_R((G^2 - 2MK)^{p/2})\,\psi \;=\; \left(\frac{\partial^2}{\partial x^2} - 2\Lambda(M)\frac{\partial}{\partial t}\right)^{p/2} \psi \;=\; 0\,, \quad \text{for (i),} \tag{12}$$

$$\pi_R(G^p)\,\psi \;=\; \frac{\partial^p}{\partial x^p}\,\psi \;=\; 0\,. \quad \text{for (ii).} \tag{13}$$

We see from (12) that the heat/Schrödinger equation in (1+1)-dimensional spacetime is the first member of the hierarchy. Furthermore, it is not difficult to verify that the singular vectors (10) and (11) give intertwiners for the representations of $\hat{\mathfrak{s}}(1)$ given by the left action. More specifically, the differential operators in (12) and (13) are intertwiners for the representations (8) with lowest weights $\Lambda(D)$ and $\Lambda(D) + p$.

5 Higher dimensional cases

The procedure in the previous section has a straightforward extension to higher dimensional spacetime. We give the results for (2+1) and (3+1)-dimensional cases in this section. Detailed analysis will be given in a separate publication[6]. We remark that the singular vectors of $\hat{\mathfrak{s}}(n)$ ($n = 2, 3$) for lowest weight modules are obtained by Mrugalla[7] and we shall use his results.

In (2+1)-dimensional spacetime, the generator $J_0 = iJ_{12}$ of $so(2)$ appears in the Cartan part. The singular vectors have different form for the massive ($\Lambda(M) \neq 0$) and for the massless ($\Lambda(M) = 0$) cases. If $\Lambda(M) \neq 0$, the singular vectors exist for arbitrary $\Lambda(J_0)$ and $\Lambda(D) = \frac{1}{2}(4-p)$ with p even

$$v_s^p = (G_1^2 + G_2^2 - 2MK)^{p/2}\, v_0\ . \tag{14}$$

This gives the hierarchy of $\hat{\mathfrak{s}}(2)$-invariant equations

$$\left(\frac{\partial^2}{\partial x_1^2} + \frac{\partial^2}{\partial x_2^2} - 2\Lambda(M)\frac{\partial}{\partial t}\right)^{p/2} \psi \;=\; 0\,. \tag{15}$$

It is also verified that the differential operators in (15) are intertwiners for the representations with the lowest weights $(\Lambda(D), \Lambda(J_0))$ and $(\Lambda(D) + p, \Lambda(J_0))$.

In (3+1)-dimensional spacetime, we use the triangular decomposition of $so(3)$, however, the Cartan part remains the same as (2+1)-dimensional one.

The singular vectors for the massive case ($\Lambda(M) \neq 0$) are classified into three types according to the value of $\Lambda(D)$ and $\Lambda(J_0)$. The most interesting case may be $\Lambda(D) = \frac{1}{2}(5-p)$ with p positive even integer and $\Lambda(J_0)$ is arbitrary. The singular vectors for this case are given by

$$v_s^p = \left(\sum_{a=1}^{3} G_a^2 - 2MK \right)^{p/2} v_0 \ , \tag{16}$$

and this gives the hierarchy of invariant equations of the form

$$\left(\frac{\partial^2}{\partial x_1^2} + \frac{\partial^2}{\partial x_2^2} + \frac{\partial^2}{\partial x_3^2} - 2\Lambda(M)\frac{\partial}{\partial t} \right)^{p/2} \psi = 0. \tag{17}$$

The differential operators in (17) are intertwiners for the representations with lowest weights $(\Lambda(D), \Lambda(J_0))$ and $(\Lambda(D)+p, \Lambda(J_0))$.

In summary, we have constructed $\hat{\mathfrak{s}}(n)$-invariant intertwining operators for $n = 1, 2, 3$ by applying the method for semisimple Lie algebras. We obtained hierarchies of invariant equations and observed that the heat/Schrödinger equation is the first member of the hierarchy.

Acknowledgments

N. A. was supported by the Ministry of Education, Culture, Sport, Science and Technology, Japan.

References

1. V. K. Dobrev, *Rep. Math. Phys.* **25**, 159 (1988).
2. V. K. Dobrev, *J. Phys. A* **27**, 4841 (1994).
3. V. K. Dobrev, Habilitationschrifts, Technische Universität Clausthal (1994).
4. A. O. Barut and R. Rączka, *Theory of Group Representations and Applications*, PWN, Warszawa, (1980).
5. V. K. Dobrev, H.-D. Doebner and Ch. Mrugalla, *Rep. Math. Phys.* **39**, 201 (1997).
6. N. Aizawa, V. K. Dobrev and H.-D. Doebner, in preparation.
7. Ch. Mrugalla, PhD Thesis, Technische Universität Clausthal (1997).

SYMMETRIES AND REDUCED SYSTEMS OF BETHE EQUATIONS

V.A.ANDREEV

P.N.Lebedev Physical Institute, Russian Academy of Sciences, Leninskii Pr. 53, Moscow 117924, Russia
e-mail: andr@sci.lebedev.ru

It is shown that the structure of equations for the Bethe ansatz parameters is closely connected with the algebraic structure of the model. The Hamiltonians of Gaudin's models have the form of the sum of products of operators of two algebras. The original Bethe equations are connected only will the form of these algebras. But it turns out, that if one considers the concrete representations of these algebras, the structure of corresponding equations can be simplified. We will call them reduced Bethe equations. This system of Bethe equations is found for three-boson interaction in explicit form.

1 Introduction

Bethe ansatz plays an important role in the theory of integrable quantum models [1-4]. For the first time it has been applied to the lattice models such as isotropic and anisotropic spin chain [1,2] and Hubbard model [5,6]. It was also applied to the quantum versions of sine-Gordon [7,8] and nonlinear Schrödinger equations [9].

The other class of models was introduced by Gaudin [10]. The possibility of functional Bethe ansatz for them was shown by Sklyamin [11]. These models have been investigated in many works [12-15]. The algebraic Bethe ansatz was applied by Jurĉo to the Gaudin's models related to nonlinear quantum optics [16]. They describe such processes as second harmonic generation, three-boson interaction, the Dike model [17] and some cases of four-boson interaction.

The dynamics of three-boson interaction was investigated in our work [18]. It turned out, that in this case the Bethe ansatz formalism has some specific properties distinguishihg the three-boson interaction from other models. The matter is that the eigenvectors of Hamiltonian in this approach have the form

$$|\lambda_1, \ldots, \lambda_N\rangle = (N!)^{-1/2} B(\lambda_1) \cdots B(\lambda_N)|0\rangle. \tag{1}$$

Here $|0\rangle$ is a pseudovacuum vector and $B(\lambda_i)$ are some specific operators depending on the parameters λ_i. In the most models vectors (1), corresponding to different sets $\{\lambda_1, \ldots, \lambda_N\}$, $\{\lambda'_1, \ldots, \lambda'_N\}$, are linearly independent. This condition together with the demand of (1) to be the eigenvector for Hamiltonian gives us the equations for the parameters $\{\lambda_i,\ i = 1, \ldots, N\}$. And in fact, in the spin chain models, in quantum sine-Gordon and nonlinear Schrödinger equations these vectors are linearly independent [19-21], but in the case of three-boson interaction it is not so. In this model eigenvectors (1) with different $\{\lambda_i, i = 1, \ldots, N\}$ can be linearly dependent or even coincide. Therefore the admissible parameters defining some eigenvector (1) occupy some surface S in space C^N. Equations of these surfaces have been found in Ref.[22]. In this work we will show how the existence of such dependencies can change the form of equations for the Bethe ansatz parameters. In more exact terms the structure of equations for the Bethe ansatz parameters is found to be more closely connected with the algebraic structure of the model. The Hamiltonians of Gaudin's models have the form of the sum of products of operators of two algebras. The original Bethe equations are connected only will the form of these algebras. But it turns out, that if one considers the concrete representations of these algebras, the structure of corresponding equations can be simplified. We will call them reduced Bethe equations.

2 Algebraic Bethe ansatz scheme

According to the usual Bethe ansatz scheme in terms of Gaudin models [16] one must consider the 2×2 matrix $L(\lambda)$ with operator matrix elements

$$L(\lambda) = \begin{pmatrix} A(\lambda) & B(\lambda) \\ C(\lambda) & D(\lambda) \end{pmatrix}. \tag{2}$$

This matrix must be a solution of the classical Yang-Baxter equation

$$\Big[L(\lambda) \bigotimes I,\, I \bigotimes L(\mu)\Big] + \Big[r(\lambda - \mu), L(\lambda) \bigotimes I + I \bigotimes L(\mu)\Big] = 0. \tag{3}$$

Here $I = \begin{pmatrix} 1 & 0 \\ 0 & 1 \end{pmatrix}$ and $r(\lambda)$ is 4×4 matrix

$$r(\lambda) = \begin{pmatrix} f(\lambda) & 0 & 0 & 0 \\ 0 & 0 & g(\lambda) & 0 \\ 0 & g(\lambda) & 0 & 0 \\ 0 & 0 & 0 & f(\lambda) \end{pmatrix}. \tag{4}$$

It follows from equation (3) that elements of matrix (2) must satisfy the following commutation relations

$$[A(\lambda), B(\mu)] = g(\lambda - \mu)B(\lambda) - f(\lambda - \mu)B(\mu), \tag{5.1}$$

$$[A(\lambda), C(\mu)] = f(\lambda - \mu)C(\mu) - g(\lambda - \mu)C(\lambda), \tag{5.2}$$

$$[D(\lambda), B(\mu)] = f(\lambda - \mu)B(\mu) - g(\lambda - \mu)B(\lambda), \tag{5.3}$$

$$[D(\lambda), C(\mu)] = g(\lambda-\mu)C(\lambda) - f(\lambda-\mu)C(\mu), \qquad (5.4)$$

$$[B(\lambda), C(\mu)] = g(\lambda-\mu)\big(A(\lambda) - A(\mu) - D(\lambda) + D(\mu)\big), \qquad (5.5)$$

$$[A(\lambda), A(\mu)] = [B(\lambda), B(\mu)] = [C(\lambda), C(\mu)] =$$

$$= [D(\lambda), D(\mu)] = [A(\lambda), D(\mu)] = 0 \qquad (5.6).$$

The Hamiltonian of a three-boson interaction has the form

$$H = H_0 + gH_I = \omega_1 a_1^+ a_1 + \omega_2 a_2^+ a_2 + \omega_3 a_3^+ a_3 + g(a_1^+ a_2 a_3 + a_3^+ a_2^+ a_1), \quad (6)$$

$$\omega_1 = \omega_2 + \omega_3.$$

In this case the elements of matrix $L(\lambda)$ can be taken as

$$A(\lambda) = -\frac{1}{2}\lambda - \frac{1}{\lambda}S_3,\ B(\lambda) = a_3^+ + \frac{1}{\lambda}S_+,\ C(\lambda) = a_3 + \frac{1}{\lambda}S_-,\ D(\lambda) = -A(\lambda). \qquad (7)$$

Here the 2-boson realization of $SU(2)$ algebra is used

$$S_+ = a_1^+ a_2, \quad S_- = a_2^+ a_1, \quad S_3 = \frac{1}{2}(a_1^+ a_1 - a_2^+ a_2). \qquad (8)$$

It is easy to verify that if
$$f(\lambda) = g(\lambda) = \frac{1}{\lambda} \qquad (9)$$

operators (7) satisfy commutation relations (5). The eigenfunctions of Hamiltonian (6) can be constructed in form (1) with the pseudovacuum vector

$$|0\rangle_M = |0, M, 0\rangle. \qquad (10)$$

This vector satisfies the equations

$$C(\lambda)|0\rangle_M = 0, \qquad A(\lambda)|0\rangle_M = a(\lambda)|0\rangle_M, \qquad a(\lambda) = \frac{1}{2\lambda}(M - \lambda^2). \quad (11)$$

Now we will find the equations for the parameters $\{\lambda_i, i = 1, \dots, N\}$. For this aim one must consider the trace of the matrix $L^2(\lambda)$.

$$T(\lambda) = \frac{1}{2}TrL^2(\lambda). \qquad (12)$$

This operator has the important property

$$[T(\lambda), T(\mu)] = 0, \qquad (13)$$

$$[T(\lambda), B(\mu)] = g(\lambda-\mu)\big(A(\mu)B(\lambda) - B(\lambda)D(\mu)\big)+ \qquad (14)$$

$$+f(\lambda-\mu)\big(B(\mu)D(\mu)-A(\lambda)B(\mu)\big).$$

The pseudovacuum vector (10) is the eigenvector for operator (12)

$$T(\lambda)|0\rangle_M = t_{M,0}(\lambda)|0\rangle_M,\ \ t_{M,0}(\lambda)=\frac{1}{2}\big(a^2(\lambda)+d^2(\lambda)+a'(\lambda)-d'(\lambda)\big). \quad (15)$$

In the case of three-boson interaction, i.e. for (7),(8),(11) we have

$$T(\lambda)=\frac{1}{4}\lambda^2+(a_3^+a_3+\frac{1}{2})+S_3+\frac{1}{\lambda^2}\big(S_3^2+\frac{1}{2}(S_+S_-+S_-S_+)\big)+$$
$$+\frac{1}{\lambda}(a_3S_++a_3^+S_-), \quad (16)$$

$$t_{M,0}(\lambda)=\frac{1}{4}\lambda^2-\frac{M}{2}+\frac{1}{2}+\frac{1}{\lambda^2}\frac{M}{2}\left(\frac{M}{2}+1\right).$$

It is obvious that the operator $T(\lambda)$ commutes both with the whole Hamiltonian (6) and with its interacting part H_I

$$[T(\lambda),H]=[T(\lambda),H_I]=0 \quad (17)$$

and
$$H_I = res\ T(\lambda), \qquad \lambda\to 0. \quad (18)$$

We are lookihg for the eigenvectors of the operator $T(\lambda)$ in the form (1)

$$\Psi_{M,N}(\lambda_1,\ldots,\lambda_N)\equiv|\lambda_1,\ldots,\lambda_N\rangle_M=(N!)^{-\frac{1}{2}}B(\lambda_1)\cdots B(\lambda_N)|0\rangle_M. \quad (19)$$

Acting on it by operator (12) and using the commutation relations (5) and (14) one can find the equation [16]

$$T(\lambda)|\lambda_1,\ldots,\lambda_N\rangle_M= \quad (20)$$
$$=\left\{\sum_{i=1}^N f^2(\lambda-\lambda_i)-g^2(\lambda-\lambda_i)+\sum_{j\neq i}f(\lambda-\lambda_i)f(\lambda-\lambda_j)+t_{M,o}(\lambda)+\right.$$
$$\left.+\sum_{i=1}^N f(\lambda-\lambda_i)\big(d(\lambda)-a(\lambda)\big)\right\}|\lambda_1,\ldots,\lambda_N\rangle_M+$$
$$+\sum_{i=1}^N g(\lambda-\lambda_i)\Big\{a(\lambda_i)-d(\lambda_i)-2\sum_{j\neq i}f(\lambda_i-\lambda_j)\Big\}|\lambda_1,\ldots,\lambda_{i-1},\lambda,\lambda_{i+1},\ldots,\lambda_N\rangle_M.$$

As we want from vector (19) to be the eigenvector of operator (12), it follows from this equation that the additional terms in right-hand part of (20), containing vectors

$$\Psi^i_{M,N} = |\lambda_1, \ldots, \lambda_{i-1}, \lambda_0, \lambda_{i+1}, \ldots, \lambda_N\rangle, \tag{21}$$

must be equal to zero. If all these vectors are linearly independent and ortogonal to vector (19) we have the system

$$a(\lambda_i) - d(\lambda_i) - 2\sum_{j\neq i} f(\lambda_i - \lambda_j) = 0, \qquad i = 1, \ldots, N. \tag{22}$$

In the case of (9),(11) this system takes the form

$$\lambda_i + \sum_{j\neq i} \frac{2}{\lambda_i - \lambda_j} = \frac{M}{\lambda_i}, \qquad i = 1, \ldots, N. \tag{23}$$

This very system (23) is usually regarded as the system of Bethe equations for the three-boson interaction, see for example [16]. It is clear from (20) that the conditions (23) are necessary and sufficient for vector (19) to be the eigenvector of the operator (16) if only all vectors (21) are linearly independent, but in the case of three-boson interaction these vectors are linearly independent only if $N \leq M$, but if $N > M$ vectors (19),(21) are linearly dependent. As the result of this fact the conditions (23) can be weakened and replaced by the another system. Now we will find these equations.

3 Reduced system of Bethe equations

The explicit form of vector (19) was found for the three-boson interaction in [18].

$$\Psi_{M,N}(\lambda_1, \ldots, \lambda_N) = \tag{24}$$

$$= \sum_{k=0}^{\min(N,M)} \left(\frac{k!M!(N-k)!}{(M-k)!N!}\right)^{\frac{1}{2}} \Big(\sum_{i_1<\cdots<i_k} (\lambda_{i_1}\cdots\lambda_{i_k})^{-1}\Big)|k, M-k, N-k\rangle.$$

We see that for every $N > M$ this sum contains $(M+1)$ members, therefore among $(N+1)$ vectors (19),(21) only $(M+1)$ are linearly independent. Let us construct the set of linearly independent vectors. As the first member of this set we will take vector (24), the other M vectors have the form

$$\Psi_l \equiv |0, M, N\rangle - \left(\frac{(M-l)!N!}{l!M!(N-l)!}\right)^{\frac{1}{2}} \Big(\sum_{i_1<\cdots<i_l} (\lambda_{i_1}\ldots\lambda_{i_l})^{-1}\Big)^{-1} |l, M-l, N-l\rangle.$$

$$l = 1, \ldots, M. \tag{25}$$

All vectors (25) are linearly independent and ortogonal to the vector (24). Let us calculate the scalar product

$$c_l^j = \left\langle \Psi_l \mid \Psi_{M,N}^j \right\rangle = 1 - \Big(\sum_{i_1 < \cdots < i_l}^{j} (\lambda_{i_1} \ldots \lambda_{i_l})^{-1} \Big) \Big(\sum_{i_1 < \cdots < i_l} (\lambda_{i_1} \ldots \lambda_{i_l})^{-1} \Big)^{-1}. \tag{26}$$

in the sum $\sum^J$ parameter λ_j is replaced by the parameter λ.
Substituting in (20) vectors (25) instead of vectors (21) one gets the system

$$\sum_{i=1}^{N} g(\lambda - \lambda_i) \Big\{ a(\lambda_i) - d(\lambda_i) - 2 \sum_{j \neq i} f(\lambda_i - \lambda_j) \Big\} c_l^i = 0, \quad l = 1, \ldots, M. \tag{27}$$

Taking into account formulas (9), (11), system (27) can be transformed into the system

$$\sum_{i=1}^{N} (\lambda_i)^{-l} \Big(\lambda_i - \frac{M}{\lambda_i} + 2 \sum_{j \neq i} \frac{1}{\lambda_i - \lambda_j} \Big) = 0, \qquad l = 1, \ldots, M. \tag{28}$$

Thus, if $M < N$ one can use instead of usual system (23) the reduced system (28). System (23) contains N equations and N unknown parameters λ_i. Its solutions are points in N-dimensional complex space. System (28) contains M equations for N unknown parameters, its solutions are $(N - M)$-dimensional complex surfaces in N-dimensional space of parameters. Each surface consists of M connected components. All points from the same connected component give the same vector (24), so

$$\Psi_{M,N}(\lambda_1, \ldots, \lambda_N) \equiv \Psi_{M,N}(\mu_1, \ldots, \mu_N),$$

if $\{\lambda_i\}$ and $\{\mu_i\}$ belong to the same connected component. Equations (23) give us the specific points from each component. For $M = 1,\ N = 1, 2, \ldots, 14$ these points have been evaluated in [18]. Thus we see that the usual Bethe equations (23) give the correct set of parameters, but this set is not unique. Note, that equations (28), in contrast to (23), permit solutions containing equal $\lambda_i = \lambda_j$.

References

1. L.D. Faddeev, *Soviet Sci. Review* **C1**, 107 (1980).
2. P.P. Kulish and E.K. Sklyanin, *Lecture Notes in Physies* **151**, 61 (1981).

3. P.P. Bogoliubov, A.G. Izergin and V.E. Korepin, *Lecture Notes in Physics* **242**, 220 (1985), also *Quantum Inverse Scattering Method and Correlation Functions*, Cambridge Univ. Press (1993).
4. H.J. De Vega, *Int. J. Mod. Phys.* **4**, 2371 (1989).
5. B.S. Shastry, *J. Stat. Phys.* **50** 57 (1988).
6. M. Wadati, E.Olmedilla and Y. Akutsu, *J. Phys. Soc. Jap.* **56**, 1340 (1987).
7. A.G. Izergin and V.E.Korepin, *Lett. Math. Phys.* **5**, 199 (1981).
8. E.K. Sklyanin, *Nucl. Phys.B* **326**, 719 (1989).
9. P.P. Kulish, *Lett. Math. Phys.* **5**, 191 (1981).
10. Gaudin, *Note CEA* **1559**, 1 (1972); *CEN-Saclay* **1559**, 2 (1973.
11. E.K. Sklyanin, *Zap. Nauchn. Sem. LOMI* **164**, 151 (1987).
12. V.B. Kuznetsov, *J. Math. Phys.* **33**, 3240 (1992).
13. A.J. Mcfarlane, *Phys. Lett. B* **268**, 71 (1994), *ibid.*, 3261 (1994).
14. D.R.D. Scott, *J. Math. Phys.* **35**, 5831 (1994).
15. B. Jurcŏ, *J. Math. Phys.* **30**, 1289;1739 (1990).
16. V.I. Rupasov, *ZhETF* **83**, 1711 (1982), *ibid.* **171**, 145 (1992).
17. A.R. Its, A.G. Izergin, V.E.Korepin and N.A.Slavnov, *Int. J. Mod. Phys. B* **4**, 1003 (1990).
18. V.E. Korepin and N.A. Slavnov, *Commun. Math. Phys.* **136**, 633 (1991).
19. A.G.Choudhury and A.R.Chowdhyry, *Phys. Lett. A* **220**, 205 (1996).
20. V.A. Andreev and O.A. Ivanova, *Phys. Lett. A* **216**, 76 (1996).

CONSTRUCTION OF WAVELETS BASIS FOR THE FIBONACCI CHAIN VIA THE SPLINE FUNCTIONS OF VARIOUS ORDERS

MIROSLAV ANDRLE

Laboratoire de Physique Théorique de la Matière Condensée, Université Paris 7 – Denis Diderot, 75251 Paris Cedex 05, France
and
Department of Mathematics, Faculty of Nuclear Sciences and Physical Engineering, Czech Technical University in Prague, Trojanova 13, 120 00, Prague 2, Czech Republic
E-mail: andrle@ccr.jussieu.fr

We present here wavelets of class $C^n(\mathbb{R})$ living on a sequence of aperiodic discretization of $\mathbb{R}$, known as the Fibonacci chain, constructed via the splines functions. The construction method is based on an algorithm published by G. Bernau. Corresponding multiresolution analysis is defined and numerical examples of linear scaling functions and wavelets are shown.

1 Introduction

Wavelets are functions generated obviously from one basic function by dilations and translations. They are used as analyzing tools in harmonic analysis, signal analysis, for numerical solving of differential equations etc. A particulary interesting was discovery of orthonormal bases of wavelets. It means functions $\psi(x) \in L^2(\mathbb{R})$, such that the family of functions

$$\psi_{j,k}(x) = 2^{j/2}\psi(2^j x - k), \quad \text{where } j,k \in \mathbb{Z}, \tag{1}$$

forms an orthonormal basis for $L^2(\mathbb{R})$. Wavelets are usually constructed using the multiresolution analysis.

In previous articles [1,2,3] we presented a basic definition of multiresolution analysis for an infinite sequence $\cdots \subset \Lambda/\tau^{2(j-1)} \subset \Lambda/\tau^{2j} \subset \Lambda/\tau^{2(j+1)} \subset \cdots$ of aperiodic discretization of $\mathbb{R}$ and the elementary Haar example (piece-wise constant functions) was introduced. Starting from the Haar case we would like to explore smoother wavelets, wavelets of class C^n, $n \geq 0$ with compact supports. Let us remark that we don't demand the orthogonality conditions because it takes rather strong constraints for the construction. We will follow a wavelet construction published by G. Bernuau in his PhD work[4] (see also [5]). Let us briefly introduce the general settings of the Bernuau's construction. At first we say some basic definitions.

Definition 1 *$\Lambda \subset \mathbb{R}$ is a Delone point set in the real line if and only if Λ is* uniformly discrete *(the distances between any pair of points in Λ are greater than a fixed $r > 0$) and* relatively dense *(there exists $R > 0$ such that $\mathbb{R}$ is covered by intervals of length $2R$ centered at points of Λ).*
We will demand for our method two additional properties of Delone sets, the set Λ has to be

1. self-similar: there exists a number $\theta > 1$ (*inflation factor*) such that
$$\theta\Lambda \subset \Lambda, \tag{2}$$
2. locally finite: for all $R > 0$, the family of finite point sets
$$\{(\Lambda - \lambda) \cap (-R, R)\}_{\lambda\in\Lambda} \tag{3}$$
is finite. One also says that Λ has locally finite complexity, and this means that local environments of points in Λ are not different in infinite fashions.

Typically, such sets Λ are mathematical models for one-dimensional structures having a long-range order, like quasicrystals. Our aim in this paper is to construct a Riesz basis of $L^2(\mathbb{R})$ in the form:
$$\theta^j \psi_\lambda(\theta^j x - \lambda),\ j \in \mathbb{Z},\ \lambda \in \Lambda, \tag{4}$$
for which the set of *mother wavelets* $\{\psi_\lambda(x),\ \lambda \in \Lambda\}$ is finite and the simplest possible.

2 Space of spline functions for multiresolution analysis

2.1 The definition of the space $V_0^{(s)}(\Lambda)$

Any Delone set Λ determines a space of splines of order s in the following way
Definition 2 *Let $s \geq 2$. Then $V_0^{(s)}(\Lambda)$ is the closed subspace of $L^2(\mathbb{R})$ defined by $V_0^{(s)}(\Lambda) = \left\{f(x) \in L^2(\mathbb{R}) \mid \frac{\mathrm{d}^s}{\mathrm{d}x^s} f(x) = \sum_{\lambda\in\Lambda} a_\lambda \delta_\lambda\right\}$.*
Let us rewrite Λ as $\Lambda = \{\lambda_n\}_{n\in\mathbb{Z}}$, where $\cdots\lambda_{n-1} < \lambda_n < \lambda_{n+1}\cdots$. Then an equivalent definition is $V_0^{(s)}(\Lambda) = \{f \in C^{s-2},\ f \in L^2(\mathbb{R}),\ f|_{[\lambda_n,\lambda_{n+1}]}$ is a polynomial of degree $\leq s-1\}$. So $V_0^{(s)}(\Lambda)$ is the space of splines of order s with nodes in Λ. Let us now recall a classical result about the existence of a Riesz basis for $V_0^{(s)}(\Lambda)$ by Schumaker[6].
Theorem 1 *For all Delone sets $\Lambda = \{\lambda_n\}_{n\in\mathbb{Z}} \subset \mathbb{R}$ and for all $s \geq 2$, there exists a Riesz basis $\{B_\lambda^{(s)}, \lambda \in \Lambda\}$ of $V_0^{(s)}(\Lambda)$. The function $B_\lambda^{(s)}$ (called B-spline) is the unique function in $V_0^{(s)}(\Lambda)$ satisfying the following conditions:*

(i) supp $B_\lambda^{(s)} = [\lambda, \lambda']$, *where* $\lambda' \in \Lambda$.
(ii) *The interval* (λ, λ') *contains exactly* $s-1$ *points of* Λ.
(iii) $\int_{\mathbb{R}} B_\lambda^{(s)} = \frac{\lambda'-\lambda}{s}$.

Note that (i) and (ii) give the exact information about the (compact) support of $B_\lambda^{(s)}$ whilst (iii) is just a normalization condition.

2.2 Construction of the Riesz basis of $V_0^{(s)}(\Lambda)$

The construction of $B_\lambda^{(s)}(x)$ can be done using several methods:

- by recurrence [7], with the use of the formula for all $\lambda_n \in \Lambda$:

$$B_{\lambda_n}^{(1)}(x) = \begin{cases} 1 & \text{if } \lambda_n \le x \le \lambda_{n+1}, \\ 0 & \text{otherwise,} \end{cases} \tag{5}$$

$$B_{\lambda_n}^{(s)}(x) = \omega_{s,n}(x) B_{\lambda_n}^{(s-1)}(x) + (1-\omega_{s,n+1}(x))\, B_{\lambda_{n+1}}^{(s-1)}(x), \text{ for } s \ge 2 \tag{6}$$

$$\text{with } \omega_{s,n}(x) = \begin{cases} (x-\lambda_n)/(\lambda_{n+s-1}-\lambda_n) & \text{if } \lambda_n < \lambda_{n+s-1}, \\ 0 & \text{otherwise,} \end{cases}$$

- by the condition of minimal support which means that $\forall f \in V_0^{(s)}(\Lambda)$,

$$\text{supp } f \subset \text{supp } B_\lambda^{(s)} \Longrightarrow f \propto B_\lambda^{(s)}, \tag{7}$$

- by Fourier transform, see [4]

Proposition 1 *Let* $\Lambda \subset \mathbb{R}$ *be a Delone set of finite type. Then the B-splines of order* s *based on* Λ *are of the form* $B_\lambda^{(s)}(x) = \phi_\lambda(x-\lambda)$, $\lambda \in \Lambda$, *where the set* $\{\phi_\lambda(x),\ \lambda \in \Lambda\}$ *is a finite set of functions with compact support.*

This proposition results from the construction of functions $B_\lambda^{(s)}$ by Fourier transformation (see proof in [4]). In the finite type case of Delone sets, it is possible to partition the indexing set $\mathbb{Z}$ for Λ into a finite set of equivalence classes $\bar{n}$, $\mathbb{Z} = \cup_{\bar{n}=\bar{0}}^{\bar{q}} \bar{n}$ where class $\bar{n}$ is defined by

$$\bar{n} = \left\{ n' \in \mathbb{Z} \mid \forall x\ B_{\lambda_{n'}}^{(s)}(x+\lambda_{n'}) = B_{\lambda_n}^{(s)}(x+\lambda_n) \right\}.$$

For a given s, the point set Λ is partitioned into $\Lambda = \cup_{\bar{n}=\bar{0}}^{\bar{q}} \Lambda_{\bar{n}}$ with $\Lambda_{\bar{n}} = \{\lambda_k,\ k \in \bar{n}\}$. Each class $\bar{n}$ is associated with the function $\phi_{\bar{n}}(x) \equiv \phi_{\lambda_k}(x) = B_{\lambda_k}^{(s)}(x+\lambda_k)$, $k \in \bar{n}$. Following this way, the space $V_0^{(s)}(\Lambda)$ decomposes into the direct sum

$$V_0^{(s)}(\Lambda) = \bigoplus_{\bar{n}=\bar{0}}^{\bar{q}} V_{0,\bar{n}}, \tag{8}$$

where $V_{0,\bar{n}}$ is the closure of the linear span of the functions $\phi_{\bar{n}}(x-\lambda_k)$, $k \in \bar{n}$.

3 Multiresolution analysis

Let Λ be a Delone set of finite type and self-similar with inflation factor $\theta > 1$. $(\theta\Lambda \subset \Lambda)$. Changing the scale allows us to define subspaces $V_j^{(s)}(\Lambda),\ j \in \mathbb{Z}$.

Definition 3 $V_j^{(s)}(\Lambda) = \left\{ f(x) \in L^2(\mathbb{R}) \mid \frac{\mathrm{d}^s}{\mathrm{d}x^s} f(x) = \sum_{\lambda\in\Lambda} a_\lambda \delta_{\theta^{-j}\lambda} \right\}$.

For this sequence of nested subspaces the following proposition holds.

Proposition 2 *The sequence of subspaces $(V_j^{(s)}(\Lambda))_{j\in\mathbb{Z}}$ is a multiresolution analysis of $L^2(\mathbb{R})$, i.e.*

(i) *for any $j \in \mathbb{Z}$, $V_j^{(s)}(\Lambda)$ is a closed subspace of $L^2(\mathbb{R})$,*

(ii) $\cdots \subset V_{-1}^{(s)}(\Lambda) \subset V_0^{(s)}(\Lambda) \subset V_1^{(s)}(\Lambda) \subset \cdots$,

(iii) $\bigcup_{j\in\mathbb{Z}} V_j^{(s)}(\Lambda)$ *is dense in $L^2(\mathbb{R})$,*

(iv) $\bigcap_{j\in\mathbb{Z}} V_j^{(s)}(\Lambda) = \{0\}$,

(v) $f(x) \in V_j^{(s)}(\Lambda)$ *if and only if* $f(\theta^{-j}x) \in V_0^{(s)}(\Lambda)$,

(vi) *there exists a finite number of functions $\varphi_{\bar{n}}(x) \in V_0^{(s)}$, called* scaling functions *such that $\{\varphi_{\bar{n}}(x - \lambda_n)\}_{n\in\bar{n},\, \bar{0}\leq\bar{n}\leq\bar{q}}$ is a Riesz basis in $V_0^{(s)}$.*

4 Construction of spline wavelet basis

The whole wavelet basis could be obtained by a standard method. We will find the orthogonal complement of $V_{-1}^{(s)}(\Lambda)$ in $V_0^{(s)}(\Lambda)$:

$$V_0^{(s)}(\Lambda) = V_{-1}^{(s)}(\Lambda) \oplus_\perp W_{-1}^{(s)}(\Lambda) \text{ and } V_{j+1}^{(s)}(\Lambda) = V_j^{(s)}(\Lambda) \oplus_\perp W_j^{(s)}(\Lambda). \tag{9}$$

for general j. Then it is not difficult to see that

$$L^2(\mathbb{R}) = \bigoplus_{-\infty}^{\infty} W_j^{(s)}(\Lambda). \tag{10}$$

This represents our type of basis presented in (4). An important step in the characterization of Riesz basis of $W_{-1}^{(s)}(\Lambda)$ is the following theorem.

Theorem 2 *Let $\Lambda \in \mathbb{R}$ be a Delone set of finite type, self-similar with the factor $\theta > 1$. Then $\forall s \geq 2$ there exists a Riesz basis of $W_{-1}^{(s)}(\Lambda)$ of the form*

$$\left\{ \psi_{\lambda_n}^{(s)}(x - \lambda_n),\ \lambda_{n+1} \in \Lambda \setminus \theta\Lambda \right\}, \quad \text{where } \left\{ \psi_{\lambda_n}^{(s)}(x),\ \lambda_{n+1} \in \Lambda \setminus \theta\Lambda \right\}$$

is a finite set of functions.

The proof goes through a sequence of intermediate results which we won't mention here. A characterization of $W_{-1}^{(s)}$ is given by the following proposition. Let us recall $H^s(\mathbb{R})$ denotes the Sobolev space of functions $f \in L^2(\mathbb{R})$ such that $\frac{\mathrm{d}^\alpha}{\mathrm{d}x^\alpha} f$, with $\alpha \leq s$, is a element of $L^2(\mathbb{R})$.

Proposition 3 *Let us denote by $H^s(\mathbb{R})$ the Sobolev space of order s. Then, for $s \geq 2$, $W_{-1}^{(s)}$ is the space of sth derivatives of elements in $H^{2s}(\mathbb{R})$ which vanish on $\theta\Lambda$:*

$$W_{-1}^{(s)}(\Lambda) \equiv \frac{\mathrm{d}^s}{\mathrm{d}x^s} K_{2s} = \left\{ \frac{\mathrm{d}^s}{\mathrm{d}x^s} h,\ h \in H^{2s}(\mathbb{R}) \mid \forall \lambda \in \theta\Lambda,\ h(\lambda) = 0 \right\}.$$

Using this proposition we know where to find the functions from $W_{-1}^{(s)}$.

5 The Fibonacci chain

The Fibonacci chain, a discretization of $\mathbb{R}$, is the set which could be generated by the two-letter tiling of the real line. This tiling is based on the substitution

$$\varsigma : \begin{cases} \mathrm{L} \to \mathrm{LLS}, \\ \mathrm{S} \to \mathrm{LS}, \end{cases} \tag{11}$$

for which the substitution matrix has eigenvalues $\tau^2 = (3+\sqrt{5})/2$ and $1/\tau^2 = (3-\sqrt{5})/2$. Therefore, the Pisot algebraic integer τ^2 is the scaling factor of the substitution ς. Let us associate to the letters L and S the tiles of length 1 and $1/\tau$ respectively. Starting from the origin with L on the right and with S on the left, we apply the substitution ς^∞. We will get the point set Λ as the set of the nodes of this tiling on the real line. The first pieces of this tiling around the origin are shown in the picture below. In algebraic

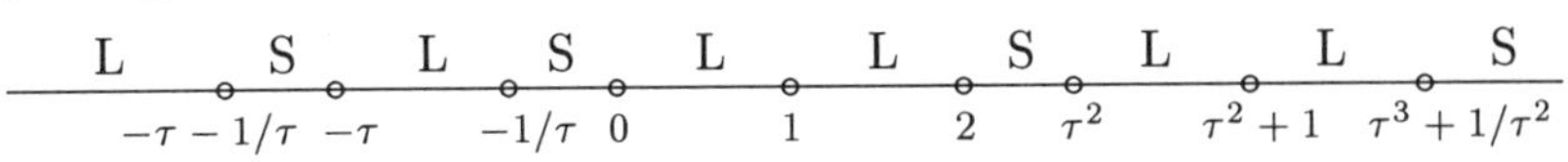

Figure 1. Fibonacci chain around the point 0.

terms, $\Lambda = \{\lambda = m + n\tau;\ m, n \in \mathbb{Z} \mid \lambda' = m - n\tau^{-1} \in [0, \tau^2)\}$, where λ' is the Galois conjugate of λ. Let us note that the set Λ is a set of finite type, self-similar with inflation factor τ^2 ($\tau^2\Lambda \subset \Lambda$) and for Λ we have for any $s \in \mathbb{N}$ decomposition $\Lambda = \cup_{i=-s}^{0} \Lambda_{\Omega_i}$ where Λ_{Ω_i} is the set of points of the beginning of s-letters admissible world Ω_i (sequences of L and S). There are only $s+1$ words build up from s letters (the substitution is sturmian with the minimal complexity). We can easily find all this admissible words. The only simple rule is that the begin of the word Ω_i is the point λ_i if we denote $\Lambda = \{\lambda_n\}_{n \in \mathbb{Z}}$, where $\cdots \lambda_{n-1} < \lambda_n < \lambda_{n+1} \cdots$ and $\lambda_0 = 0$.

6 Splines for the Fibonacci chain

In this case the inflation factor is $\theta = \tau^2$. Accordingly to the Def. 2 we have $V_0^{(s)}(\Lambda) = \left\{ f \in C^{s-2}, f \in L^2(\mathbb{R}),\ f|_{[\lambda_n, \lambda_{n+1}]} \text{ is a polynom. of degree} \leq s-1 \right\},$

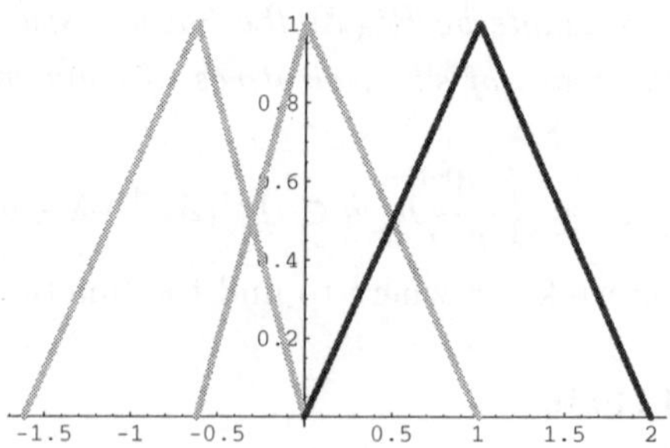

Figure 2. Linear scaling functions $\phi_{\mathrm{LS}}(x+\tau)$, $\phi_{\mathrm{SL}}(x+1/\tau)$ and $\phi_{\mathrm{LL}}(x)$.

for $s \geq 2$. The Riesz basis of $V_0^{(s)}(\Lambda)$ is $s+1$ functions $\phi_{\Omega_{-s}}, \ldots, \phi_{\Omega_0}$, where Ω_i, $i = -s, \ldots, 0$ are all admissible s-letters words. Let consider that the support corresponding to the word Ω_i is the interval $[\lambda_i, \lambda_{i+s}]$. Then we can construct all $s+1$ functions $\phi_{\Omega_i} \equiv \phi_i$ through the conditions below:

- $\phi_i(x - \lambda_i) = 0$ if $x \notin (\lambda_i, \lambda_{i+s})$,
- $\phi_i(x - \lambda_i)|_{[\lambda_{i+k}, \lambda_{i+k+1}]}$ is a polynomial of the degree $s-1$ for $k \leq s-1$,
- $\phi_i^{(l)}(x - \lambda_i)$, $0 \leq l \leq s-2$ are continuous functions also in the points λ_{i+k}, $0 \leq k < s$,
- $\int_{\mathbb{R}} \phi_i(x)\,\mathrm{d}x = \frac{\lambda_{i+s} - \lambda_i}{s}$.

6.1 *Splines of order $s = 2$ for the Fibonacci chain*

From the figure 1 and also the substitution rules (11) we can see that in this case there are just 3 admissible words LL, LS and SL. So the Riesz basis of $V_0^{(2)}(\Lambda)$ can be build by admissible translations of these functions (see their graphs on the figure 2):

$$\phi_{\mathrm{LL}}(x) = \begin{cases} x & x \in [0,1] \\ 2-x & x \in [1,2] \end{cases} \tag{12}$$

$$\phi_{\mathrm{LS}}(x-1) = \begin{cases} x-1 & x \in [1,2] \\ \tau^2 - \tau(x-1) & x \in [2,\tau^2] \end{cases} \tag{13}$$

$$\phi_{\mathrm{SL}}(x+1/\tau) = \begin{cases} \tau(x+1/\tau) & x \in [-1/\tau, 0] \\ \tau - (x+1/\tau) & x \in [0,1] \end{cases} \tag{14}$$

The Riesz basis of $V_0^{(2)}(\Lambda)$ is the set of functions

$$\{\phi_{\mathrm{LL}}(x-\lambda_{\mathrm{LL}}),\ \phi_{\mathrm{LS}}(x-\lambda_{\mathrm{LS}}),\ \phi_{\mathrm{SL}}(x-\lambda_{\mathrm{SL}})\}_{\lambda_{\mathrm{LL}}\in\Lambda_{\mathrm{LL}},\lambda_{\mathrm{LS}}\in\Lambda_{\mathrm{LS}},\lambda_{\mathrm{SL}}\in\Lambda_{\mathrm{SL}}}.$$

According to the proposition 3 we will construct the wavelet basis of $W_{-1}^{(2)}(\Lambda)$. The supports of *mother wavelets* was found LLSLS, LSLSLL, LSLLS, LLSLL.

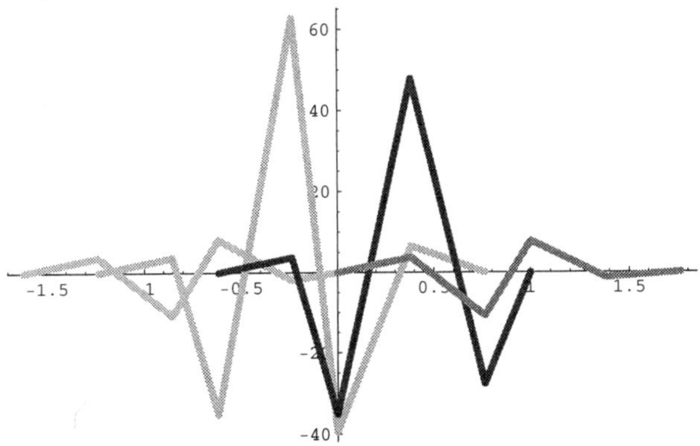

Figure 3. The *mother wavelets* of basis of $W_0^{(2)}(\Lambda)$

7 Conclusions

There has been presented a general method of construction spline wavelets corresponding to the Fibonacci chain. As we said in the beginning out wavelets are not orthogonal, so if we want to do some signal analysis we need to find the biorthogonal wavelet basis. How to construct corresponding biortogonal basis is our next aim.

Acknowledgment

I'am grateful to my professor J.P. Gazeau for his big patience and help.

References

1. M. Andrle, *Czech Journal of Physics* **51**, 285 (2001).
2. M Andrle, in *The sixth international school of theoretical physics "Symmetry and structural properties of condensed matter"*, ed. T Lulek, World Scientific, Singapore (2001).
3. M. Andrle *et al.*,*J. Phys A: Math. Gen.* **33**, 47 (2000).

4. G. Bernuau, *Propriétés spectrales et géométriques des quasicristaux. Ondelettes adaptées aux quasicristaux*, PhD thesis, Ceremade, Université Paris IX Dauphine, France, (1998).
5. G. Bernuau, *J. Math. Phys.* **39**, 4213 (1998).
6. L.L. Schumaker, *Spline Functions: Basic Theory,* Wiley, New-York, (1981).
7. J.J. Risler, *Mathématiques pour la CAO*, Masson, Paris (1991).

ALGEBRAIC MODEL OF MOLECULAR THERMODYNAMICS

MAIA ANGELOVA

School of Computing and Mathematics
University of Northumbria, Newcastle upon Tyne, UK GB-NE1 8ST

A. FRANK

Instituto de Ciencias Nucleares and Centro de Ciencias Físicas
UNAM, A.P. 70-543, Mexico, D.F., 04510 Mexico

A molecular algebraic model is applied to the analysis of thermodynamic properties of molecules. The local anharmonic effects are described by anharmonic bosons associated with the $SU(2)$ algebra. The vibrational high-temperature partition function and the related thermodynamic potentials are derived in terms of the parameters of the model. Quantum analogues of anharmonic bosons, q-bosons, are used to describe anharmonic properties of molecules. A physical interpretation of a linear quantum deformation associated with the model is given.

1 Introduction

The algebraic models used in molecular physics [1,2] combine Lie algebraic techniques, describing the interatomic interactions, with discrete symmetry techniques associated with the local symmetry of the molecules . The algebraic model [1] exploits the isomorphism of $SU(2)$ algebra and the one-dimensional Morse oscillator. The anharmonic effects are described by anharmonic boson operators [1], $\hat{b}$, $\hat{b}^\dagger$ and $\hat{v}$, which satisfy the commutation relations,

$$\left[\hat{b}, \hat{v}\right] = \hat{b}, \qquad \left[\hat{b}^\dagger, \hat{v}\right] = -\hat{b}^\dagger, \qquad \left[\hat{b}, \hat{b}^\dagger\right] = 1 - \frac{2\hat{v}}{N} \tag{1}$$

where $\hat{v}$ is the Morse phonon operator with an eigenvalue v. N is the total number of bosons fixed by the potential shape, and v is the number of quanta in the oscillator, $v = 1, 2, \ldots, \left[\frac{N}{2}\right]$. The value of N is dependent on the depth D and the width d of the Morse potential well [3],

$$N + 1 = \left(\frac{8\mu D d^2}{\hbar^2}\right)^{\frac{1}{2}} \tag{2}$$

where μ is the mass of the Morse oscillator. The harmonic limit is obtained when $N \to \infty$, in which case $\left[\hat{b}, \hat{b}^\dagger\right] \to 1$ giving the usual boson commutation relations.

The Morse Hamiltonian can be written in terms of $\hat{b}$ and $\hat{b}^{\dagger}$,

$$\mathcal{H}_M \sim \frac{1}{2}\left(\hat{b}\hat{b}^{\dagger} + \hat{b}^{\dagger}\hat{b}\right) \tag{3}$$

which corresponds to vibrational energies

$$\varepsilon_v = \hbar\omega_0\left(v + \frac{1}{2} - \frac{v^2}{N}\right), \; v = 1, 2, \ldots, \left[\frac{N}{2}\right] \tag{4}$$

where ω_0 is the harmonic oscillator frequency. The Morse potential leads to a deformation of the harmonic oscillator algebra. For an infinite potential depth, $N \to \infty$, the Morse potential cannot be distinguished from the harmonic potential.

The anharmonic model provides a systematic procedure for studying vibrational excitations in a simple form by describing the stretching and bending modes in a unified scheme based on $U(2)$ algebras which incorporate the anharmonicity at the local level [3-7]. In this paper we apply the algebraic approach to the vibrational high-temperature thermodynamics of diatomic molecules. We study the partition function [8] and the basic thermodynamic properties in terms of the parameters of the model. The idea of critical temperature is introduced. We show that the linear quantum deformation [9] is related to the fixed number of anharmonic bosons and the shape of the anharmonic potential. The results are applied to the diatomic molecule of HCl.

2 Vibrational Partition Function

The vibrational partition function of a diatomic anharmonic molecule is

$$Z_N = \sum_{v=0}^{[N/2]} e^{-\beta\varepsilon_v} \tag{5}$$

where $\beta = 1/k_B T$, the vibrational energies ε_v are given by equation (4) and N is the fixed total number of anharmonic bosons. Introducing new parameters, $\alpha = \frac{\beta\hbar\omega_0}{2}$, $N_0 = \left[\frac{N}{2}\right]$ and $l = N_0 - v$, the exact value of vibrational partition function can be written as,

$$Z_N = e^{-\alpha}\sum_{l=0}^{N_0} e^{-\frac{\alpha}{N_0}\left(N_0^2 - l^2\right)}. \tag{6}$$

At high temperatures T, for N_0 large and α small, the sum can be replaced by the integral, which can be evaluated exactly in terms of the error function,

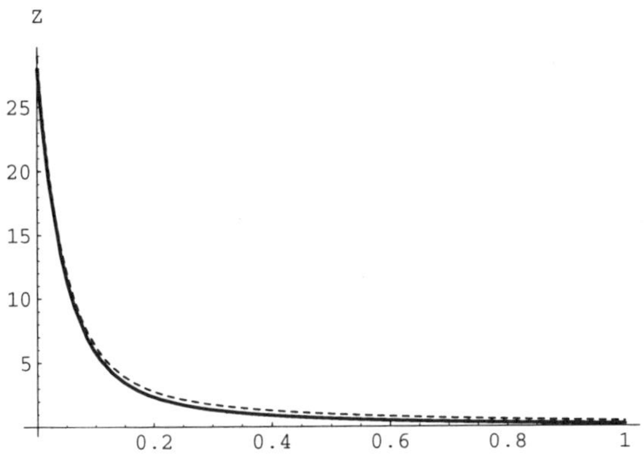

Figure 1. Vibrational partition function Z_{56} as a function of α

$\mathrm{erf}\,i\big(\sqrt{\alpha N_0}\big)$ (as defined in [10]),

$$Z_N = \frac{1}{2}\sqrt{\frac{N_0\pi}{\alpha}}e^{-\alpha(N_0+1)}\mathrm{erf}\,i\left(\sqrt{\alpha N_0}\right). \tag{7}$$

Equation (7) represents the high-temperature value of the vibrational partition function in the Morse-like spectrum. When $N_0 \to \infty$, we obtain [8] the classic result for the harmonic vibrational partition function of a diatomic molecule at high temperatures. The partition function [11] is evaluated using an approximation of the classical density of states.

Using the values of the harmonic and anharmonic constants [12] for the zero lines of the molecule of $^1H^{35}Cl$, and the relations between these constants and the parameters of the model [4], we obtain $N = 56$, $N_0 = 28$. Figure 1 represents the dependence of the partition function on the parameter α in the high-temperature region, $0 < \alpha \leq 1$. The partition function Z_{56} from equation (7) is given with thick line. The exact partition function from equation (6) is given for comparison (dashed line). It is clear that the integral approximation (7) is in a very good agreement with the exact representation (6) and does not change the value and appearance of the partition function.

3 Molecular Thermodynamics

The mean vibrational energy is given by

$$U_N = -\frac{\partial}{\partial\beta}\ln Z_N = -\frac{\hbar\omega_0}{2Z_N}\frac{\partial Z_N}{\partial\alpha}. \tag{8}$$

Taking into account equation (7), we obtain the following expression for the mean vibrational energy in terms of the parameters of the model,

$$U_N = \frac{\hbar\omega_0}{2}\left(1 + N_0 + \frac{1}{2\alpha} - \sqrt{\frac{N_0}{\alpha\pi}}\frac{e^{\alpha N_0}}{\mathrm{erf}\,i\left(\sqrt{\alpha N_0}\right)}\right). \tag{9}$$

The vibrational part of the specific heat is,

$$C_N = \frac{\partial U_N}{\partial T} = -\frac{\hbar\omega_0}{2k_B T^2}\frac{\partial U_N}{\partial \alpha}. \tag{10}$$

Using equation (9), we obtain the specific heat in terms of α and N_0,

$$C_N = \frac{k_B}{2} + k_B\sqrt{\frac{\alpha N_0}{\pi}}\frac{e^{\alpha N_0}}{\mathrm{erf}\,i\left(\sqrt{\alpha N_0}\right)}\left(\alpha N_0 - \frac{1}{2} - \sqrt{\frac{\alpha N_0}{\pi}}\frac{e^{\alpha N_0}}{\mathrm{erf}\,i\left(\sqrt{\alpha N_0}\right)}\right) \tag{11}$$

It is clear from the relation (11) that all anharmonic contributions to the vibrational part of the partition function depend on the temperature. When $N_0 \to \infty$, the harmonic limit of the model gives the vibrational specific heat of a diatomic molecule at very high temperatures, $C_\infty \sim k_B$.

Figure 2 represents the dependence of the vibrational specific heat, $2C_{56}/k_B$, on the parameter α for the molecule $^1\mathrm{H}^{35}\mathrm{Cl}$ (thick line) in the high-temperature region. For comparison the harmonic limit $2C_\infty/k_B$ is also given (dashed line). The graph shows that the specific heat has a maximum

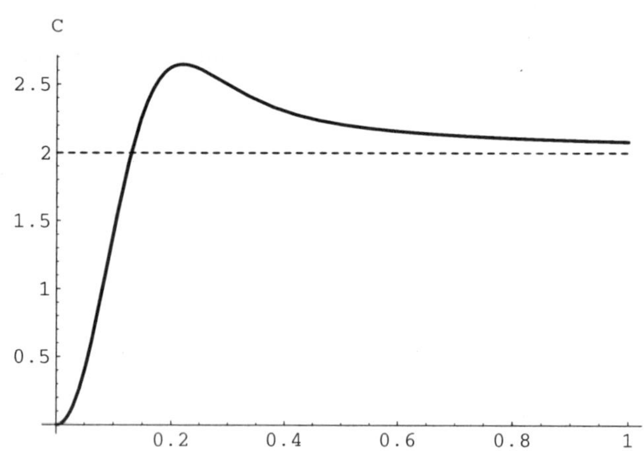

Figure 2. Vibrational specific heat $2C_{56}/k_B$ as a function of α.

for a value of the parameter α_C which corresponds to a critical temperature, T_C. For $^1Na^{35}Cl$, the critical temperature is $T_C = 8895$K. The anomaly of the specific heat is a result of the fixed number of the anharmonic bosons in the algebraic model.

4 Anharmonic Bosons

The mean vibrational energy in the anharmonic model can be written in terms of mean number $\langle \nu_N \rangle$ of anharmonic quanta, each with energy $\hbar\omega_0$,

$$U_N = \hbar\omega_0 \left(\langle \nu_N \rangle + \frac{1}{2} \right) \tag{12}$$

Substituting U_N by equation (9), we obtain the high-temperature value for the mean number of anharmonic bosons,

$$\langle \nu_N \rangle = \frac{N_0}{2} + \frac{1}{4\alpha} - \sqrt{\frac{N_0}{4\pi\alpha}} \frac{e^{\alpha N_0}}{\operatorname{erf} i \left(\sqrt{\alpha N_0} \right)} \tag{13}$$

In our paper [9], we have shown that the anharmonic bosons $\hat{b}$ and $\hat{b}^\dagger$, can be obtained as an approximation of the q-bosons [13,14], which enter the Heisenberg-Weyl q-algebra HW_q by the following commutation relations:

$$[a, a^\dagger] = q^{\hat{n}} \ , \quad [\hat{n}, a] = -a \ , \quad [\hat{n}, a^\dagger] = a^\dagger \tag{14}$$

Here, the deformation parameter q is in general a complex number. As shown in [9], the anharmonic commutation relations (1) can be recovered for real values of the deformation q close to 1, $q < 1$, and a linear expansion of q in terms of a parameter p,

$$q^{\hat{n}} = 1 - \frac{\hat{n}}{p} \ . \tag{15}$$

where $p \equiv 1/(1-q)$. If we substitute the parameter p with $N/2$, $\hat{n}$ with $\hat{v}$ in (15), and the creation and annihilation operators $a, a^\dagger$, with $b, b^\dagger$ in (14), the $SU(2)$ anharmonic relations (1) are recovered. The form (1) of the $SU(2)$ commutation relations can be considered as a deformation of the usual (harmonic oscillator) commutation relations, with a deformation parameter $N = 2p$.

This gives a physical interpretation of the the quantum deformation. The quantum deformation parameter p is the fixed number N_0 of the anharmonic bosons in the oscillator. The quantum deformation is also related to the shape of the Morse potential well (2).

Now, we can substitute $N_0 = p$ in the expressions for the partition function (7), mean energy (9), specific heat (11) and mean number of anharmonic bosons (13), and obtain the thermodynamic properties of diatomic molecules in terms of the deformation parameter p. For the molecule $^1\mathrm{Na}^{35}\mathrm{Cl}$, $p = 28$, which gives $q = 27/28$.

5 Conclusion

We have applied the algebraic model to those thermodynamic properties of diatomic molecules which at high temperature strongly depend on its anharmonicity. We have introduced the idea of critical temperature, characterising the anharmonic vibrations. We have shown that a linear quantum deformation, related to the shape of the Morse potential, can be associated with the anharmonic bosons. The model can be further generalised for the thermodynamics of polyatomic molecules, solids and plasma.

References

1. A. Frank A and P. Van Isacker,*Algebraic Methods in Molecular and Nuclear Structure Physics* , John Wiley (1994) .
2. F. Iachello and R. Levine, *Algebraic Theory of Molecules* , OUP (1995).
3. A. Frank, R. Lemus, R. Bijker, F. Pérez-Bernal and J.M. Arias, *Annals of Physics* **252**, 211 (1996).
4. F. Iachello, *Chem. Phys. Lett.* **78**, 581 (1981).
5. A. Frank, R. Lemus, M. Carvajal, C. Jung and E. Ziemniak, *Chem. Phys. Lett.* ***308***, 91 (1999).
6. F. Pérez-Bernal and J.M. Arias, A. Frank, R. Lemus and R. Bijker, *J. Mol. Spectrosc.* **184**, 1 (1997).
7. S. Oss, *Adv. Chem. Phys.* *43*, 455 (1996).
8. Maia Angelova and A. Frank. Anharmonic Algebraic Model of Thermodynamic Properties of Diatomic Molecules, preprint cond-mat/*0105554*.
9. M. Angelova, V. Dobrev and A. Frank, *J Phys A***34**, L503 (2001), preprint cond-mat/*0105193*.
10. *Handbook of Mathematical Functions*, Edited by M. Abramowitz and I.A. Stegun (National Bureau of Standards 1964).
11. D. Kusnetzov, *J. Chem. Phys.* **101** , 2289 (1994).
12. G. Herzberg, *Molecular Spectra and Molecular Structure*, volume II, Van Nostrand (1945).
13. L. C. Biedenharn, *J. Phys. A: Math. Gen.* **22**, L873 (1989).
14. A. J. Macfarlane, *J. Phys. A: Math. Gen.* **22**, 4581 (1989).

INDUCED REPRESENTATIONS OF BICROSSPRODUCT ALGEBRAS

OSCAR ARRATIA
Departamento de Matemática Aplicada a la Ingeniería, Universidad de Valladolid. E-47011, Valladolid, Spain.
E-mail: oscarr@wmatem.eis.uva.es

MARIANO A. DEL OLMO
Departamento de Física Teórica, Universidad de Valladolid, E-47011, Valladolid, Spain
E-mail: olmo@fta.uva.es

We present a method to construct representations of quantum algebras having the structure of bicrossproduct. This method is, in some sense, the quantum counterpart of the Mackey method for Lie groups with structure of semidirect product. As an example we consider the quantum κ–Galilei algebra.

1 Introduction

In the last years we have developed in a series of works [1,2,3] a procedure to construct induced representations of quantum algebras. Our method is algebraic and makes use of objects like modules and comodules and concepts as duality (or pairing). The representations of a Hopf algebra $U_q(\mathfrak{g})$ are obtained after dual bases of it and its dual $Fun_q(G)$ are known. We have applied this procedure to obtain the induced representations of quantum kinematical algebras in $(1+1)$ dimensions. Based on the fact that this kind of quantum Hopf algebras also has a structure of bicrossproduct (which is in some sense the quantum counterpart of the structure of semidirect product of groups) we have developed a theory of induced representations of quantum bicrossproduct algebras.

In the following we will consider quantum Hopf algebras having a bicrossproduct structure $H = \mathcal{K} \triangleright\!\!\blacktriangleleft \mathcal{L}$, such that $\mathcal{K}$ is cocommutative and $\mathcal{L}$ commutative [4]. We will construct the representations induced by 'real' characters of the commutative sector $\mathcal{L}$. We will see that the solution of this problem can be reduced to the study of certain dynamical systems which present, in general, a nonlinear action. Other authors have written about the same subject like [5,6,7,8].

We will denote the different actions involved by the following symbols (or their symmetric for the corresponding right actions and coactions): $\triangleright$ ($\blacktriangleleft$) (bicrossproduct actions (coactions)), $\vdash$ (induced and inducting representations) and $\succ$ ($\prec$) (regular actions (coactions)).

Note that we will use multi-index notation [3]: let A be an algebra generated by the elements $(a_1, a_2, \ldots, a_r)$ such that the ordered monomials $a^n := a_1^{n_1} a_2^{n_2} \cdots a_r^{n_r} \in A$, $n = (n_1, n_2, \ldots, n_r) \in \mathbb{N}^r$, form a basis of the linear space underlying to A. An arbitrary product of generators of A is written in a normal ordering if it is expressed in terms of the (ordered) basis $(a^n)_{n \in \mathbb{N}^r}$. The unit of A, 1_A, is denoted by a^0 ($0 \in \mathbb{N}^n$). Multi-factorials and multi-deltas are defined by $l! = \prod_{i=1}^n l_i!$ and $\delta_l^m = \prod_{i=1}^n \delta_{l_i}^{m_i}$.

2 Induced representations

The main results about the theory of induced representation of quantum bicrossproduct algebras will be summarized in this section (see [9] for a detailed description).

Let us suppose that the subalgebras $\mathcal{K}$ and $\mathcal{L}$ of H are finite generated by the sets $\{k_i\}_{i=1}^r$ (the k_i's are primitive) and $\{l_i\}_{i=1}^s$, respectively, and $(k_n)_{n \in \mathbb{N}^r}$ and $(l_m)_{m \in \mathbb{N}^s}$ are bases of the vector spaces underlying to $\mathcal{K}$ and $\mathcal{L}$. The dual algebras (in the sense of a left and right nondegenerate pairing $\langle .,. \rangle$ that we will simply call nondegenerate pairing [10]) of $\mathcal{K}$ and $\mathcal{L}$ ($\mathcal{K}^*$, $\mathcal{L}^*$) have dual systems to those of $\mathcal{K}$ and $\mathcal{L}$ with analogue properties to them. Two dual algebras and their nondegenerate pairing determine a 'nondegenerate triplet'. Duality between H and $H^* = \mathcal{K}^* \blacktriangleright\!\!\triangleleft \mathcal{L}^*$ is given by $\langle k_m l_n, \kappa^{m'} \lambda^{n'} \rangle = m! n!\, \delta_m^{m'} \delta_n^{n'}$.

Dual bases and $*$–structures over bicrossproduct Hopf algebras can be constructed when the corresponding bases and $*$–structures of the bicrossproduct factors are known [9]. If $H = \mathcal{K} \triangleright\!\!\blacktriangleleft \mathcal{L}$ and $(\mathcal{K}, \mathcal{K}^*, \langle \cdot, \cdot \rangle_1)$ and $(\mathcal{L}, \mathcal{L}^*, \langle \cdot, \cdot \rangle_2)$ are nondegenerate triplets, then $\langle kl, \kappa\lambda \rangle = \langle k, \kappa \rangle_1 \langle l, \lambda \rangle_2$ defines a nondegenerate pairing between H and H^*. If (k_m), (κ_m) are dual bases for $\mathcal{K}$, $\mathcal{K}^*$ and (l_n), (λ_n) for $\mathcal{L}$, $\mathcal{L}^*$, then $(k_m l_n)$ and $(\kappa^m \lambda^n)$ are dual bases for H and H^*.

Theorem 2.1. *Let $H = \mathcal{K} \triangleright\!\!\blacktriangleleft \mathcal{L}$ be a bicrossproduct Hopf algebra with $\mathcal{K}$ and $\mathcal{L}$ equipped with $*$–structures verifying $(l \triangleleft k)^* = l^* \triangleleft S(k)^*$. Then a $*$–structure on the algebra sector of H is determined by $(kl)^* = l^* k^*$, $k \in K$, $l \in L$.*

It can be proved that the carrier space, $\mathbb{C}^\uparrow$, of the representation of H induced by the character $a \in \mathbb{C}^s$ of $\mathcal{L}$

$$1 \dashv l_n = a_n, \qquad n \in \mathbb{N}^s, \tag{1}$$

is isomorphic to $\mathcal{K}^*$ and is constituted by the elements of the form $\kappa\psi$, where $\kappa \in \mathcal{K}^*$ and $\psi = e^{a_1 \lambda_1} e^{a_2 \lambda_2} \cdots e^{a_s \lambda_s}$. The induced action of the elements h of

H is given by

$$f \dashv h = \sum_{m \in \mathbb{N}^r} \kappa^m \langle h \frac{k_m}{m!}, f \rangle \psi, \qquad f \in \mathbb{C}^{\dagger}. \tag{2}$$

Since $\mathcal{L}$ is commutative, it can be identified with the algebra of functions $F(\mathbb{R}^s)$ by the morphism $\mathcal{L} \to F(\mathbb{R}^s)$, $(l \mapsto \tilde{l})$, mapping the generators of $\mathcal{L}$ into the canonical projections

$$\tilde{l}_j(x) = x_j, \qquad \forall x = (x_1, x_2, \ldots, x_s) \in \mathbb{R}^s, \quad j = 1, 2, \ldots, s.$$

This identification allows to chose a $*$–structure keeping invariant the generators of $\mathcal{L}$ by $l_j^* = l_j$, $(j = 1, 2, \ldots, s)$. So, the characters of $\mathcal{L}$ (1) compatible with the $*$–structure are real and we can write $1 \dashv l = \tilde{l}(a)$ with $a \in \mathbb{R}^n$.

The right action of $\mathcal{K}$ on $\mathcal{L}$ can be also carried to $F(\mathbb{R}^s)$. The generators of $\mathcal{K}$ act by derivations on the $\mathcal{K}$–module algebra of $\mathcal{K} \triangleright\!\!\blacktriangleleft \mathcal{L}$ inducing vector fields, X_i, on $\mathbb{R}^s$ by

$$X_i \tilde{l} = \widetilde{l \triangleleft k_i}, \qquad i = 1, 2, \ldots, r.$$

The associated flow to X_i, $\Phi_i : \mathbb{R} \times \mathbb{R}^s \to \mathbb{R}^s$, is given by

$$(X_i f)(x) = (D f_{x, \Phi_i})(0), \tag{3}$$

where $f_{x,\Phi_i}(t) = f \circ \Phi_i^t(x)$ and D is the derivative operator.

The explicit action of the generators of $\mathcal{K}$ and $\mathcal{L}$ in the induced representation determined by the action (2) and with carrier space $\mathcal{K}^*$ is given by

$$\kappa \dashv k_i = \kappa \prec k_i \ , \qquad \kappa \dashv l_j = \kappa \ \hat{l}_j \circ \Phi_{(\kappa_1, \kappa_2, \ldots, \kappa_r)}(a) \ , \tag{4}$$

where $i \in \{1, \ldots, r\}$, $j \in \{1, \ldots, s\}$, the symbol $\prec$ denotes the regular action of $\mathcal{K}$ on $\mathcal{K}^*$ and $\Phi_{(\kappa_1, \kappa_2, \ldots, \kappa_r)} = \Phi_r^{\kappa_r} \circ \cdots \circ \Phi_2^{\kappa_2} \circ \Phi_1^{\kappa_1}$.

The relation between representations and modules allows us to reformulate the induction procedure [3,9]. The four regular H–modules associated to a Hopf algebra H and its dual H^*, $(H, \prec, H)$, $(H^*, \succ, H)$, $(H, \succ, H)$ and $(H^*, \prec, H)$, will be use to describe the actions as well as the carrier spaces involved in the induced representations. The modules $(H^*, \succ, H)$ and $(H^*, \prec, H)$ are called co-spaces since they can be considered as an algebraic generalization of the concept of G–space.

The bicrossproduct Hopf algebras, that we consider, may be factorized as $H = U(\mathfrak{k}) \triangleright\!\!\blacktriangleleft F(L)$, $H^* = F(K) \blacktriangleright\!\!\triangleleft U(\mathfrak{l})$ where K and L are Lie groups with associated Lie algebras $\mathfrak{k}$ and $\mathfrak{l}$, respectively.

For an effective description of the regular modules it is essential to use elements of H and H^* like $k\lambda \in H$, $(k \in K, \ \lambda \in F(L))$ and

$\kappa l \in H^*$, $(\kappa \in F(K),\ l \in L)$, respectively, instead of the standard bases of ordered monomials. The action on each of the four regular H–modules is:

$$\begin{array}{lll} (H,\prec,H) : & (k\lambda) \prec k' = kk'(\lambda \triangleleft k'), & (k\lambda) \prec \lambda' = k\lambda\lambda'; \\ (H^*,\succ,H) : & k' \succ (\kappa l) = (k' \succ \kappa)(k' \triangleright l), & \lambda' \succ (\kappa l) = \lambda'(l)\kappa l; \\ (H,\succ,H) : & k' \succ (k\lambda) = k'k\lambda, & \lambda' \succ (k\lambda) = k(\lambda' \triangleleft k)\lambda; \\ (H^*,\prec,H) : & (\kappa l) \prec k' = (\kappa \prec k')l, & (\kappa l) \prec \lambda' = \kappa(\lambda' \circ \hat{l})l; \end{array} \tag{5}$$

where $\hat{l}$ is the map, $K \to L$ $(k \mapsto k \triangleright l)$, projecting the group K on the orbit passing through $l \in L$ such that $\langle l^{(1)}, \lambda\rangle l^{(2)} = \lambda \circ \hat{l}$, for all $k, k' \in K$, $\lambda, \lambda' \in F(L)$, $\kappa \in F(K)$ and $l \in L$ (the coaction associated to the structure of $U(\mathfrak{l})$ as right $F(K)$–comodule has been used, i.e., $l \blacktriangleleft = l^{(1)} \otimes l^{(2)} \in U(\mathfrak{l}) \otimes F(K)$).

The following theorem summarizes the induction procedure for bicrossproduct algebras.

Theorem 2.2. *Let us consider an element $l \in L$ supporting a global action of the group K. The carrier space, $\mathbb{C}^\uparrow$, of the representation of H induced by the character determined by l is the set of elements of H^* of the form*

$$\kappa l, \qquad \kappa \in F(K).$$

The spaces $\mathbb{C}^\uparrow$ and $F(K)$ are isomorphic $(\kappa \mapsto \kappa l)$. The action induced by the elements of the form $k \in K$ and $\lambda \in F(L)$ on the space $F(K)$ is

$$\kappa \dashv k = \kappa \prec k, \qquad \kappa \dashv \lambda = \kappa(\lambda \circ \hat{l}).$$

The modules induced by l and $k \triangleright l$ are isomorphic. Thus, the induction algorithm establishes a correspondence between the space of orbits L/K and the set of equivalence classes of representations.

If the group K is compact the induced representation is unitary in the space $L^2(K)$, of square-integrable functions with respect to the right invariant Haar measure, when the $*$–structure given by Theorem 2.1 is considered.

3 Quantum κ–Galilei algebra

The quantum κ–Galilei algebra in $(1+1)$ dimensions, $U_\kappa(\mathfrak{g}(1,1))$, has the following Lie commutators and coproduct [11]

$$\begin{array}{lll} [H,K] = -P, & [P,K] = \frac{P^2}{2\kappa}, & [H,P] = 0; \\ \Delta H = H \otimes 1 + 1 \otimes H, & \Delta X = X \otimes 1 + e^{-H/\kappa} \otimes X, & X \in \{P, K\}. \end{array}$$

It has a bicrossproduct structure $U_\kappa(\mathfrak{g}(1,1)) = \mathcal{K} \triangleright\!\!\blacktriangleleft U_\kappa(\mathfrak{t}_2)$, where $U_\kappa(\mathfrak{t}_2)$ is the commutative and non-cocommutative Hopf subalgebra spanned by P and

H, and $\mathcal{K}$ is the commutative and cocommutative Hopf algebra generated by K. The right action of $\mathcal{K}$ on $U_\kappa(\mathfrak{t}_2)$ is given by

$$P \triangleleft K = [P, K] = \frac{P^2}{2\kappa}, \qquad H \triangleleft K = [H, K] = -P, \tag{6}$$

and the left coaction of $U_\kappa(\mathfrak{t}_2)$ over the generator of $\mathcal{K}$ is

$$K \blacktriangleleft= e^{-H/\kappa} \otimes K.$$

The dual algebra has also a bicrossproduct structure $F_\kappa(G(1,1)) = \mathcal{K}^* \blacktriangleright\!\!\triangleleft \mathcal{L}^*$, with $\mathcal{K}^*$ generated by v and $\mathcal{L}^*$ by x and t.

The pairing between $U_\kappa(\mathfrak{g}(1,1))$ and $F_\kappa(G(1,1))$ is given by $\langle K^m P^n H^p, v^q x^r t^s \rangle = m!n!p!\, \delta_q^m \delta_r^n \delta_s^p$.

The bicrossproduct structure of $U_\kappa(\mathfrak{g}(1,1))$ may be interpreted as

$$U_\kappa(\mathfrak{g}(1,1)) = U(\mathfrak{V}) \triangleright\!\!\blacktriangleleft F(T_{\kappa,2}),$$

where $T_{\kappa,2}$ is the group defined by $(a', b')(a, b) = (a' + e^{-b'/\kappa} a, b' + b)$. The functions P and H, defined by $P(a, b) = a$, $H(a, b) = b$, determine a global chart on $T_{\kappa,2}$.

The action of K on the $U(\mathfrak{V})$–module algebra $F(T_{\kappa,2})$ is given by (6). So, the induced vector field is

$$X_K = \frac{P^2}{2\kappa}\frac{\partial}{\partial P} - P\frac{\partial}{\partial H}.$$

The integral curves in the region $a < 0$ are

$$a(s) = \frac{-1}{c_1 + \frac{s}{2\kappa}}, \qquad b(s) = 2\kappa \ln(c_1 + \frac{s}{2\kappa}) + c_2.$$

The flow associated to X_K is

$$\Phi^s(a, b) = (\frac{a}{1 - \frac{sa}{2\kappa}}, b + 2\kappa \ln(1 - \frac{sa}{2\kappa})).$$

The action of $\mathfrak{V}$ on $T_{\kappa,2}$ is given by $e^{sK} \triangleright (a, b) = \Phi^s(a, b)$ and is not global. Under it $T_{\kappa,2}$ is decomposed in two strata: i) each orbit is a point of the form $(0, b)$, and ii) the rest of the space is foliated by 1–dimensional orbits.

The elements of $F_\kappa(G(1,1)$ can be written as $\phi(a, b)$ with $\phi \in F_\kappa(\mathfrak{V})$ and $(a, b) \in T_{\kappa,2}$. The action on the regular co-space $(F_\kappa(G(1,1)), \prec, U_\kappa(\mathfrak{g}(1,1)))$ is obtained applying (5)

$$\begin{aligned}
(\phi(a,b)) \prec e^{sK} &= \phi(e^{sK} \cdot)(a, b), \\
(\phi(a,b)) \prec P &= \phi \tfrac{a}{1 - \frac{av}{2\kappa}}(a, b), \\
(\phi(a,b)) \prec H &= \phi(b + 2\kappa \ln(1 - \tfrac{av}{2\kappa}))(a, b),
\end{aligned}$$

with $\phi \in F(\mathfrak{V})$, $(a,b) \in T_{\kappa,2}$ and the dot stands for the argument of the function $\phi = \phi(\,\cdot\,)$.

According to Theorem 2.2 each element $(a,b) \in T_{\kappa,2}$ induces a representation given by

$$\begin{aligned} \phi \dashv e^{sK} &= \phi(e^{sK}\,\cdot\,), \\ \phi \dashv P &= \phi_{\frac{a}{1-\frac{av}{2\kappa}}}, \\ \phi \dashv H &= \phi(b + 2\kappa \ln(1 - \tfrac{av}{2\kappa})). \end{aligned} \tag{7}$$

Using the local coordinates v, x, t as generator elements for $F_\kappa(G(1,1))$ we easily find that $v^m \dashv K = m v^{m-1}$. So, we can rewrite that $\phi \dashv K = \phi'(v)$, with ϕ belonging to the space of formal series of v. With this change (7) coincides with the result obtained in [1,3] by other method. This representation is reducible and a decomposition of it can be found in [1].

Acknowledgments

This work has been partially supported by DGES of the Ministerio of Educación and Cultura of Spain under project PB98-0360, and the Junta de Castilla y León (Spain).

References

1. O. Arratia and M.A. del Olmo in *Anales de Física, Monografías*, vol. 5 (Ciemat/RSEF, Madrid, 1998); math.QA/0110265.
2. O. Arratia and M.A. del Olmo in *Publicaciones de la Real Sociedad Matemática Española*, vol. 1 (RSME, Madrid, 2000); math.QA/0110266.
3. O. Arratia and M.A. del Olmo, *J. Math. Phys.* **41**, 4817 (2000).
4. S. Majid, *Foundations of quantum group theory* (Cambridge Univ. Press, Cambridge, 1995).
5. A. González-Ruiz and L.A. Ibort, *Phys. Lett.* **B296**, 104 (1992).
6. V.K. Dobrev, *J. Phys. A* **27**, 4841 and 6633 (1994).
7. F. Bonechi, R. Giachetti, E. Sorace and M. Tarlini, *Lett. Math. Phys.* **43**, 309 (1998).
8. N. Ciccoli, *Induction of quantum group representations*, math/9804138.
9. O. Arratia and M.A. del Olmo, *Representations of quantum bicrossproduct algebras.* Preprint, UVA 2001; math.QA/0110275.
10. V. Chari and A. Pressley, *A guide to quantum groups*, (Cambridge Univ. Press, Cambridge 1994).
11. J. C. Pérez Bueno, Ph.D. Thesis (Univ. de Valencia, Valencia, 1998).

SEPARATION OF DISCRETE VARIABLES IN THE 2-DIM FINITE OSCILLATOR[a]

NATIG M. ATAKISHIYEV,[b] GEORGE S. POGOSYAN,[c]
LUIS EDGAR VICENT AND KURT BERNARDO WOLF
Centro de Ciencias Físicas
Universidad Nacional Autónoma de México
Apartado Postal 48-3, Cuernavaca 62251, Morelos, México
E-mail: bwolf@fis.unam.mx

The finite oscillator in two dimensions is a system whose dynamical algebra is so(4); it has a discrete, finite configuration space whose points can be arranged following cartesian or polar coordinates. Its wavefunctions satisfy 'Schrödinger' difference equations; in the cartesian model they are Wigner d-functions involving Kravchuk polynomials, while in the radial model they are su(2) Clebsch-Gordan functions containing dual Hahn polynomials. An su(2) symmetry algebra and group can be imported from the ordinary oscillator; coherent states exist, and a covariant Wigner function on a compact phase space can be formulated.

1 Introduction

In this contribution we study the finite oscillator in two dimensions, in cartesian and in polar coordinates. A finite oscillator has a finite number of discrete energy levels, and is defined over a finite number of points in configuration space. This is the first instance that we know of where a finite Hamiltonian system can be separated in two distinct arrays: one conforms to cartesian coordinates, with $(N+1)^2$ points (q_x, q_y) in the range $q_x, q_y \in \{-\frac{1}{2}N, -\frac{1}{2}N+1, \ldots, \frac{1}{2}N\}$ ($0 \leq N$ integer),[1] while the other also has $(N+1)^2$ points, but distributed by polar coordinates, on circles of integer radii $\rho \in \{0, 1, \ldots, N\}$, with $2\rho+1$ points on each circle.[2]

Both finite oscillator models are Hamiltonian systems with unitary evolution, support coherent states, have a compact phase space, and admit a Wigner function which is covariant with the dynamical group of the system. In this Section we spell out the **axioms** for the D-dim harmonic oscillator. Section 2 will review the 1-dim finite oscillator which is based on the dynamical algebra $\mathrm{u}(1) \oplus \mathrm{su}(2)$, and applies to a planar optical waveguide of $N+1$ sensors; for

[a]Work supported by project DGAPA–UNAM IN112300 *Optica Matemática*. G.S.P. acknowledges CONACYT for Cátedra Patrimonial Nivel II, and Russian Foundation for Basic Research, grant RFBR N00–02–81023. L.E.V. thanks CONACYT Graduate Scholarship #150666.
[b]Instituto de Matemáticas, UNAM, Apartado Postal 273-3, 62210 Cuernavaca, México
[c]Permanent address: Laboratory of Theoretical Physics, Joint Institute for Nuclear Research, Dubna, Russia

energy $\nu + \frac{1}{2}$ ($\nu \in \{0, 1, \ldots, N\}$), its wavefunctions are $d^{N/2}_{\nu-N,q}(\frac{1}{2}\pi)$ and involve Kravchuk polynomials. Section 3 will contain a brief account of the 2-dim finite **cartesian** oscillator, based on the dynamical algebra $\mathrm{u}(1) \oplus \mathrm{su}(2)_x \oplus \mathrm{su}(2)_y$, where the SU(2) symmetry has been imported to form states of angular momentum m on a square array of sensor points. In Section 4 we introduce the 2-dim finite **radial** oscillator, based on the same dynamical algebra, but reduced with respect to a (noncanonical) $\mathrm{so}(4) \subset \mathrm{so}(3)$ chain; its wavefunctions are the su(2) Clebsch-Gordan coefficients $C^{\frac{1}{2}N,\frac{1}{2}N,\rho}_{\frac{1}{2}(m+\nu),\frac{1}{2}(m-\nu),m}$.

The **OSCILLATOR AXIOMS** used to define the common well-known quantum oscillator, serve equally well to define finite oscillator models.

1: There exists a **position** subalgebra, of algebraic basis $\vec{Q} \stackrel{\text{def}}{=} \mathrm{span}\,\{Q_x, Q_y, \ldots, Q_D\}$. Spectra of operators in this subalgebra are the position coordinates of the system.

2: There exists a **Hamiltonian** operator, self-adjoint and compact generator of time evolution, which satisfies the Newton-Lie ⇔ Hamilton-Lie equations:

$$[H,[H,\vec{Q}]] = \vec{Q} \quad \Longleftrightarrow \quad \begin{cases} [H,\vec{Q}] \stackrel{\text{def}}{=} -i\vec{P}, \\ [H,\vec{P}] = \ \ i\vec{Q}, \end{cases}$$

so there exists a **momentum** subalgebra, defined by the first Hamilton equation, $\vec{P} \stackrel{\text{def}}{=} \mathrm{span}\,\{P_x, P_y, \ldots, P_D\}$. The second Hamilton equation contains the oscillator dynamics.

3: There exists an **angular momemtum** subalgebra $\mathrm{so}(D) = \mathrm{span}\,\{M_{\alpha,\beta}\}$ commuting with H, and under which $\vec{Q}$, and hence $\vec{P}$, transform as D-vectors:

$$\begin{aligned} [M_{\alpha,\beta}, Q_\gamma] &= -i(\delta_{\beta,\gamma}Q_\alpha - \delta_{\alpha,\gamma}Q_\beta), \\ [M_{\alpha,\beta}, P_\gamma] &= -i(\delta_{\beta,\gamma}P_\alpha - \delta_{\alpha,\gamma}P_\beta), \end{aligned} \qquad [M_{\alpha,\beta}, H] = 0.$$

4: Under Lie brackets, all operators close into a **Lie algebra**.

These axioms allow for several solutions. The ordinary quantum oscillator model is obtained with the Heisenberg-Weyl algebra W_D for $\vec{Q}$, $\vec{P}$, and $\hat{1}$, plus angular momentom $\Lambda_{j,k} = Q_jP_k - Q_kP_j$ and Hamiltonian $H = \frac{1}{2}(\vec{P}^2 + \vec{Q}^2)$. The **finite** oscillator models satisfy these axioms with the algebra so(D). Replacing 'Lie algebras' in the last axiom for 'q-algebras' also yields q-oscillator solutions.[3]

2 One-dimensional finite oscillator

In the 1-dim case there is no angular momentum, but only one position operator Q, one momentum operator P, and the Hamiltonian H. These three operators are posited to close into an su(2) algebra span $\{\Lambda_{j,k}\}$, $1 \le j < k \le 3$, with the following identification of the generators:

$$\begin{array}{|c|c|}\hline \Lambda_{1,2} & \Lambda_{1,3} \\ \hline \multicolumn{1}{c|}{} & \Lambda_{2,3} \\ \cline{2-2} \end{array} = \begin{array}{|c|c|}\hline J_3 & -J_2 \\ \hline \multicolumn{1}{c|}{} & J_1 \\ \cline{2-2} \end{array} = \begin{array}{|c|c|}\hline H-j-\frac{1}{2} & P \\ \hline \multicolumn{1}{c|}{} & Q \\ \cline{2-2} \end{array}$$

in the unitary irrep representation labelled by $j \stackrel{\text{def}}{=} \frac{1}{2}N \in \{0, \frac{1}{2}, 1, \ldots\}$. There, the Casimir operator is a multiple of unity, $J^2 \stackrel{\text{def}}{=} J_1^2 + J_2^2 + J_3^2 = j(j+1)\hat{1}$, and the number j is used to displace the spectrum of H to $\nu + \frac{1}{2}$, $\nu \in \{0, 1, \ldots, N\}$. We write $E_J(E_J + 1) \stackrel{\text{def}}{=} J^2$ and use the central operator $E_J = j\hat{1}$, generator of an u(1) algebra, to specify the representation.

The identification ($\leftrightarrow$) of physical observables in the finite oscillator with the generators in the algebra and their spectra is thus:

$$\begin{array}{llll} \text{position:} & \longleftrightarrow Q = J_1, & & q \in \{-j, -j+1, \ldots, j\}, \\ -\text{momentum:} & \longleftrightarrow -P = J_2, & & p \in \{-j, -j+1, \ldots, j\}, \\ \text{Hamiltonian:} & \longleftrightarrow & H = J_3 + j + \frac{1}{2}, & h \in \{\frac{1}{2}, \frac{3}{2}, \ldots, 2j + \frac{1}{2}\}; \\ \text{mode number:} & & H - \frac{1}{2} = J_3 + j, & \nu \in \{0, 1, \ldots, N\}. \end{array}$$

The su(2) Lie brackets of the finite oscillator dynamical algebra are thus

$$[H, P] = iQ, \quad [H, Q] = -iP, \quad [Q, P] = i(H - j - \tfrac{1}{2}).$$

The first two are common and incontestable; the third is a 'nonstandard' commutator which is characteristic of the finite oscillator model.

States of definite position and energy are then defined by

$$\begin{array}{rlrl} E_J\,|N;q\rangle_1 &= \frac{1}{2}N\,|N;q\rangle_1, & E_J\,|N;\nu\rangle_{\text{H}} &= \frac{1}{2}N\,|N;\nu\rangle_{\text{H}}, \\ Q\,|N;q\rangle_1 &= q\,|N;q\rangle_1, & H\,|N;\nu\rangle_{\text{H}} &= (\nu + \frac{1}{2})\,|N;\nu\rangle_{\text{H}}, \end{array}$$

and their overlap are the 1-dim finite oscillator **wavefunctions**,

$$\begin{aligned} \Phi_\nu^{(N)}(q) &\stackrel{\text{def}}{=} {}_1\langle N; q | N; \nu\rangle_{\text{H}} = {}_{\text{H}}\langle N; q + j | e^{+i\frac{\pi}{2}J_2}\,|N;\nu\rangle_{\text{H}} \\ &= d^j_{q,\nu-j}(-\tfrac{1}{2}\pi) = d^j_{\nu-j,q}(\tfrac{1}{2}\pi) = (-1)^{\nu-j-q}\, d^j_{j-\nu,-q}(\tfrac{1}{2}\pi) \\ &= \frac{(-1)^\nu}{2^j}\sqrt{\binom{2j}{\nu}\binom{2j}{j+q}}\,K_\nu(j+q; \tfrac{1}{2}, 2j), \end{aligned}$$

where $K_\nu(x; \frac{1}{2}, N) \stackrel{\text{def}}{=} {}_2F_1(-\nu, -\frac{1}{2}N - x; -N; 2)$ are the symmetric **Kravchuk** polynomials of degree ν in x. Plots of the Kravchuk functions appear in Refs.

1 and 5; for $\nu \ll N$ they reproduce quite faithfully the well-known Hermite functions on the finite interval $(-\frac{1}{2}N-1, \frac{1}{2}N+1)$. When $\nu \approx \frac{1}{2}N$ however, the wavefunction zig-zags every two points with an amplitude which increases a bit at the ends of the interval, and for $\nu \approx N$ it alternates in sign the $\nu \approx 0$-functions, due to the factor $(-1)^q$ in the before-last line of the previous formula; $\Phi_N^{(N)}(q)$ is the state of highest energy in the finite oscillator or shallow optical waveguide.

The time evolution of the finite oscillator is generated by H. In the common oscillator this is a phase $\exp(i\frac{1}{4}\pi\alpha)$ times the fractional Fourier transform $\mathcal{F}^\alpha$ of power $\alpha = 2t/\pi$. In the finite oscillator, we define the fractional Fourier-*Kravchuk* operator and the summation kernel of the **fractional Fourier-Kravchuk transform** to be

$$\mathcal{K}^\alpha \stackrel{\text{def}}{=} \exp[-i\tfrac{1}{2}\pi\alpha\,(J_3+j)] = e^{i\frac{1}{4}\pi\alpha}\exp(-i\tfrac{1}{2}\pi\alpha\,H),$$

$$K_{q,q'}^{(N,\alpha)} \stackrel{\text{def}}{=} {}_1\langle N,q|\mathcal{K}^\alpha\,|N,q'\rangle_1 = i^{q'-q}\,d_{q,q'}^{\frac{1}{2}N}(\tfrac{1}{2}\pi\alpha).$$

In the contraction limit $N \to \infty$, this returns the ordinary Fourier kernel and the oscillator time evolution.[6]

3 The 2-dim cartesian oscillator model

The direct product of two finite oscillators with the same number of points in the x- and y-directions is a 2-dim cartesian oscillator. Its dynamical algebra is $\mathrm{u}(1)\oplus\mathrm{su}(2)_x\oplus\mathrm{su}(2)_y = \mathrm{u}(1)\oplus\mathrm{so}(4)$, with the common central u(1) generator E_J specifying the size $j_x = j_y = \frac{1}{2}N$ of the square array of points. In image processing, the cartesian 2-dim oscillator is a square, pixellated screen.

In a restricted sense, the Cartesian oscillator can be said to be somewhat trivial —the eigenfunctions of $H \stackrel{\text{def}}{=} H_x + H_y$ are just the product of two 1-dim oscillators, $\Phi^{(N)}_{\nu_x,\nu_y}(q_x,q_y) = \Phi^{(N)}_{\nu_x}(q_x)\,\Phi^{(N)}_{\nu_y}(q_y)$, with total mode number $\nu = \nu_x+\nu_y$. But the model raises an interesting question: where is the famous SU(2) symmetry of the common quantum oscillator? *Rotations* are apparently impossible. In the common oscillator, states of the same total ν form su(2) multiplets, but in the finite case they do **not**. However, we take the states with the same total ν and linearly combine the finite states *as if* they were a multiplet; then it can be shown [1] that the x- and y-Fourier-Kravchuk transforms, and these rotations, *do* close into a group SU(2). In image processing this is a *bona fide* group of unitary (hence precisely invertible) transformations, which approximates the continuous transformations, including rotations, of a square pixellated image.

4 The 2-dim radial oscillator model

The same symmetric $(N+1)^2$-dimensional representation of the same dynamical group is used with the following identification between so(4) generators:

$$\begin{array}{|c|c|c|}\hline \Lambda_{1,2} & \Lambda_{1,3} & \Lambda_{1,4} \\ \hline \multicolumn{1}{c|}{} & \Lambda_{2,3} & \Lambda_{2,4} \\ \cline{2-3} \multicolumn{1}{c}{} & \multicolumn{1}{c|}{} & \Lambda_{3,4} \\ \cline{3-3} \end{array} \;=\; \begin{array}{|c|c|c|}\hline J & P_x & P_y \\ \hline \multicolumn{1}{c|}{} & Q_x & Q_y \\ \cline{2-3} \multicolumn{1}{c}{} & \multicolumn{1}{c|}{} & M \\ \cline{3-3} \end{array} \;=\; \begin{array}{|c|c|c|}\hline S_3 & -S_2 & S_1 \\ \hline \multicolumn{1}{c|}{} & R_1 & R_2 \\ \cline{2-3} \multicolumn{1}{c}{} & \multicolumn{1}{c|}{} & R_3 \\ \cline{3-3} \end{array}$$

Here, $J \stackrel{\text{def}}{=} H - j - \frac{1}{2}$ is the (displaced) Hamiltonian, M is the single angular momentum generator, and in the last pattern we indicate the so(3) position subalgebra by $\{R_1, R_2, R_3\} = \{Q_x, Q_y, M\}$. The Casimir operator $\vec{R}^2$ has eigenvalues $\rho(\rho+1)$, $\rho \in \{0, 1, \ldots, N\}$, and we interpret ρ as the radius. The model implies now the 'nonstandard' commutators $[Q_x, P_x] = iJ = [Q_y, P_y]$ as before, but now also $[Q_x, Q_y] = iM = [P_x, P_y]$. There are arguments in optics and image processing to consider this as an expected result.

The overlap between the eigenstates of the subalgebra chains $\mathrm{so}(4) \supset \mathrm{so}(2)_{\mathrm{J}} \oplus \mathrm{so}(2)_{\mathrm{M}}$ and $\mathrm{so}(4) \supset \mathrm{so}(3)_{\mathrm{R}} \supset \mathrm{so}(2)_{\mathrm{M}}$,[d] yields the finite oscillator radial wavefunctions for $|m| \le \rho \le N$ integers, $m_a \stackrel{\text{def}}{=} \frac{1}{2}(\nu+m)$ and $m_b \stackrel{\text{def}}{=} \frac{1}{2}(\nu-m)$:

$$\begin{aligned}(-1)^\rho \Phi^{(N)}_{\nu,m}(\rho) &\stackrel{\text{def}}{=} {}_{\mathrm{R}}\langle N;\rho,m|N;\nu,m\rangle_{\mathrm{JM}} = C^{\frac{1}{2}N,\frac{1}{2}N,\rho}_{\frac{1}{2}(m+\nu),\frac{1}{2}(m-\nu),m} \\ &= \frac{N!}{m!}\sqrt{\frac{(2\rho+1)}{(N-\rho)!\,(N+\rho+1)!}\,\frac{(\rho+m)!}{(\rho-m)!}\,\frac{(j+m_b)!\,(j-m_b)!}{(j+m_a)!\,(j-m_a)!}} \\ &\qquad \times(-1)^{j-m_a}\,R_{j-m_a}\Big(\rho(\rho+1);m,-m,N\Big),\end{aligned}$$

with dual *Hahn* polynomials $R_n(\rho(\rho+1);m,m',N) \stackrel{\text{def}}{=} {}_3F_2\left({-n,\,-\rho,\,\rho+m+m'+1 \atop m+1,\,-N}\Big|1\right)$ These are finite, unitary approximants to the well-known radial harmonic oscillator wavefunctions with Laguerre functions, which are obtained in the contraction limit $N \to \infty$.

Finally, from the states $|N;\nu,m\rangle_{\mathrm{JM}}$ of radius ρ, the discrete $(2\rho+1)$-dim Fourier transform produces the radius-and-angle states

$$|N;\rho,\theta_k\rangle_\circ \stackrel{\text{def}}{=} \frac{1}{\sqrt{2\rho+1}}\sum_{m=-\rho}^{\rho} e^{im\theta_k}\,|N;\rho,m\rangle_{\mathrm{R}}, \qquad \begin{array}{l}\theta_k = 2\pi k/N, \\ k \in \{-\rho,-\rho+1,\ldots,\rho\}.\end{array}$$

[d] We are careful to relate the position subalgebra to the Gel'fand-Tsetlin canonical chain.

The discrete surface element in polar coordinates is $\Delta\sigma \stackrel{\text{def}}{=} \rho\,\Delta\rho\,\Delta\theta_k$; with $\Delta\rho = 1$ and $\Delta\theta_k = \theta_k - \theta_{k-1} = 2\pi/(2\rho+1)$, so the density of sensor points or pixels is, $\pi\rho/(\rho+\frac{1}{2}) \to \pi$.

Fractional Hankel-Hahn transforms are the elements of the 2-dim Fourier operator $\mathcal{K}_x^\alpha\,\mathcal{K}_y^\alpha$ (which is diagonal in m) between the radial position kets. Its summation kernel is $H_m^{(N,\alpha)}(\rho,\rho') = {}_{\text{R}}\langle N;\rho,m|\mathcal{F}^\alpha\,|N;\rho',m\rangle_{\text{R}} \stackrel{\text{def}}{=} d^{N,0}_{\rho,m,\rho'}(\frac{1}{2}\pi\alpha)$, where the last expression is the SO(4) d-function. These Hankel-Hahn transforms are finite-N unitary approximations to the Hankel integral transforms. This is of interest in sonar acoustics, where Hankel transforms must be evaluated and integrated numerically. The traditional interpolation logic can be thus replaced by a unitary (hence invertible) finite transform.

5 Further developments

In theoretical physics, the understanding of systems whose wavefunctions are Wigner or Bargmann d's, Clebsch-Gordan's or higher recoupling coefficients such as Raccah's, has some value. The importation of symmetries from the continuum to the discrete, opens the way to possible discretizations of quantum systems, where symplecticity becomes unitarity. In finite optics these systems are shallow cylindrical waveguides with a finite number of sensors, where coherent states may live, and where measurement theory (uncertainty relations in spot size) can take a place.

For signal processing, we have square and round pixellated screens, whose images can be unitarily mapped onto each other, and linearly Fourier-transformed or rotated. Nonsymmetric representations of so(4) will lead to rectangular and elliptic pixellated screens. Special functions such as the discrete polynomials of Kravchuk, Meixner, and Hahn are given physical life. Their second-order difference equations, in the contraction limit $N \to \infty$, become the Schrödinger differential equations.[6] In this way, we achieve a separation of discrete variables into (here, two) coordinate systems.

1. N.M. Atakishiyev, G.S. Pogosyan, L.E. Vicent and K.B. Wolf, I, *J.Phys.A*, accepted 2001.
2. N.M. Atakishiyev, G.S. Pogosyan, L.E. Vicent and K.B. Wolf, II, *J.Phys.A*, accepted 2001.
3. M. Arık, N.M. Atakishiyev and K.B. Wolf, *J. Phys. A* **32**, L371 (1999).
4. N.M. Atakishiyev and K.B. Wolf, *J. Opt. Soc. Am. A* **14**, 1467 (1997).
5. N.M. Atakishiyev, L.E. Vicent and K.B. Wolf, *J. Comp. Appl. Math.* **107**, 73 (1999).
6. N.M. Atakishiyev, G.S. Pogosyan and K.B. Wolf, in these Proceedings.

CONTRACTIONS OF THE FINITE OSCILLATOR

NATIG M. ATAKISHIYEV,[a] GEORGE S. POGOSYAN,[b]
AND KURT BERNARDO WOLF
Centro de Ciencias Físicas
Universidad Nacional Autónoma de México
Apartado Postal 48–3, Cuernavaca 62251, Morelos, México
E-mail: bwolf@fis.unam.mx

The spectrum-generating algebra of the finite oscillator model is so(3) [or su(2)], whose generators are the position, momentum and mode number operators. The spectra thus consist of a finite number of equally-spaced values. We examine the contraction of this model to the ordinary quantum oscillator as the number and density of points increases. This is done for the algebra and for the wave functions.

1 Introduction

The finite oscillator model uses the well-known theory of angular momentum to propose a new physical interpretation for the generators of the algebra so(3) [or su(2)] in the role of observables of position, momentum and energy [1]. Because the algebra is compact, the spectra of these observables is discrete and finite [2]. In this contribution we investigate the contractions of the finite oscillator model to the common quantum oscillator system.

Lie algebra contractions were introduced by Inönü and Wigner [3] in 1953. Contractions are interpreted as singular limits of representations of Lie algebras, and correspondingly generalized for their wave functions and operators. Many authors, among them Talman [4], have considered contractions ($\longrightarrow$) beside so(3) $\longrightarrow$ e(2), in particular so(3) $\longrightarrow \mathcal{W}_1$ to the Heisenberg-Weyl algebra. A new contraction procedure, called analytic contraction, was presented recently [5]. Following these methods, the contraction parameter appears in the operators of the algebra, in the eigenvalues and eigenfunctions, and not only in the structure constants.

In Section 2 we present the essentials of the finite oscillator model, its operators, eigenvalues and eigenfunctions given with Kravchuk polynomials. A special role is played by the Wigner 'little-d' functions of argument $\frac{1}{2}\pi$. The contraction of so(3) [or su(2)] to the Heisenberg-Weyl algebra $\mathcal{W}_1$ is addressed in Section 3. In Section 4 we contract the finite oscillator Kravchuk functions to the Hermite functions of the ordinary quantum oscillator. The derivation is

[a]Instituto de Matemáticas, UNAM, Apartado Postal 273-3, 62210 Cuernavaca, México
[b]Permanent address: Laboratory of Theoretical Physics, Joint Institute for Nuclear Research, Dubna, Russia

new in the sense that it does not rely on the limit of the recursion relations, but proceeds directly from function properties. It turns out that the ${}_2F_1$-hypergeometric representation is not suited for this limit, but a recent relation between Wigner little-d and ${}_3F_2$-functions [6] permits the direct proof reported here. The contraction limit of dynamical to the oscillator algebra H_4 of four generators is considered in Section 5.

2 One-dimensional finite oscillator

Consider the following realization of so(3) [or su(2)] by difference operators [7]:

$$J_1 = s, \qquad J_2 = -\frac{i}{2}\left[\alpha^\ell(s)\, e^{-\partial_s} - \alpha^\ell(-s)\, e^{\partial_s}\right],$$
$$J_3 = -\frac{1}{2}\left[\alpha^\ell(s)\, e^{-\partial_s} + \alpha^\ell(-s)\, e^{\partial_s}\right],$$

where $\alpha^\ell(s) = \sqrt{(s+\ell)(\ell+1-s)} = \alpha^\ell(1-s)$, with $\ell \geq 0$ and s are integers or half-integers such that $\ell \leq s \leq \ell \geq 0$, and the difference operators act on functions as $e^{a\partial_s} f(s) = f(s+a)$, for any real constant a. In this representation of the algebra, the Casimir operator is a number,

$$J^2 = J_1^2 + J_2^2 + J_3^2 = \ell(\ell+1),$$

which means that the representation of su(2) is irreducible, that the spectra of the generators consist of $2\ell+1$ points, and that the label ℓ characterizes the 'size' of the finite system. The eigenbasis of the two commuting operators $\{J^2, J_3\}$ is in fact determined by only one operator, J_3, so we can define the eigenbasis by

$$J_3\, \psi_m^\ell(s) = m\, \psi_m^\ell(s), \qquad J_\pm\, \psi_m^\ell(s) = \sqrt{(\ell \pm m + 1)(\ell \mp m)}\, \psi_{m\pm1}^\ell(s).$$

To find the difference equation for the function $\psi_m^\ell(s)$, we use J_3 to write

$$-2m\, \psi_m^\ell(s) = \sqrt{(\ell-s)(\ell+s+1)}\, \psi_m^\ell(s+1) + \sqrt{(\ell+s)(\ell-s+1)}\, \psi_m^\ell(s-1).$$

Multiplying by a square root of binomial coefficients, one can transform this into the Kravchuk difference equation [1], so that the solution is

$$\psi_m^\ell(s) = \frac{(-1)^{\ell+m}}{2^\ell}\sqrt{\binom{2\ell}{\ell+s}\binom{2\ell}{\ell+m}}\, K_{\ell+m}(\ell+s; \tfrac{1}{2}, 2\ell). \qquad (1)$$

The Kravchuk polynomials are connected to the Wigner d-functions of argument $\frac{1}{2}\pi$:

$$d^{\ell}_{m,m'}(\tfrac{1}{2}\pi) = \frac{(-1)^{\ell-m}}{2^{\ell}}\sqrt{\binom{2\ell}{\ell+m}\binom{2\ell}{\ell+m'}}\,K_{\ell+m}(\ell+m';\tfrac{1}{2},2\ell).$$

The finite oscillator wave functions are therefore

$$\psi^{\ell}_{m}(s) = (-1)^{2\ell+m-s}\,d^{\ell}_{m,s}(\tfrac{1}{2}\pi).$$

Using the addition theorem for the $d^{\ell}_{m,m'}(\frac{1}{2}\pi)$ functions,

$$\sum_{m''=-\ell}^{\ell} e^{-im''\phi}\,d^{\ell}_{m,m''}(\tfrac{1}{2}\pi)\,d^{\ell}_{m'',m'}(\tfrac{1}{2}\pi) = e^{-im\frac{1}{2}\pi}\,d^{\ell}_{m,m'}(\pi-\phi)\,e^{-im'\frac{1}{2}\pi},$$

and their symmetry properties, we obtain the orthogonality and completeness of the finite oscillator wave functions. Also from here we find the eigenfunctions of operators J_1 and J_2, which we denote by $\psi^{\ell}_{m'}(s)$ and $\psi^{\ell}_{m''}(s)$ respectively:

$$\psi^{\ell}_{m'}(s) = \sum_{m=-\ell}^{\ell} D^{\ell}_{m,m'}(0,\tfrac{1}{2}\pi,\tfrac{1}{2}\pi)\psi^{\ell}_{m}(s) = e^{-im'\frac{1}{2}\pi}\,\delta_{m',s}$$

$$\psi^{\ell}_{m''}(s) = \sum_{m=-\ell}^{\ell} D^{\ell}_{m,m''}(\tfrac{1}{2}\pi,\tfrac{1}{2}\pi,0)\psi^{\ell}_{m}(s) = e^{i(m''-s)\frac{1}{2}\pi}\,d^{\ell}_{m'',s}(\tfrac{1}{2}\pi)$$

We see thus that the group SO(3) [or SU(2)] is the **noninvariance**, or **dynamical**, group of the finite oscillator. When we introduce the quantum number $n = \ell + m$ and diagonal **Hamiltonian** operator,

$$\mathcal{H} = J_3 + \ell + \tfrac{1}{2},$$

we find that the following 'Schrödinger' difference equation is satisfied:

$$\mathcal{H}\,\psi^{\ell}_{n}(s) = (n+\tfrac{1}{2})\,\psi^{\ell}_{n}(s), \qquad n = 0,1,2....2\ell.$$

3 Contractions of the $so(3)$ [or $su(2)$] algebra

We shall use $\ell^{-1/2}$ as the contraction parameter, and for convenience we introduce the new variable $\xi = s/\sqrt{\ell}$. Then, the generators of the group SO(3) [or SU(2)] can be expressed as

$$\begin{aligned}
&\textit{position} \quad Q = \frac{J_1}{\sqrt{\ell}} = \xi, \\
&\textit{momentum} \quad P = \frac{J_2}{\sqrt{\ell}} = -\frac{i}{2\sqrt{\ell}}\left[\alpha^\ell(\xi)\, e^{-\frac{1}{\sqrt{\ell}}\partial_\xi} - \alpha^\ell(-\xi)\, e^{\frac{1}{\sqrt{\ell}}\partial_\xi}\right], \\
&\textit{`unit'} \quad E = -\frac{J_3}{\ell} = \frac{1}{2\ell}\left[\alpha^\ell(\xi)\, e^{-\frac{1}{\sqrt{\ell}}\partial_\xi} + \alpha^\ell(-\xi)\, e^{\frac{1}{\sqrt{\ell}}\partial_\xi}\right],
\end{aligned}$$

and the commutation relations are:

$$[Q,\, P] = -iE, \qquad [P,\, E] = -i\frac{Q}{\ell}, \qquad [Q,\, E] = -i\frac{P}{\ell}.$$

Hence, for $\ell \to \infty$, we have so(3) $\longrightarrow \mathcal{W}_1$ with $E = \mathit{1}$,

$$[Q,\, P] = -i\mathit{1}, \qquad [P,\, \mathit{1}] = [Q,\, \mathit{1}] = 0.$$

Thus the 'position' operator Q contracts to the position ξ, the momentum operator P to $P = i\partial_\xi$.

4 Contractions of the wave functions

We now prove that the Kravchuk *functions* (1) contract to the ordinary quantum harmonic oscillator wave functions, commonly called Hermite functions, which involve Hermite polynomials. It is easy to see that the expression of the Kravchuk functions $\psi^\ell_m(s)$ in terms of hypergeometric functions ${}_2F_1$, is ineligible for the contraction limit $\ell \to \infty$. To make this contraction we will use the expression of the Wigner d-functions in terms of the ${}_3F_2$ functions. We have

$$\psi^\ell_m(s) = \frac{(-1)^{s+\ell+m/2}}{\sqrt{\pi}\,\ell!}\sqrt{(\ell+s)!(\ell-s)!}$$

$$\times \begin{cases} \sqrt{\frac{\Gamma((\ell+m+1)/2)\Gamma((\ell-m+1)/2)}{\Gamma(1+(\ell+m)/2)\Gamma(1+(\ell-m)/2)}}\, {}_3F_2\left(\left. {-s, s, (\ell+m+1)/2 \atop \frac{1}{2}, \ell+1} \right| 1\right), & \ell+m \text{ even}, \\ \frac{2is}{\ell+1}\sqrt{\frac{\Gamma(1+(\ell+m)/2)\Gamma(1+(\ell-m)/2)}{\Gamma((\ell+m+1)/2)\Gamma((\ell-m+1)/2)}}\, {}_3F_2\left(\left. {-s, s, 1+(\ell+m)/2 \atop \frac{3}{2}, \ell+2} \right| 1\right), & \ell+m \text{ odd}. \end{cases}$$

Now in the contraction limit $\ell \to \infty$, $m \to -\infty$ such that $\ell + m = n$ is fixed and $s \sim \sqrt{\ell}\,\xi$, we get

$$\lim_{\ell\to\infty} \ell^{1/4}\, \psi^\ell_m(s) = \frac{2^{n/2} e^{\xi^2}}{\sqrt{\sqrt{\pi} n!}} \begin{cases} \frac{\Gamma((n+1)/2)}{\Gamma(1/2)}\, {}_1F_1((n+1)/2; \tfrac{1}{2}; -\xi^2), & n \text{ even}, \\ 2i\xi\frac{\Gamma(n/2+1)}{\Gamma(1/2)}\, {}_1F_1(n/2+1; \tfrac{3}{2}; -\xi^2), & n \text{ odd}. \end{cases}$$

Comparing this with the expression for the Hermite functions, we finally obtain

$$\lim_{\ell\to\infty} \ell^{1/4}\, \psi^{\ell}_{m}(s) = \psi_n(\xi) = \frac{e^{-\xi^2/2}}{\sqrt{\sqrt{\pi}2^n n!}} H_n(\xi).$$

To compute the contraction limit for the eigenfunctions of the momentum operator J_2, denoted $\psi^{\ell}_{m''}(s)$, we can use Meller's formula for the Hermite polynomials[8]:

$$\sum_{n=0}^{\infty} \frac{z^n}{2^n\, n!} H_n(x)\, H_n(y) = \frac{1}{\sqrt{1-z^2}} \exp\left(\frac{2xyz - (x^2+y^2)z^2}{1-z^2}\right).$$

Now from the relation between the Hermite functions and the Hermite polynomials and the generating function for two Hermite polynomials, it is easy to prove that choosing the value $z = i$, we arrive at the Fourier transform kernel between the position and the momentum representations of the continuous system

$$\sum_{n=0}^{\infty} i^n\, \psi_n(\xi_1)\, \psi_n(\xi_2) = \frac{e^{i\xi_1\xi_2}}{\sqrt{2\pi}}.$$

Consider now the contraction limit $\ell \to \infty$, $m \to -\infty$ such that $\ell + m = n$ is fixed, and $s \sim \xi\sqrt{\ell}$, $m'' \sim p\sqrt{\ell}$. Then we get

$$\lim_{\ell\to\infty} (-1)^{\ell/2}\, \ell^{1/2}\, \psi^{\ell}_{m''}(s) = \lim_{\ell\to\infty} (-1)^{\ell/2}\, \ell^{1/2}\, d^{\ell}_{m'',s}(\tfrac{1}{2}\pi) = \frac{e^{ip\xi}}{\sqrt{2\pi}}.$$

Thus the eigenfunctions of the momentum operator J_2 contract to the eigenfunctions of the ordinary quantum momentum P. This asymptotic formula for the Wigner d-function is new.

5 The contraction of $\mathrm{U}(2) \longrightarrow \mathcal{H}_4$ to the oscillator group

We can form the algebra u(2) with the previous dynamical algebra su(2) of generators $J_\pm = \frac{1}{\sqrt{2}}(J_1 \pm iJ_2)$ and J_3, and adding the central generator $L = \ell 1$ which commutes with the previous three. Then we perform the following linear transformation of the basis:

$$\begin{pmatrix} A_+ \\ A_- \\ A_3 \\ A_0 \end{pmatrix} = \begin{pmatrix} 1/\sqrt{2\ell} & 0 & 0 & 0 \\ 0 & 1/\sqrt{2\ell} & 0 & 0 \\ 0 & 0 & 1 & 1 \\ 0 & 0 & 0 & 1/\ell \end{pmatrix} \begin{pmatrix} J_+ \\ J_- \\ J_3 \\ L \end{pmatrix}.$$

Then the commutation relations for new generators are:

$$[A_3, A_\pm] = \pm i A_\pm, \ [A_+ A_-] = i\frac{1}{\ell} A_3 - iA_0, \ [A_\pm, A_0] = [A_3, A_0] = 0$$

In the limit $\ell \to \infty$, the commutation relation transforms into those of the well-known oscillator algebra $\mathcal{H}_4$ with $A_0 \equiv \mathbf{1}$, namely

$$[A_3, A_\pm] = \pm i A_\pm, \quad [A_+ A_-] = -iA_0, \quad [A_\pm, A_0] = 0 = [A_3, A_0].$$

For the generators thus we get

$$\lim_{\ell\to\infty} A_+ = \tfrac{1}{\sqrt{2}}(Q - iP), \qquad \lim_{\ell\to\infty} A_- = \tfrac{1}{\sqrt{2}}(Q + iP),$$

$$\lim_{\ell\to\infty} A_3 = \lim_{\ell\to\infty} (J_3 + L) = \tfrac{1}{2}(Q^2 + P^2).$$

Finally note that the Heisenberg-Weyl algebra $\mathcal{W}_1$ with the generators $(Q, P, 1)$ is a subalgebra of $\mathcal{H}_4$.

Acknowledgements

We thank the support of the Dirección General de Asuntos del Personal Académico, Universidad Nacional Autónoma de México (DGAPA–UNAM) by the grant IN112300 *Optica Matemática.* G.S.P. acknowledges the Consejo Nacional de Ciencia y Tecnología (México) for a Cátedra Patrimonial Nivel II, and Russian Foundation for Basic Research, grant RFBR N00–02–81023.

1. M. Krawtchouk, *C.R. Acad. Sci. Paris* **189**, 620 (1929).
2. N.M. Atakishiyev and K.B. Wolf, *J. Opt. Soc. Am. A* **14**, 1467 (1997).
3. E. Inönü and E.P. Wigner, *Proc. Nat. Acad. Sci. (US)* **39**, 510 (1953).
4. J.D. Talman, *Special Functions —a Group Theoretic Approach*, W.A. Benjamin, Inc., New York (1968).
5. A.A. Izmest'ev, G.S. Pogosyan, A.N. Sissakian and P. Winternitz, *J. Phys. A* **29**, 5940 (1996); A.A. Izmest'ev, G.S. Pogosyan, A.N. Sissakian and P. Winternitz, *J. Math. Phys.* **40**, 1549 (1999).
6. A.A. Izmest'ev, G.S. Pogosyan, A.N. Sissakian and P. Winternitz, *J. Phys. A* **34**, 521 (2001).
7. N.M. Atakishiyev and S.K. Suslov, *Teor. Math. Phys.* **85**, 1055 (1991).
8. A. Erdélyi, W. Magnus, F. Oberhettinger, and F.G. Tricomi, *Higher Transcendental Functions*, Vol. 2, McGraw-Hill, New York (1953).

A GLIMPSE AT MATHEMATICAL DIFFRACTION THEORY

MICHAEL BAAKE
Institut für Mathematik, Universität Greifswald
Jahnstr. 15 a, 17487 Greifswald, Germany
Email: mbaake@uni-greifswald.de

Mathematical diffraction theory is concerned with the analysis of the diffraction measure of a translation bounded complex measure ω. It emerges as the Fourier transform of the autocorrelation measure of ω. The mathematically rigorous approach has produced a number of interesting results in the context of perfect and random systems, some of which are summarized here.

1 Introduction

Diffraction experiments are widely used to gain information on the atomic structure of solids. Neutron and X-ray diffraction share the property that fundamental aspects of the diffraction image can be understood very well in single scattering theory together with Born's approximation, while electron diffraction needs more sophistication, see Ref. 7 for details. In the former case, on which we will focus here, the diffraction image is then described as the Fourier transform of the autocorrelation of the original distribution of scatterers. In this situation, important information is lost, namely the relative phases. Consequently, the inverse problem of structure determination has, in general, no unique solution, unless extra information is available from other sources of knowledge.

All this can be phrased in mathematically rigorous terms, but hardly anybody seemed really interested in it – until quasicrystals were discovered in the early 1980's. They are solids which, like crystals, show sharp Bragg diffraction images in X-ray diffraction, yet with maliciously perfect point symmetries which are crystallographically forbidden – at least in dimensions ≤ 3. As turned out quickly, they can be formally described as projections of portions (strips) of lattices in higher dimensions, and this can account both for their symmetries and for their perfect Bragg diffraction spectra. However, they also leave a certain degree of uneasiness towards the obviously not sufficiently well understood possibilities in diffraction theory.

This was the starting point of the rigorous development of mathematical diffraction theory in a paper by Hof [9] which was later also extended to the treatment of high temperature.[10] The setting is as follows. The structure under investigation is described as a translation bounded complex measure ω

on n-dimensional Euclidean space, whose autocorrelation, γ_ω is obtained as a volume-weighted convolution,

$$\gamma_\omega \;=\; \lim_{r\to\infty} \frac{\omega|_r * \tilde{\omega}|_r}{\mathrm{vol}(B_r)} \tag{1}$$

where B_r is the (open) ball of radius r around 0, $\omega|_r$ denotes the restriction of ω to B_r, and $\tilde{}$ stands for taking the complex conjugate of the origin-inverted measure. It is not even clear in general whether this limit exists, but if not, the right hand side certainly has points of accumulation which can then be obtained as limits along a discrete subsequence of averaging balls. We will thus assume, from now on, that the limit in (1) exists.

If so, the autocorrelation is a positive definite measure, and as such automatically possesses a Fourier transform, denoted by $\hat{\gamma}_\omega$, which is a positive measure by Bochner's theorem, compare Ref. 6 for background material. Furthermore, $\hat{\gamma}_\omega$ has a unique decomposition into three parts,

$$\hat{\gamma}_\omega \;=\; \left(\hat{\gamma}_\omega\right)_{pp} + \left(\hat{\gamma}_\omega\right)_{sc} + \left(\hat{\gamma}_\omega\right)_{ac}, \tag{2}$$

where *pp*, *sc* and *ac* as usual stand for pure point, singular continuous and absolutely continuous, with the ordinary Lebesgue measure as reference. The pure point (absolutely continuous) part corresponds to the Bragg spectrum (to diffuse scattering), while the term 'singular continuous' does not appear in the standard crystallographic literature.

In what follows, we will summarize a few aspects of this setting and give some pointers for further reading.

2 Pure point diffraction

One important class of structures are those with pure Bragg spectrum, i.e. those where $\hat{\gamma}_\omega$ is a pure point measure. This is certainly the case if the measure ω is crystallographic, i.e. periodic with a full lattice Γ of periods. So, $\omega = h * \delta_\Gamma$ where h is the density (or measure) in a fundamental domain and δ_Γ is the uniform lattice Dirac comb. With this, the autocorrelation reads $\gamma_\omega = (h * \tilde{h})\,\mathrm{dens}(\Gamma)\,\delta_\Gamma$, where $\mathrm{dens}(\Gamma)$ is the density of the lattice, and the diffraction measure is

$$\hat{\gamma}_\omega \;=\; |\hat{h}|^2 \left(\mathrm{dens}(\Gamma)\right)^2 \delta_{\Gamma^*} \tag{3}$$

with $\Gamma^* = \{u \mid uv \in \mathbb{Z} \text{ for all } v \in \Gamma\}$ the dual lattice. This rests upon Poisson's summation formula, $\hat{\delta}_\Gamma = \mathrm{dens}(\Gamma)\delta_{\Gamma^*}$, and reconfirms that the diffraction spectrum of a perfect crystal is pure point.

Other by now well-known examples of pure point diffractive systems include perfect quasicrystals as described by the projection method. More precisely, given a cut and project scheme,[14] a generic, regular model set based on it will produce a pure point diffraction spectrum. Here, a subset of a lattice is projected which lies in an incommensurately oriented strip with constant cross section. The latter is called window and lives in the so-called internal space, while the complement, which the pattern is projected to, is termed direct or physical space. The diffraction result was first proved by Hof [9] for Euclidean internal spaces with polytopal windows. It was later extended by various people to include more general internal spaces, and finally proved in full generality by Schlottmann.[15] This fuelled the speculation that basically only model sets have this property, but the set of visible lattice points [5] proved otherwise. In fact, model sets only form a relatively sparse set of examples with pure point diffraction.

After some effort into solving this puzzle, the answer uses a different box of tools and relates diffraction properties to almost periodicity of the autocorrelation measure. Partly based on results of Gil de Lamadrid and Argabright, it is shown in Ref. 4 that a translation bounded measure ω has pure point diffraction if and only if its autocorrelation γ_ω is strongly almost periodic as a measure. This criterion can actually be checked in many cases, and it also provides an independent proof of Schlottmann's theorem as a by-product. The result also shows that the diffraction problem really consists of two steps: one from ω to its autocorrelation, γ_ω, and one from there to the Fourier transform, $\hat{\gamma}_\omega$. The second step is ultimately easy because it is at least one-to-one, even though many open questions are still to be settled. So, let us take a look at the first step, which lies at the heart of the phase reconstruction problem.

3 The homometry problem

Two different measures can share the same autocorrelation, as is obvious from the limit in (1), in which case they are called *homometric*. In particular, modifying a given measure by a *finite* measure has no effect on the limit, i.e. on γ_ω. But much worse operations are possible, which makes the equivalence relation defined by homometry a rather nasty one. To show how bad the situation is, we recall a result from Ref. 8. There, the binary Rudin-Shapiro sequence, which is based on a substitution rule and thus has entropy 0, is realized as a Dirac comb along the integers. It has the same autocorrelation as the generic Bernoulli Dirac comb with occupation probability 1/2 per site, which is truly random and has maximal entropy, log(2). The homometry class also contains an example for each intermediate value of the entropy. This shows how bad the inverse problem really is, unless further information is available about the structure, or a general principle is invoked to select a 'reasonable' representative from the homometry class, e.g. by maximizing the entropy.

Another example is discussed in Ref. 1. A subset S of a lattice is homometric with its complement set, if both have the same density – no matter whether the subset itself is periodic, aperiodic or random. So, one can in particular construct homometric pairs for all possible types of spectra, pure and mixed. Note that the pure point part which is due to the underlying lattice can be made extinct by simply subtracting a multiple of the uniform lattice Dirac comb. This way, one can even realize purely singular continuous spectra, e.g. based upon substitution rules of Thue-Morse type.

4 Diffraction of lattice subsets

A subset of a lattice Γ inherits a good deal of the lattice structure, no matter how unusual the selection of lattice points may be. In particular, if we once again make the (rather weak) assumption that a given subset S has a unique autocorrelation, the diffraction measure $\hat{\gamma}_\omega$ of the Dirac comb δ_S is periodic, with the dual lattice Γ^* as lattice of periods.[1] What is more, the autocorrelation admits a representation in the form $\gamma_\omega = \Phi \cdot \delta_\Gamma$ with a Lipschitz continuous function Φ which interpolates the autocorrelation coefficients between the discrete lattice points. It can actually be chosen to have an extension to an entire function, which is a bit surprising.

In connection with this, the diffraction measure takes the form

$$\hat{\gamma}_\omega = \varrho * \delta_{\Gamma^*} \tag{4}$$

where ϱ is a finite, positive measure concentrated on a true fundamental domain of the dual lattice, see Ref. 1 for details. This means that the analysis of the spectral type can be done with ϱ, which makes it considerably easier.

This result, like many mentioned so far, can be generalized substantially, e.g. to weighted lattice subsets or to the setting of lattices in locally compact Abelian groups that are σ-compact. On the other hand, it has a wide range of applications. First of all, the above mentioned set of visible lattice points falls into the class, so must have a diffraction of the convolution form (4), as was previously derived explicitly.[5] Another class of interesting point sets are those obtained from lattice substitution systems which have recently received a lot of attention.[13] Finally, all lattice gases, with or without interaction, are covered. This gives a unified explanation for the periodicity observed in the diffraction of these stochastic systems.

5 Diffraction from stochastic point sets

Let us finally take a closer look at systems with some degree of randomness, either due to a lattice gas structure or due to an underlying random tiling approach. Bernoulli subsets of lattices and model sets are well understood, even in terms of elementary methods from stochastics, see Ref. 3 and references therein. Recently, this approach has been generalized considerably, with more sophisticated machinery, to cover stochastic selections from rather general Delone sets.[11] The outcome proves the folklore claim that uncorrelated random removal of scatterers has two effects, namely reducing the overall intensity of the diffraction of the fully occupied set, without changing the relative intensities, and adding a white noise type constant diffuse background.

More interesting would be a detailed understanding of the spectra of stochastic systems with interactions. Though some qualitative features are known, quantitative results are rare, i.e. basically restricted to systems in one dimension or to systems that can be mapped to exactly solvable models of statistical mechanics, see Ref. 2 and the literature given there. A particularly interesting question concerns the precise spectral nature of random tilings of the plane, both with crystallographic and non-crystallographic symmetries (which is essentially meant as symmetry on average here). While typical dimer based lattice random tilings have mixed spectra with pure point and absolutely continuous components,[2] some quasicrystal related random tilings seem to have singular continuous parts. There are very convincing scaling arguments and numerical simulations in favour of this observation, but no proof is available at present. This is an open challenge, together with the investigation how robust and hence relevant such a singular continuous diffraction spectrum would be for real world crystallography.

Acknowledgments

This summary grew out of discussions and cooperations with a number of colleagues, in particular Moritz Höffe, Robert V. Moody and Martin Schlottmann, whose input is gratefully acknowledged. I thank Uwe Grimm for reading the manuscript and suggesting a number of improvements.

References

1. M. Baake, Diffraction of weighted lattice subsets, Preprint (2000); math.MG/0008063.
2. M. Baake and M. Höffe, Diffraction of random tilings: some rigorous results, J. Stat. Phys. **99**, 219–261 (2000); math-ph/9904005.
3. M. Baake and R. V. Moody, Diffractive point sets with entropy, J. Phys. A: Math. Gen. **31**, 9023–9039 (1998); math-ph/9809002.
4. M. Baake and R. V. Moody, Weighted Dirac combs with pure point diffraction, Preprint (2001).
5. M. Baake, R. V. Moody and P. A. B. Pleasants, Diffraction from visible lattice points and k-th power free integers, Discr. Math. **221**, 3–42 (2000); math.MG/9906132.
6. C. Berg and G. Forst, *Potential Theory on Locally Compact Abelian Groups*, Springer, Berlin (1975).
7. J. M. Cowley, *Diffraction Physics*, 3rd ed., North-Holland, Amsterdam (1995).
8. M. Höffe and M. Baake, Surprises in diffuse scattering, Z. Kristallographie **215**, 373–376 (2000); math-ph/9911014.
9. A. Hof, On diffraction by aperiodic structures, Commun. Math. Phys. **169**, 25–43 (1995).
10. A. Hof, Diffraction of aperiodic structures at high temperatures, J. Phys. A: Math. Gen. **28**, 57–62 (1995).
11. C. Külske, Universal bounds on the selfaveraging of random diffraction measures, WIAS-preprint 676 (2001); math-ph/0109005.
12. J. C. Lagarias, Mathematical quasicrystals and the problem of diffraction, in: *Directions in Mathematical Quasicrystals*, eds. M. Baake and R. V. Moody, CRM Monograph Series, vol. 13, Amer. Math. Soc., Providence, RI (2000), pp. 61–93.
13. J.-Y. Lee and R. V. Moody, Lattice substitution systems and model sets, Discr. Comput. Geom. **25**, 173–201 (2001).
14. R. V. Moody, Model sets: A survey, in: *From Quasicrystals to More Complex Systems*, eds. F. Axel, F. Dénoyer and J. P. Gazeau, EDP Sciences, Les Ulis, and Springer, Berlin (2000), pp. 145–166.
15. M. Schlottmann, Generalized model sets and dynamical systems, in: *Directions in Mathematical Quasicrystals*, eds. M. Baake and R. V. Moody, CRM Monograph Series, vol. 13, Amer. Math. Soc., Providence, RI (2000), pp. 143–159.

EXACTLY SOLVABLE HAMILTONIANS CONNECTED WITH $\mathrm{su}_q(2)$ AND QUANTUM OPTICAL MODELS

A. BALLESTEROS

Departamento de Física, Universidad de Burgos,
Pza. Misael Bañuelos, E-09001 Burgos, Spain
E-mail: angelb@ubu.es

S.M. CHUMAKOV

Departamento de Física, Universidad de Guadalajara,
Revolución 1500, 44420, Guadalajara, Jalisco, México
E-mail: sergei@cencar.udg.mx

The $\mathrm{su}_q(2)$ algebra is considered as a natural dynamical symmetry algebra for some quantum optical models. This allows us to discuss spectrum and dynamical properties of these models. We use the Wigner quasiprobabilty function on the sphere to visualize quantum nonlinear dynamics.

1 Introduction

Let a, b, c be the boson operators of three different modes of the radiation field. The trilinear Hamiltonian

$$H = \omega_a a^\dagger a + \omega_b b^\dagger b + \omega_c c^\dagger c + g(a^\dagger b^\dagger c + c^\dagger ba) \tag{1}$$

describes (under certain conditions) nonlinear quantum optical processes as frequency conversion, Raman and Brillouin scattering, see, *e.g.* [1]. Through the Jordan-Schwinger transformation, this model is algebraically equivalent to the resonant interaction of a system of two-level atoms with a single mode of the quantized radiation field ("the Dicke model", see [2] and references therein):

$$H = H_0 + H_D = \omega a^\dagger a + \omega_{at} S_z + g(a^\dagger S_+ + S_- a). \tag{2}$$

Here S_k are collective atomic operators. Exact solutions for the Hamiltonians (1),(2) does not exist[3]. In [2] a perturbative approach was proposed based on the approximate su(2) dynamical symmetry of the models (1),(2). The Hamiltonian (2) comutes with the excitation number operator $N = a^\dagger a + S_z + A/2$ (where $A = 2l$ is the number of atoms). In every subspace with given value of N the interaction Hamiltonian H_D has a tridiagonal form, with the following non-zero matrix elements (we restrict ourselves with the exact resonance, $\omega = \omega_{at}$ and with the case $N > A$):

$$A_m \equiv \left[H_D^N\right]_{m,m-1} = \left[H_D^N\right]_{m-1,m} = \sqrt{(l+m)(l-m+1)(N-l+m)}, \tag{3}$$

$$m = -l+1, -l+2, \ldots, l.$$

Under certain dynamical conditions (in either a strong ($N \gg A$) or weak ($N \ll A$) field domain) the matrix (3) can be written as the $2J_x = J_+ + J_-$ generator of the corresponding irreducible representation of su(2) plus small corrections. In this way, the perturbative approach to the spectrum and dynamical properties of the models (1),(2) can be developed [2].

Recently we have shown[4] that the $\mathrm{su}_q(2)$ quantum algebra can be also used to define a Hamiltonian of the type H_D^N, whose eigenvalues and eigenvectors can be analytically found. The spectrum of the $\mathrm{su}_q(2)$ model is essentially anharmonic, and can efficiently serve as the 0-th order approximation to the Dicke model. Here we show that the corresponding perturbation theory can be used for the cases of the strongest nonlinearity of (1),(2), where the su(2) perturbation theory does not work. We also discuss some dynamical properties of the deformed $\mathrm{su}_q(2)$ model.

2 A tridiagonal Hamiltonian defined on $\mathrm{su}_q(2)$

Let the operators J_z , $J_\pm$ generate the quantum algebra $\mathrm{su}_q(2)$(see, e.g. [5,6,7]), with the coproduct,

$$\Delta(J_\pm) = J_\pm \otimes q^{-J_z} + q^{J_z} \otimes J_\pm\,, \quad \Delta(J_z) = J_z \otimes 1 + 1 \otimes J_z\,.$$

Deformed commutation rules consistent with the previous map are given in the form:

$$[J_z, J_\pm] = \pm\, J_\pm, \qquad [J_+, J_-] = [2\,J_z],$$

where $[x] := \frac{q^x - q^{-x}}{q - q^{-1}}$ and $q = e^{z/2}$. We shall assume that q is not a root of unity. Let us introduce, firstly, the "bare" basis of eigenvectors of J_z,

$$2J_z|l,m\rangle = 2m|l,m\rangle\,. \tag{4}$$

In this basis, the $(2\,l+1)$-dimensional irreducible representation of $\mathrm{su}_q(2)$ is given by (4) and

$$J_\pm|l,m\rangle = \sqrt{[l \mp m][l \pm m+1]}|l,m+1\rangle.$$

Let us consider the following $\mathrm{su}_q(2)$ operator as a Hamiltonian:

$$H_q = q^{J_z/2}\,(J_+ + J_-)\,q^{J_z/2}\,. \tag{5}$$

The $(2\,l+1)$-dimensional representation of H takes a tridiagonal form similar to (3) with

$$A_m(q) = q^{m-1/2}\sqrt{[l{+}m][l{-}m{+}1]}\,. \tag{6}$$

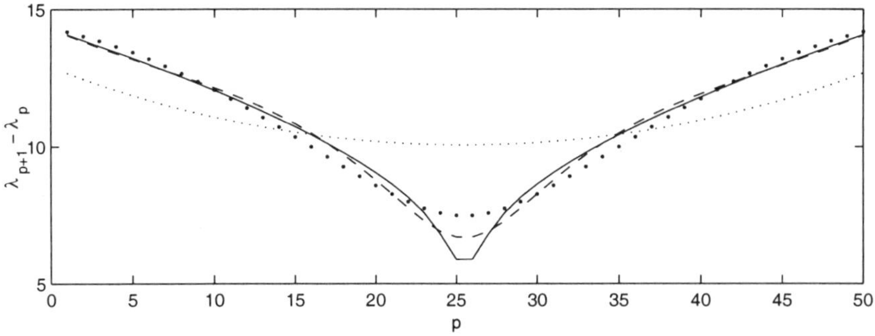

Figure 1. Dicke transition frequencies: —— exact, ··· 0-th, --- 1-st, – – 2-nd orders.

It can be proven (by induction and using the coproduct to construct higher dimensional representations from the three-dimensional one) that the spectrum of this operator for a given l is just

$$H_q|\underline{l,m}\rangle = [2m]|\underline{l,m}\rangle, \quad m = -l, \ldots, l.$$

Eigenvectors $|\underline{l,m}\rangle$ can be found in [4] (see also [8]). Note, that one can recover the common $\overline{\mathrm{su}(2)}$ results when $q \to 1$; then $H_q \to 2\,J_x$, the su(2)-generator used in [2] as the 0-th order approximation to the Dicke model Hamiltonian (2). The operator (5) has the following advantages: *(a)* one has an additional parameter q for better choice of the 0-th order approximation; *(b)* its spectrum is anharmonic from the beginning. On the other hand, the $\mathrm{su}_q(2)$environment is as convenient for the analytic calculations as the su(2) environment.

3 Second harmonics generation through $\mathrm{su}_q(2)$

Let us now consider the second harmonics generation which is a particular case of (3) when $N = A$. This is the most nonlinear case of (1),(2). As we have shown in [4], Eqs. (3) and (6) reveal similar behaviour and one can approximate the Hamiltonian H_D^N as follows,

$$H_0 = \Omega H_q, \quad \Omega = \frac{4(A+1)^{3/2}}{\sqrt{27}[A+1]},$$
$$A \log q = \frac{3}{2} \log \frac{\sqrt{5}-1}{2} \approx -0.7218. \tag{7}$$

(Note that the only parameter Ω was available in the case of su(2) perturbation theory [2].) We present the results for the approximation of transition

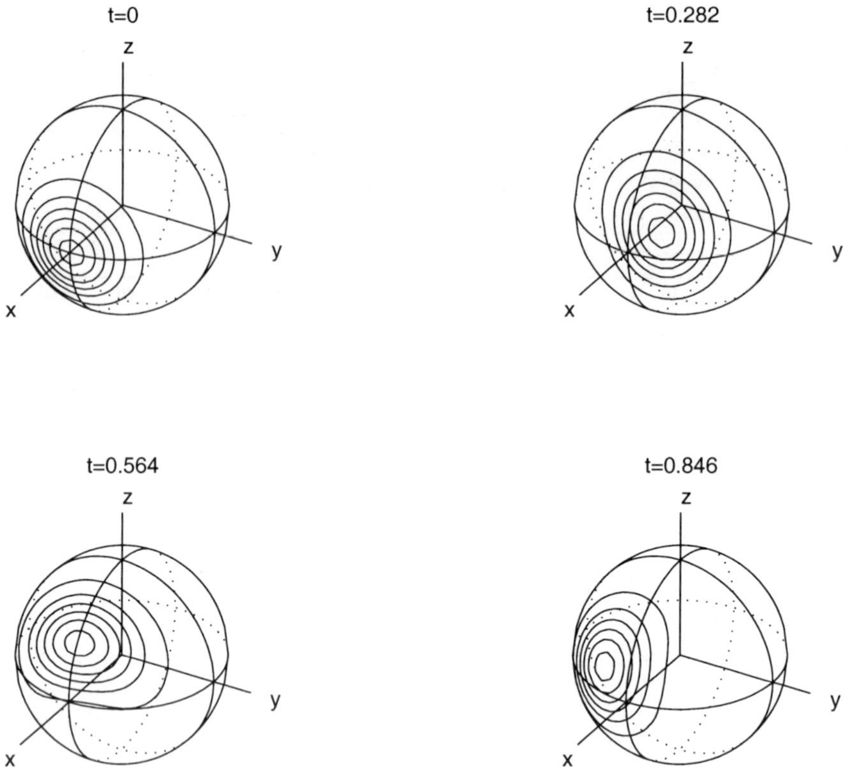

Figure 2. Wigner function dynamics, $S_x|\text{in}\rangle = A|\text{in}\rangle$.

frequencies, $\lambda_{p+1} - \lambda_p$ for the second harmonics generation in Fig. 1. (Here, λ_p, $p = 0, \ldots, A$ are eigenvalues). The exact numeric solution together with the results of the 0-th, 1-st and 2-nd order approximations are shown for the case $N = A = 50$. Note that the 1-st and 2-nd order approximations capture the typical singularity of transition frequencies at point $A/2$, the feature which is usually difficult to reproduce analytically.

4 Dynamical properties of H_q

Here we discuss some dynamical properties of the 0-th order approximation (5) which also reflect the properties of the exact Dicke Hamiltonian (2). The sphere $\mathcal{S}_2$ is a natural phase space for these models. We use the Stratonovich-

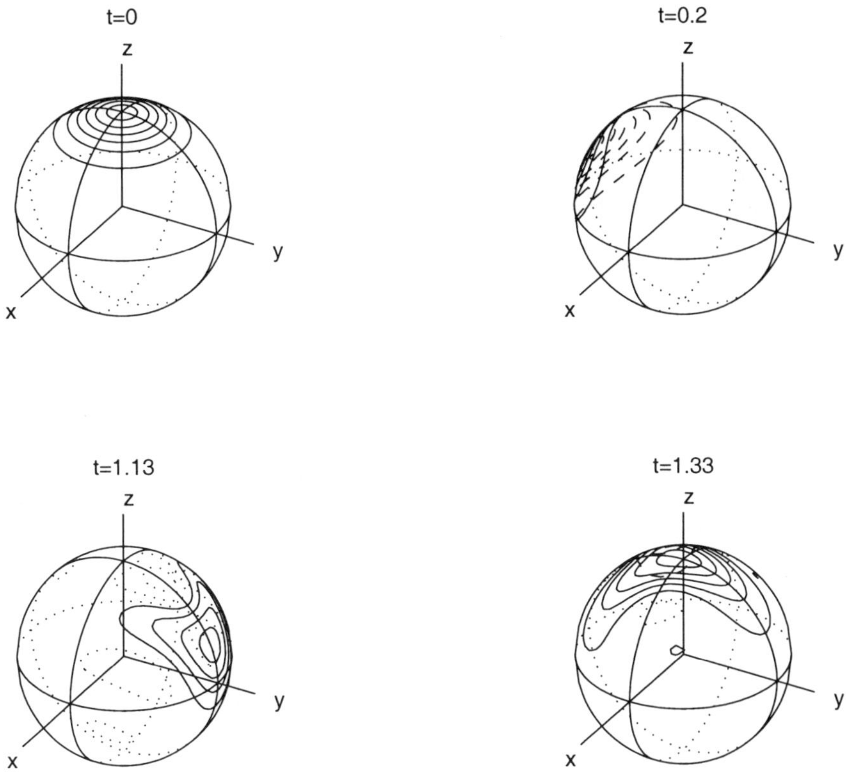

Figure 3. Wigner function dynamics, $S_z|\text{in}\rangle = A|\text{in}\rangle$.

Wigner function on the sphere (see [9] and, e.g., [10]) to visualize quantum dynamics. It is clear from the classical counterpart of Eq. (5) (which can be chosen as the Stratonovich-Wigner symbol of the operator H_q) that there are two stable points:

$$\theta_{1,2} = \arccos\left[1 - \frac{\log\left(1/2 + q^{2A}/2\right)}{A\log q}\right], \quad \phi_1 = 0, \quad \phi_2 = \pi.$$

In particular, for the value (7) of the parameter q we have $\theta_{1,2} = \arccos(1/3)$.

The motion in the small vicinity of a stable point has a semiclassical nature, as it can be seen in Fig. 2, where the evolution of the initial su(2) coherent state (CS) $|\theta = \pi/2, \phi = 0\rangle$ (along the x-direction) under the Hamiltonian H_q, Eqs. (5),(7) is presented for the case $A = 10$ (using level plots of the Stratonovich-Wigner function on the sphere). Note, that the revolution period round the stable point is ≈ 1.13, in this case. Here the evolution leads to a "soft" deformation of the shape of the initial CS and negative values of the Wigner function does not appear, for sufficiently long times. This is typical for the semiclassical region.

We consider the evolution of the initial CS $|\theta = 0\rangle$ (at the north pole) in Fig. 3. Here, the revolution period is ≈ 1.3. This is an example of the dynamics far from stable points, which shows strong deformation of the shape of the initial CS. Here, negative values of the Wigner function appear even for short times (the small dotted cloused contours in plots for $t = 1.13$ and $t = 1.33$). This is a signature of nonlinear quantum dynamics.

Acknowledgments

We thank Dr. A.B. Klimov for interesting discussions. SMC is grateful to the University of Burgos for hospitality.

References

1. *Coherence and Statistics of Photons and Atoms*, ed. J. Peřina (John Wiley and Sons, New York, 2001).
2. S. Chumakov and M. Kozierowski, *Quant. Semiclas. Optics* **8**, 775 (1996).
3. V.P. Karassiov and A.B. Klimov, *Phys. Lett. A* **189**, 43 (1994).
4. A. Ballesteros and S. Chumakov, *J. Phys. A* **32**, 6261 (1999).
5. P. Kulish and N. Reshetikhin *Zap. Nauch. Sem. LOMI* **101**, 101 (1981).
6. V.G. Drinfel'd in *Quantum Groups*, (Proceedings of the International Congress of Mathematics, MRSI Berkeley, 1986).
7. M. Jimbo, *Lett. Math. Phys* **10**, 63 (1985).
8. N.M. Atakishiyev and P. Winternitz, *J. Phys. A* **33**, 5303 (2000).
9. R.L. Stratonovich, *Sov. Phys. JETP* **31**, 1012 (1956).
10. A.B. Klimov and S.M. Chumakov, *JOSA A* **17**, 2315 (2000).

DYNAMICAL COALGEBRA SYMMETRY AND INTEGRABILITY

A. BALLESTEROS, F.J. HERRANZ

Departamento de Física, Universidad de Burgos, 09001-Burgos, Spain
E-mail: angelb@ubu.es, fjherranz@ubu.es

O. RAGNISCO

Dipartimento di Fisica, Terza Universitá di Roma, Via della Vasca Navale 84, 00146-Roma, Italy
E-mail: ragnisco@fis.uniroma3.it

A procedure for the systematic construction of integrable systems from symplectic realizations of a Poisson coalgebra with Casimir element is presented. Several examples of Hamiltonians with (either undeformed or 'quantum') coalgebra symmetry are given, and their Liouville integrability is discussed.

1 Poisson coalgebras and integrability

The relevance of Poisson coalgebras as the 'hidden' symmetries underlying the integrability properties of many families of finite-dimensional Hamiltonian systems has been recently outlined. [1,2,3,4,5,6] The aim of this contribution is to provide a summary of a large set of Hamiltonians with such a coalgebra symmetry and to comment on their Liouville integrability.

Let us summarize the general construction. [2] Let (A, Δ) be a Poisson coalgebra with generators X_i ($i = 1, \ldots, l$), Casimir function $\mathcal{C}(X_1, \ldots, X_l)$ and coassociative coproduct

$$\Delta \equiv \Delta^{(2)} : A \to A \otimes A, \tag{1}$$

which is a Poisson map with respect to the usual Poisson bracket on $A \otimes A$:

$$\{X_i \otimes X_j, X_r \otimes X_s\}_{A\otimes A} = \{X_i, X_r\}_A \otimes X_j X_s + X_i X_r \otimes \{X_j, X_s\}_A. \tag{2}$$

The j-th coproduct map $\Delta^{(j)}(X_i)$ of the generators of the algebra

$$\Delta^{(j)} : A \to A \otimes A \otimes \ldots^{j)} \otimes A, \tag{3}$$

can be defined by applying recursively the coproduct $\Delta^{(2)}$ in the form

$$\Delta^{(j)} := (id \otimes id \otimes \ldots^{j-2)} \otimes id \otimes \Delta^{(2)}) \circ \Delta^{(j-1)}. \tag{4}$$

Such an induction ensures that $\Delta^{(j)}$ is also a Poisson map.

Table 1. Functions obtained by applying the coproduct map.

X_1	X_2	$\ldots$	X_l	$\mathcal{C}$
$\Delta^{(2)}(X_1)$	$\Delta^{(2)}(X_2)$	$\ldots$	$\Delta^{(2)}(X_l)$	$\Delta^{(2)}(\mathcal{C})$
$\Delta^{(3)}(X_1)$	$\Delta^{(3)}(X_2)$	$\ldots$	$\Delta^{(3)}(X_l)$	$\Delta^{(3)}(\mathcal{C})$
$\vdots$	$\vdots$		$\vdots$	$\vdots$
$\Delta^{(N)}(X_1)$	$\Delta^{(N)}(X_2)$	$\ldots$	$\Delta^{(N)}(X_l)$	$\Delta^{(N)}(\mathcal{C})$

In this way, we can construct the set of functions shown in Table 1. ¿From them, given a smooth function $\mathcal{H}(X_1,\ldots,X_l)$, the N-sites Hamiltonian is defined as the N-th coproduct of $\mathcal{H}$:

$$H^{(N)} := \Delta^{(N)}(\mathcal{H}(X_1,\ldots,X_l)) = \mathcal{H}(\Delta^{(N)}(X_1),\ldots,\Delta^{(N)}(X_l)). \tag{5}$$

By construction, the $(N-1)$ functions $(m = 2,\ldots,N)$

$$C^{(m)} := \Delta^{(m)}(\mathcal{C}(X_1,\ldots,X_l)) = \mathcal{C}(\Delta^{(m)}(X_1),\ldots,\Delta^{(m)}(X_l)), \tag{6}$$

Poisson-commute with the Hamiltonian

$$\left\{C^{(m)}, H^{(N)}\right\}_{A\otimes A\otimes\ldots^{N)}\otimes A} = 0 \qquad m = 2,\ldots,N. \tag{7}$$

Moreover, all these integrals of the motion are in involution:

$$\left\{C^{(m)}, C^{(n)}\right\}_{A\otimes A\otimes\ldots^{N)}\otimes A} = 0 \qquad m,n = 2,\ldots,N. \tag{8}$$

Therefore, once a (non-degenerate) symplectic realization $A \to C^\infty(q,p)$ is given, the Hamiltonian $H^{(N)}$ will be a function of N canonical pairs (q_i,p_i) and is –by construction– completely integrable, since the functional independence of the $C^{(m)}$ functions is guaranteed (the m-th integral $C^{(m)}$ only depends on the first m pairs (q_i,p_i) of canonical coordinates).

The most natural option is to consider Hamiltonians $\mathcal{H}(X_1,\ldots,X_l)$ defined on Poisson–Lie algebras, with the (primitive) coalgebra structure

$$\Delta(X_i) = X_i \otimes 1 + 1 \otimes X_i. \tag{9}$$

On the other hand, the Poisson analogues of quantum algebras and groups [7] are also (deformed) coalgebras (A_z,Δ_z) (where z is the deformation parameter). Consequently, any function of the generators of a given 'quantum' Poisson algebra (with Casimir element C_z) will provide, under an appropriate (deformed) symplectic representation, an integrable deformation of the Hamiltonian defined on the coalgebra (A,Δ). Some examples of these deformed systems will also be given.

2 Hamiltonians with $sl(2)$ coalgebra symmetry

The Calogero-Gaudin (CG) Hamiltonian [8,9]

$$H^{(N)} = \sum_{i<j}^{N} 2\, p_i\, p_j\, (1 - \cos(q_i - q_j)), \tag{10}$$

was proven to have $sl(2)$ coalgebra symmetry [2,1] and its constants of the motion can be identified with the coproducts of the $sl(2)$ Casimir. The more general system

$$H^{(N)} = \left(\sum_{i=1}^{N} p_i\right)^2 + \frac{1}{2} \sum_{i,j=1}^{N} p_i\, p_j ((\kappa_1 + \kappa_2) \cos(q_i - q_j) - (\kappa_1 - \kappa_2) \cos(q_i + q_j)). \tag{11}$$

(where κ_i are free parameters) was also proven to be completely integrable. [2]

In general, the modification of the symplectic realization drastically changes the 'shape' of the Hamiltonian. For instance, by using the Gelfan'd-Dyson symplectic map the CG system reads [6]

$$H^{(N)} = \sum_{i<j}^{N} \{-\, p_i\, p_j\, (q_i - q_j)^2 - b\, (p_i - p_j)\, (q_i - q_j)\} + \frac{b^2}{4} N^2, \tag{12}$$

where b is a constant. On the other hand, the $N = 2$ rational Calogero-Moser Hamiltonian [10] has also $sl(2)$ coalgebra symmetry, [6] although the $N > 2$ case is still an open problem.

The following new integrable generalization of the N-dimensional Smorodinsky-Winternitz (SW) Hamiltonian is also $sl(2)$ coalgebra-invariant [3]

$$H^{(N)} = \sum_{i=1}^{N} \left(p_i^2 + \frac{c_i}{q_i^2}\right) + \mathcal{F}\left(\sum_{i=1}^{N} q_i^2\right) \tag{13}$$

where $\mathcal{F}$ is an arbitrary smooth function. In fact, the $sl(2)$ coalgebra provides a common set of integrals of the motion for any choice of $\mathcal{F}$.

3 An integrable deformation of the CG system from $sl_q(2)$

Given a Hamiltonian with certain coalgebra symmetry, any deformation of the underlying coalgebra provides an integrable deformation of the initial Hamiltonian. This fact has been shown for the first time by constructing the following deformed CG system: [1]

$$H_z^{(N)} = \sum_{i<j}^{N} 2\, \pi_i\, \pi_j\, (1 - \cos(q_i - q_j)), \tag{14}$$

where

$$\pi_k = 2\,\frac{\sinh(\frac{z}{2}p_k)}{z}\prod_{i=1}^{k-1} e^{-\frac{z}{2}p_i}\prod_{j=k+1}^{N} e^{\frac{z}{2}p_j}. \tag{15}$$

The corresponding constants of the motion come from the (deformed) coproduct of the (deformed) Casimir of $sl_q(2)$ (the standard deformation of $sl(2)$ with $q = e^z$) and, as expected, in the limit $z \to 0$ we recover the 'classical' CG system given in Eq. (10). We remark that the quantum mechanical version of this deformed CG system has been explicitly solved [11] and a 'twisted' version of the Gaudin magnet has already been introduced [4] through a quantum deformation of $gl(2)$. On the other hand, an integrable deformation of the $N = 2$ SW system has also been obtained [6] by using the same procedure.

4 Non-standard integrable deformations from $sl_h(2)$

Another integrable deformation of the generalized SW Hamiltonian

$$H_h^{(N)} = \sum_{i=1}^{N}\left(\frac{\sinh hq_i^2}{hq_i^2}p_i^2 + \frac{hc_i}{\sinh hq_i^2}\right)e^{hK_i^{(N)}(q^2)} + \mathcal{F}\left(\sum_{i=1}^{N} q_i^2\right), \tag{16}$$

(where the K-functions are defined by $K_i^{(m)}(x) = -\sum_{k=1}^{i-1} x_k + \sum_{l=i+1}^{m} x_l$) has been obtained [3] by making use of the non-standard deformation $sl_h(2)$. [12] Note that when all the constants c_i vanish, Eq. (16) provides an integrable deformation of the motion of a particle in an N-dimensional Euclidean space under the action of an arbitrary central potential. We also stress the long range nature of the 'interaction terms' appearing in Eqs. (14) and (16).

5 Ruijsenaars-Schneider like systems and $P_z(1+1)$ coalgebra

Another interesting example of coalgebra-invariant system is the following analogue [2] of the Ruijsenaars-Schneider model: [13]

$$H_z^{(N)} = \sum_{i=1}^{N}\cosh\theta_i \,\exp\left(-\frac{z}{2}(\sum_{j=1}^{i-1} q_j) + \frac{z}{2}(\sum_{k=i+1}^{N} q_k)\right), \tag{17}$$

where (q_i, θ_i) are canonically conjugate coordinates. This completely integrable Hamiltonian was obtained by using the Poisson analogue of a quantum deformation of the (1+1) dimensional Poincaré algebra. [14]

6 The oscillator coalgebra h_4 and related systems

The Hamiltonian [15]

$$H^{(N)} = (\lambda + \mu) \sum_{i=1}^{N} p_i + 2\,\mu \sum_{i<j}^{N} \sqrt{p_i\, p_j}\, \cosh(q_i - q_j), \tag{18}$$

can be proven to be coalgebra invariant [2] under the harmonic oscillator coalgebra h_4. The quantum version of this Hamiltonian has also been solved [16] and the system turns out to be equivalent to a chain of coupled oscillators. [16,17]

7 Two-photon Poisson coalgebras and quasi-integrability

The Poisson version of the two-photon/Schrödinger algebra h_6 [18,19] provides a huge family of Hamiltonians with coalgebra symmetry, namely

$$H^{(N)} = \mathcal{H}\left(f_N^{(N)}, f_M^{(N)}, f_{A_+}^{(N)}, f_{A_-}^{(N)}, f_{B_+}^{(N)}, f_{B_-}^{(N)}\right), \tag{19}$$

where

$$\begin{aligned} f_{A_+}^{(N)} &= \sum_{i=1}^{N} p_i \qquad & f_{A_-}^{(N)} &= \sum_{i=1}^{N} m_i q_i \qquad & f_N^{(N)} &= \sum_{i=1}^{N} \left(q_i p_i - \frac{m_i}{2}\right) \\ f_{B_+}^{(N)} &= \sum_{i=1}^{N} \frac{p_i^2}{m_i} \qquad & f_{B_-}^{(N)} &= \sum_{i=1}^{N} m_i q_i^2 \qquad & f_M^{(N)} &= \sum_{i=1}^{N} m_i . \end{aligned} \tag{20}$$

However, the symplectic realization given by Eq. (20) introduces a slight degeneracy, in the sense that the integral of the motion that comes from $\Delta^{(2)}(\mathcal{C})$ is left. As a consequence, for an arbitrary choice of the function $\mathcal{H}$ the Hamiltonian (19) is quasi-integrable (it has $(N-2)$ constants of the motion). [5]

Note that many interesting types of N-dimensional Hamiltonians can be obtained as certain choices for $\mathcal{H}$ in Eq. (19) (natural systems with polynomial potentials, Fokker-Planck Hamiltonians, static electromagnetic systems, Drach systems, etc.). [20] The superintegrability of some of these physically relevant Hamiltonians can be recovered when the function $\mathcal{H}$ is defined on a subalgebra of h_6, since additional constants can be obtained as the coproducts of the Casimir of the subalgebra. But such subalgebra cases do not exhaust the set of superintegrable systems included in Eq. (19). For instance,

$$H^{(N)} = \frac{1}{g}\left(\sum_{i=1}^{N} \frac{p_i^2}{2m_i} + \beta_1 \sum_{i=1}^{N} m_i q_i + \beta_2 \sum_{i=1}^{N} m_i q_i^2 + \beta_3\right), \tag{21}$$

with $g = \alpha_1 \sum_{i=1}^{N} m_i q_i + \alpha_2 \sum_{i=1}^{N} m_i q_i^2 + \alpha_3$ turns out to be superintegrable. [20]

Finally, we mention that –by following the same approach– new integrable systems have been defined on quantum Heisenberg algebras, [6] classical and quantum Galilei algebras, [20,21] classical and quantum $gl(2)$ algebras, [4,20] quantum h_6 algebras, [22] $so(3,2)$ Poisson coalgebras [6] and Poisson-Lie groups. [23]

Acknowledgments

A.B. and F.J.H. have been partially supported by Junta de Castilla y León (Project CO2/399) and by Ministerio de Ciencia y Tecnología (Project BFM2000-1055), Spain. O.R. has been partially supported by INFN and by MURST (PRIN 'Geometry and Integrability'), Italy.

References

1. A. Ballesteros *et al.*, *Czech. J. Phys.* **46**, 1153 (1996).
2. A. Ballesteros and O. Ragnisco, *J. Phys. A: Math. Gen.* **31**, 3791 (1998).
3. A. Ballesteros and F.J. Herranz, *J. Phys. A: Math. Gen.* **32**, 8851 (1999).
4. A. Ballesteros, *Int. J. Mod. Phys. B* **13**, 2903 (1999).
5. A. Ballesteros and F.J. Herranz, *J. Nonlinear Math. Phys.* **8** Suppl., 18 (2001).
6. A. Ballesteros and O. Ragnisco, *J. Math. Phys*, in press (2001).
7. V. Chari and A. Pressley, *A Guide to Quantum Groups* (Cambridge University Press, Cambridge, 1994).
8. F. Calogero, *Phys. Lett. A* **201**, 306 (1995).
9. M. Gaudin, *La Fonction d'Onde de Bethe* (Masson, Paris, 1983).
10. F. Calogero, *J. Math. Phys.* **12**, 419 (1971).
11. F. Musso and O. Ragnisco, *J. Math. Phys* **41**, 7386 (2000).
12. C. Ohn, *Lett. Math. Phys.* **25**, 85 (1992).
13. S.N.M. Ruijsenaars and H. Schneider, *Ann. Phys.* **170**, 370 (1986).
14. L. Vaksman and L.I. Korogodskii, *Sov. Math. Dokl.* **39**, 173 (1989).
15. F. Calogero, *J. Math. Phys.* **36**, 9 (1995).
16. F. Calogero and J.F. Van Diejen, *Phys. Lett. A* **205**, 143 (1995).
17. V. Karimipour, *J. Math. Phys.* **38**, 1577 (1997).
18. W.M. Zhang *et al.*, *Rev. Mod. Phys.* **62**, 867 (1990).
19. G. Burdet *et al.*, *J. Math. Phys.* **19**, 1758 (1978).
20. A. Ballesteros and F.J. Herranz, in preparation (2001).
21. A. Ballesteros *et al.*, *J. Phys. A: Math. Gen.* **33**, 3431 (2000).
22. A. Ballesteros and F.J. Herranz, *Czech. J. Phys.* **50**, 1239 (2000).
23. J. Grabowski *et al.*, *Mod. Phys. Lett. A* **14**, 2109 (1999).

$\pi\rho\omega$ VERTEX IN NUCLEAR MATTER

ANNA BARAN, AGNIESZKA BIENIEK, AND WOJCIECH BRONIOWSKI
The H. Niewodniczański Institute of Nuclear Physics, ul. Radzikowskiego 152
PL-31342 Krakow, Poland

Medium modifications of the $\pi\omega\rho$ vertex are analyzed in context of the $\omega \to \pi^0\gamma^*$ and $\rho \to \pi\gamma^*$ decays in nuclear matter. A relativistic hadronic model with mesons, nucleons, and $\Delta(1232)$ isobars is applied. A substantial increase of the widths for the decays $\omega \to \pi^0\gamma^*$ and $\rho \to \pi\gamma^*$ is found for photon virtualities in the range $0.3 - 0.6$GeV. This enhancement has a direct importance for the description of dilepton yields obtained in relativistic heavy-ion collisions.

The experience gathered over the past years led to the conviction that the properties of hadrons in medium are substantially modified [1,2,3,4,5,6,7,8,9,10,11,12,13,14,15,16,17,18,19,20]. This view is supported by numerous studies of hadronic Green functions. Only a few investigations of the in-medium hadronic *couplings* exist [21,22,23,24,25,26,27]. For instance, it has been found that the value of the $\rho\pi\pi$ coupling is considerably enhanced. In Ref. [28], which is the basis for this talk, we have examined the *in-medium* $\pi\omega\rho$ coupling. This vertex enters the decays $\omega \to \pi^0\gamma^*$ and $\rho \to \pi\gamma^*$, which are important for description of the dilepton production in relativistic heavy-ion collisions [29,30]. The existing calculations [31,32,33] have always used the unmodified (vacuum) value of the $\pi\omega\rho$ coupling. We have found that the medium effects can significantly alter the input of such calculations, thereby modifying the predicted dilepton production and creating hopes for a resolution of the long-standing problem of the experimental dilepton enhancement in the range $M_{l^+l^-} \sim 0.3 - 0.5$GeV.

In our study we use a typical hadronic model with meson, nucleon and Δ degrees of freedom, and for simplicity work at leading baryon density and at zero temperature. In addition, we constrain ourselves to the case where the decaying particle is at rest with respect to the medium. The effects of the relative motion turn out to be not very large [34]. In the vacuum the $\pi\omega\rho$ coupling is

$$-iV_{\pi\omega^\mu\rho^\nu} = i\frac{g_{\pi\omega\rho}}{F_\pi}\epsilon^{\mu\nu pQ}, \tag{1}$$

where we have used the notation $\epsilon^{\mu\nu pQ} = \epsilon^{\mu\nu\alpha\beta}p_\alpha Q_\beta$, and $F_\pi = 93$MeV is the pion decay constant. In our convention Q is the incoming momentum of the pion, p is the outgoing momentum of the ρ, and $q = Q - p$ is the outgoing momentum of the ω. The vector mesons are, in general, virtual. For the

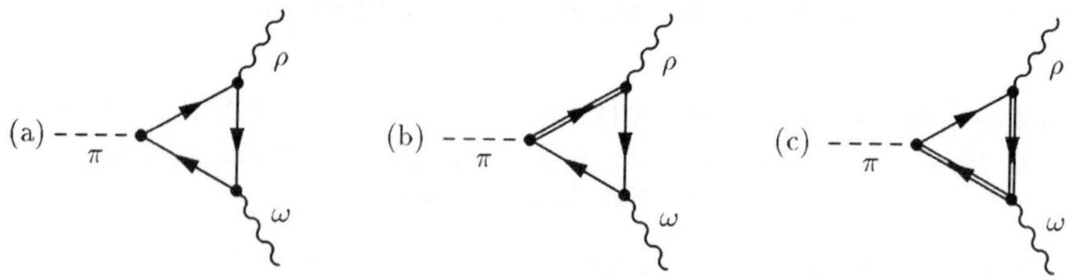

Figure 1. Diagrams included in our calculation (crossed ones not shown). Wavy lines denotes the ρ or ω, dashed lines the pions, solid lines the in-medium nucleon, and double lines the Δ resonance.

kinematics where all external momenta vanish, the vector meson dominance models gives $g_{\pi\to\omega\rho} = -\frac{g_\rho g_\omega}{e^2} g_{\pi\gamma\gamma}$, with $g_{\pi\gamma\gamma} = \frac{e^2}{4\pi^2 F_\pi}$. The nucleon propagator in nuclear matter is decomposed into the *free* and *density* parts in the usual way [4]:

$$iS(k) = iS_F(k) + iS_D(k) = i(\gamma^\mu k_\mu + m_N)[\frac{1}{k^2 - m_N^2 + i\varepsilon} + \frac{i\pi}{E_k}\delta(k_0 - E_k)\theta(k_F - |k|)]. \quad (2)$$

Here m_N denotes the nucleon mass, $E_k = \sqrt{m_N^2 + k^2}$, and k_F is the Fermi momentum. The Rarita-Schwinger propagator [35,36] for the spin 3/2 $\Delta(1232)$ particle has the form

$$iS_\Delta^{\mu\nu}(k) = i\frac{\gamma^\mu k_\mu + M_\Delta}{k^2 - M_\Delta^2}(-g^{\mu\nu} + \frac{1}{3}\gamma^\mu\gamma^\nu + \frac{2k^\mu k^\nu}{3M_\Delta^2} + \frac{\gamma^\mu k^\nu - \gamma^\nu k^\mu}{3M_\Delta}). \quad (3)$$

The effects of the non-zero width of the Δ are incorporated by replacing M_Δ by the complex mass $M_\Delta - i\Gamma/2$.

The diagrams used in our study are shown in Fig. 1. The density effects are produced when exactly one of the nucleon lines in each of the our diagrams of Fig. 1 involves the nucleon *density* propagator, S_D. The diagrams with more than one S_D vanish for kinematic reasons. On the other hand, diagrams with no S_D describe the vacuum polarization effects, not considered in the present study devoted to the Fermi sea effect only.

The meson-baryon vertices needed in our analysis have the standard form

$$
\begin{aligned}
-iV_{\omega^\mu NN} &= ig_\omega \gamma^\mu, \\
-iV_{\rho_b^\mu NN} &= i\frac{g_\rho}{2}(\gamma^\mu - \frac{i\kappa_\rho}{2m_N}\sigma^{\mu\nu}p_\nu)\tau^b, \\
-iV_{\pi^a NN} &= \frac{g_A}{2F_\pi}\not{Q}\gamma^5\tau^a.
\end{aligned} \tag{4}
$$

For the interactions involving the Δ resonance we use

$$
\begin{aligned}
-iV_{N\Delta^\alpha\pi^a} &= g_{N\Delta\pi}Q^\alpha T^a, \\
-iV_{N\Delta^\alpha\rho_b^\mu} &= ig_{N\Delta\rho}(\not{p}\gamma^5 g^{\alpha\mu} - \gamma^\mu\gamma^5 p^\alpha)T^b, \\
-iV_{\Delta^\alpha\Delta^\beta\omega^\mu} &= -ig_\omega(\gamma^\alpha\gamma^\mu\gamma^\beta - \gamma^\beta g^{\alpha\mu} - \gamma^\alpha g^{\beta\mu} + \gamma^\mu g^{\alpha\beta}).
\end{aligned} \tag{5}
$$

Above T^a is the standard isospin $\frac{1}{2} \to \frac{3}{2}$ transition matrix [37,36,38,39,40]. We choose the following values for the physical parameters: $g_\omega = 10.4$, $g_\rho = 5.2$, $g_A = 1.26$, $g_{N\Delta\pi} = 2.12/m_\pi$, $g_{N\Delta\rho} = 2.12\sqrt{2}/m_\pi$, $\kappa_\rho = 3.7$.

The $\pi\omega\rho$ vertex has the following tensor structure in the nuclear medium moving with the four-velocity u:

$$
\begin{aligned}
A^{\mu\nu} = {} & A_1\varepsilon^{\mu\nu pQ} + A_2\varepsilon^{\mu\nu uQ} + A_3\varepsilon^{\mu\nu pu} + A_4\varepsilon^{\mu upQ}p^\nu \\
& + A_5\varepsilon^{\mu upQ}Q^\nu + A_6\varepsilon^{\mu upQ}u^\nu + A_7\varepsilon^{\nu upQ}p^\mu + A_8\varepsilon^{\nu upQ}Q^\mu + A_9\varepsilon^{\nu upQ}u^\mu.
\end{aligned} \tag{6}
$$

This structure, restricted by the Lorentz invariance and parity, is more involved than in the vacuum case (1) due to the availability of the additional four-vector u. The leading-density calculation of diagrams of Fig. 1 is very simple, and follows the steps of Ref. [27]. First, the trace factors are evaluated, then k^μ is substituted by $m_N u^\mu$, with $u^\mu = (1,0,0,0)$ in the rest frame of the medium, and, finally, the loop integration is replaced by $\int \frac{d^4k}{(2\pi)^4}\Theta(k_f - |\vec{k}|)\delta(k_0 - E_k) \to \frac{1}{8\pi}\rho_B$, where ρ_B denotes the baryon density [27].

For the pion is at rest with respect to the medium, the four-vectors Q and u are parallel, and then $A^{\mu\nu} = (A_1 + A_3/m_\pi)\varepsilon^{\mu\nu pQ}$. We introduce the notation $A^{\mu\nu} = i\frac{e^2}{g_\rho g_\omega}\left(\frac{g_{\pi\rho\omega}}{F_\pi} + \frac{\rho_B}{F_\pi}B\right)\epsilon^{\nu\mu pQ}$, which puts the medium effects in B. Our results below will be presented relative to the vacuum value $g_{\rm vac} = g_{\pi\rho\omega}$, hence we define $g_{\rm eff} \equiv g_{\pi\rho\omega} + \rho_0 B$.

For the processes $\omega \to \pi^0\gamma^*$ and $\rho^a \to \pi^a\gamma^*$, where ω and ρ are on mass shell, we use the following vacuum values of the coupling constants: $g_{\omega\to\rho\pi} = -1.13$ and $g_{\rho\to\omega\pi} = -0.76$, obtained from the experimental partial decay width $\Gamma_{\omega\to\pi^0\gamma} = 717\text{keV}$ and $\Gamma_{\rho^a\to\pi^a\gamma} = 79\text{keV}$. These coupling constants differ from each other, as well as from $g_{\pi\to\omega\rho} = -\frac{g_\rho g_\omega}{e^2}g_{\pi\gamma\gamma} = -1.36$ inferred

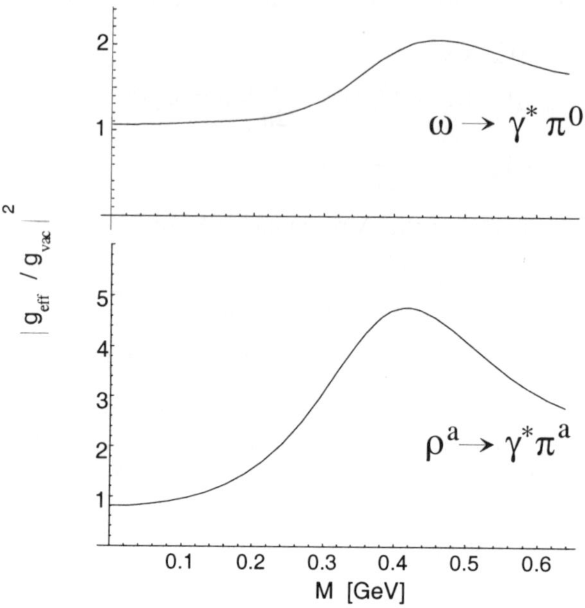

Figure 2. Top: $|g_{\rm eff}/g_{\rm vac}|^2$ for the decay $\omega \to \gamma^*\pi^0$ at the nuclear saturation density, plotted as a function of the virtuality of the photon, M. Bottom: the same for the decay ρ.

from the π^0 decay. This difference is a manifestation of form-factor effects related to virtuality of vector mesons.

The numerical results are shown in Fig. 2. At low values of M the effective coupling constant remains practically unchanged. At higher values of M, above 0.2GeV, the value of $g_{\rm eff}$ is much larger than in the vacuum: for the square of $g_{\rm eff}$ we have an enhancement by a factor of ~ 2 for the ω, and ~ 5 for the ρ around $M = 0.4$GeV. These values strongly indicate the possibility of large medium effects in the Dalitz decays of vector mesons. The enhancement of Fig. 2 directly affects the calculations of the dilepton yields in relativistic heavy-ion collisions. The theoretical output from the Dalitz decays of vector mesons becomes enhanced in the region of $0.3 - 0.6$GeV, which is exactly where the existing calculations have serious problems in supplying enough strength to explain the CERES and HELIOS experiments [29,30]. The dilepton production rate from the Dalitz decays of a vector meson is given by the formula [32]

$$\frac{dN_{l^+l^-}}{d^4x\,dM^2} = \int \frac{d^3p_v}{(2\pi)^3 E_v} f(p_v) \frac{3m_v}{\pi M^3} \Gamma_{v\to\pi^0\gamma^*} \Gamma_{\gamma^*\to l^+l^-} \tag{7}$$

where v relates to the vector meson, and $f(p_v)$ is the distribution function. The width $\Gamma_{v\to\pi^0\gamma^*}$ depends on the baryon density, ρ_B, which in turn, depends on the space-time point x. Clearly, an increased value of $\Gamma_{\omega\to\pi^0\gamma^*}$ results in an increased dilepton yield. A more exact estimate requires a detailed model of the evolution of the fireball, and is left for a further study [34]. According to our analysis (*cf.* Fig. 3), at $M \approx 400$ MeV we have $\Gamma_{\omega\to\pi^0\gamma^*} \approx 900$keV and $\Gamma_{\rho^a\to\pi^a\gamma^*} \approx 250$keV. However, in a thermalized system due to isospin degeneracy the ρ mesons are three times more numerous, hence the effect of the ρ is comparable in size to the effect of the ω. The Dalitz yields from the π^0 decays are not altered by the medium for a double reason: first, virtually all pions decay outside of the fireball due to their long lifetime, second, the width for this decay is practically unaltered by the medium [28].

Acknowledgements

Research supported by the Polish State Committee for Scientific Research grant 2P03B09419.

References

1. G. E. Brown and M. Rho, *Phys. Rev. Lett.* **66**, 2720 (1991).
2. L. S. Celenza, A. Pantziris, C. M. Shakin, and W.-D. Sun, *Phys. Rev. C* **45**, 2015 (1992).
3. B. D. Serot and J. D. Walecka, *Adv. Nucl. Phys.* **16**, 1 (1986).
4. S. A. Chin, *Ann. Phys. (NY)* **108**, 301 (1977).
5. T. Hatsuda, H. Shiomi, and H. Kuwabara, *Prog. Theor. Phys.* **95**, 1009 (1996).
6. H.-C. Jean, J. Piekarewicz, and A. G. Williams, *Phys. Rev. C* **49**, 1981 (1994).
7. M. Herrmann, B. L. Friman, and W. Noerenberg, *Nucl. Phys. A* **560**, 411 (1993).
8. B. Friman and H. J. Pirner, *Nucl. Phys. A* **617**, 496 (1997).
9. R. Rapp, G. Chanfray, and J. Wambach, *Nucl. Phys. A* **617**, 472 (1997).
10. S. Leupold and U. Mosel, *Prog. Part. Nucl. Phys.* **42**, 221 (1999).

11. S. Gao, C. Gale, C. Ernst, H. Stocker, and W. Greiner, *Nucl. Phys. A* **661**, 518 (1999).
12. T. Hatsuda and S. H. Lee, *Phys. Rev. C* **46**, R34 (1993).
13. S.-H. Lee, *Nucl. Phys. A* **670**, 119 (2000).
14. U. Vogl and W. Weise, *Prog. Part. Nucl. Phys.* **27**, 195 (1991).
15. F. Klingl, N. Kaiser, and W. Weise, *Nucl. Phys. A* **624**, 527 (1997).
16. G. Chanfray, *Nucl. Phys. A* **685**, 328 (2001).
17. W. Broniowski and B. Hiller, *Phys. Lett. B* **392**, 267 (1997).
18. V. L. Eletsky, B. L. Ioffe, and J. I. Kapusta, *Eur. J. Phys. A* **3**, 381 (1998).
19. B. Friman, *Acta Phys. Polon. B* **29**, 3195 (1998).
20. M. Lutz, B. Friman, and G. Wolf, *Nucl. Phys. A* **661**, 526 (1999).
21. C. Song and V. Koch, *Phys. Rev. C* **54**, 3218 (1996).
22. W. Broniowski, W. Florkowski, and B. Hiller, *Acta Phys. Polon. B* **30**, 1079 (1999).
23. W. Broniowski, W. Florkowski, and B. Hiller, *Eur. Phys. J. A* **7**, 287 (2000).
24. M. Urban, M. Buballa, R. Rapp, and J. Wambach, *Nucl. Phys. A* **641**, 433 (1998).
25. M. Urban, M. Buballa, R. Rapp, and J. Wambach, *Nucl. Phys. A* **673**, 357 (2000).
26. M. K. Banerjee and J. A. Tjon, *Phys. Rev. C* **56**, 497 (1997).
27. W. Broniowski, W. Florkowski, and B. Hiller, nucl-th/0103027.
28. A. Bieniek, A. Baran, and W. Broniowski, nucl-th/0106053.
29. CERES Collab., G. Agakichiev *et al.*, *Phys. Rev. Lett.* **75**, 1272 (1995).
30. HELIOS/3 Collab., M. Masera *et al.*, *Nucl. Phys. A* **590**, 93 (1995).
31. G. Q. Li, C. M. Ko, and G. E. Brown, *Nucl. Phys. A* **606**, 568 (1996).
32. P. Koch, *Z. Phys. C* **57**, 283 (1993).
33. R. Rapp and J. Wambach, *Adv. Nucl. Phys.* **25**, 1 (2000).
34. A. Bieniek and W. Broniowski, to be published.
35. W. Rarita and J. Schwinger, *Phys. Rev.* **60**, 61 (1941).
36. M. Benmerrouche, R. M. Davidson, and N. C. Mukhopadhyay, *Phys. Rev. C* **39**, 2339 (1989).
37. T. O. E. Ericson and W. Weise, *Pions and nuclei*, Clarendon Press, Oxford (1988).
38. T. R. Hemmert, B. R. Holstein, and J. Kambor, *J. Phys. G* **24**, 1831 (1998).
39. V. Pascalutsa and R. Timmermans, *Phys. Rev. C* **60**, 042201 (1999).
40. H. Haberzettl, nucl-th/9812043.

UNUSUAL EIGENVALUE PROBLEMS

CARL M. BENDER
Physics Department, Washington University,
St. Louis, MO 63130, USA
E-mail: cmb@howdy.wustl.edu

In this talk I discuss a new class of exactly solvable differential-equation eigenvalue problems $-y''(x) + x^{2N+2}y(x) = x^N Ey(x)$ $(N = -1, 0, 1, 2, 3, \ldots)$ on the interval $-\infty < x < \infty$. These problems can be solved exactly for the eigenvalues E and corresponding eigenfunctions $y(x)$. The eigenvalues are all integers and the eigenfunctions are confluent hypergeometric functions. The eigenfunctions can be rewritten as products of polynomials and functions that decay exponentially as $x \to \pm\infty$. For odd N the polynomials obtained in this way are new and interesting classes of orthogonal polynomials. For example, when $N = 1$, the eigenfunctions are orthogonal polynomials in x^3 multiplying Airy functions of x^2. The properties of the polynomials for all N are described.

1 Introduction

Let us consider the class of differential-equation eigenvalue problems

$$-y''(x) + x^{2N+2}y(x) = x^N Ey(x) \qquad (N = -1, 0, 1, 2, 3, \ldots) \tag{1}$$

on the interval $-\infty < x < \infty$. The eigenfunction $y(x)$ obeys the boundary conditions that $y(x)$ vanish exponentially rapidly as $x \to \pm\infty$. For each integer $N \geq -1$, we can solve these eigenvalue problems exactly. The eigenvalues are all integers and the associated eigenfunctions are confluent hypergeometric functions. All the eigenfunctions for each value of N can be written as the product of a polynomial and a given function that vanishes exponentially for large $|x|$. The classes of polynomials obtained in this way are orthogonal and for odd N are apparently new. The details are given in a forthcoming paper.[1]

The eigenvalue problem (1) arises in many contexts. In classical physics a perturbative technique called boundary-layer theory has been developed to find approximate solutions to boundary-value problems of the form

$$\epsilon w''(x) + a(x)w'(x) + b(x)w(x) = 0, \qquad w(-1) = A,\ w(1) = B, \tag{2}$$

where $\epsilon > 0$ is small. Equations of this sort arise in fluid-flow problems in various geometries. Perturbative treatments of this equation are usually quite straightforward.[2] However, a particularly difficult special case of (2) may occur when there is a *resonant* internal boundary layer. Suppose that $a(0) = 0$, so that there is an internal boundary layer at $x = 0$. Suppose also that near

$x = 0$, $a(x) \sim \alpha x$ and $b(x) \sim \beta$. Then, near $x = 0$ the differential equation in (2) is approximated by $\epsilon w''(x) + \alpha x w'(x) + \beta w(x) = 0$. The change of variables $w(x) = e^{-\alpha x^2/(4\epsilon)} y(x)$ converts this equation to Schrödinger form: $-y''(x) + \frac{\alpha^2}{4\epsilon^2} x^2 y(x) = \frac{2\beta-\alpha}{2\epsilon} y(x)$. Apart from a scaling, this equation is the $N = 0$ case of (1). It describes the quantum harmonic oscillator, and eigenvalues occur when the parameters α and β satisfy $\beta = (n+1)\alpha \quad (n = 0, 1, 2, 3, \ldots)$. When the parameters α and β are related in this fashion, the internal boundary layer is said to be *resonant*. Unlike conventional boundary layers, a resonant boundary layer is not narrow; that is, its thickness is not small when ϵ is small. As a result it is difficult to treat the resonant case using ordinary boundary-layer methods [2].

In general, if $a(x) \sim \alpha x^{N+1}$ and $b(x) \sim \beta x^N$ when x is near 0, we obtain the differential equation in (4). This equation becomes (1) upon the change of variables in (3). Thus, the eigenvalue problem (1) characterizes the *general* resonant case in the theory of internal boundary layers. The special case of the harmonic oscillator discussed above occurs when $N = 0$.

The eigenvalue problem (1) also arises in supersymmetric quantum mechanics and quasi-exactly solvable models. It appears, for example, in recent work of Voros [3] and Dorey *et al* [4]. The same equation was also considered by Znojil [5] but with boundary conditions imposed on a semi-infinite interval. The problem (1) may be viewed as an inverse approach to quasi-exact solvability. Usually, in this field one tries to find potentials for which a finite number of eigenvalues of the spectrum can be calculated exactly, while the rest of the spectrum remains analytically intractable. The problem in (1) is to construct potentials $V(x)$ for which there is an exactly zero eigenvalue. The zero eigenvalue may or may not be the ground-state energy of the potential $V(x) = x^{2N+2} - x^N E$ that has been constructed. We will see that the case of odd-integer N is much more interesting than the even-N case. The eigenvalue problem (1) is especially interesting because, as we will see, leading-order WKB theory (physical optics) gives the exact spectrum E for all odd N and almost the exact answer for even N.

The plan of this talk is to give the exact solution to the eigenvalue problem (1) in Sec. 2. In Sec. 3 we examine the eigenfunctions for even-integer N. For this case the eigenspectrum is positive and the nth eigenfunction has definite parity. This eigenfunction has the form of a polynomial of degree n and of argument x^{N+2} multiplied by the exponential $\exp(-x^{N+2})$, which decays as $x \to \pm\infty$. The polynomials are generalized Laguerre polynomials. The polynomials for even n form an orthogonal set and the polynomials for odd n form a different orthogonal set. In Sec. 4, we study the eigenfunctions for odd-integer N. For this case the spectrum of eigenvalues E ranges from $-\infty$ to ∞ and the

nth eigenfunction does not exhibit definite parity. For each N the nth eigenfunction has the general form $xA_N(x^{N+1})P_n(x^{N+2}) + A'_N(x^{N+1})Q_n(x^{N+2})$. Here, $P_n(z)$ and $Q_n(z)$ are polynomials of degree n that satisfy the same recursion relation but have different initial conditions. The functions $A_N(z)$ are independent of n and are generalized Airy functions that obey the differential equation $A''_N(z) = z^{2/(N+1)}A_N(z)$. When $N = 1$, the function $A_1(z)$ is the conventional Airy function $\mathrm{Ai}(z)$.

We emphasize that the eigenvalues that are obtained here are not the energies of a conventional Schrödinger equation. However, the eigenfunctions might well be useful for solving some conventional Schrödinger equations. For example, it might be useful to express the eigenfunctions of the anharmonic oscillator, $-y''(x) + x^4y(x) = Ey(x)$ as linear combinations of the eigenfunctions discussed here. It may well be advantageous to expand the function $y(x)$ as $P(x)\mathrm{Ai}(x^2) + Q(x)\mathrm{Ai}'(x^2)$, where P and Q are series in powers of x.

2 Exact Solution of the Eigenvalue Problem

To find the eigenvalues of the differential equation (1) we convert it to a confluent hypergeometric equation and impose the boundary conditions. We let

$$y(x) = e^{-\frac{1}{N+2}x^{N+2}}w(x). \tag{3}$$

The function $w(x)$ then satisfies the differential equation

$$w''(x) - 2x^{N+1}w'(x) + \beta x^N w(x) = 0, \tag{4}$$

where $\beta \equiv E - N - 1$.

Consider first the case for which N is even. There are two linearly independent solutions to (4):

$$w(x) = {}_1F_1\left(-\frac{\beta}{2N+4}, 1 - \frac{1}{N+2}; \frac{2}{N+2}x^{N+2}\right), \tag{5}$$

$$w(x) = x\,{}_1F_1\left(\frac{1}{N+2} - \frac{\beta}{2N+4}, 1 + \frac{1}{N+2}; \frac{2}{N+2}x^{N+2}\right). \tag{6}$$

When the first parameter of the confluent hypergeometric function is a negative integer, its Taylor series truncates to a Laguerre polynomial:

$${}_1F_1(-n, c; t) = n!\,\frac{\Gamma(c)}{\Gamma(c-n)}\,\mathrm{L}_n^{(c-1)}(t). \tag{7}$$

Thus, for the solution in (5) we obtain $\frac{\beta_n}{2N+4} = n \quad (n = 0, 1, 2, 3, \ldots)$, or

$$E_n = 2n(N+2) + N + 1 \qquad (n = 0, 1, 2, 3, \ldots). \tag{8}$$

The corresponding eigenfunctions have even parity. For the solution in (6) we obtain $-\frac{1}{N+2} + \frac{\beta_n}{2N+4} = n \quad (n = 0, 1, 2, 3, \ldots)$, or

$$E_n = 2n(N+2) + N + 3 \qquad (n = 0, 1, 2, 3, \ldots). \tag{9}$$

The corresponding eigenfunctions have odd parity.

When N is odd, the solution to (4) that is well behaved as $x \to \infty$ is a particular linear combination of the solutions in (5) and (6) known as a confluent hypergeometric function of the second kind:

$$\begin{aligned} w(x) = &\frac{\Gamma\left(\frac{1}{N+2}\right)}{\Gamma\left(\frac{2-\beta}{2N+4}\right)} {}_1F_1\left(-\frac{\beta}{2N+4}, \frac{N+1}{N+2}; \frac{2}{N+2}x^{N+2}\right) \\ &+ \frac{\Gamma\left(-\frac{1}{N+2}\right)x}{\Gamma\left(-\frac{\beta}{2N+4}\right)} {}_1F_1\left(\frac{2-\beta}{2N+4}, \frac{N+3}{N+2}; \frac{2}{N+2}x^{N+2}\right). \end{aligned} \tag{10}$$

As $x \to +\infty$, the function $y(x)$ in (3) vanishes exponentially. However, as $x \to -\infty$, the exponential factor in (3) grows and

$$\begin{aligned} y(x) \sim\ & e^{-\frac{1}{N+2}x^{N+2}} \left(-\frac{2}{N+2}x^{N+2}\right)^{\frac{\beta}{2N+4}} \\ &\times \left[\frac{\Gamma\left(\frac{1}{N+2}\right)\Gamma\left(\frac{N+1}{N+2}\right)}{\Gamma\left(\frac{2-\beta}{2N+4}\right)\Gamma\left(\frac{2N+2+\beta}{2N+4}\right)} - \frac{\Gamma\left(-\frac{1}{N+2}\right)\Gamma\left(\frac{N+3}{N+2}\right)}{\Gamma\left(-\frac{\beta}{2N+4}\right)\Gamma\left(\frac{2N+2+\beta}{2N+4}\right)}\right] \\ &\times \left[1 + \mathrm{O}\left(\frac{1}{|x|^{N+2}}\right)\right]. \end{aligned} \tag{11}$$

Because $\exp\left(-\frac{1}{N+2}x^{N+2}\right)$ grows exponentially as $x \to -\infty$, the only way to satisfy the boundary condition as $x \to -\infty$ is for the expression in square brackets to vanish. This condition gives the result

$$E_n = (2n+1)(N+2) \qquad (n = 0, \pm1, \pm2, \pm3, \ldots). \tag{12}$$

Leading-order WKB (physical optics) analysis of (1) gives the exact eigenvalues for odd N and almost the exact eigenvalues when N is even. Consider the two-turning-point Schrödinger equation boundary-value problem $-y''(x)+Q(x)y(x)=0,\quad y(\pm\infty)=0$. Ordinarily, $Q(x)=V(x)-E$, where $V(x)$ is the potential and E is the energy. In the WKB approximation a solution exists if

$$\int_A^B dx\,\sqrt{-Q(x)}=\left(n+\frac{1}{2}\right)\pi \qquad (n=0,1,2,3,\ldots), \tag{13}$$

where the turning points A and B satisfy $Q(A)=Q(B)=0$. Applying the quantization condition (13) to (1), where $Q(x)=x^{2N+2}-Ex^N$, we obtain $\int_0^B dx\,\sqrt{Ex^N-x^{2N+2}}=\left(n+\frac{1}{2}\right)\pi \quad (n=0,1,2,3,\ldots)$, where we assume without loss of generality that E is positive. The turning point B satisfies $B^{N+2}=E$. This integral can be evaluated exactly as a beta function and we get $E=(2n+1)(N+2)$. This is precisely the result in (12) for N odd. Also, it is nearly the results for even N in (8) and (9), which combine to give $E=(2n+1)(N+2)\pm 1$.

When N is even, the eigenvalues of (1) are positive but for odd N the eigenvalues are both positive and negative. This is reminiscent of the Klein-Gordon equation for bosons, which has positive-energy states only, and the Dirac equation for fermions, which has positive-energy states (electrons) and negative-energy states (positrons or holes).

3 Eigenvalue Problem for Even N

In this and the next section we describe the two cases N odd and N even in depth. When N is even, the eigenvalues are positive and the eigenfunctions have either even or odd parity. The even-parity eigenfunctions are

$$y_{2n}(x)=e^{-\frac{1}{N+2}x^{N+2}}\mathrm{L}_n^{\left(-\frac{1}{N+2}\right)}\left(\frac{2}{N+2}x^{N+2}\right) \qquad (n=0,1,2,3,\ldots) \tag{14}$$

and the corresponding eigenvalues are

$$E_{2n}=2n(N+2)+N+1, \tag{15}$$

where $\mathrm{L}_n^{(\alpha)}$ is the generalized Laguerre polynomial. The odd-parity eigenfunctions are

$$y_{2n+1}(x)=e^{-\frac{1}{N+2}x^{N+2}}x\,\mathrm{L}_n^{\left(\frac{1}{N+2}\right)}\left(\frac{2}{N+2}x^{N+2}\right) \qquad (n=0,1,2,3,\ldots) \tag{16}$$

and the corresponding eigenvalues are $E_{2n+1}=2n(N+2)+N+3$.

The eigenfunctions have the form of a decaying exponential multiplying a polynomial. For the even-parity solutions we can write the polynomial as a monic[a] polynomial $p_n(z)$ in the variable $z = 4x^{N+2}$:

$$p_n(z) = (-1)^n n!\,(2N+4)^n\,\mathrm{L}_n^{\left(-\frac{1}{N+2}\right)}\left(\frac{z}{2N+4}\right) \tag{17}$$

and for the odd-parity solutions we have

$$q_n(z) = (-1)^n n!\,(2N+4)^n\,\mathrm{L}_n^{\left(\frac{1}{N+2}\right)}\left(\frac{z}{2N+4}\right). \tag{18}$$

These polynomials satisfy similar recurrence relations

$$\begin{aligned} p_{n+1}(z) &= (z - 4Nn - 2N - 8n - 2)p_n(z) \\ &\quad -4(N+2)n(Nn+2n-1)p_{n-1}(z), \\ q_{n+1}(z) &= (z - 4Nn - 2N - 8n - 6)q_n(z) \\ &\quad -4(N+2)n(Nn+2n+1)q_{n-1}(z), \end{aligned} \tag{19}$$

where the initial conditions are $p_0(z) = 1$, $p_1(z) = z - 2N - 2$, $q_0(z) = 1$, $q_1(z) = z - 2N - 6$. The differential equations and differential relations for $p_n(z)$ and $q_n(z)$ are similar:

$$\begin{aligned} (2N+4)zp_n''(z) + (2N+2-z)p_n'(z) + np_n(z) &= 0, \\ (2N+4)zq_n''(z) + (2N+6-z)q_n'(z) + nq_n(z) &= 0, \end{aligned} \tag{20}$$

$$\begin{aligned} (2N+4)zp_n'(z) + p_{n+1}(z) + (2Nn + 2N + 4n + 2 - z)p_n(z) &= 0, \\ (2N+4)zq_n'(z) + q_{n+1}(z) + (2Nn + 2N + 4n + 6 - z)q_n(z) &= 0. \end{aligned} \tag{21}$$

The generating functions for these polynomials are also quite similar:

$$\begin{aligned} G_p(z,t) &\equiv \sum_{n=0}^{\infty} \frac{(-1)^n t^n}{n!} p_n(z) = [1-(2N+4)t]^{-1+\frac{1}{N+2}}\, e^{\frac{zt}{(2N+4)t-1}}, \\ G_q(z,t) &\equiv \sum_{n=0}^{\infty} \frac{(-1)^n t^n}{n!} q_n(z) = [1-(2N+4)t]^{-1-\frac{1}{N+2}}\, e^{\frac{zt}{(2N+4)t-1}}. \end{aligned} \tag{22}$$

The polynomials $p_n(z)$ and $q_n(z)$ are separately orthogonal:

$$\int_0^\infty dz\, w_p(z) p_m(z) p_n(z) = (2N+4)^{2n} n!\, \frac{\Gamma\left(n + \frac{N+1}{N+2}\right)}{\Gamma\left(\frac{N+1}{N+2}\right)}\, \delta_{m,n},$$

[a] The term *monic* means that the coefficient of the highest power in the polynomial is 1.

$$\int_0^\infty dz\, w_q(z) q_m(z) q_n(z) = (2N+4)^{2n} n! \frac{\Gamma\left(n+\frac{N+3}{N+2}\right)}{\Gamma\left(\frac{N+3}{N+2}\right)} \delta_{m,n}, \tag{23}$$

where the weight functions $w_p(z)$ and $w_q(z)$ are given by

$$\begin{aligned} w_p(z) &= e^{-\frac{z}{2N+4}} z^{-\frac{1}{N+2}} (2N+4)^{-\frac{N+1}{N+2}} \Big/ \Gamma\left(\frac{N+1}{N+2}\right), \\ w_q(z) &= e^{-\frac{z}{2N+4}} z^{\frac{1}{N+2}} (2N+4)^{-\frac{N+3}{N+2}} \Big/ \Gamma\left(\frac{N+3}{N+2}\right). \end{aligned} \tag{24}$$

The moments of these weight functions are

$$\begin{aligned} a_n^{(p)} &\equiv \int_0^\infty dz\, w_p(z) z^n = (2N+4)^n \Gamma\left(n+\frac{N+1}{N+2}\right) \Big/ \Gamma\left(\frac{N+1}{N+2}\right), \\ a_n^{(q)} &\equiv \int_0^\infty dz\, w_q(z) z^n = (2N+4)^n \Gamma\left(n+\frac{N+3}{N+2}\right) \Big/ \Gamma\left(\frac{N+3}{N+2}\right). \end{aligned} \tag{25}$$

The (divergent) power series constructed from these moments have continued fraction expansions in which the coefficients are all integers:

$$\sum_{n=0}^\infty a_n^{(p)} t^n = \cfrac{1}{1-\cfrac{[2(N+2)-2]t}{1-\cfrac{[2(N+2)]t}{1-\cfrac{[4(N+2)-2]t}{1-\cfrac{[4(N+2)]t}{1-\cfrac{[6(N+2)-2]t}{1-\cfrac{[6(N+2)]t}{1-\cdots}}}}}}},$$

$$\sum_{n=0}^\infty a_n^{(q)} t^n = \cfrac{1}{1-\cfrac{[2(N+2)+2]t}{1-\cfrac{[2(N+2)]t}{1-\cfrac{[4(N+2)+2]t}{1-\cfrac{[4(N+2)]t}{1-\cfrac{[6(N+2)+2]t}{1-\cfrac{[6(N+2)]t}{1-\cdots}}}}}}}. \tag{26}$$

Special Case $N = 0$: The Harmonic Oscillator. For this case the eigenvalues in (15) are $E = 1, 3, 5, 7, \ldots$ and $p_n(z)$ and $xq_n(z)$ in (17) and (18) coalesce to become the Hermite polynomials $\mathrm{H}_n(x)$:

$$\begin{aligned} p_n(z) &= \mathrm{H}_{2n}(x) = (-1)^n 2^{2n} n!\, \mathrm{L}_n^{(-1/2)}(x^2), \\ x q_n(z) &= \mathrm{H}_{2n+1}(x) = (-1)^n 2^{2n+1} n!\, x \mathrm{L}_n^{(1/2)}(x^2). \end{aligned} \tag{27}$$

Special Case $N = 2$. Here, the eigenvalues are $E = 3,\ 5,\ 11,\ 13,\ 19,\ 21,\ \ldots$ and the first few monic polynomials $p_n(x)$ and $q_n(x)$ in (17) and (18) are

$$\begin{aligned} p_0(z) &= 1, \\ p_1(z) &= z - 6, \\ p_2(z) &= z^2 - 28z + 84, \\ p_3(z) &= z^3 - 66z^2 + 924z - 1848, \\ q_0(z) &= 1, \\ q_1(z) &= z - 10, \\ q_2(z) &= z^2 - 36z + 180, \\ q_3(z) &= z^3 - 78z^2 + 1404z - 4680. \end{aligned} \tag{28}$$

4 Eigenvalue Problem for Odd N

For this case the eigenvalues are

$$E_n = (2n+1)(N+2) \qquad (n = 0, \pm 1, \pm 2, \pm 3, \ldots) \tag{29}$$

and the corresponding eigenfunctions $y_n(x)$, which do not have definite parity, are confluent hypergeometric functions of the second kind:

$$y_n(x) = e^{-\frac{1}{N+2}x^{N+2}}\, \mathrm{U}\left(-n - \frac{1}{2N+4}, 1 - \frac{1}{N+2}; \frac{2}{N+2}x^{N+2}\right). \tag{30}$$

Note that the boundary condition as $x \to +\infty$ is already satisfied and the quantization comes from requiring that $y(x) \to 0$ as $x \to -\infty$.

For each N the eigenfunctions $y_n(x)$ in (30) can be expressed in terms of what we call *generalized Airy functions* $A_N(x)$ combined with polynomials P_n and Q_n as follows:[b]

$$\begin{aligned} y_{-n-1}(x) = 2^{-\frac{N+1}{2N+4}}\, x A_N &\left[\left(2^{-\frac{N+1}{2N+4}}\, x\right)^{N+1}\right] P_n(4x^{N+2}) \\ &+ A_N'\left[\left(2^{-\frac{N+1}{2N+4}}\, x\right)^{N+1}\right] Q_n(4x^{N+2}) \qquad (n \geq 0) \end{aligned} \tag{31}$$

[b] To avoid confusion, for odd N we use upper-case notation P_n and Q_n to represent the polynomials; we use lower-case notation p_n and q_n to represent the polynomials associated with even N.

and $y_n(x) = y_{-n-1}(-x) \qquad (n \geq 0)$.

We define the *generalized Airy function* $A_N(x)$ as the solution to

$$A''_N(x) = x^{\frac{2}{N+1}} A_N(x) \tag{32}$$

that decays exponentially as $x \to +\infty$. When $N = 1$, the $A_1(x)$ is just the conventional Airy function $\mathrm{Ai}(x)$. We can express $A_N(x)$ in terms of the associated Bessel function $K_\nu(z)$:

$$A_N(x) = \frac{1}{2\pi}[4(N+1)]^{\frac{N+1}{2N+4}} \sqrt{\frac{x}{N+2}} K_{\frac{N+1}{2N+4}} \left(\frac{N+1}{N+2} x^{\frac{N+2}{N+1}} \right). \tag{33}$$

With this choice the function $A_N(x)$ is normalized so that

$$\int_0^\infty dx\, A_N(x) = \frac{1}{\pi}\Gamma\left(\frac{N+1}{N+2}\right)\Gamma\left(\frac{N+1}{2N+4}\right) 2^{-\frac{N+7}{2N+4}} (N+1)^{\frac{1}{N+2}} (N+2)^{-\frac{3}{2N+4}}. \tag{34}$$

This gives the standard result $\int_0^\infty dx\, \mathrm{Ai}(x) = \frac{1}{3}$ when $N = 1$.

The polynomials $P_n(z)$ and $Q_n(z)$, where $z = 4x^{N+2}$, satisfy the *same* recursion relation

$$\begin{aligned} P_{n+1}(z) &= [z + (2N+4)(2n+1)]P_n(z) - [(2Nn+4n)^2 - 1]P_{n-1}(z), \\ Q_{n+1}(z) &= [z + (2N+4)(2n+1)]Q_n(z) - [(2Nn+4n)^2 - 1]Q_{n-1}(z), \end{aligned} \tag{35}$$

but have different initial conditions $P_0(z) = 1$, $P_1(z) = z + 2N + 5$, $Q_0(z) = 1$, $Q_1(z) = z + 2N + 3$.

The polynomials $P_n(z)$ and $Q_n(z)$ satisfy *coupled* differential equations

$$\begin{aligned} &4(N+2)zP''_n(z) + 4(N+3)P'_n(z) + 2zQ'_n(z) + Q_n(z) = (2n+1)P_n(z), \\ &4(N+2)zQ''_n(z) + 4(N+1)Q'_n(z) + 2zP'_n(z) + P_n(z) \\ &\quad = (2n+1)Q_n(z), \end{aligned} \tag{36}$$

and *coupled* differential relations

$$\begin{aligned} 4(N+2)zP'_n(z) &= 2P_{n+1}(z) - zQ_n(z) - [4(N+2)(n+1) + 2 + z]P_n(z), \\ 4(N+2)zQ'_n(z) &= 2Q_{n+1}(z) - zP_n(z) \\ &\quad -[4(N+2)(n+1) - 2 + z]Q_n(z). \end{aligned} \tag{37}$$

The generating functions for these polynomials are

$$\begin{aligned} G_P(z,t) &\equiv \sum_{n=0}^{\infty} \frac{t^n}{n!} P_n(z) \\ &= [1 - (2N+4)t]^{-\frac{2N+5}{2N+4}} \frac{{}_1F_1\left(\frac{2N+5}{2N+4}, \frac{N+3}{N+2}; \frac{z}{(2N+4)(1-2Nt-4t)}\right)}{{}_1F_1\left(\frac{2N+5}{2N+4}, \frac{N+3}{N+2}; \frac{z}{2N+4}\right)}, \end{aligned} \tag{38}$$

$$G_Q(z,t) \equiv \sum_{n=0}^{\infty} \frac{t^n}{n!} Q_n(z)$$

$$= [1-(2N+4)t]^{-\frac{2N+3}{2N+4}} \frac{{}_1F_1\left(\frac{2N+3}{2N+4}, \frac{N+1}{N+2}; \frac{z}{(2N+4)(1-2Nt-4t)}\right)}{{}_1F_1\left(\frac{2N+3}{2N+4}, \frac{N+1}{N+2}; \frac{z}{2N+4}\right)}. \quad (39)$$

The polynomials $P_n(z)$ and $Q_n(z)$ obey identical-looking orthogonality and normalization conditions

$$\int_{-\infty}^{\infty} dx\, W_P(x) P_m(x) P_n(x) = \sin \frac{\pi}{2N+4} (2N+4)^{2n+1} \Gamma\left(n + \frac{2N+3}{2N+4}\right) \times\Gamma\left(n + \frac{2N+5}{2N+4}\right) \frac{\delta_{m,n}}{\pi},$$

$$\int_{-\infty}^{\infty} dx\, W_Q(x) Q_m(x) Q_n(x) = \sin \frac{\pi}{2N+4} (2N+4)^{2n+1} \Gamma\left(n + \frac{2N+3}{2N+4}\right) \times\Gamma\left(n + \frac{2N+5}{2N+4}\right) \frac{\delta_{m,n}}{\pi}. \quad (40)$$

The weight functions $W_P(x)$ and $W_Q(x)$ are real and positive and are expressible as principal-part integrals:

$$W_P(x) = \int_{-\infty}^{x} ds\, \mathcal{P} \int_{-\infty}^{\infty} \frac{(4N+8)dt}{(4N+8)t-s} \ln\left[\sqrt{\frac{-2t}{\pi}}\, e^{-t} K_{\frac{N+3}{2N+4}}(-t)\right],$$

$$W_Q(x) = \int_{-\infty}^{x} ds\, \mathcal{P} \int_{-\infty}^{\infty} \frac{(4N+8)dt}{(4N+8)t-s} \ln\left[\sqrt{\frac{-2t}{\pi}}\, e^{-t} K_{\frac{N+1}{2N+4}}(-t)\right], \quad (41)$$

where $\mathcal{P}$ indicates principal-part integration and the integral is performed on the sheet for which $-1 \equiv e^{-i\pi}$.

The moments of the weight functions $W_P(x)$ and $W_Q(x)$ are given by $a_n^{(P)} = \int_{-\infty}^{\infty} dx\, x^n W_P(x)$ and $a_n^{(Q)} = \int_{-\infty}^{\infty} dx\, x^n W_Q(x)$. The divergent power series constructed from these moments have continued-fraction expansions whose coefficients are integers:

$$\sum_{n=0}^{\infty} a_n^{(P)} t^n = \cfrac{1}{1-\cfrac{[2(N+2)+1]t}{1-\cfrac{[2(N+2)-1]t}{1-\cfrac{[4(N+2)+1]t}{1-\cfrac{[4(N+2)-1]t}{1-\cfrac{[6(N+2)+1]t}{1-\cfrac{[6(N+2)-1]t}{1-\cdots}}}}}}},$$

$$\sum_{n=0}^{\infty} a_n^{(Q)} t^n = \cfrac{1}{1-\cfrac{[2(N+2)-1]t}{1-\cfrac{[2(N+2)+1]t}{1-\cfrac{[4(N+2)-1]t}{1-\cfrac{[4(N+2)+1]t}{1-\cfrac{[6(N+2)-1]t}{1-\cfrac{[6(N+2)+1]t}{1-\cdots}}}}}}}. \tag{42}$$

Special case $N = -1$. For this case Eq. (32) is not valid. However, the formula for the eigenvalues E in (29) is still valid and $E = \pm 1, \ \pm 3, \ \pm 5, \ \pm 7, \ \ldots$. The eigenfunctions $y_n(x)$ in (30) are now *Bateman functions* $k_{E_n}(x)$:

$$y_n(x) = k_{E_n}(x) \equiv \frac{2}{\pi} \int_0^{\pi/2} d\theta \, \cos(x \tan\theta - E_n\theta). \tag{43}$$

Aside from a multiplicative constant, this solution can be written in terms of associated Bessel functions combined with polynomials. For negative eigenvalues $y_{-n-1}(x) = xK_0(x)P_n(4x) + xK_0'(x)Q_n(4x) \quad (n \geq 0)$, and for positive eigenvalues $y_n(x) \equiv y_{-n-1}(-x) \quad (n \geq 0)$. Note that the eigenfunctions are finite at the origin but that there is a branch cut. For definiteness, we take the branch cut to run up the positive imaginary-x axis.

In terms of $z = 4x$ the first few polynomials $P_n(z)$ and $Q_n(z)$ are

$$\begin{aligned}
P_0(z) &= 1,\\
P_1(z) &= z + 3,\\
P_2(z) &= z^2 + 9z + 15,\\
P_3(z) &= z^3 + 19z^2 + 90z + 105,\\
Q_0(z) &= 1,
\end{aligned}$$

$$\begin{aligned} Q_1(z) &= z+1, \\ Q_2(z) &= z^2+7z+3, \\ Q_3(z) &= z^3+17z^2+58z+15. \end{aligned} \tag{44}$$

The polynomials $P_n(x)$ and $Q_n(x)$ satisfy the recursion relations (35), the coupled second-order differential equations (36), and the coupled differential relations (37) with $N=-1$. The generating functions for $P_n(z)$ and $Q_n(z)$ are expressed in terms of Bateman functions k_ν:

$$\begin{aligned} G_P(z,t) &\equiv \sum_{n=0}^{\infty} \frac{t^n}{n!} P_n(z) = (1-2t)^{-3/2} \frac{e^{\frac{z}{4(1-2t)}}\, k_{-3}\left(\frac{z}{4(1-2t)}\right)}{e^{z/4}\, k_{-3}(z/4)} \\ G_Q(z,t) &\equiv \sum_{n=0}^{\infty} \frac{t^n}{n!} Q_n(z) = (1-2t)^{-1/2} \frac{e^{\frac{z}{4(1-2t)}}\, k_{-1}\left(\frac{z}{4(1-2t)}\right)}{e^{z/4}\, k_{-1}(z/4)}. \end{aligned} \tag{45}$$

The polynomials $P_n(z)$ and $Q_n(z)$ obey identical-looking orthogonality and normalization conditions

$$\begin{aligned} \int_{-\infty}^{\infty} dx\, W_P(x) P_m(x) P_n(x) &= (2n-1)!!\,(2n+1)!!\,\delta_{m,n}, \\ \int_{-\infty}^{\infty} dx\, W_Q(x) Q_m(x) Q_n(x) &= (2n-1)!!\,(2n+1)!!\,\delta_{m,n}, \end{aligned} \tag{46}$$

where $(-1)!!=1$. Note that the weight functions $W_P(x)$ and $W_Q(x)$ are *real* and *positive* and are expressible as principal-part integrals:

$$\begin{aligned} W_P(x) &= \int_{-\infty}^{x} ds\, \mathcal{P} \int_{-\infty}^{\infty} \frac{dt}{t-s} \ln\left[\sqrt{\frac{-t}{2\pi}}\, e^{-t/4}\, K_1(-t/4)\right], \\ W_Q(x) &= \int_{-\infty}^{x} ds\, \mathcal{P} \int_{-\infty}^{\infty} \frac{dt}{t-s} \ln\left[\sqrt{\frac{-t}{2\pi}}\, e^{-t/4}\, K_0(-t/4)\right]. \end{aligned} \tag{47}$$

The moments of the weight functions $W_P(x)$ and $W_Q(x)$ give rise to the lovely continued-fraction expansions

$$\sum_{n=0}^{\infty} a_n^{(P)} t^n = \cfrac{1}{1-\cfrac{3t}{1-\cfrac{t}{1-\cfrac{5t}{1-\cfrac{3t}{1-\cfrac{7t}{1-\cfrac{5t}{1-\cdots}}}}}}},$$

$$\sum_{n=0}^{\infty} a_n^{(Q)} t^n = \cfrac{1}{1-\cfrac{t}{1-\cfrac{3t}{1-\cfrac{3t}{1-\cfrac{5t}{1-\cfrac{5t}{1-\cfrac{7t}{1-\cdots}}}}}}}. \tag{48}$$

Special case $N = 1$ (Airy functions). here, the eigenvalues E in (29) are $E = \pm 3, \ \pm 9, \ \pm 15, \ \pm 21, \ \ldots$ and the eigenfunctions $y_n(x)$ in (30) are written in terms of Airy functions combined with polynomials. For negative eigenvalues we have

$$y_{-n-1}(x) = 2^{-1/3}\, x\, \mathrm{Ai}(2^{-2/3}x^2)P_n(4x^3) + \mathrm{Ai}'(2^{-2/3}x^2)Q_n(4x^3) \qquad (n \geq 0), \tag{49}$$

and for positive eigenvalues we have $y_n(x) \equiv y_{-n-1}(-x) \quad (n \geq 0)$. The polynomials P_n and Q_n are functions of the variable $z = 4x^3$. The first few such polynomials are

$$\begin{aligned}
P_0(z) &= 1, \\
P_1(z) &= z + 7, \\
P_2(z) &= z^2 + 25z + 91, \\
P_3(z) &= z^3 + 55z^2 + 698z + 1729, \\
Q_0(z) &= 1, \\
Q_1(z) &= z + 5, \\
Q_2(z) &= z^2 + 23z + 55, \\
Q_3(z) &= z^3 + 53z^2 + 602z + 935.
\end{aligned} \tag{50}$$

The polynomials $P_n(x)$ and $Q_n(x)$ satisfy the recursion relations (35), the coupled second-order differential equations (36), and the coupled differential relations (37) with $N = 1$. The generating functions $G_P(z,t)$ and $G_Q(z,t)$ and the integral representations of the weight functions W_P and W_Q are obtained by setting $N = 1$ in (38), (39), and (41). The polynomials $P_n(z)$ and $Q_n(z)$ obey identical-looking orthogonality and normalization conditions

$$\begin{aligned}
\int_{-\infty}^{\infty} dz\, W_P(z) P_m(z) P_n(z) &= \frac{3}{\pi} 6^{2n} \Gamma\left(n + \frac{5}{6}\right) \Gamma\left(n + \frac{7}{6}\right) \delta_{m,n} \\
\int_{-\infty}^{\infty} dz\, W_Q(z) Q_m(z) Q_n(z) &= \frac{3}{\pi} 6^{2n} \Gamma\left(n + \frac{5}{6}\right) \Gamma\left(n + \frac{7}{6}\right) \delta_{m,n}
\end{aligned} \tag{51}$$

and the moments of $W_P(x)$ and $W_Q(x)$ have the continued-fraction expansions

$$\sum_{n=0}^{\infty} a_n^{(P)} t^n = \cfrac{1}{1-\cfrac{7t}{1-\cfrac{5t}{1-\cfrac{13t}{1-\cfrac{11t}{1-\cfrac{19t}{1-\cfrac{17t}{1-\cdots}}}}}}},$$

$$\sum_{n=0}^{\infty} a_n^{(Q)} t^n = \cfrac{1}{1-\cfrac{5t}{1-\cfrac{7t}{1-\cfrac{11t}{1-\cfrac{13t}{1-\cfrac{17t}{1-\cfrac{19t}{1-\cdots}}}}}}}. \quad (52)$$

There is an interesting connection between the moments $a_n^{(Q)}$ and the combinatorial numbers $C_{2n}^{[3]}$, which represent the sum of the symmetry numbers of the $2n$-vertex connected vacuum graphs in a ϕ^3 quantum field theory [6]: $a_n^{(Q)} = 6nC_{2n}^{[3]}4^n$.

Acknowledgments

I thank my coauthor and graduate student Q. Wang for his contribution to this research and I thank A. Voros and P. Dorey for interesting discussions. This work was supported by the U.S. Department of Energy.

References

1. C. M. Bender and Q. Wang, *J. Phys. A: Math. Gen.* (in press).
2. C. M. Bender and S. A. Orszag, *Advanced Mathematical Methods for Scientists and Engineers*, McGraw-Hill, New York (1978), Chap. 6.
3. A. Voros, *J. Phys. A* **33**, 7423 (2000).
4. P. Dorey, C. Dunning, and R. Tatao, hep-th/0104119.
5. M. Znojil, *J. Math. Phys.* **38**, 5087 (1997).
6. C. M. Bender and K. A. Milton, *J. Math. Phys.* **35**, 364 (1994).

ELECTROWEAK INTERACTION, FERMIONS AND VECTOR BOSONS FROM GAUGE AND SPACE-TIME SYMMETRY UNIFICATION

J. BESPROSVANY

Instituto de Física, Universidad Nacional Autónoma de México, Apartado Postal 20-364, México 01000, D. F., MÉXICO

E-mail: bespro@fenix.fisica.unam.mx

Unification ideas lead to the consideration of a generalized Dirac equation with boson and fermion solutions. These fields naturally support an algebra and inner product which lead to vertices among them. Symmetries in the equation can be associated with flavor and gauge groups. Extended versions predict isospin and hypercharge $SU(2)_L \times U(1)$ symmetries, their vector carriers, two-flavor charged and chargeless leptons, and scalar particles. A mass term produces breaking of the symmetry to an electromagnetic $U(1)$, a Weinberg's angle with $sin^2(\theta_W) = .25$, and respective coupling constants $g = 1/\sqrt{2} \approx .707$ and $g' = 1/\sqrt{6} \approx .408$.

1 Introduction

Although the accepted theory of elementary particles, the standard model (SM), is successful in describing their behavior, it is phenomenological; a justification for its input is still required at a more fundamental level, as evidenced by the large number of experimental parameters it needs. We also lack information on the origin of the particular groups $SU(3) \times SU(2)_L \times U(1)$ which define the interactions, and on why the isospin acts only on a given chirality. Neither is clear the origin of the spectrum of fermions, the three generations, the masses and mixing angles, and the gauge-group fundamental representations in which they appear. The Higgs particle is a useful mathematical device but we lack a more basic reason for its presence.

The idea of unification has proved to be a powerful tool in physical research, finding links between originally assumed separated phenomena, and also predicting new ones. For example, by assuming the SM interactions have a shared origin in the same unified group[1] some clues can be obtained on the fermion representations in the SM and also relations among the coupling constants. On another plane, the Kaluza-Klein idea, originally used in the framework of general relativity, has been applied to understand interactions by relating them to additional spatial dimensions. While both spin and configuration spaces are equally necessary elements for the description of particles, the possibility of relating interactions to extended spin spaces has not been explored.

In this contribution we consider an equation with fermion and boson solutions sharing the same solution space. We use an interaction prescription that emerges naturally among the field solutions, with the allowed interactions determined by the symmetries present. We consider an extension of the equation at 5+1 dimensions, which predicts an $SU(2)_L \times U(1)$ symmetry with corresponding gauge fields and leptons with correct quantum numbers.

2 Generalized Dirac equation

We depart from the equation

$$\gamma_0(i\partial_\mu\gamma^\mu - M)\Psi = 0, \tag{1}$$

where, we now assume Ψ represents a matrix. Eq. 1 contains four conditions over four spinors in a 4×4 matrix. There are, then, additional possible ones to further classify Ψ. The transformations and symmetry operations on the Dirac operator $(i\partial_\mu\gamma^\mu - M) \to U(i\partial_\mu\gamma^\mu - M)U^{-1}$ induce the lhs of the transformation

$$\Psi \to U\Psi U^\dagger. \tag{2}$$

Also, Ψ is postulated to transform as indicated on the rhs. The latter transformation is consistent with the additional equation

$$\Psi\gamma_0(-i \overleftarrow{\partial_\mu} \gamma^\mu - M) = 0 \tag{3}$$

(the Dirac operator transforming accordingly), which will not necessarily be assumed for Ψ, but used as reference to classify the solutions of Eq. 1.[a]

Ψ can be understood to be constructed from a tensor product of the usual configuration (or momentum) space and two spinors with expansion $\sum_{i,j} a_{ij}|w_i\rangle\langle w_j|$. The resulting tensor product of operators leads the corresponding generators of the Poincaré algebra in transformation U to be formed of two spin operators, acting on each side of Ψ, and derivative ones acting only once, from either side. The expansion of Ψ implies it Lorentz transforms as a scalar, a vector, or an antisymmetric tensor. In fact, it can be shown[3] certain choices of symmetry operators allow for a fermion interpretation of some solutions too.

If A, B are solutions, the matrix product $C = AB$, defines an algebra whose elements may or may not be solutions. The inner product of A, B is

[a]The Bargmann-Wigner equations[2] contain Eq. 1 but set the different second condition as $\Psi^*\gamma_0(-i \overleftarrow{\partial_\mu} \gamma^\mu - M) = 0$.

naturally defined by

$$\langle A|B\rangle = tr(A^\dagger B), \tag{4}$$

as expected for an extended space of tensor product of spinors. A trace over the coordinates is also implied.

The conjugated fields $\Psi^\dagger$ satisfy the hermitian conjugate Dirac equation $0 = \Psi^\dagger\gamma_0(-i \overleftarrow{\partial}_\mu \gamma^\mu - M)$, with corresponding equation to Eq. 3. The form of transformation 2 then becomes also valid for these fields. In fact, it is by taking also account of these fields that we can span the function space within the 32-dimensional complex 4×4 matrices. Then, we shall extend our space of solutions by considering also combinations of fields $A + B^\dagger$. Flavor spin-1/2 and boson solutions are obtained which define a vertex and coupling constant[3].

3 5+1 Dimensional Extension

The simplest generalization of the above model is to consider the six-dimensional Clifford algebra, (the $d=5$ lives also in a 4×4 space), composed of 64 8×8 matrices. To describe it we use the quaternion-like objects 1, I, J, K and now generalized 8×8 matrices γ_μ (and γ_5). A 6-d Clifford algebra $\{\gamma'_\mu, \gamma'_\nu\} = g_{\mu\nu}$ can be formed with the 4-d elements $\gamma'_\mu = \gamma_\mu$, $\mu = 0, 1, 3$, $\gamma'_2 = I\gamma_2$, and the 4-d scalars $\gamma'_5 = J\gamma_2$, $\gamma'_6 = K\gamma_2$. Altogether, the scalars (and pseudo-) are 1, I, $J\gamma_2$, $K\gamma_2$, γ_5, $I\gamma_5$, $J\gamma_2\gamma_5$, $K\gamma_2\gamma_5$. From these, $I\gamma_5$ commutes with the rest. Excluding it and the identity, the remaining six elements generate an $SO(4)$ algebra, or equivalently, an $SU(2)\times SU(2)$ algebra. The eight scalars have a Cartan algebra of dimension four, for which we can take the basis 1, I, γ_5, $I\gamma_5$. The only projector operator L that can be constructed with these, which describes both fermions and bosons through the Lorentz generator $J^L_{\mu\nu} = LJ'_{\mu\nu}$, with $J'_{\mu\nu} = i(x_\mu\partial_\nu - x_\nu\partial_\mu) + \frac{1}{2}\sigma'_{\mu\nu}$, and $\sigma'_{\mu\nu} = \frac{i}{2}[\gamma'_\mu, \gamma'_\nu]$, allows for a non-abelian group symmetry, and permits parity to be a good quantum number for some solutions is $L = \frac{3}{4} - \frac{1}{4}(I + \gamma_5 + I\gamma_5) = 1 - \frac{1}{4}(1+I)(1+I\gamma_5)$; this can be interpreted as the lepton number. The equation

$$iL\gamma_0\partial^\mu\gamma'_\mu\Psi = 0,\ \ \mu = 0, ..., 3, \tag{5}$$

is invariant under $J^L_{\mu\nu}$ and allows for a two-flavor symmetry to be defined. It is also invariant under the following scalars: the hypercharge $Y = -1 + 1/2(I + \gamma_5)$, and the isospin $SU(2)_L$ with generators $I_1 = \frac{1}{4}(1 - I\gamma_5)J\gamma_2$, $I_2 = \frac{1}{4}(1 - I\gamma_5)K\gamma_2$, $I_3 = \frac{1}{4}(1 - I\gamma_5)I$; this identification of Y and I_i comes from the correct vertices and quantum numbers of the leptons, as argued below.

This choice of scalars leads, as expected, to a restricted set of solutions. The massless positive solutions are shown schematically on Table 1 and they include $V - A$ and $V+A$ vectors, with spin components $\tilde{B}_\mu = \frac{1}{4\sqrt{2}}(1-I)(1+I\gamma_5)\gamma_0\gamma_\mu$, $B_\mu = \frac{1}{4}(1-I\gamma_5)\gamma_0\gamma_\mu$, which amount to eight degrees of freedom (df); isospin $V - A$ vectors $W^i_\mu = I_i B_\mu$ with twelve df; also left and right-handed antisymmetric tensors and scalars $n_{L,R}$, $v_{L,R}$ in an isospin doublet, amounting to eight df, and with their antiparticles sixteen df. These add up to thirty-six bosons. We also obtain spin 1/2 left-handed particles in an isospin doublet $(\nu, l^-)_L$ with $Y = -1$ and a right-handed isospin singlet l^-_R with $Y = -2$; for example, $l^-_R(k) = \frac{1}{8}(1+I\gamma_5)(J\gamma_2 - iK\gamma_2)\gamma_0(\gamma_1 + iI\gamma_2)$ is a right-handed spin component with these quantum numbers and given flavor. The choice of Y can also be deduced from other requirements[3] related to gauge symmetry. Taking account of antiparticles and the two flavors, we have twenty-four fermion df. The reason for not having altogether sixty-four active df is the four inert df, defining the flavor, projected by $1 - L$, and which are not connected to the Hamiltonian (top-left matrix in Table 1).

Table 1: Arrangement of solutions in a $6-d$ 8×8 matrix model, with each box occupying a 2×2 matrix.

	e^+_L	$\bar{\nu}_R$	l^+_R
e^-_R	$\tilde{B}_\mu$	n_R	v_R
ν_L	n_L	B_μ, W^0_μ	W^1_μ, W^2_μ
l^-_L	v_L	W^1_μ, W^2_μ	B_μ, W^0_μ

In seeking a massive extension of Eq. 5, we expect all the hermitian combinations of the scalar terms $M_i\gamma_0$ to be scalars with respect to $J'_{\mu\nu}$. However, if we also demand that they be scalars with respect to $J^L_{\mu\nu}$, then the choices are reduced. Now, the only non-trivial scalar that commutes with all $M_i\gamma_0$ terms is L. Nevertheless, if we relax this condition we obtain in addition that precisely and only

$$Q = I_3 + \frac{1}{2}Y \tag{6}$$

commutes with $M_3\gamma_0$ and $M_4\gamma_0$, $M_3 = -(M/2)J\gamma_2(1+\gamma_5)$, $M_4 = (M/2)$ $K\gamma_2(1+\gamma_5)$, and M is the mass constant. As Q is the electric charge we deduce the electromagnetic $U(1)_{em}$ remains a symmetry while the hypercharge and isospin are broken. We stress that Q is deduced, rather than being imposed, as the only additional symmetry consistent with massive terms.

4 Relation to physical fields

The addition of a mass term to Eq. 5, say, $M_3\gamma_0$ modifies the vector and fermion fields obtained, which can be classified with the help of the discrete symmetries. The vectors include non-axial massive terms given by $A_\mu = \frac{1}{2}Q\gamma_0\gamma_\mu$, and their orthogonal combination Z_μ. A_μ can be represented as a mixture of two chargeless and massless components. On one hand, the B_μ and $\tilde{B}_\mu$ terms form $B_\mu = \frac{1}{2\sqrt{3}}Y\gamma_0\gamma_\mu$, that is, the hypercharge carriers. Thus, we obtain another argument to set Y whose origin is in the way we arrive at the expression for Q in Eq. 6. On the other hand, we can extract the chargeless vector isospin triplet component $W^3_\mu = I_3\gamma_0\gamma_\mu$. ¿From the expressions for Q, A_μ, and W^3_μ we easily obtain

$$A_\mu = \frac{1}{2}W^3_\mu + \frac{\sqrt{3}}{2}B_\mu. \qquad (7)$$

The value of Weinberg's angle θ_W is derived immediately from this equation by making an analogy with the new fields obtained in the SM, after application of the Higgs mechanism[4]. The photon there has the form $A_\mu = \frac{1}{\sqrt{g^2+g'^2}}(gB_\mu + g'W^0_\mu)$, where g and g' are respectively the isospin and hypercharge coupling constants. We obtain $\frac{g'}{g} = \frac{1}{\sqrt{3}}$. As in the SM $tan(\theta_W) = \frac{g'}{g}$ we find $sin^2(\theta_W) = .25$.

5 Vector fermion-current vertices

The spin 1/2 components form a charge $q = 1$ (from Q) massive Dirac particle l (left-handed part l_L) and a $q = 0$ particle ν, which remains massless. This is as occurs in the SM for charged leptons and neutrinos, to be identified respectively with l and ν. It is possible to derive the SM vertices describing the coupling of the fermions to the vectors obtained from the solutions, using the above arguments.

In addition, the vertices give information on the coupling constants g' and g. g can be obtained, e.g., from the coupling of the massive charged vectors W^+_μ and the charged current, defined by the neutrino and the charged leptons wave functions. These are $g = 1$, $g' = 1/\sqrt{3}$. The next Clifford algebra model at 7+1 dimensions, assuming irreducible representations for the gauge groups, predicts coupling constants of

$$g = 1/\sqrt{2} \approx .707 \quad g' = 1/\sqrt{6} \approx .408 \ . \tag{8}$$

It is a consistency feature of the theory that these values are in accordance with θ_W. In addition, these values are to be compared with the experimentally measured ones at energies of the mass of the Z_μ particle, which is where the breakdown of the $SU(2)_L \times U(1)_Y$ symmetry occurs. These are $g'_{exp} \approx .35$, $g_{exp} \approx .65$, and $sin^2(\theta_{W exp}) \approx .23$. The obtained predictions are unique because other representations with possible similar properties will not predict, for example, the same coupling constants. Indeed, the embedding of $5+1$ $|\Psi_i\rangle_{5+1}$ solutions in $7+1$ d, implies their mapping into 7+1 double irreducible representations $\frac{1}{\sqrt{2}}(|\Psi_1\rangle_{5+1} + |\Psi_2\rangle_{5+1})$, amounting to a $\frac{1}{\sqrt{2}}$ factor in the coupling obtained from the trace in Eq. 4, with respect to the lower-dimensional algebra. The theory thus presented succeeds in reproducing many aspects of the electroweak sector of the SM. The close connection between the results hence derived and the physical particles' phenomenology makes plausible the idea that, as with the spin, the gauge vector and matter fields, and their interactions originate in the structure of space-time.

Acknowledgments

The author acknowledges support from DGAPA-UNAM through project IN11860(

References

1. H. Georgi and S. L. Glashow, *Phys. Rev. Lett.* **32**, 438 (1974).
2. V. Bargmann and E. P. Wigner, *Proc. Nat. Acad. Sci. (USA)* **34**, 211 (1948).
3. J. Besprosvany, *Int. Jour. Theor. Phys.* **39**, 2797 (2000).
4. S. Glashow, *Nucl. Phys.* **22**, 579 (1961); S. Weinberg, *Phys. Rev. Lett.* **19**, 1264 (1961); A. Salam in W. Svartholm (Ed.), *Elementary Particle Theory,* Almquist and Wiskell, Stockholm (1968).

IN SEARCH OF ALTERNATIVES TO HIGGS FIELDS

I.L.BOGOLUBSKY, A.A.BOGOLUBSKAYA
Joint Institute for Nuclear Research, LIT,
141980, Dubna, Moscow region, Russia
Email: bogolubs@cv.jinr.ru

Field equations of S^2 sigma model ("the $A3$ model") with spontaneously broken $Z(2)$ symmetry are presented for $(D+1)$-dimensional space-time. The $A3$ model is an extension of the sine-Gordon equation (SGE) and supports kink-like $U(1)$ charged solitons which are generalization of neutral solitons of the SGE. Natural question arises: is the $A3$ model completely integrable in (1+1)-dimensional space-time? The Lorentz-invariant scalar $A3$ field can be viewed as promising alternative to Higgs field.

1 Introduction

Despite the impressive successes of the Standard Model of EW interactions, the scalar sector of the theory requires further investigation as the Higgs scalar particles have not yet been found. It seems promising to search for possible alternatives to Higgs fields among scalar fields possessing mathematical beauty, e.g., among scalar fields taking values on manifolds with continuous and discrete symmetries, which can be spontaneously broken. Given growing interest to superstring models it seems productive to focus on the models which give rise to stable localized (e.g., soliton and vortex) solutions in low-dimensional $((1+1)$ and $(2+1))$ models. Among aesthetically appealing models of special interest are completely integrable ones (for example, the sine-Gordon equation (SGE) model).

Below we shall discuss periodic in time (hence, electrically charged) kink and antikink solutions of the easy-axis Heisenberg antiferromagnet (AFM) model in quasi-classical approximation, which we shall call for brevity "the $A3$ model" ("Anisotropic 3–component"). We shall also consider the zero-curvature representation of the anisotropic N-component sigma model which includes the $A3$ model as a particular case.

2 The $A3$ model (the easy-axis AFM sigma model)

Lagrangian and Hamiltonian densities of the $A3$ field model in standard field theory notation are as follows (recall that $s_a s_a = 1$):

$$\mathcal{L} = \frac{1}{2}[\partial_\mu s_a \partial^\mu s_a + (s_3^2 - 1)] = \frac{1}{2}[(\partial_0 s_a)^2 - (\partial_1 s_a)^2 + (s_3^2 - 1)], \qquad (1)$$

$$\mathcal{H} = \frac{1}{2}[(\partial_0 s_a)^2 + (\partial_1 s_a)^2 + (1 - s_3^2)], \qquad (2)$$

here $a = 1, 2, 3$, $\mu = 0, 1, ..., D$ ($D = 1$ in the present paper), $\partial_0 = \frac{\partial}{\partial t}$, $\partial_1 = \frac{\partial}{\partial x}$, $\partial_\mu \partial^\mu = \partial_0^2 - \partial_1^2 = \frac{\partial^2}{\partial t^2} - \frac{\partial^2}{\partial x^2}$, and summation over repeated indices μ, a is meant. The Euler–Lagrange equations have the form:

$$\partial_\mu \partial^\mu s_i + (\partial_\mu s_a \partial^\mu s_a) s_i - s_3(\delta_{i3} - s_i s_3) = 0, \qquad i = 1, 2, 3. \qquad (3)$$

The $A3$ model emerges, in particular, when going over to the continuous limit in lattice models of the Heisenberg antiferromagnet (AFM) [1] and ferroelectrics [2], which have the easy–axis anisotropy. This model is Lorentz-invariant; it possesses $U(1) \times Z(2)$ internal symmetry with two equivalent vacua on its field manifold S^2. $Z(2)$ symmetry of the $A3$ model is spontaneously broken both on vacuum and kink solutions. So this model having its roots in condensed matter physics seems to be of interest as a field theory model.

It is also instructive to note that the $A3$ model is the extension to $(1+1)$–dimensional (x, t) space–time of $(0+1)$–dimensional equations of C.Neumann problem, describing motion of a point particle on the surface of the unit sphere S^2 with quadratic potential set on it [3]. It is known that C.Neumann problem is completely integrable (see, e.g., [4] and Refs. therein). Further, the $A3$ model is the extension of completely integrable sine–Gordon equation. Thus the natural question arises: is the $A3$ model completely integrable?

3 Generalization of SGE and charged kinks

Using representation of components s_i of the unit isovector $\mathbf{s}(\mathbf{x}, t)$ in terms of angular variables θ, ϕ on the sphere S^2,

$$s_1 = \sin\theta \cos\phi, \quad s_2 = \sin\theta \sin\phi, \quad s_3 = \cos\theta, \qquad (4)$$

we derive from (1), (2) expressions for $\mathcal{L}$ and $\mathcal{H}$ in terms of θ, ϕ

$$\mathcal{L} = \frac{1}{2}[\partial_\mu \theta \partial^\mu \theta + \sin^2\theta \partial_\mu \phi \partial^\mu \phi - \sin^2\theta] =$$

$$= \frac{1}{2}\left\{\left(\frac{\partial\theta}{\partial t}\right)^2 - \left(\frac{\partial\theta}{\partial x}\right)^2 + \sin^2\theta\left[\left(\frac{\partial\phi}{\partial t}\right)^2 - \left(\frac{\partial\phi}{\partial x}\right)^2\right] - \sin^2\theta\right\}, \qquad (5)$$

$$\mathcal{H}=\frac{1}{2}\left\{\left(\frac{\partial\theta}{\partial t}\right)^2+\left(\frac{\partial\theta}{\partial x}\right)^2+\sin^2\theta\left[\left(\frac{\partial\phi}{\partial t}\right)^2+\left(\frac{\partial\phi}{\partial x}\right)^2\right]+\sin^2\theta\right\}. \tag{6}$$

The dynamic system defined by relations (1), (3) possesses the following densities of momentum $\mathcal{P}_x$ and the third component of "isospin" $\mathcal{S}_3$ ("charge")

$$\mathcal{P}_x=-\frac{\partial s_a}{\partial t}\frac{\partial s_a}{\partial x}, \qquad \mathcal{S}_3=s_1\frac{\partial s_2}{\partial t}-s_2\frac{\partial s_1}{\partial t}. \tag{7}$$

The Noether invariants of the system (1), (3) can be calculated by means of integration of the densities $\mathcal{H},\mathcal{P},\mathcal{S}_3$:

$$H=\int\mathcal{H}dx, \qquad P_x=\int\mathcal{P}_x dx, \qquad S_3=\int\mathcal{S}_3 dx. \tag{8}$$

In particular

$$S_3=2\int\sin^2\theta\frac{\partial\phi}{\partial t}dx. \tag{9}$$

Euler-Lagrange equations which follow from (5) are (cf. [1,5]):

$$2\partial_\mu\partial^\mu\theta+\sin 2\theta(1-\partial_\mu\partial^\mu\phi)=$$

$$=2\left(\frac{\partial^2\theta}{\partial t^2}-\frac{\partial^2\theta}{\partial x^2}\right)+\sin 2\theta\left[1-\left(\frac{\partial\phi}{\partial t}\right)^2+\left(\frac{\partial\phi}{\partial x}\right)^2\right]=0, \tag{10}$$

$$2\cos\theta\partial_\mu\theta\partial^\mu\phi+\sin\theta\partial_\mu\partial^\mu\phi=$$

$$=2\cos\theta\left(\frac{\partial\theta}{\partial t}\frac{\partial\phi}{\partial t}-\frac{\partial\theta}{\partial x}\frac{\partial\phi}{\partial x}\right)+\sin\theta\left(\frac{\partial^2\phi}{\partial t^2}-\frac{\partial^2\phi}{\partial x^2}\right)=0. \tag{11}$$

Notice that by setting $\phi=0$ (i.e., $s_2=0$) we get the famous completely integrable SGE [6-11] for the 2θ variable:

$$\frac{\partial^2(2\theta)}{\partial t^2}-\frac{\partial^2(2\theta)}{\partial x^2}+\sin 2\theta=0. \tag{12}$$

Making substitution $z=u\exp(i\phi)$, $u=\tan\frac{\theta}{2}$ and setting $\phi=0$, we find once again the SGE for $\theta(x,t)$. Equations (10), (11) can be rewritten in the form (cf. [12])

$$(1+z^*z)\partial_\mu\partial^\mu z-2z^*\partial_\mu z\partial^\mu z+(1-z^*z)z=$$

$$= (1 + z^*z)\left(\frac{\partial^2 z}{\partial t^2} - \frac{\partial^2 z}{\partial x^2}\right) - 2z^*\left[\left(\frac{\partial z}{\partial t}\right)^2 - \left(\frac{\partial z}{\partial x}\right)^2\right] + (1 - z^*z)z = 0. \quad (13)$$

Equation (13) may be viewed as complexification of the SGE. It possesses soliton solutions which prove to be direct generalizations of the SGE solitons. Zero velocity soliton of the model can be easily found in terms of z (denote it $z_0(x,t)$):

$$z_0(x,t) = \exp\left[\pm(1-\omega^2)^{\frac{1}{2}}x\right]\exp(i\omega t + i\phi_0), \quad |\omega| < 1, \quad (14)$$

in terms of θ, ϕ:

$$\theta(x,t) = 2\arctan\left\{\exp\left[\pm(1-\omega^2)^{\frac{1}{2}}x\right]\right\}, \quad \phi(x,t) = \phi_0 + \omega t, \quad \phi_0 = const, \quad (15)$$

and finally by using (4), (15) in terms of $s_i(x,t), \quad i = 1,2,3$.

Making Lorentz transformation

$$x \to \gamma(x - vt), \quad t \to \gamma(t - vx), \quad \gamma = (1 - v^2)^{-\frac{1}{2}}, \quad (16)$$

we can find from (14), (15) the soliton moving at velocity $v, -1 < v < +1$. To illustrate this, we write it down in $z(x,t)$ terms:

$$z_v(x,t) = z_0[\gamma(x-vt), \gamma(t-vx)] = \exp[\pm P\gamma(x-vt)]\exp[i\omega\gamma(t-vx) + i\phi_0] =$$

$$= \exp(p_0 t - p_1 x + i\phi_0) = \exp(p_\mu x_\mu + i\phi_0), \qquad P = (1-\omega^2)^{\frac{1}{2}}, \quad (17)$$

where we use notation

$$p_0 = i\omega\gamma \pm \gamma v P, \qquad p_1 = i\omega\gamma v \mp \gamma P. \quad (18)$$

The values p_0 and p_1 are related by the identity:

$$p_\mu p^\mu = p_0^2 - p_1^2 = -1. \quad (19)$$

In the next section we describe some results of the search for zero-curvature representation for a more general class of models, containing the $A3$ model as a special case.

4 Zero-curvature representation of anisotropic N-component sigma models

Consider N-component field $s_a(x,t)$, $s_a s_a = 1$, $a, b = 1, 2, ..., N$, with the Lagrangian density

$$\mathcal{L} = \frac{1}{2}\left\{(\partial_\mu s_a \partial^\mu s_a) + J_a \delta_{ab}[s_b(x,t)s_a(x,t) - s_b(\infty,t)s_a(\infty,t)]\right\} =$$

$$= \frac{1}{2}\left[(\partial_\mu s_a \partial^\mu s_a) + (\mathbf{J}\mathbf{s}, \mathbf{s}) - \mathbf{J}_N\right], \tag{20}$$

here $\mathbf{J}$ is the diagonal matrix, $\mathbf{J} = diag(J_1, J_2, ..., J_N)$; we assume that $J_1 \leq J_2 \leq ... < J_N$, δ_{ab} is the Kronecker symbol.

The Euler-Lagrange equation derived from (20) is

$$\left(\frac{\partial^2 s_i}{\partial t^2} - \frac{\partial^2 s_i}{\partial x^2}\right) - s_k\left(\frac{\partial^2 s_k}{\partial t^2} - \frac{\partial^2 s_k}{\partial x^2}\right) s_i + s_i\left(\sum_{k=1}^{N} J_k s_k s_k - J_i\right) = 0. \tag{21}$$

Note that (3) describing the easy-axis $A3$ field arises from (21)in the special case $N = 3$, $J_1 = J_2 < J_3$ after simple scaling transformation of x, t : $x' = (\Delta J)^{\frac{1}{2}} x$, $t' = (\Delta J)^{\frac{1}{2}} t$, $\Delta J = J_3 - J_1$.

We succeeded to find the zero-curvature representation of (21)

$$\frac{\partial U}{\partial t} - \frac{\partial V}{\partial x} + [U, V] = 0, \tag{22}$$

where

$$\mathbf{U} = \begin{pmatrix} \mathbf{U}_{11} & \mathbf{U}_{12} \\ \mathbf{U}_{12} & \mathbf{U}_{11} \end{pmatrix}, \qquad \mathbf{V} = \begin{pmatrix} \mathbf{V}_{11} & \mathbf{V}_{12} \\ \mathbf{V}_{12} & \mathbf{V}_{11} \end{pmatrix}, \tag{23}$$

$$\mathbf{U}_{11} = a_1(\lambda)\mathbf{l}_0 + b_1(\lambda)\mathbf{l}_1 + c_1(\lambda)\mathbf{J} + d_1(\lambda)\mathbf{N},$$

$$\mathbf{U}_{12} = e_1(\lambda)\mathbf{l}_0 + f_1(\lambda)\mathbf{l}_1 + q_1(\lambda)\mathbf{J} + h_1(\lambda)\mathbf{N},$$

$$\mathbf{V}_{11} = a_2(\lambda)\mathbf{l}_0 + b_2(\lambda)\mathbf{l}_1 + c_2(\lambda)\mathbf{J} + d_2(\lambda)\mathbf{N},$$

$$\mathbf{V}_{12} = e_2(\lambda)\mathbf{l}_0 + f_2(\lambda)\mathbf{l}_1 + q_2(\lambda)\mathbf{J} + h_2(\lambda)\mathbf{N}. \tag{24}$$

with functions $a_1(\lambda), ..., h_2(\lambda)$ to be found from equation set

$$b_1 = a_2 = b_1 a_2 - a_1 b_2 + f_1 e_2 - e_1 f_2,$$

$$a_1 = b_2 = d_1 c_2 - c_1 d_2 + h_1 q_2 - q_1 h_2, \qquad a_1(\lambda) \neq 0,$$

$$c_1 a_2 - a_1 c_2 + q_1 e_2 - e_1 q_2 = 0,$$

$$c_1 b_2 - b_1 c_2 + q_1 f_2 - f_1 q_2 = 0,$$

$$d_1 a_2 - a_1 d_2 + h_1 e_2 - e_1 h_2 - \frac{d_1}{2} = 0,$$

$$d_1 b_2 - b_1 d_2 + h_1 f_2 - f_1 h_2 + \frac{d_2}{2} = 0,$$

$$f_1 = e_2 = f_1 a_2 - e_1 b_2 + b_1 e_2 - a_1 f_2,$$

$$e_1 = f_2 = h_1 c_2 - q_1 d_2 + d_1 q_2 - c_1 h_2, \qquad e_1(\lambda) \neq 0,$$

$$q_1 a_2 - e_1 c_2 + c_1 e_2 - a_1 q_2 = 0,$$

$$q_1 b_2 - f_1 c_2 + c_1 f_2 - b_1 q_2 = 0,$$

$$h_1 a_2 - e_1 d_2 + d_1 e_2 - a_1 h_2 - \frac{h_1}{2} = 0,$$

$$h_1 b_2 - f_1 d_2 + d_1 f_2 - b_1 h_2 - \frac{h_2}{2} = 0, \tag{25}$$

We have not yet found the solutions $a_1(\lambda), ..., h_2(\lambda)$ to (25), the investigation of this problem and the search for the higher (non-Noether) integrals of motion of (3), which are known [6-11] to be attributes of completely integrable systems, are in progress; the issue of integrability of the $A3$ model requires further investigation.

5 Interaction of the A3 and Maxwell fields; 2D solitonic strings-vortices

The global $U(1)$ symmetry of the $A3$ model can be made local by switching on "minimal" interaction of the $A3$ field with Maxwell field [13]. The gauge-invariant Lagrangian density of the resulting "$A3M$ model" reads:

$$\mathcal{L} = \eta^2 \left(\bar{\mathcal{D}}_\mu s_- \mathcal{D}^\mu s_+ + \partial_\mu s_3 \partial^\mu s_3 \right) - V(s_a) - \frac{1}{4} F^2_{\mu\nu},$$

$$\bar{\mathcal{D}}_\mu = \partial_\mu + igA_\mu, \quad \mathcal{D}_\mu = \partial_\mu - igA_\mu,$$

$$s_+ = s_1 + is_2, \quad s_- = s_1 - is_2,$$

$$F_{\mu\nu} = \partial_\mu A_\nu - \partial_\nu A_\mu, \quad V(s_a) = \beta^2(1 - s_3^2), \tag{26}$$

where β^2, η^2 are constants, g is a coupling constant, $\mu, \nu = 0, 1, ..., D$, and summation over repeated indices μ, ν is meant. It is straightforward to rewrite this Lagrangian density in the equivalent form

$$\mathcal{L} = \eta^2 (\partial_\mu s_a)^2 - V(s_a) - \frac{1}{4} F^2_{\mu\nu} + 2g\eta^2 A_\mu (s_2 \partial^\mu s_1 - s_1 \partial^\mu s_2) + g^2 \eta^2 (s_1^2 + s_2^2) A_\mu A^\mu. \tag{27}$$

By rescaling of x_μ Euler-Lagrange equations of the model are transformed to dimensionless form with the only parameter $p = \beta^2 g^{-2} \eta^{-4}$. These equations possess exponentially localized solutions for $p \in (0, p_{cr}], p_{cr} \approx 0.4$ with their energy $E(p)$ monotonically growing with p increase [13]; they are dynamically *stable*, and can be viewed as *soliton* analogues of Nielsen-Olesen strings-defects [14] in abelian Higgs model [15]. For given topological charge and all p, $0 < p < p_{cr}$, the energy $E(p)$ of the $A3M$ solitonic strings-vortices *does not exceed* [13] the energy of Belavin-Polyakov solitons E_{BP} in *isotropic* Heisenberg sigma model [16] (despite positive definiteness of $V(s_a)$ and $\frac{1}{4}F^2_{\mu\nu}$ terms in (27)!). Perhaps the most amazing fact is that $E(p) \to E_{BP}$ for $p \to p_{cr}$ [13].

One can conclude that the $A3$ field has interesting mathematical properties and can be considered as possible alternative to Higgs scalar field.

References

1. A.M.Kosevich, B.A.Ivanov and A.S.Kovalev, *Nonlinear magnetization waves. Dynamic and topological solitons*, Naukova Dumka, Kiev (1983) [in Russian]; *Phys. Rep.* **194**, 1 (1994).
2. J.Pouget, G.A.Maugin, *Phys. Rev.* **B30**, 5306 (1984).
3. C.Neumann, *J. Reine und Angew. Math.*, 1859, **56**, S, 46 (1859).
4. B.A.Dubrovin, *Uspekhi Matem. Nauk* **36**, 11 (1981).
5. I.V.Baryakhtar, B.A.Ivanov, *Zh. Eksp. Teor. Fiz.* **85**, 328 (1983).
6. V.E.Zakharov, S.V.Manakov, S.P.Novikov, L.P.Pitaevsky, *Soliton theory: Inverse scattering method* (ed. S.P.Novikov), Nauka, Moscow (1980).
7. G.L.Lamb, *Elements of soliton theory*, John Wiley, New York (1980).
8. M.J.Ablowitz, H.Segur, *Solitons and the inverse scattering transform*, SIAM, Philadelphia (1981).
9. R.K.Dodd, J.C.Eilbeck, J.D.Gibbon, H.C.Morris, *Solitons and nonlinear wave equations*, Academic Press Inc., London (1984).
10. A.C.Newell, *Solitons in mathematics and physics*, SIAM, Philadelphia (1985).
11. L.D.Faddeev, L.A.Takhtajan, *Hamiltonian methods in the theory of solitons*, Springer, Berlin (1987).
12. R.Leese, *Nucl. Phys. B* **336**, 283 (1991).
13. I.L.Bogolubsky, A.A.Bogolubskaya, *Phys.Lett. B* **395**, 269 (1997).
14. H.B.Nielsen and P.Olesen, *Nucl. Phys. B* **61**, 45 (1973) .
15. P.Higgs, *Phys. Lett.* **12**, 132 (1964).
16. A.A.Belavin and A.M.Polyakov, *Pis'ma v Zh.E.T.F.* **22**, 503 (1975); [*JETP Lett.* **22**, 245 (1975)].

Q-PLANE WAVE SOLUTIONS OF Q-MAXWELL EQUATIONS

V.K. DOBREV

School of Computing and Mathematics,
University of Northumbria,
Newcastle upon Tyne NE1 8ST, UK
vladimir.dobrev@unn.ac.uk
and
Institute of Nuclear Research and Nuclear Energy,
Bulgarian Academy of Sciences,
72 Tsarigradsko Chaussee, 1784 Sofia, Bulgaria
dobrev@inrne.bas.bg
(permanent address)

S.T. PETROV

Institute of Nuclear Research and Nuclear Energy,
Bulgarian Academy of Sciences,
72 Tsarigradsko Chaussee, 1784 Sofia, Bulgaria
spetrov@inrne.bas.bg

We give solutions of the quantum conformal deformations of the full Maxwell equations in terms of deformations of the plane wave.

1 Introduction

One of the purposes of quantum deformations is to provide an alternative of the regularization procedures of quantum field theory. Applied to Minkowski space-time the quantum deformations approach is also an alternative to Connes' noncommutative geometry[1]. The first problem to tackle in a noncommutative deformed setting is to analyze the behaviour of the wave equations analogues. Here we continue the study of hierarchies of deformed equations derived[2,3] with the use of quantum conformal symmetry. Actually, we study a q-deformation of the full Maxwell equations. We give solutions of the q-Maxwell equations using a deformation of the plane wave[4] which is a formal power series in the noncommutative coordinates of q-Minkowski space-time and four-momenta. For the latter deformations we use the one[2] related to a deformation of the conformal group which is not the case with the other known examples[5,6,7] .

2 Preliminaries

First we introduce new Minkowski variables:

$$x_\pm \equiv x_0 \pm x_3 , \qquad v \equiv x_1 - ix_2 , \qquad \bar{v} \equiv x_1 + ix_2 , \tag{1}$$

which, (unlike the x_μ), have definite group–theoretical interpretation[2]) as part of a six-dimensional coset of the conformal group $SU(2,2)$. In terms of these variables, e.g., the d'Alembert equation is:

$$\Box \varphi = (\partial_- \partial_+ - \partial_v \partial_{\bar{v}}) \varphi = 0 . \tag{2}$$

In the q-deformed case we use the noncommutative q-Minkowski space-time[2] which is given by the following commutation relations (with $\lambda \equiv q - q^{-1}$):

$$x_\pm v = q^{\pm 1} v x_\pm , \quad x_\pm \bar{v} = q^{\pm 1} \bar{v} x_\pm , \quad x_+ x_- - x_- x_+ = \lambda v \bar{v} , \quad \bar{v} v = v \bar{v} , \tag{3}$$

and the deformation parameter being a phase: $|q| = 1$. Relations (3) are preserved by the anti-linear anti-involution ω :

$$\omega(x_\pm) = x_\pm , \quad \omega(v) = \bar{v} , \quad \omega(q) = \bar{q} = q^{-1} , \quad (\omega(\lambda) = -\lambda) . \tag{4}$$

The solution spaces consist of formal power series in the q-Minkowski coordinates:

$$\varphi = \sum_{j,n,\ell,m \in \mathbb{Z}_+} \mu_{jn\ell m} \, \varphi_{jn\ell m} , \quad \varphi_{jn\ell m} = v^j \, x_-^n \, x_+^\ell \, \bar{v}^m . \tag{5}$$

The solution spaces (5) are representation spaces of the quantum algebra $U_q(sl(4))$. For the latter we use the rational basis of Jimbo[8]. The action of $U_q(sl(4))$ on $\varphi_{jn\ell m}$ was given earlier[9]. Because of the conjugation ω we are actually working with the conformal quantum algebra which is a deformation of $U(su(2,2))$. (We note that there is another basis[10] in the conjugated ordering: $\bar{v}, x_+, x_-, v$.)

Further we suppose that q is not a nontrivial root of unity.

In order to write our q-deformed equations in compact form it is necessary to introduce some additional operators. We first define the operators:

$$\begin{aligned} \hat{M}_\kappa^\pm \, \varphi &= \textstyle\sum_{j,n,\ell,m \in \mathbb{Z}_+} \mu_{jn\ell m} \, \hat{M}_\kappa^\pm \, \varphi_{jn\ell m} , \qquad \kappa = \pm, v, \bar{v} , \\ T_\kappa^\pm \, \varphi &= \textstyle\sum_{j,n,\ell,m \in \mathbb{Z}_+} \mu_{jn\ell m} \, T_\kappa^\pm \, \varphi_{jn\ell m} , \qquad \kappa = \pm, v, \bar{v} , \end{aligned} \tag{6}$$

and $\hat{M}_+^\pm$, $\hat{M}_-^\pm$, $\hat{M}_v^\pm$, $\hat{M}_{\bar{v}}^\pm$, resp., acts on $\varphi_{jn\ell m}$ by changing by ± 1 the value of j, n, ℓ, m, resp., while $T_+^\pm$, $T_-^\pm$, $T_v^\pm$, $T_{\bar{v}}^\pm$, resp., acts on $\varphi_{jn\ell m}$ by multiplication by $q^{\pm j}, q^{\pm n}, q^{\pm \ell}, q^{\pm m}$, resp. Now we can define the q-difference operators:

$$\hat{\mathcal{D}}_\kappa \varphi = \frac{1}{\lambda} \hat{M}_\kappa^{-1} \left(T_\kappa - T_\kappa^{-1}\right) \varphi . \tag{7}$$

Note that when $q \to 1$ then $\hat{\mathcal{D}}_\kappa \to \partial_k$. Using (6) and (7) the q-d'Alembert equation may be written as[3] :

$$\left(q \hat{\mathcal{D}}_- \hat{\mathcal{D}}_+ T_v T_{\bar{v}} - \hat{\mathcal{D}}_v \hat{\mathcal{D}}_{\bar{v}}\right) T_v T_- T_+ T_{\bar{v}} \varphi = 0 , \tag{8}$$

Note that when $q \to 1$ equation (8) goes to (2). Note that the operators in (6), (7), (8) for different variables commute, i.e., we have passed to commuting variables. However, keeping the normal ordering it is straightforward to pass back to noncommuting variables.

3 Solutions of the q-Maxwell equations

We consider the quantum conformal deformation[2] of Maxwell's equations as part of Maxwell's hierarchy of equations. The equations of the hierarchy are[2]:

$$_qI_n^+ \, _qF_n^+ = {}_qJ^n , \qquad _qI_n^- \, _qF_n^- = {}_qJ^n \tag{9}$$

where the operators are[2] :

$$_qI_n^+ = \tfrac{1}{2} \left(\left(q\hat{\mathcal{D}}_v + \hat{M}_{\bar{z}} \hat{\mathcal{D}}_+ (T_- T_v)^{-1} T_{\bar{v}} \right) T_- [n+2-N_z]_q - q^{-n-2} \times \right.$$

$$\left. \times \left(\hat{\mathcal{D}}_- T_- + q^{-1} \hat{M}_{\bar{z}} \hat{\mathcal{D}}_{\bar{v}} - \lambda \hat{M}_v \hat{M}_{\bar{z}} \hat{\mathcal{D}}_- \hat{\mathcal{D}}_+ T_{\bar{v}} \right) T_-^{-1} \hat{\mathcal{D}}_z \right) T_+ T_v T_z T_{\bar{z}}^{-1} , \tag{10}$$

$$_qI_n^- = \tfrac{1}{2} \left(\hat{\mathcal{D}}_{\bar{v}} + q\hat{M}_z \hat{\mathcal{D}}_+ T_{\bar{v}} T_- T_v^{-1} - q\lambda \hat{M}_v \hat{\mathcal{D}}_- \hat{\mathcal{D}}_+ T_{\bar{v}} \right) T_{\bar{v}} \times$$

$$\times [n+2-N_{\bar{z}}]_q - \tfrac{1}{2} q^{n+3} \left(\hat{\mathcal{D}}_- + q\hat{M}_z \hat{\mathcal{D}}_v T_- \right) \hat{\mathcal{D}}_{\bar{z}} T_- T_{\bar{v}} . \tag{11}$$

Maxwell's equations $\partial^\mu F_{\mu\nu} = J_\nu$, $\epsilon_{\mu\nu\rho\sigma} \partial^\mu F^{\rho\sigma} = 0$ are obtained from (9) for $n = 0$, $q = 1$, substituting the fixed helicity constituents $F^\pm$ by: $F^+ = z^2(F_1^+ + iF_2^+) - 2zF_3^+ - (F_1^+ - iF_2^+)$, $F^- = \bar{z}^2(F_1^- - iF_2^-) - 2\bar{z}F_3^- - (F_1^- + iF_2^-)$, $F_k^\pm = F_{k0} \pm \frac{i}{2}\varepsilon_{k\ell m} F_{\ell m} = E_k \pm iH_k$, $J^0 = \bar{z}z(J_0 + J_3) + z(J_1 + iJ_2) + \bar{z}(J_1 - iJ_2) + (J_0 - J_3)$, and then comparing the coefficients of the resulting polynomials in z and $\bar{z}$.

We shall use the fact that Maxwell's equations belong also to another hierarchy[3] for which we know solutions in terms of deformations of the plane wave. In the present paper we shall use the first deformation of the plane wave[4] (more general deformations were given later[11]) :

$$\exp_q(k,x) = \sum_{s=0}^{\infty} \frac{1}{[s]_q!}\, f_s \,, \tag{12}$$

$$[s]_q! \equiv [s]_q[s-1]_q \cdots [1]_q \,, \quad [0]_q! \equiv 1 \,, \quad [n]_q \equiv \frac{q^n - q^{-n}}{q - q^{-1}} \,,$$

$$f_s = \beta^s \sum_{a,b,n \in \mathbb{Z}_+} \frac{(-1)^{s-a-b}\, q^{n(s-2a-2b+2n) \,+\, a(s-a-1) \,+\, b(-s+a+b+1)}}{\Gamma_q(a-n+1)\Gamma_q(b-n+1)\Gamma_q(s-a-b+n+1)[n]_q!} \times$$
$$\times\, k_v^{s-a-b+n} k_-^{b-n} k_+^{a-n} k_{\bar{v}}^{n} v^n x_-^{a-n} x_+^{b-n} \bar{v}^{s-a-b+n} \,, \tag{13}$$

$$(\beta^s)^{-1} = \sum_{p=0}^{s} \frac{q^{(s-p)(p-1)+p}}{[p]_q!\,[s-p]_q!} \,,$$

where the momentum components $(k_v, k_-, k_+, k_{\bar{v}})$ are supposed to be non-commutative between themselves (obeying the same rules (3) as the q-Minkowski coordinates), and commutative with the coordinates. Further, Γ_q is q-deformation of the Γ-function, of which here we use only the properties: $\Gamma_q(p) = [p-1]_q!$ for $p \in \mathbb{N}$, $1/\Gamma_q(p) = 0$ for $p \in \mathbb{Z}_-$; $P_s(a,b)$ is a polynomial in a, b. Note that $(f_s)|_{q=1} = (k\cdot x)^s$ and thus $(\exp_q(k,x))|_{q=1} = \exp(k\cdot x)$. This q-plane wave has some properties analogous to the classical one but is not an exponent or q-exponent[12]. This is enabled also by the fact (true also for $q = 1$) that solving the equations may be done in terms of the components f_s . Each f_s satisfies the q-d'Alembert equation (8) on the momentum q-cone:

$$\mathcal{L}_q^k \equiv k_- k_+ - q^{-1} k_v k_{\bar{v}} = k_+ k_- - q\, k_v k_{\bar{v}} = 0 \,. \tag{14}$$

The solutions of (9) for $n = 0$ in the homogeneous case ($J = 0$) are:

$$\hat{F}^{\pm} \doteq \left({}_qF_0^{\pm}\right)_{J=0} = \sum_{m,s=0}^{\infty} \frac{1}{[s]_q!}\, \hat{F}^{\pm}_{ms}(k)\, f_s \,, \tag{15}$$

$$\hat{F}^{+}_{ms}(k) = \sum_{i=0}^{m} \Bigg(\sum_{j=0}^{m-i} p_{ij}^{ms1}\, k_v^i k_-^{m-i-j} k_{\bar{v}}^j\, (k_v - q^{s+6} z k_-)(k_v - q^{s+3} z k_-) +$$
$$+\, p_i^{ms2}\, k_v^i k_{\bar{v}}^{m-i}\, (k_v - q^{s+6} z k_-)(k_+ - q^{s+3} z k_{\bar{v}}) +$$
$$+ \sum_{j=0}^{m-i} p_{ij}^{ms3}\, k_v^i k_+^{m-i-j} k_{\bar{v}}^j\, (k_+ - q^{s+6} z k_{\bar{v}})(k_+ - q^{s+3} z k_{\bar{v}}) \Bigg) \,, \tag{16}$$

$$\hat{F}^-_{ms}(k) = \sum_{i=0}^{m} \Big(\sum_{j=0}^{m-i} r^{ms1}_{ij} \, k^i_v k^{m-i-j}_- k^j_{\bar{v}} \, (k_{\bar{v}} - q^{-1}\bar{z}k_-)(k_{\bar{v}} - \bar{z}k_-) \, +$$

$$+ \, r^{ms2}_i \, k^i_v k^{m-i}_{\bar{v}} \, (k_+ - q^{-1}\bar{z}k_v)(k_{\bar{v}} - \bar{z}k_-) \, +$$

$$+ \sum_{j=0}^{m-i} r^{ms3}_{ij} \, k^i_v k^{m-i-j}_+ k^j_{\bar{v}} \, (k_+ - q^{-1}\bar{z}k_v)(k_+ - \bar{z}k_v) \Big) \, , \tag{17}$$

where $p^{msa}_{i(j)}, r^{msa}_{i(j)}$ are independent constants. The check that these are solutions is done for commutative Minkowski coordinates and noncommutative momenta on the q-cone. The terms with $m = 0$ of the solutions (15), (17), (18) were obtained earlier[13] (later they were generalized using more general q-plane waves[11]). The solution (17) can be written in terms of the deformed plane wave if we suppose that the $r^{msa}_{i(j)}$ for different s coincide: $r^{msa}_{i(j)} = r^{ma}_{i(j)}$. Then we have:

$$\hat{F}^- = \sum_{m=0}^{\infty} \hat{F}^-_m(k) \, \exp_q(k,x) \, , \quad \hat{F}^-_m(k) = \hat{F}^-_{ms}(k) \, . \tag{18}$$

If we use the conjugated basis mentioned after (5) then the roles of the solutions $\hat{F}^-$ and $\hat{F}^+$ will be exchanged. The asymmetry between $\hat{F}^-$ and $\hat{F}^+$ may be avoided[14] in a Connes-Lott type model[15].

In the inhomogeneous case the solutions of (9) for $n = 0$ are:

$${}_qJ^0 = \bar{z}zJ_+ + zJ_v + \bar{z}J_{\bar{v}} + J_- \, , \tag{19}$$

$$J_\kappa = \sum_{m,s=o}^{\infty} \frac{1}{[s]_q!} \, J^{ms}_\kappa(k) \, f_{s-1} \, , \quad \kappa = \pm, v, \bar{v} \, , \tag{20}$$

$$\begin{aligned}
J^{ms}_v(k) &= K^s_m(k) \, k_{\bar{v}} \, , \\
J^{ms}_-(k) &= -q^{-s-2} K^s_m(k) \, k_+ \, , \\
J^{ms}_+(k) &= -K^s_m(k) \, k_- \, , \\
J^{ms}_{\bar{v}}(k) &= q^{-s-2} K^s_m(k) \, k_v \, , \\
K^s_m(k) &\doteq \gamma^s_v k^{m+1}_v + \gamma^s_- k^{m+1}_- + \gamma^s_+ k^{m+1}_+ + \gamma^s_{\bar{v}} k^{m+1}_{\bar{v}} \, ,
\end{aligned} \tag{21}$$

$${}_qF^\pm_0 = \tilde{F}^\pm + \hat{F}^\pm \, , \tag{22}$$

$$\tilde{F}^\pm = \sum_{m,s=0}^{\infty} \frac{1}{[s]_q!} \, \tilde{F}^\pm_{ms}(k) \, f_s \, , \tag{23}$$

$$
\begin{aligned}
\tilde{F}^{+}_{ms}(k) &= 2d_s q^{-s}\big((q^{-s-5}\gamma^s_- k^m_- + z\gamma^s_v k^m_v)(k_v - q^{s+3}zk_-) \; + \\
&\quad + (q^{-s-5}\gamma^s_{\bar{v}} k^m_{\bar{v}} + z\gamma^s_+ k^m_+)(k_+ - q^{s+3}k_{\bar{v}})\big) \; , \\
\tilde{F}^{-}_{ms}(k) &= 2d_s q^{-2s-2}\big((\gamma^s_- k^m_- + q^{-2}\bar{z}\gamma^s_{\bar{v}} k^m_{\bar{v}})(k_{\bar{v}} - \bar{z}k_-) \; + \\
&\quad + (\gamma^s_- k^m_v + q^{-2}\bar{z}\gamma^s_+ k^m_+)(k_+ - \bar{z}k_v)\big) \; ,
\end{aligned}
$$

where $d_s = \beta^s/\beta^{s+1}$. As in the homogeneous case we can not make $\tilde{F}^{+}_{ms}(k)$ independent of s. We can make either $\tilde{F}^{-}_{ms}(k)$ or $J^{ms}_{\kappa}(k)$ (but not both) independent of s by choosing $\gamma^s_\kappa \sim d_s^{-1}$ or $\gamma^s_\kappa = \gamma_\kappa$, respectively. These issues will be discussed in more detail elsewhere.

References

1. A. Connes, *Noncommutative Geometry*, Academic Press (1994).
2. V.K. Dobrev, *Phys. Lett. B* **341**, 133 (1994) & **346**, 427 (1995).
3. V.K. Dobrev, *J. Phys. A* **28**, 7135 (1995).
4. V.K. Dobrev and B.S. Kostadinov, *Phys. Lett. B* **439**, 337 (1998).
5. U. Carow-Watamura et al, *Zeit. f. Phys. C* **48**, 159 (1990).
6. W.B. Schmidke, J. Wess and B. Zumino, *Zeit. f. Phys. C* **52**, 471 (1991).
7. S. Majid, *J. Math. Phys.* **32**, 3246 (1991).
8. M. Jimbo, *Lett. Math. Phys.* **10**, 63 (1985); *Lett. Math. Phys.* **11**, 247 (1986).
9. V.K. Dobrev, *J. Phys. A* **27**, 4841 & 6633 (1994).
10. V.K. Dobrev, R.I. Gushterski, S.T. Petrov and B.S. Zlatev, Invited talk by V.K.D. at III International Workshop "Classical and Quantum Integrable Systems" (Yerevan, 1998), Proc., eds. L.G. Mardoyan, G.S. Pogosyan and A.N. Sissakian, Publ. Dept. of JINR, Dubna (1998), p. 39.
11. V.K. Dobrev, S.T. Petrov and B.S. Zlatev, Talk by V.K.D. at PDE2000 Conference (Clausthal, Germany), Proc. *Partial Differential Equations and Spectral Theory*, ed. M. Demuth et al, *Operator Theory: Advances and Applications, Vol. 126*, Birkhäuser, Berlin (2001), p. 113.
12. G. Gaspar and M. Rahman, *Basic Hypergeometric Series*, Cambridge U. Press (1990).
13. V.K. Dobrev and B.S. Zlatev, *Czech. J. Phys.* **50**, 53 (2000).
14. V.K. Dobrev, S.T. Petrov and B.S. Zlatev, Helicity Asymmetry of q-Plane Wave Solutions of q-Maxwell Equations, *Physics of Atomic Nuclei*, to appear; UNN-SCM-M-00-12 (2000).
15. A. Connes and J. Lott, *Nucl. Phys. B (Proc. Suppl.)* **18**, 29 (1991).

SYMMETRIES AND WAVELETS

I.M. DREMIN
P.N.Lebedev Physical Institute, 119991 Moscow, Russia

Wavelet analysis of different patterns reveals some symmetries and singularities otherwise hidden in the pattern. Its general methods are briefly reviewed. Examples from turbulence, cavitation, Cherenkov gluon emission and quark-gluon jets structure in QCD are considered.

Symmetries are crucial in the surrounding us world and therefore lie at the very heart of any theoretical construction. Symmetries can be either global or local. The invariance under the global gauge transformation requires a definite symmetry between particles and antiparticles. More important consequences appear from the requirement of the local gauge invariance. Both in QED and QCD it defines the interaction lagrangian and, consequently, the symmetries of main equations. In classical and quantum physics the symmetries of equations and of initial (boundary) conditions determine the symmetries of the corresponding patterns observed. Thus the problem of the local analysis of these patterns should be solved.

One of the most popular methods is to use the Fourier transform. The set of functions (sine, cosine and imaginary exponents) is such that all of them are widespread along their definition axis. Thus this transform is a global one. With its help one can get the knowledge of frequencies (scales) important in the problem considered but can not learn about their locations. In other words, this transform is suited for stationary (homogeneous) processes. For non-stationary (inhomogeneous) processes one should apply a local transform. With the above set of functions, it becomes possible only if one uses the so-called windowed Fourier transform. Namely, the analyzed function is multiplied by an auxiliary function steeply decreasing (or equal to zero) outside some finite range (window) of its variables. Sometimes, this leads to uncontrollable consequences.

The only way to control such transform is to build up a complete set of orthogonal functions with the compact support. This program is realized with the help of wavelets [1,2,3]. They allow for the direct and inverse local transform to be done. Thus one gets the local scale (frequency) characteristics (fluctuations) of the analyzed process within any required resolution. Moreover, this resolution is self-adjustable in contrast to fixed windows in the Fourier transform. It means that wavelets automatically construct the so-called Heisenberg windows adjusted to the local properties of the considered function. To resolve its fast variation with high frequencies they admit small time-window but large frequency-window while for low frequencies the wide time-window and narrow frequency-window are used, i.e., the width of the window is proportional to its mean value. That is why they are called the relative band-width filters.

Wavelets originate from the functional equation which relates the so-called scaling function with its shifted and translated version. The coefficients h_k of this linear equation define in a unique way the explicit form of the wavelet which is not used directly, however. The wavelet transform coefficients can be explicitly obtained from the iterative equations containing h_k. This procedure is called the fast wavelet transform. It allows for the complete decomposition of the considered process at any resolution level and does not require any integrartion. Computer calculations are done quickly.

For those dealing with quantum field theory, I'd like to mention also that wavelet analysis actually is very close to the renormalization group approach (see, e.g., [4]) because it uses the translated versions of the same function. These features must be further exploited.

The equation for the scaling function $\varphi(x)$ is

$$\varphi(x) = \sqrt{2} \sum_{k=0}^{2M-1} h_k \varphi(2x - k) \tag{1}$$

with the dyadic dilation 2, integer translation k and the coefficients h_k determined from conditions of orthogonality (both mutual and to some polynomials) after one chooses their number $2M$. If the scaling function is known, one can form a "mother wavelet " (or a basic wavelet) $\psi(x)$ according to

$$\psi(x) = \sqrt{2} \sum_{k=0}^{2M-1} g_k \varphi(2x - k), \tag{2}$$

where

$$g_k = (-1)^k h_{2M-k-1}. \tag{3}$$

The dilated and translated versions of the scaling function φ and the "mother wavelet" ψ

$$\varphi_{j,k} = 2^{j/2} \varphi(2^j x - k), \tag{4}$$

$$\psi_{j,k} = 2^{j/2} \psi(2^j x - k) \tag{5}$$

form the orthonormal basis.

One may decompose any function f of $L^2(R)$ at any resolution level j_n in a series

$$f = \sum_k s_{j_n,k}\varphi_{j_n,k} + \sum_{j\geq j_n,k} d_{j,k}\psi_{j,k}. \tag{6}$$

The wavelet coefficients $s_{j,k}$ and $d_{j,k}$ can be calculated as

$$s_{j,k} = \int dx f(x)\varphi_{j,k}(x), \tag{7}$$

$$d_{j,k} = \int dx f(x)\psi_{j,k}(x). \tag{8}$$

However, in practice their values are determined from the fast wavelet transform. In general, one can get the iterative formulas of the fast wavelet transform

$$s_{j+1,k} = \sum_m h_m s_{j,2k+m}, \tag{9}$$

$$d_{j+1,k} = \sum_m g_m s_{j,2k+m} \tag{10}$$

where

$$s_{0,k} = \int dx f(x)\varphi(x-k). \tag{11}$$

These equations yield fast algorithms (the so-called pyramid algorithms) for computing the wavelet coefficients, asking now just for $O(N)$ operations to be done. Starting from $s_{0,k}$, one computes all other coefficients provided the coefficients h_m, g_m are known. The explicit shape of the wavelet is not used. The wavelet coefficients show the strength of fluctuations of f at the location k and the scale j.

The behavoir of wavelet coefficients is closely related to the structure of the local singularities of the function f and therefore can provide the knowledge which class of functions it belongs to. Also, the fractal properties of the analyzed set can be determined from the partition function $Z_q(j)$. Namely, let us consider the sum Z_q of the q-th moments of the coefficients of the wavelet transform at various scales j

$$Z_q(j) = \sum_k |d_{j,k}|^q, \tag{12}$$

where the sum is over the maxima of $|d_{j,k}|$. Then it was shown that for a fractal signal this sum should behave as

$$Z_q(j) \propto 2^{j[\tau(q)+\frac{q}{2}]}, \tag{13}$$

i.e.,

$$\log Z_q(j) \propto j[\tau(q) + \frac{q}{2}]. \tag{14}$$

Thus the necessary condition for a signal to possess fractal properties is the linear dependence of $\log Z_q(j)$ on the level number j. If this requirement is fulfilled the dependence of τ on q shows whether the signal is monofractal or multifractal. Monofractal signals are characterized by a single dimension and, therefore, by a linear dependence of τ on q, whereas multifractal ones are described by a set of such dimensions, i.e., by non-linear functions $\tau(q)$. Monofractal signals are homogeneous, in the sense that they have the same scaling properties throughout the entire signal. Multifractal signals, on the other hand, can be decomposed into many subsets characterized by different local dimensions, quantified by a weight function. The wavelet transform removes lowest polynomial trends that could cause the traditional box-counting techniques to fail in quantifying the local scaling of the signal. The function $\tau(q)$ can be considered as a scale-independent measure of the fractal signal. It can be further related to the Renyi dimensions, Hurst and Hölder exponents (for more detail, see Ref. [5]).

The wavelet transform can be also used for representation of differential operators, for solving the differential equations etc.

Let me describe the wavelet application to the analysis of some symmetrical patterns. Many such patterns are observed, e.g., on the water surface [6] (see Figs. 1, 2). Macroscopically they are described by some order parameters. The wavelet decomposition of the observed pattern allows to ascribe the definite scale at any location within the pattern and extract wave vector information from the pattern. In particular, the regions with scales within the chosen range can be separated. E.g., the stability regions with large scales are shown in Fig. 3 in white. Thus the transition from microscopic pattern to its macroscopic description becomes feasible. The location of some defects (broken symmetry) is easily recognized as well.

Another interesting effect is the turbulence. Its wavelet analysis reveals the fractal properties of the velocity field. Closer to the topic of this conference are the similar patterns observed in jet production in e^+e^--annihilation at high energies (for the review see [5,7]). The partonic cascade structure with "jets inside jets inside jets" leads to the mono- and multifractal distribution of the created particles within the available phase space (see [5]). The fluctuations increase in smaller phase space bins, and, correspondingly, the factorial moments of the multiplicity distributions should increase linearly (for monofractals) for smaller bins and flatten off for multifractals. This is demonstrated in Fig. 4. It is explained in QCD [8,9,10] as a consequence of the asymptotic freedom behavior of the coupling strength.

The fractal properties have been also discovered by the wavelet analysis in the distributions of the heartbeat intervals. One would be inclined to consider this process as a completely periodic stationary one. In fact, it has been found that it possesses the fractal properties and the distribution moments (correlations) at different scales behave in a different way for healthy and diseased patients [11,12]. The results of the wavelet analysis were proposed as a first clinically significant measure of heart disease.

The functions somewhat similar, at first sight, to those for heartbeat intervals are obtained for the pressure variation in gas turbine compressors. However their wavelet analysis revealed absence of fractal properties with some signature of a singular behavior. More important discovery is that the behavior of the dispersion of the wavelet coefficients at some resolution level can be used as a precursor of the extremely dangerous effect [13]. It is initiated by some instability (singularity), reminds the cavitation and is called the stall. This instability results in the complete damage of the engine and an aircraft crash.

Coming back to particle interactions, let me mention the important field of wavelet application to pattern recognition in multiparticle production. One can ask what kind of patterns are formed within the available phase space by particles created in very high energy interactions. This problem becomes experimentally feasible, e.g., in AA-collisions at RHIC where more than 4000 charged particles are on average produced in a single event at c.m. energy 130 GeV/c. In a search for the ring-like structure implied by ideas on Cherenkov gluon radiation [14], the wavelet analysis of several events of central Pb-Pb collisions at l.s. energy 158 GeV/c with more than 1000 charged secondaries was done. Some events possessing such a symmetry in the long-range correlations have been found [15] (see Fig. 4). This analysis can be extended to search for other symmetry patterns as well.

In conclusion, I'd like to stress how powerful is the method of the wavelet analysis to search for symmetries as is seen already from the above examples. However, its use is at the very beginning now and further results are coming.

Fig. 1

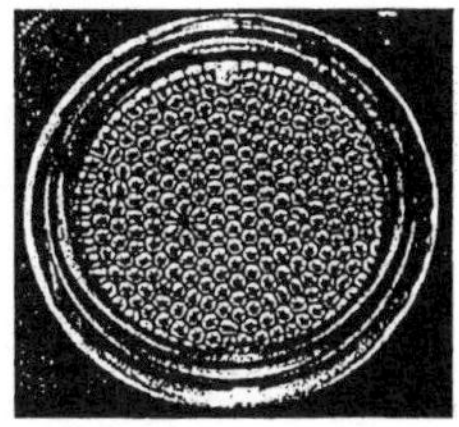

Fig. 2

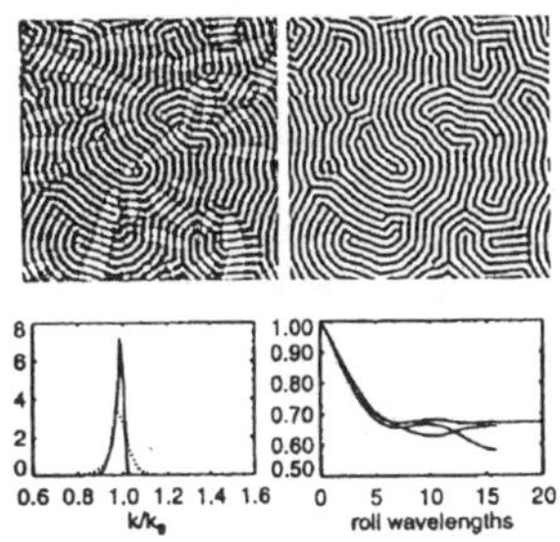

Fig. 3

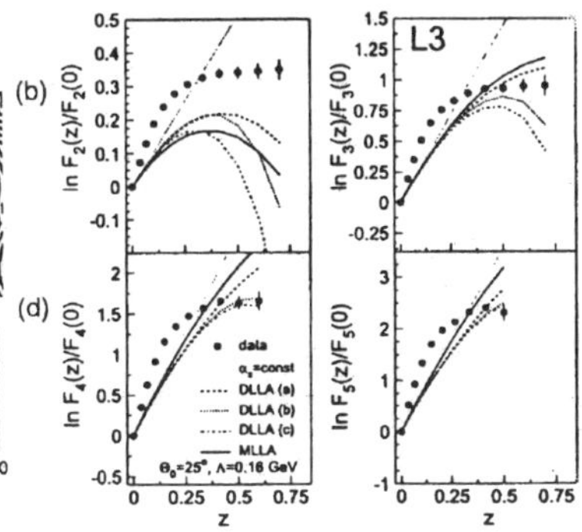

Fig. 4

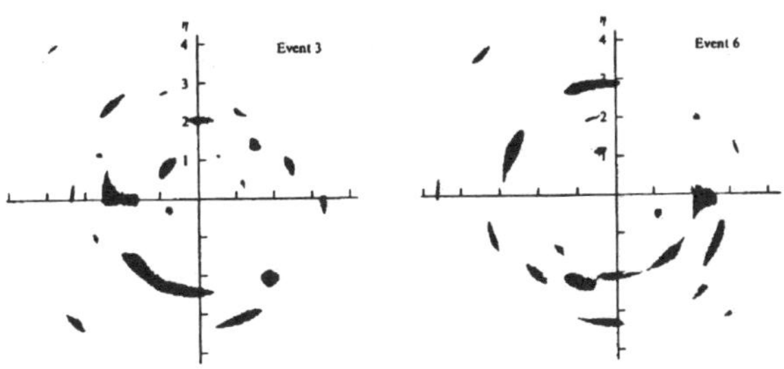

Fig. 5

Figure 1. The pattern in experimental convection (from [16]).

Figure 2. The symmetry and imperfections in a hexagonal Benard convection cell (from [17]).

Figure 3. The original pattern (b) is wavelet analyzed and (a) the regions of high wave vectors ($k > k_B$) are marked in white, (c) the wave-number histogram computed with the wavelet algorithm (solid line) is more precise than that with the Fourier transform (dotted line); k_B is the wave number separating regions with straight rolls stable ($k > k_B$) and unstable to bending, (d) the correlations of the pattern along various cross sections are shown to be similar (from [6]).

Figure 4. The factorial moments F_q of multiplicity distributions in e^+e^--collisions at Z^0 energy show the mono- and multifractal behavior (linear and curved parts of the lines, correspondingly) of particles pattern in the available phase space (from [18]). Increasing z corresponds to smaller bins. Experimental data are shown by dots. Different lines correspond to different calculations within some approximations of QCD.

Figure 5. The ring-like structure of some events of Pb-Pb interaction at 158 GeV/c (from [15]) revealed by the wavelet analysis demonstrates the symmetry otherwise hidden in a huge background.

References

1. Meyer Y., *Wavelets: Algorithms and Applications*, Philadelphia: SIAM, (1993).
2. Daubechies I., *Ten Lectures on Wavelets*, Philadelphia: SIAM, (1991).
3. Dremin I. M., Ivanov A. V. and Nechitailo V. A.. *UFN* **171**, 465 (2001); *Physics-Uspekhi* **44**, (May 2001).
4. Feshbach H., *MIT-CTP-2192*, (1993).
5. De Wolf E., Dremin I. M., and Kittel W. *Phys. Rep.* **270**, 1 (1996).
6. Bowman C., and Newell A. C., *Rev. Mod. Phys.* **70**, 289 (1998).
7. Dremin I. M. and Gary J. W., *Phys. Rep.* **349**, 301 (2001).
8. Dokshitzer Yu. L., and Dremin I. M., *Nucl. Phys.* **B 402**, 139 (1993).
9. Ochs W., and Wosiek J., *Phys. Lett.* **B 289**, 159 (1992); *ibid.* **B 304**, 144 (1993).
10. Brax Ph., Meunier J. L., and Peschanski R., *Z. Phys.* **C 62**, 649 (1994).
11. Thurner S., Feurstein M. C., and Teich M. C., *Phys. Rev. Lett.* **80**, 1544 (1998).
12. Amaral L. A. N., Goldberger A. L., Ivanov P. C., *et al.*, *Phys. Rev. Lett.* **81**, 2388 (1998).
13. Dremin I. M., Furletov V. I., Ivanov O. V., *et al*, *Control Engineering Practice* (2001), (to be published).
14. Dremin I. M., *JETP Lett* **30**, 140 (1979); *Yad Fiz* **33** 1357 (1981).
15. Dremin I. M., Ivanov O. V., Kalinin S. A., *et al.* *Phys Lett* **B 499**, 97 (2001).
16. Assenheimer M., and Steinberg V., *Nature* **367**, 345 (1994)
17. Koschmieder E. L., *Adv Chem Phys* **26**, 177 (1974).
18. L3 Collaboration, Acciari M., *et al.*, *Phys Lett* **B 428**, 186 (1998).

ON INTERPRETATIONS AND CONSTRUCTIONS OF CLASSICAL DYNAMICAL R-MATRICES

L. FEHÉR AND A. GÁBOR

Department of Theoretical Physics, University of Szeged
Tisza Lajos krt 84-86, H-6720 Szeged, Hungary
E-mail: lfeher@sol.cc.u-szeged.hu

In this note we complement recent results on the exchange r-matrices appearing in the chiral WZNW model by providing a direct, purely finite-dimensional description of the relationship between the monodromy dependent 2-form that enters the chiral WZNW symplectic form and the exchange r-matrix that governs the corresponding Poisson brackets. We also develop the special case in which the exchange r-matrix becomes the 'canonical' solution of the classical dynamical Yang-Baxter equation on an arbitrary self-dual Lie algebra.

1 Introduction

Let G be a connected (real or complex) Lie group whose Lie algebra $\mathcal{G}$ is self-dual in the sense of admitting a nondegenerate invariant scalar product $\langle\ ,\ \rangle$. The corresponding chiral WZNW phase space consists of G-valued quasiperiodic fields on the real line, $g(x+2\pi) = g(x)M$ ($\forall x \in \mathbf{R}$), where $M \in G$ is the 'monodromy matrix'. After restricting M to some open submanifold $\check{G} \subset G$ (see below), this phase space is equipped [1,2] with a Poisson structure that can be symbolically described as follows:

$$\kappa\,\{g(x) \overset{\otimes}{,} g(y)\} = (g(x)\otimes g(y))\left(\hat{r}(M) + \frac{1}{2}\hat{I}\,\mathrm{sign}\,(y-x)\right),\quad 0 < x, y < 2\pi \qquad (1)$$

where $\hat{r} : \check{G} \to \mathcal{G}\otimes\mathcal{G}$, that is $\hat{r}(M) = r^{\alpha\beta}(M) T_\alpha \otimes T_\beta$, is an *antisymmetric* 'exchange r-matrix', and $\hat{I} = T_\alpha \otimes T^\alpha$ using dual bases $\{T_\alpha\}$, $\{T^\beta\}$ of $\mathcal{G}$; κ is some constant. The Jacobi identity of the Poisson bracket (1) is equivalent to a certain dynamical generalization of the (modified) classical Yang-Baxter equation on the exchange r-matrix, which we call the 'G-CDYBE'. If for any function ψ on G we introduce

$$(\mathcal{L}_\alpha\psi)(M) := \frac{d}{dt}\psi(e^{tT_\alpha}M)\Big|_{t=0}, \qquad (\mathcal{R}_\alpha\psi)(M) := \frac{d}{dt}\psi(Me^{tT_\alpha})\Big|_{t=0}, \qquad (2)$$

then the G-CDYBE [2] is given as follows:

$$[\hat{r}_{12}, \hat{r}_{23}] + T_1^\alpha\big(\frac{1}{2}\mathcal{D}_\alpha^+ + r_\alpha{}^\beta\mathcal{D}_\beta^-\big)\hat{r}_{23} + \text{cycl. perm.} = -\frac{1}{4}\hat{f}. \qquad (3)$$

Here $\mathcal{D}^{\pm}_{\alpha} := \mathcal{R}_{\alpha} \pm \mathcal{L}_{\alpha}$, the cyclic permutations act on the three tensorial factors as usual, and $\hat{f} = f_{\alpha\beta\gamma} T^{\alpha} \otimes T^{\beta} \otimes T^{\gamma}$ with the structure constants of $\mathcal{G}$. The Poisson structure (1) arises by inverting [2] a (weak) symplectic form on the chiral WZNW phase space, which is defined [1] with the aid of a 2-form ρ on $\check{G}$ whose exterior derivative coincides with the restriction of the canonical 3-form on G. In this context, the restriction to some $\check{G} \subset G$ is necessary since the canonical 3-form on G is closed but not exact. In the next section we provide a self-contained *finite-dimensional* characterization of the relationship between the 2-form ρ and the exchange r-matrix $\hat{r}$. This relationship has previously been established [2] by indirect arguments relying on the properties of the *infinite-dimensional* chiral WZNW phase space.

2 The 2-form ρ and the exchange r-matrix

For any (real or complex) manifold Q, denote by $\mathcal{F}(G,Q)$ the set of (smooth or holomorphic) maps from G to Q, and let $\mathcal{F}(G)$ be the set of (smooth or holomorphic) scalar functions on G. Let $\bar{M} \in \mathcal{F}(G, \mathrm{End}(\mathcal{G}))$ be the map

$$\bar{M} : M \mapsto \mathrm{Ad}\, M \qquad \forall M \in G. \tag{4}$$

With any $X \in \mathcal{F}(G,\mathcal{G})$ associate the vector field $\mathcal{R}_X$ on G by the definition

$$(\mathcal{R}_X \psi)(M) := \frac{d}{dt}\psi(M e^{tX(M)})\Big|_{t=0}, \qquad \forall M \in G, \quad \psi \in \mathcal{F}(G). \tag{5}$$

Let $\check{G}$ be an open domain in G and choose a function $r \in \mathcal{F}(\check{G}, \mathrm{End}(\mathcal{G}))$. Define $r_{\pm} := r \pm \frac{1}{2} I$ and introduce $\mathcal{A} \in \mathcal{F}(\check{G}, \mathrm{End}(\mathcal{G}))$ as

$$\mathcal{A} := (\bar{M}^{-1} \circ r_{+} - r_{-}). \tag{6}$$

With these notations, the G-CDYBE (3) can be rewritten as follows:

$$\langle [r\xi, r\eta], \zeta\rangle - \langle \mathcal{R}_{\mathcal{A}(\xi)} r\eta, \zeta\rangle + \text{ c.p.} = -\frac{1}{4}\langle [\xi,\eta],\zeta\rangle \qquad \forall \xi,\eta,\zeta \in \mathcal{G}, \tag{7}$$

where the cyclic permutations act on ξ, η, ζ.

Consider the canonical 3-form Φ on G,

$$\Phi(\mathcal{R}_X, \mathcal{R}_Y, \mathcal{R}_Z) := \frac{1}{2}\langle [X,Y], Z\rangle, \tag{8}$$

where $[X,Y](M) := [X(M), Y(M)]$. Let $q \in \mathcal{F}(\check{G}, \mathrm{End}(\mathcal{G}))$ be a map for which $q(M)$ is antisymmetric with respect to $\langle\ ,\ \rangle$ for any $M \in \check{G}$. With such a map q associate a 2-form ρ on $\check{G}$ by

$$\rho(\mathcal{R}_X, \mathcal{R}_Y) = \langle X, qY\rangle \qquad \forall X, Y \in \mathcal{F}(\check{G}, \mathcal{G}). \tag{9}$$

Proposition 1. *Suppose that $d\rho = \Phi$ on an open submanifold $\check{G} \subset G$ and*

$$\det\left(q_+(M) - q_-(M) \circ \mathrm{Ad}\,(M^{-1})\right) \neq 0 \qquad \forall M \in \check{G}, \tag{10}$$

where $q_\pm := q \pm \frac{1}{2}I$. Then

$$r := \frac{1}{2}\left(q_+ - q_- \circ \bar{M}^{-1}\right)^{-1} \circ \left(q_+ + q_- \circ \bar{M}^{-1}\right) \tag{11}$$

solves the G-CDYBE (7) on $\check{G}$.

Remark. In fact, under the assumption (10) formula (11) gives the unique solution of the factorization equation

$$q_+ \circ r_- = q_- \circ \bar{M}^{-1} \circ r_+ \tag{12}$$

for r, and one also has the equivalent formula

$$r_- = -q_- \circ \left(q_- - \bar{M} \circ q_+\right)^{-1}. \tag{13}$$

Proof of proposition 1. We start by noting that, as a special case of a standard identity between the Lie and the exterior derivatives, any 2-form ρ on G (or $\check{G}$) satisfies

$$(d\rho)(\mathcal{R}_X, \mathcal{R}_Y, \mathcal{R}_Z) = \mathcal{R}_X(\rho(\mathcal{R}_Y, \mathcal{R}_Z)) - \rho([\mathcal{R}_X, \mathcal{R}_Y], \mathcal{R}_Z) + \text{c.p.}, \tag{14}$$

where the cyclic permutations act on X, Y, Z. The Lie bracket of the vector fields $\mathcal{R}_X, \mathcal{R}_Y$ is given by

$$[\mathcal{R}_X, \mathcal{R}_Y] = \mathcal{R}_{\mathcal{B}(X,Y)} \quad \text{with} \quad \mathcal{B}(X,Y) = [X,Y] + \mathcal{R}_X Y - \mathcal{R}_Y X. \tag{15}$$

By using the Leibniz rule to evaluate the first term of (14) and inserting (15) into (14), (9) implies that

$$(d\rho)(\mathcal{R}_X, \mathcal{R}_Y, \mathcal{R}_Z) = \langle q[X,Y], Z\rangle - \langle (\mathcal{R}_X q)Y, Z\rangle + \text{c.p.} \tag{16}$$

Now taking into account the assumption $d\rho = \Phi$, we obtain the identity

$$-\langle (\mathcal{R}_X q)Y, Z\rangle + \text{c.p} = \langle [X,Y], qZ\rangle + \frac{1}{6}\langle [X,Y], Z\rangle + \text{c.p.} \tag{17}$$

$\forall X, Y, Z \in \mathcal{F}(\check{G}, \mathcal{G})$. Then a further important identity is

$$\langle \mathcal{R}_{\mathcal{A}(\xi)} r\eta, \zeta\rangle = \langle (\mathcal{R}_{\mathcal{A}(\xi)} q)\mathcal{A}(\eta), \mathcal{A}(\zeta)\rangle - \langle (\bar{M} \circ r_- - r_+)\xi, [r_+\eta, r_+\zeta]\rangle \tag{18}$$

$\forall \xi, \eta, \zeta \in \mathcal{G}$. One can derive this by calculating the derivative $\mathcal{R}_{\mathcal{A}(\xi)} r$ from the formula (13), and using many times (12), the invariance of $\langle\ ,\ \rangle$, and the antisymmetry of r and q. The identities (17), (18) allow us to express the derivative term in the G-CDYBE (7) in terms of non-derivative terms containing r. More precisely, we need one more identity to do this,

$$\langle[\mathcal{A}(\xi),\mathcal{A}(\eta)],q\mathcal{A}(\zeta)\rangle=\frac{1}{2}\langle[\mathcal{A}(\xi),\mathcal{A}(\eta)],\mathcal{A}(\zeta)\rangle+\langle[\mathcal{A}(\xi),\mathcal{A}(\eta)],r_-\zeta\rangle,\qquad(19)$$

which follows from the definition of $\mathcal{A}$ (6) and (13). With the last three identities in hand, the G-CDYBE (7) can be verified straightforwardly as

$$\begin{aligned}
\langle[r\xi,r\eta],\zeta\rangle-\langle\mathcal{R}_{\mathcal{A}(\xi)}r\eta,\zeta\rangle+\text{c.p.}&=\langle[r\xi,r\eta],\zeta\rangle-\langle(\mathcal{R}_{\mathcal{A}(\xi)}q)\mathcal{A}(\eta),\mathcal{A}(\zeta)\rangle\\
&\quad+\langle(\bar{M}\circ r_--r_+)\xi,[r_+\eta,r_+\zeta]\rangle+\text{c.p.}\\
=\langle[r\xi,r\eta],\zeta\rangle&+\langle(\bar{M}\circ r_--r_+)\xi,[r_+\eta,r_+\zeta]\rangle\\
&+\langle[\mathcal{A}(\xi),\mathcal{A}(\eta)],q\mathcal{A}(\zeta)\rangle+\frac{1}{6}\langle[\mathcal{A}(\xi),\mathcal{A}(\eta)],\mathcal{A}(\zeta)\rangle+\text{c.p.}\\
=\langle[r\xi,r\eta],\zeta\rangle&+\langle[\bar{M}^{-1}\circ r_+\xi,\bar{M}^{-1}\circ r_+\eta],r_-\zeta\rangle\\
&-\langle[\bar{M}^{-1}\circ r_+\xi,\bar{M}^{-1}\circ r_+\eta],\bar{M}^{-1}\circ r_+\zeta\rangle\\
&+\langle[\mathcal{A}(\xi),\mathcal{A}(\eta)],r_-\zeta\rangle+\frac{2}{3}\langle[\mathcal{A}(\xi),\mathcal{A}(\eta)],\mathcal{A}(\zeta)\rangle+\text{c.p.}\\
=\langle[r\xi,r\eta],\zeta\rangle&-\frac{1}{3}\langle[\bar{M}^{-1}\circ r_+\xi,\bar{M}^{-1}\circ r_+\eta],\bar{M}^{-1}\circ r_+\zeta\rangle\\
&+\frac{1}{3}\langle[r_-\xi,r_-\eta],r_-\zeta\rangle+\text{c.p.}=-\frac{1}{4}\langle[\xi,\eta],\zeta\rangle,
\end{aligned}\qquad(20)$$

which proves the proposition. *Q.E.D.*

Proposition 2. *Suppose that $r\in\mathcal{F}(\check{G},\mathrm{End}(\mathcal{G}))$ solves the G-CDYBE (7) on an open submanifold $\check{G}\subset G$ and*

$$\det\big(\mathrm{Ad}\,(M^{-1})\circ r_+(M)-r_-(M)\big)\neq0\qquad\forall M\in\check{G}.\qquad(21)$$

Then the 2-form ρ defined by (9) with

$$q:=\frac{1}{2}\left(\bar{M}^{-1}\circ r_++r_-\right)\circ\left(\bar{M}^{-1}\circ r_+-r_-\right)^{-1}\qquad(22)$$

satisfies $d\rho=\Phi$ on $\check{G}$.

Proof. Consider the equivalent formula of the q

$$q_-=r_-\circ\left(\bar{M}^{-1}\circ r_+-r_-\right)^{-1}=r_-\circ\mathcal{A}^{-1},\qquad(23)$$

where $\mathcal{A}$ (6) is invertible due to (21). By taking the derivative of this equation one can show that relation (18) is valid in this case as well. By combining

(18) with (7) we obtain that

$$
\begin{aligned}
&\langle(\mathcal{R}_{\mathcal{A}(\xi)}q)\mathcal{A}(\eta),\mathcal{A}(\zeta)\rangle + \text{c.p.} = \\
&= \langle[r\xi, r\eta],\zeta\rangle + \frac{1}{12}\langle[\xi,\eta],\zeta\rangle - \langle(r_+ - \bar{M}\circ r_-)\xi,[r_+\eta, r_+\zeta]\rangle + \text{c.p.} \\
&= \langle q[\mathcal{A}(\xi),\mathcal{A}(\eta)],\mathcal{A}(\zeta)\rangle - \frac{1}{6}\langle[\mathcal{A}(\xi),\mathcal{A}(\eta)],\mathcal{A}(\zeta)\rangle + \text{c.p.}
\end{aligned} \tag{24}
$$

$\forall \xi,\eta,\zeta\in\mathcal{G}$. The second equality here is derived by using several times (23) and the corresponding equation for q_+, i.e., $q_+ = \bar{M}^{-1}\circ r_+\circ\mathcal{A}^{-1}$. By (16) and (8), this allows us to conclude that

$$
(d\rho)(\mathcal{R}_{\mathcal{A}(\xi)},\mathcal{R}_{\mathcal{A}(\eta)},\mathcal{R}_{\mathcal{A}(\zeta)}) = \Phi(\mathcal{R}_{\mathcal{A}(\xi)},\mathcal{R}_{\mathcal{A}(\eta)},\mathcal{R}_{\mathcal{A}(\zeta)}). \tag{25}
$$

Since the correspondence $\xi\mapsto\mathcal{A}(\xi)$ is invertible due to (21), this implies that $d\rho=\Phi$ holds. *Q.E.D.*

3 Recovering a canonical solution of the CDYBE

Let us now suppose that $\check{G}$ is diffeomorphic to a domain $\check{\mathcal{G}}\subset\mathcal{G}$ by the exponential parametrization, whereby we write $\check{G}\ni M = e^{\omega}$ with $\omega\in\check{\mathcal{G}}$. Then choose [2] the 2-form ρ on $\check{G}\simeq\check{\mathcal{G}}$ to be

$$
\rho(\omega) := -\frac{1}{2}\int_0^1 dx\langle d\omega \overset{\wedge}{,} de^{x\omega}e^{-x\omega}\rangle. \tag{26}
$$

It is not difficult to check that $d\rho=\Phi$ holds and one can also calculate the explicit form of the operator $q(\omega)$ that corresponds to ρ in (26) by means of (9). Let $\mathcal{Q}$ and $\mathcal{R}$ denote the complex meromorphic functions

$$
\mathcal{Q}(z) = \frac{2z + e^{-z} - e^{z}}{2(e^z-1)(1-e^{-z})}, \qquad \mathcal{R}(z) := \frac{1}{2}\coth\frac{z}{2} - \frac{1}{z}, \tag{27}
$$

which are regular around $z=0$. In fact, (26) implies the formula

$$
q(\omega) = \mathcal{Q}(\operatorname{ad}\omega), \tag{28}
$$

where the right hand side is defined by means of the power series expansion of the function $\mathcal{Q}$ around $z=0$ if ω is near to the origin. By inserting this into equation (12) that determines r in terms of q, one finds that the exchange r-matrix associated with the 2-form ρ in (26) is given by

$$
r(\omega) = \mathcal{R}(\operatorname{ad}\omega). \tag{29}
$$

Consider the holomorphic complex function

$$
h : z \mapsto \frac{e^z - 1}{z}, \tag{30}
$$

and recall that for a curve $t \mapsto A(t)$ of finite-dimensional linear operators

$$\frac{de^{\pm A(t)}}{dt} = \pm e^{\pm A(t)} h(\mp \mathrm{ad}_{A(t)})(\dot{A}(t)), \qquad \dot{A}(t) := \frac{dA(t)}{dt}. \tag{31}$$

The right hand side of this equation is defined by means of the Taylor expansion of $h(z)$ around $z = 0$, and $\mathrm{ad}_{A(t)}(\dot{A}(t)) = [A(t), \dot{A}(t)]$. By using relation (31) it is not difficult to derive (28). Relation (31) also implies that the r-matrix in (29) satisfies the following identity:

$$\left(\frac{1}{2}\mathcal{D}_\alpha^+ + r_\alpha{}^\beta \mathcal{D}_\beta^-\right) = \frac{\partial}{\partial \omega^\alpha}, \tag{32}$$

where we use that $M = e^\omega$ and $\mathcal{D}_\alpha^\pm = (\mathcal{R}_\alpha \pm \mathcal{L}_\alpha)$ are given by (2). Incidentally, requiring (32) for *the ansatz* $r(\omega) = \mathcal{R}(\mathrm{ad}\,\omega)$ with some odd function $\mathcal{R}(z)$ that is holomorphic around 0 leads uniquely to the function $\mathcal{R}(z)$ in (27).

As a consequence of proposition 1, we know that $r(\omega) = \mathcal{R}(\mathrm{ad}\,\omega)$ with (27) satisfies the G-CDYBE (3). We now notice from (32) that *for this particular r-matrix* (3) becomes the standard CDYBE [3] on the Lie algebra $\mathcal{G}$:

$$\left[\hat{r}_{12}, \hat{r}_{23}\right] + T_1^\alpha \frac{\partial}{\partial \omega^\alpha} \hat{r}_{23} + \text{cycl. perm.} = -\frac{1}{4}\hat{f}. \tag{33}$$

We call the r-matrix in (29) the 'canonical' solution of the CDYBE (33) on $\mathcal{G}$ since many solutions of the CDYBE with respect to subalgebras $\mathcal{H} \subset \mathcal{G}$ can be derived from it by applying Dirac reduction [4] to the dynamical variable ω. The above proof of the statement that the canonical r-matrix solves (33) was extracted from the study of the WZNW model, where it arose in a natural manner [2]. For different proofs of the same result, and other applications of the canonical r-matrix and its reductions, we refer to the literature [3,5,6].

Acknowledgments

This investigation was supported in part by the Hungarian Scientific Research Fund (OTKA) under T034170 and M028418.

References

1. F. Falceto and K. Gawędzki, *J. Geom. Phys.* **11**, 251 (1993).
2. J. Balog, L. Fehér and L. Palla, *Nucl. Phys.* B **568**, 503 (2000).
3. P. Etingof and A. Varchenko, *Commun. Math. Phys.* **192**, 77 (1998).
4. L. Fehér, A. Gábor and B.G. Pusztai, *J. Phys.* A **34**, 7235 (2001).
5. A. Alekseev and E. Meinrenken, *Invent. Math.* **139**, 135 (2000).
6. B.G. Pusztai and L. Fehér, preprint *math.QA/0109082*.

DECOUPLING OF TRANSLATIONS FROM HOMOGENEOUS TRANSFORMATIONS IN INHOMOGENEOUS QUANTUM GROUPS

GAETANO FIORE

Dip. di Matematica e Applicazioni, V. Claudio 21, 80125 Napoli, Italy;
I.N.F.N., Sezione di Napoli, Complesso MSA, V. Cintia, 80126 Napoli, Italy
E-mail: gaetano.fiore@na.infn.it

We briefly report on our result that, if there exists a realization of a Hopf algebra H in a H-module algebra $\mathcal{A}$, then their cross-product is equal to the product of $\mathcal{A}$ itself with a subalgebra isomorphic to H and *commuting* with $\mathcal{A}$. We illustrate its application to the Euclidean quantum groups in $N \geq 3$ dimensions.

1 Introduction and main results

Let H be a Hopf algebra and $\mathcal{A}$ a unital H-module algebra. Here for the sake of brevity we stick to consider the case of a right H-module algebra. Then by definition there exists a bilinear map $\triangleleft : (a,g) \in \mathcal{A} \times H \to a \triangleleft g \in \mathcal{A}$, called a right action, such that

$$a \triangleleft (gg') = (a \triangleleft g) \triangleleft g' \tag{1}$$

$$(aa') \triangleleft g = (a \triangleleft g_{(1)})\,(a' \triangleleft g_{(2)}). \tag{2}$$

We use a Sweedler-type notation with suppressed summation sign for the coproduct $\Delta(g)$ of g, namely the short-hand notation $\Delta(g) = g_{(1)} \otimes g_{(2)}$. The cross-product algebra $\mathcal{A} \rtimes H$ is $H \otimes \mathcal{A}$ as a vector space, whereas the product of two elements is given by

$$(g \otimes a)(g' \otimes a') = gg'_{(1)} \otimes (a \triangleleft g'_{(2)})a' \tag{3}$$

for any $a, a' \in \mathcal{A}$, $g, g' \in H$. To simplify the notation, in the sequel we denote $g \otimes a$ by ga and omit either unit $\mathbf{1}_{\mathcal{A}}, \mathbf{1}_H$ whenever multiplied by non-unit elements; consequently, for $g = \mathbf{1}_H$, $a' = \mathbf{1}_{\mathcal{A}}$ (3) takes the form

$$ag' = g'_{(1)}\,(a \triangleleft g'_{(2)}). \tag{4}$$

$\mathcal{A} \rtimes H$ itself is a H-module algebra under the right action $\triangleleft$ of H if we extend the latter on the elements of H as the adjoint action,

$$h \triangleleft g = Sg_{(1)} h g_{(2)}, \qquad\qquad g, h \in H. \tag{5}$$

One very important situation in which one encounters cross products is in inhomogeneous (quantum) groups, that are Hopf algebras whose algebras are cross-products $\mathcal{A}\rtimes H$, where H is the Hopf subalgebra of "homogeneous transformations" and $\mathcal{A}$ the (braided) Hopf subalgebra of "translations".

In Ref. [3] we have found a sufficient condition for $\mathcal{A}\rtimes H$ to be equal to a product $\mathcal{A}H'$, where $H' \subset H\ltimes\mathcal{A}$ is a subalgebra isomorphic to H and *commuting* with $\mathcal{A}$: it is sufficient to assume that there exists an algebra map

$$\tilde{\varphi} : \mathcal{A}\rtimes H \to \mathcal{A} \tag{6}$$

acting as the identity on the subalgebra $\mathcal{A}$, namely for any $a \in \mathcal{A}$

$$\tilde{\varphi}(a) = a. \tag{7}$$

Here we briefly recall this result, as well as a similar one requiring a weaker assumption, namely that H admits a Gauss decomposition into two Hopf subalgebras H^+, H^- for each of which analogous maps $\tilde{\varphi}^+, \tilde{\varphi}^-$ exist. In the last section we illustrate the application of these general results to a class of inhomogeneous quantum groups, the Euclidean quantum groups in $N \geq 3$ dimensions, for which $\mathcal{A}$ is ("the algebra of functions on") the quantum Euclidean N-dimensional space; the corresponding maps $\tilde{\varphi}, \tilde{\varphi}^+, \tilde{\varphi}^-$ have been determined in [1] and further studied in [4]. [Note that no such maps can exist for the (undeformed) Euclidean algebra, for which $\mathcal{A} \equiv$the algebra of functions on $\mathbf{R}^N$, which is abelian]. Other applications include $U_q\mathbf{g}$ -covariant Heisenberg (or Clifford) algebras and the q-deformed 2-dimensional fuzzy sphere $S^2_{q,M}$ [5].

Let $\tilde{\mathcal{C}}$ be the commutant of $\mathcal{A}$ within $\mathcal{A}\rtimes H$, i.e. the subalgebra

$$\tilde{\mathcal{C}} := \{c \in \mathcal{A}\rtimes H \mid [c,a] = 0 \quad \forall a \in \mathcal{A}\}. \tag{8}$$

Clearly $\tilde{\mathcal{C}}$ contains the center $\mathcal{Z}(\mathcal{A})$ of $\mathcal{A}$. Let $\tilde{\zeta} : H \to \mathcal{A}\rtimes H$ be the map defined by

$$\tilde{\zeta}(g) := g_{(1)})\tilde{\varphi}(Sg_{(2)}). \tag{9}$$

Note that if we apply $\tilde{\varphi}$ to $\tilde{\zeta}$ and recall that $\tilde{\varphi}$ is both a homomorphism and idempotent we find $\tilde{\varphi} \circ \tilde{\zeta} = \varepsilon$. Now, $\varepsilon(g)$ is a complex number times $\mathbf{1}_{\mathcal{A}}$ and therefore trivially commutes with $\mathcal{A}$. Since by definition the commutation relations between $\mathcal{A}$ and either H or $\tilde{\varphi}(H)$ are the same, we expect that also $\tilde{\zeta}(g)$ commutes with $\mathcal{A}$. This is confirmed by

Theorem 1 [3] *Under the above assumptions the map $\tilde{\zeta}$ is an injective algebra homomorphism $\tilde{\zeta} : H \to \tilde{\mathcal{C}}$; moreover $\tilde{\mathcal{C}} = \tilde{\zeta}(H)\,\mathcal{Z}(\mathcal{A})$ and $\mathcal{A}\rtimes H = \tilde{\zeta}(H)\,\mathcal{A}$. If, in particular $\mathcal{Z}(\mathcal{A}) = \mathbf{C}$, then $\tilde{\mathcal{C}} = \zeta(H)$ and $\tilde{\zeta} : H \leftrightarrow \tilde{\mathcal{C}}$ is an algebra isomorphism. Moreover, the center of the cross-product $\mathcal{A}\rtimes H$ is given by $\mathcal{Z}(\mathcal{A}\rtimes H) = \mathcal{Z}(\mathcal{A})\,\tilde{\zeta}\,(\mathcal{Z}(H))$. Finally, if $H_c, \mathcal{A}_c$ are maximal abelian subalgebras of H and $\mathcal{A}$ respectively, then $\mathcal{A}_c\,\tilde{\zeta}(H_c)$ is a maximal abelian subalgebra of $\mathcal{A}\rtimes H$.*

In other words, the subalgebra H' looked for will be obtained by setting $H' := \tilde{\zeta}(H)$. For these reasons we shall call $\tilde{\zeta}$, as well as the other maps $\tilde{\zeta}^{\pm}$ which we shall introduce below, *decoupling maps.*

To introduce the latter we just need a weaker assumption, namely that, instead of a map $\tilde{\varphi}$, we just have at our disposal two homomorphisms $\tilde{\varphi}^{+}, \tilde{\varphi}^{-}$

$$\tilde{\varphi}^{\pm} : H^{\pm} \ltimes \mathcal{A} \to \mathcal{A} \tag{10}$$

fulfilling (7), where H^{+}, H^{-} denote two Hopf subalgebras of H such that Gauss decompositions $H = H^{+}H^{-} = H^{-}H^{+}$ hold. (The typical case is when $H = U_q\mathbf{g}$ and H^{+}, H^{-} denote its positive and negative Borel subalgebras.) Then Theorem 1 will apply separately to $\mathcal{A}{>\!\!\!\triangleleft} H^{+}$ and $\mathcal{A}{>\!\!\!\triangleleft} H^{-}$, if we define corresponding maps $\tilde{\zeta}^{\pm} : H^{\pm} \to \mathcal{A}$ by

$$\tilde{\zeta}^{\pm}(g) := g_{(1)}\tilde{\varphi}^{\pm}(Sg_{(2)}), \tag{11}$$

where $g \in H^{\pm}$ respectively. What can we say about the whole $H \ltimes \mathcal{A}$?

Theorem 2 [3] *Under the above assumptions formulae (11) define injective algebra homomorphisms $\tilde{\zeta}^{\pm} : H^{\pm} \to \tilde{\mathcal{C}}$. Moreover,*

$$\tilde{\mathcal{C}} = \tilde{\zeta}^{+}(H^{+})\, \tilde{\zeta}^{-}(H^{-})\, \mathcal{Z}(\mathcal{A}) = \tilde{\zeta}^{-}(H^{-})\, \tilde{\zeta}^{+}(H^{+})\, \mathcal{Z}(\mathcal{A}) \tag{12}$$

and

$$\mathcal{A}{>\!\!\!\triangleleft} H = \tilde{\zeta}^{+}(H^{+})\, \tilde{\zeta}^{-}(H^{-})\, \mathcal{A} = \tilde{\zeta}^{-}(H^{-})\, \tilde{\zeta}^{+}(H^{+})\, \mathcal{A}. \tag{13}$$

In particular, if $\mathcal{Z}(\mathcal{A}) = \mathbf{C}$, then $\tilde{\mathcal{C}} = \tilde{\zeta}^{+}(H^{+})\tilde{\zeta}^{-}(H^{-}) = \tilde{\zeta}^{-}(H^{-})\tilde{\zeta}^{+}(H^{+})$. Furthermore, any element c of the center $\mathcal{Z}(\mathcal{A}{>\!\!\!\triangleleft} H)$ of the cross-product $\mathcal{A}{>\!\!\!\triangleleft} H$ can be expressed in the form

$$c = \tilde{\zeta}^{+}\left(c^{(1)}\right) \tilde{\zeta}^{-}\left(c^{(2)}\right) c^{(3)}, \tag{14}$$

where $c^{(1)} \otimes c^{(2)} \otimes c^{(3)} \in H^{+} \otimes H^{-} \otimes \mathcal{Z}(\mathcal{A})$ and $c^{(1)}c^{(2)} \otimes c^{(3)} \in \mathcal{Z}(H) \otimes \mathcal{Z}(\mathcal{A})$; vice versa any such object c is an element of $\mathcal{Z}(\mathcal{A}{>\!\!\!\triangleleft} H)$. Finally, if $H_c \subset H^{+} \cap H^{-}$ and $\mathcal{A}_c$ are maximal abelian subalgebras of H and $\mathcal{A}$ respectively, then $\mathcal{A}_c\, \tilde{\zeta}^{+}(H_c)$ (as well as $\mathcal{A}_c\, \tilde{\zeta}^{-}(H_c)$) is a maximal abelian subalgebra of $\mathcal{A}{>\!\!\!\triangleleft} H$.

As a consequence of this theorem, for any $g^{+} \in H^{+}$, $g^{-} \in H^{-}$ there exists a sum $c^{(1)} \otimes c^{(2)} \otimes c^{(3)} \in \mathcal{Z}(\mathcal{A}) \otimes H^{-} \otimes H^{+}$ (depending on g^{+}, g^{-}) such that

$$\tilde{\zeta}^{+}(g^{+})\tilde{\zeta}^{-}(g^{-}) = c^{(1)}\tilde{\zeta}^{-}(c^{(2)})\tilde{\zeta}^{+}(c^{(3)}). \tag{15}$$

These will be the "commutation relations" between elements of $\tilde{\zeta}^{+}(H^{+})$ and $\tilde{\zeta}^{-}(H^{-})$. Their form will depend on the specific algebras considered.

Assume that H is a Hopf $*$-algebra and $\mathcal{A}$ a H-module $*$-algebra. Then, as known, these two $*$-structures can be glued in a unique one to make $\mathcal{A}{>\!\!\!\triangleleft} H$

a $*$-algebra itself. What can we say about the behaviour of the decoupling maps under the latter $*$-structure?

Proposition 1 [3] *If* $\tilde{\varphi}: \mathcal{A} \rtimes H \to \mathcal{A}$ *is a* $*$*-homomorphism, then also the map* $\tilde{\zeta}: H \to \tilde{\mathcal{C}}$ *is. If* $\tilde{\varphi}^{\pm}$ *are* $*$*-homomorphism, then also the map* $\tilde{\zeta}^{\pm}: H^{\pm} \to \tilde{\mathcal{C}}$ *are. If* $\tilde{\varphi}^{\pm}$ *fulfill the condition* $\tilde{\varphi}^{\pm}(\alpha^*) = [\tilde{\varphi}^{\mp}(\alpha)]^*$ *for any* $\alpha \in H^{\mp} \ltimes \mathcal{A}$*, then* $\tilde{\zeta}^{\pm}$ *fulfill*

$$\tilde{\zeta}^{\pm}(g^*) = [\tilde{\zeta}^{\mp}(g)]^*, \qquad g \in H^{\mp}. \tag{16}$$

2 Application to the Euclidean quantum group $\mathbf{R}_q^N \rtimes U_q so(N)$

As an algebra $\mathcal{A}$ we shall consider a slight extension of the quantum Euclidean space $\mathbf{R}_q^N$ [2] (the $U_q so(N)$-covariant quantum space), i.e. of the unital associative algebra generated by p^i fulfilling the relations

$$\mathcal{P}_a{}^{ij}_{hk} p^h p^k = 0, \tag{17}$$

where $\mathcal{P}_a$ denotes the q-deformed antisymmetric projector (29), and the indices take the values $i = -n, \ldots, -1, 0, 1, \ldots n$ for N odd, and $i = -n, \ldots, -1, 1, \ldots n$ for N even; here $n := \left[\frac{N}{2}\right]$ is the rank of $so(N)$, . The multiplet (p^i) carries the fundamental N-dim (or vector) representation ρ of $U_q so(N)$: for any $g \in U_q so(N)$

$$p^i \triangleleft g = \rho^i_j(g) p^j. \tag{18}$$

As a set of generators of $H \equiv U_q so(N)$ it is convenient to introduce the FRT generators [2] $\mathcal{L}^{+i}{}_j, \mathcal{L}^{-i}{}_j$, together with the square roots of the diagonal elements $\mathcal{L}^{+j}{}_j, \mathcal{L}^{-j}{}_j$. The FRT generators and the braid matrix $\hat{R}$ are related to the so-called universal R-matrix $\mathcal{R} \in H \otimes H$ by

$$\mathcal{L}^{+a}{}_l := \mathcal{R}^{(1)} \rho^a_l(\mathcal{R}^{(2)}) \qquad \mathcal{L}^{-a}{}_l := \rho^a_l(\mathcal{R}^{-1(1)}) \mathcal{R}^{-1(2)} \tag{19}$$

and $\hat{R}^{ij}_{hk} = \rho^j_h(\mathcal{L}^{+i}{}_k) = (\rho^j_h \otimes \rho^i_k)(\mathcal{R})$. Since in our conventions $\mathcal{R} \in H^+ \otimes H^-$ (H^+, H^- denote the positive, negative Borel subalgebras) we see that $\mathcal{L}^{+a}{}_l \in H^+$ and $\mathcal{L}^{-a}{}_l \in H^-$.

To define $\tilde{\varphi}$ or $\tilde{\varphi}^{\pm}$ one [1] slightly enlarges $\mathbf{R}_q^N$ as follows. One introduces some new generators $\sqrt{P_a}$, with $1 \leq a \leq \frac{N}{2}$, together with their inverses $(\sqrt{P_a})^{-1}$, requiring that

$$P_a^2 = \sum_{h=-a}^{a} p^h p_h = \sum_{h,k=-a}^{a} g_{hk} p^h p^k. \tag{20}$$

In the previous equation g_{hk} denotes the 'metric matrix' of $SO_q(N)$:

$$g_{ij} = g^{ij} = q^{-\rho_i}\delta_{i,-j}. \tag{21}$$

It is a $SO_q(N)$-isotropic tensor and is a deformation of the ordinary Euclidean metric. We have introduced the notation [2] $(\rho_i) = (n-\frac{1}{2},\ldots,\frac{1}{2},0,-\frac{1}{2},\ldots,\frac{1}{2}-n)$ for N odd, $(n-1,\ldots,0,0,\ldots,1-n)$ for N even. In the sequel we shall call P_n^2 also P^2. Moreover for odd N we add also $\sqrt{p^0}$ and its inverse as new generators. The commutation relations involving these new generators can be fixed consistently, in particular one finds that

$$\sqrt{P_a}p^i = p^i\sqrt{P_a} \times \begin{cases} 1 \text{ if } |i| \leq a \\ \sqrt{q} \text{ if } i > a \\ 1/\sqrt{q} \text{ if } i < a \end{cases} \tag{22}$$

The center of $\mathbf{R}_q^N$ is generated by $\sqrt{P}$ and, only in the case of even N, by $\sqrt{p^1/p^{-1}}$ and its inverse $\sqrt{p^{-1}/p^1}$. In the case of even N one needs to include also the FRT generator $\mathcal{L}^{-}{}^{1}_{1} = \mathcal{L}^{+}{}^{-1}_{-1}$ and its inverse $\mathcal{L}^{+}{}^{1}_{1} = \mathcal{L}^{-}{}^{-1}_{-1}$ [which are generators of $U_qso(N)$ belonging to the natural Cartan subalgebra] among the generators of $\mathcal{A}$. They satisfy nontrivial commutation relations with the generators of $\mathcal{A}$, and the standard FRT relations with the rest of $U_qso(N)$. As a consequence, $\sqrt{p^{\pm 1}/p^{\mp 1}}$ are eliminated from the center of $\mathcal{A}$ (in fact they do not q-commute with $\mathcal{L}^{\pm}{}^{1}_{1}$).

The homomorphisms [1] $\tilde{\varphi}^{\pm} : \mathcal{A} \bowtie U_q^{\pm}so(N) \rightarrow \mathcal{A}$ take the simplest and most compact expression on the FRT generators of $U_q^{\pm}so(N)$. Let us introduce the short-hand notation $[A,B]_x = AB - xBA$. The images of $\tilde{\varphi}^{-}$ on the FRT generators read

$$\tilde{\varphi}^{-}(\mathcal{L}^{-}{}^{i}_{j}) = g^{ih}[\mu_h, p^k]_q g_{kj}, \tag{23}$$

$$\tilde{\varphi}^{+}(\mathcal{L}^{+}{}^{i}_{j}) = g^{ih}[\bar{\mu}_h, p^k]_{q^{-1}} g_{kj}, \tag{24}$$

where

$$\begin{array}{lll} \mu_0 = \gamma_0(p^0)^{-1} & \bar{\mu}_0 = \bar{\gamma}_0(p^0)^{-1} & \text{for } N \text{ odd}, \\ \mu_{\pm 1} = \gamma_{\pm 1}(p^{\pm 1})^{-1}\mathcal{L}^{\pm}{}^{1}_{1} & \bar{\mu}_{\pm 1} = \bar{\gamma}_{\pm 1}(p^{\pm 1})^{-1}\mathcal{L}^{\mp}{}^{1}_{1} & \text{for } N \text{ even}, \\ \mu_a = \gamma_a P_{|a|}^{-1}P_{|a|-1}^{-1}p^{-a} & \bar{\mu}_a = \bar{\gamma}_a P_{|a|}^{-1}P_{|a|-1}^{-1}p^{-a} & \text{otherwise}, \end{array} \tag{25}$$

and $\gamma_a, \bar{\gamma}_a \in \mathbf{C}$ are normalization constants fulfilling the conditions

$$\begin{array}{lll} \gamma_0 = -q^{-\frac{1}{2}}h^{-1} & \bar{\gamma}_0 = q^{\frac{1}{2}}h^{-1} & \text{for } N \text{ odd}, \\ \gamma_{\pm 1} = -k^{-1} & \bar{\gamma}_{\pm 1} = k^{-1} & \text{for } N \text{ even}, \\ \gamma_1\gamma_{-1} = -q^{-1}h^{-2} & \bar{\gamma}_1\bar{\gamma}_{-1} = -qh^{-2} & \text{for } N \text{ odd}, \\ \gamma_a\gamma_{-a} = -q^{-1}k^{-2}\omega_a\omega_{a-1} & \bar{\gamma}_a\bar{\gamma}_{-a} = -qk^{-2}\omega_a\omega_{a-1} & \text{for } a > 1. \end{array} \tag{26}$$

Here $h := q^{\frac{1}{2}} - q^{-\frac{1}{2}}$, $k := q - q^{-1}$, $\omega_a := (q^{\rho_a} + q^{-\rho_a})$. One can choose the γ_a's, $\bar{\gamma}_a$'s: 1) only for odd N, in such a way that $\tilde{\varphi}^-, \tilde{\varphi}^+$ can be "glued" into a unique homomorphism φ of the type (6), (7); 2) for $|q| = 1$ or $q \in \mathbf{R}^+$ respectively, in such a way that either condition necessary for the application of Proposition 1 is fulfilled. From definition (9), using (30), (23), (24), we find

$$\begin{aligned} \tilde{\zeta}^-(\mathcal{L}^-{}^i_j) &= \mathcal{L}^-{}^i_h \tilde{\varphi}^-(S\mathcal{L}^-{}^h_j) = \mathcal{L}^-{}^i_h[\mu_h, p^i]_q, \\ \tilde{\zeta}^+(\mathcal{L}^+{}^i_j) &= \mathcal{L}^+{}^i_h \tilde{\varphi}^+(S\mathcal{L}^+{}^h_j) = \mathcal{L}^+{}^i_h[\bar{\mu}_h, p^i]_{q^{-1}}. \end{aligned} \tag{27}$$

In Ref. [3] we have determined the commutation relations (15) between them.

Appendix

The braid matrix of $U_q so(N)$ is given by

$$\hat{R} = q \sum_{i \neq 0} e^i_i \otimes e^i_i + \sum_{\substack{i \neq j, -j \\ \text{or } i=j=0}} e^j_i \otimes e^i_j + q^{-1} \sum_{i \neq 0} e^{-i}_i \otimes e^i_{-i} \tag{28}$$

$$+ k\Big(\sum_{i<j} e^i_i \otimes e^j_j - \sum_{i<j} q^{-\rho_i + \rho_j} e^{-j}_i \otimes e^j_{-i}\Big).$$

It admits the orthogonal projector decomposition

$$\hat{R} = q\mathcal{P}_s - q^{-1}\mathcal{P}_a + q^{1-N}\mathcal{P}_t; \tag{29}$$

$\mathcal{P}_a, \mathcal{P}_t, \mathcal{P}_s$ are q-deformed antisymmetric, trace, trace-free symmetric projectors. The coproduct and antipode of the FRT generators are given by

$$\Delta(\mathcal{L}^{\pm}{}^i_j) = \mathcal{L}^{\pm}{}^i_h \otimes \mathcal{L}^{\pm}{}^h_j, \qquad S\mathcal{L}^{\mp}{}^j_i = g_{ih} \mathcal{L}^{\mp}{}^h_k g^{kj}. \tag{30}$$

References

1. B. L. Cerchiai, G. Fiore, J. Madore, *Commun. Math. Phys.* **217**, 521 (2001).
2. L.D. Faddeev, N.Y. Reshetikhin, L. Takhtadjan, *Algebra i Analiz* **1**, 178 (1989), translated from Russian in *Leningrad Math. J.* **1**, 193 (1990).
3. G. Fiore, *On the Decoupling of the Homogeneous and Inhomogeneous Parts in Inhomogeneous Quantum Groups*, Preprint 00-31 DMA, Università di Napoli, DSF/3-2001 and math.QA/0101218.
4. G. Fiore, H. Steinacker and J. Wess, "Unbraiding the braided tensor product", math/0007174.
5. H. Grosse, J. Madore, H. Steinacker, "Field Theory on the q–deformed Fuzzy Sphere", hep-th/0005273.

SYMMETRIES IN THE UNIVERSE AT LARGE SCALES

P. FLIN[1,2], K. BAJAN[3], W. GODLOWSKI[4]

1. Bogoliubov Laboratory of Theoretical Physics, Joint Institute for Nuclear Research, Dubna, Moscow Region 141980, Russia

2. Pedagogical University, Institute of Physics, 25-406 Kielce, ul.Swietokrzyska 15, Poland

3. H. Niewodniczanski Institute of Nuclear Physics, 31-342 Krakow, ul.Radzikowskiego 152, Poland

4. Astronomical Observatory of the Jagiellonian University, 30-244 Krakow, ul. Orla 171, Poland

1 Introduction

The existence of symmetry in the microworld manifests itselfs, among other things, through quantisation of some parameters describing physical quantities. There are no a priori counterindications as to the existence of quantised physical quantities in the macroworld. It should be stressed that such effects as wave and particle duality, characteristic for the microworld, were observed also outside the microworld, when interference of fulerons was detected. In the large-scale Universe the search for regularities is connected with testing whether radial velocities of galaxies can admit arbitrary values, or else some regular patterns regarded as periodisation or quantisation of galaxy redshifts are observed.

The discussion of redshift quantisation started with the work of Tifft. In the Coma cluster of galaxies[1], optically determined redshifts of member objects can be grouped into bands with the differences of 72.46 km $\cdot$ s^{-1}. A repeated analysis based on more accurate, i.e. HI, data confirmed this periodicity. The result was preceded by the study of the distribution of galaxy redhifts versus the nuclear magnitudes in the Coma cluster core (Tifft[2] and references therein), where strong redshift periodicities of 220 km $\cdot$ s^{-1} were observed. The existence of a band structure in the on nuclear magnitude-redshift diagram was reported by Nanni et al.[3] on the basis of better nuclear magnitude and redshift determination. However, in order to find the effect of quantisation it was necessary to apply also the correction connected with the solar galactocentric movement to heliocentric galaxy redhifts.

A few years later Tifft and Cocke[4] generalised their claim for all galaxies. There exists a global periodicity, equal to 36.3 km $\cdot$ s^{-1} for broad-line galaxies and 24.2 km $\cdot$ s^{-1} for the narrow-line ones. The latter value was not confirmed afterwards, so presently the value of about 36 (or 72) km $\cdot$ s^{-1} is considered as the basic one.

2 Discretisation of redshift differences in paired galaxies

Tifft[5] showed that redshift differences for double galaxies have a statistically significant grouping near multiples of 72 km $\cdot$ s^{-1}. Afterwards, the initial sample of galaxies was enlarged and objects with more accurate redshift determination ($\sigma \leq 25$ km $\cdot$ s^{-1}) considered (Tifft[6,7]). The periodicity of 72 km $\cdot$ s^{-1} was observed only for such well determined redshifts. Greater uncertainties blurred the observed periodicity. The way in which statistical analysis had been performed was criticised by Newman, Haynes & Terzian[8]. Cocke and Tifft[9] re-addressed this problem, concluding that the approach was statistically correct. Schneider and Salpeter[10] used the sample of 107 isolated binary galaxies with very reliable 21 cm redshifts. They did not find the multiplicity of 144, 216 ... km $\cdot$ s^{-1}, suggested by Tifft, but confirmed the excess in velocity differences near 72 km $\cdot$ s^{-1}.

3 Redshift quantisation in the Local Group

Considering 40 members of the Local Group, Rudnicki et al.[11] claimed the existence, at 95% significance level, of redshift periodisation in the sample considered, however, without precise periodisation value. They concluded that the uncertainties in the published redshift determination do not allow one to make more precise statements.

It is worthwhile noting that redshift discretisation in the Local Group concerns not only galaxies, but also globular clusters. Further analysis of the problem (Godlowski et al.[12]) showed that strict quantisation is excluded, but a weak evidence of period P of about 36 (or 24) km $\cdot$ s^{-1} is observed.

4 Redshift periodicity in the Local Supercluster

Guthrie and Napier[13] extracted 89 non-Virgo spirals with galactocentric redshifts $\leq$ 1000 km $\cdot$ s^{-1} from the database, and quoted redshift accuracy $\sigma \leq 4$ km $\cdot$ s^{-1}. An evidence of periodicity of about 37.5 km $\cdot$ s^{-1} was noted. Afterwards they[14] repeated their analysis using 247 spirals with galactocentric radial velocities $\leq$ 2600 km $\cdot$ s^{-1}, finding strong periodisation with P $\sim$ 37.6 km $\cdot$ s^{-1}. Their paper precisely describes the procedure of Power

Spectrum Analysis, the hypothesis tested, as well as the determination of the solar vector in respect to the Galaxy centre.

5 Large-scale periodisation

Broadhurst et al.[15] found high peaks separated by regular spacing, strongly indicating the existence of a 128 $h^{-1} \cdot$ Mpc (where h^{-1} is the Hubble constant in units of 100 km $\cdot$ s^{-1}) period in the redshift distribution. This result, based on the narrow pencil-beam method, was confirmed by other researchers who, using the whole sky distribution of Abell galaxy clusters, found a 120 Mpc periodicity (Einasto et al.[16]).

The Voronoi foam being a model describing galaxy distribution in sheets, filaments and clusters surrounding voids, was applied to study the possibility of explaining this periodicity. Van de Weygaert[17] was able to explain the effect, but he concluded that the periodicity, even in the cellular Universe, is not a common phenomenon. A similar conclusion was reached by Ikeuchi and Turner[18], who stressed that the quasi-periodic pattern is a smaller fluctuation in a Voronoi tessellation, than 2σ only. An other simple possibility is the existence of a coherent peculiar velocity field with the amplitude of the order of 10^{-3} and the period of 128 $h^{-1} \cdot$ Mpc (Hill, Steinhardt and Turner[19]). Such an amplitude is similar to the peculiar velocity field observed in our vicinity and connected with the so-called Great Attractor. The oscillating Universe (Morikawa[20]) or the change of the values in the values fundamental constants (Hill, Steinhardt and Turner[21]) were also invoked as a possible explanation for this periodicity. Nevertheless, as pointed out by Kirshner[22], several redshift surveys indicate the presence of 100-Mpc scale structures.

6 Quasar redshift quantisation

This phenomenon is probably the oldest one, because the first claims were published already in the late 1960's. Cowan[23] reported an effect with a period of 0.1666 in redshift $z = \Delta\lambda/\lambda$. The result was broadly discussed and existing data, with an increased number of quasars considered, were re-analyzed. The Power Spectrum Analysis was implemented for periodicity search, which allows one to determine precisely the statistical significance of the effects (Lake & Roeder[24]). Some further claims were made that the periodicity is 0.089 in $\log(1+z)$, which corresponds to a factor of 1.227 between the values of two consecutive maxima in the $1+z$ redshift distribution. Kjaergaard[25] carefully discussed the manner in which quasars are found, and concluded that the strong selection effects influence the observed redshift distribution, causing the existence of several peaks. Thus, according to his conclusions, the observed peaks are due to selection effects.

7 Conclusions

There are several effects which are called redhift quantisations. The scale of these effects is drastically different, from almost local values to a substantial part of the Universe. Clear interpretation exists only in the case of 100-Mpc periodisation, which is probably due to the large-scale structure, yet its origin is not easily explained. The discretisation of quasar redshifts seems to be an artifact of selection effects. In the case of 36 (or 72) km $\cdot$ s^{-1} periodisation, which is of the main interest for the advocates of redshift quantisation, our oppinion is that the effect is not shown convincingly. It is not quite clear why from several thousands of galaxies in the Local Supercluster only about 250 objects (i.e., less than 4%) are taken into account, which can cause a serious bias. It will be interesting to compare the real differences between various galaxy redshift determination, not just addressing the formal errors quoted by the authors. This point was also stressed by Guthrie and Napier[14].

If this alleged effect is confimed, it will open completely new areas of investigation, also unlocking the doors to the so-called 'new physics'. Beforehand, however, a detailed analysis of observational errors and applied statistical methods should be perfomed. Papers presenting the results obtained so far clearly show the weak and the strong points, which will be fruitful in further work. Moreover, the investigations of redshift distribution in other superclusters will be of great interest. Due to the accummulation of precise redshift measurements over the last years, a new re-discussion of redshift discretisation is possible, hopefully helping to solve this bizzare problem.

References

1. W.G. Tifft, *ApJ* **206**, 38 (1976).
2. W.G. Tifft, *The formation and Dynamics of Galaxies (J.R. Shakeshaft ed)*, D. Reider, Dordrecht, p.243 (1974).
3. D. Nanni, G. Pittella, D. Trevese, A. Vignato, *Astron. Astrophys.* **95**, 188 (1981).
4. W.G. Tifft, W.J. Cocke, *ApJ* **287**, 492 (1984).
5. W.G. Tifft, *ApJ* **236**, 70 (1980).
6. W.G. Tifft, *ApJ* **257**, 442 (1982).
7. W.G. Tifft, *ApJ* **262**, 44 (1982).
8. W.I. Newman, M.P. Haynes, Y. Terzian, *ApJ* **344**, 111 (1989).
9. W.J. Cocke, W.G. Tifft, *ApJ* **368**, 383 (1991).

10. S.S. Schneider, E.E. Salpater, *ApJ* **385**, 32 (1992).
11. K. Rudnicki, W. Godlowski, A. Magdziarz, In: ed. K. Rudnicki, *Gravitation, Electromagnetism and Cosmology: Toward a New Synthesis*, Montral, Apeiron (2001).
12. W. Godlowski, K. Bajan, P. Flin, in preparation (2001).
13. B.N.G. Guthrie, W.M. Napier, MNRAS 253, 533 (1991).
14. B.N.G. Guthrie, W.M. Napier, *Astron. Astrophys.* **310**, 353 (1996).
15. T.J. Broadhurst, R.S. Ellis, D.C. Koo, A.S. Szaley, *Nature* **343**, 726 (1990).
16. J. Einasto, M. Einasto, S. Gottloeber, V. Mueller, V. Saar, A.A. Starobinsky, D. Tucker, H. Andernacht, P. Frisch, *Nature* **385**, 139 (1997).
17. R. van de Weygaert, *Astron. Astrophys.* **249**, 159 (1991).
18. S. Ikeuchi, E.L. Turner, MNRAS 250, 519 (1991).
19. C.T. Hill, P.J. Steinhardt, M.S. Turner, *ApJ* **366**, L57 (1991).
20. M. Morikawa, *ApJ* **362**, L37 (1990).
21. C.T. Hill, P.J. Steinhardt, M.S. Turner, *Phys. Lett.* B **252**, 343 (1990).
22. R. Kirshner, *Nature* **385**, 122 (1997).
23. C.L. Cowan, *ApJ* **154**, L5 (1968).
24. R.G. Lake, R.C Roeder, RASC Jour. 66, 111 (1972).
25. P. Kjaergaard, *Physica Scripta* **17**, 347 (1978).

FROM GEOMETRY TO QUANTUM COMPUTATION

KAZUYUKI FUJII
Department of Mathematical Sciences
Yokohama City University, Yokohama 236-0027, JAPAN

The aim of this paper is to introduce our idea of Holonomic Quantum Computation (Computer). Our model is based on both harmonic oscillators and non–linear quantum optics, not on spins of usual quantum computation and our method is moreover completely geometrical.

1 Introduction

After the breakthrough by P. Shor [1] there has been remarkable progress in Quantum Computer or Computation (QC briefly). This discovery had a great influence on scientists. This drived not only theoreticians to finding other quantum algorithms, but also experimentalists to building quantum computers. See [2] and [3] in outline.

Gauge Theories had been widely recognized as the basis in quantum field theory. Therefore it is very natural to intend to include gauge theories in QC $\cdots$ a construction of "gauge theoretical" or "geometric" quantum computation. The merit of geometric method of QC may be strong for the influence from the environment.

In [4] and [5] Zanardi and Rasetti proposed an attractive idea $\cdots$ Holonomic Quantum Computation $\cdots$ using the non-abelian Berry phase (quantum holonomy). They gave a few simple but interesting examples to explain their idea. To make their works more mathematical and rigorous the author has given the mathematical reinforcement to their works, [7], [8], [9] and [10]. But his works are still not sufficient.

In this talk we will introduce Holonomic Quantum Computation and discuss some problems to be solved.

2 Mathematical Preliminaries

We start with mathematical preliminaries. Let $\mathcal{H}$ be a separable Hilbert space over $\mathbf{C}$. For $m \in \mathbf{N}$, we set

$$St_m(\mathcal{H}) \equiv \left\{ V = (v_1, \cdots, v_m) \in \mathcal{H} \times \cdots \times \mathcal{H} | V^\dagger V = 1_m \right\} , \tag{1}$$

$(V^\dagger V = 1_m \Longleftrightarrow \langle v_i | v_j \rangle = \delta_{ij})$ where 1_m is a unit matrix in $M(m, \mathbf{C})$. This is called a (universal) Stiefel manifold. Note that the unitary group $U(m)$ acts on $St_m(\mathcal{H})$ from the right :

$$St_m(\mathcal{H}) \times U(m) \to St_m(\mathcal{H}) \ : \ (V,a) \mapsto Va. \tag{2}$$

Next we define a (universal) Grassmann manifold

$$Gr_m(\mathcal{H}) \equiv \{X \in M(\mathcal{H}) | X^2 = X, X^\dagger = X \text{ and } \mathrm{tr}X = m\} \ , \tag{3}$$

where $M(\mathcal{H})$ denotes a space of all bounded linear operators on $\mathcal{H}$. Then we have a projection

$$\pi : St_m(\mathcal{H}) \to Gr_m(\mathcal{H}) \ , \quad \pi(V) \equiv VV^\dagger \tag{4}$$

compatible with the action (2). Now the set $\{U(m), St_m(\mathcal{H}), \pi, Gr_m(\mathcal{H})\}$ is called a (universal) principal $U(m)$ bundle, see [7]. We set

$$E_m(\mathcal{H}) \equiv \{(X,v) \in Gr_m(\mathcal{H}) \times \mathcal{H} | Xv = v\} \ . \tag{5}$$

Then we have also a projection

$$\pi : E_m(\mathcal{H}) \to Gr_m(\mathcal{H}) \ , \quad \pi((X,v)) \equiv X \ . \tag{6}$$

The set $\{\mathbf{C}^m, E_m(\mathcal{H}), \pi, Gr_m(\mathcal{H})\}$ is called a (universal) m-th vector bundle. This vector bundle is one associated with the (universal) principal $U(m)$ bundle above.

Next let M be a finite or infinite dimensional differentiable manifold and the map $P : M \to Gr_m(\mathcal{H})$ be given (called a projector). Using this P we can consider pullback bundles over M :

$$\{U(m), \widetilde{St}, \pi_{\widetilde{St}}, M\} \equiv P^*\{U(m), St_m(\mathcal{H}), \pi, Gr_m(\mathcal{H})\} \ , \tag{7}$$

$$\{\mathbf{C}^m, \widetilde{E}, \pi_{\widetilde{E}}, M\} \equiv P^*\{\mathbf{C}^m, E_m(\mathcal{H}), \pi, Gr_m(\mathcal{H})\} \ . \tag{8}$$

Let $\mathcal{M}$ be a parameter space and we denote by λ its element. Let λ_0 be a fixed reference point of $\mathcal{M}$. Let H_λ be a family of Hamiltonians parameterized by $\mathcal{M}$ which act on a Fock space $\mathcal{H}$. We set $H_0 = H_{\lambda_0}$ for simplicity and assume that this has a m–**fold degenerate** vacuum $H_0 v_j = \mathbf{0}$, $j = 1 \sim m$. These v_j's form a m-dimensional vector space. We may assume that $\langle v_i | v_j \rangle = \delta_{ij}$. Then $(v_1, \cdots, v_m) \in St_m(\mathcal{H})$ and $F_0 \equiv \{\sum_{j=1}^m x_j v_j | x_j \in \mathbf{C}\} \cong \mathbf{C}^m$. Namely, F_0 is the vector space associated with o.n.basis $\{v_1, \cdots, v_m\}$.

Next we assume for simplicity that a family of unitary operators parameterized by $\mathcal{M}$

$$W : \mathcal{M} \to U(\mathcal{H}), \quad W(\lambda_0) = \mathrm{id}. \tag{9}$$

is given and H_λ above is given by the following isospectral family

$$H_\lambda \equiv W(\lambda) H_0 W(\lambda)^{-1}. \tag{10}$$

In this case there is no level crossing of eigenvalues. Making use of $W(\lambda)$ we can define a projector

$$P : \mathcal{M} \to Gr_m(\mathcal{H}), \quad P(\lambda) \equiv W(\lambda)(\textstyle\sum_{j=1}^{m} v_j v_j^\dagger)W(\lambda)^{-1} \tag{11}$$

and have the pullback bundles over $\mathcal{M}$ $\{U(m), \widetilde{St}, \pi_{\widetilde{St}}, \mathcal{M}\}$, $\{\mathbf{C}^m, \widetilde{E}, \pi_{\widetilde{E}}, \mathcal{M}\}$. For the latter we set $|vac\rangle = (v_1, \cdots, v_m)$. In this case a canonical connection form $\mathcal{A}$ of $\{U(m), \widetilde{St}, \pi_{\widetilde{St}}, \mathcal{M}\}$ is given by

$$\mathcal{A} = \langle vac|W(\lambda)^{-1} dW(\lambda)|vac\rangle, \tag{12}$$

where d is a differential form on $\mathcal{M}$, and its curvature form by $\mathcal{F} \equiv d\mathcal{A} + \mathcal{A}\wedge\mathcal{A}$. Let γ be a loop in $\mathcal{M}$ at λ_0., $\gamma : [0,1] \to \mathcal{M}, \gamma(0) = \gamma(1)$. For this γ a holonomy operator $\Gamma_{\mathcal{A}}$ is defined :

$$\Gamma_{\mathcal{A}}(\gamma) = \mathcal{P}exp\left\{\oint_\gamma \mathcal{A}\right\} \in U(m), \tag{13}$$

where $\mathcal{P}$ means path-ordered. This acts on the fiber F_0 at λ_0 of the vector bundle $\{\mathbf{C}^m, \widetilde{E}, \pi_{\widetilde{E}}, M\}$ as follows : $\mathbf{x} \to \Gamma_{\mathcal{A}}(\gamma)\mathbf{x}$. The holonomy group $Hol(\mathcal{A})$ is in general subgroup of $U(m)$. In the case of $Hol(\mathcal{A}) = U(m)$, $\mathcal{A}$ is called irreducible. In the Holonomic Quantum Computation we take

$$\begin{aligned} &\text{Encoding of Information} \Longrightarrow \mathbf{x} \in F_0, \\ &\text{Processing of Information} \Longrightarrow \Gamma_{\mathcal{A}}(\gamma) : \mathbf{x} \to \Gamma_{\mathcal{A}}(\gamma)\mathbf{x}. \end{aligned} \tag{14}$$

3 Holonomic Quantum Computation

We apply the results of last section to Quantum Optics and discuss (optical) Holonomic Quantum Computation proposed by [4] and [6]. Let $a(a^\dagger)$ be the annihilation (creation) operator of the harmonic oscillator. If we set $N \equiv a^\dagger a$ (: number operator), then

$$[N, a^\dagger] = a^\dagger \ , \ [N, a] = -a \ , \ [a, a^\dagger] = 1 \ . \tag{15}$$

Let $\mathcal{H}$ be a Fock space generated by a and $a^\dagger$, and $\{|n\rangle | n \in \mathbf{N} \cup \{0\}\}$ be its basis. The actions of a and $a^\dagger$ on $\mathcal{H}$ are given by $a|n\rangle = \sqrt{n}|n-1\rangle$, $a^\dagger|n\rangle = \sqrt{n+1}|n+1\rangle$, where $|0\rangle$ is a vacuum ($a|0\rangle = 0$). In the following we treat coherent operators and squeezed operators.

$$\text{Coherent Operator} \quad D(\alpha) = \exp\left(\alpha a^\dagger - \bar{\alpha} a\right) \quad \text{for } \alpha \in \mathbf{C}, \tag{16}$$

$$\text{Squeezed Operator} \quad S(\beta) = \exp\{(\beta(a^\dagger)^2 - \bar{\beta}a^2)/2\} \quad \text{for } \beta \in \mathbf{C}. \tag{17}$$

For the details see [7]. Next we consider the system of n–harmonic oscillators. If we set

$$a_i = 1 \otimes \cdots \otimes 1 \otimes a \otimes 1 \otimes \cdots \otimes 1, \quad {a_i}^\dagger = 1 \otimes \cdots \otimes 1 \otimes a^\dagger \otimes 1 \otimes \cdots \otimes 1, \tag{18}$$

for $1 \le i \le n$, then it is easy to see $[a_i, a_j] = [{a_i}^\dagger, {a_j}^\dagger] = 0$, $[a_i, {a_j}^\dagger] = \delta_{ij}$. We also denote by $N_i = {a_i}^\dagger a_i$ $(1 \le i \le n)$ number operators.

3.1 Two–Qubit Case

Since we want to consider coherent states based on Lie algebras $su(2)$ and $su(1,1)$, we make use of Schwinger's boson method, see [12] and [13]. Namely if we set

$$su(2): \quad J_+ = a_1^\dagger a_2, \; J_- = a_2^\dagger a_1, \; J_3 = (a_1^\dagger a_1 - a_2^\dagger a_2)/2, \tag{19}$$

$$su(1,1): \quad K_+ = a_1^\dagger a_2^\dagger, \; K_- = a_2 a_1, \; K_3 = (a_1^\dagger a_1 + a_2^\dagger a_2 + 1)/2, \tag{20}$$

then we have

$$su(2): \quad [J_3, J_+] = J_+, \; [J_3, J_-] = -J_-, \; [J_+, J_-] = 2J_3, \tag{21}$$

$$su(1,1): \quad [K_3, K_+] = K_+, \; [K_3, K_-] = -K_-, \; [K_+, K_-] = -2K_3. \tag{22}$$

In the following we treat unitary coherent operators based on Lie algebras $su(2)$ and $su(1,1)$.

$$U(\lambda) = \exp\left(\lambda J_+ - \bar{\lambda} J_-\right) \quad \text{for } \lambda \in \mathbf{C}, \tag{23}$$

$$V(\mu) = \exp\left(\mu K_+ - \bar{\mu} K_-\right) \quad \text{for } \mu \in \mathbf{C}. \tag{24}$$

For the details of $U(\lambda)$ and $V(\mu)$ see [11] and [12].

Let H_0 be a Hamiltonian with nonlinear interaction produced by a Kerr medium., that is $H_0 = \mathrm{X}N(N-1)$, where X is a certain constant, see [6]. The eigenvectors of H_0 corresponding to 0 is $\{|0\rangle, |1\rangle\}$, so its eigenspace is $\mathrm{Vect}\,\{|0\rangle, |1\rangle\} \cong \mathbf{C}^2$. The space $\mathrm{Vect}\,\{|0\rangle, |1\rangle\}$ is called 1-qubit (**qu**antum **bit**) space. Since we are considering the system of two particles, the Hamiltonian that we will treat is

$$H_0 = \mathrm{X}N_1(N_1 - 1) + \mathrm{X}N_2(N_2 - 1). \tag{25}$$

The eigenspace of 0 of this Hamiltonian becomes therefore

$$F_0 = \mathrm{Vect}\,\{|0\rangle, |1\rangle\} \otimes \mathrm{Vect}\,\{|0\rangle, |1\rangle\} = \mathrm{Vect}\,\{|0,0\rangle, |0,1\rangle, |1,0\rangle, |1,1\rangle\} \cong \mathbf{C}^4.$$

We set $|vac\rangle = (|0,0\rangle, |0,1\rangle, |1,0\rangle, |1,1\rangle)$. Next we consider the following isospectral family of H_0 :

$$H_{(\alpha_1,\beta_1,\lambda,\mu,\alpha_2,\beta_2)} = W(\alpha_1,\beta_1,\lambda,\mu,\alpha_2,\beta_2) H_0 W(\alpha_1,\beta_1,\lambda,\mu,\alpha_2,\beta_2)^{-1}, \tag{26}$$

$$W(\alpha_1,\beta_1,\lambda,\mu,\alpha_2,\beta_2) = W_1(\alpha_1,\beta_1) O_{12}(\lambda,\mu) W_2(\alpha_2,\beta_2). \tag{27}$$

where

$$O_{12}(\lambda,\mu)=U(\lambda)V(\mu),\quad W_j(\alpha_j,\beta_j)=D_j(\alpha_j)S_j(\beta_j)\quad \text{for}\quad j=1,2. \tag{28}$$

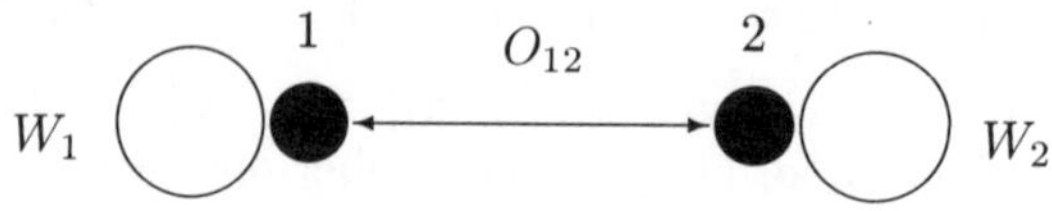

In this case

$$\mathcal{M}=\left\{(\alpha_1,\beta_1,\lambda,\mu,\alpha_2,\beta_2)\in \mathbf{C}^6\right\},\quad \text{and} \tag{29}$$

$$\mathcal{A}=\langle vac|W^{-1}dW|vac\rangle, \tag{30}$$

where d is the exterior differential form on $\mathcal{M}$ (29). We want to caluculate (30), but the calculation is not so easy. See [10] for the details.
Problem Is the connection form irreducible in $U(4)$?
Our analysis in [10] shows that the holonomy group generated by $\mathcal{A}$ may be $SU(4)$ not $U(4)$. To obtain $U(4)$ a sophisticated trick $\cdots$ higher dimensional holonomies [15] $\cdots$ may be necessary. A further study is needed.

3.2 N–Qubit Case

A reference Hamiltonian is in this case

$$H_0=\mathrm{X}\textstyle\sum_{i=1}^{n}N_i(N_i-1) \tag{31}$$

and the eigenspace to 0–eigenvalue (n–qubit) becomes

$$F_0=\mathrm{Vect}\{|0,\cdots,0,0\rangle,|0,\cdots,0,1\rangle,\cdots,|1,\cdots,1,0\rangle,|1,\cdots,1,1\rangle\}\cong \mathbf{C}^{2^n}$$

and set $|vac\rangle=(|0,\cdots,0,0\rangle,|0,\cdots,0,1\rangle,\cdots,|1,\cdots,1,0\rangle,|1,\cdots,1,1\rangle)$.
The $u(n)$–algebra is defined by

generators : $\{E_{ij}\mid 1\le i,j\le n\}$, relations : $[E_{ij},E_{kl}]=\delta_{jk}E_{il}-\delta_{li}E_{kj}$,

and $\{E_{in}\mid 1\le i\le n-1\}$ is a Weyl basis. Boson representation of $u(n)$–algebra is well–known to be

$$E_{ij}={a_i}^{\dagger}a_j\quad 1\le i,j\le n\ . \tag{32}$$

The $u(n-1,1)$–algebra is also defined by

generators : $\{E_{ij}\mid 1\le i,j\le n\}$, relations : $[E_{ij},E_{kl}]=\eta_{jk}E_{il}-\eta_{li}E_{kj}$,

where $\eta=\mathrm{diag}(1,\cdots,1,-1)$, and $\{E_{in}\mid 1\le i\le n-1\}$ is a Weyl basis. Boson representation of $u(n-1,1)$–algebra is given by

$$E_{ij}={a_i}^{\dagger}a_j\quad 1\le i,j\le n-1\ ,\quad E_{nn}={a_n}^{\dagger}a_n+1$$
$$E_{in}={a_i}^{\dagger}{a_n}^{\dagger}\ ,\quad E_{ni}=a_na_i\quad 1\le i\le n-1\ . \tag{33}$$

A family of Hamiltonians that we treat is

$$H = WH_0W^{-1}\ , \qquad W = \prod_{j=1}^{n} W_{jn} \tag{34}$$

and W_{jn} is

$$W_{jn}(\alpha_j, \beta_j, \lambda_j, \mu_j) = W_j(\alpha_j, \beta_j)O_{jn}(\lambda_j, \mu_j), \quad \text{and} \tag{35}$$

$$W_j(\alpha_j, \beta_j) = D_j(\alpha_j)S_j(\beta_j), \quad O_{jn}(\lambda_j, \mu_j) = U_j(\lambda_j)V_j(\mu_j), \tag{36}$$

and for $1 \le j \le n-1$

$$U_j(\lambda_j) = \exp(\lambda_j {a_j}^{\dagger} a_n - \bar{\lambda}_j {a_n}^{\dagger} a_j), \quad V_j(\mu_j) = \exp(\mu_j {a_j}^{\dagger} {a_n}^{\dagger} - \bar{\mu}_j a_n a_j) \tag{37}$$

making use of each Weyl basis. We note that $O_{nn} = 1$. In this case

$$\mathcal{M} = \{(\cdots, \alpha_j, \beta_j, \lambda_j, \mu_j, \alpha_{j+1}, \beta_{j+1}, \cdots)\} \cong \mathbf{C}^{4n-2}, \quad \text{and} \tag{38}$$

$$\mathcal{A} = \langle vac|W^{-1}dW|vac\rangle\ , \tag{39}$$

where d is the exterior differential form on $\mathcal{M}$ (38). Then
Problem Is the connection form irreducible in $U(2^n)$?

References

1. P. W. Shor, *SIAM J. Computing*, **26**, 1484 (1997).
2. A. Steane, *Rept. Prog. Phys.*, **61**, 117 (1998).
3. K. Fujii, quant-ph/0103011.
4. P. Zanardi and M. Rasetti, *Phys. Lett. A* **264**, 94 (1999).
5. J. Pachos, P. Zanardi and M. Rasetti, *Phys. Rev. A* **61**, 010305R (2000).
6. J. Pachos and S. Chountasis, *Phys. Rev. A* **62** 052318 (2000).
7. K. Fujii, *J. Math. Phys.* **41**, 4406 (2000).
8. K. Fujii, to appear in *Rept. Math. Phys.* **48** (2001), quant-ph/0004102.
9. K. Fujii, quant-ph/0005129.
10. K. Fujii, quant-ph/00101102.
11. A. Perelomov, *Generalized Coherent States and Their Applications*, Springer Verlag (1986).
12. K. Funahashi, T. Kashiwa, S. Sakoda and K. Fujii, *J. Math. Phys.* **36**, 3232 (1995).
13. K. Funahashi, T. Kashiwa, S. Sakoda and K. Fujii, *J. Math. Phys.* **36**, 4590 (1995).
14. J. Pachos and P. Zanardi, quant-ph/0007110.
15. O. Alvarez, L. A. Ferreira and J. S. Guillen, *Nucl. Phys. B* **529**, 689 (1998).

PARTITION FUNCTION ZEROS OF APERIODIC ISING MODELS

UWE GRIMM

Applied Mathematics Department, Faculty of Mathematics and Computing, The Open University, Walton Hall, Milton Keynes MK7 6AA, UK
E-mail: u.g.grimm@open.ac.uk

PRZEMYSŁAW REPETOWICZ

Department of Mathematics, Heriot-Watt University, Riccarton, Edinburgh EH14 4AS, UK
E-mail: p.repetowicz@ma.hw.ac.uk

We consider Ising models defined on periodic approximants of aperiodic graphs. The model contains only a single coupling constant and no magnetic field, so the aperiodicity is entirely given by the different local environments of neighbours in the aperiodic graph. In this case, the partition function zeros in the temperature variable, also known as the Fisher zeros, can be calculated by diagonalisation of finite matrices. We present the partition function zero patterns for periodic approximants of the Penrose and the Ammann-Beenker tiling, and derive precise estimates of the critical temperatures.

1 Introduction

The Ising model is the paradigm for a second-order phase transition in two-dimensional models of statistical mechanics. Without an external magnetic field, the partition function of the two-dimensional Ising model can be calculated explicitly for regular lattices. However, if the Ising model is defined on an aperiodic graph, for instance on the Penrose tiling (PT), this is no longer the case. Therefore, the influence of an aperiodic order in the underlying structure on the phase transition of the Ising model has been investigated by various means, including numerical simulations, series expansions, and zeros of the partition function; an overview on the results and a comprehensive lists of references on the subject can be found in recent review articles.[1,2]

The most general prediction stems from heuristic scaling arguments, adapted from a relevance criterion for disordered Ising models.[2] It yields an inequality involving a characteristic exponent that describes the fluctuations in the disordered or aperiodically ordered system, the correlation critical exponent ν of the pure system, and the space dimension of the fluctuation. Planar quasiperiodic graphs obtained by cut-and-project methods[3] have low fluctuations, because they are flat sections through higher-dimensional peri-

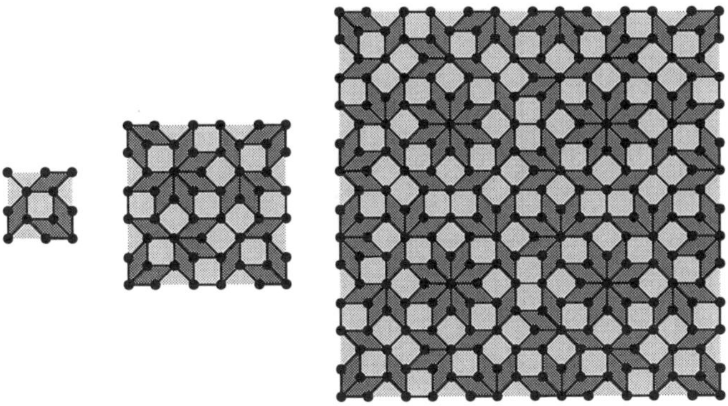

Figure 1. The first three periodic approximants ($m = 1, 2, 3$) of the Ammann-Beenker tiling.

odic lattices. Therefore, the aperiodicity is expected to be irrelevant in these cases, which conforms with the results obtained for specific examples, notably the PT.

Here, we consider partition function zeros obtained for periodic approximants of the rhombic PT and the Ammann-Beenker tiling (ABT). The unit cells of the first three approximants for the latter are shown in Fig. 1.

2 The Ising model

We place Ising spins $\sigma_{j,k} \in \{-1, 1\}$ on the vertices of the periodic approximants. Spins interact by a nearest-neighbour interaction J. The energy of a configuration $\sigma = \{\sigma_j \mid 1 \leq j \leq M\} \in \{-1, 1\}^M$ on a finite graph with M vertices is

$$E(\sigma) = -\sum_{\langle j,k \rangle} J \sigma_j \sigma_k, \tag{1}$$

where the summation is performed over all pairs $\langle j, k \rangle$ of neighbouring spins, i.e., those located on vertices that are connected by an edge. The corresponding partition function is the sum on all configurations

$$Z(\beta) = \sum_{\sigma} \exp[-\beta E(\sigma)], \tag{2}$$

where β is the inverse temperature.

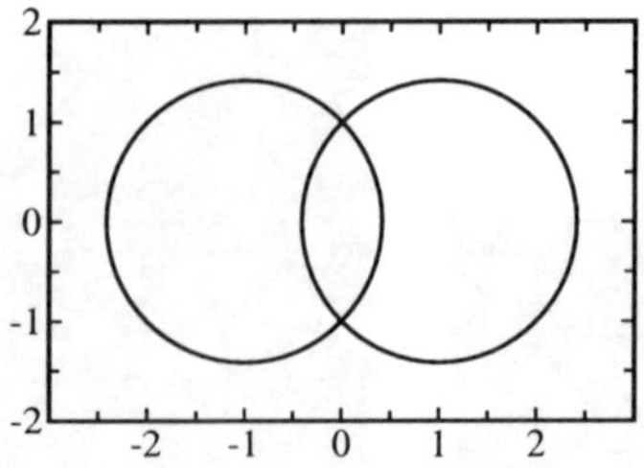

Figure 2. Partition function zeros of the square lattice Ising model.

3 Partition function zeros

We are interested in the pattern of zeros of this function in the complex temperature plane. In the thermodynamic limit $M \to \infty$, the zeros accumulate, filling curves or areas in the complex plane. These separate different regions of analyticity of the free energy, which is essentially the logarithm of the partition function, and hence correspond to phase transitions.

The zero pattern for the square lattice case is shown in Fig. 2, which displays the zeros in the complex variable $z = \exp(2\beta J)$. It consists of two circles. The two intersections at $z = \sqrt{2}+1$ and $z = \sqrt{2}-1$ with the positive real axis correspond to the ferromagnetic and antiferromagnetic critical points $w_\mathrm{c} = \tanh(\beta_\mathrm{c} J) = \pm\sqrt{2}$, respectively.

4 Results and conclusions

The partition function zeros of the Ising model on any periodic planar graph with N spins in a unit cell can be calculated by diagonalising a two-parameter family of $4N \times 4N$ matrices.[4,5] The powerful tool behind this method is the Kac-Ward determinant or Pfaffian method for calculating the Ising model partition function on general planar graphs.[4] Thus, this method allows to compute partition function zeros for Ising models on infinite periodic graphs with an arbitrary numerical precision.

The result for the first approximants of the PT and the ABT are shown in Figs. 3 and 4. These show accumulated zeros obtained by diagonalising the Kac-Ward matrix for various values of the two parameters, again in the complex variable $z = \exp(2\beta J)$. The corresponding values of the critical temperature are listed in Table 1. These yield precise estimates for the critical temperature for the aperiodic tilings, the errors were estimated on the basis of the apparent convergence of the first few terms.

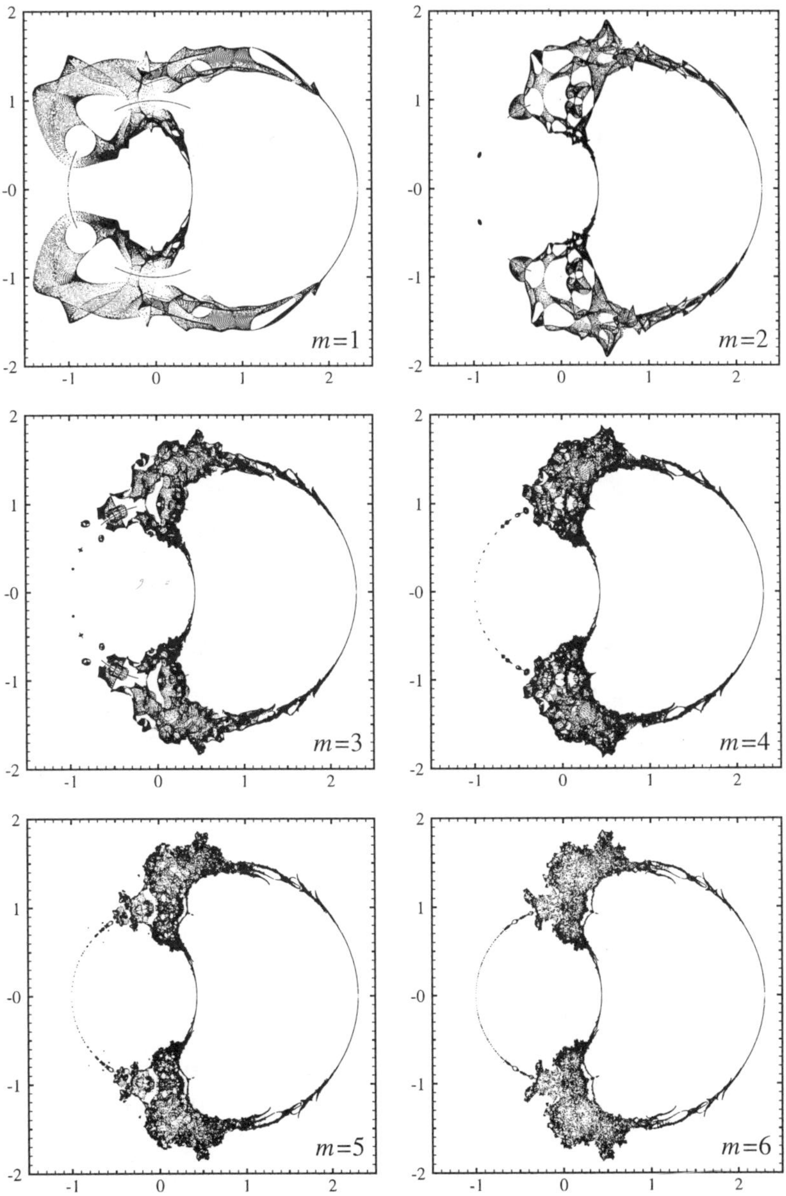

Figure 3. Part of partition function zeros for periodic approximants of the Penrose tiling.

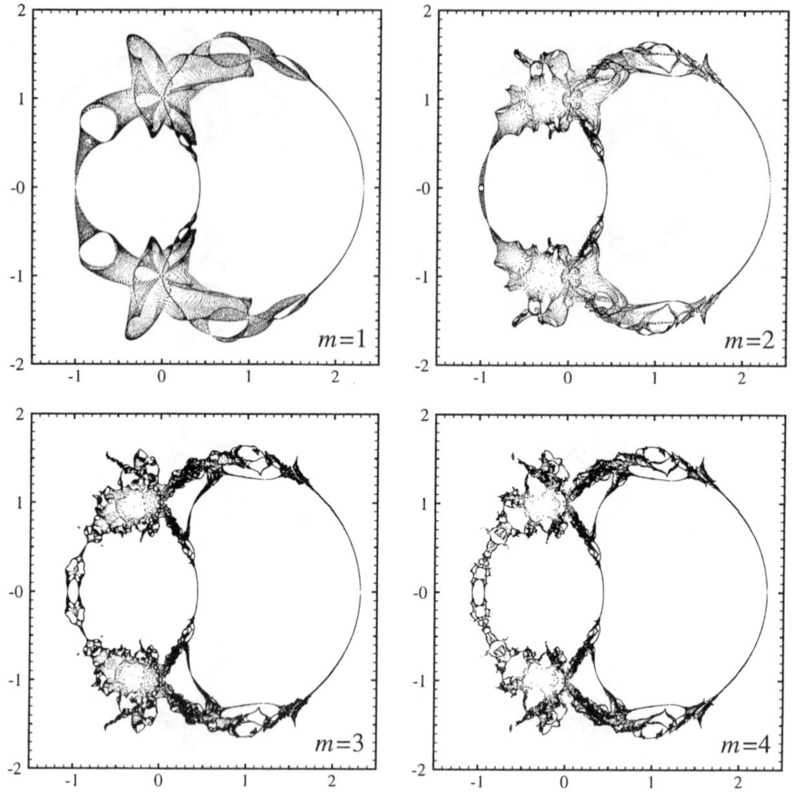

Figure 4. Partition function zeros for periodic approximants of the Ammann-Beenker tiling.

The zero patterns, besides the trivial symmetry under complex conjugation of z, still show the $z \leftrightarrow z^{-1}$ symmetry that is a consequence of the bipartiteness of the tilings. This also implies that the ferromagnetic and antiferromagnetic phase transitions are of the same type. However, some of the symmetry of the square lattice case, corresponding to the self-duality of the square-lattice Ising model, is lost. In fact, the dual graphs of the aperiodic graphs under consideration contain triangles, hence are not bipartite. The antiferromagnetic model on these dual graphs will suffer frustration. This may be responsible for the rather complicated, presumably fractal or fractally bounded zero patterns, particularly further away from the positive real

Table 1. Critical temperatures $w_c = \tanh(\beta_c J)$ for periodic approximants of the PT and the ABT. Here, m labels the approximants with N spins per unit cell.

m	Penrose tiling		Ammann-Beenker tiling	
	N	w_c	N	w_c
1	11	0.401 440 380	7	0.396 850 570
2	29	0.395 411 099	41	0.396 003 524
3	76	0.395 082 894	239	0.395 985 346
4	199	0.394 554 945	1393	0.395 984 811
5	521	0.394 523 576	8119	0.395 984 795
6	1364	0.394 454 880	47321	0.395 984 795
7	3571	0.394 451 035		
8	9349	0.394 441 450		
9	24476	0.394 439 826		
10	64079	0.394 439 319		
∞	∞	0.394 439(1)	∞	0.395 984 79(1)

axis. The zeros around the two phase transition points still lie on a circle intersecting the real axis orthogonally, consistent with the fact that the phase transition belongs to the same universality class as on the square lattice.

References

1. U. Grimm and M. Baake, Aperiodic Ising models, in *The Mathematics of Long-Range Aperiodic Order*, ed. R.V. Moody, Kluwer, Dordrecht (1997), p. 199.
2. U. Grimm, Aperiodicity and disorder - does it matter? in *Computational Statistical Physics — From Billiards to Monte-Carlo*, eds. K.-H. Hoffmann and M. Schreiber, Springer, Berlin (2001).
3. U. Grimm and M. Schreiber, Aperiodic Tilings on the Computer, in *Quasicrystals*, eds. J.-B. Suck, M. Schreiber and P. Häussler, Springer, Berlin (2002).
4. P. Repetowicz, U. Grimm and M. Schreiber, *J. Phys. A: Math. Gen.* **32**, 4397 (1999).
5. P. Repetowicz, U. Grimm and M. Schreiber, *Mat. Sci. Eng. A* **294–296**, 638–641 (2000).

CAYLEY–KLEIN CONTRACTIONS OF ORTHOSYMPLECTIC SUPERALGEBRAS

N. A. GROMOV, I. V. KOSTYKOV, V. V. KURATOV

Department of Mathematics, Syktyvkar Branch of IMM UrD RAS, Chernova st., 3a, Syktyvkar, 167982, Russia

E-mail: gromov@dm.komisc.ru

We define a class of orthosymplectic superalgebras $osp(m;j|2n;\omega)$ which may be obtained from $osp(m|2n)$ by contractions and analytic continuations in a similar way as the orthogonal and the symplectic Cayley-Klein algebras are obtained from the corresponding classical ones. Contractions of $osp(1|2)$ and $osp(3|2)$ are regarded as an examples.

Since their discovery [1 2 3] in 1971 the supersymmetry is used in different physical theories such as Kaluza–Klein supergravity [4], supersymmetric field theories of the Wess–Zumino type [5], massless higher-spin field theories [6]. Recently the secret theory [7] (or S-theory) that includes superstring theory and its super p-brane and D-brane [8] generalizations was discussed. All these theories are build algebraically with the help of some superalgebra in their base. In this work we wish to present a wide class of Cayley–Klein (CK) superalgebras which may be used for constructions of different sypersymmetric models.

1 $osp(m|2n)$ superalgebra

Let $e_{IJ} \in M_{m+2n}$ satisfying $(e_{IJ})_{KL} = \delta_{IL}\delta_{JK}$ are elementary matrices. One defines the following graded matrix

$$G = \left(\begin{array}{c|cc} I_m & \multicolumn{2}{c}{0} \\ \hline 0 & 0 & I_n \\ & -I_n & 0 \end{array}\right)$$

where I_m, I_n are identity matrices. Let $i, j, \ldots = 1, \ldots, m$, $\bar{i}, \bar{j} = m+1, \ldots, m+2n$. The generators of the orthosymplectic superalgebra $osp(m|2n)$ are given by

$$E_{ij} = -E_{ji} = \sum_k (G_{ik}e_{kj} - G_{jk}e_{ki}), \quad E_{\bar{i}\bar{j}} = E_{\bar{j}\bar{i}} = \sum_{\bar{k}} (G_{\bar{i}\bar{k}}e_{\bar{k}\bar{j}} + G_{\bar{j}\bar{k}}e_{\bar{k}\bar{i}}),$$

$$E_{i\bar{j}} = E_{\bar{j}i} = \sum_k G_{ik}e_{k\bar{j}} + \sum_{\bar{k}} G_{\bar{j}\bar{k}}e_{\bar{k}i}, \tag{1}$$

where the even (bosonic) E_{ij} generate the $so(m)$ part, the even (bosonic) $E_{\bar{i}\bar{j}}$ generate the $sp(2n)$ part and the rest $E_{i\bar{j}}$ are the odd (fermionic) generators of superalgebra. They satisfy the following (super) commutation relations

$$[E_{ij}, E_{kl}] = G_{jk}E_{il} + G_{il}E_{jk} - G_{ik}E_{jl} - G_{jl}E_{ik},$$

$$[E_{\bar{i}\bar{j}}, E_{\bar{k}\bar{l}}] = -G_{\bar{j}\bar{k}}E_{\bar{i}\bar{l}} - G_{\bar{i}\bar{l}}E_{\bar{j}\bar{k}} - G_{\bar{j}\bar{l}}E_{\bar{i}\bar{k}} - G_{\bar{i}\bar{k}}E_{\bar{j}\bar{l}},$$

$$[E_{ij}, E_{k\bar{l}}] = G_{jk}E_{i\bar{l}} - G_{ik}E_{j\bar{l}}, \quad [E_{i\bar{j}}, E_{\bar{k}\bar{l}}] = -G_{\bar{j}\bar{k}}E_{i\bar{l}} - G_{\bar{j}\bar{l}}E_{i\bar{k}},$$

$$[E_{ij}, E_{\bar{k}\bar{l}}] = 0, \quad \{E_{i\bar{j}}, E_{k\bar{l}}\} = G_{ik}E_{\bar{j}\bar{l}} - G_{\bar{j}\bar{l}}E_{ik}. \tag{2}$$

In the matrix form $osp(m|2n) = \{M \in M_{m+2n} | M^{st}G + GM = 0\}$. If the matrix M has the following form: $M = \sum_{i,j} a_{ij}E_{ij} + \sum_{\bar{i},\bar{j}} b_{\bar{i}\bar{j}}E_{\bar{i}\bar{j}} + \sum_{i\bar{j}} \mu_{i\bar{j}}E_{i\bar{j}}$, with $a_{ij}, b_{\bar{i}\bar{j}} \in \mathbf{R}$ or $\mathbf{C}$ and $\mu_{i\bar{j}}$ as the odd nilpotent elements of Grassmann algebra: $\mu_{i\bar{j}}^2 = 0$, $\mu_{i\bar{j}}\mu_{i'\bar{j}'} = -\mu_{i'\bar{j}'}\mu_{i\bar{j}}$, then the corresponding supergroup $Osp(m|2n)$ is obtained by the exponential map $\mathcal{M} = \exp M$ and act on (super)vector space by matrix multiplication $\mathcal{X}' = \mathcal{M}\mathcal{X}$, where $\mathcal{X}^t = (x|\theta)^t$, x is a n–dimentsional even vector and θ is a $2m$–dimensional odd vector with odd Grassmann elements. The form $inv = \sum_{i=1}^n x_i^2 + 2\sum_{k=1}^m \theta_{+k}\theta_{-k} = x^2 + 2\theta^2$ is invariant under this action of orthosymplectic supergroup.

2 Cayley-Klein orthogonal and symplectic algebras

Orthogonal $so(m)$ and symplectic $sp(2n)$ algebras are even subalgebras of $osp(m|2n)$. On the other hand both of these algebras may be contracted and analytically continued to the set of CK orthogonal (and symplectic) algebras. CK group $SO(m;j)$ is defined as the set of transformations of vector space $\mathbf{R}_m(j)$, which preserve the form $x^2(j) = x_1^2 + \sum_{k=2}^m [1,k]^2 x_k^2$, where $[i,k] = \prod_{p=\min(i,k)}^{\max(i,k)-1} j_p$, $[i,i] = 1$, each parameter $j_k = 1, \iota_k, i$, where ι_k are nilpotent $\iota_k^2 = 0$, commutative $\iota_k\iota_p = \iota_p\iota_k \neq 0$ generators of Pimenov algebra $P(\iota)$. Cartesian components of vector $x(j) \in \mathbf{R}_m(j)$ are $x^t(j) = (x_1, j_1x_2, \ldots, [1,m]x_m)^t$, as it is easily follows from $x^2(j)$. For $m \times m$ matrix $g(j) \in SO(m;j)$ the transformation $g(j) : \mathbf{R}_m(j) \to \mathbf{R}_m(j)$ means that the vector $x'(j) = g(j)x(j)$ has exactly the same distribution of parameters j among its components as $x(j)$. This requirement give an opportunity to obtain the distribution of parameter j among elements of matrix $g(j)$, i.e. to build the fundamental representation of CK group $SO(m;j)$ starting from the quadratic form. It is remarkable that the same distribution of the paramcters j is hold for CK Lie algebra $so(m;j)$, namely $A_{ik} = [i,k]a_{ik}$, for $A \in so(m;j)$.

CK symplectic group $Sp(2n;\omega)$ is defined as the set of transformations of $\mathbf{R}_n(\omega) \times \mathbf{R}_n(\omega)$, which preserve the bilinear form $S(\omega) = S_1 + \sum_{k=2}^n (1,k)^2 S_k$, where $S_k(y,z) = y_k z_{n+k} - y_{n+k} z_k$, $(i,k) = \prod_{p=\min(i,k)}^{\max(i,k)-1} \omega_k$, $(i,i) = 1$, $\omega_k = 1, \xi_k, i$, $\xi_k^2 = 0$, $\xi_k \xi_p = \xi_p \xi_k$. The distribution of parameters ω_k among matrix elements of the fundamental representation $M(\omega) = \begin{pmatrix} H(\omega) & E(\omega) \\ F(\omega) & -H^t(\omega) \end{pmatrix}$ of the CK symplectic algebra $sp(2n;\omega)$ may be obtained as for orthogonal CK algebras and is as follows: $B_{ik} = (i,k)b_{ik}$, where $B = H(\omega), E(\omega), F(\omega)$.

3 CK orthosymplectic superalgebras $osp(m;j|2n;\omega)$

We shall define these superalgebras starting with the invariant form

$$inv = u^2 \sum_{k=1}^{m} [1,k]^2 x_k^2 + v^2 2 \sum_{k=m+1}^{m+n} (1, \hat{\bar{k}} - m)^2 \theta_k \theta_{-k} \equiv u^2 x^2(j) + v^2 2\theta^2(\omega), \quad (3)$$

$\hat{\bar{k}} = \bar{k} - m$, $\bar{k} = m+1, \ldots, m+n$; $\hat{\bar{k}} = \bar{k} - 2m$, $\bar{k} = m+n+1, \ldots, m+2n$, which is the natural unification of CK orthogonal and symplectic forms. The distributions of contraction parameters j, ω among matrix elements of the fundamental representation of $osp(m;j|2n;\omega)$ and transformations of the generators (1) are obtained in a standard CK manner and are as follows:

$$E_{ik} = [i,k] E_{ik}^*, \quad E_{\bar{i}\bar{k}} = (\hat{\bar{i}}, \hat{\bar{k}}) E_{\bar{i}\bar{k}}, \quad E_{i\bar{k}} = u[1,i] v(1,\hat{\bar{k}}) E_{i\bar{k}}^*, \quad (4)$$

where E^* are generators (1) of the starting superalgebra $osp(m|2n)$. The transformed generators are subject of the (super) commutation relations:

$$[E_{ij}, E_{kl}] = [i,j][k,l] \left(\frac{G_{jk} E_{il}}{[i,l]} + \frac{G_{il} E_{jk}}{[j,k]} - \frac{G_{ik} E_{jl}}{[j,l]} - \frac{G_{jl} E_{ik}}{[i,k]} \right),$$

$$[E_{\bar{i}\bar{j}}, E_{\bar{k}\bar{l}}] = -(\hat{\bar{i}}, \hat{\bar{j}})(\hat{\bar{k}}, \hat{\bar{l}}) \left(\frac{G_{\bar{j}\bar{k}} E_{\bar{i}\bar{l}}}{(\hat{\bar{i}}, \hat{\bar{l}})} + \frac{G_{\bar{i}\bar{l}} E_{\bar{j}\bar{k}}}{(\hat{\bar{j}}, \hat{\bar{k}})} + \frac{G_{\bar{i}\bar{k}} E_{\bar{j}\bar{l}}}{[\hat{\bar{j}}, \hat{\bar{l}})} + \frac{G_{\bar{j}\bar{l}} E_{\bar{i}\bar{k}}}{[\hat{\bar{i}}, \hat{\bar{k}})} \right),$$

$$[E_{ij}, E_{\bar{k}\bar{l}}] = 0, \quad [E_{ij}, E_{k\bar{l}}] = [i,j][1,k] \left(\frac{G_{jk} E_{i\bar{l}}}{[1,i]} - \frac{G_{ik} E_{j\bar{l}}}{[1,j]} \right),$$

$$[E_{i\bar{j}}, E_{\bar{k}\bar{l}}] = -(1, \hat{\bar{j}})(\hat{\bar{k}}, \hat{\bar{l}}) \left(\frac{G_{\bar{j}\bar{k}} E_{i\bar{l}}}{(1, \hat{\bar{l}})} + \frac{G_{\bar{j}\bar{l}} E_{i\bar{k}}}{(1, \hat{\bar{k}})} \right),$$

$$\{E_{i\bar{j}}, E_{k\bar{l}}\} = u^2 v^2 [1,i](1,\hat{\bar{j}})[1,k](1,\hat{\bar{l}}) \left(\frac{G_{ik} E_{\bar{j}\bar{l}}}{(\hat{\bar{j}}, \hat{\bar{l}})} - \frac{G_{\bar{j}\bar{l}} E_{ik}}{[i,k]} \right). \quad (5)$$

For $u = \iota$ or $v = \iota$, $\iota^2 = 0$ superalgebra $osp(m|2n)$ is contracted to inhomogeneous superalgebra, which is semidirect sum $\{E_{i\bar{j}}\} \ni (so(m) \bigoplus sp(2n))$, with all anticommutators of the odd generators equal to zero $\{E_{i\bar{j}}, E_{k\bar{p}}\} = 0$.

4 Examples

4.1 Kinematical contractions of $osp(1|2)$

This contractions was described in detail in [9] both on the level of (super) commutation relations and as subrepresentations of the fundamental matrix representations of $osp(3|2)$. The isomorphism of low dimensional Lie algebras $sp(2)$ and $so(3)$ are essentially used for contraction $osp(1|2)$ to $(1+1)$ dimensional Poincare and Galilei superalgebras. This case is not included in general CK contractions of the previous section and we give here the fundamental 3×3 representations of $(1+1)$ Poincare and Galilei superalgebras which was absent in [9]. For this purpose we need to introduce the algebra $A_4(\xi)$, which is free generated by ξ_1, ξ_2, where $\xi_1\xi_2 = \xi_2\xi_1$, $\xi_1^4 = \xi_2^4 = 0$. If one take the basis $X_{23}^* = \frac{i}{2}E_{23}$, $X_{12}^* = \frac{i}{4}(E_{33}+E_{22})$, $X_{13}^* = \frac{1}{4}(E_{33}-E_{22})$, $Q_+^* = E_{12}$, $Q_-^* = E_{13}$, and one transform the generators as follows

$$X_{12} = \omega_1^2 X_{12}^*,\ X_{23} = \omega_2^2 X_{23}^*,\ X_{13} = \omega_1^2\omega_2^2 X_{13}^*,\ Q_\pm = \omega_1\omega_2 Q_\pm^*, \tag{6}$$

where 3×3 matrix E are given by (1), each parameter $\omega_k = 1, \xi_k, i$, $k = 1, 2$, then the (super) commutation relations of $osp(1|2;\omega)$ may be written in the form

$$[X_{12}, X_{13}] = \omega_1^4 X_{23},\ [X_{13}, X_{23}] = \omega_2^4 X_{13},\ [X_{23}, X_{12}] = X_{13},$$
$$[X_{12}, Q_\pm] = \pm\frac{i}{2}\omega_1^2 Q_\mp,\ [X_{13}, Q_\pm] = \frac{1}{2}\omega_1^2\omega_2^2 Q_\mp, [X_{23}, Q_\pm] = \pm\frac{i}{2}\omega_2^2 Q_\mp,$$
$$\{Q_+, Q_-\} = -2i\omega_1^2 X_{12},\ \{Q_+, Q_+\} = -2(X_{13} + i\omega_2^2 X_{12}),$$
$$\{Q_-, Q_-\} = 2(X_{13} - i\omega_2^2 X_{12}), \tag{7}$$

which is coincided with the super commutators (4.51) in [9] for the standard contractions of $osp(1|2)$. Our designations of generators and contraction parameters are connected with corresponding of [9] as follows: $\omega_k = \epsilon_k, X_{12} = K_{21} = H$, $X_{13} = K_{20} = P$, $X_{23} = K_{01} = K$. The slight differences in a structure constant are due to the use of complex $so(3;\mathbf{C})$ instead of its anti de Sitter real form $so(2,1)$. The $(1+1)$ super Poincare algebra is obtained for $\omega_1 = \xi_1, \omega_2 = 1$ (compare with (4.12) in [9]) and $(1+1)$ super Galilei algebra is given by (7) for $\omega_1 = \xi_1, \omega_2 = \xi_2$ (compare with (4.39) in [9]). The commutators $[A, B] = \xi^k C$, $k = 1, 2, 3$, are regarded as zero, i.e. $[A, B] = 0$.

The Grassmann-hull [10] $M(\omega) = 2\alpha X_{23} + 2\beta X_{12} + 2\gamma X_{13} + \mu Q_+ + \nu Q_-$ of $osp(1|2;\omega)$ is represented by the matrix

$$M(\omega) = \left(\begin{array}{c|cc} 0 & \omega_1\omega_2\mu & \omega_1\omega_2\nu \\ \hline -\omega_1\omega_2\nu & -i\omega_2^2\alpha & -\omega_1^2(i\beta + \omega_2^2\gamma) \\ \omega_1\omega_2\mu & \omega_1^2(i\beta - \omega_2^2\gamma) & i\omega_2^2\alpha \end{array}\right),$$

where μ, ν are odd grassmannian elements: $\mu^2 = \nu^2 = 0$, $\mu\nu = -\nu\mu$. For this simplest case it is possible to find the corresponding supergroup $Osp(1|2;\omega)$ explicitly[11], namely

$$\mathcal{M}(\omega) = \exp M(\omega) = I + \frac{\sinh u}{u} M(\omega) + \frac{\cosh u - 1}{u^2} M^2(\omega) +$$

$$+\omega_1^2\omega_2^2 \frac{2(1-\cosh u) + u\sinh u}{u^2}\mu\nu A + \omega_1^2\omega_2^2 \frac{u\cosh u - \sinh u}{u^3}\mu\nu B(\omega),$$

where $u^2 = \omega_1^4(\beta^2 + \omega_2^4\gamma^2) - \omega_2^4\alpha^2$,

$$M^2(\omega) = \left(\begin{array}{c|cc} -2\omega_1^2\omega_2^2\mu\nu & -i\omega_2^2\alpha\mu + \omega_1^2(i\beta - \omega_2^2\gamma)\nu & i\omega_2^2\alpha\nu - \omega_1^2(i\beta + \omega_2^2\gamma)\mu \\ \hline i\omega_2^2\alpha\nu - \omega_1^2(i\beta + \omega_2^2\gamma)\mu & u^2 + \omega_1^2\omega_2^2\mu\nu & 0 \\ i\omega_2^2\alpha\mu - \omega_1^2(i\beta - \omega_2^2\gamma)\nu & 0 & u^2 + \omega_1^2\omega_2^2\mu\nu \end{array}\right),$$

$$A = \left(\begin{array}{c|cc} 0 & 0 & 0 \\ \hline 0 & 1 & 0 \\ 0 & 0 & 1 \end{array}\right), \quad B(\omega) = \left(\begin{array}{c|cc} 0 & 0 & 0 \\ \hline 0 & -i\omega_2^2\alpha & -\omega_1^2(i\beta + \omega_2^2\gamma) \\ 0 & \omega_1^2(i\beta - \omega_2^2\gamma) & i\omega_2^2\alpha \end{array}\right).$$

In the case of superalgebra $osp(1|4)$ the isomorphism $sp(4) \cong so(5)$ is used in [9] and the standard kinematical contractions of $osp(1|4)$ to $(3+1)$ super Poincare and $(3+1)$ super Galilei are regarded for abstract generators and for embedding in the fundamental representations of $osp(5|4)$.

4.2 CK contractions of $osp(3|2)$

This superalgebra has $so(3)$ as even subalgebra therefore their contractions to the kinematical $(1+1)$ Poincare, Newton and Galilei superalgebras may be fulfilled according to general CK scheme of the first section. But unlike of two odd generators of $osp(1|2)$ the superalgebra $osp(3|2)$ has six odd generators. In the basis $X_{ik} = E_{ki}$, $k, i = 1, 2, 3$, $F = \frac{1}{2}E_{44}$, $E = -\frac{1}{2}E_{55}$, $H = -E_{45}$, $Q_k = E_{k4}$, $Q_{-k} = E_{k5}$ the generators are affected by the contraction coefficients j_1, j_2 in the following way

$$X_{ik} \to [i,k]X_{ik}, \quad Q_{\pm k} \to [1,k]Q_{\pm k} \tag{8}$$

and H, F, E are remained unchanged. Then superalgebra $osp(3; j|2)$ is given by

$$[X_{12}, X_{13}] = j_1^2 X_{23}, \quad [X_{13}, X_{23}] = j_2^2 X_{12}, \quad [X_{23}, X_{12}] = X_{13},$$

$$[H,E]=2E,\quad [H,F]=-2F,\quad [E,F]=H,$$
$$[X_{ik},Q_{\pm i}]=Q_{\pm k},\quad [X_{ik},Q_{\pm k}]=-[i,k]^2Q^2_{\pm i},\ \ i<k,$$
$$[H,Q_{\pm k}]=\mp Q_{\pm k},\quad [E,Q_k]=-Q_{-k},\quad [F,Q_{-k}]=-Q_k,$$
$$\{Q_k,Q_k\}=[1,k]^2F,\quad \{Q_{-k},Q_{-k}\}=-[1,k]^2E,$$
$$\{Q_k,Q_{-k}\}=-[1,k]^2H,\quad \{Q_{\pm i},Q_{\mp k}\}=\pm[1,k]^2X_{ik}.$$

The non-minimal Poincare superalgebra is obtained for $j_1=\iota_1$, $j_2=i$ and has the structure of the semidirect sum $T \ni (\{X_{23}\}\oplus osp(1|2))$, where abelian $T=\{X_{12},X_{13},Q_{\pm 2},Q_{\pm 3}\}$ and $osp(1|2)=\{H,E,F,Q_{\pm 1}\}$. The Newton superalgebra $osp(3;\iota_2|2)=T_2 \ni osp(2|2)$, where $T_2=\{X_{13},X_{23},Q_{\pm 3}\}$ and $osp(2|2)$ is generated by $X_{12},H,E,F,Q_{\pm 1},Q_{\pm 2}$. Finally the non-minimal Galilei superalgebra may be presented as semidirect sums $osp(3;\iota_1,\iota_2|2)=(T \ni \{X_{23}\}) \ni osp(1|2)=T \ni (\{X_{23}\}\oplus osp(1|2))$.

Acknowledgments

NG would like to thank Mariano del Olmo for sending the copy of paper[9]. This work was supported by Russian Foundation for Basic Research under Project 01-01-96433.

References

1. Yu. A. Golfand and E. P. Likhtman, *JETP. Lett.* **13**, 452 (1971).
2. D. V. Volkov and V. P. Akulov, *JETP Lett.* **16**, 621 (1972).
3. J. Wess and B. Zumino, *Nucl. Phys.* **B70**, 139 (1974).
4. P. West, *Introduction to supersymmetry and supergravity*, World Scientific, Singapore (1986).
5. B. W. Keck , *J. Phys. A: Math. Gen.* **8**, 1819 (1975).
6. M. A. Vasiliev, *Phys. Lett. B* **243**, 378 (1990)
7. I. Bars, hep-th/9608061, (1996).
8. S. Bellucci, E. Ivanov and S. Krivonos, *Nucl. Phys. B (Proc. Suppl.)* **102&103**, 26 (2001).
9. V. Hussin, J. Negro and M.A. del Olmo, *J. Phys. A: Math. Gen.* **32**, 5097 (1999).
10. H. Boseck, *Seminar Sophus Lie* **1**, 109 (1991).
11. V. Hussin and L.M. Nieto, *Preprint* CRM–1863 (1993).

DISTINGUISHED BASES OF $sl(n, C)$ AND THEIR SYMMETRIES

M. HAVLÍČEK, E. PELANTOVÁ,
Department of Mathematics, Czech Technical University, Trojanova 13, CZ – 110 00 Prague 1, Czech Republic
E-mail: edita.pelantova@fjfi.cvut.cz

J. PATERA
Centre de Recherches Mathématiques, Université de Montréal, C.P. 6128 – Centre Ville, Montréal H3C 3J7, Canada
E-mail: patera@crm.umontreal.ca

J. TOLAR
Department of Physics, Czech Technical University, Břehová 7, CZ – 115 19 Prague 1, Czech Republic
E-mail: jiri.tolar@fjfi.cvut.cz

Fine gradings of simple Lie algebras determine their distinguished bases. Certain subgroups of automorphisms yield the symmetries of these bases. Complete description of the bases and their symmetries is given for $sl(2, C)$ and $sl(3, C)$.

1 Introduction

Gradings of simple Lie algebras L over C or R are basic structural properties of each L [1]. Their fine gradings, recently classified [2,3], yield special distinguished bases in L. Of course, the study of possible symmetries of gradings is of interest — not only as the symmetries of the bases but also as the symmetries of e.g. the system of quadratic equations determining the contractions which preserve the given grading [4,5]. In this contribution we consider the classical Lie algebras $A_n = sl(n+1, C)$ and describe our specific results for $sl(2, C)$ and $sl(3, C)$ [5].

The decomposition $\Gamma : \ L = \bigoplus_{i\in I} L_i$ is called a *grading* if, for any pair of indices $i, j \in I$, there exists an index $k \in I$ such that $0 \neq [L_i, L_j] \subseteq L_k$. If $\Gamma : L = \bigoplus_{i\in I} L_i$ is a grading and g is an automorphism of L, then $\tilde{\Gamma} : L = \bigoplus_{i\in I} g(L_i)$ is also a grading. Two such gradings are called *equivalent.* A grading $\Gamma : L = \bigoplus_{i\in I} L_i$ is a *refinement* of the grading $\overline{\Gamma} : L = \bigoplus_{i\in J} \overline{L}_j$ if for any $i \in I$ there exists $j \in J$ such that $L_i \subseteq \overline{L}_j$. A grading which cannot be properly refined is called *fine.*

To construct a grading of L one can use a diagonalizable automorphism g from $Aut\, L$. It is easy to see that the resulting decomposition of L into eigenspaces of g is a grading. Furthermore, if one takes a set of commuting automorphisms from $Aut\,L$, and decomposes L according to all automorphisms into their common eigenspaces, a finer grading of L is obtained in general. This procedure can be followed until a maximal Abelian group of diagonalizable automorphisms — a **MAD–group** — of L is obtained. MAD-groups are important for the theory of fine gradings of simple Lie algebras. The basic fact was proven [1]:

> If L is a simple Lie algebra over C, then a grading Γ of L is fine if and only if there exists a MAD-group $\mathcal{G}$ such that Γ is the decomposition into eigenspaces of automorphisms from $\mathcal{G}$. Further, two fine gradings Γ_1 and Γ_2 are equivalent if and only if the corresponding MAD-groups $\mathcal{G}_1$ and $\mathcal{G}_2$ are conjugate in $Aut\,L$. Moreover, if L is simple, the index set I can be embedded in a finite Abelian group G; if the multiplication in G is written additively, then $0 \neq [L_i, L_j] \subseteq L_{i+j}$.

Hence any fine grading of a simple Lie algebra over C is the decomposition Γ of L into eigenspaces of automorphisms belonging to some MAD-group $\mathcal{G} \subset Aut\,L$. For any subspace L_i of the fine grading and any element $g \in \mathcal{G}$ we have $g\,L_i = L_i$, i.e. the elements of a MAD-group preserve each subspace of the corresponding grading ($\mathcal{G}$-grading). Here we are interested in symmetries of a given grading, i.e. in automorphisms of L which preserve a $\mathcal{G}$-grading of L but not each of its subspaces separately. In other words, we are looking for those elements of $Aut\,L$ which permute the grading subspaces.

We can see that they are just the elements of the *normalizer* $\mathcal{N}(\mathcal{G})$ of the MAD-group $\mathcal{G}$ in $Aut\,L$. Namely, the normalizer $\mathcal{N}(\mathcal{G})$ is defined as the set $\{h \in Aut\,L \mid h^{-1}\mathcal{G}h \subset \mathcal{G}\}$. Let $h \in \mathcal{N}(\mathcal{G})$ and L_i be a subspace of the fine grading $\bigoplus_{i\in I} L_i$ corresponding to the MAD-group $\mathcal{G}$. Since $h^{-1}fh \in \mathcal{G}$ for arbitrarily chosen $f \in \mathcal{G}$, we can find $g \in \mathcal{G}$ such that $h^{-1}fh = g$. Applying this automorphism to L_i and using that $gL_i = L_i$, we obtain $f(hL_i) = hL_i$, i.e. hL_i is an eigenspace for any automorphism $f \in \mathcal{G}$, which means that $hL_i = L_j$ for some index $j \in I$. Any $h \in \mathcal{N}(\mathcal{G})$ thus generates some permutation π on grading indices $\pi : I \to I$.

The complete classification of MAD-groups for classical Lie algebras over C is given in [2]. Let us recall the particular case of the Lie algebras $sl(n, C)$ we are interested in. The group of automorphisms $Aut\,sl(n, C)$ consists of inner automorphisms $Ad_A X$ and the outer ones $Out_A X$:

$$\begin{aligned} Ad_A X &:= A^{-1}XA, & A \in SL(n,C) = \{B \in C^{n\times n}\,, \det B = 1\} \\ Out_A X &:= -(A^{-1}XA)^T\,, & A \in SL(n,C) \end{aligned}$$

Sections 2 and 3 are devoted to $L = sl(2, C)$ and $L = sl(3, C)$, respectively. For each grading we give the explicit form of: the grading group $\mathcal{G}$, the index group G, the $\mathcal{G}$-grading decomposition Γ, the matrix form of the distinguished basis $\mathcal{B}$ and the group $\mathcal{N}(\mathcal{G})/\mathcal{G}$ of non–trivial symmetries of the basis.

2 The two distinguished bases of $sl(2,C)$ and their symmetries

Aut $sl(2,C)$ has two non–conjugate MAD–groups [2], and therefore two inequivalent fine gradings. Note that both MAD-groups are formed by inner automorphisms only. Let us list them.

$$\mathcal{G}_1 = \{Ad_A \mid A = diag(\alpha_1,\alpha_2), \alpha_1,\alpha_2 \in C^*\}$$
$$G_1 = Z_3$$
$$\Gamma_1 : \ sl(2,C) = L_{-1} \oplus L_0 \oplus L_{+1}$$
$$\mathcal{B}_1 = \{\begin{pmatrix}0&0\\1&0\end{pmatrix}, \begin{pmatrix}1&0\\0&-1\end{pmatrix}, \begin{pmatrix}0&1\\0&0\end{pmatrix}\}$$
$\mathcal{N}(\mathcal{G}_1)/\mathcal{G}_1 = S_2$ has 2 elements generated by Ad_P, $P = \begin{pmatrix}0&1\\1&0\end{pmatrix}$.

$$\mathcal{G}_2 = \left\{Ad_{P^jQ^k} \mid P = \begin{pmatrix}0&1\\1&0\end{pmatrix}, Q = \begin{pmatrix}1&0\\0&-1\end{pmatrix}, j,k = 0,1\right\}$$
$$G_2 = Z_2 \times Z_2$$
$$\Gamma_2 : \ sl(2,C) = L_{01} \oplus L_{10} \oplus L_{11}$$
$$\mathcal{B}_2 = \{\begin{pmatrix}1&0\\0&-1\end{pmatrix}, \begin{pmatrix}0&1\\1&0\end{pmatrix}, \begin{pmatrix}0&1\\-1&0\end{pmatrix}\}$$
$\mathcal{N}(\mathcal{G}_2)/\mathcal{G}_2 = SL(2,Z_2)$ has 6 elements generated by Ad_S, Ad_D,
$S = \begin{pmatrix}1&1\\1&-1\end{pmatrix}$, $D = \begin{pmatrix}1&0\\0&-i\end{pmatrix}$.

3 The four distinguished bases of $sl(3,C)$ and their symmetries

Aut $sl(3,C)$ has four non-conjugate MAD-groups [2], and therefore four inequivalent fine gradings. They are listed below.

$$\mathcal{G}_1 = \{Ad_A \mid A = diag(\alpha_1,\alpha_2,\alpha_3), \alpha_1,\alpha_2,\alpha_3 \in C^*\}$$
$$G_1 = Z_3 \times Z_3$$
$$\Gamma_1 : \ sl(3,C) = L_{00} \oplus L_{10} \oplus L_{01} \oplus L_{11} \oplus L_{-1-1} \oplus L_{0-1} \oplus L_{-10}$$
$$\mathcal{B}_1 = \{\begin{pmatrix}a&0&0\\0&b&0\\0&0&-a-b\end{pmatrix}, \begin{pmatrix}0&1&0\\0&0&0\\0&0&0\end{pmatrix}, \begin{pmatrix}0&0&0\\0&0&1\\0&0&0\end{pmatrix}, \begin{pmatrix}0&0&1\\0&0&0\\0&0&0\end{pmatrix}, \begin{pmatrix}0&0&0\\0&0&0\\1&0&0\end{pmatrix},$$

$$\begin{pmatrix}0&0&0\\0&0&0\\0&1&0\end{pmatrix}, \begin{pmatrix}0&0&0\\1&0&0\\0&0&0\end{pmatrix}\}$$

$\mathcal{N}(\mathcal{G}_1)/\mathcal{G}_1 = Z_2 \times S_3$ has 12 elements generated by Out_I, Ad_P, Ad_B,

$$P = \begin{pmatrix}0&1&0\\0&0&1\\1&0&0\end{pmatrix}, \; B = \begin{pmatrix}1&0&0\\0&0&1\\0&1&0\end{pmatrix}.$$

$\mathcal{G}_2 = \{Ad_A \mid A = diag(\varepsilon_1, \varepsilon_2, \varepsilon_3),\ \varepsilon_i = \pm 1\} \cup \{Out_A \mid A = diag(\varepsilon_1, \varepsilon_2, \varepsilon_3),\ \varepsilon_i = \pm 1\}$

$G_2 = Z_2 \times Z_2 \times Z_2$

$\Gamma_2: \ sl(3, C) = L_{001} \oplus L_{111} \oplus L_{101} \oplus L_{011} \oplus L_{110} \oplus L_{010} \oplus L_{100}$

$$\mathcal{B}_2 = \{\begin{pmatrix}a&0&0\\0&b&0\\0&0&-a-b\end{pmatrix}, \begin{pmatrix}0&1&0\\1&0&0\\0&0&0\end{pmatrix}, \begin{pmatrix}0&0&1\\0&0&0\\1&0&0\end{pmatrix}, \begin{pmatrix}0&0&0\\0&0&1\\0&1&0\end{pmatrix}, \begin{pmatrix}0&-1&0\\1&0&0\\0&0&0\end{pmatrix},$$
$$\begin{pmatrix}0&0&0\\0&0&-1\\0&1&0\end{pmatrix}, \begin{pmatrix}0&0&-1\\0&0&0\\1&0&0\end{pmatrix}\}$$

$\mathcal{N}(\mathcal{G}_2)/\mathcal{G}_2 = D \times S_3$ has 18 elements generated by Ad_H, Ad_P, Ad_B,

$$H = \begin{pmatrix}1&0&0\\0&i&0\\0&0&i\end{pmatrix}.$$

$\mathcal{G}_3 = \{Ad_A \mid A = diag(\varepsilon, \alpha, \alpha^{-1}),\ \varepsilon = \pm 1, \alpha \in C^*\} \cup$
$$\cup \left\{ Out_A \mid A = \begin{pmatrix}\varepsilon&0&0\\0&0&\alpha\\0&\alpha^{-1}&0\end{pmatrix},\ \varepsilon = \pm 1, \alpha \in C^* \right\}$$

$G_3 = Z_8$

$\Gamma_3: \ sl(3, C) = L_0 \oplus L_1 \oplus L_2 \oplus L_3 \oplus L_4 \oplus L_5 \oplus L_6 \oplus L_7$

$$\mathcal{B}_3 = \{\begin{pmatrix}0&0&0\\0&1&0\\0&0&-1\end{pmatrix}, \begin{pmatrix}0&1&0\\0&0&0\\-1&0&0\end{pmatrix}, \begin{pmatrix}0&0&0\\0&0&1\\0&0&0\end{pmatrix}, \begin{pmatrix}0&0&1\\1&0&0\\0&0&0\end{pmatrix}, \begin{pmatrix}2&0&0\\0&-1&0\\0&0&-1\end{pmatrix},$$
$$\begin{pmatrix}0&1&0\\0&0&0\\1&0&0\end{pmatrix}, \begin{pmatrix}0&0&0\\0&0&0\\0&1&0\end{pmatrix}, \begin{pmatrix}0&1&0\\-1&0&0\\0&0&0\end{pmatrix}\}$$

$\mathcal{N}(\mathcal{G}_3)/\mathcal{G}_3 = Z_2 \times Z_2$ has 4 elements generated by Ad_H, Ad_B.

$$\mathcal{G}_4 = \left\{ Ad_{P^jQ^k} \mid P = \begin{pmatrix} 0\,1\,0 \\ 0\,0\,1 \\ 1\,0\,0 \end{pmatrix},\ Q = \begin{pmatrix} 1 & 0 & 0 \\ 0 & \omega & 0 \\ 0 & 0 & \omega^2 \end{pmatrix},\ \omega = e^{i\frac{2\pi}{3}},\ j,k = 0,1,2 \right\}$$

$G_4 = Z_3 \times Z_3$

$\Gamma_4 :\ sl(3,C) = L_{10} \oplus L_{20} \oplus L_{01} \oplus L_{02} \oplus L_{11} \oplus L_{21} \oplus L_{12} \oplus L_{22}$

$\mathcal{B}_4 = \{P, P^2, Q, Q^2, PQ, P^2Q, PQ^2, P^2Q^2\}$

$\mathcal{N}(\mathcal{G}_4)/\mathcal{G}_4 = Z_2 \times SL(2, Z_3)$ has 48 elements generated by Out_I, Ad_S, Ad_D,

$$S_{ij} = \omega^{-ij},\ D = \begin{pmatrix} 1\,0\,0 \\ 0\,1\,0 \\ 0\,0\,\omega \end{pmatrix}.$$

Note that $\mathcal{G}_1$ and $\mathcal{G}_4$ are MAD-groups formed by inner automorphisms only. The group $\mathcal{G}_1$ is infinite. It is the maximal torus of $sl(3,C)$. The group $\mathcal{G}_4$ is finite and consists of 9 elements; it has been called the Pauli group. The groups $\mathcal{G}_2$ and $\mathcal{G}_3$ contain outer automorphisms. There are 8 elements in $\mathcal{G}_2$. The group $\mathcal{G}_3$ is infinite, containing a 1–parameter subgroup of the torus.

Acknowledgments

M.H., E.P. and J.T. gratefully acknowledge the support of the Ministry of Education of Czech Republic under the contract MSM 210000018. J.P. has been partially supported by the National Science and Engineering Research Council of Canada.

References

1. J. Patera, H. Zassenhaus, *J. Math. Phys.* **29**, 665 (1988)
2. M. Havlíček, J. Patera, E. Pelantová, *Linear Algebra & Its Appl.* **277**, 97 (1998)
3. M. Havlíček, J. Patera, E. Pelantová, *Linear Algebra & Its Appl.* **314**, 1 (2000)
4. M. de Montigny, J. Patera, *J. Phys. A: Math. Gen.* **24**, 525 (1991)
5. M. Havlíček, J. Patera, E. Pelantová, J. Tolar, Automorphisms of the fine grading of $sl(n,C)$ associated with the generalized Pauli matrices, to appear in *J. Math. Phys.*

QUANTUM EFFECTS IN MATTER-WAVE DIFFRACTION

GERHARD C. HEGERFELDT AND THORSTEN KÖHLER

Institut für Theoretische Physik
Universität Göttingen, Germany
email: hegerf@theorie.physik.uni-goettingen.de

Advances in micro-technology of the last years have made it possible to carry optics textbooks experiments over to atomic and molecular beams, such as diffraction by a double slit or transmission grating. The usual wave-optical approach gives a good qualitative description of these experiments. However, small deviations therefrom and sophisticated quantum mechanics yield new surprising insights on the size of particles and on their interaction with surfaces.

1 Introduction

The wave nature of subatomic particles was postulated by de Broglie in 1923 and this idea suffices to explain many diffraction experiments. Indeed, some people have argued that for matter diffraction a good optics book is sufficient. The aim of this contribution is to show that in many cases this is not so, and that full quantum mechanics may be necessary to describe and evaluate recent more sophisticated experiments. First of all it is clear from the statistical interpretation of quantum mechanics that a diffraction picture is build up slowly from individual particles each of which contributes a single dot on the screen. A single particle does not give an interference picture, only the complete particle beam does. This shows the wave property belongs to the beam and not the single particle.

Interestingly, for atoms the simple but fundamental double slit experiment has been just a thought experiment for a long time. This is due to their small de Broglie wavelengths $\lambda = H/p$, with p the particle moment. For usual beam velocities o a few hundred meters per seconds this is only about 1 Å. Therefore very small slit widths and distances are needed to obtain observable diffraction angles. Only the recent advances in micro-technology have made atomic diffraction experiments possible.

Let us first consider wave optics and a transmission grating of period d with N slits of widths s. If a classical wave passes through a grating one observes behind it and outside the original direction an intensity with characteristic directional modulation. For a perpendicularly incident plane wave the

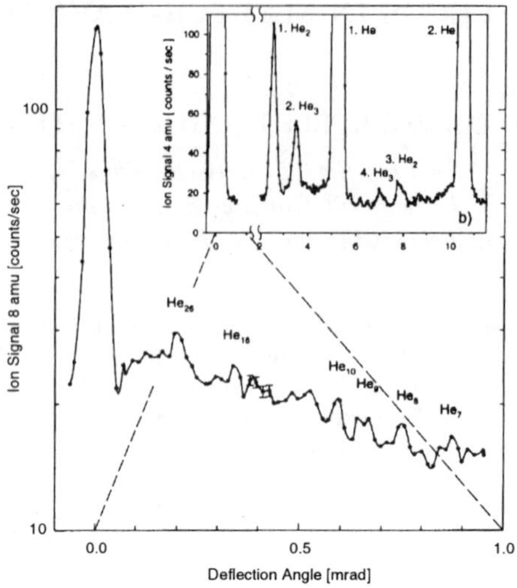

Figure 1: Diffraction pattern of helium atoms and molecules up to He_{26} and thereby the first definite detection of these exotic molecules (cf. Refs. 3 and 4).

diffracted intensity is given by

$$I(\theta) \propto \left(\frac{\sin(\frac{1}{2}Nd\,k\sin\theta)}{\sin(\frac{1}{2}d\,k\sin\theta)}\right)^2 \times \left(\frac{\sin(\frac{1}{2}s\,k\sin\theta)}{\frac{1}{2}s\,k\sin\theta}\right)^2 . \tag{1}$$

The first factor is the so called grating function, with narrow maxima of height Nd at the angles

$$\sin\vartheta_n = n\lambda/d \quad (n = 0, \pm1, \pm2, \cdots) , \tag{2}$$

and these maxima together with the second factor, the so called slit function, give the intensity

$$I_n \equiv I(\theta_n) \propto \frac{\sin^2(n\pi s_0/d)}{(n\pi s_0/d)^2} . \tag{3}$$

For many purposes this simple wave-optical approach gives a good description of matter diffraction also. However, in some cases effects from the full quantum theory may be important so that the simple wave picture is no longer appropriate. This will now be explained for recent experiments.

The typical experimental setup for atomic and molecular diffraction consists of a beam of particles with very small velocity distibution which pass through a transmission grating or double slit [1,2,3]. The grating used by the Toennies group in Göttingen has a period of $d = 100$ nm. A beautiful diffraction pattern for a helium beam is shown in Fig. 1. In the inset in the upper right hand corner one observes the first helium diffraction order and left of it at half the angle another small maximum. The latter provided the first direct evidence of the exotic helium molecule $^4He_2$. In the atomic beam there can be helium clusters, all moving with the same velocity. Therefore their de Broglie wavelengths and their diffraction angles are inversely proportional to their mass. The main part of the figure shows diffraction maxima of higher clusters up to He_{26}. The Zeilinger group in Vienna recently observed diffraction of the fullerenes C_{60} and C_{70} [5].

Deviations from the simple wave-optical diffraction theory are expected to occur [6] due to

- the inner structure of the particles and van der Waals potentials
- the spatial extent of the particles
- the breakup of weakly bound molecules.

Are these expected deviations just "dirt effects" or do they contain surprises with useful information? To investigate this question one needs the full quantum theory.

2 Quantum Theory of Matter-Wave Diffraction

First of all one has to realize that matter diffraction off a grating is not a classical wave phenomenon but a quantum mechanical *scattering* problem. The diffraction is not caused by the slits but by scattering of the particles off the grating bars. This is depicted in Fig. 2. In addition to the reflective particle-surface interaction one also has to take a attractive van der Waals surface interaction into account.

Starting from the Schrödinger equation and using Faddeev scattering theory in the formulation of Alt, Grassberger and Sandhas [7,8] we have obtained a general expression for the diffraction intensity I_n of an extended molecule in the form [9]

$$I_n \propto e^{-(2\pi n\sigma/d)^2} \left[\frac{\sin^2(n\pi s_{\rm eff}/d) + \sinh^2(n\pi\delta/d)}{(n\pi s_{\rm eff}/d)^2 + (n\pi\delta/d)^2} \right] \tag{4}$$

where $s_{\rm eff}$ denotes an effective slit width which is smaller than s. The term δ diminishes the contrast and the exponential term takes into account that the number of molecules in the individual diffraction orders may decrease due to

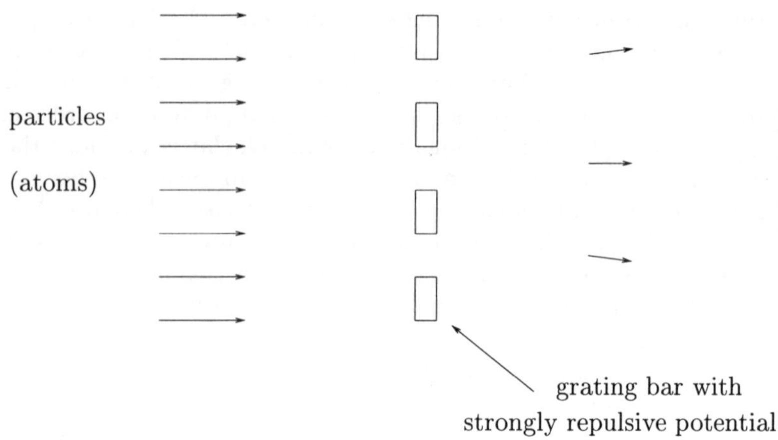

Figure 2: Matter diffraction arises from scattering off the grating bars.

breakups at the bars and that small variations in the bar widths may occur. All these parameters may depend on the particle-grating interaction, on the spatial extent of the particles and on their velocity.

2.1 *Surface Effects*

During the passage through a slit an atom experiences an additional attractive surface van der Waals potential $V = -C_3/l^3$ where l denotes the distance from the surface of a grating bar and where C_3 depends on the particle species. Quite recently we have found the as yet unpublished result that for rare gas atoms s_{eff} behaves as $s_{\text{eff}} \propto 1/\sqrt{v}$ where v is the particle velocity. Therefore we have plotted $s_{\text{eff}}(v)$ obtained from experimental diffraction patterns for variable helium beam velocities as a function of $1/\sqrt{v}$ in Fig. 3. From the slope one can determine C_3. The intersection with the ordinate axis yields the true slit width. The result agrees with an alternative procedure we have used before [9]. The method is so sensitive that the geometrical trapeze form of the grating base has to be taken into account.

For the evaluation of these measurements one does indeed need quantum mechanics but qualitatively the difference to wave-optics is not so pronounced since an optical grating with a dielectric coating can produce similar effects. This however is no longer true in the following example.

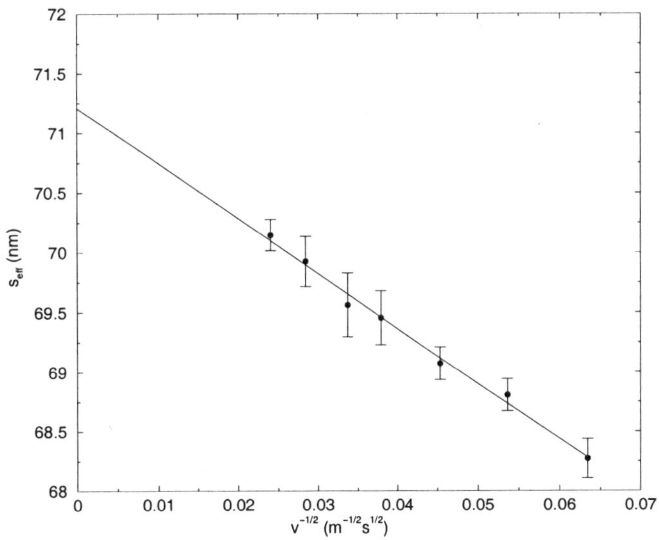

Figure 3: Plot of s_{eff} over $1/\sqrt{v}$ for helium. From the slope of the straight line one can calculate C_3. The ordinate intersection gives the true slit width of $s = 71.2$ nm.

2.2 Size of the Helium Dimer

The main difference to wave-optics occurs for excitations of higher levels [10] as for the helium trimer He_3 and for size and breakup effects [6]. The latter two effects are particularly interesting for the exotic He_2 which is fifty times larger than a hydrogen molecule and whose binding energy is a hundred million times smaller than that of an electron in a hydrogen atom. Therefore He_2 is extremely fragile and thus very difficult to investigate by conventional methods. But at small diffraction angles most helium dimers arrive at the detectors unbroken so that deviations of the diffraction pattern from the predictions of wave-optics can be measured and analyzed.

In Ref. [11] experimental dimer diffraction intensities up to 7th order have been fitted to the expression I_n from Eq. (4) and s_{eff} was determined. By quantum mechanical multi-channel scattering theory we obtained a relation between $s_{\text{eff}}(v)$ and the inter-nuclear distance $\langle r \rangle$. A simplified result is $\langle r \rangle / 2 = s_{\text{eff}}^{\text{He}}(v) - s_{\text{eff}}^{^4\text{He}_2}(v)$, which gives $< r > \approx 50$ Å. The more precise theory yields $\langle r \rangle = 52 \pm 4$ Å.

3 Conclusions

These examples show that refinements of a simple textbook experiment on matter diffraction can lead to new quantum mechanical applications in atomic and molecular physics. In particular we have discussed

- its use as a quantum mass spectrometer
- exploitation of quantum effects for
  - particle surface interaction
  - influence of the particle size on the diffraction pattern: size determination
  - detection of excited energy levels for He_3 (work in progress)

As a further development we mention the setup of an atom interferometer which, in a first approximation, can be understood similarly as in wave-optics, but to describe finer effects one needs full quantum mechanics.

References

1. D. W. Keith, M. L. Schattenburg, H. I. Smith, D. E. Pritchard, *Phys. Rev. Lett.* **61**, 1580 (1988).
2. O. Carnal and J. Mlynek, *Phys. Rev. Lett.* **66**, 2689 (1991).
3. W. Schöllkopf, J. P. Toennies, *Science* **266**, 1345 (1994).
4. W. Schöllkopf, Doctoral Dissertation, University of Göttingen (1998).
5. M. Arndt, O. Nairz, J. Vos-Andreae, C. Keller, G. van der Zouv, and A. Zeilinger, *Nature* (London), **401**, 680 (1999).
6. G. C. Hegerfeldt, T. Köhler, *Phys. Rev. A* **57**, 2021 (1998); *Phys. Rev. A* **61**, 023606 (2000).
7. E. O. Alt, P. Grassberger, and W. Sandhas, *Nucl. Phys. B* **2**, 167 (1967).
8. G. C. Hegerfeldt, T. Köhler, *Few Body Systems Suppl.* **10**, 263 (1999).
9. R. E. Grisenti, W. Schöllkopf, J. P. Toennies, G. C. Hegerfeldt, T. Köhler, *Phys. Rev. Lett.* **83**, 1755 (1999).
10. G. C. Hegerfeldt, T. Köhler, *Phys. Rev. Lett.* **84**, 3215 (2000).
11. R. E. Grisenti, W. Schöllkopf, J. P. Toennies, G. C. Hegerfeldt, T. Köhler, M. Stoll, *Phys. Rev. Lett.* **85**, 2284 (2000).

TIME-TYPE QUANTUM DEFORMATION OF THE CONFORMAL ALGEBRA SO(4,2)

FRANCISCO J. HERRANZ

Departamento de Física, Universidad de Burgos, 09006 Burgos, Spain
E-mail: fjherranz@ubu.es

A new non-standard quantum deformation of $so(4,2)$, the conformal algebra of the $(3+1)$-dimensional Minkowskian spacetime, is explicitly constructed by starting from the non-standard (or Jordanian) classical r matrix of $sl(2,\mathbb{R})$. Since the deformation parameter has the dimension of a time, this quantum $so(4,2)$ algebra is called 'time-type'. The corresponding Hopf structure, universal R matrix and differential-difference realization are presented. The latter allows us to show, in a natural way, that this quantum algebra is the symmetry algebra of a time discretization of the $(3+1)$-dimensional wave equation on a uniform lattice. The space-type deformation counterpart is also pointed out.

1 Introduction

The $so(4,2)$ Lie algebra of the group of conformal transformations of the $(3+1)$D Minkowskian spacetime is spanned by the generators of rotations J_i, time translations P_0, space translations P_i, boosts K_i, special conformal transformations C_0, C_i, and dilations D $(i = 1,2,3)$. Throughout the paper sum over repeated indices will be assumed, $i, j, k = 1,2,3$, and $\mu, \nu = 0,1,2,3$. The Lie brackets of $so(4,2)$ are given by:

$$\begin{array}{lll}
[J_i, J_j] = \varepsilon_{ijk} J_k & [J_i, K_j] = \varepsilon_{ijk} K_k & [K_i, K_j] = -\varepsilon_{ijk} J_k \\
[J_i, P_j] = \varepsilon_{ijk} P_k & [J_i, P_0] = 0 & [P_\mu, P_\nu] = 0 \\
[J_i, C_j] = \varepsilon_{ijk} C_k & [J_i, C_0] = 0 & [C_\mu, C_\nu] = 0 \\
[K_i, P_0] = P_i & [K_i, P_j] = \delta_{ij} P_0 & [P_0, C_i] = 2K_i \\
[K_i, C_0] = C_i & [K_i, C_j] = \delta_{ij} C_0 & [C_0, P_i] = 2K_i \\
[P_0, C_0] = -2D & [P_i, C_j] = 2\delta_{ij} D - 2\varepsilon_{ijk} J_k & [D, K_i] = 0 \\
[D, P_\mu] = P_\mu & [D, C_\mu] = -C_\mu & [D, J_i] = 0,
\end{array} \qquad (1)$$

where ε_{ijk} is the completely antisymmetric unit tensor with $\varepsilon_{123} = 1$. Any generator with three components will be denoted as $\mathbf{X} = (X_1, X_2, X_3)$.

In this conformal framework it is worth recalling that $so(4,2)$ has two remarkable Lie subalgebras: $\{\mathbf{J}, \mathbf{K}, \mathbf{P}, P_0\}$ that generate the Poincaré algebra $\mathcal{P}$ of isometries of the $(3+1)$D Minkowskian spacetime, and $\{\mathbf{J}, \mathbf{K}, \mathbf{P}, P_0, D\}$ that span the Weyl–Poincaré subalgebra $\mathcal{WP}$, that is, the similitude algebra of the spacetime (isometries plus dilations).

The aim of this contribution is to present a new quantum deformation of $so(4,2)$ of non-standard or triangular type, [1,2] denoted $U_\tau(so(4,2))$, which has the following properties: (i) It is simpler and more manageable than other known non-standard quantum conformal algebras. [3,4] (ii) By construction, generic properties/structures known for low dimensional cases (such as $U_\tau(so(2,2))$ [5]) can directly be extended to $U_\tau(so(4,2))$ (this also happens for Drinfeld–Jimbo deformations). [6] (iii) By means of a differential-difference realization, this quantum algebra is shown to be the symmetry algebra of a time discretization of the wave equation on a uniform lattice. (iv) The deformation parameter has dimension of time and is identified with the time lattice constant. (v) $U_\tau(so(4,2))$ has a well defined non-relativistic limit giving rise to the quantum conformal algebra of the Galilean spacetime. In particular, the Hopf structure of $U_\tau(so(4,2))$ is given in Sec. 2 and its role as a quantum conformal algebra of time discrete symmetries is developed in Secs. 3 and 4.

2 Time-type quantum deformation

Let us consider the non-standard classical r matrix [1,2] of $sl(2,\mathbb{R}) \simeq so(2,1)$ written in a conformal basis $\{J_3, J_+, J_-\} \equiv \{D, P_0, C_0\}$:

$$r = -\tau D \wedge P_0, \tag{2}$$

where τ is the deformation parameter. As in the $so(2,2)$ case, [5] we consider Eq. (2) as the r matrix for the whole $so(4,2)$ algebra; hence we find the following cocommutator δ defining the Lie bialgebra $(so(4,2), \delta(r))$:

$$\begin{array}{lll} \delta(P_0) = 0 & \delta(P_i) = \tau P_i \wedge P_0 & \delta(J_i) = 0 \\ \delta(K_i) = -\tau D \wedge P_i & \delta(D) = -\tau D \wedge P_0 & \\ \delta(C_0) = -\tau C_0 \wedge P_0 & \delta(C_i) = -\tau C_i \wedge P_0 + 2\tau D \wedge K_i . & \end{array} \tag{3}$$

Note that $\mathcal{WP}$ appears as a Lie sub-bialgebra: $(\mathcal{WP}, \delta(r)) \subset (so(4,2), \delta(r))$. Next the quantum deformation, $U_\tau(so(4,2))$, is obtained in three steps. First, we deduce the coproduct Δ (for which Eq. (3) is the linear term):

$$\begin{aligned} \Delta(P_0) &= 1 \otimes P_0 + P_0 \otimes 1 \\ \Delta(P_i) &= 1 \otimes P_i + P_i \otimes \mathrm{e}^{\tau P_0} \\ \Delta(J_i) &= 1 \otimes J_i + J_i \otimes 1 \\ \Delta(K_i) &= 1 \otimes K_i + K_i \otimes 1 - \tau D \otimes \mathrm{e}^{-\tau P_0} P_i \\ \Delta(D) &= 1 \otimes D + D \otimes \mathrm{e}^{-\tau P_0} \\ \Delta(C_0) &= 1 \otimes C_0 + C_0 \otimes \mathrm{e}^{-\tau P_0} \\ \Delta(C_i) &= 1 \otimes C_i + C_i \otimes \mathrm{e}^{-\tau P_0} + 2\tau D \otimes \mathrm{e}^{-\tau P_0} K_i \\ &\qquad -\tau^2 (D^2 + D) \otimes \mathrm{e}^{-2\tau P_0} P_i . \end{aligned} \tag{4}$$

Second, we find the compatible commutation rules for the Weyl–Poincaré Hopf subalgebra $U_\tau(\mathcal{WP}) \subset U_\tau(so(4,2))$:

$$
\begin{array}{lll}
[J_i, J_j] = \varepsilon_{ijk} J_k & [J_i, K_j] = \varepsilon_{ijk} K_k & [K_i, K_j] = -\varepsilon_{ijk} J_k \\
[J_i, P_j] = \varepsilon_{ijk} P_k & [J_i, P_0] = 0 & [P_\mu, P_\nu] = 0 \\
[K_i, P_0] = \mathrm{e}^{-\tau P_0} P_i & [K_i, P_j] = \delta_{ij} \dfrac{\mathrm{e}^{\tau P_0} - 1}{\tau} & [D, P_0] = \dfrac{1 - \mathrm{e}^{-\tau P_0}}{\tau} \\
[D, P_i] = P_i & [D, K_i] = 0 & [D, J_i] = 0,
\end{array}
\tag{5}
$$

and, finally, the commutators involving special conformal transformations:

$$
\begin{array}{ll}
[J_i, C_j] = \varepsilon_{ijk} C_k & [J_i, C_0] = 0 \\
[C_i, C_j] = 0 & [C_0, C_i] = -\tau(DC_i + C_i D) \\
[K_i, C_0] = C_i & [K_i, C_j] = \delta_{ij}(C_0 - \tau D^2) \\
[P_i, C_j] = 2\delta_{ij} D - 2\varepsilon_{ijk} J_k & [C_0, P_i] = 2K_i + \tau(DP_i + P_i D) \\
[P_0, C_0] = -2D & [P_0, C_i] = \mathrm{e}^{-\tau P_0} K_i + K_i\, \mathrm{e}^{-\tau P_0} \\
[D, C_i] = -C_i & [D, C_0] = -C_0 + \tau D^2 .
\end{array}
\tag{6}
$$

We remark that, by construction, some relevant subalgebras of $so(4,2)$ become into Hopf subalgebras of $U_\tau(so(4,2))$ after deformation. In particular, besides $\mathcal{WP}$, we find the following embedding of Hopf subalgebras:

$$
\begin{array}{ll}
U_\tau(so(2,1)) & \{D, P_0, C_0\} \\
\cap & \cap \\
U_\tau(so(2,2)) & \{D, P_0, P_1, C_0, C_1, K_1\} \\
\cap & \cap \\
U_\tau(so(3,2)) & \{D, P_0, P_i, C_0, C_i, K_i, J_3\} \quad i = 1,2 \\
\cap & \cap \\
U_\tau(so(4,2)) & \{D, P_0, \mathbf{P}, C_0, \mathbf{C}, \mathbf{K}, \mathbf{J}\}
\end{array}
\tag{7}
$$

All of them share the same classical r matrix (2). The chain (7) has an important consequence: properties previously known for a low dimensional case can directly be extended to higher dimensions. A first application is provided by the universal R matrix of $U_\tau(sl(2,\mathbb{R})) \simeq U_\tau(so(2,1)) \subset U_\tau(so(4,2))$: [7]

$$
\mathcal{R} = \exp\{\tau P_0 \otimes D\} \exp\{-\tau D \otimes P_0\}, \tag{8}
$$

that fulfils the quantum Yang–Baxter equation and the property

$$
\mathcal{R}\Delta(X)\mathcal{R}^{-1} = \sigma \circ \Delta(X) \qquad X \in \{D, P_0, C_0\}, \tag{9}
$$

where $\sigma(X \otimes Y) = Y \otimes X$. It can be proven that Eq. (8) also gives the universal R matrix for all the quantum algebras arising in the sequence (7) including, obviously, $U_\tau(so(4,2))$. In the next Section, we shall see another application: the extension of the known time discretization of the $(1+1)$D wave equation [5] associated to $U_\tau(so(2,2))$ to the present $(3+1)$D case.

3 Discrete time symmetries

As is well known, $so(4,2)$ is the Lie symmetry algebra [8] of the $(3+1)$D wave equation (or massless Klein–Gordon equation). In what follows we show that $U_\tau(so(4,2))$ is the symmetry algebra of a time discretization of the wave equation on a uniform lattice. Let us consider the spacetime coordinates $(x^\mu) \equiv (x^0, \mathbf{x}) \equiv (x^0, x^1, x^2, x^3)$, a differential operator $E = E(x^\mu, \partial_\mu)$ acting on a function $\Phi(x^\mu)$ $(\partial_\mu \equiv \partial/\partial x^\mu)$, and the differential equation:

$$E(x^\mu, \partial_\mu)\Phi(x^\mu) = 0. \tag{10}$$

An operator $\mathcal{O}$ is a symmetry of the Eq. (10) if $\mathcal{O}$ transforms solutions into solutions:

$$E\,\mathcal{O} = \mathcal{Q}\,E \qquad \text{or} \qquad [E, \mathcal{O}] = \mathcal{Q}'\,E, \tag{11}$$

where $\mathcal{Q}$ is another operator and $\mathcal{Q}' = \mathcal{Q} - \mathcal{O}$.

Now we consider a differential-difference realization of the deformed commutators (5) and (6):

$$\begin{array}{lll} J_1 = x^3\partial_2 - x^2\partial_3 & J_2 = x^1\partial_3 - x^3\partial_1 & J_3 = x^2\partial_1 - x^1\partial_2 \\ K_i = -x^i\Delta_0 - x^0T_0^{-1}\partial_i & P_\mu = \partial_\mu & D = -x^0T_0^{-1}\Delta_0 - x^i\partial_i - 1 \\ \multicolumn{3}{l}{C_0 = \left(\mathbf{x}^2 + (x^0)^2T_0^{-1}\right)\Delta_0 + 2x^0\left(x^i\partial_i + 1\right) + \tau\left(x^i\partial_i + 1\right)^2} \\ \multicolumn{3}{l}{C_i = \left(\mathbf{x}^2 - (x^0)^2T_0^{-2}\right)\partial_i - 2x^i\left(x^0T_0^{-1}\Delta_0 + x^j\partial_j + 1\right) + \tau x^0T_0^{-2}\partial_i,} \end{array} \tag{12}$$

where $\mathbf{x}^2 = (x^1)^2 + (x^2)^2 + (x^3)^2$, $T_0 = \mathrm{e}^{\tau\partial_0}$ is a shift operator and $\Delta_0 = (T_0 - 1)/\tau$ is a discrete derivative in the time direction acting on $\Phi(x^\mu)$ as

$$T_0\Phi(x^0, \mathbf{x}) = \Phi(x^0 + \tau, \mathbf{x}) \qquad \Delta_0\Phi(x^0, \mathbf{x}) = \frac{\Phi(x^0 + \tau, \mathbf{x}) - \Phi(x^0, \mathbf{x})}{\tau}. \tag{13}$$

If we now take as the operator E the deformed Casimir invariant E_τ of the Poincaré subalgebra $\mathcal{P}$ (which is not a Hopf subalgebra) given by

$$E_\tau = P_1^2 + P_2^2 + P_3^2 - \left(\frac{\mathrm{e}^{\tau P_0} - 1}{\tau}\right)^2, \tag{14}$$

and we introduce the realization (12), we find a time discretization of the $(3+1)$D wave equation in the form

$$(\partial_1^2 + \partial_2^2 + \partial_3^2 - \Delta_0^2)\Phi(x^\mu) = 0. \tag{15}$$

Therefore the deformation parameter τ is the time lattice constant on the uniform lattice, while the space coordinates $\mathbf{x}$ remain as continuous variables.

We conclude that the quantum algebra $U_\tau(so(4,2))$ is the symmetry algebra of the Eq. (15) since the generators (12) are symmetry operators fulfilling

$$\begin{array}{ll} [E_\tau, X] = 0 \quad X \in \{\mathbf{J}, \mathbf{K}, \mathbf{P}, P_0\} & [E_\tau, D] = -2E_\tau \\ {[E_\tau, C_i] = -4x^i E_\tau} & [E_\tau, C_0] = 4(x^0 + \tau x^i \partial_i + 2\tau) E_\tau . \end{array} \tag{16}$$

If we apply the limit $\tau \to 0$ in all the previous expressions, we recover the usual differential operators, continuous wave equation and Lie symmetry algebra.

4 Twist map

The non-standard nature of $U_\tau(so(4,2))$ ensures that its Hopf structure can be transformed into another one with classical commutation rules but with deformed coproduct [1,2] by means of a non-linear change of basis, that is, a twist map. Once more, such a map is a direct generalization of the already known one for the Hopf subalgebra $U_\tau(sl(2,\mathbb{R}))$, [9,10] which reads

$$\mathcal{P}_0 = \frac{e^{\tau P_0} - 1}{\tau} \qquad \mathcal{C}_0 = C_0 - \tau D^2, \tag{17}$$

with the remaining generators unchanged. If we apply the twist map to the the commutators (5) and (6) of $U_\tau(so(4,2))$, we recover the classical ones (1), while the coproduct (4) is transformed into:

$$\begin{aligned} &\Delta(\mathcal{P}_\mu) = 1 \otimes \mathcal{P}_\mu + \mathcal{P}_\mu \otimes 1 + \tau \mathcal{P}_\mu \otimes \mathcal{P}_0 \qquad \Delta(\mathcal{J}_i) = 1 \otimes \mathcal{J}_i + \mathcal{J}_i \otimes 1 \\ &\Delta(\mathcal{K}_i) = 1 \otimes \mathcal{K}_i + \mathcal{K}_i \otimes 1 - \tau \mathcal{D} \otimes \frac{\mathcal{P}_i}{1 + \tau \mathcal{P}_0} \\ &\Delta(\mathcal{D}) = 1 \otimes \mathcal{D} + \mathcal{D} \otimes \frac{1}{1 + \tau \mathcal{P}_0} \\ &\Delta(\mathcal{C}_0) = 1 \otimes \mathcal{C}_0 + \mathcal{C}_0 \otimes \frac{1}{1 + \tau \mathcal{P}_0} - 2\tau \mathcal{D} \otimes \frac{1}{1 + \tau \mathcal{P}_0} \mathcal{D} \\ &\qquad + \tau(\mathcal{D}^2 + \mathcal{D}) \otimes \frac{\tau \mathcal{P}_0}{(1 + \tau \mathcal{P}_0)^2} \\ &\Delta(\mathcal{C}_i) = 1 \otimes \mathcal{C}_i + \mathcal{C}_i \otimes \frac{1}{1 + \tau \mathcal{P}_0} + 2\tau \mathcal{D} \otimes \frac{1}{1 + \tau \mathcal{P}_0} \mathcal{K}_i \\ &\qquad - \tau^2 (\mathcal{D}^2 + \mathcal{D}) \otimes \frac{\mathcal{P}_i}{(1 + \tau \mathcal{P}_0)^2} . \end{aligned} \tag{18}$$

On the other hand, the discretization of the wave equation (15) remains invariant under the twist map, while the differential-difference realization (12) and relations (16) are transformed into

$$\begin{array}{l} \mathcal{P}_0 = \Delta_0 \qquad [E, \mathcal{P}_0] = 0 \qquad [E, \mathcal{C}_0] = 4x^0\, T_0^{-1} E \\ \mathcal{C}_0 = \left(\mathbf{x}^2 + (x^0)^2 T_0^{-2}\right) \Delta_0 + 2x^0 T_0^{-1} \left(x^i \partial_i + 1\right) - \tau x^0 T_0^{-2} \Delta_0 , \end{array} \tag{19}$$

where the omitted expressions are unchanged and E is the (undeformed) second-order Poincaré Casimir. In this context, recall that the twist for a different non-standard quantum $so(3,2)$ algebra [4] was studied by Aizawa [11], together with a discrete analogue of the (2 + 1)D Klein–Gordon equation.

To end with, it is worth pointing out that a similar construction for a space-type quantum deformation of $so(4,2)$ can be developed by starting from another $sl(2,\mathbb{R})$ subalgebra now spanned by $\{D, P_1, C_1\}$. Hence the initial object generating such a quantum algebra $U_\sigma(so(4,2))$ would be the classical r matrix given by

$$r = -\sigma D \wedge P_1, \tag{20}$$

where σ is the deformation parameter. [12] In this case, σ would be interpreted as a space lattice constant in the x^1 direction, and the remaining space and time coordinates (x^2, x^3, x^0) would be preserved as continuous variables. Further to this second possibility, the non-relativistic contraction of these quantum algebras leading to (time-type and space-type) quantum deformations of the conformal algebra of the Galilean spacetime can also been carried out. All these results will be presented elsewhere.

Acknowledgments

This work has been partially supported by the Ministerio de Ciencia y Tecnología (Project BFM2000-1055) and by JCyL, Spain (Project CO2/399).

References

1. V.G. Drinfeld, in *Proc. Int. Cong. Math. Berkeley 1986*, ed. A.V. Gleason, AMS, Providence (1987).
2. V.G. Drinfeld, *Leningrad Math. J.* **1**, 1419 (1990).
3. J. Lukierski *et al*, *Phys. Lett.* B **371**, 215 (1996).
4. F.J. Herranz, *J. Phys. A: Math. Gen.* **30**, 6123 (1997).
5. F.J. Herranz, *J. Phys. A: Math. Gen.* **33**, 8217 (2000).
6. V.K. Dobrev, *J. Phys. A: Math. Gen.* **26**, 1317 (1993).
7. A. Ballesteros and F.J. Herranz, *J. Phys. A: Math. Gen.* **29**, L311 (1996).
8. A.O. Barut and B. Xu, *Phys. Lett. A* **82**, 218 (1981).
9. B. Abdesselam *et al*, *Mod. Phys. Lett. A* **14**, 765 (1999).
10. A. Ballesteros *et al*, *J. Phys. A: Math. Gen.* **33**, 4859 (2000).
11. Aizawa N, *Rep. Math. Phys.* to appear (math.QA/0008055).
12. F.J. Herranz *et al*, *Czech. J. Phys.* **51**, 321 (2001).

q-LAPLACE OPERATOR RELATED TO THE QUANTUM UNITARY GROUP

N. Z. IORGOV, A. U. KLIMYK

Bogolyubov Institute for Theoretical Physics, Kiev 03143, Ukraine
E-mail: aklimyk@bitp.kiev.ua

The aim of this paper is to study q-Laplace operator and the space of q-harmonic polynomials on the quantum complex vector space generated by q-commuting elements z_i, w_i, $i = 1, 2, \cdots, n$, on which the quantum group $GL_q(n)$ acts.

Everywhere below we suppose that q is not a root of unity. The Drinfeld–Jimbo quantum algebra $U_q(\mathrm{gl}_n)$ is generated by the elements $k_i \equiv q^{h_i}$, $k_i^{-1} \equiv q^{-h_i}$, e_j, f_j, $i = 1, 2, \cdots, n$, $j = 1, 2, \cdots, n-1$, satisfying Serre relations and the relations $k_i e_j k_i^{-1} = q^{a_{ij}} e_j$, $k_i f_j k_i^{-1} = q^{-a_{ij}} f_j$,

$$[e_i, f_j] \equiv e_i f_j - f_j e_i = \delta_{ij} \frac{k_i k_{i+1}^{-1} - k_i^{-1} k_{i+1}}{q - q^{-1}},$$

where $a_{ii} = 1$, $a_{i,i-1} = -1$ and $a_{ij} = 0$ otherwise. The Hopf algebra structure on $U_q(\mathrm{gl}_n)$ is given by formulas (1.29) in [1].

The algebra $U_q(\mathrm{gl}_n)$ acts on the q-analogue of the complex vector space which is determined by the algebra of polynomials $\mathcal{A} \equiv \mathbb{C}[z_1, \cdots, z_n, w_1, \cdots, w_n]$ in elements $z_1, z_2, \cdots, z_n, w_1, w_2 \cdots, w_n$ satisfying the defining relations

$$z_i z_j = q z_j z_i, \quad w_i w_j = q^{-1} w_j w_i, \quad i < j, \qquad w_j z_i = q z_i w_j, \quad i \neq j, \tag{1}$$

$$w_k z_k = z_k w_k + (1 - q^2)(z_1 w_1 + \cdots + z_{k-1} w_{k-1}). \tag{2}$$

The set of all monomials

$$z_1^{r_1} z_2^{r_2} \cdots z_n^{r_n} w_1^{s_1} w_2^{s_2} \cdots w_n^{s_n}, \qquad r_j, s_j = 0, 1, 2, \cdots, \tag{3}$$

form a basis of the algebra $\mathcal{A}$. The set

$$w_1^{r_1} w_2^{r_2} \cdots w_n^{r_n} z_1^{s_1} z_2^{s_2} \cdots z_n^{s_n}, \qquad r_j, s_j = 0, 1, 2, \cdots, \tag{4}$$

also form a basis of this algebra.

The vector space of the algebra $\mathcal{A}$ can be represented as a direct sum of the vector subspaces $\mathcal{A}_{m,m'}$ consisting of homogeneous polynomials of homogeneity degree m in $z_1, z_2, \cdots, z_n$ and of homogeneity degree m' in $w_1, w_2, \cdots, w_n$, $m, m' = 0, 1, 2, \cdots$: $\mathcal{A} = \bigoplus_{m,m'=0}^{\infty} \mathcal{A}_{m,m'}$.

We can define an action of the algebra $U_q(\mathrm{gl}_n)$ on the vector space $\mathcal{A}$. For this we determine an action of $U_q(\mathrm{gl}_n)$ on each subspace $\mathcal{A}_{m,m'}$. To determine this action we give the action of $U_q(\mathrm{gl}_n)$ on z_j and w_j by the formulas

$$k_i \triangleright z_j = q^{\delta_{ij}} z_j, \qquad e_i \triangleright z_j = \delta_{j,i+1} z_{j-1}, \qquad f_i \triangleright z_j = \delta_{j,i} z_{j+1}, \tag{5}$$

$$k_i \triangleright w_j = q^{-\delta_{ij}} w_j, \qquad e_i \triangleright w_j = -\delta_{j,i} q^{-1} w_{j+1}, \qquad f_i \triangleright w_j = -\delta_{j,i+1} q w_{j-1} \tag{6}$$

and extend this action to $\mathcal{A}_{m,m'}$ by using linearity and the comultiplication (see, for example, [1]).

This action of $U_q(\mathrm{gl}_n)$ on the vector space $\mathcal{A}_{m,m'}$ determines a representation of $U_q(\mathrm{gl}_n)$ on this space. We denote it by $L_{m,m'}$. These representations determine the representation of $U_q(\mathrm{gl}_n)$ on the space $\mathcal{A}$, which is denoted by L.

We define on $\mathcal{A}$ the q-differentiations ∂_i and $\bar{\partial}_i$ as the linear operators acting as $\partial_i p = 0$ ($\bar{\partial}_i p' = 0$) on monomials p of the form (3) not containing z_i (on monomials p' of the form (4) not containing w_i) and as

$$\partial_i = \breve{z}_i^{-1} \frac{\gamma_i - \gamma_i^{-1}}{q - q^{-1}}, \qquad \bar{\partial}_i = \breve{w}_i^{-1} \frac{\bar{\gamma}_i - \bar{\gamma}_i^{-1}}{q - q^{-1}},$$

in other cases, where $\breve{z}_i$ and $\breve{w}_i$ are the linear operators defined by

$$\breve{z}_i(z_1^{r_1} z_2^{r_2} \cdots z_n^{r_n} w_1^{s_1} w_2^{s_2} \cdots w_n^{s_n}) = z_1^{r_1} z_2^{r_2} \cdots z_n^{r_n} z_i w_1^{s_1} w_2^{s_2} \cdots w_n^{s_n},$$

$$\breve{w}_i(w_1^{r_1} w_2^{r_2} \cdots w_n^{r_n} z_1^{s_1} z_2^{s_2} \cdots z_n^{s_n}) = w_1^{r_1} w_2^{r_2} \cdots w_n^{r_n} w_i z_1^{s_1} z_2^{s_2} \cdots z_n^{s_n},$$

and γ_i, $\bar{\gamma}_i$ are the linear operators given by

$$\gamma_i(z_1^{r_1} z_2^{r_2} \cdots z_n^{r_n} w_1^{s_1} w_2^{s_2} \cdots w_n^{s_n}) = q^{r_i} z_1^{r_1} z_2^{r_2} \cdots z_n^{r_n} w_1^{s_1} w_2^{s_2} \cdots w_n^{s_n},$$

$$\bar{\gamma}_i(w_1^{r_1} w_2^{r_2} \cdots w_n^{r_n} z_1^{s_1} z_2^{s_2} \cdots z_n^{s_n}) = q^{r_i} w_1^{r_1} w_2^{r_2} \cdots w_n^{r_n} z_1^{s_1} z_2^{s_2} \cdots z_n^{s_n}.$$

Note, a similar differential calculus on the algebra $\mathcal{A}$ was considered in [2].

The element

$$Q = \sum_{i=1}^{n} z_i w_i = \sum_{i=1}^{n} q^{2(n-i)} w_i z_i \in \mathcal{A}_{1,1}$$

of the algebra $\mathcal{A}$ is called the *squared q-radius*. It is checked by a direct computation that *Q is invariant with respect to the representation $L_{1,1}$*: $L(k_i^{\pm 1})Q = Q$, $L(e_j)Q = 0$, $L(f_j)Q = 0$. Similarly, the element $Q^k \in \mathcal{A}_{k,k}$ is invariant with respect to the representation $L_{k,k}$. The squared q-radius Q belongs to the center of the algebra $\mathcal{A}$. We shall also use the elements $Q_j = \sum_{i=1}^{j} z_i w_i = \sum_{i=1}^{j} q^{2(j-i)} w_i z_i$, which are squared q-radiuses for the corresponding subalgebras of the algebra $\mathcal{A}$, generated by $z_1, w_1, \cdots, z_j, w_j$. They satisfy the relations $Q_jQ_i = Q_iQ_j$.

The operator

$$\Delta_q \equiv \Delta = \partial_1\bar{\partial}_1 + \partial_2\bar{\partial}_2 + \cdots + \partial_n\bar{\partial}_n. \tag{7}$$

on $\mathcal{A}$ is called the *q-Laplace operator* on the quantum complex vector space.

To the element Q there corresponds the operator

$$\hat{Q} = \hat{z}_1\hat{w}_1 + \hat{z}_2\hat{w}_2 + \cdots + \hat{z}_n\hat{w}_n$$

on $\mathcal{A}$, where $\hat{z}_j$ and $\hat{w}_j$ are operators of left multiplications by z_j and w_j, respectively.

The following relations are proved by a direct calculations:

$$\Delta\hat{Q}^k - \hat{Q}^k\Delta = q^{n-1}\hat{Q}^{k-1}[k]\{q^{k+n-1}\gamma\bar{\gamma}\}, \tag{8}$$

$$\Delta(Q^k) = q^{n-1}Q^{k-1}[k][k+n-1], \tag{9}$$

where $\gamma = \gamma_1\gamma_2\cdots\gamma_n$, $\bar{\gamma} = \bar{\gamma}_1\bar{\gamma}_2\cdots\bar{\gamma}_n$ and

$$[a] = \frac{q^a - q^{-a}}{q - q^{-1}}, \qquad \{a\} = \frac{a - a^{-1}}{q - q^{-1}}.$$

If $p \in \mathcal{A}_{mm'}$, then $\gamma p = q^m p$ and $\bar{\gamma} p = q^{m'} p$.

A polynomial $p \in \mathcal{A}$ is called *q-harmonic* if $\Delta p = 0$. The linear subspace of $\mathcal{A}$ consisting of all q-harmonic polynomials is denoted by $\mathcal{H}$. Let $\mathcal{H}_{m,m'} = \mathcal{A}_{m,m'} \cap \mathcal{H}$. Then $\mathcal{H} = \bigoplus_{m,m'=0}^{\infty} \mathcal{H}_{m,m'}$.

Proposition 1. *The space $\mathcal{A}_{m,m'}$ can be represented in the form of the direct sum*

$$\mathcal{A}_{m,m'} = \mathcal{H}_{m,m'} \oplus Q\mathcal{A}_{m-1,m'-1} = \bigoplus_{j=0}^{\min(m,m')} Q^j\mathcal{H}_{m-j,m'-j}. \tag{10}$$

It follows from Proposition 1 that *if $p \in \mathcal{H}_{m,m'}$, then p cannot be represented as $p = Q^kp'$, $k \neq 0$, with some polynomial p'.*

Remark. If $n = 1$, then $\mathcal{A}$ consists of all polynomials in commuting elements z_1 and w_1. In this case, the space $\mathcal{H}$ of q-harmonic polynomials has a basis consisting of the polynomials $1, z_1^k, w_1^k$, $k = 1, 2, \cdots$.

It follows from Proposition 1 that $\mathcal{A} \simeq \mathbb{C}[Q] \otimes \mathcal{H}$. This decomposition is a *q-analogue of the theorem on separation of variables* for Lie groups in an abstract form. We have

$$\mathcal{A} \simeq \mathbb{C}[Q] \otimes \mathcal{H} \simeq \mathbb{C}[Q] \otimes \bigoplus_{m\geq 0}\bigoplus_{m'\geq 0} \mathcal{H}_{m,m'} = \bigoplus_{m\geq 0}\bigoplus_{m'\geq 0} (\mathbb{C}[Q] \otimes \mathcal{H}_{m,m'}).$$

Proposition 2. *The q-Laplace operator Δ and the operator $\hat{Q}$ commute with all operators of the representation L of $U_q(\mathrm{gl}_n)$.*

It follows from Proposition 2 that the subspace $\mathcal{H}_{m,m'}$ is invariant with respect to the representation $L_{m,m'}$ of the algebra $U_q(\mathrm{gl}_n)$. We denote the restriction of this representation to $\mathcal{H}_{m,m'}$ by $T_{m,m'}$. Since $\hat{Q}$ commutes with action of $U_q(\mathrm{gl}_n)$, it follows from (10) that $L_{m,m'} = \bigoplus_{j=0}^{\min(m,m')} T_{m-j,m'-j}$.

Proposition 3. *The representations $T_{m-j,m'-j}$ of $U_q(\mathrm{gl}_n)$ are irreducible and have highest weights $(m-j, 0, \cdots, 0, -m'+j)$, respectively.*

We denote by $\mathcal{A}^{U_q(\mathrm{gl}_n)}$ the space of elements of $\mathcal{A}$ consisting of invariant elements with respect to the action of $U_q(\mathrm{gl}_n)$.

Proposition 4. *We have $\mathcal{A}^{U_q(\mathrm{gl}_n)} = \mathbb{C}[Q]$ and $\mathcal{A}^{U_q(\mathrm{gl}_{n-1})} = \mathbb{C}[Q, z_n, w_n]$.*

We construct the projector $\mathbf{H}_{m,m'} : \mathcal{A}_{m,m'} = \mathcal{H}_{m,m'} \oplus Q\mathcal{A}_{m-1,m'-1} \to \mathcal{H}_{m,m'}$. It has the form

$$\mathbf{H}_{m,m'}p = \sum_{k=0}^{\min(m,m')} \frac{(-1)^k}{q^{(n-1)k}} \frac{[m+m'+n-k-2]!}{[k]![m+m'+n-2]!} \hat{Q}^k \Delta^k p, \quad p \in \mathcal{A}_{m,m'}. \tag{11}$$

where $[s]! = [s][s-1]\cdots[1]$ for $s \neq 0$ and $[0]! = 1$. Note that $\mathbf{H}^2_{m,m'} = \mathbf{H}_{m,m'}$ and $\mathbf{H}_{m,m'}$ commutes with the action of $U_q(\mathrm{gl}_n)$.

A polynomial φ of the space $\mathcal{H}_{m,m'}$ is called *zonal* if it is invariant with respect to operators $L_{m,m'}(X)$, $X \in U_q(\mathrm{gl}_{n-1})$. Zonal polynomials are constructed by using the projectors $\mathbf{H}_{m,m'}$. Namely, we take the polynomial $z_n^m w_n^{m'}$ (invariant with respect to $U_q(\mathrm{gl}_{n-1})$) and act by $\mathbf{H}_{m,m'}$ on it. The resulting polynomial is $U_q(\mathrm{gl}_{n-1})$-invariant. In this way we obtain the following result.

Proposition 5. *The subspace of zonal polynomials in $\mathcal{H}_{m,m'}$ is one-dimensional. Up to a constant, a zonal polynomial of $\mathcal{H}_{m,m'}$ is given by*

$$\varphi'_{m,m'} = z_n^{m-m'} Q^{m'} \sum_{s=0}^{m'} \frac{(q^{-2m'};q^2)_s (q^{2(m+n-1)};q^2)_s}{(q^{2(n-1)};q^2)_s (q^2;q^2)_s} \frac{Q^s_{n-1}}{Q^s} q^{2s}$$

if $m \geq m'$ and by the formula

$$\varphi'_{m,m'} = Q^m \sum_{s=0}^{m} \frac{(q^{-2m};q^2)_s (q^{2(m'+n-1)};q^2)_s}{(q^{2(n-1)};q^2)_s (q^2;q^2)_s} \frac{Q^s_{n-1}}{Q^s} q^{2s} w_n^{m'-m},$$

if $m \leq m'$, where the expression $(a;q)_s$ is defined as

$$(a;q)_s = (1-a)(1-qa)(1-q^2a)\cdots(1-q^{s-1}a),$$

The zonal polynomials from Proposition 5 can be written in the terms of the little q-Jacobi polynomials as

$$\varphi'_{m,m'} = Q^{m'} z_n^{m-m'} P_{m'}^{(n-2,m-m')}(Q_{n-1}/Q;\ q^2)$$

if $m \geq m'$ and as

$$\varphi'_{m,m'} = Q^{m} P_{m}^{(n-2,m'-m)}(Q_{n-1}/Q;\ q^2) w_n^{m'-m}$$

if $m \leq m'$. Definition of these polynomials see in [3].

The following proposition describes important property of q-harmonic polynomials which is used for studying separation of variables for q-Laplace operator in the same way as in [4].

Proposition 6. *Let s and s' be integers such that $0 \leq s \leq m$ and $0 \leq s' \leq m'$. Let $h_{s,s'}(\mathbf{z}', \mathbf{w}')$ be a homogeneous harmonic polynomial of degree s in $\mathbf{z}' = (z_1, z_2, \cdots, z_{n-1})$ and of degree s' in $\mathbf{w}' = (w_1, w_2, \cdots, w_{n-1})$. Then for $z_n^{m-s} w_n^{m'-s'} h_{s,s'}(\mathbf{z}', \mathbf{w}') \in \mathcal{A}_{m,m'}$ we have*

$$\mathbf{H}_{m,m'}(z_n^{m-s} w_n^{m'-s'} h_{s,s'}(\mathbf{z}', \mathbf{w}')) = z_n^{m-s-m'+s'} Q^{m'-s'} d_{ss'}^{mm'} h_{s,s'}(\mathbf{z}', \mathbf{w}')),$$

where $m - s \geq m' - s'$,

$$d_{ss'}^{mm'} = q^{-2(m-s)(m'-s')} \frac{(q^{-2m'-2s-2n+4}; q^2)_{m'-s'}}{(q^{-2m-2m'-2n+4}; q^2)_{m'-s'}} \times$$

$$\times {}_2\varphi_1(q^{-2(m'-s')}, q^{2(m+s'+n-1)}; q^{2(s+s'+n-1)};\ q^2;\ q^2 Q_{n-1}/Q),$$

and

$$\mathbf{H}_{m,m'}(z_n^{m-s} w_n^{m'-s'} h_{s,s'}(\mathbf{z}', \mathbf{w}')) = Q^{m-s} d_{ss'}^{mm'} w_n^{m'-s'-m+s} h_{s,s'}(\mathbf{z}', \mathbf{w}')),$$

where $m - s \leq m' - s'$,

$$d_{ss'}^{mm'} = q^{-2(m-s)(m'-s')} \frac{(q^{-2m-2s'-2n+4}; q^2)_{m-s}}{(q^{-2m-2m'-2n+4}; q^2)_{m-s}} \times$$

$$\times {}_2\varphi_1(q^{-2(m-s)}, q^{2(m'+s+n-1)}; q^{2(s+s'+n-1)};\ q^2;\ q^2 Q_{n-1}/Q).$$

The unital associative algebra $\mathcal{F}(S_{q,n-1}^{\mathbb{C}})$ generated by the elements $z_1, \cdots, z_n, w_1, \cdots, w_n$ satisfying the relations (1), (2) and the relation $z_1 w_1 + \cdots + z_n w_n = 1$ is called the *algebra of functions on the quantum sphere* $S_{q,n-1}^{\mathbb{C}}$. It is clear that $\mathcal{F}(S_{q,n-1}^{\mathbb{C}}) = \mathcal{A}/\mathcal{I}$, where $\mathcal{I}$ is the two-sided ideal of $\mathcal{A}$ generated by the element $\sum_{i=1}^{n} z_i w_i - 1$. We denote by τ the canonical algebra homomorphism $\tau : \mathcal{A} \to \mathcal{A}/\mathcal{I}$. This homomorphism is called the *restriction* of polynomials of $\mathcal{A}$ onto the quantum sphere $S_{q,n-1}^{\mathbb{C}}$.

Proposition 7. *The homomorphism τ is a one-to-one mapping on the spaces $\mathcal{H}_{m,m'}$, that is, the restriction of an homogeneous q-harmonic polynomial to the sphere $S^{\mathbb{C}}_{q,n-1}$ determines uniquely this polynomial.*

It is clear that $\tau\mathcal{A} \equiv \tau(\mathbb{C}[Q] \otimes \mathcal{H}) = \hat{\mathcal{H}} \equiv \bigoplus_{m,m'=0}^{\infty} \hat{\mathcal{H}}_{m,m'}$, where $\hat{\mathcal{H}}_{m,m'}$ is the set of polynomials from $\mathcal{H}_{m,m'}$ restricted to the quantum sphere $S^{\mathbb{C}}_{q,n-1}$.

Proposition 6 allows us to determine a scalar product on the space $\mathcal{H}$. For this, we use the invariant functional h on the quantum sphere defined in [1], section 4.2. A scalar product on $\mathcal{H}$ is given by

$$\langle p_1, p_2 \rangle = h((\tau p_1)(\tau p_2)^*).$$

We have $\mathcal{H}_{mm'} \perp \mathcal{H}_{rr'}$ if $(m, m') \neq (r, r')$

The formulas $ke = q^2 ek$, $kf = q^{-2} fk$ and $ef - fe = \{k\}$ determine the quantum algebra $U_q(\mathrm{sl}_2)$. This algebra acts on the algebra $\mathbb{C}[z_1, \cdots z_n, w_1 \cdots, w_n]$. This action is uniquely determined by the formulas

$$\omega(e) = q^{n-1} Q, \quad \omega(f) = -\Delta, \quad \omega(k) = q^n \gamma\bar{\gamma}.$$

The homomorphism $\omega : U_q(\mathrm{sl}_2) \to \mathcal{L}(\mathcal{A})$, where $\mathcal{L}(\mathcal{A})$ is the space of linear operators on $\mathcal{A}$, gives a representation of $U_q(\mathrm{sl}_2)$.

Proposition 8. *The action of $U_q(\mathrm{sl}_2)$ on $\mathcal{A}$ commutes with the action of $U_q(\mathrm{gl}_n)$ on $\mathcal{A}$.*

This proposition means that $U_q(\mathrm{sl}_2)$ and $U_q(\mathrm{gl}_n)$ constitute a dual pair on $\mathcal{A}$.

Acknowledgments

The research contained in this paper was supported in part by Award No. UP1-2115 of the Civilian Research and Development Foundation for the Independent States of the Former Soviet Union (CRDF). The first author (NZI) was also supported by INTAS Grant No.2000-334.

References

1. M. Noumi, M. Yamada, and K. Mimachi, *Japan J. Math.* **19**, 31 (1993).
2. W. Pusz and S. L. Woronowicz, *Rep. Math. Phys.* **27**, 231 (1989).
3. G. Gasper and M. Rahman, *Basic Hypergeometric Functions* (Cambridge Univ. Press, Cambridge, 1991).
4. N. Iorgov and A. Klimyk, *J. Math. Phys.* **42**, 1326 (2001).

CONTRACTION AND INTERBASES EXPANSIONS ON N-SPHERE

A.A. IZMEST'EV, G.S. POGOSYAN,[a], A.N. SISSAKIAN
Joint Institute for Nuclear Research, Dubna, Moscow Region, 141980, Russia

P. WINTERNITZ
Centre de recherches mathématiques, Université de Montréal, C. P. 6128, succ. Centre Ville, Montréal, Québec, H3C 3J7, CANADA

Lie algebra contractions from o(n+1) to e(n) are used to construct the asymptotic limit of the interbasis expansions between two general subgroup chains $O(n+1) \supset O(n_\alpha+n_\beta)\otimes O(n_\gamma)$ and $O(n+1) \supset O(n_\alpha)\otimes O(n_\gamma+n_\beta)$, where $n+1 = n_\alpha+n_\gamma+n_\beta$. It provides the asymptotic formula of Racah coefficients for $SU(2)$ and $SU(1,1)$ groups.

1 Introduction

In a series of papers [1,2,3,4] we have presented a **new aspect** of the theory of Lie group and Lie algebra contractions: *the relation between separable coordinates systems in curved and flat spaces, related by the contraction of their isometry group.*

In article [1] we considered the two-dimensional sphere S_2. The contractions we use are analytical ones: **the radius of the sphere** R is built into the infinitesimal operators and into the sets of commuting operators, not only into the structure constants. The contractions can be viewed as singular changes of bases, as was the case of the original Inönü-Wigner [5] contractions. For $R \to \infty$ the sphere $S_n \sim O(n+1)/O(n)$ goes into the Euclidean space $E_n \sim E(n)/O(n)$.

In paper [3] the dimension of the space was arbitrary, but only the simplest types of coordinates were considered, **namely subgroup type coordinates**. Furthemore, we introduced a *graphical method* of connecting subgroup–type coordinate systems on the sphere S_n (characterized by tree diagrams) and on the Euclidean space E_n (characterized by cluster diagrams) and gave rules relating the contraction limit $R \to \infty$ of the coordinates, invariant operators, eigenvalues and basis functions.

Vilenkin, Kuznetsov and Smorodinsky [6] (see also [7,8]) developed a graphical method, the "method of trees" to describe subgroup type coordinates on S_n. The corresponding separated eigenfunctions are hyperspherical functions [5-8]. Their relation to subgroup chains and subgroup diagrams was analyzed in

[a]Permanent address: International Center for Advanced Studies, Yerevan State University, Alex Manougian st.1, Yerevan, 375049, Armenia

paper[3], as were their contractions to subgroup type separated basis functions for the groups $E(n)$.

In many-body theories it is often necessary to expand one type of hyperspherical functions in terms of other ones. The expansion coefficients have been called T-coefficients, or overlap functions[9].

The purpose of this contribution is to investigate the $R \to \infty$ contraction limit of the **interbases expansions and overlap functions** for the different spherical and hyperspherical functions on S_n.

A convenient way of calculating the T-coefficients corresponding to a transformation from one tree to another is to introduce a sequence of "intermediate" trees, each differing from the previous one by the transplantation of exactly one branch from one side of a branching point to the other. The T-matrix will be factorized into a product of "elementary T-matrices", corresponding to such elementary transformations. Each elementary T-matrix can be represented by a tree-type diagram consisting of a cell with three ends, each of which can be either open or closed. Therefore there are eight inequivalent diagrams of this type. The T-coefficients for all 8 types of elementary transformations were calculated by Kil'dyushov[9]. They were expressed in terms of generalized hypergeometric functions of argument $x = 1$: ${}_2F_1(1)$, ${}_3F_2(1)$, ${}_4F_3(1)$, Wigner D-functions, or Clebsch-Gordon and Racah coefficients for positive discrete series of representations of the group $SU(1,1)$[7,10].

The T-coefficients representing the general transformation corresponding to the diagram on Fig.1a have the form[9]

$$\left\| \begin{matrix} j_1\, j_2\, j_3 \\ \\ j_{12}\, j\, j_{23} \end{matrix} \right\| = \frac{\sqrt{(2j_{12}+1)(2j_{23}+1)}}{\Gamma(2j_2+2)} \sqrt{\frac{\Gamma(j_2+j_3+j_{23}+2)}{\Gamma(j_{23}-j_2-j_3)}}$$

$$\times \sqrt{\frac{\Gamma(j_2-j_3+j_{23}+1)\Gamma(j-j_{12}-j_3)\Gamma(j_1+j_{12}+j_2+2)}{\Gamma(j_{23}-j_2+j_3+1)\Gamma(j-j_{12}+j_3+1)\Gamma(j_1+j_{12}-j_2+1)}}$$

$$\times \sqrt{\frac{\Gamma(j+j_{12}-j_3+1)\Gamma(j_{12}-j_1+j_2+1)\Gamma(j+j_1+j_{23}+2)}{\Gamma(j+j_{12}+j_3+2)\Gamma(j_{12}-j_1-j_2)\Gamma(j-j_1-j_2+j_3)}}$$

$$\times \sqrt{\frac{\Gamma(j+j_1-j_{23}+1)\Gamma(j-j_1-j_2+j_3)\Gamma(j-j_1-j_2+j_3)}{\Gamma(j+j_1+j_2-j_3+2)\Gamma(j-j_1-j_{23})\Gamma(j-j_1+j_{23}+1)}}$$

$$\times {}_4F_3 \left\{ \begin{matrix} -j_{12}+j_1+j_2+1, j_{12}+j_1+j_2+2, j_{23}+j_2-j_3+1, j_2-j_3-j_{23} \\ \\ 2j_2+2, j+j_1+j_2-j_3+2, j_1+j_2-j_3-j+1 \end{matrix} \middle| 1 \right\}. \quad (1)$$

Here $j_i(j_{ik})$ label the discrete representations of $SU(1,1)$. It was pointed out in[7] that if the number $j_i = -1/4$, or -3/4 is associated with an end point, then one obtains values of the T-matrix for transitions in cells with open ends. The transition matrices for all 8 types of cells can be obtained in this manner.

2 Contractions of the Interbasis Expansion

Let us have two wave functions $\Psi^{j_1,j_2,j_3}_{j,j_{12}}$ and $\Psi^{j_1,j_2,j_3}_{j,j_{23}}$, which are solutions of the Helmholtz equation on S_n and correspond to the two subgroup chains $O(n+1) \supset O(n_\alpha + n_\beta) \otimes O(n_\gamma)$ and $O(n+1) \supset O(n_\alpha) \otimes O(n_\gamma + n_\beta)$, where $n+1 = n_\alpha + n_\gamma + n_\beta$. The transformations between these two functions corresponds to the recoupling of the angular momentum $\{j_1, j_2, j_3, j_{12}, j\}$ and $\{j_1, j_2, j_3, j_{23}, j\}$. The overlap functions expressed in terms of Racah coefficients and the corresponding interbases expansion is

$$\Psi^{j_1,j_2,j_3}_{j,j_{12}}(\theta_1', \theta_2') = \sum_{j_{23}=j_2+j_3+1}^{j-j_1-1} \left\| \begin{matrix} j_1 \; j_2 \; j_3 \\ \\ j_{12} \; j \; j_{23} \end{matrix} \right\| \Psi^{j_1,j_2,j_3}_{j,j_{23}}(\theta_1, \theta_2), \tag{2}$$

where the angles are related as follows

$$\cos\theta_1 = \cos\theta_1' \cos\theta_2' \qquad \cot\theta_2 = \cot\theta_1' \sin\theta_2',$$

The wave functions in eq. (2) can be constructed following the rules presented in papers[6,7] (see also[4]) and have the form

$$\Psi^{j_1,j_2,j_3}_{j,j_{12}}(\theta_1', \theta_2') = \sqrt{\frac{(2j+1)(2j_{12}+1)\Gamma(j_1+j_2+j_{12}+2)\Gamma(j+j_3+j_{12}+2)}{\Gamma(j_{12}+j_1-j_2+1)\Gamma(j_{12}-j_1+j_2+1)\Gamma(j-j_{12}+j_3+1)}}$$

$$\times\sqrt{\frac{\Gamma(j_{12}-j_1-j_2)\Gamma(j-j_{12}-j_3)}{\Gamma(j+j_{12}-j_3+1)}}\,(\sin\theta_2')^{2j_2+1}(\cos\theta_2')^{2j_1+1} \tag{3}$$

$$\times(\sin\theta_1')^{2j_3+1}(\cos\theta_1')^{2j_{12}+1}\,P^{(2j_2+1,2j_1+1)}_{j_{12}-j_1-j_2-1}(\cos 2\theta_2')P^{(2j_3+1,2j_{12}+1)}_{j-j_{12}-j_3-1}(\cos 2\theta_1')$$

$$\Psi^{j_1,j_2,j_3}_{j,j_{23}}(\theta_1, \theta_2) = \sqrt{\frac{(2j+1)(2j_{23}+1)\Gamma(j+j_1+j_{23}+2)\Gamma(j_2+j_3+j_{23}+2)}{\Gamma(j_{23}-j_3+j_2+1)\Gamma(j_{23}-j_2+j_3+1)\Gamma(j+j_{23}-j_3+1)}}$$

$$\times\sqrt{\frac{\Gamma(j_{23}-j_3-j_2)\Gamma(j-j_1-j_{23})}{\Gamma(j-j_{23}+j_3+1)}}\,(\sin\theta_2)^{2j_3+1}(\cos\theta_2)^{2j_2+1} \tag{4}$$

$$\times (\sin\theta_1)^{2j_{23}+1} (\cos\theta_1)^{2j_1+1}\, P^{(2j_3+1,2j_2+1)}_{j_{23}-j_3-j_2-1} (\cos 2\theta_2) P^{(2j_{23}+1,2j_1+1)}_{j-j_{23}-j_1-1} (\cos 2\theta_1)$$

here $P_n^{(\alpha,\,\beta)}(x)$ are Jacobi polynomials.

Consider the contraction limit $R \to \infty$ in the wave functions (3)-(4) and overlap functions (1). For large R we put

$$j \sim kR, \quad j_{12} \sim pR, \quad j_1 \sim qR, \quad \theta_1' \sim \frac{r_{j_2}}{R}, \quad \theta_2' \sim \frac{r_{j_2}}{R}, \quad \theta_1 \sim \frac{r_{j_{23}}}{R}, \tag{5}$$

where $r_{j_{23}} = \sqrt{r_{j_2}^2 + r_{j_3}^2}$, $k_{j_2}^2 = p^2 - q^2$, $k_{j_3}^2 = k^2 - p^2$ and $k_{j_{23}}^2 = k_{j_2}^2 + k_{j_3}^2$. Taking into account the formula [3,4]

$$\lim_{R\to\infty} \frac{\Gamma(j - j_{23} - j_1 - 1)}{\Gamma(j + j_{23} - j_1)}\, P^{(2j_{23}+1,2j_1+1)}_{j-j_{23}-j_1-1}(\cos 2\theta_1) = 2^{2j_{23}+1}\, \frac{J_{2j_{23}+1}(k_{j_{23}}\, r_{j_{23}})}{(k_{j_{23}} r_{j_{23}})^{2j_{23}+1}},$$

where $J_\mu(x)$ is Bessel function, we have

$$\begin{aligned} \lim_{R\to\infty} \Psi^{j_1,j_2,j_3}_{j,j_{12}}(\theta_1', \theta_2') &= \Phi^{j_2,j_3}_{kk_{j_2}k_{j_3}}(r_{j_2}, r_{j_3}) \\ &= 2\sqrt{kp}\, J_{2j_2+1}(k_{j_2} r_{j_2}) J_{2j_3+1}(k_{j_3} r_{j_3}), \end{aligned} \tag{6}$$

$$\begin{aligned} \lim_{R\to\infty} \Psi^{j_1,j_2,j_3}_{j,j_{23}}(\theta_1, \theta_2) &= \Phi^{j_{23},j_2,j_3}_{kk_{j_{23}}}(r_{j_{23}}, \theta_2) = 2\sqrt{k}\, J_{2j_{23}+1}(k_{j_{23}}\, r_{j_{23}}) \\ &\times \sqrt{\frac{(2j_{23}+1)\Gamma(j_{23}+j_2+j_3+2)\Gamma(j_{23}-j_2-j_3)}{\Gamma(j_{23}-j_2+j_3+1)\Gamma(j_{23}+j_2-j_3+1)}} \\ &\times (\sin\theta_2)^{2j_3+1}(\cos\theta_2)^{2j_2+1}\, P^{(2j_3+1,2j_2+1)}_{j_{23}-j_3-j_2-1}(\cos 2\theta_2). \end{aligned} \tag{7}$$

Using the asymptotic formulas for ${}_4F_3$ functions, we obtain

$$\begin{aligned} \lim_{R\to\infty} \left\| \begin{matrix} j_1\, j_2\, j_3 \\ \\ j_{12}\, j\, j_{23} \end{matrix} \right\| &= \frac{\sqrt{(2j_{23}+1)}}{\Gamma(2j_{23}+2)} \sqrt{\frac{\Gamma(j_2+j_3+j_{23}+2)\Gamma(j_2-j_3+j_{23}+1)}{\Gamma(j_{23}-j_2-j_3)\Gamma(j_{23}-j_2+j_3+1)}} \\ &\times \sqrt{\frac{2p}{k^2-q^2}} \left(\frac{p^2-q^2}{k^2-q^2}\right)^{j_2+\frac{1}{2}} \left(\frac{k^2-p^2}{k^2-q^2}\right)^{j_3+\frac{1}{2}} \\ &\times {}_2F_1\left(-j_3 - j_{23} + j_2, -j_3 + j_{23} + j_2 + 1; 2j_2 + 2; \frac{p^2-q^2}{k^2-q^2}\right) \end{aligned}$$

Using the transformation

$${}_2F_1(a,\ b;\ c;\ z) = (1-z)^{c-a-b} \cdot {}_2F_1(c-a,\ c-b;\ c;\ z)$$

we finally get that

$$\lim_{R\to\infty} \left\| \begin{matrix} j_1\, j_2\, j_3 \\ \\ j_{12}\, j\, j_{23} \end{matrix} \right\| = \sqrt{\frac{p^2(2j_{23}+1)\Gamma(j_2-j_3+j_{23}+1)\Gamma(j_2+j_3+j_{23}+2)}{(k^2-q^2)\Gamma(j_{23}-j_2-j_3)\Gamma(j_{23}-j_2+j_3+1)}}$$

$$\times (\cos\phi)^{2j_2+1}(\sin\phi)^{2j_3+1} \cdot P^{(2j_2+1,\; 2j_3+1)}_{j_{23}-j_2-j_3-1}(\cos 2\phi), \qquad (8)$$

where

$$\cos\phi = \left(\frac{p^2-q^2}{k^2-q^2}\right)^{1/2}, \quad \sin\phi = \left(\frac{k^2-p^2}{k^2-q^2}\right)^{1/2}.$$

Taking the contraction limit $R \to \infty$ on both sides of the interbases expansion (2) and using the formulas (6)-(7) and (8), we have ($\theta_2 \equiv \theta$)

$$\Phi^{j_2,j_3}_{kk_{j_2}k_{j_3}}(r_{j_2}, r_{j_3}) = \sum_{j_{23}=j_2+j_3+1}^{\infty} W^{j_{23},j_2,j_3}_{kk_{j_2}k_{j_3}}\; \Phi^{j_{23},j_2,j_3}_{kk_{j_{23}}}(r_{j_{23}}, \theta_2) \qquad (9)$$

or in explicit form

$$J_{2j_2+1}(z\cos\theta\cos\phi)\, J_{2j_3+1}(z\sin\theta\sin\phi) = (\sin\phi\sin\theta)^{2j_3+1}\; (\cos\phi\cos\theta)^{2j_2+1}$$

$$\times \sum_{j_{23}=j_2+j_3+1}^{\infty} \frac{(-1)^{j_{23}-j_2-j_3-1}}{z}\; \frac{(2j_{23}+1)\Gamma(j_{23}+j_2+j_3+2)\Gamma(j_{23}-j_2-j_3)}{\Gamma(j_{23}-j_2+j_3+1)\Gamma(j_{23}+j_2-j_3+1)}$$

$$\times P^{(2j_3+1,2j_2+1)}_{j_{23}-j_2-j_3-1}(\cos 2\phi)\; P^{(2j_3+1,\,2j_2+1)}_{j_{23}-j_2-j_3-1}(\cos 2\theta) J_{2j_{23}+1}(z) \qquad (10)$$

where $z \equiv k_{j_{23}} r_{j_{23}}$ and

$$r_{j_2} = r_{j_{23}}\cos\theta, \qquad r_{j_3} = r_{j_{23}}\sin\theta,$$

$$k_{j_2} = k_{j_{23}}\cos\phi, \qquad k_{j_3} = k_{j_{23}}\sin\phi,$$

The last expansion is equivalent to well–known formulas in the theory of Bessel functions[11], namely expansions of the product of two Bessel functions in terms of the product of one Bessel function and two Jacobi polynomials.

The presented procedure of contraction is illustrated on Fig.1. The vertical arrows correspond to the contraction limit (5) from the S_n to E_n. The left tree in Fig.1a contracts to bihyperspherical coordinates; the second to hyperspherical ones. The contraction of the coefficients T is given by eq. (8). This is an asymptotic formula for the Racah coefficients, where the three momenta satisfy $j, j_1, j_{12} \to \infty$. The interbases expansions (9) between two E_n cluster diagrams are obtained from the interbases expansions (2) for the S_n tree diagrams. We thus obtain relations between the bihyperspherical and hyperspherical solutions of the Helmholtz equation on E_n.

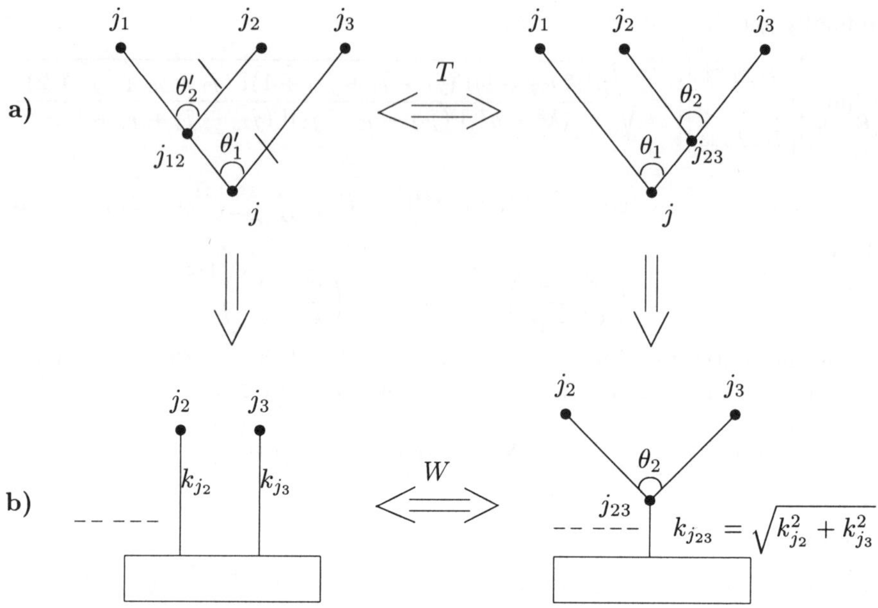

Figure 1: Contractions in interbasis expansion for a general transformation.

References

1. A.A.Izmest'ev, G.S.Pogosyan, A.N.Sissakian and P.Winternitz, *J.Phys. A: Math. Gen.* **29**, 5940-5962 (1996).
2. A.A.Izmest'ev, G.S.Pogosyan, A.N.Sissakian and P.Winternitz, *Inter. J. Mod.Phys.* **A12**, 53-61 (1997).
3. A.A.Izmest'ev, G.S.Pogosyan, A.N.Sissakian and P.Winternitz, *J.Math.Phys.* **40**, 1549-1573 (1999).
4. A.A.Izmest'ev, G.S.Pogosyan, A.N.Sissakian and P.Winternitz, *J.Phys. A: Math. Gen.* **34**, 521-554 (2001).
5. E.Inönü and E.P.Wigner, *Proc. Nat. Acad. Sci. (US)* **39**, 510-524 (1953)
6. N.Ya.Vilenkin, G.I.Kuznetsov and Ya.A.Smorodinskii, *Sov. J. Nucl. Phys.* **2**, 645-655 (1965).
7. G.I.Kuznetsov, S.S.Moskalyuk, Yu.F.Smirnov and V.P.Shelest, *Graph-*

ical Theory of Representations of Orthogonal Groups and its Physical Applications (Kiev: Naukova Dumka 1999) (in Russian).

8. N.Ya.Vilenkin, A.U.Klimyk, *Representations of Lie Groups and Special Functions*, (Dordrecht: Kluwer, 1991).
9. M.S.Kildyushov, *Sov. J. Nucl. Phys.* **15** 113-123 (1972).
10. V.A.Knyr, P.P.Pepiraite and Yu.F.Smirnov, *Sov.J.Nucl.Phys.* **22** 554 (1975)
11. G.Bateman, A.Erdelyi, *Higher Transcendental Functions* (MC Graw-Hill Book Company, INC. New York-Toronto-London 1953)

QUANTIZED SYMMETRIZATIONS AND $\star$-PRODUCTS

HANS PLESNER JAKOBSEN
Mathematics Institute
Universitetsparken 5, DK-2100 Copenhagen Ø, Denmark
E-mail: jakobsen@math.ku.dk

The interrelation between certain quadratic algebras occurring in quantized enveloping algebras on the one hand side, and Poisson structures and deformation quantization on the other, is discussed. It is shown that there are two different methods of constructing $\star$-products available. Further implications in the direction of quantized wave- and Dirac-operators are investigated.

1 Content

The purpose of this article is to make precise some of the results we have reported in talks in Bedlevo (Sep. 2000) and Krakow (July 2001). We have pointed out there that for any algebra having a PBW-type basis, it is possible to construct a $\star$-product. At the conference this has also been made pertinent by J. Wess in his talk and R. Twarock also uses Bergman's Diamond Lemma [1]. On the other hand, in our article [7] we have shown, in a work that falls naturally in the line of investigations of Procesi and de Concini [2], and also of the Diamond Lemma, that it is possible to construct (apparently) another $\star$-product for a class of quadratic algebras. In this connection we wish to stress, in view of Kontsevich's result [10], the importance of determining the gauge group. For all of the known cases, these algebras are generated by certain elements of the quantized enveloping algebra corresponding to a hermitian symmetric space, and actually, a close inspection - and fine tuning - of the construction reveals that the results extend to both the quantized Borel subalgebra as well as the full quantized enveloping algebra. Here, the result of Levendorskii [12] ,[13] is also needed. Even an arbitrary (classical) Lie algebra falls under this and we rediscover here the Kontsevich $\star$-product [10]. The $\star$-products have direct implications for the quantized Dirac-operator - an operator we want to act on "classical" functions on a "classical" (indeed, linearly flat) space.

2 The first $\star$-products; Algebras with PBW-type bases

Let A be an iterated twisted Ore extension, or, more generally, an algebra over a field $\mathbb{F}$ with a PBW-type basis $x^\alpha = x_1^{\alpha_1} x_2^{\alpha_2} \ldots x_n^{\alpha_n}$. Let

$$x^\alpha \cdot x^\beta = \sum_\gamma a^\gamma_{\alpha,\beta} x^\gamma$$

be the expansion in the basis x^α of the product of x^α and x^β in A. Then

$$z^\alpha \star_1 z^\beta = \sum_\gamma a^\gamma_{\alpha,\beta} z^\gamma$$

defines an associative product which we may interpret as a product in $\mathbb{F}[z_1, z_2, \ldots, z_n]$.

3 $\star$-products arising from a certain class of algebras

Consider an algebra $\mathcal{A}$ generated by (linearly independent) elements $X_1, \ldots, X_N$. For each $i = 1, \ldots, N$ let $\mathcal{A}_i$ denote the algebra generated by $X_1, \ldots, X_i$. We assume that the defining relations are of the form:

(Rel) If $i > j$ then $X_i X_j = b_{ij} X_j X_i + p_{ij}$
with p_{ij} of "lower order".

Let V denote the N-dimensional complex vector space spanned by the elements $X_1, \ldots, X_N$, let $T = T(V)$ denote the tensor algebra over V, and let I_R denote the ideal in T generated by elements $X_i X_j - (b_{ij} X_j X_i + p_{ij})$. Then

$$\mathcal{A} := T/I_R. \tag{1}$$

Remark 3.1 *The last condition could e.g. be of the form $p_{ij} \in \mathcal{A}_{i-1}$ The algebra is quadratic if each p_{ij} is quadratic. Originally, this was the most interesting case and it should be noted that there are extremely many examples of this situation in contemporary mathematical physics. To each quantized hermitian symmetric space there is such an algebra. The quantized enveloping algebra is, by Levendorskii's result* [12,13]*, another example. We think of the relation as $X_i X_j = b_{ij} X_j X_i$ with "an error term p_{ij}" and the decisive feature of p_{ij} is that it should be of lower order in some appropriate sense. With this in mind it may be seen that even a classical Lie algebra with $XY = YX + [X, Y]$ fits into the framework and is covered by the results to come.*

We may at first think of (Rel) as a reduction system with the reductions

$$(X_i X_j, b_{ij} X_j X_i + p_{ij}) \quad (\text{ for all } i > j). \tag{2}$$

The essential assumption (EA) below is needed to avoid situations where, due to some special cancellations, a sum of elements in I_R might add up to an element which strictly precedes all the summands in the order. Specifically, assume for any element $u \in I_R$

$$\begin{aligned}\text{(EA):} \qquad & u \in \mathrm{Span}\{\mathrm{a} \cdot (\mathrm{X_i X_j} - \mathrm{b_{ij} X_j X_i} - \mathrm{p_{ij}}) \cdot \mathrm{b} \mid \\ & a, b \in T \text{ and } a \cdot (X_i X_j) \cdot b \leq_l u\}.\end{aligned}$$

It can be seen that the requirements for the Diamond Lemma [1] to be applicable to our situation indeed are satisfied. Thus,

Proposition 3.2 *All elements of T are reduction unique. Moreover, the set $\{X_1^{i_1} \cdots X_N^{i_N} \mid 1_1, \ldots, i_n \in \mathbb{N}_0\}$ is a basis for $\mathcal{A}$ and $\mathcal{A}$ is a domain and is in fact an iterated twisted polynomial (Ore) algebra. In particular, the assumptions of Procesi and de Concini [2] are satisfied. Conversely, one can recover our situation from theirs.*

Remark 3.3 *It would be interesting to classify all quadratic algebras that satisfy this reduction assumption (EA). It is clearly a quite strong assumption, on the order of complication of e.g. the Jacobi Identity in the enveloping algebra.*

4 Projections and quantized symmetries.

The $\star$-product of Section 2 only assumed the existence of a PBW-type basis. In this section we indicate briefly how an algebra $\mathcal{A}$ as in Section 3 gives rise to yet another kind of $\star$-product:

Definition 4.1 *We define a linear map $\Sigma : V \otimes V \longrightarrow V \otimes V$ by*

$$\Sigma(X_i \otimes X_j) = b_{ij} X_j X_i + p_{ij} \text{ if } i > j, \tag{3}$$

$$\Sigma(X_j \otimes X_i) = (b_{ij})^{-1}(X_i X_j - p_{ij}) \text{ if } i > j, \text{ and} \tag{4}$$

$$\Sigma(X_i \otimes X_i) = X_i \otimes X_i \text{ for all } i = 1, \ldots, N. \tag{5}$$

Observe that p_{ij} may be a higher – or lower – order tensor. From now on, we assume that $\forall i, j : b_{ij} = q^{\alpha_{ij}}$ where q is a non-zero complex number which is not a root of unity. Recall that the associated quasi-polynomial algebra is the quadratic algebra $\overline{\mathcal{A}}$, generated by elements $\overline{x}_1, \ldots, \overline{x}_N$ with relations $\overline{x}_i \overline{x}_j = q^{\alpha_{ij}} \overline{x}_j \overline{x}_i$.

Definition 4.2 *For $i \in \mathbb{N}, \sigma_i$ denotes the linear map $T \longrightarrow T$ given by*

$$\sigma_i(v_1 \otimes \ldots \otimes v_{i-1} \otimes v_i \otimes v_{i+1} \otimes \cdots \otimes v_n) = \tag{6}$$

$$v_1 \otimes \ldots \otimes v_{i-1} \otimes \Sigma(v_i \otimes v_{i+1}) \otimes \cdots \otimes v_n \text{ and} \tag{7}$$

$$\sigma_i(v_1 \otimes \ldots \otimes v_r) = v_1 \otimes \ldots \otimes v_r \text{ if } r \leq i. \tag{8}$$

Finally, we define operators $\overline{\sigma}_i$, $T \longrightarrow T$, *analogously, but based on maps* $\overline{\Sigma}$ *in which all* p_{ij} *are zero.*

Lemma 4.3 *For each* $i \in \mathbb{N}$, $\overline{\sigma}_i\ \overline{\sigma}_{i+1}\ \overline{\sigma}_i = \overline{\sigma}_{i+1}\ \overline{\sigma}_i\ \overline{\sigma}_{i+1}$, *and hence* $\overline{\sigma}_1, \ldots, \overline{\sigma}_{n-1}$ *define a representation, called* **quasi-permutation**, *of the symmetric group* S_n *on* T^n.

Lemma 4.4 *For each* $i \in \mathbb{N}$, $\sigma_i = \overline{\sigma}_i$ *modulo lower order.*

Lemma 4.5 *The following hold: 1) For each* $i \in \mathbb{N}$, σ_i *preserves* I_R. *2) For each* $i \in \mathbb{N}$, *if for* $u \in T : \overline{\sigma}_i(u) = u$, *then* $\sigma_i(u) = u$. *3)* $\sigma_i\sigma_{i+1}\sigma_i = \sigma_{i+1}\sigma_i\sigma_{i+1}$ *modulo* I_R *or modulo lower order terms.*

We wish to introduce an analogue of the usual symmetrization map on T. Let us first consider the representation of S_n described in Lemma 4.3. For any $\sigma \in S_n$ we denote the resulting operator on T^n as $\overline{\sigma}$ and we set

$$P_{\text{quasi-sym}} = \frac{1}{n!} \sum_{\sigma \in S_n} \overline{\sigma} \qquad (\textbf{ quasi-symmetrization}). \tag{9}$$

Lemma 4.6

$$\forall i : \qquad \overline{\sigma}_i \cdot P_{quasi\text{-}sym} = P_{quasi\text{-}sym} \cdot \overline{\sigma}_i = P_{quasi\text{-}sym}.$$

We set $S_{\text{quasi-sym}} = P_{\text{quasi-sym}}(T)$. As a vector space, this is clearly equivalent to $\overline{\mathcal{A}}$.

We next want to define a similar operator on T^n with respect to the σ_i's. The problem is, of course, that we do not have a bona fide representation. In spite of this we proceed by defining for each $\sigma \in S_n$ an operator $\widehat{\sigma} = \sigma_{i_1}\sigma_{i_2}\cdots\sigma_{i_r}$ if $\sigma = s_{i_1}s_{i_2}\cdots s_{i_r}$, where $s_j, j = 1, \ldots, n-1$, denotes the elementary transpositions in S_n and we set, for each such set of decompositions of elements,

$$P(n) = \frac{1}{n!} \sum_{\sigma \in S_n} \widehat{\sigma}. \tag{10}$$

Definition/Proposition 4.7 *Set*

$$\mathcal{P}^{q\text{-}sym} = \lim_{N \longrightarrow \infty} (\lim_{n \longrightarrow \infty} P(n))^N. \tag{11}$$

Then $\mathcal{P}^{q\text{-}sym}$ *is a well-defined projection satisfying* $\mathcal{P}^{q\text{-}sym}(I_R) = 0$. *The image of* $\mathcal{P}^{q\text{-}sym}$ *is equal to the image of* $\mathcal{P}_{quasi\text{-}sym}$, *i.e.* $\overline{\mathcal{A}}$. *Moreover,* $\forall i : \mathcal{P}^{q\text{-}sym} \circ \sigma_i = \mathcal{P}^{q\text{-}sym}$.

Proposition 4.8

$$(\star) \quad \forall r, s : (I_r \otimes \mathcal{P}_k^{q\text{-}sym} \otimes I_s)\mathcal{P}^{q\text{-}sym} = \mathcal{P}^{q\text{-}sym},$$

$$(\star\star) \quad \forall r, s : \mathcal{P}^{q\text{-}sym}(I_r \otimes \mathcal{P}_k^{q\text{-}sym} \otimes I_s) = \mathcal{P}^{q\text{-}sym}.$$

If $u, v \in S_{\text{quasi-sym}}$ (construed as a subset of T) we define

$$u \star_2 v = \mathcal{P}^{q\text{-sym}}(u \otimes v). \tag{12}$$

Remark 4.9 *For a classical Lie algebra it follows, using* [11] *and* [15]*, that* $\star_2$ *is the star product one gets from Kontsevich's general construction* [10].

Remark 4.10 *For the purely quadratic algebras, the Poisson structure is also quadratic. It will be interesting to relate the general* $\star_2$*-product to that of Kontsevich. Furthermore, Shklyarov* [14] *has related the algebraic structures more directly to the co-product* Δ *and it remains to be seen if there is a direct way of relating* $\star_2$ *to that.*

Let $\mathcal{B}_q^+$ denote the Borel part of $\mathcal{U}_q$ and let $\mathcal{B}_q^-$ be defined analogously. Let χ_λ be a 1-dimensional representation of $\mathcal{B}_q^+$ corresponding to the weight λ. Let $H^0(\chi_\lambda) = \{f : \mathcal{U}_q \mapsto \mathbb{C} \mid f(ub^+) = \chi_\lambda^{-1}(b^+)f(u)$ and $f \in \text{FIN}(\mathcal{U}_q(\mathfrak{k})\}$ with the action $(x \cdot f)(u) = f(S(x)u)$, S being the antipode. The condition FIN denotes the usual finiteness condition. This is the quantized analogue of a holomorphically induced representation. Covariant (quantized) differential operators [3,4,5,6] arise in this context. Indeed, any element $u_0 \in \mathcal{U}_q$ may in some sense be viewed as an intertwiner through the map $f \mapsto f_{u_0}$ $f_{u_0}(u) = f(u \cdot u_0)$, but to be an intertwiner, there are severe restrictions on u_0 (and χ).

4.1 Translation into ordinary functions; quadratic algebras

We now specialize to the quantized hermitian symmetric spaces - in particular to the one corresponding to the conformal algebra.

Here, $\mathcal{B}_q^- = \mathcal{N}_q^- \otimes \mathbf{C}[K_1^{\pm 1}, \ldots, K_r^{\pm 1}]$ and $\mathcal{N}_q^- = \mathcal{A}_q^- \mathcal{U}_q(\mathfrak{k}_q^-)$. It is a very interesting question to decide between which of these spaces (which all fit into the general framework) should be used, but for now we choose $\mathcal{A}_q^-$: For $su_q(2,2)$ it is generated by w_1, w_2, w_3, w_4 with the usual relations:

$$\begin{aligned} w_1w_2 &= qw_2w_1, w_1w_3 = qw_3w_1, w_3w_4 = qw_4w_3, \\ w_2w_4 &= qw_4w_2, w_2w_3 = w_3w_3, \\ w_1w_4 - w_4w_1 &= (q - q^{-1})w_2w_3. \end{aligned}$$

The elements of $H^0(\chi_\lambda)$ may be viewed as (vector valued) functions on $\mathcal{A}_q^-$, so let us just work with polynomials on the classical space $\mathfrak{p}_0^-$ (identifiable with $\mathfrak{p}_0^+$. For the conformal algebra this is simply polynomials on $\mathbb{C}^4$. Let $z^\gamma = z^{\gamma_1}z^{\gamma_2}z^{\gamma_3}z^{\gamma_4}$, let w^γ in $\mathcal{A}_q^- \subseteq \mathcal{U}_q^-$ be defined analogously, and let B denote the map

$$z^\gamma \mapsto \frac{z^\gamma}{[\gamma]!_q} w^\gamma.$$

We shall not argue here for inserting the factor $[\gamma]!_q$ "for good measure" but proceed to define a linear map S as the composite $S = P_{\text{quasi-sym}} \circ S_{classical}$ where $S_{classical}$ is the classical symmetrization map.

We now basically have two maps B, S from polynomials on $\mathbb{C}^4$ into $\mathcal{U}_q^-$. If $\circ$ denotes the product in $\mathcal{U}_q^-$, then the two $\star$-products are given as follows:

$$x_1 \star_1 x_2 = (x_1, x_2)_B := B^{-1}\left((Bx_1) \circ (Bx_2)\right)$$
$$x_1 \star_2 x_2 = (x_1, x_2)_S := S^{-1}\left((Sx_1) \circ (Sx_2)\right)$$

If $R := B^{-1}S$ then $(x_1, x_2)_S = R^{-1}\left((Rx_1, Rx_2)_B\right)$ - but this map is presumably not "geometrical" (a gauge-equivalence). Both maps B, S take first order expressions into first order expressions. We may then translate left multiplication in $\mathcal{U}_q^-$ into first order differential operators:

$$\left(\partial_x^B p\right)(y) := p(y \star_1 x). \quad \text{Likewise,}$$
$$\left(\partial_x^S p\right)(y) := p(y \star_2 x)$$
$$= \left(\partial_x^B (p(R^{-1}(\cdot)))\right)(Ry)$$

Of course, any $\star$-product with these properties might be used. Even a rescaling of the basis will in some sense (c.f. below) yield another. For now, we end the article by showing how the map B may be used to define the differential operators occurring in the quantized Dirac operator and we discuss possible interpretations of these, thereby continuing the investigation in [9].

The natural first order differential operators.

Consider operators of the form $((w_0^R)^\dagger \cdot f)(w^\gamma) = f(w^\gamma w_0)$. Presently, w_0 will be either w_1, w_2, w_3, w_4, or $w_1 w_4 - q w_2 w_3$. Transformed into operators on functions on $\mathbb{C}^4$ via (4.1) these become q-differential operators expressible in terms of, among other things, the usual q-differential operators on $\mathbb{C}^1$ as well as scaling operators, that is

$$(\left[\frac{\partial}{\partial x}\right]_q \psi)(x) = \frac{f(xq) - f(xq^{-1})}{x(q - q^{-1})}, \text{ and}$$
$$K_i(z_1^{\alpha_1} z_2^{\alpha_2} z_3^{\alpha_3} z_4^{\alpha_4}) = q^{-\alpha_i} z_1^{\alpha_1} z_2^{\alpha_2} z_3^{\alpha_3} z_4^{\alpha_4} \quad i = 1, 2, 3, 4.$$

In terms of un-quantized operators, if $q = e^\hbar$ then $K_i = e^{-\hbar S_i} = \sum_{n=0}^\infty \frac{(-\hbar S_i)^n}{n!}$, where $S_i = z_i \frac{\partial}{\partial z_i}$. Also observe that $K_i = \left[\frac{\partial}{\partial z_i}\right]_q z_i -$

$qz_i \left[\frac{\partial}{\partial z_i}\right]_q$. We obtain:

$$(w_4^R)^\dagger = \left[\frac{\partial}{\partial z_4}\right]_q, (w_2^R)^\dagger = K_4 \left[\frac{\partial}{\partial z_2}\right]_q, (w_3^R)^\dagger = K_4 \left[\frac{\partial}{\partial z_3}\right]_q, \text{ and}$$

$$(w_1^R)^\dagger = K_2 K_3 K_4^2 \left[\frac{\partial}{\partial z_1}\right]_q + z_4(1-q^{-2})K_4 \Box_q, \text{ where}$$

$$\Box_q \overset{Def.}{=} \left((w_1 w_4 - q w_2 w_3)^R\right)^\dagger = K_2 K_3 \left[\frac{\partial}{\partial z_1}\right]_q \left[\frac{\partial}{\partial z_4}\right]_q - \left[\frac{\partial}{\partial z_2}\right]_q \left[\frac{\partial}{\partial z_3}\right]_q.$$

The last defined operator is of course nothing else but the quantized wave operator.

If, instead, we use the classical $\gamma!$ we get quite similar formulas but with ordinary differentiations - except for the $\Box_q$. In this connection observe that the map $\Gamma : z^\gamma \mapsto \frac{[\gamma]_q!}{\gamma!}$ intertwines $\frac{\partial}{\partial z}$ and $\left[\frac{\partial}{\partial z}\right]_q$!

We may rewrite the equation $w_1 w_4 - w_4 w_1 = (q - q^{-1}) w_2 w_3$ as $q w_4 w_1 - q^{-1} w_1 w_4 = (q - q^{-1}) \det_q$ where $\det_q = w_1 w_4 - q w_2 w_3$ generates the center of $\mathcal{A}_q^-$. Here, this becomes

$$q(w_4^R)^\dagger (w_1^R)^\dagger - q^{-1}(w_1^R)^\dagger (w_4^R)^\dagger = (q - q^{-1})\Box_q. \tag{13}$$

Since $\Box_q = c \cdot I$, with $c \in \mathbb{C}$ is any irreducible representation, this has the flavor of a quantized Heisenberg algebra.

Along a similar vein,

$$(w_1^R)^\dagger \mapsto K_2 K_3 K_4^2 \left[\frac{\partial}{\partial z_1}\right]_q + c \cdot z_4(1-q^{-2})K_4.$$

which points towards a covariant derivative; an interpretation which may be further supported since the K_i's may be removed by a change of generators.

Another way of seeing the same phenomenon is by letting

$$G(f) = \left(f, \Box_q f, \Box_q^2 f, \Box_q^3 f, \ldots\right)$$

and B the matrix whose $i,(i+1)$ entry is $z_4(1-q^{-2})K_4$ and zeros everywhere else.

Then we have

$$(K_2 K_3 K_4^2 \left[\frac{\partial}{\partial z_1}\right]_q + B)G(f) = G\left(\left(K_2 K_3 K_4^2 \left[\frac{\partial}{\partial z_1}\right]_q + z_4(1-q^{-2})K_4 \Box_q\right) f\right)$$

This version is also well behaved with respect to the other generators.

Observe: $\left[\frac{\partial}{\partial z}\right]_q (zK) - q^{-2}(zK)\left[\frac{\partial}{\partial z}\right]_q = I$, i.e. $\left[\frac{\partial}{\partial z}\right]_q$ and zK are "quantized conjugate operators".

References

1. G.M. Bergman, *Adv. in Math.* **29**, 178 (1978).
2. C. de Concini and C. Procesi, *Quantum groups.* Lecture notes in mathematics **1565**, 31, Springer (1993).
3. V.K. Dobrev, *J. Phys. A* **27**, 4841, 6633 (1994), & **28**, 7135 (1995).
4. V.K. Dobrev, *Phys. Lett. B* **341**, 133 (1994) & **346**, 427 (1995).
5. M. Harris and H.P. Jakobsen *Math. Ann.* **259**, 227 (1983).
6. H.P. Jakobsen, *Ann. scient. Èc. Norm. Sup.* **18**, 421 (1985).
7. H.P. Jakobsen, Q-differential operators, preprint math.QA/9907009.
8. H.P. Jakobsen and Jøndrup, Quantized Rank R Matrices, J. Alg., to appear.
9. H.P. Jakobsen, *Czech. J. Phys.* **50**, 1265 (2000).
10. M. Kontsevich, Deformation quantization of Poisson manifolds I, preprint math.QA/9709040.
11. V. Kathotia, Kontsevich universal formula for the quantization and the Campbell-Baker-Hausdorff formula, preprint math.QA/9811174.
12. S.Z. Levendorskii, A short proof of the PBW-theorem for quantized enveloping algebras, preprint 1991.
13. S.Z. Levendorskii and Y.S. Soibelman, *Comm. Math. Phys.* **139**, 141-170 (1991).
14. D.L. Shklyarov, in *Lectures on q-Analogues of Cartan Domains and Harish-Chandra Modules*, Ed. L. Vaksman, preprint, Kharkov (2001).
15. B. Shoikhet, Vanishing of Konsetvich integrals of the wheels, preprint math.QA/0007080.

UNIQUENESS RESULTS BY COVARIANCE IN COVARIANT QUANTUM MECHANICS

JOSEF JANYŠKA

Department of Mathematics, Masaryk University
Janáčkovo nám 2a, 662 95 Brno, Czech Republic
E-mail: janyska@math.muni.cz

MARCO MODUGNO

Department of Applied Mathematics, Florence University
Via S. Marta 3, 50139 Florence, Italy
E-mail: modugno@dma.unifi.it

We show that the only requirement of general covariance determines essentially the quantum operators associated with a classical quantisable function and the Schrödinger operator. Our framework is the covariant quantum mechanics of a scalar quantum particle in a curved spacetime which is fibred over absolute time and equipped with given spacelike metric, gravitational field and electromagnetic field. In particular, in the flat case, we recover essentially the standard operators.

Introduction

Since the very beginning of quantum mechanics, the quantum operators associated with classical quantisable functions and the Schrödinger operator have been approached in several ways; indeed, the related literature is huge.

The most usual approaches to quantum operators are based on Hamiltonian techniques, such as the Dirac's canonical quantisation and its refinement provided by geometric quantisation. Analogously, the standard approaches to the Schrödinger operator are based on the quantisation of the classical Hamiltonian.

On the other hand, there are several theories involved, in some respects, with different formulations of these operators (see, for instance, [1,2] and further references therein).

In this paper, we claim that the quantum operators associated with a classical quantisable function and the Schrödinger operator can be essentially determined by the only requirement of general covariance, in a reasonable setting with given exterior fundamental fields and under weak hypotheses of the order of the operators.

These results are obtained in the framework of "covariant quantum mechanics" proposed by A. Jadczyk and M. Modugno [3,4] and further developed in cooperation with other authors (see, for instance, [5,6] and references therein). Some aspects of this formulation have been discussed in the previous sessions of the meeting "Lie Theory and Its Applications in Physics" [7]. In a few words, we consider a classical spacetime, which is fibred over absolute time and is equipped with given Riemannian spacelike metric, gravitational field and electromagnetic field. Partially similar settings have been considered by several authors; in particular, we are indebted to C. Duval and K. Künzle (see, for instance, [2]). The above fundamental fields yield, in a covariant way, a cosymplectic 2–form. Then, we consider a complex bundle over spacetime equipped with a Hermitian metric and a universal, Hermitian connection, whose curvature is proportional to the cosymplectic 2–form. These objects, yield, in a covariant way, good candidates for the quantum operators, through an isomorphism between the Lie algebra of Hermitian quantum vector fields and a distinguished Lie algebra of classical functions. Moreover, these objects yield, in a covariant way, a good candidate for the Schrödinger operator, equivalently, via quantum covariant differentials and via a quantum Lagrangian.

Then, a natural question arises whether we could obtain, in a covariant way, further candidates for quantum operators and Schrödinger operator. In this paper, we answer this question showing that there are essentially no further covariant solutions of the problem. Even more, the covariance determines directly these operators, in a pure geometric way, independently of any Lagrangian or Hamiltonian approaches!

1 Covariant quantum mechanics

Let us summarise briefly the setting of our model for covariant quantum mechanics. Further details can be found in [4,6,7].

We assume the following "positive 1-dimensional semi-vector spaces" over $I\!R^+$ as fundamental unit spaces: the space $\mathbb{T}$ of *time intervals*, the space $\mathbb{L}$ of *lengths* and the space $\mathbb{M}$ of *masses*. Moreover, we assume the *Planck constant* to be an element $\hbar \in \mathbb{T}^* \otimes \mathbb{L}^2 \otimes \mathbb{M}$.

Time is an an affine space of dimension 1 associated with the vector space $\mathbb{T} \otimes I\!R$.

Spacetime is an oriented curved manifold $\boldsymbol{E}$ of dimension 4, which is equipped with:

- a *time fibring* $t : \boldsymbol{E} \to \boldsymbol{T}$,
- a scaled Riemannian *spacelike metric* $g : \boldsymbol{E} \to \mathbb{L}^2 \otimes S^2 V^* \boldsymbol{E}$,
- a *gravitational field*, constituted by a linear connection $K^\natural$, such that $\nabla^\natural dt = 0$ and $\nabla^\natural g = 0$, and whose curvature fulfills the identity $R^{\natural i}{}_\lambda{}^j{}_\mu =$

$R^{\natural j}{}_{\mu}{}^{i}{}_{\lambda}$,

- an *electromagnetic field* constituted by a closed 2–form $f : \boldsymbol{E} \to (\mathbb{L} \otimes \mathbb{M})^{1/2} \otimes \Lambda^2 T^* \boldsymbol{E}$.

With reference to a particle with mass m and charge q, we define the *re–scaled* spacelike metric $G := \frac{m}{\hbar}\, g : \boldsymbol{E} \to \mathbb{T} \otimes S^2 V^* \boldsymbol{E}$ and electromagnetic field $F := \frac{q}{\hbar}\, f : \boldsymbol{E} \to \Lambda^2 T^* \boldsymbol{E}$.

Then, G, $K^\natural$ and F yield the *total connection* $K := K^\natural + (dt \otimes \widehat{F} + \widehat{F} \otimes dt)$ of $\boldsymbol{E}$ (where $\widehat{F} := G^\sharp(F)$), which fulfills the same properties of $K^\natural$.

Moreover, g yields a *spacelike volume form* η and a *spacetime volume form* $\upsilon = dt \wedge \eta$. Then, we obtain the *spacetime divergence* $\mathrm{div}_\upsilon X$, for each spacetime vector field X, and the *spacelike divergence* $\mathrm{div}_\eta X$, for each projectable spacetime vector field X.

We denote by $(x^\lambda) = (x^0, x^i)$ the *spacetime charts*, adapted to the time fibring the affine structure of time and a time unit of measurement $u^0 \in \mathbb{T}$.

We assume the first jet space $J_1 \boldsymbol{E}$ as classical *phase space*. We denote by (x^λ, x^i_0) the *phase charts* induced by a spacetime chart. We obtain, in a covariant way, the *contact map* $\mathrm{d} = u^0 \otimes (\partial_0 + x^i_0\, \partial_i)$.

Then, G and K yield, in a covariant way, a *cosymplectic 2–form* $\Omega : J_1 \boldsymbol{E} \to \Lambda^2 T^* \boldsymbol{E}$.

The Reeb *2nd order connection* γ associated with Ω yields the generalised *Newton law* for classical motions effected by the given gravitational and electromagnetic fields.

The closed 2–form Ω admits locally a *horizontal potential* Θ, defined up to a closed spacetime 1–form. The horizontal phase 1–form $\mathcal{L}[\Theta] := \mathrm{d} \lrcorner \Theta$ is the *classical Lagrangian*; moreover, for each observer o, the horizontal phase 1–form $\mathcal{H}[\Theta, o] := -o \lrcorner \Theta$ is the observed *classical Hamiltonian*, the vertical phase 1–form $\mathcal{P}[\Theta, o] := \nu[o] \lrcorner \Theta$ is the observed *classical momentum*, and the spacetime 1–form $A[\Theta, o] := o^* \Theta$ is the observed *classical potential*. The Euler–Lagrange equation associated with $\mathcal{L}[\Theta]$ turns out to be just the Newton law.

A function of the type $f = \frac{1}{2}\, f^0\, G^0_{ij}\, x^i_0\, x^j_0 + f^i\, G^0_{ij}\, x^j_0 + \overset{o}{f}$, where $f^0, f^i, \overset{o}{f} \in \mathrm{map}(\boldsymbol{E}, I\!\!R)$ is called *special quadratic* and $f'' := f^0\, u_0 \in \mathrm{map}(\boldsymbol{E}, \mathbb{T} \otimes I\!\!R)$ is called the "time component" of f.

The sheaf of special quadratic functions $\mathrm{spec}(J_1 \boldsymbol{E}, I\!\!R)$ is a Lie algebra through the bracket $[\![f, g]\!] := \{f, g\} + \gamma(f'').g - \gamma(g'').f$.

The conditions $\partial_i f^0 = 0$, or $\partial_\lambda f^0 = 0$, or $f^0 = 0$, or $f^\lambda = 0$, yield, respectively, the Lie subalgebras of *projectable*, *fine*, *affine* and *spacetime* functions $\mathrm{spec}(J_1 \boldsymbol{E}, I\!\!R) \supset \mathrm{proj}(J_1 \boldsymbol{E}, I\!\!R) \supset \mathrm{fine}(J_1 \boldsymbol{E}, I\!\!R) \supset \mathrm{aff}(J_1 \boldsymbol{E}, I\!\!R) \supset \mathrm{map}(\boldsymbol{E}, I\!\!R)$.

In particular, the spacetime coordinates, the momentum and the Hamiltonian are special quadratic functions.

The map $\text{spec}(J_1\boldsymbol{E}, I\!R) \to \text{sec}(\boldsymbol{E}, T\boldsymbol{E}) : f \mapsto X[f] = f^0\,\partial_0 - f^i\,\partial_i$ is a covariant morphism of Lie algebras.

The *quantum bundle* is a 1–dimensional complex bundle $\boldsymbol{Q} \to \boldsymbol{E}$, which is equipped with:

- a *quantum metric*, which is a Hermitian metric valued in the space of spacelike volume forms $\mathrm{h} : \boldsymbol{Q} \to \mathbb{C} \otimes \Lambda^3 V^*\boldsymbol{E}$,

- a *quantum connection*, which is a "universal", Hermitian connection of the *extended quantum bundle* $\boldsymbol{Q}^\uparrow := J_1\boldsymbol{E} \underset{\boldsymbol{E}}{\times} \boldsymbol{Q} \to J_1\boldsymbol{E}$, whose curvature is $R[\mathrm{Q}^\uparrow] = -2\,\Omega \otimes \mathbb{I}^\uparrow$, where $\mathbb{I}^\uparrow$ is the Liouville vector field of $\boldsymbol{Q}^\uparrow$.

We recall that the Planck constant $\hbar$ has been incorporated into Ω through G.

We refer to *quantum charts* (x^λ, z), where (x^λ) is a spacetime chart and z is the complex fibre coordinate induced by a *quantum basis* (b) normalised as $\mathrm{h}(\mathrm{b}, \mathrm{b}) = \eta$.

The coordinate expression of the quantum connection is $\mathrm{Q}^\uparrow{}_0 = -\mathrm{i}\,\mathcal{H}_0$, $\mathrm{Q}^\uparrow{}_i = -\mathrm{i}\,\mathcal{P}_i$ and $\mathrm{Q}^\uparrow{}^0_i = 0$, where the gauge of the classical functions $\mathcal{H}_0$ and $\mathcal{P}_i$ is locally determined by the chosen quantum basis.

For each $f \in \text{spec}(J_1\boldsymbol{E}, I\!R)$, we obtain the *quantum vector field* defined by the covariant formula $Y[f] = f^0\,\partial_0 - f^i\,\partial_i + \mathrm{i}\,(\overset{o}{f} + A_0\,f^0 - A_i\,f^i)\,\mathbb{I}$, where $A_\lambda\,d^\lambda$ is the classical potential determined by the quantum connection and the chosen quantum chart. The map $f \mapsto Y[f]$ is a Lie algebra isomorphism.

The *Hermitian quantum vector fields* are classified by the projectable functions through the covariant formula $H[f] = Y[f] - \frac{1}{2}\,\text{div}_\eta X[f]\,\mathbb{I}$.

These vector fields act as Lie derivatives on quantum sections.

A criterion of projectability yields in a covariant way the *quantum Lagrangian* $\mathrm{L}_{(k)} = \frac{1}{4}\,\big((z\,\bar{z}_0 - \bar{z}\,z_0) + \mathrm{i}\,2\,A_0\,\bar{z}\,z - \mathrm{i}\,G^{ij}_0\,(z_i\,\bar{z}_j + A_i\,A_j\,\bar{z}\,z) + A^i_0\,(\bar{z}\,z_i - z\,\bar{z}_i) + 4\,k\,\mathrm{i}\,r_0\,\bar{z}\,z\big)\upsilon^0$, where r is the scalar spacelike curvature induced by G and $k \in I\!R$.

The Euler–Lagrange operator associated with $\mathrm{L}_{(k)}$ yields, in a covariant way, the *Schrödinger operator* $\mathrm{S}_{(k)} = \big(\overset{o}{\nabla}_0 + \frac{1}{2}\,(\text{div}_\eta\,o)_0 - \frac{1}{2}\,\mathrm{i}\,(\overset{o}{\Delta}_0 + k\,r_0)\big)\,u^0$, where $\overset{o}{\nabla}$ is the the observed quantum covariant differential, $\text{div}_\eta\,o$ is the divergence of the observer, $\overset{o}{\Delta}$ is the observed quantum Laplacian; indeed the above combination turns out to be observer independent.

Then, we define the (infinite dimensional) *functional quantum bundle* $\boldsymbol{H} \to \boldsymbol{T}$ to be the fibred set over $\boldsymbol{T}$, whose fibres are constituted by the compact support smooth sections, at fixed time, of the quantum bundle. The quantum metric h equips the functional quantum bundle with a fibred *pre-Hilbert metric* $\langle \, | \, \rangle$. Then, a true Hilbert bundle can be obtained by a completion procedure.

The Hermitian quantum vector fields $H[f]$ (via Lie derivatives) and the Schrödinger operator $\mathsf{S}_{(k)}$ yield, in a covariant way, symmetric operators on the fibres of the functional quantum bundle. Moreover, the Schrödinger operator can be regarded as a Hermitian operator of the functional quantum bundle (actually, as the covariant differential associated with a Hermitian connection).

2 Uniqueness results by covariance

Let us consider the complex linear fibred automorphisms $\Phi : \boldsymbol{Q} \to \boldsymbol{Q}$ projectable on fibred automorphisms $\phi : \boldsymbol{E} \to \boldsymbol{E}$. The tangent prolongations of the Φ,s and ϕ,s, at fixed points of $\boldsymbol{E}$, yield the group bundles $\mathcal{G}(\boldsymbol{Q})$ and $\mathcal{G}(\boldsymbol{E})$ over $\boldsymbol{E}$. The fibred surjection $\mathcal{G}(\boldsymbol{Q}) \to \mathcal{G}(\boldsymbol{E})$ is a fibred central extension. Moreover, we have fibred actions of $\mathcal{G}(\boldsymbol{E})$ on $\boldsymbol{E} \to \boldsymbol{T}$ and of $\mathcal{G}(\boldsymbol{Q})$ on $\boldsymbol{Q} \to \boldsymbol{E}$ and on all other bundles we have derived from them.

Moreover, we consider the trivial group bundle $\mathcal{G}(\mathbb{T}) := \boldsymbol{E} \times I\!R^+$ of changes of time scales. We have a fibred action of $\mathcal{G}(\mathbb{T})$ on time scaled bundles such as $\mathbb{T}^* \otimes T\boldsymbol{E}$ and $\mathbb{T}^* \otimes \boldsymbol{Q}$.

First, let us classify the covariant operators of Schrödinger type.

Theorem 2.1. *All 2nd order operators $\mathsf{O}(\Psi) : \boldsymbol{E} \to \mathbb{T}^* \otimes \boldsymbol{Q}$, which depend on dt, G, K, h, $\mathsf{Q}^\uparrow$ and are covariant with respect to $\mathcal{G}(\boldsymbol{Q})$ and $\mathcal{G}(\mathbb{T})$, are of the type*

$$\mathsf{O}(\Psi) = \alpha \, \mathsf{S}_{(0)}(\Psi) + \beta \, r \, \Psi \,, \quad \textit{with} \quad \alpha, \beta \in \mathbb{C} \,.$$

Moreover, if we add the condition that the operators $\mathsf{O}(\Psi)$ yield Hermitian operators on the functional quantum bundle, then they are of the type

$$\mathsf{O}(\Psi) = \mathsf{S}_{(k)}(\Psi) \,. \qquad \square$$

Next, let us classify the covariant operators of quantum Lagrangian type.

Theorem 2.2. *All 2nd order operators $\mathsf{L}(\Psi) : \boldsymbol{E} \to \mathrm{i}\, \Lambda^4 T^*\boldsymbol{E}$, which depend on dt, G, K, h, $\mathsf{Q}^\uparrow$ and are covariant with respect to $\mathcal{G}(\boldsymbol{Q})$ and $\mathcal{G}(\mathbb{T})$, are of the type*

$$\begin{aligned} \mathsf{L}[\Psi] &= a \, \mathsf{L}_{(k)}(\Psi) + b \, \tfrac{1}{2} \, dt \wedge \big(\mathsf{h}(\Psi, \mathsf{S}_{(0)}[\Psi]) - \mathsf{h}(\mathsf{S}_{(0)}[\Psi], \Psi)\big) \\ &\quad + c \, \mathrm{i} \, \tfrac{1}{2} \, dt \wedge \big(\mathsf{h}(\Psi, \mathsf{S}_{(0)}[\Psi]) + \mathsf{h}(\mathsf{S}_{(0)}[\Psi], \Psi)\big) \,, \quad a, b, c \in I\!R \,. \qquad \square \end{aligned}$$

The Euler–Lagrange operators associated with the above Lagrangians are just the above Schrödinger operators. Moreover, we observe that the 2nd order terms in the above Lagrangians are not relevant because they yield again the Schrödinger operator.

Eventually, let us classify the covariant operators associated with special quadratic functions.

Theorem 2.3. *All 2nd order operators $\mathsf{O}[f](\Psi) : \boldsymbol{E} \to \boldsymbol{Q}$, which depend on dt, G, K, $\hbar$, $\mathrm{Q}^{\uparrow}$, depend linearly on a special quadratic function f and are covariant with respect to $\mathcal{G}(\boldsymbol{Q})$ and $\mathcal{G}(\mathbb{T})$, are of the type*

$$\begin{aligned}\mathsf{O}[f](\Psi) = \big(&\alpha_1\, f'' \lrcorner S + \alpha_2\, L_{Y[f]} + \alpha_3\, L_{Y[\mathrm{d}.f'']} \\ &+ \alpha_4\, \operatorname{div}_\upsilon X[f] + \alpha_5\, \operatorname{div}_\eta X[\mathrm{d}.f''] + \alpha_6\, f'' \lrcorner r\big)\,(\Psi)\,,\end{aligned}$$

where $\alpha_1, \dots, \alpha_6 \in \mathbb{C}$.

Moreover, if we add the condition $\mathsf{O}[1](\Psi) = \Psi$, then we obtain $a_2 = \mathrm{i}$.

Furthermore, if we add the condition that the operators $\mathsf{O}[f](\Psi)$ yield operators $\widehat{\mathsf{O}}[f]$ acting on the fibres of the functional quantum bundle, then we obtain $a_1 = -\mathrm{i}$.

Eventually, if we add the condition that the induced operators $\widehat{\mathsf{O}}[f]$ are symmetric, then the special functions must be projectable and, for such functions, the corresponding operators are of the type

$$\begin{aligned}\mathsf{O}[f](\Psi) = \big(&\mathrm{i}\,(L_{H[f]} - f'' \lrcorner S_{(k)}) + a\, \operatorname{div}_\upsilon X[f] \\ &+ b\,\mathrm{i}\, L_{H[\mathrm{d}.f'']} + c\, \operatorname{div}_\eta X[\mathrm{d}.f'']\big)\,(\Psi)\,,\end{aligned}$$

where $a, b, c, k \in \mathbb{R}$. □

In order to fix the undetermined coefficients $a, b, c, k \in \mathbb{R}$ we need further requirements.

Corollary 2.4. *All operators as above associated, respectively, with $f \in \operatorname{fine}(\boldsymbol{E})$, $f \in \operatorname{aff}(\boldsymbol{E})$, $f \in \operatorname{map}(\boldsymbol{E}, \mathbb{R})$, are of the type*

$$\begin{aligned}
&\mathsf{O}[f](\Psi) = \big(\mathrm{i}\,(L_{H[f]} - f'' \lrcorner S_{(k)}) + a\, \operatorname{div}_\upsilon X[f]\big)\,(\Psi)\,, && \textit{for} \quad f \in \operatorname{fine}(J_1\boldsymbol{E}, \mathbb{R})\,,\\
&\mathsf{O}[f](\Psi) = \big(\mathrm{i}\, L_{H[f]} + a\, \operatorname{div}_\eta X[f]\big)\,(\Psi)\,, && \textit{for} \quad f \in \operatorname{aff}(J_1\boldsymbol{E}, \mathbb{R})\,,\\
&\mathsf{O}[f](\Psi) = f\,\Psi\,, && \textit{for} \quad f \in \operatorname{map}(\boldsymbol{E}, \mathbb{R})\,. \square
\end{aligned}$$

Example 2.5. All operators as above associated with the classical Hamiltonian $\mathcal{H}_0 = \frac{1}{2}\, G^0_{ij}\, x^i_0\, x^j_0 - A_0$ and the classical momentum $\mathcal{P}_i = G^0_{ij}\, x^j_0 + A_i$ are, respectively, of the type

$$\mathrm{O}[\mathcal{H}_0] = -(\tfrac{1}{2}\,\overset{o}{\Delta}_0 + \tfrac{1}{2}\,k\,r_0 - A_0) + a\,\frac{\partial_0\sqrt{|g|}}{\sqrt{|g|}}\,,$$

$$\mathrm{O}[\mathcal{P}_j] = -\mathrm{i}\,(\partial_j + \tfrac{1}{2}\,\frac{\partial_j\sqrt{|g|}}{\sqrt{|g|}}) - a\,\frac{\partial_i\sqrt{|g|}}{\sqrt{|g|}}\,. \qquad \square$$

Let us sketch briefly the proofs of the above Theorems.

Proof. The proofs can be achieved by the following steps:

- in virtue of the "orbit reduction theorem" [8] and by taking into account the covariance with respect to $\mathcal{G}(\boldsymbol{Q})$, we can express the operators through the curvatures of the connections and the covariant derivatives of the sections up to 2nd order;

- the "fundamental identities" postulated for the fields $dt, G, K, \mathrm{h}, \mathrm{Q}^{\uparrow}$ make some of the above objects vanishing, or express some of them through the others;

- in virtue of the "homogeneous function theorem" [8] and by taking into account the covariance with respect to the subbundle of $\mathcal{G}(\boldsymbol{Q})$ constituted by homotheties, we can express the operators in terms of polynomials;

- in virtue of the "metric covariant function" [8], by taking into account the covariance with respect to $\mathcal{G}(\mathbb{T})$ and counting the covariant and contravariant indices occurring in the previous expression of the operators, we can compute the only contractions allowed and get the result. $\square$

We have assumed no distinguished time scale. But, if we suppose that the quantum system be involved with a distinguished $\tau \in \mathbb{T}$, then we could not require the covariance with respect to $\mathcal{G}(\mathbb{T})$ and we would obtain several additional solutions of our classification problems. For instance, for the Schrödinger operator we would obtain the additional covariant non–linear term

$$\mathrm{O}'(\Psi) = \tau^{1/2}\,(\hbar/m)^{3/2}\,\frac{\mathrm{h}}{\eta}(\Psi,\Psi)\,\Psi\,.$$

Acknowledgements

This research has been supported by University of Florence, GNFM of INDAM, the Ministry of Educations of the Czech Republic under the project MSM 143100009 and the grant of the GA ČR No. 201/99/0296.

Thanks are due to H.-D. Doebner for stimulating discussions.

References

1. H.-D. Doebner, G. A. Goldin, P. Nattermann, *J. Math. Phys.* **40**, 49 (1999); C. Ghirardi, A. Rimini, T. Weber, *Phys. Rev.* D **34**, 470 (1986); K. Kuchař, *Phys. Rev.* D **22**, 1285 (1980); A. Kyprianidis, *Phys. Rep.* **155**, 1 (1987); W. M. Tulczyjew, *J. Geom. Phys.* **2**, 93 (1985).
2. C. Duval, H. P. Künzle, *G.R.G.* **16**, 333 (1984).
3. A. Jadczyk, M. Modugno, in *Differential geometric methods in theoretical physics*, C. N. Yang, M. L. Ge and X. W. Zhou Eds., World Scientific, Singapore, 543–556, (1992).
4. A. Jadczyk, M. Modugno, *Galilei general relativistic quantum mechanics*, report, Dept. of Appl. Math., Univ. of Florence, (1994).
5. D. Canarutto, A. Jadczyk, M. Modugno, *Rep. on Math. Phys.* **36**, 95 (1995); J. Janyška, *Rendiconti di Matematica, S. VII (Roma)* **15**, 457 (1995); J. Janyška, in *The Proceedings of the Winter School "Geometry and Topology"* (Srní, 2000), Suppl. ai Rendiconti del Circolo Matematico di Palermo, Serie II, **66**, 117, (2001); J. Janyška, M. Modugno, *Covariant pre-quantum operators*, preprint 2001; M. Modugno, R. Vitolo, in *Gravitation, electromagnetism and geometrical structures*, G. Ferrarese Ed., Pitagora Editrice Bologna, 237, (1996); D. Saller, R. Vitolo, *J. Math. Phys.* **41**, 6824 (2000); R. Vitolo, *Annales de l'Institut Henri Poincaré* **64**, 177 (1996); R. Vitolo, in *Proc. of the XII It. Conf. on G.R. and Grav. Phys.*, Roma, 1996; World Scientific, Singapore; R. Vitolo, *Ann. Inst. 'H. Poinc.* **70**, 239 (1999).
6. A. Jadczyk, J. Janyška, M. Modugno, in *Geometria, Física-Matemática e outros Ensaios*, Homenagem a António Ribeiro Gomes, A. S. Alves, F. J. Craveiro de Carvalho and J. A. Pereira da Silva Eds., 253–313, Coimbra (1998); J. Janyška, M. Modugno, D. Saller, in *Recent Developments in General Relativity*, Genova, September 2000, World Scientific, Singapore; J. Janyška, M. Modugno, *Covariant Quantum Mechanics*, book in preparation (2001).
7. M. Modugno, in *Lie Theory and Its Applications in Physics*, August 1997, Clausthal, Eds. H.-D. Doebner, V. K. Dobrev and J. Hilgert, World Scientific, Singapore (1998), p.183; M. Modugno, C. Tejero Prieto, R. Vitolo, in *Lie Theory and Its Applications in Physics*, July 1999, Clausthal, Germany, H.-D. Doebner, V.K. Dobrev and J. Hilgert Eds., World Scientific, Singapore, (2000), p.155.
8. I. Kolář, P. Michor, J. Slovák, *Natural operators in differential geometry*, Springer-Verlag, Berlin (1993).
9. J. Sniatycki, *Geometric quantization and quantum mechanics*, Springer, New York (1980).
10. N. Woodhouse, *Geometric quantization*, Clarendon Press, Oxford, 2nd Edit. (1992).

WAVES AND SOLITONS IN BORN-INFELD ELECTRODYNAMICS

RICHARD KERNER

L.P.T.L. - Tour 22, 4-ème étage, Boıte 142
Université Paris-VI, 4, Place Jussieu, 75005 Paris, France

We recall first the exceptional propagation properties of classical Born-Infeld theory discovered by G. Boillat in 1970 ([3]). Then we display some ancient and more recent static solutions of solitonic type, and discuss their properties

1 Introduction - a short glance at the history

Since the discovery of the electron by J.J.Thomson in 1899, physicists tried to develop models of finite energy charge concentrations that could describe the elementary electric charge. One of the ideas was to use non-linear generalizations of Maxwell's theory, deviating from it only at very short distances and very strong fields in order to ensure a cut-off and to avoid singularity at $r \to 0$.

G. Mie ([1]) was first to introduce such a model, based on the assumption that the electric field $\mathbf{E}$ can not exceed the limiting value $\mathbf{E_0}$, and that the repulsive force should be proportional to the expression

$$\mathbf{F} \sim \frac{\mathbf{E}}{\sqrt{1 - \frac{\mathbf{E}^2}{\mathbf{E}_0^2}}} \tag{1}$$

In this model it was possible to find a nonsingular solution with finite energy and charge, and with the field $\mathbf{E}$ falling off as r^{-2} at great distances, but this solution was not covariant with respect to the Lorentz transformations.

In 1932 and in 1934 Born and Infeld have published by now celebrated version of non-linear electrodynamics, in whihc they proposed the following Lorentz-invariant lagrangian:

$$\mathcal{L} = \beta^2 \left[\sqrt{\det \left(\delta^\mu_\lambda + \beta^{-1} \, F^\mu_{\ \lambda} \right)} \right] \tag{2}$$

The constant β appears for dimensional reasons, and plays the same rôle here as the limiting value of the electric field in G. Mie's non-linear electrodynamics.

When expressed in terms of two invariants of Maxwell's tensor,

$$P = \frac{1}{4} \, F_{\mu\nu} \, F^{\mu\nu} \quad \text{and} \qquad S = \frac{1}{4} \, F_{\mu\nu} \, \tilde{F}^{\mu\nu}\,, \quad \text{with} \quad \tilde{F}^{\mu\nu} = \frac{1}{2} \, \epsilon^{\mu\nu\lambda\rho} \, F_{\lambda\rho}$$

this lagrangian can be written explicitly as

$$\mathcal{L}_{BI} = \beta^2 \left[1 - \sqrt{1 + 2P - S^2} \right] \tag{3}$$

or as $$\mathcal{L}_{BI} = \beta^2 \left[1 - \sqrt{1 + \frac{1}{2\beta^2} (\mathbf{B}^2 - \mathbf{E}^2) - \frac{1}{16\beta^4} (\mathbf{E} \cdot \mathbf{B})^2} \right] \tag{4}$$

With the advent of Quantum Mechanics and Dirac's equation for the electron, the interest in classical models of charged particles has considerably faded. However, in 1970 G. Boillat ([3] considered the Born-Infeld electrodynamics as an example of non-linear theory in order to study its propagation nproperties. Starting with the most general non-linear theory derived from an arbitrary Lagrangian depending on two Lorentz invariants of Maxwell's tensor, $\mathcal{L}\,(P, S)$, he discovered that among all such non-linear theories, the Born-Infeld electrodynamics is the only one ensuring the absence of bi-refringence, i.e. propagation along a single light-cone, and the absence of shock waves. In this respect the Born-Infeld theory is unique (except for another singular and unphysical Lagrangian $\mathcal{L} = P/S$). A beautiful discussion of these properties can be found in I. Bialynicki-Birula's paper ([4]). Let us remind very shortly how the bi-refringence phenomenon may occur in non-linear theories.

From the mathematical point of view, these theories are based on systems of second order partial differential equations, linear in highest derivatives, with coefficients which depend only on the fields (but not on their derivatives). The systems of this type can be reduced to a set of differential equations of first order via introduction of auxiliary fields, which are the independent linear combinations of the first partial derivatives of functions corresponding to the degrees of freedom of our system.

The differential system can be represented by means of a matrix whose entries contain the operators of partial derivation or multiplicative coefficients, acting on a vector-column representing auxiliary fields. If the vector-column $\mathbf{u}$ with the fields ψ, χ_i, E_i and B_i contains N elements, then let us denote by $\mathcal{A}$ the NxN matrix containing the partial derivatives and by $\mathcal{B}$ the NxN matrix containing the multiplicative factors. Then the field equations can be written as:

$$\mathcal{A}^{\mu}(\mathbf{u})\partial_{\mu}\mathbf{u} + \mathcal{B}(\mathbf{u})\mathbf{u} = 0\,. \tag{5}$$

If the hypersurface defined by the implicit equation

$$\Sigma\,(x^{\mu}) \,= 0\,. \tag{6}$$

is a surface of discontinuity, then the first derivatives of fields are discontinuous across this surface, whereas the fields themselves are continuous. So, when applied to the *discontinuities* across the hypersurface (6), the equation (5) reduces to

$$(\mathcal{A}^\mu \, \Sigma_\mu) \, \delta_1 \mathbf{u} = 0 \,, \tag{7}$$

where $\Sigma_\mu \equiv \partial_\mu \Sigma$, and $\delta_1 \mathbf{u}$ denotes the discontinuity of the first derivative across Σ, $\delta_1 \mathbf{u} \equiv \partial \mathbf{u}/\partial \Sigma|_+ - \partial \mathbf{u}/\partial \Sigma|_-$. By definition, for a characteristic surface one has $\delta_1 \mathbf{u} \neq 0$, therefore, in order for (7) to hold, one must have

$$\det \left(\mathcal{A}^\mu \, \Sigma_\mu \right) = 0 \,, \tag{8}$$

on the surface of discontinuity. The characteristic equation (8) determines the surface whose generic equation is $H(x, \Sigma_\mu) = 0$, with H a homogeneous function of order N in Σ_μ. The Born-Infeld theory turns out to be *completely exceptional* since it obeys the corresponding condition of [3], namely $\delta_0 H \equiv H|_+ - H|_- = 0$.

Let us give an illustration of this principle on the simplest case: the scalar field wave equation in a two-dimensional space-time (t, x):

$$\partial_0^2 \, \phi - \partial_x^2 \, \phi = 0 \,. \tag{9}$$

(Partial derivatives in Lorentz indices, $(0, x, y, z)$ or $(0, 1, 2, 3)$, will be denoted by ∂_0, ∂_x), etc.. According to the prescription, we can use as auxiliary fields ψ and χ the first derivatives of the scalar field ϕ, $\partial_0 \phi = \psi$ and $\partial_x \phi = \chi$. Then by definition, the first derivatives of auxiliary fields are not independent, because we have, as $\partial_0 \, (\partial_x \, \phi) = \partial_x \, (\partial_0 \, \phi)$, automatically $\partial_0 \chi - \partial_x \psi = 0$. On the other hand the dynamical equation (9) can be written as $\partial_0 \psi - \partial_x \chi = 0$. In the matrix notation of (7) these two equations can be combined to yield

$$\begin{pmatrix} 0 & 1 \\ 1 & 0 \end{pmatrix} \partial_0 \begin{pmatrix} \psi \\ \chi \end{pmatrix} + \begin{pmatrix} -1 & 0 \\ 0 & -1 \end{pmatrix} \partial_x \begin{pmatrix} \psi \\ \chi \end{pmatrix} = \begin{pmatrix} 0 \\ 0 \end{pmatrix} . \tag{10}$$

We then find

$$\mathcal{A}^\mu \Sigma_\mu = \begin{pmatrix} -\Sigma_x & \Sigma_0 \\ \Sigma_0 & -\Sigma_x \end{pmatrix} , \tag{11}$$

and the characteristic equation $\det(\mathcal{A}^\mu \Sigma_\mu) = 0$ can be written as

$${\Sigma_0}^2 - {\Sigma_x}^2 = 0 \,, \tag{12}$$

where $\Sigma_0 \equiv \partial_0 \Sigma$ and $\Sigma_x \equiv \partial_x \Sigma$. The last equation defines the characteristic surfaces $\Sigma(t, x)$, which in this case are the light-cones in two space-time dimensions.

The same technique can be easily applied to the electromagnetic Maxwellian field. We have to solve a 6×6 matrix, because we have now six independent combinations of its first derivatives (the fields **E** and **B**) appearing in the first-order Maxwell's equations. As we know, the characteristic surfaces in four dimensions are given by $\Sigma_{,\mu}\,\Sigma^{,\mu} = {\Sigma_0}^2 - {\Sigma_x}^2 - {\Sigma_y}^2 - {\Sigma_z}^2 = 0$.

The same is true for the Born-Infeld non-linear electrodynamics. A more exhaustive discussion of the propagation properties of various non-linear generalizations of the electromagnetism can be found in recent papers ([5], [6]).

An entirely new and unexpected impulse for the revival of interest in the Born-Infeld electrodynamics, and in its non-abelian generalizations, came from recent developments of String and Membrane theories. The string lagrangian in (4 + D) dimensions, which defines a minimal surface in a (4+D)-dimensional Minkowskian space-time, is in fact a generalization of geoddesic equation for a point-like particle.

Consider a two-dimensional surface with cylindrical topology, parametrized with one time-like and one space-like parameter, τ and σ, respectively, and embedded in a (4+D)-dimensional space-time. The embedding functions will be denoted by $X^\mu\,(\tau,\sigma)$, or equivalentlu, as $X^\mu\,(\xi^a)$, with $\mu, \nu = 0, 1, 2, ...3 + D$, and $A, B, .. = 1, 2$ so that $\xi^1 = \tau$, $\xi^2 = \sigma$.

The exterior space-time metric $g_{\mu\lambda}$ induces the internal metric of the world-sheet spanned by the string,

$$G_{AB} = g_{\mu\lambda}\,\partial_a X^\mu\,\partial_b X^\lambda \tag{13}$$

Let $h^{AB}\,(\xi^c)$ be an arbitrary metric on the world-sheet; the variational principle introduced first by A. Polyakov reads then

$$\delta\, S = -\frac{1}{4\pi\alpha'}\,\delta \int\!\!\int \sqrt{-h}\, h^{AB}\, G_{AB}\, \mathrm{d}\tau \mathrm{d}\sigma\; = 0 \tag{14}$$

Under the independent variations δx^μ and $\delta\, h^{AB}$ one gets the following equations:

$$G_{AB} - \frac{1}{2}\, h_{AB} \Big(h^{CD}\, G_{CD} \Big) = 0 \tag{15}$$

$$h^{AB} \left[\nabla_A \nabla_B x^\mu + \Gamma^\mu_{\lambda\rho}\, \partial_A x^\lambda \partial_B x^\rho \right] = 0. \tag{16}$$

After dimensional reduction from 11 to 10 dimensions, auxiliary fields A_μ and ϕ do appear, and the total lagrangian takes on the form that contains the

Born-Infeld lagrangian ([14,15,16]).

$$\mathcal{L} = \frac{1}{2}\,\mathcal{D}_\mu \Phi \mathcal{D}^\mu \Phi^* + \beta^2\,(1-\mathcal{R}) - \frac{\lambda}{2}\,(\Phi^*\,\Phi - v^2)^2 + \frac{1}{16\pi G}\,R \qquad (17)$$

with Φ denoting scalar field, R the Riemann curvature scalar, and $\mathcal{R}$ given by

$$\mathcal{R} = \sqrt{1 + \frac{1}{2\beta^2} F^a_{\mu\nu} F^{\mu\nu}_a - \frac{1}{16\beta^4} (F^a_{\mu\nu} \tilde{F}^{\mu\nu}_a)^2}. \qquad (18)$$

For dimensional reductions onto lower dimensions, the non-abelian generalizations of this lagrangian are naturally produced.
In a pure Yang-Mills theory in flat space-time, with the usual Lagrangian density $\mathcal{L}_{YM} = -\frac{1}{4} g_A B\, F^A_{\mu\nu}\, F^{B\,\mu\nu}$ there are no finite energy static non-singular solutions describing a charged soliton. This fact can be explained by the conformal invariance of the theory, and the tracelesness of the energy-momentum tensor,

$$T^\mu{}_\mu = -T_{00} + \sum_{i=1}^{3} T_{ii} = 0, \qquad (19)$$

Given the positivity of energy, i.e.$T_{00} > 0$, this means that the sum of principal pressures is positive, too, $\sum T_{ii} > 0$, which leads to the conclusion that the Yang-Mills "matter" is naturally subjectsd to repulsive forces only.

The presence of the Higgs field breaks the conformal invariance, which leads to the existence of 't Hooft and Prasad-Sommerfield magnetic monopoles. In what follows, we are interested in soliton-like solutions arising in other non-linear theories, including non-abelian versions of Yang-Mills theories, which are no more conformally invariant.

2 Non-linear electrodynamics from the Kaluza-Klein theory

An interesting non-linear generalization of electrodynamics derived from the Kaluza-Klein theory in five dimensions has been proposed in ([8], [9]). It is based on the addition of the Gauss-Bonnet term, $R_{ABCD}\, R^{ABCD} - 4\, R_{AB}\, R^{AB} + R^2$, which in five dimensions is not a topological invariant, leading to non-trivial equations of motion of second order when added to the Einstein-Hilbert Lagrangian.

In a flat space-time and without the scalar field the Kaluza-Klein metric is

$$g_{AB} = \begin{pmatrix} g_{\mu\nu} + A_\mu A_\nu & A_\mu \\ A_\nu & 1 \end{pmatrix}, \qquad (20)$$

where $A, B = 0, 1, 2, 3, 5$ and $\mu, \nu = 0, 1, 2, 3$ (or $\mu, \nu = 0, x, y, z$ following the covention we have been using). The full Lagrangian is taken to be (see [8,9]):

$$\mathcal{L} = R + \gamma\,(R_{ABCD}\,R^{ABCD} - 4\,R_{AB}\,R^{AB} + R^2)\,, \tag{21}$$

with γ being a certain dimensional parameter characterizing the strength of the non-linearity. When expressed in four dimensions in terms of the Maxwell tensor, it becomes

$$\mathcal{L} = -\frac{1}{4}F_{\mu\nu}\,F^{\mu\nu} - \frac{3\gamma}{16}\left[(F_{\mu\nu}F^{\mu\nu})^2 - 2\,(F_{\mu\lambda}F_{\nu\rho}F^{\mu\nu}F^{\lambda\rho})\right]. \tag{22}$$

In terms of the invariants P and S this Lagrangian is given by $\mathcal{L} = 2P + \frac{3\gamma}{2}S^2$, which for the choice $\gamma = -\frac{2}{3}$ yields essentially the square of the Born-Infeld Lagrangian. The equations of motion are:

$$F_{\lambda\rho,\mu} + F_{\rho\mu,\lambda} + F_{\mu\lambda,\rho} = 0\,, \tag{23}$$

which correspond to the Bianchi identities and are geometrical equations valid independently of the Lagrangian chosen, and the dynamical equations resulting from the variational principle,

$$[F^{\lambda\rho} - \frac{3\gamma}{2}\,(F_{\mu\nu}F^{\mu\nu})F^{\lambda\rho} + \frac{3\gamma}{2}F_{\mu\nu}F^{\lambda\mu}F^{\rho\nu}]_{,\lambda} = 0\,. \tag{24}$$

The Lagrangian (22) is particularly simple when expressed in more familiar terms with the fields $\mathbf{E}$ and $\mathbf{B}$:

$$\mathcal{L} = \frac{1}{2}(\mathbf{B}^2 - \mathbf{E}^2) + \frac{3\gamma}{2}\,(\mathbf{E}\cdot\mathbf{B})^2 \tag{25}$$

The equations of motion also display a clear physical meaning when expressed in terms of $\mathbf{E}$ and $\mathbf{B}$. The equation (24) becomes

$$\mathbf{div\,B} = 0, \qquad \mathbf{rot\,E} = -\partial_0\mathbf{B}, \tag{26}$$

whereas the equations (23) become

$$\mathbf{div\,E} = -3\gamma\,\mathbf{B}\cdot\mathbf{grad}\,(\mathbf{E}\cdot\mathbf{B})$$

$$\mathbf{rot\,B} = \partial_0\mathbf{E} + 3\,\gamma\left[\mathbf{B}\partial_0(\mathbf{E}\cdot\mathbf{B}) - \mathbf{E}\times\mathbf{grad}(\mathbf{E}\cdot\mathbf{B})\right]. \tag{27}$$

which show how the density of charge and the current are created by the non-linearity of the field: indeed, we can introduce

$$\rho = -3\gamma\,\mathbf{B}\cdot\mathbf{grad}\,(\mathbf{E}\cdot\mathbf{B}) \quad \text{and} \quad \mathbf{j} = 3\,\gamma\left[\mathbf{B}\partial_0(\mathbf{E}\cdot\mathbf{B}) - \mathbf{E}\times\mathbf{grad}(\mathbf{E}\cdot\mathbf{B})\right] \tag{28}$$

which satisfy the continuity equation

$$\partial_0 \rho + \mathbf{div}\,\mathbf{j} = 0 \tag{29}$$

The Poynting vector conserves its form known from the Maxwellian theory, but the energy density is modified:

$$\mathbf{S} = \mathbf{E} \times \mathbf{B}, \qquad \mathcal{E} = \frac{1}{2}\left(\mathbf{E}^2 + \mathbf{B}^2\right) + \frac{3\gamma}{2}\left(\mathbf{E}\cdot\mathbf{B}\right)^2, \tag{30}$$

with the continuity equation resuming the energy conservation satisfied by virtue of the equations of motion:

$$\partial_0 \mathcal{E} + \mathbf{divS} = 0\,. \tag{31}$$

It can be easily proved that there is birefringence in this theory. One wave propagates in a Maxwellian way, the other possible wave solution propagates differently; in fact, it is delayed (see [5] for details).

The properties of possible stationary axisymmetric solutions, endowed with non-vanishing charge, intrinsic kinetic and magnetic moments, have been discussed in [8,9]. In the theory based on the Gauss-Bonnet term in 5 dimensions, one can try to find axially-symmetric configurations displaying both finite electric charge and finite magnetic moment; also, a kinetic momentum can be expected, parallel to the magnetic moment.

In cylindric coordinates ρ, ϕ, z we expect the induced current density to be aligned on the $\mathbf{e}_\phi$-vector of the local frame, giving a current density circulating around the z-axis; the fields $\mathbf{E}$ and $\mathbf{B}$ should be contained in the $\rho - z$ planes orthogonal to $\mathbf{e}_\phi$. Recalling the fact that the lines of strength of $\mathbf{B}$ must be closed, the best description of this configuration can be obtained using the toroidal curvilinear coordinates (μ, η, ϕ) defined as follows in terms of cylindric coordinates:

$$\rho = \frac{a \cosh \mu}{\cosh \mu - \cos \eta}, \qquad z = \frac{a \sin \eta}{\cosh \mu - \cos \eta}, \qquad \phi.$$

with $0 \leq \phi < 2\pi\,, \quad 0 \leq \eta < 2\pi\,, \quad 0 \leq \mu \leq \infty$. The coordinate lines of ϕ are concentric circles in the $(z = 0)$-plane, while the coordinate lines of the variable η are excentric tori concentrating around the circle $\rho = a$. We shall suppose that the lines of force of the magnetic field coincide with the coordinate curves given by $\phi =$ Const. and $\mu =$ Const. The configuration we seek can be written as:

$$\mathbf{E} = E_\rho\, \mathbf{e}_\rho + E_z\, \mathbf{e}_z = E_\mu \mathbf{e}_\mu + E_\eta\, \mathbf{e}_\eta; \tag{32}$$

$$\mathbf{B} = B_\rho\, \mathbf{e}_\rho + B_z\, \mathbf{e}_z = B_\eta\, \mathbf{e}_\eta; \qquad \mathbf{j}_{ind} = j_{ind}\, \mathbf{e}_\phi. \tag{33}$$

It can be also shown that the whole problem can be reduced to determining just two unknown functions of the variables (μ, η), because $\mathbf{B} = rot\mathbf{A}$ and $B_\mu = 0$, and because here $\mathbf{E} = -gradV$, we have $\mathbf{A} = A(\mu, \eta)\,\mathbf{e}_\phi$, and $V = V(\mu, \eta)$.

Approximate solutions of this form have been found in ([8], [9]); here we shall only remind their essential features. At great distances, the fields **E** and **B** behave as if they were generated by a finite charge Q and a finite magnetic dipole **m** :

$$\mathbf{E}_\infty \simeq \frac{Q\,\mathbf{r}}{4\pi r^3}\,, \qquad \mathbf{B}_\infty \simeq \frac{\mathbf{m} \wedge \mathbf{r}}{4\pi r^3} \tag{34}$$

The charge is concentrated around the circle $\rho = a$ and "smeared" in its vicinity; if it is chosen to be positive, there is a little "halo" of negative charge density farther away, imitating the vacuum polarization effect. The charge density's fall-off is vary rapid, behaving at short distances as r^{-9}, and then falling off exponentially; the same concerns the density of induced currenh $\mathbf{j}_{ind}$ which falls off as r^{-8}. The induced current behaves as if it were produced by the charge density rotating around the z-axis with the speed of light.

Another interesting feature of this solution is its $Z_2 \times Z_2$ symmetry. Indeed, any such solution displaying the total energy (mass) $\mathcal{E}$, the total charge Q, magnetic momentum **m** and the total spin **sigma** is followed by three similar solutions with the same energy, but either with the same charge, but with the spin and magnetic momentum in the opposite direction (both "down"), or another couple of solutions having the opposite charge, and spin and magnetic moment up or down, but always opposite to each other - just like with what we know about the electron and the positron. The following table shows the properties of the four solutions:

Fields	*Energy*	*Charge*	**m**	*Spin*
E, **B**	$\mathcal{E}$	Q	**m**	s
E, −**B**	$\mathcal{E}$	Q	−**m**	−s
−**E**, **B**	$\mathcal{E}$	$-Q$	**m**	−s
−**E**, −**B**	$\mathcal{E}$	$-Q$	−**m**	s

Tab 1. The symmetry properties of four solutions.

Unfortunately, these solutions present a mild singularity on the circle $\rho = a$, which can not be avoided. Its presence can be proved by using Poincaré's lemma; the details can be found in ([8], [9]).

3 An SU(2)-based non-abelian generalization of Born-Infeld theory

The superstring theory gives rise to one important modification of the standard Yang-Mills quadratic Lagrangian suggesting the action of the Born-Infeld (BI) type [14,15,16]. Because this modification breaks the scale invariance, the natural question arises whether in the Born–Infeld–Yang–Mills (BIYM) theory the non-existence of classical particle-like solutions can be overruled. Although a mere scale invariance breaking, being a necessary condition, by no means guarantees the existence of particle-like solutions, a detailed study ([10] has shown that the $SU(2)$ BIYM classical glueballs indeed do exist.

Non–Abelian generalization of the Born–Infeld action presents an ambiguity in specifying how the trace over the the matrix–valued fields is performed in order to define the Lagrangian. Here we adopt the version with the ordinary trace which leads to a simple closed form for the action. In fact, another trace prescription is favored in the superstring context, namely, the *symmetrized* trace [14], but so far the explicit Lagrangian with such trace is known only perturbatively [17]. Since for our purposes the full non-perturbative Lagrangian is needed we consider the model with ordinary trace. But both numerical evidence and analytical arguments show that particle-like solutions should persist in the theory with symmetrized trace, too.

The BIYM action with the ordinary trace looks like a straightforward generalisation of the corresponding $U(1)$ action in the "square root" form

$$S = \frac{\beta^2}{4\pi} \int (1 - \mathcal{R})\, d^4x, \tag{35}$$

where

$$\mathcal{R} = \sqrt{1 + \frac{1}{2\beta^2} F^a_{\mu\nu} F^{\mu\nu}_a - \frac{1}{16\beta^4} (F^a_{\mu\nu} \tilde{F}^{\mu\nu}_a)^2}. \tag{36}$$

It is easy to see that the BI non-linearity breaks the conformal symmetry ensuring the non-zero trace of the stress–energy tensor

$$T^\mu_\mu = \mathcal{R}^{-1} \left[4\beta^2 (1 - \mathcal{R}) - F^a_{\mu\nu} F^{\mu\nu}_a\right] \neq 0. \tag{37}$$

This quantity vanishes in the limit $\beta \to \infty$ when the theory reduces to the standard one. For the Yang-Mills field we assume the usual monopole ansatz

$$A^a_0 = 0, \quad A^a_i = \epsilon_{aik} \frac{n^k}{r} (1 - w(r)), \tag{38}$$

where $n^k = x^k/r$, $r = (x^2 + y^2 + z^2)^{1/2}$, and $w(r)$ is the real-valued function. After the integration over the sphere in (35) one obtains a two-dimensional action from which β can be eliminated by the coordinate rescaling $\sqrt{\beta}t \to t$, $\sqrt{\beta}r \to r$. The following static action results then :

$$S = \int L dr, \quad L = r^2(1 - \mathcal{R}), \tag{39}$$

with

$$\mathcal{R} = \sqrt{1 + 2\frac{w'^2}{r^2} + \frac{(1-w^2)^2}{r^4}}, \tag{40}$$

where prime denotes the derivative with respect to r. It is worth noticing that the non-linearity arises here because of the non-linear dependence of the tensor $F^a_{\mu\nu}$ on the potentials A^b_μ. The corresponding equation of motion reads

$$\left(\frac{w'}{\mathcal{R}}\right)' = \frac{w(w^2-1)}{r^2\mathcal{R}}. \tag{41}$$

A trivial solution $w \equiv 0$ corresponds to the pointlike magnetic BI-monopole with the unit magnetic charge (embedded $U(1)$ solution). In the Born–Infeld theory it has a finite self-energy [18]. For time-independent configurations the energy density is equal to minus the Lagrangian, so the total energy (mass) is given by the integral

$$M = \int_0^\infty (\mathcal{R} - 1) r^2 dr. \tag{42}$$

For $w \equiv 0$ one finds

$$M = \int \left(\sqrt{r^4+1} - r^2\right) dr = \frac{\pi^{3/2}}{3\Gamma(3/4)^2} \approx 1.23604978. \tag{43}$$

Looking now for the essentially non–Abelian solutions of finite mass, we observe that in order to assure the convergence of the integral (42) the quantity $\mathcal{R} - 1$ must fall down faster than r^{-3} as $r \to \infty$. Thus, far from the core the BI corrections have to vanish and the Eq.(41) should reduce to the ordinary Yang-Mills equation, equivalent to the following two-dimensional autonomous system [19,20,21]:

$$\dot{w} = u, \quad \dot{u} = u + (w^2 - 1)w, \tag{44}$$

where a dot denotes the derivative with respect to $\tau = \ln r$. This dynamical system has three non-degenerate stationary points ($u = 0, w = 0, \pm 1$), from which $u = w = 0$ is a focus, while two others $u = 0$, $w = \pm 1$ are saddle points with eigenvalues $\lambda = -1$ and $\lambda = 2$. The separatices along the directions

$\lambda = -1$ start at infinity and after passing through the saddle points go to the focus with the eigenvalues $\lambda = (1 \pm i\sqrt{3})/2$.

It has been proved in ([10]) that *the only finite-energy configurations with non-vanishing magnetic charge are the embedded U(1) BI-monopoles.* Indeed, such solutions should have asymptotically $w = 0$, which does not correspond to bounded solutions unless $w \equiv 0$. The remaining possibility is $w = \pm 1, \dot{w} = 0$ asymptotically, which corresponds to zero magnetic charge. Coming back to r-variable one finds from (41)

$$w = \pm 1 + \frac{c}{r} + O(r^{-2}), \tag{45}$$

where c is a free parameter. This gives a convergent integral (42) as $r \to \infty$. The two values $w = \pm 1$ correspond to two neighboring topologically distinct Yang-Mills vacua.

Now consider local solutions near the origin $r = 0$. For convergence of the total energy (42), w should tend to a finite limit as $r \to 0$. Then using the Eq.(41) one finds that the only allowed limiting values are $w = \pm 1$ again. In view of the symmetry of (44) under reflection $w \to \pm w$, one can take without loss of generality $w(0) = 1$. Then the following Taylor expansion can be checked to satisfy the Eq.(44):

$$w = 1 - br^2 + \frac{b^2(44b^2+3)}{10(4b^2+1)} r^4 + O(r^6), \tag{46}$$

with b being (the only) free parameter.

As $r \to 0$, the function $\mathcal{R}$ tends to a finite value

$$\mathcal{R} = \mathcal{R}_0 + O(r^2), \qquad \mathcal{R}_0 = 1 + 12b^2. \tag{47}$$

therefore it is not a solution of the initial system (42). What remains to be done is to find appropriate values of constant b leading to smooth finite-energy solutions by gluing together the two asymptotic solutions between 0 and ∞.

It has been proved in ([10] that *any regular solution of the* Eq.(41) *belongs to the one-parameter family of local solutions* (46) *near the origin.*

It follows that the global finite energy solution starting with (46) should meet some solution from the family (45) at infinity. Since both these local solutions are non–generic, one can at best match them for some discrete values of parameters. This technique has been used first in ([20])

For some precisely tuned value of b the solution will remain a monotonous function of τ reaching the value -1 at infinity (Fig.1). This happens for $b_1 = 12.7463$.

By a similar reasoning one can show that for another fine-tuned value $b_2 > b_1$ the integral curve $w(\tau)$ which has a minimum in the lower part of the strip and then becomes positive will be stabilized by the friction term in the upper half of the strip $[-1, 1]$ and tend to $w = 1$. This solution will have two nodes. Continuing this process we obtain the increasing sequence of parameter values b_n for which the solutions remain entirely within the strip $[-1, 1]$ tending asymptotically to $(-1)^n$. The lower values b_n found numerically are given in Tab.2.

n	b	M
1	1.27463×10^1	1.13559
2	8.87397×10^2	1.21424
3	1.87079×10^4	1.23281
4	1.27455×10^6	1.23547
5	2.65030×10^7	1.23595

Tab 2. Parameters b, M for first five solutions.

References

1. G. Mie, *Annalen der Physik*, **37**, 511 (1912).
2. M. Born, L. Infeld, *Nature* **132**, 970 (1932); also, *Proc. Roy. Soc.* **A 144**, 425 (1934).
3. G. Boillat, *Journal of Mathematical Physics*, **11**, 941 (1970).
4. I. Bialynicki-Birula, in J. Lopuszanski's *Festschrift*, *Quantum Theory of Particles and Fields"*, Eds. B. Jancewicz and J. Lukierski, World Scientific, Singapore (1983), p.31.
5. J.P.S. Lemos and R. Kerner, *Gravitation and Cosmology*, **6**, 49 (2000).
6. G.W. Gibbons and C. Herdeiro, *Phys.Rev.D* **63**, 064006 (2001).
7. F. Müller-Hoissen, *Physics Letters* **156** B, 315 (1985).
8. R. Kerner, *Comptes Rendus Acad. Sci. Paris*, **304**, série 2, 621 (1987)
9. R. Kerner, in *Infinite dimensional Lie algebras and Quantum Field Theory, Proceedings of the Varna Summer School, Bulgaria, 1987*, H.D. Doebner, J.D. Hennig and T.D. Palev eds., World Scientific (1988), p.53.
10. D.V. Gal'tsov, R. Kerner, *Phys. Rev. Letters,* **84** (26), 5955 (2000).
11. D.V. Gal'tsov and M.S. Volkov, *Phys. Lett. B*, **273**, 255 (1991).
12. D. Sudarsky and R.M. Wald, *Phys. Rev. D*, **46**, 1453 (1992).
13. M.S. Volkov and D.V. Gal'tsov, *Phys. Reports*, **319**, 1 (1999); hep-th/9810070.
14. A.Tseytlin, *Nucl. Phys. B*, **501** , 41 (1997).
15. J.P. Gauntlett, J. Gomis and P.K. Townsend, *JHEP*, **01**, 003 (1998).
16. D. Brecher and M.J. Perry, *Nucl. Phys. B*, **527**, 121 (1998).

17. N. Grandi, E.F. Moreno and F.A. Shaposhnik, Monopoles in non-Abelian Dirac-Born-Infeld theory, hep-th/9901073.
18. G.W. Gibbons, Wormholes on the World Volume: Born-Infeld particles and Dirichlet p-branes, hep-th/9801106.
19. D.S. Chernavskii and R. Kerner, *Journ. of Math. Phys.*, **9**, 287 (1978).
20. R. Kerner, *Phys. Rev. D* **19**, 1243 (1979).
21. P. Breitenlohner, P. Forgacs, and D. Maison, *Comm. Math. Phys.*, **163**, 141 (1994).

QUASIPERIODIC PATTERNS IN FARADAY INSTABILITY AND ADAPTED BETA-LATTICE

A. D. EL KHARRAT, J. P. GAZEAU, M. GUIDY, S. METENS
LPTMC
Box 7020, 2 Place Jussieu 75251, CEDEX 05, Paris, FRANCE
E-mail: kharrat@ccr.jussieu.fr

It has recently been shown that quasicrystalline structures of interest in physics (namely the 5 and 10-fold, 8-fold and 12-fold structures) are subsets of specific quasiperiodic point sets, named beta-lattices. Beta-lattices are eligible labelling frames for quasiperiodic sets. They are based on discrete sequences related to the irrational number $\beta > 1$ as lattices are based on integers. In this article we show how it is possible to embed a 12-fold quasiperiodic pattern observed in the hydrodynamic Faraday instability in an adapted beta-lattice.

Introduction

The notion of quasiperiodicity has been mainly considered in solid state sciences and quasicrystals. This is however too restrictive, because it is possible to encounter quasiperiodicity in other scientific fields such as dynamical systems, dissipative systems [1], hydrodynamics and biology [2]. In this article we focus on the hydrodynamic instability of Faraday : Fauve and Edwards have shown experimentally [3] that this instability can display quasiperiodic patterns.

Back to the field of solid state science it has been shown that the geometric models of atomic sites of quasicrystals, as well as their diffraction patterns beyond a certain cut-off, are embedded in a quasiperiodic adapted grid, named beta-lattice, which we shall denote by β-lattice from now on [5]. Beta-lattices can roughly be described as quasiperiodic "millimeter paper", enabling one to label sites in quasiperiodic structures. They are spatially self-similar and they have space inversion symmetry. Their construction is free of arbitrariness (as in the case of mathematical quasicrystals). Above all, β-lattices share with lattices the property of counting systems on the same level as integers [5]. Beta-lattices were used successfully by R. Krejcar to perform Monte-Carlo simulations for finding ground states for two-dimensional quasicrystals as shown in [6].

In the present article we consider β-lattices adapted to 12-fold symmetry, since we focus our attention on a model for the Faraday instability displaying 12-fold quasiperiodic patterns. We ask ourselves if the concepts and techniques developed in [5], in the physics of quasicrystals in solid, can be applied in the context of out of equilibrium systems displaying quasiperiodic patterns. This work is a preliminary one to numerical simulations and more profound mathematical demonstrations based on β-lattices.

The present article is organized as follows. In the first section we present briefly the experimental Faraday instability, and in the second section we present a theoretical model for Faraday patterns proposed by Lifshitz and Petrich [4]. In the third section we give some insight on why β-lattices can be used as labelling frames for quasicrystals, and in the fourth and last section we show how to embed the quasiperiodic structure obtained by the theoretical model in the adapted β-lattice.

1 The Hydrodynamic Instability of Faraday

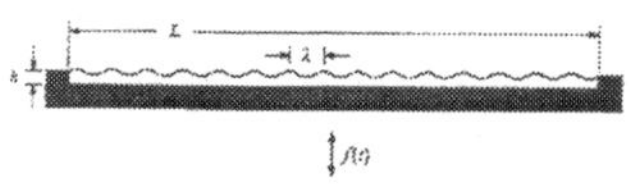

Figure 1. Experimental device.

Faraday hydrodynamic instability is the parametric excitation of surface waves via vertical oscillations of a flat bottomed-container filled with liquid. As a result, patterns appear on the surface. It is exactly what happens when a cup of water is put on a table in a train : the up and down motion of the train destabilizes the surface of the liquid and makes it oscillate. Very nice circular patterns then appear.

Fauve and Edwards [3] did the following experience: they excited the fluid with a two-frequency forcing:

$$f(t) = a\left[\cos(\chi)\cos(m\omega t) + \sin(\chi)\cos(n\omega t + \phi)\right] . \tag{1}$$

a is the amplitude of the forcing, χ is an amplitude parameter, and ϕ is a phase. The coefficients m and n are even and odd integers. This type of forcing generates 2 wave vectors q_1 and q_2. In the case of $m = 4$ and $n = 5$ q_1 and q_2 are forming a $\pi/6$ angle and are in incommensurate ratio:

$$\frac{q_1}{q_2} = \sqrt{2+\sqrt{3}} .$$

They used viscous fluid in low depth in order to get rid of border effects.

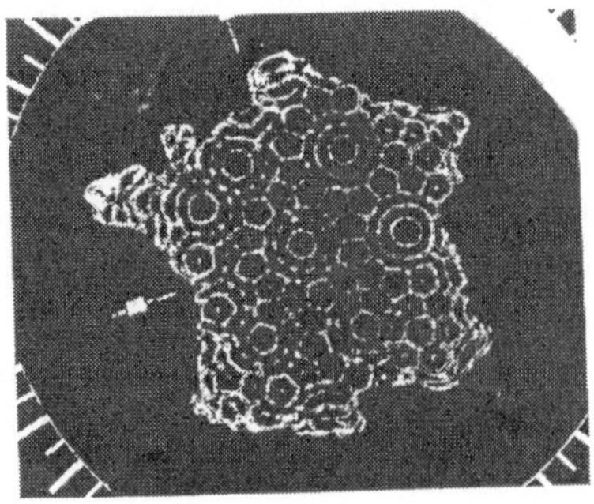

Figure 2. Fauve and Edwards observed a 12-fold quasiperiodic pattern free of border effects as one can see on this figure.

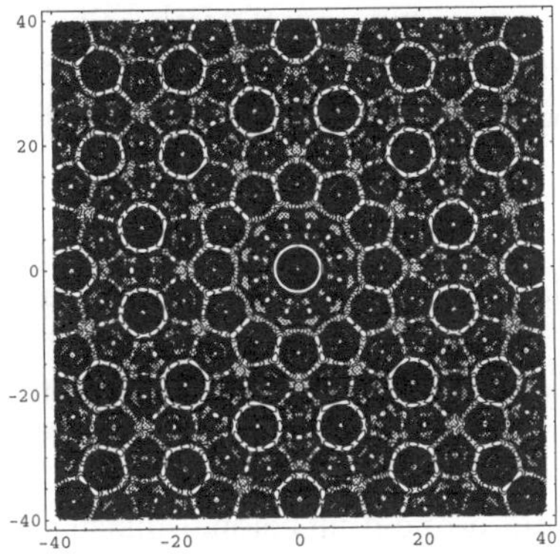

Figure 3. Numerical result obtained by weakly non-linear treatment of the Lifshitz-Petrich model.

2 Lifshitz and Petrich Model

The simplest model for the Faraday instability was coined by Lifshitz and Petrich. The starting point to this model is the Swift-Hohenberg equation, in which they introduced a double diffusive term to model the double frequency forcing, and a quadratic term to model the breaking of the inversion symmetry of the surface occurring under such a forcing. The equation they obtained is the following:

$$\partial_t \zeta = \epsilon\, \zeta - c\left(\nabla^2 + q_1^2\right)^2 \left(\nabla^2 + q_2^2\right)^2 \zeta + \alpha\, \zeta^2 - \zeta^3\,, \tag{2}$$

the ratio q_1/q_2 being equal to $\sqrt{2+\sqrt{3}}$ as previously stated.

Modelling the Faraday waves in such a way is in fact quite astute: although the original Swift-Hohenberg equation has nothing to do with the Faraday instability, Lifshitz and Petrich managed to make a very handy model out of it by introducing the minimal physics required : two wave numbers in an irrational ratio and a breaking of symmetry quadratic term. Yet it is not possible to derive equation (2) from the Navier-Stokes fundamental equation for incompressible viscous liquids.

We solved equation (2) by weakly non-linear treatment, around the bi-criticality [a], and discarded all harmonics of q_1 and q_2 by adiabatic elimination.

Furthermore we can say that if one integrates formally (2), one obtains what is called the Lyapunov functional of the equation, which can be considered as the free energy of the equation. For a certain set of parameters the 12-fold quasiperiodic pattern is the global minimum of the functional, therefore it is globally stable.

3 Beta-Lattice Adapted to 12-Fold Symmetry

In this section we shall not give any demonstration on β-lattices. Instead we shall give some insight on why β-lattices are handy labelling frames for quasicrystals.

It is possible to build a one-dimensional quasicrystal with the aid of the irrational number $\beta = 2+\sqrt{3}$. Let $\mathbb{Z}^2$ be the periodic square lattice. Let E be the straight line of irrational slope $-(2+\sqrt{3}) = -\beta$, and E' the line of slope $-1/(2+\sqrt{3}) = -1/\beta$. A model set Λ is a set obtained by projecting all the points of $\mathbb{Z}^2$ that are lying within a stripe parallel to E and of arbitrary width, onto E and along E'. Λ is also called a "cut and project" set. This construction is very easily generalized for 2 and 3 dimensional quasiperiodic point sets, projecting selected points of $\mathbb{Z}^4$ and $\mathbb{Z}^6$ respectively on $\mathbb{R}^2$ and $\mathbb{R}^3$.

It is possible to prove that one-dimensional cut and project sets, such as Λ, are best labelled by the so-called set of β-integers, denoted by $\mathbb{Z}_\beta$. Yet in the case of β-integers and β-lattices adapted to 12-fold symmetry, it is necessary to take the decorated set [b] $\widetilde{\mathbb{Z}}_\beta = \mathbb{Z}_\beta + \{0, \pm 1/\beta\}$. Beta-integers are the real numbers that are polynomial when developed [c] in powers of β. It is

[a] In this model, the two wave numbers are unstable for the same critical amplitude.

[b] In the case of 5-fold, 10-fold and 8-fold symmetry their is no decoration needed.

[c] The development of real numbers in powers of β is given here by the greedy algorithm.

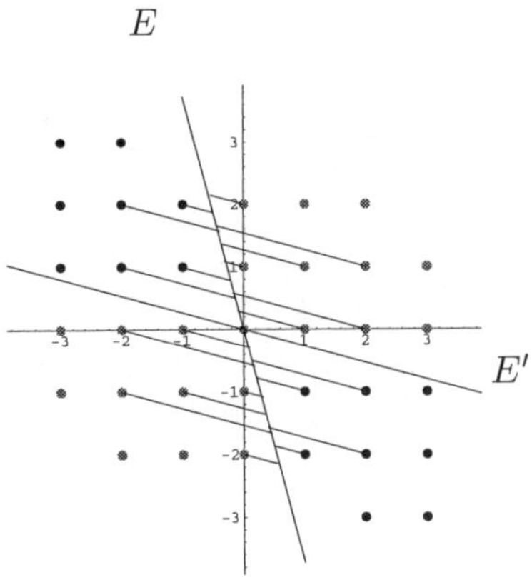

Figure 4. Specific points of $\mathbb{Z}^2$ are projected onto E and along E', to be the β-integers.

possible to obtain $\mathbb{Z}_\beta$, and it's decoration, by projecting specific nodes of $\mathbb{Z}^2$ onto E and along E' as shown on figure 4. The set of points of $\mathbb{Z}^2$ defining Λ is a subset of the set of points of $\mathbb{Z}^2$ defining $\widetilde{\mathbb{Z}}_\beta$. Note that their is no arbitrariness in the construction of $\widetilde{\mathbb{Z}}_\beta$.

4 Embedding of the Faraday Quasiperiodic Pattern in the Adapted β-lattice

Our aim here is to embed the structure formed by the peaks of the quasiperiodic patterned obtained by the Lifshitz-Petrich model into a two-dimensional β-lattice, with $\beta = 2 + \sqrt{3}$. We search systematically the local maxima of the surface by two methods : a **continuous method** consisting in finding the points of the surface such that the modulus of the surface gradient $|\overrightarrow{\nabla}\zeta| = 0$ and the hessian matrix is definite negative, and a **discrete method** consisting in calculating the surface gradient in the points of the β-lattice and retaining only the points where the gradient is null and the hessian matrix is definite negative. In fact the two methods don't match perfectly, yet most of the points match. We notice that the Faraday surface is composed of a macrostructure, made of sharp peaks, and a decoration of the macrostructure. This decoration is made of undulating circular ridges surrounding the peaks of the macrostructure. The peaks of the macrostructure reach lower heights than the ridges. We can then retain the maxima of the macrostructure by operating a cut-off below a certain hight. The result of this operation is displayed on figure 5.

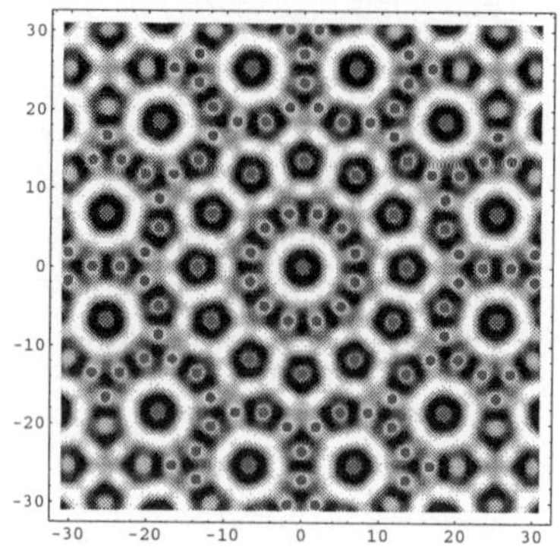

Figure 5. The macrostructure, marked here with dots, is a subset of the β-lattice.

Now the structures obtained by the continuous method and by the discrete method match, which means that the peaks of the macrostructure form a subset of the β-lattice since all the points obtained by the discrete method are within the β-lattice.

Acknowledgments

We are greatly indebt to J.C. Levy, D. Mercier and S. Fauve whose advices were priceless.

References

1. G. Dewel, M. Bachir, P. Borckmans and S.Metens, Motifs en Super-Réseaux et Structures Quasicristallines dans les Systèmes Bistables, *C.R.A.S.P. serie IIb*, to be published, (2001).
2. Y. Couder, S.Douady, in *Statphys 19*, Editor Hao Bailin, World Scientific, Singapore (1995).
3. W.S. Edwards, S. Fauve, *J. Fluid. Mech.* **278**, 123 (1994).
4. R. Lifshitz and D.M. Petrich, *Phys. Rev. Lett.* **79**, 1261 (1997).
5. Ch. Frougny, J.P. Gazeau and R. Krejcar (1999), Additive and Multiplicative Properties of Point Sets based on Beta-integers, *Theor. Comp. Sci.*, to appear, (2002).
6. R. Krejcar, Beta-quasi-lattices for Quasicrystals, Ph.D. Thesis, Université de Paris 7 - Denis Diderot, UFR de Phys. et Phys. Théorique (2000).
7. D. Mercier, J.-C. S. Levy, *Phys. Rev. B* **27**, 1293 (1983).

ON THE SU(2) WIGNER FUNCTION DYNAMICS

A.B. KLIMOV AND S.M. CHUMAKOV

Departamento de Física, Universidad de Guadalajara,
Revolución 1500, 44420, Guadalajara, Jalisco, México
E-mail: klimov@cencar.udg.mx, sergei@cencar.udg.mx

We study the quantum dynamics of the SU(2) quasiprobability distribution ("Wigner function") for the simple nonlinear Hamiltonian (finite analog of the Kerr medium, $H = S_z^2$). The quasiclassical approximation for the Wigner function and corresponding evolution of mean values are considered and compared with the exact and classical solutions.

1 Introduction

The advantage of the phase-space description of a quantum system in terms of Wigner quasiprobability function consists in that reflecting all quantum features, the Wigner function keeps the closest correspondence with its classical counterpart [1,2]. To provide the phase space description for spin systems Stratonovich [3] in 1956 introduced the quasiprobability distribution function on the sphere $(\theta, \phi) \in \mathcal{S}_2$ (see also [4,5,6]). This function is naturally related to the $SU(2)$ dynamical group and we will later call it the $SU(2)$ Wigner function. This function was proved to be very useful to visualize nonclassical properties of a collection of two-level atoms [7] and polarization optics [8].

The $SU(2)$ Wigner function is defined as follows,

$$W_\rho(\theta, \phi) = \mathrm{Tr}\left(\rho \hat{w}(\theta, \phi)\right), \tag{1}$$

where ρ is the density matrix and $\hat{w}(\theta, \phi)$ is the Wigner operator

$$\hat{w}(\theta, \phi) = \frac{2\sqrt{\pi}}{\sqrt{2S+1}} \sum_{L=0}^{2S} \sum_{M=-L}^{L} Y_{L,M}^*(\theta, \phi)\, \hat{T}_{L,M}^{(S)}, \tag{2}$$

such that $\mathrm{Tr}\hat{w}(\theta, \phi) = 1$, $\quad (2S{+}1)/4\pi \int_{\mathcal{S}_2} d\Omega \hat{w}(\theta, \phi) = I$. Here we use spherical harmonics $Y_{L,M}(\theta, \phi) = Y_{L,-M}^*(\theta, \phi)$ and irreducible tensor operators $\hat{T}_{L,M}^{(S)}$ (*cf.* [9], Eq. 2.4(6)),

$$\hat{T}_{L,M}^{(S)} = \sqrt{\frac{2L+1}{2S+1}} \sum_{m,m'=-S}^{S} C_{S,m;L,M}^{S,m'} \, |S, m'\rangle \, \langle S, m|, \tag{3}$$

here $C_{S,m;L,M}^{S,m'}$ are the Clebsch-Gordan coefficients which couple two representations of spin S and L $(0 \le L \le 2S)$ to a total spin S. The function $W_\rho(\theta, \phi)$

is covariant under rotations and provides the overlap relation

$$\frac{2S+1}{4\pi}\int_{S_2} d\Omega W_\rho(\theta,\phi)\, W_A(\theta,\phi) = \mathrm{Tr}\,(\rho A)\,. \tag{4}$$

Here $d\Omega = \sin\theta\, d\theta\, d\phi$ is the invariant measure on the sphere and $W_A(\theta,\phi) = \mathrm{Tr}\left[\hat{A}\hat{w}(\theta,\phi)\right]$ is the Wigner symbol of the operator A. The density matrix can be reconstructed from the Wigner function (1) through the relation

$$\rho = \frac{2S+1}{4\pi}\int_{S_2} d\Omega\, \hat{w}(\theta,\phi) W_\rho(\theta,\phi). \tag{5}$$

In this paper we will consider the dynamics of the $SU(2)$ Wigner function for the simplest nontrivial example of the finite-level analog of the Kerr Medium [10,11,12]. We will show that the leading order of the expansion in the inverse powers of $2S+1$ (the representation dimension) leads to the classical evolution of the initial distribution on the sphere. It allows us to calculate mean values of the spin operators and gives the results which are drastically better then the "naive" solution of the Heisenberg equations of motion with decoupled correlators. On the other hand, the quantum self-interference phenomena (like Shcrödinger cats [10,11]) are clearly beyond the scope of this semiclassical approximation. In the rest of paper we will consider only integer values of S, which corresponds to the $SO(3)$ group rather then $SU(2)$.

2 Wigner function dynamics

2.1 Linear dynamics

Let us consider the dynamics of the Wigner function $W_\rho(\theta,\phi)$ under the action of a Hamiltonian from the universal enveloping algebra of $su(2)$. First of all we note that the evolution under a linear Hamiltonian

$$H = \omega_0 S_z + g_1 S_x + g_2 S_y \tag{6}$$

is equivalent to a rotation round some axis:

$$H \to H_d = U^\dagger H U = \omega S_z, \quad \omega = \sqrt{\omega_0^2 + g_1^2 + g_2^2}, \tag{7}$$

where U is the $SU(2)$-rotation. Substituting the density matrix in terms of Wigner function (5) to the equation of motion, $i\partial_t\rho = [H,\rho]$ we obtain

$$i\int_{S_2} d\Omega\, \hat{w}(\theta,\phi)\partial_t W_\rho(\theta,\phi) = \int_{S_2} d\Omega\, [H, \hat{w}(\theta,\phi)]\, W_\rho(\theta,\phi). \tag{8}$$

Taking into account the commutation relations $\left[S_z, \hat{T}^{(S)}_{L,M}\right] = M\hat{T}^{(S)}_{L,M}$, we get, $[S_z, \hat{w}(\theta,\phi)] = -i\partial_\phi \hat{w}(\theta,\phi)$. Replacing this equation into (8) and integrating by parts we obtain the following equation of motion for the Wigner function

$$\partial_t W_\rho(\theta,\phi) = -\omega\partial_\phi W_\rho(\theta,\phi).$$

Its solution is $W_\rho(\theta,\phi|t) = W_\rho(\theta,\phi-\omega t|t=0)$, which, of course, corresponds to the covariance propriety of the Wigner function[4].

2.2 Kerr dynamics

Now let us consider the simplest non-linear Hamiltonian

$$H = \chi S_z^2, \tag{9}$$

which gives the simplest example when the quantum dynamics differs essentially from the corresponding classical one [10]. It leads to generation of squeezed atomic states [11] and atomic Shcrödinger cats [12].

To find an approximate dynamical equation for the Wigner function we use the expansion in powers of $\varepsilon = (2S+1)^{-1}$ (the inverse representation dimension). Taking into account that

$$\left[S_z^2, \hat{w}(\theta,\phi)\right] = \{S_z, [S_z, \hat{w}(\theta,\phi)]\}_+ = i\partial_\phi\{S_z, \hat{w}(\theta,\phi)\}_+, \tag{10}$$

where $\{.,.\}_+$ stands for anticommutator, calculating $\{S_z, \hat{w}(\theta,\phi)\}_+$, substituting into (8) and neglecting terms of order ε (the zero-order approximation) we arrive at the quasiclassical equation for the distribution function,

$$\partial_t W_\rho(\theta,\phi) = -\frac{\chi}{\varepsilon}\cos\theta\partial_\phi W_\rho(\theta,\phi), \tag{11}$$

with the solution,

$$W_\rho(t) = W_\rho\left(\theta, \phi - \frac{\chi t}{\varepsilon}\cos\theta \middle| t=0\right) \tag{12}$$

This quasiclassical Wigner function describes well the evolution for times $\chi t \preceq 1$. Note that Eq. (11) corresponds to the classical evolution equation that involves the Poisson brackets on the sphere (expressed in terms of the angles θ,ϕ, see e.g. [15]) between the classical Hamiltonian function and the classical distribution function.

The diffusion-like terms of order $O(\varepsilon)$ neglected in (11) describe quantum corrections to the quasiclassical motion in analogy to the quasiclassical expansion of the Moyal equation [1]. They are responsible for the formation of Shcrödinger cats on the sphere.

To obtain approximate forms of the Wigner functions for different initial distributions we will use the following integral representation of the Wigner operator (2)

$$\hat{w}(\theta,\phi)=\int_0^{2\pi} d\omega e^{-i\omega\,\vec{n}\cdot\vec{S}}f(\omega) \tag{13}$$

where the weight function $f(\omega)$ in the limit case of large representation dimensions, $S\gg 1$ takes the following asymptotic form [13] (see also [14]),

$$f(\omega)\to(-1)^S\left[\delta(\omega-\pi)-\frac{i}{S}\frac{\partial}{\partial\omega}\delta(\omega-\pi)\right],\quad S\to\infty, \tag{14}$$

The Wigner function of an eigenstate $|k,S\rangle$ of the operator S_z in the limit $S\gg 1$, $|k|\sim S$, can be approximated as follows,

$$W_k(\theta,\phi)\simeq(-1)^{k+S}d^S_{kk}(2\theta)\left[1+\frac{k}{S}\cos\theta\right],$$

where $d^S_{kk}(\vartheta)=\langle k,S|\exp\left(i\vartheta S_y\right)|k,S\rangle$ is the d-function. These states are invariant under the evolution (9). In particular, the Wigner function of the state $|S,S\rangle$ takes the form

$$W_S(\theta,\phi)\simeq\cos^{2S}\theta\,[1+\cos\theta]\,.$$

On the other hand, starting from the atomic coherent state located along the x-direction, $|\theta=\pi/2,\phi=0\rangle$, $S_x|\pi/2,0\rangle=(S/2)|\pi/2,0\rangle$, we get ($S\gg 1$),

$$W_\rho(\theta,\phi|t=0)\simeq(\sin\theta\cos\phi)^{2S-1}\left(1+\sin\theta\cos\phi\right),$$

which evolves as

$$W_\rho(\theta,\phi|t)\simeq g(\theta,\phi|t)^{2S-1}\left[1+g(\theta,\phi|t)\right], \tag{15}$$

where $g(\theta,\phi|t)=\sin\theta\cos(\phi-\varepsilon^{-1}\chi t\cos\theta)$. In the same manner, we can find evolution of the Wigner function for the initial superposition of atomic coherent states. For instance, for the state

$$|\psi\rangle=\frac{1}{\sqrt{2}}\left(|\theta=\frac{\pi}{2},\phi=0\rangle+|\theta=\frac{\pi}{2},\phi=\pi\rangle\right),$$

(the Schrodinger cat state on the sphere[12]), the Wigner function takes the form

$$W_\rho\left(\theta,\phi|t\right)=g(\theta,\phi|t)^{2S}-\mathrm{Re}\left(g(\theta,\phi+\pi/2|t)+i\cos\theta\right)^{2S}$$

2.3 Evolution of average values

Using the overlap relation (4) and the semiclassical equation (12) we can determine the evolution of average values:

$$\langle A\rangle = \frac{2S+1}{4\pi}\int_{S_2} d\Omega W_A(\theta,\phi)W_\rho(\theta,\phi). \tag{16}$$

The Wigner symbols of S_j, $j = x, y, z$ are

$$W_{S_j}(\theta,\phi) = \sqrt{S\,(S+1)}n_j, \quad \vec{n} = (\cos\phi\sin\theta,\, \sin\phi\sin\theta,\, \cos\theta),$$

giving, for the initial coherent state $|\pi/2, 0\rangle$, the following average values:

$$\begin{aligned}\langle S_x(t)\rangle &= \sqrt{S(S+1)}\Gamma(S+3/2)\left(\frac{2}{z}\right)^{S+1/2} J_{S+1/2}(z), \qquad (17)\\ \langle S_z(t)\rangle &= \text{const}, \quad \langle S_y(t)\rangle = 0, \quad z = \chi t\,(2S+1)\,.\end{aligned}$$

This result is to be compared with the exact quantum solution

$$\langle S_x(t)\rangle_{ex} = S\cos^{2S-1}\chi t. \tag{18}$$

Eqs. (17) and (18) are in a very good agreement for times $\chi t \sim 1$. We stress that the quasiclassical evolution equation for the Wigner function does not lead to the classical equations of motion for the average values. Indeed, decoupling the correlators in the Heisenberg equations ("parametric approximation") one arrives at the classical equations,

$$\frac{d}{dt}\langle S_z\rangle = 0, \quad \frac{d}{dt}\langle S_x\rangle = -2\chi\langle S_y\rangle\langle S_z\rangle, \quad \frac{d}{dt}\langle S_y\rangle = 2\chi\langle S_x\rangle\langle S_z\rangle.$$

(This system can also be obtained form the classical Hamiltonian $H = \chi S_z^2$ with the Poisson brackets $\{S_i, S_j\} = S_k$, $i, j, k = x, y, z$.) The solution is

$$\begin{aligned}\langle S_x(t)\rangle_{cl} &= \langle S_x(0)\rangle\cos 2\langle S_z(0)\rangle\chi t - \langle S_y(0)\rangle\sin 2\langle S_z(0)\rangle\chi t,\\ \langle S_y(t)\rangle_{cl} &= \langle S_x(0)\rangle\sin 2\langle S_z(0)\rangle\chi t + \langle S_y(0)\rangle\cos 2\langle S_z(0)\rangle\chi t. \qquad (19)\end{aligned}$$

This "classical" solution diverges from the exact one (18) even for very small times. Indeed, the Tailor expansion of Eq. (18) gives, $\langle S_x(t)\rangle_{ex} \approx S - S(S-1/2)(\chi t)^2 + O(t^4)$, while Eq. (19) for this initial state gives a constant, $\langle S_x(t)\rangle_{cl} = S$. On the other hand, Eq. (19) can be rewritten in terms of spherical angles, $S_z = S\cos\theta$, $S_x = S\sin\theta\cos\phi$, $S_y = S\sin\theta\sin\phi$,

$$\theta = \text{const}, \quad \phi(t) = \phi(0) + 2S\chi t\cos\theta.$$

Thus, the quasiclassical evolution of the Wigner function (12) corresponds to the motion of every point of the initial distribution along the classical trajectory on the sphere (compare with the Heisenberg-Weyl case[16,17,18]).

In summary, we have considered the evolution of the $SU(2)$ Wigner quasiprobability function on the sphere under the action of the simplest nonlinear Hamiltonian - the "finite Kerr medium". We obtained the exact equation of motion for the Wigner function and solved it for the case of large representation dimensions. In this "quasiclassical" limit different parts of the initial distribution rotate with different velocities (depending of the angle θ) which leads to a deformation of the initial distribution (without self-interference). This quasiclassical Wigner function leads to the results which are essentially different from the solution of the classical equations of motion for mean values (parametric approximation). The quasiclassical Wigner function describes well the system dynamics up to times $\chi t \sim 1$ (while the parametric approximation fails for these times).

References

1. J.E. Moyal, *Proc. Cambridge Philos. Soc.* **45**, 99 (1949).
2. E.P. Wigner, *Phys. Rev.* **40**, 749 (1932).
3. R.L. Stratonovich, *Sov. Phys. JETP* **31**, 1012 (1956).
4. J.C. Várilly and J.M. Gracia-Bondía, *Annals of Physics* **190**, 107 (1989).
5. C. Brif and A. Mann, *Phys. Rev. A* **59**, 971 (1999).
6. G.S. Agarwal, *Phys. Rev. A* **24**, 2889 (1981).
7. J.P. Dowling, G.S. Agarwal and W.P. Schleich, *Phys. Rev. A* **49**, 4101 (1994).
8. V.P. Karassiov, *Bull. Lebedev Phys. Inst. (Russian Academy of Sience)* N9, 34 (1999).
9. D.A. Varshalovich, A.N. Moskalev and V.K. Khersonskiĭ, *Quantum Theory of Angular Momentum*, World Scientific, Singapore (1988).
10. S.M. Chumakov, A. Frank and K.B. Wolf, *Phys. Rev. A* **60**, 1817 (1999).
11. M. Kitagawa and M. Ueda, *Phys. Rev. A* **47**, 5138 (1993).
12. G.S. Agarwal, R.R. Puri and R.P. Singh, *Phys. Rev. A* **56**, 2249 (1997).
13. A.B. Klimov and S.M. Chumakov, *JOSA A* **17**, 2315 (2000).
14. J.-P. Amiet and S. Weigert, *Phys. Rev. A*, (2000).
15. A. Perelomov, *Generalized Coherent states and their applications,* Springer-Verlag, Berlin (1986).
16. G.J. Milburn, *Phys. Rev. A* **33**, 674 (1986).
17. A.L. Rivera *et al*, *Phys. Rev. A* **55**, 876 (1997).
18. G. Drobný, A. Bandilla, and I. Jex, *Phys. Rev. A* **55**, 78 (1997).

VI HILBERT'S PROBLEM, INFINITE LIE GROUPS, AND RELATIVISTIC PHYSICS

N. P. KONOPLEVA
Bogoliubov Laboratory of Theoretical Physics, JINR,
Dubna, 141980, Russia
E-mail: nelly@thsun1.jinr.dubna

Just hundred years ago in 1900 D.Hilbert formulated 23 problems which, in his opinion, the mathematicians of XX century would have to solve. Among them the sixth problem pointed to necessity to state the mathematical formulation of the axioms of physics. As particular case of this problem Hilbert considered the possibility of the physical axioms construction according to the model of the axioms of geometry. Thus the sixth Hilbert's problem contained the problem of physics geometrization. For all XX century long this problem formed the strategies of scientific researces in theoretical physics and in some new mathematical topics, especially in geometry. Appearence of the special and general relativity as well as the geometrical gauge field theory can be regarded as consequent stages in the sixth Hilbert's problem solution. According to these physical theories the corresponding new geometries appeared: Minkowskian 4D-geometry (for SR), Riemannian 4D-geometry and its Cartan's formulation by tetrads (for GR), and after all, the fibre bundle space geometry which is the base of the geometrical gauge field theory. The gauge field theory is successful in explanation of phenomena of particle physics and gravity. Now the problem of today consists in application of the geometrical gauge field theory for relativistic nuclear physics.

1 Contents

2 Introduction. Physics axiomatization: What for?

Research of 23 problems from different mathematical topics. which D.Hilbert formulated, could. in his opinion, stimulate further science progress. Among them the sixth problem was named "Mathematical formulation of physical axioms". At first sight it seems unexpected and misunderstood that Hilbert put pure mathematicians attention to the problem of physics axiomatization.

Those were the years when the physicists put experiments in first place in science investigation, and physics axioms were really not exist. But Maxwell already allowed that a theory of new physical phenomena could be under some conditions constructed before full experimental investigation of them has been fulfiled. Maxwell considered the theoretical physics a mathematics. In his turn Hilbert regarded geometry as physics branch. The connection of physics and geometry was established by Kant in XVIII century.

What is physics axiomatization necessary for?

Axiomatization of science ensure trustworthiness of its predictions. It is known that before Euclidean times geometry was experimental science and formed as a sum of experimentally fixed facts and knowledge about extension properties of bodies. Thanks to its axiomatization geometry became the strict mathematical discipline. Euclidean axiom system permitted to obtain many geometrical relations following formal logic rules from the first principles and without use of experiments on each stage. Axiomatization of geometry make trustworthy its theoretical predictions. Therefore when one will experimentally detect that triangle angle sum differs from 2π he will look for mistakes in his measurements but have no doubts in Euclidean geometry. In this case the experiment conditions have to proposed corresponding to just Euclidean geometry.

If physics axiomatization will be done its theoretical predictions will be more trustworthy also. Then some test of statement truth will appear within the theory without experiments. Such theory would be so much trustworthy that it can be used for engineering calculations and correct predictions for the future. Only the theory which is fundamental in such meaning can be the base of the further science progress.

3 VI Hilbert's problem formulation and its role in science knowledge

VI Hilbert's problem formulation begins with note about investigations on geometry foundations are closely tied with the problem of axiomatization of those physical disciplines where mathematics is already now playing the leading role. In the first place it concerns to probability theory and mechanics.

Then Hilbert explains: "To construct the physical axioms according to the model of axioms of geometry, one must first try to encompass the largest possible class of physical phenomena by means of a small number of axioms and then, by adding each subsequent axiom, to arrive at more special theories after which there may arise a classification principle which can make use of the deep theory of infinite Lie groups of transformations. Moreover, as it is done in geometry, the mathematician must bear in mind not only the facts of actual reality, but also all the logically possible theories, and must be particularly careful to obtain the most complete survey of the totality of consequences which follow from the adopted systematization. "

Thus VI Hilbert's problem is not a concrete narrow mathematical one demanding the concrete answer, but it points the way of development for many fields of knowledge basing on the mathematical methods. The Hilbert's problem talks about science investigation strategy in nature study of XX century and close connection between geometry and physics. Just this strategy led to active use of group theory in physics, whereas S.Lie and F.Klein were first who applied group theory in geometry. From this strategy the new fundamental physical theories, as special and general relativity, quantum mechanics, quantum and gauge field theories arose. Among them SR, GR and classical gauge field theory have now purely geometrical forms. That is, relativistic physics of XX century arose thanks to the fact that from the century beginning it was found the right direction to science development and effective methods were used. VI Hilbert's problem solved just this task. It did really stimulate science development in XX century.

However, it must be noted that Hilbert's hopes were realized step by step. When probabilty theory was axiomatized it was found that it made use not Lie groups but other algebraic methods. Special relativity, quantum mechanics and elementary particle theory make use a classification principle being based on finite Lie group theory, but it is no concern of infinite Lie groups. The classification principle, which is a talk of VI Hilbert's problem, did only arise in a gauge field theory. This principle classifies not elementary particles but all the possible interactions between them.

The gauge field theory is fundamental unified theory of all existing in nature interactions including nuclear forces. At the same time this theory can have purely geometrical form like Einstein's gravity theory. But in this case the geometry being in use is more general than Riemannian 4D one. Thus the geometrical theory of nuclear forces and geometrical nuclear physics must exist. They will be new relativistic theories of XXI century.

The process of solution of VI Hilbert's problem stimulated development not only physics but geometry also. New geometries arose simultaneously with new physical theories. Just 4D Minkowski geometry arose as geometrical interpretation of SR. For GR 4D pseudoriemannian geometry was developed. Cartan's tetrad formulation of Riemannian geometry arose thanks to necessity to unify the Riemannian geometry conceptions and Lie-Klein's axiomatics of geometry using Lie group theory. Generalization of Cartan's geometrical approach led to a new geometry of fibre bundle spaces. It was in 50-60th of

XX century simultaneously with appearance of the gauge field theory. Fibre bundle space geometry permitted to geometrize the gauge field theory. Thus the most part of modern physics can be at once translated into geometrical language.

4 Relativistic physics: close-range action hypothesis and geometry

The foundation of relativistic physics is two basic ideas: symmetry principles and close-range action hypothesis. As it arose, the first relativistic physical theory (Maxwell's electrodynamics) did not use symmetry principles. The role of these principles for construction of physical theory was understood later. But close-range action hypothesis was basic in Maxwell theory. Just this idea about an interaction as a process propagating from point to point differs in essence Maxwell's electrodynamics from Newtonian mechanics. All following relativistic physical theories use the close-range action hypothesis also.

In mechanics without gravity separated from one another bodies do not interact between themselves. They are moving in vacuum which does not act to its motion. Interaction does only arise under collisions, i.e., under intersections of trajectories. The only known interaction in mechanics is gravity. But it transmits momentally to any distances and so is a long-distance interaction. Newton wrote that the process of gradual propagation of interaction would be more adequate to physics, but he did not know any equations describing such process. Moreover Newton noted that he proposed the idea of absolute infinite space enveloping the whole universe because he did not know how to describe the great number of real finite frames of reference used by observers.

Electrodynamics of Maxwell and SR settled the Newton's questions. SR is in essence the first physical theory unified mechanics and electrodynamics. Maxwell equations describe the propagation process of electromagnetic interaction between bodies. The velocity of this propagation occurred finite and equal to light velocity c. Therefore the electromagnetic interaction realizes the close-range action hypothesis.

Can gravity be interaction of close-range action kind and simultaneously described by geometry of space-time? For that a space-time geometry must not be given globally in a whole universe. In this case the space-time as a whole and nonlocal characteristics (for example, length) are only determined from point to point. This is Riemannian geometrical point of view. In general Riemannian spaces do not possess any degree of homogeneity. The invariants of Riemannian geometry are differential invariants of the group of arbitrary continuous transformations of the coordinates. The group of these general covariant transformations has no any invariants like invariants of finite Lie groups describing Klein spaces. Therefore it does not lead to ordinary conservation laws. Consequently any classification like the elementary particle classification taking place in homogeneous spaces of Minkowski or Galiley is not exist in Riemannian space. We meet with the same problems when a global internal symmetry becomes local one as it happens in the gauge field theory.

The solution was found thanks to Cartan's formulation of Riemannian geometry. The essense of the matter consist in following. 4D Riemannian space-time can be regarded as possessing the same symmetry properties that Minkowski space-time but only in the neighbourhood of each its point. In fact, a Riemannian space can be represented as a manifold whose "points" are Minkowski spaces, these being "interlinked" by Ricci or Christoffel connection coefficients. The geometrical concept of connection coefficients corresponds in physics to the gravitation interaction. If in a similar way we consider a 4D manifold whose points are homogeneous spaces of the representations of internal symmetry group, we obtain an example of a fibre bundle space. The connection coefficients introduced in it correspond to vector-potentials of gauge fields, or multiplets of vector mesons. This geometrical interpretation of gauge fields allows us to consider the trajectories of particles interacting with gauge field as free trajectories in a fiber bundle space. Thus in the description of any interactions mediated by some gauge field, we can get rid of the concept of force and make the theory of such interactions purely geometrical as in GR. This is also true for electrodynamics.

These results were obtained by N.P.Konopleva in 1965/67 and reported in Dubna on International Seminar on Vector Mesons and Electromagnetic interactions in 1969. The geometry of fibre bundle spaces was developed in 50-60th years by mathematicians of Moscow and Kazan state universities and by their schools. It was at the same time when the gauge field theory was formed by Sakurai, Yang and Mills, Utiyama, De Witt, L.Faddeev and V.Popov and others.

An important property of local symmetries is the existance of identity relations between the extremals and their derivatives. These identities can be expressed in the form of conservation laws which are strong, i.e., are satisfied independently of the specific form of the Lagrangian and the equations of motion. Integration of such conservation laws yields invariants having topological sense.

The basis of the theory of gauge fields comprises symmetry principles and close-range action hypothesis, which converts global symmetries into local one. The principle of local gauge invariance reflects a deep relationship between the universality of the various interactions, conservation of the vector currents, and the existence of the interactions themselves. This principle determines the form of all interactions, irrespective of their physical nature, and thereby opens the way to the construction of a consistent unified theory of the elementary particle interactions. At the same time, the principle of local gauge invariance, like Einstein's general principle of relativity, gives the theory such form, which admits a purely geometrical interpretation. As a result, it becomes possible to generalize Einstein's idea that the geometry of space-time must determined by the physical processes and have a dynamical character. Such geometry effectively reflects the influence on a distinguished test particle of all the remaining matter in the world. The possibility to classify elementary particles in terms of the invariants of the finite Lie groups arise again, but this classification becomes local one.

5 Summary: is the geometrical nuclear physics possible?

The geometrical description of interactions makes it possible to axiomatize physics, at least those its topics which are described by the gauge field theory. In this theory different kinds of interactions possess the different local gauge symmetry groups. Such groups, just as general covariant group of coordinate transformations in GR, belong to infinite Lie's groups. So, in the gauge field theory the classification principle arise which use a deep theory of the infinite Lie groups, as it predicted Hilbert in his VI problem.

Now the geometrical picture corresponds to the transmission of interaction through a medium. Therefore gauge models have naturally led to the notion of the vacuum as a medium. In description of vacuum in the gauge field theory GR occurs very useful. It can be used for development of nonperturbative quantization methods.

The classical theory of gauge fields is rapidly developing. The nonlinearity of the classical equations of non-Abelian gauge fields has given birth to a new industry among theoreticians. This is the study of particle-like solutions of these equations (solitons, instantons, kinks, monopoles and vortices). Particle-like solutions possess a new type of charge - topological charge, which one can attempt to assosiate with the new quantum numbers, that characterize the elementary particles. Threfore the gauge field theory puts the relations between quantum and classical physics to a new way.

The gauge field theory describes all interactions known in our days. It is very effective in elementary particle physics, gravity and condensed matter physics. But nuclear physics makes weakly use the gauge fild theory. The nuclear forces are described by the gauge field theory as three kinds of forces:

1. universal nuclear forces associated with $U(1)$-symmetry (Sakurai, model of vector dominance);

2. weak nuclear forces associated with $SU(2)$-symmetry (Yang-Mills fields);
3. strong nuclear forces associated with $SU(3)$-symmetry (quantum chromodynamics).

While Sakurai model led to the discovery of the vector mesons ρ, ω, ϕ, $SU(2)$-symmetry led to discovery of the W- and Z-bosons, $SU(3)$- symmetry was responsible for the discovery of the Ω-resonance and the appearence of new models of hadrons known as quark models.

Thus the results of the gauge theory of nuclear forces apply to elementary particle physics but do not for the present concern to a nuclear structure. The problem of XXI century is to construct the nuclear theory on the base of idears of the gauge field theory. Such nuclear theory will be relativistic and can be done purely geometrical.

Acknowledgments

It is a pleasure to thank the Organizers of the International Symposium QTS-2 and BLTP of JINR administration for their support of this investigation.

References

1. D. Hilbert, *Die Grundlagen der Physik, Nach. von der Kön. Ges. der Wiss. zu Göttingen, Math.-Phys. Kl.* **Heft 3**, (1915), pp.395-407.
2. A. Poincaré, *Derniéres Pensées*, Paris (1913).
3. A. Einstein, *Sitz. Preuss. Akad. Wiss.* **2**, 1111 (1916).
4. *Proc. of the Int. Sem. on Vector Mesons and Electromagn. Interactions*, ed. A.M.Baldin, JINR, Dubna (1969).
5. N. P. Konopleva, V. N. Popov, *Gauge fields*, Harwood Academic Publishers, Chur-London-New York (1981).
6. N.P.Konopleva, *Geometrical description of interactions*, Habilitation Thesis, Lebedev Institute of Physics, Moscow (1969).
7. N.P.Konopleva, Instantons and gravity, JINR Comm., E2-96-459, Dubna (1996).
8. N.P.Konopleva, Hyperbolical instantons and gauge fields vacuum structure, JINR Comm., E2-98-388, Dubna (1998).
9. N.P.Konopleva, VI Hilbert's problem and Lie infinite groups, JINR Comm., E2-99-301, Dubna (1999).

ON THE UNCERTAINTY RELATIONS AND SQUEEZED STATES FOR A QUANTUM PARTICLE ON A CIRCLE

K. KOWALSKI[1], J. REMBIELIŃSKI[1,2]

[1]Department of Theoretical Physics, University of Łódź, ul. Pomorska 149/153, 90-236 Łódź, Poland,

[2]College of Computer Science, ul. Rzgowska 17A, 93-008 Łódź, Poland

E-mail: kowalski@krysia.uni.lodz.pl

The uncertainty relations for a quantum particle on a circle are introduced minimized by the corresponding coherent states. The sqeezed states for the quantum mechanics on a circle and the related generalized uncertainty relations are identified as well.

An extended version of this paper will be sent in the near future to arXiv:quant-ph. In spite of its importance and long history[1-8] the problem of finding the uncertainty relations for the quantum mechanics on a circle still remains open. In this work we introduce the new uncertainty relations referring to the very recently found coherent states for a quantum particle on a circle [9,10]. We also identify the squeezed states for the quantum mechanics on a circle and discuss them in the context of the uncertainty relations.

Consider a free particle on a circle S^1. We assume for simplicity that the particle has unit mass and it moves in a unit circle. The angle operator $\hat{\varphi}$ and the angular momentum operator $\hat{J}$ obey the following commutation relation:

$$[\hat{\varphi}, \hat{J}] = \mathrm{i}, \tag{1}$$

where we set $\hbar = 1$. It can be demonstrated that the commutator (1) is defined only on the zero vector. The better candidate for representing the position of a quantum particle on a circle S^1 than $\hat{\varphi}$ is the unitary operator U of the form

$$U = e^{i\hat{\varphi}}, \tag{2}$$

satisfying

$$[\hat{J}, U] = U. \tag{3}$$

Consider the eigenvectors $|j\rangle$ of the angular momentum operator $\hat{J}$

$$\hat{J}|j\rangle = j|j\rangle. \tag{4}$$

The operators U and $U^\dagger$ act on the vectors $|j\rangle$ as follows:

$$U|j\rangle = |j+1\rangle, \qquad U^\dagger|j\rangle = |j-1\rangle. \tag{5}$$

The time-reversal invariance of the algebra (3) implies that the eigenvalues of the operator $\hat{J}$ can be only integer or half-integer [9]. In this work we restrict, for brevity, to the case with integer j. Of course, the eigenvectors of the angular momentum satisfy the orthogonality and completeness conditions

$$\langle j|k\rangle = \delta_{jk}, \tag{6}$$

$$\sum_{j=-\infty}^{\infty} |j\rangle\langle j| = I. \tag{7}$$

We now recall the basic properties of the coherent states for a quantum particle on a circle. These states can be introduced by [9]

$$Z|\xi\rangle = \xi|\xi\rangle, \tag{8}$$

where $Z = e^{-\hat{J}+\frac{1}{2}}U$, and the complex number $\xi = e^{-l+\mathrm{i}\varphi}$ parametrizes the cylinder which is the classical phase space for the particle moving in a circle. The projection of the vectors $|\xi\rangle$ onto the basis vectors $|j\rangle$ is

$$\langle j|\xi\rangle = \xi^{-j}e^{-\frac{j^2}{2}}. \tag{9}$$

The coherent states are not orthogonal. We have

$$\langle \xi|\eta\rangle = \sum_{j=-\infty}^{\infty} (\xi^*\eta)^{-j}e^{-j^2} = \theta_3(\tfrac{\mathrm{i}}{2\pi}\ln \xi^*\eta|\tfrac{\mathrm{i}}{\pi}), \tag{10}$$

where θ_3 is the Jacobi theta-function defined as

$$\theta_3(v|\tau) = \sum_{n=-\infty}^{\infty} q^{n^2}(e^{\mathrm{i}\pi v})^{2n}, \tag{11}$$

where $q = e^{\mathrm{i}\pi\tau}$ and $\operatorname{Im}\tau > 0$.

The most important feature of the discussed coherent states is that they are as close as possible to the classical phase space. Namely, the expectation value of the angular momentum in the coherent states satisfies

$$\langle \hat{J}\rangle_\xi = \frac{\langle \xi|\hat{J}|\xi\rangle}{\langle \xi|\xi\rangle} \approx l, \tag{12}$$

where the approximation is very good and we have the *exact* equality in the case of l integer or half-integer. Thus, it turns out that the parameter l labelling he coherent states can be interpreted as the classical angular momentum. Furthermore, the relative expectation value $\langle U\rangle_\xi/\langle U\rangle_1 :=$

$\langle\xi|U|\xi\rangle/\langle 1|U|1\rangle$, which is a natural candidate to describe the average position on a circle, obeys

$$\frac{\langle U\rangle_\xi}{\langle U\rangle_1} \approx e^{i\phi}, \tag{13}$$

where the approximation is very good. It thus appears that the parameter φ can be really regarded as the classical angle.

We now introduce the uncertainty relations for a quantum particle on a circle. On using the relations

$$\langle e^{-2\hat{J}}\rangle_\xi = e^{1-2l}, \qquad \langle e^{2\hat{J}}\rangle_\xi = e^{1+2l}, \tag{14}$$

and

$$\langle U^2\rangle_\xi = \frac{\langle\xi|U^2|\xi\rangle}{\langle\xi|\xi\rangle} = e^{-1}e^{2i\varphi}, \tag{15}$$

we arrive at the following remarkable identity:

$$\langle e^{-2\hat{J}}\rangle_\xi\langle e^{2\hat{J}}\rangle_\xi = \frac{1}{|\langle U^2\rangle_\xi|^2}. \tag{16}$$

We now define the uncertainty of the angular momentum

$$\Delta^2_\phi(\hat{J}) := \frac{1}{4}\ln\left(\langle e^{-2\hat{J}}\rangle_\phi\langle e^{2\hat{J}}\rangle_\phi\right), \tag{17}$$

and the uncertainty of the angle

$$\Delta^2_\phi(\hat{\varphi}) := \frac{1}{4}\ln\frac{1}{|\langle U^2\rangle_\phi|^2}. \tag{18}$$

It can be easily demonstrated that both uncertainties (17) and (18) are non-negative. Further, using the cumulant expansion

$$\langle e^X\rangle = \exp(\langle\!\langle X\rangle\!\rangle + \tfrac{1}{2!}\langle\!\langle X^2\rangle\!\rangle + \tfrac{1}{3!}\langle\!\langle X^3\rangle\!\rangle + \ldots), \tag{19}$$

where $\langle\!\langle X^n\rangle\!\rangle$, $n = 1, 2, \ldots$, are the cumulants such that

$$\begin{aligned}\langle\!\langle X\rangle\!\rangle &= \langle X\rangle,\\ \langle\!\langle X^2\rangle\!\rangle &= \langle X^2\rangle - \langle X\rangle^2,\\ \langle\!\langle X^3\rangle\!\rangle &= \langle X^3\rangle - 3\langle X^2\rangle\langle X\rangle + 2\langle X\rangle^3,\end{aligned} \tag{20}$$

and so on, we find that in the first approximation neglecting the cumulants of order four and higher, the uncertainties (17) and (18) coincide with the usual variances of the angular momentum and angle, respectively. Now, we have

$$\Delta^2_\xi(\hat{J}) = \Delta^2_\xi(\hat{\varphi}) = \tfrac{1}{2}. \tag{21}$$

Hence,

$$\Delta_\xi^2(\hat{J}) + \Delta_\xi^2(\hat{\varphi}) = 1. \tag{22}$$

The form of (22) and experience with the standard coherent states for a particle on a real line indicate the following uncertainty relations for a quantum particle on a circle:

$$\Delta^2(\hat{J}) + \Delta^2(\hat{\varphi}) \geq 1. \tag{23}$$

These uncertainty relations are supported numerical simulations.

We finally study the squeezed states for the quantum mechanics on a circle. Motivated by the fact that the eigenvectors of the annihilation operators

$$a(s) = e^{-s\frac{\hat{p}^2}{2}}\hat{q}e^{s\frac{\hat{p}^2}{2}} = \hat{q} + \mathrm{i}s\hat{p}, \tag{24}$$

where $\hat{q}$ and $\hat{p}$ are the standard position and momentum operators, respectively, and $s > 0$, are the standard squeezed states, we define the squeezed states $|\xi\rangle_s$ for a quantum particle on a circle as a solution of the eigenvalue equation

$$Z(s)|\xi\rangle_s = \xi|\xi\rangle_s, \tag{25}$$

where

$$Z(s) = e^{-s\frac{\hat{J}^2}{2}}Ue^{s\frac{\hat{J}^2}{2}} = e^{-s(\hat{J}-\frac{1}{2})}U. \tag{26}$$

Notice that the coherent states for a particle on a circle satisfying (8) correspond to the particular case of $s = 1$. We also remark that (24) and (26) are example of the general annihilation operators which have been very recently introduced in ref. 11. The squeezed states obey

$$\langle j|\xi\rangle_s = \xi^{-j}e^{-\frac{sj^2}{2}}, \tag{27}$$

$$_s\langle\xi|\eta\rangle_s = \sum_{j=-\infty}^{\infty}(\xi^*\eta)^{-j}e^{-sj^2} = \theta_3(\tfrac{\mathrm{i}}{2\pi}\ln\xi^*\eta|\tfrac{\mathrm{i}s}{\pi}), \tag{28}$$

where $s > 0$. Furthermore, we have the generalizations of the formulae (14) and (15) such that

$$\langle e^{-2s\hat{J}}\rangle_{\xi,s} = e^{s-2l}, \qquad \langle e^{2s\hat{J}}\rangle_{\xi,s} = e^{s+2l}, \tag{29}$$

where $\langle A\rangle_{\xi,s} = {}_s\langle\xi|A|\xi\rangle_s/{}_s\langle\xi|\xi\rangle_s$, and

$$\langle U^2\rangle_{\xi,s} = e^{-s}e^{2\mathrm{i}\varphi}. \tag{30}$$

The relations (29) and (30) lead to the following generalization of the uncertainties (17) and (18):

$$\widetilde{\Delta}^2_{\phi,s_0}(\hat{J}) = \frac{1}{4}\ln\left(\langle e^{-2s_0\hat{J}}\rangle_\phi\langle e^{2s_0\hat{J}}\rangle_\phi\right), \tag{31}$$

$$\Delta^2_{\phi,s_0}(\hat{\varphi}) \equiv \Delta^2_\phi(\hat{\varphi}). \tag{32}$$

Proceeding as with (17) and (18) we finally arrive at the generalized uncertainty relations of the form

$$\widetilde{\Delta}^2_{\phi,s_0}(\hat{J}) + \Delta^2_\phi(\hat{\varphi}) \geq s_0. \tag{33}$$

These relations are supported by the numerical simulations. We point out that, in opposition to the case of the quantum mechanics on a real line, the uncertainty relations (33), where the uncertainties are given by (31) and (32) do not enable to distinguish the coherent and the squeezed states. Moreover, the squeezed states with different s are not related by a unitary transformation as the standard squeezed states, and the problem arises of the physical interpretation of the dimensionless parameter s.

References

1. D. Judge, *Phys. Lett.* **5**, 189 (1963).
2. W.H. Luisell, *Phys. Lett.* **7**, 60 (1963).
3. P. Carruthers and M.M. Nieto, *Rev. Mod. Phys.* **40**, 411(1968).
4. I. Białynicki-Birula and J. Mycielski, *Commun. Math. Phys.* **44**, 129 (1975).
5. I. Białynicki-Birula, M. Freyberger and W. Schleich, *Physica Scripta* **T38**, 113 (1993).
6. A.S. Holevo, *Rep. Math. Phys.* **16**, 385 (1979).
7. V.A. Kostelecký and B. Tudose, *Phys. Rev. A* **53**, 1978 (1996).
8. R. Bluhm, V.A. Kostelecký and B. Tudose, *Phys. Rev. A* **52**, 2234 (1995).
9. K. Kowalski, J. Rembieliński and L.C. Papaloucas, *J. Phys. A: Math. Gen.* **29**, 4149 (1996).
10. J.A. Gonzáles and M.A. del Olmo, *J. Phys. A: Math. Gen.* **31**, 8841 (1998).
11. B.C. Hall, Coherent states on spheres, ArXiv:quant-ph/0109086.

EFFECTIVE DIPOLE-TYPE FIELD APPROACH AND THE DUAL HIGGS MODEL

G.A. KOZLOV

Bogoliubov Laboratory of Theoretical Physics,
Joint Institute for Nuclear Research,
141980 Dubna, Moscow Region, Russia
E-mail: kozlov@thsun1.jinr.ru

We consider the dual picture of the Yang-Mills theory at large distances. The dual Higgs model is reformulated in terms of two-point Wightman functions with the equations of motion involving higher derivatives.

1. The nature of confining forces in gauge theories attracts a big attention during last years. We consider the model (in four-dimensional space-time (4d)) based on the dual description of a long-distance Yang-Mills (LDY-M) theory which can provide a quark confinement in a system of static test charges. This talk follows the idea that the vacuum of quantum Yang-Mills (Y-M) theory is realized by a condensate of monopole-antimonopole pairs [1-4]. Since there are no monopoles as classical solutions with finite energy in a pure Y-M theory, it has been suggested by 't Hooft [5] going into the Abelian projection where the gauge group SU(2) is broken by a suitable gauge condition to its Abelian subgroup U(1).

It is known that the topology of the Y-M SU(N) manifold and that of its Abelian subgroup $[U(1)]^{N-1}$ are different, and new topological objects can appear in case of introducing the local gauge transformation of some gauge function in our model, eg., the field strength tensor for the gauge field $A_\mu(x)$ in quantum chromodynamics (QCD) transforms with the gauge function $\Omega(x)$ as

$$F_{\mu\nu}(x) \to F^{\Omega}_{\mu\nu}(x) = \Omega(x)\, F_{\mu\nu}(x)\, \Omega^{-1}(x) = \partial_\mu\, A^{\Omega}_\nu(x) - \partial_\nu\, A^{\Omega}_\mu(x) + \\ + ie\, [A^{\Omega}_\mu(x), A^{\Omega}_\nu(x)] - \frac{1}{i\,e}\, \Omega(x)\, [\,\partial_\mu, \partial_\nu] \Omega^{-1}(x)\ . \qquad (1)$$

The last term in (1) reflects the singular character of the above-mentioned gauge transformation. One can identify the Abelian projection by the replacement

$$F^{\Omega}_{\mu\nu}(x) \to F^{\alpha}_{\mu\nu}(x) = \partial_\mu\, A^{\alpha}_\nu(x) - \partial_\nu\, A^{\alpha}_\mu(x) - \frac{1}{i\,e}\, \Omega(x)\, [\,\partial_\mu, \partial_\nu] \Omega^{-1}(x)\ ,$$

where $A^{\Omega}_\mu \to A^{\alpha}_\mu$, while the label α reflects the Abelian world. This leads to the Dirac string and magnetic current

$$J^{mon}_{\mu}(x) = -\frac{1}{2\,i\,e}\,\epsilon_{\mu\nu\rho\sigma}\,\partial_{\nu}\,\Omega(x)[\partial_{\rho},\partial_{\sigma}]\,\Omega^{-1}(x)$$

in the Abelian gauge sector.

Analytical models of the dual QCD with monopoles were intensively investigated [6-9]. We study the Lagrangian model where the fundamental variables are an octet of dual potentials coupled minimally to three octets of monopole (Higgs) fields. The basic manifestation of the model is that it generates the equations of motion where one of them for the scalar (Higgs) field is the equation involving higher derivatives. Our aim is to apply the previous results [10-16] of Wightman functions in 4d models with equations of motion (for the scalar fields) involving the dipole-type structure at least, to the model with monopoles. The monopole fields obeying such an equation are classified by their two-point Wightman functions (TPWF). In the scheme presented in this talk, the flux distribution in the tubes formed between two heavy color charges is understood via the following statement: the Abelian monopoles are excluded from the string region while the Abelian electric flux is squeezed into the string region. In our model, we use the dual gauge field $\hat{C}^a_{\mu}(x)$ and the monopole field $\hat{B}^a_i(x)$ ($i = 1,...,N_c(N_c-1)/2$ and a=1,...,8 is a color index) which are relevant modes for infrared behaviour. The local coupling of the $\hat{B}^a_i$-field to the $\hat{C}^a_{\mu}$-field provides the mass of the dual field and, hence, a dual Meissner effect. The commutation relations, TPWF and Green's functions as well-defined distributions in the space $S(\Re^d)$ of complex Schwartz test functions on $\Re^d$, will be defined as well.

2. Let us consider the Lagrangian density (LD) L of the $U(1)\times U(1)$ dual Higgs model corresponding to the LDY-M theory [8]

$$L = 2\,Tr\left[-\frac{1}{4}\hat{F}^{\mu\nu}\hat{F}_{\mu\nu} + \frac{1}{2}\left(D_{\mu}\hat{B}_i\right)^2\right] - W\left(\hat{B}_i\right)\,, \tag{2}$$

where $\hat{F}_{\mu\nu} = \partial_{\mu}\hat{C}_{\nu} - \partial_{\nu}\hat{C}_{\mu} - ig[\hat{C}_{\mu},\hat{C}_{\nu}]$ and $D_{\mu}\hat{B}_i = \partial_{\mu}\hat{B}_i - ig\,[\hat{C}_{\mu},\hat{B}_i]$, $\hat{C}_{\mu}$ and $\hat{B}_i$ are the SU(3) matrices, g is the gauge coupling constant; $\hat{C}_{\mu} = \lambda^8\,C_{\mu}$ (λ^a is the generator of SU(3)). $\hat{F}_{\mu\nu}$ becomes the Abelian tensor with C_{μ} being the dual to an ordinary vector potential. The Higgs fields develop their vacuum expectation values (v.e.v.) $\hat{B}_{0_i}$ and the Higgs potential $W(\hat{B}_i)$ has a minimum at $\hat{B}_{0_i}$. The v.e.v. $\hat{B}_{0_i}$ produce a color monopole generating current confining the electric color flux. The interaction between a quark Q and antiquark $\bar{Q}$ is provided by $\hat{C}_{\mu}$ via a Dirac string tensor [17] $\hat{\tilde{G}}_{\mu\nu}$ having the same color structure as $\hat{C}_{\mu}$, namely $\hat{\tilde{G}}_{\mu\nu} = \lambda^8\,\tilde{G}_{\mu\nu}$. A bound state of Q and $\bar{Q}$ occurs with replacing $\hat{F}_{\mu\nu} \to \lambda^8\,G_{\mu\nu}$, where $G_{\mu\nu} = \partial_{\mu}C_{\nu} - \partial_{\nu}C_{\mu} + \tilde{G}_{\mu\nu}$.

Following [8], we choose the color structure for $\hat{B}_i$-fields as $\hat{B}_1 = \lambda_7 B_1 - \lambda_6 \bar{B}_1$, $\hat{B}_2 = -\lambda_5 B_2 + \lambda_4 \bar{B}_2$, $\hat{B}_3 = \lambda_2 B_3 - \lambda_1 \bar{B}_3$. The effective potential looks like (see also [8])

$$W(B,\bar{B},B_3) = \frac{2}{3}\lambda\{11[2(B^2+\bar{B}^2-B_0^2)^2+(B_3^2-B_0^2)^2]$$

$$+7[2(B^2+\bar{B}^2)+B_3^2-3B_0^2]^2\} ,$$

where λ is dimensionless. The LD (2) becomes

$$L(\tilde{G}_{\mu\nu}) = -\frac{1}{3}G_{\mu\nu}^2 + 4|(\partial_\mu - igC_\mu)\phi|^2 + 2(\partial_\mu\phi_3)^2 - W(\phi,\phi_3) , \qquad (3)$$

where $\phi \equiv \phi_1 = \phi_2 = B - i\bar{B}, \phi_3 = B_3$; The generating current of (3) is nothing but the monopole current confining the electric color flux $J_\mu^{mon} = (2/3)\,\partial^\nu\,G_{\mu\nu}(x)$.

Since ϕ_1 and ϕ_2 couple to C_μ in the same way [8], we choose $B(x) = b(x) + B_0, \bar{B}(x) = \bar{b}(x), B_3(x) = b_3(x) + B_0$ with $\langle B(x)\rangle_0 \neq 0$, $\langle \bar{B}(x)\rangle_0 \neq 0$, $\langle B_3(x)\rangle_0 \neq 0$. In terms of the new fields $b, \bar{b}, b_3$ the LD (3) is divided into two parts $L = L_1 + L_2$ where L_1 in the lowest order of g and λ and with the minimal weak interaction looks like

$$L_1 = -\frac{1}{3}G_{\mu\nu}^2 + 4\left[(\partial_\mu b)^2 + (\partial_\mu\bar{b})^2 + \frac{1}{2}(\partial_\mu b_3)^2\right]$$

$$+m^2\,C_\mu^2 - \frac{4}{3}\mu^2(50b^2+18b_3^2) + 8m\partial_\mu\bar{b}\,C_\mu .$$

Here, $m \equiv gB_0$ and $\mu \equiv \sqrt{2\lambda}B_0$. The equations of motion for the fields $b, \bar{b}, b_3$ and C_μ are $(\Delta^2+\mu_1^2)\,b(x) = 0$, $\Delta^2\,\bar{b}(x) + m(\partial\cdot C) = 0$, $(\Delta^2+\mu_2^2)\,b_3(x) = 0$, respectively and

$$(\Delta^2+m_1^2)\,C_\mu(x) - \partial_\mu(\partial\cdot C) + 12\,m\,\partial_\mu\bar{b}(x) - \partial^\nu\tilde{G}_{\mu\nu}(x) = 0 , \qquad (4)$$

where $\mu_1^2 = (50/3)\,\mu^2, \mu_2^2 = 12\,\mu^2, m_1^2 = 3\,m^2$. The formal solution of equation (4) looks like

$$C_\mu(x) = \alpha\,\partial^\nu\tilde{G}_{\mu\nu}(x) - \beta\,\partial_\mu\bar{b}(x) ,$$

with $\alpha \equiv (3\,m^2)^{-1}$, $\beta \equiv 4/m$. We obtain that the dual gauge field is defined via the divergence of the Dirac string tensor $\tilde{G}_{\mu\nu}(x)$ shifted by the divergence of the scalar field $\bar{b}(x)$. For large enough $\vec{x}$, the monopole field is going to its v.e.v. while $C_\mu(\vec{x}\to\infty)\to 0$ and $J_\mu^{mon}(\vec{x}\to\infty)\to 8\,m^2\,C_\mu$. It implies that in the $d = 2h$ dimensions the $\bar{b}(x)$-field obeys the equation

$$\Delta^{2h}\,\bar{b}(x) \simeq 0 , \quad h = 2,3,\dots , \qquad (5)$$

but $\Delta^2\,\bar{b}(x) \neq 0$. Here, the solutions of equation (5) obey locality, Poincare covariance and spectral conditions, and look like the dipole "ghosts" at h=2.

We define TPWF $W_h(x)$ in the d=2 h-dimensions $W_h(x) = \langle\,\bar{b}(x)\,\bar{b}(0)\,\rangle_0$ as the distribution in the Schwartz space $S'(\Re^{2h})$ of temperate distributions on $\Re^{2h}$ which obeys the equation

$$\Delta^{2\,h}\,W_h(x) = 0\ . \tag{6}$$

The general solution of (6) should be Lorentz invariant and is given in the form [10] at h=2

$$W_2(x) = a_1\,\ln\frac{l^2}{-x_\mu^2 + i\,\epsilon\,x^0} + \frac{a_2}{x_\mu^2 - i\,\epsilon\,x^0} + a_3\ , \tag{7}$$

where a_i $(i = 1,2,3)$ are the coefficients , l is an arbitrary length scale. The coefficients a_1 and a_2 in (7) can be fixed using the canonical commutation relations (CCR) $[C_\mu(x), \pi_{C_\nu}(0)]_{|_{x^0=0}} = i\,g_{\mu\nu}\,\delta^3(\vec{x})$ and $\left[\,\bar{b}(x), \pi_{\bar{b}}(0)\right]_{|_{x^0=0}} = i\,\delta^3(\vec{x})$, respectively, with $\pi_{C_\mu}(x) = -\frac{4}{3}\,G_{0\mu}(x)$ and $\pi_{\bar{b}}(x) = 8\,\left[\partial^0\,\bar{b}(x) + m\,C^0(x)\right]$. The standard commutator for the scalar field $\bar{b}$ is

$$\left[\,\bar{b}(x), \bar{b}(0)\,\right] = (2\,\pi)^2\,i\,\left[4\,a_1\,E_2(x) + a_2\,D_2(x)\right]\ ,$$

where $E_2(x) = (8\,\pi)^{-1}\,sgn(x^0)\,\theta(x^2)$ and $D_2(x) = (2\,\pi)^{-1}\,sgn(x^0)\,\delta(x^2)$ are taken into account. The direct calculation leads to $a_1 = (m^2/48\,\pi^2)$ and $a_2 = -(1/24\,\pi^2)$.

The propagator of the $\bar{b}$-field in $S'(\Re_4)$ is

$$\hat{\tau}_2(p) = weak\lim_{\tilde{\kappa}^2<<1}\frac{i}{3\,(2\,\pi)^4}\left\{m^2\left[\frac{1}{(p^2-\kappa^2+i\,\epsilon)^2} + i\,\pi^2\,\ln\frac{\kappa^2}{\tilde{\mu}^2}\,\delta_4(p)\right] - \frac{1}{2}\,\frac{1}{p^2-\kappa^2+i\,\epsilon}\right\}. \tag{8}$$

Here, κ is a parameter of representation and not an analogue of the infrared mass, $\tilde{\kappa}^2 \equiv \kappa^2/p^2$.

To define the commutation relation $[\,C_\mu(x), C_\nu(y)\,]$, let us consider the canonical conjugate pair $\{C_\mu, \pi_{C_\nu}\}$

$$\left[\frac{4}{3}C_\mu(x), \partial_\nu C_0(0) - \partial_0 C_\nu(0) - g_{0\nu}(\partial\cdot C(0)) + \Delta_{0\nu}(0)\right]_{|_{x^0=0}} = ig_{\mu\nu}\delta^3(\vec{x}), \tag{9}$$

where $\Delta_{\mu\nu}(x) = g_{\mu\nu}\,(\partial\cdot C(x)) - \tilde{G}_{\mu\nu}(x)$ tends to zero as $x \to 0$ and the Dirac string tensor $\tilde{G}_{\mu\nu}(x)$ obeys the equation $\left[\Delta^2 + (3m)^2\right]\,\tilde{G}_{\mu\nu}(x) = 0$.

Obviously, the following form of the free C_μ-field commutator:

$$[C_\mu(x), C_\nu(0)] = i g_{\mu\nu} \left[\xi\, m_1^2\, E_2(x) + c\, D_2(x) \right] \ , \tag{10}$$

ensures the CCR (9) at large x_μ^2 with both ξ and c (in (10)) being real arbitrary numbers but $\xi = \frac{3}{4} - 4\, c$.

The free dual gauge field propagator in $S'(\Re_4)$ in any local covariant gauge is given by

$$\hat{\tau}_{\mu\nu}(p) = \int d^4x\, \exp(i\, p\, x)\, \tau_{\mu\nu}(x)$$

$$= i \left[g_{\mu\nu} - \left(1 - \frac{1}{\zeta} \right) \frac{p_\mu\, p_\nu}{p^2 + i\,\epsilon} \right] \cdot \left[\xi\, m_1^2\, \hat{t}_1(p) + c\, \hat{t}_2(p) \right] \ , \tag{11}$$

where

$$\tau_{\mu\nu}(x) = \frac{i\, g_{\mu\nu}}{(4\,\pi)^2} \left[\xi\, m_1^2\, \ln(-\tilde{\mu}^2\, x_\mu^2 + i\,\epsilon) + \frac{c}{x_\mu^2 + i\,\epsilon} \right] \ ;$$

$$\hat{t}_1(p) = weak \lim_{\tilde{\kappa}^2 << 1} \left[\frac{1}{(p^2 - \kappa^2 + i\,\epsilon)^2} + i\,\pi^2\, \ln\left(\frac{\kappa^2}{\tilde{\mu}^2} \right) \delta_4(p) \right] \ ;$$

$$\hat{t}_2(p) = weak \lim_{\tilde{\kappa}^2 << 1} \frac{1}{2}\, \frac{1}{(p^2 - \kappa^2 + i\,\epsilon)} \ .$$

The gauge parameter ζ in (11) is a real number. The following requirement $(\Delta^2)^2\, \tilde{\tau}_{\mu\nu}(x) = i\, \delta_4(x)$ on Green's function $\tau_{\mu\nu}(x)$ leads to that a constant c has to be equal to zero and $\tilde{\tau}_{\mu\nu}(x) = \tau_{\mu\nu}(x)/(\xi\, m_1^2)$.

3. As for an approximate topological solution for this dual model, we fix the equations of motion

$$\partial^\nu\, G_{\mu\nu} = 6\, i\, g[\phi^*(\partial_\mu - i\, g\, C_\mu)\, \phi \ - \ \phi(\partial_\mu + i\, g\, C_\mu)\, \phi^*] \ , \tag{12}$$

$$(\partial_\mu - i\, g\, C_\mu)^2\, \phi = \frac{2}{3}\, \lambda\, (32\, B_0^2 - 25\, |\,\phi\,|^2 - 7\, \phi_3^2\,)\, \phi \ , \tag{13}$$

where $\phi(x)$ is decomposed like $\phi(x) = \frac{1}{\sqrt{2}} \exp(i\, f(x))\, [\, \chi(x) + B_0\,]$ using the new scalar variables $\chi(x)$ and $f(x)$. The equation of motion (12) transforms into the following one:

$$\partial^\nu\, G_{\mu\nu} = 6\, g\, (\chi + B_0)^2\, (g\, C_\mu - \partial_\mu\, f) \ ,$$

that means that the $\bar{b}(x)$-field is nothing but a mathematical realization of the "massive" phase $B_0 \cdot f(x)$ at large enough $\vec{x}$, i.e., $\bar{b}(x) \simeq (B_0/2)\, S(x)\, f(x)$ and $S(x) \equiv (1 + \chi(x)/B_0)^2$. Integrating out $G_{\mu\nu}$ over the 2d surface element

$\sigma^{\mu\nu}$ in the flux $\Pi = \int G_{\mu\nu}(x)\, d\sigma^{\mu\nu}$, we conclude that the phase $f(x)$ is varied by $2\pi n$ for any integer number n associated with the topological charge [18] inside the flux tube. Using the cylindrical symmetry we arrive at the field equation ($\vec{C} \to (\tilde{C}(r)/r)\vec{e_\theta}$)

$$\frac{d^2\,\tilde{C}(r)}{d\,r^2} - \frac{1}{r}\,\frac{d\,\tilde{C}(r)}{d\,r} - 3\,m^2\,[3 + 2\,S(r)]\,\tilde{C}(r) + 6\,n\,m\,B_0\,S(r) = 0$$

with the asymptotic transverse behaviour of its solution

$$\tilde{C}(r) \simeq \frac{4\,n}{7\,g} - \sqrt{\frac{\pi\,m\,r}{2\,k}}\,e^{-k\,m\,r}\left(1 + \frac{3}{8\,k\,m\,r}\right)\,, \quad k \equiv \sqrt{21}\,.$$

The field equation (13) is given by

$$\frac{d^2\,\hat{\chi}(r)}{d\,r^2} + \frac{1}{r}\,\frac{d\,\hat{\chi}(r)}{d\,r} - \left\{\left[\frac{n - g\,\tilde{C}(r)}{r}\right]^2 + \frac{25}{3}\lambda\left[2\,B_0^2 - \hat{\chi}^2(r)\right]\right\}\hat{\chi}(r) \simeq 0\,,$$

where $\hat{\chi}(r) = \chi(r) + B_0$. The profile of the color electric field in the flux tube at large r looks like

$$E_z(r) = \sqrt{\frac{\pi\,m}{2\,k\,r}}\left(k\,m - \frac{1}{2\,r}\right)e^{-k\,m\,r}\,.$$

4. We define the static potential like $P_{stat}(r) = \lim_{T\to\infty}[T^{-1}\cdot A(r)]$ and the action $A(r)$ is given by the colour source-current part of LD $L(p) = -\vec{j}^{\,\mu}_{\alpha}(-p)\,\hat{\tau}_{\mu\nu}(p)\,\vec{j}^{\,\nu}_{\alpha}(p)$ with the quark current $\vec{j}^{\,\mu}_{\alpha}(x) = \vec{Q}_\alpha\,g^{\mu 0}\,[\delta_3(\vec{x} - \vec{x_1}) - \delta_3(\vec{x} - \vec{x_2})]$. Here, $\vec{Q}_\alpha = e\,\vec{\rho}_\alpha$ is the Abelian color-electric charge of a quark while $\vec{\rho}_\alpha$ is the weight vector of the SU(3) algebra: $\rho_1 = (1/2,\,\sqrt{3}/6)$, $\rho_2 = (-1/2,\,\sqrt{3}/6)$, $\rho_3 = (0,\,-1/\sqrt{3})$ [18]; $\vec{x}_1$ and $\vec{x}_2$ are the position vectors of a quark and an antiquark, respectively.

As a consequence of the dual field propagator (11), we get (see also [14,15])

$$P_{stat}(r) = -\frac{\vec{Q}^2}{16\,\pi}\left[\xi\,m^2\,r(-12.4 + 6\,\ln\tilde{\mu}\,r) + O\left(\frac{c}{r}\right)\right]\,, \qquad (14)$$

where the string tension a in $P_{stat}(r) \sim a\,r$ emerges as $a \simeq \frac{3}{4}\,\frac{\pi}{g^2}\,m^2 = \frac{3}{4}\,\alpha_s\,m^2$ The logarithmic correction in (14) has been ignored at large distances $r \sim \tilde{\mu}^{-1}$. We obtain $a \simeq 0.18\ GeV^2$ for the mass of the dual C_μ-field $m = 0.8\ GeV$ and $\alpha_s = \pi/g^2 = e^2/(4\,\pi)$=0.37 fixed from fitting the heavy quark-antiquark spectrum [19]. The string tension $a_{fixed} = 1.0\ GeV/fm$ extracted from the Regge slope of the hadrons [20] then determines the fixed value of the mass $m_{fixed} = 0.85\ GeV$. Doing thc formal comparison, let us remind that the string tensions in papers [21] and [22] are equal to $a = (430\ MeV)^2$ and

$a = (440\ MeV)^2$, respectively. We found that for a sufficiently long string the $\sim r$-behaviour of the static potential is dominant; for a short string the singular interaction provided by the second term in (14) becomes important if the average size of the monopole is even smaller.

References

1. Y. Nambu, *Phys. Rev. D* **10**, 4262 (1974) .
2. S. Mandelstam, *Phys. Rep. C* **23**, 245 (1976); *Phys. Rev. D* **19**, 2391 (1979).
3. A.M. Polyakov, *Nucl. Phys. B* **120**, 429 (1977).
4. G.'t Hooft, *Nucl. Phys. B* **138**, 1 (1978).
5. G.'t Hooft, *Nucl. Phys. B* [FS3] **190**, 455 (1981).
6. T. Suzuki, *Prog. Theor. Phys.* **80**, 929 (1988); **81**, 752 (1988). (1989); S. Maedan and T. Suzuki, *Prog. Theor. Phys.* **81**, 229 (1989).
7. H. Suganuma, S. Sasaki and H. Toki, *Nucl. Phys. B* **435**, 207 (1995).
8. M. Baker et al., *Phys. Rev. D* **54**, 2829 (1996).
9. T. Suzuki, *Nucl. Phys. (Proc. Suppl.)* **30**, 176 (1993); M.I. Polikarpov, *Nucl. Phys. (Proc. Suppl.)* **53**, 134 (1997); A. Di Giacomo, *Prog. Theor. Phys. Suppl.* **131**, 161 (1998); M.N. Chernodub et al., *Prog. Theor. Phys. Suppl.* **131**, 309 (1998); R.W. Haymaker, *Phys. Rep.* **315**, 153 (1999).
10. R. Ferrari, *Il Nuovo Cimento A* **19**, 204 (1974).
11. D. Zwanziger, *Phys. Rev. D* **17**, 457 (1978).
12. E. D'Emilio and M. Mintchev, A gauge model with perturbative confinement in four dimensions, report IFUP, Pisa (1979).
13. E. D'Emilio and M. Mintchev, *Phys. Lett. B* **89**, 207 (1980).
14. G.A. Kozlov, S.P. Kuleshov, and R.P. Zaikov, On linear quasipotential in the framework of quantum field theory, report E2-80-352, JINR, Dubna (1980).
15. G.A. Kozlov, em IL Nuovo Cimento A **105**, 139 (1992).
16. G.A. Kozlov, *IL Nuovo Cimento A* **107**, 819, 1739 (1994).
17. P.A.M. Dirac, *Phys. Rev.* **74**, 817 (1948).
18. Y. Koma, H. Suganuma, and H. Toki, *Phys. Rev. D* **60**, 074024 (1999).
19. M. Baker, J.S. Ball and F. Zachariasen, *Phys. Rev. D* **51**, 1968 (1995).
20. K. Sailer et al., *J. Phys. G* **17**, 1005 (1991).
21. M. Baker, N. Brambilla, H.G. Dosch, and A. Vairo, *Phys. Rev. D* **58**, 034010 (1998).
22. M.N. Chernodub, *Phys. Lett. B* **474**, 73 (2000).

CP VIOLATION IN B DECAYS: THE ROLE OF FINAL-STATE INTERACTIONS

P. ŁACH AND P. ŻENCZYKOWSKI

Dept. of Theoretical Physics, Inst. of Nuclear Physics
Radzikowskiego 152, 31-342 Kraków, Poland

The main goals of the programme of studying CP violation in B decays are briefly presented. It is stressed that reliable ways of handling final-state interactions have to be found if these goals are to be achieved. On the example of $B^0 \to K\bar{K}$ decays it is shown that rescattering may strongly affect phases of decay amplitudes. The need to understand SU(3) breaking effects is also pointed out.

1 CP Violation

In 1964 [1] the long-lived neutral kaon K_L was observed to decay not only into three but also (much more rarely) two pions. Since three-pion (two-pion) states are eigenstates of the product of charge conjugation (C) and space inversion (P) operators corresponding to eigenvalues $(CP)_{3\pi} = -1$ ($(CP)_{2\pi} = +1$), CP must be violated. There are two ways in which this may occur: either 1) K_L is not an eigenstate of CP, i.e. being basically a $CP = -1$ state it contains a small admixture of the $CP = +1$ state (this is called "violation of CP through mass matrix in the $K^0\bar{K}^0$ system"), or 2) K_L has $CP = -1$, and CP is violated *during* the decay to two pions (so-called "direct CP violation"). Experimental differentiation between these alternatives was very difficult because it had to consist in identifying a tiny difference between the CP-violating (and thus already very small) decay rates $K_L \to \pi^+\pi^-$ and $K_L \to \pi^0\pi^0$. The existence of direct CP violation was confirmed only very recently.

2 CP Violation in Standard Model

From among various proposals to provide a a more fundamental "raison d'être" for CP violation, the idea of Kobayashi and Maskawa (KM) [2], accepted as the most probable one, is now included into the "Standard Model". According to this idea, CP violation may come from an unremovable phase in a 3×3 unitary matrix describing the relation between two quark bases: that of mass eigenstates and that of weak interaction eigenstates. In the 2×2 version of this matrix (the Cabibbo matrix), all such phases can be absorbed into the definitions of quark fields so that the matrix becomes real. In the 3×3 case, only five phases can be removed thus leaving four parameters: one too much for the 3×3 matrix to become orthogonal and real. The unremovable imaginary parameter appears then in interactions and leads to CP violation.

Experiment indicates that this CKM matrix is fairly close to being diagonal. In a convenient parametrization it is written as

$$\begin{bmatrix} V_{ud} & V_{us} & V_{ub} \\ V_{cd} & V_{cs} & V_{cb} \\ V_{td} & V_{ts} & V_{tb} \end{bmatrix} = \begin{bmatrix} 1-\lambda^2/2 & \lambda & A\lambda^3(\rho - i\eta) \\ -\lambda & 1-\lambda^2/2 & A\lambda^2 \\ A\lambda^3(1-\rho-i\eta) & -A\lambda^2 & 1 \end{bmatrix} \tag{1}$$

where $\lambda \approx .22$, while A, ρ, η are of order 1. In this parametrization the CP-violating parameter η appears only in the corners of the matrix where it is accompanied by λ^3. CP violation in neutral kaon system arises from box diagrams (Fig.1) with intermediate t quarks. The relevant amplitudes are proportional to $|V_{td}V_{ts}^*|^2 \propto \lambda^{10}$. Hence, CP violation in the kaon sector is tiny. Size of CP violation observed in kaon system provides constraints on the values of ρ and η, but a clearcut test of the CKM idea is not possible.

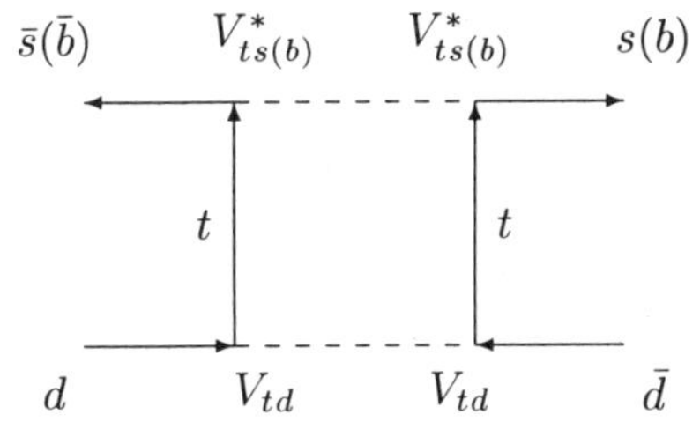

Fig. 1 Box diagram

3 B Decays as a Testing Ground

If the KM idea is correct, the matrix V should be unitary. Since this matrix constitutes a physical observable (up to reparametrizations), we should be able to test its unitarity through experiments. One way to do this is to check that

$$V_{ub}^* V_{tb} + V_{us}^* V_{ts} + V_{ud}^* V_{td} = 0 \tag{2}$$

i.e. that we are dealing with a "unitarity triangle" (Fig.2). This equation involves CKM elements V_{td} and V_{ub}, which depend on the CP-violating parameter η. Both these elements should be well testable in B decays. Indeed, the size of CP-violating box diagrams is controlled by the product $|V_{td}V_{tb}^*|^2 \propto \lambda^6$, much larger than the corresponding parameter in the kaon sector. Since box diagrams determine B^0-$\bar{B}^0$ mixing, CP-violating effects in the $B\bar{B}$ system should be much more pronounced than in $K\bar{K}$.

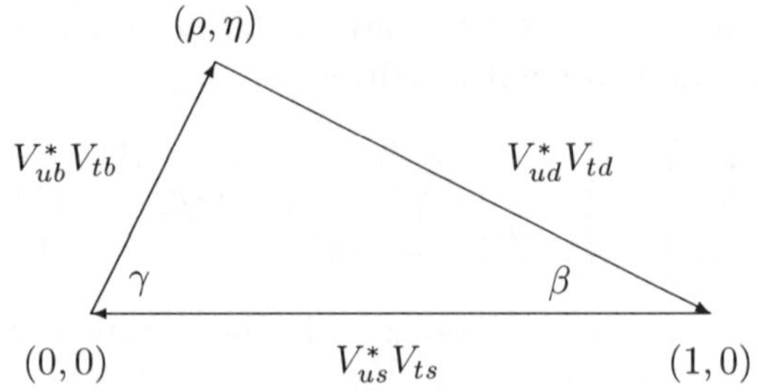

Fig.2 Unitarity triangle

The phase of CKM element V_{td}^* (angle β of the unitarity triangle) is accessible through measurement of the time-dependent CP-violating asymmetry in $B(\overline{B}) \to J/\psi K_S$, which is proportional to $\sin 2\beta$. The angle β enters here from the $B^0 - \overline{B}^0$ mixing (Fig.1). The $B(\overline{B}) \to J/\psi K_S$ decay itself introduces no other quark-level phases.

Absolute magnitude of V_{ub} controls the size of semileptonic B decays and may be extracted from data. Extraction of the phase of element V_{ub}^* (angle γ) is hampered in various ways. Analysis of $B \to \pi^+\pi^-$ was suggested as a way to determine $\beta + \gamma$. The relevant tree diagram T (Fig.3a) involves a $b \to u$ transition, thus introducing an additional phase. Together with angle β entering into consideration from the $B^0 - \overline{B}^0$ mixing, the relevant asymmetry becomes proportional to $\sin 2(\beta + \gamma)$. Unfortunately, "penguin" diagrams (Fig. 3b) complicate the situation.

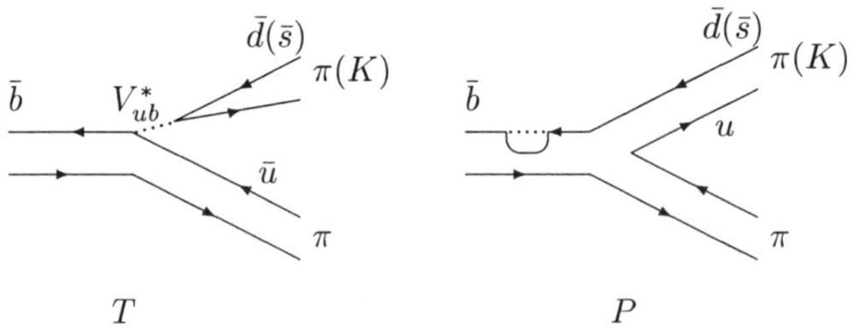

Fig. 3 a) Tree (T) and b) "penguin" (P) diagrams

4 Analyses through Interference Effects

Extraction of phases like β, γ from $B^0\bar{B}^0$ decays into CP eigenstates ($f = f_{CP}$, e.g. $\pi^+\pi^-$) is possible because relevant calculations involve superpositions of terms with opposite phases. In general, appearance of CP-violating effects requires superposition of at least two amplitudes. In decays to CP-noneigenstates $f \neq f_{CP}$ one needs interference of two amplitudes

$$A(B \to f) = |T|e^{i\delta_T}e^{i\phi_T} + |P|e^{i\delta_P}e^{i\phi_P} \tag{3}$$

with different weak phases $\phi_{T,P}$ resulting from CKM matrix and strong phases $\delta_{T,P}$ due to strong interactions. Under CP,

$$A(B \to f) \to A_{CP} \equiv A(\bar{B} \to f_{CP}) \tag{4}$$

with (strong) weak phases (not) changing sign. Consequently, the CP-asymmetry $|A|^2 - |A_{CP}|^2$ is proportional to

$$\sin(\delta_T - \delta_P)\sin(\phi_T - \phi_P) \tag{5}$$

and vanishes if $\delta_T = \delta_P$. In the above formulas T may denote the tree amplitude (Fig.3a), while P may denote the "penguin" amplitude (Fig.3b). The two amplitudes have different weak and strong phases. It is clear that extraction of the weak phases from experiment requires knowledge of the strong ones. The latter, however, are theoretically very difficult to come by. Thus, the methods proposed to extract weak phases treat the strong ones as parameters, also to be determined from experiment.

5 Understanding Final State Interactions

The following additional problem can be noticed: the angles to be extracted are defined at quark level, but they are to be determined from decays in which strong final state interactions (FSI) are possible. Since the latter may modify the formulas upon which the extraction of the parameters is based, FSI must be analysed carefully. This leads to a serious problem as it is very hard to achieve a reliable description of FSI or rescattering effects [3]. Usually, this is done with the help of models, under the assumption that some FSI are negligible. Understanding which of the rescattering effects are indeed negligible and which must be included in the analysis is a necessary prerequisite for the extraction of weak angles. In order to achieve this, theoretical studies of various forms of FSI have to be performed. One of the approaches considered is based on ideas developed in strong interaction physics in the sixties. Namely, consider the decays $B \to PP$ where P denote pseudoscalar mesons (π, K, etc.). The pair of PP mesons receding from each other may rescatter elastically or inelastically. Such processes may be described quite reliably with the help of

Reggeon exchanges which fit experimental data on $PP \to PP$ scattering very well. (A "Reggeon" is a quasi-particle exchanged between scattering mesons, equivalent to a superposition of many partial-wave amplitudes into which the full $PP \to PP$ amplitude may be resolved.) Application of these ideas to the determination of strong rescattering phases in $B \to \pi\pi, K\bar{K}$ decays was carried out in various papers, e.g.[4].

6 A Simple Example

By modifying the amplitude of B decays to CP-eigenstate $\pi^+\pi^-$, the "penguin" of Fig. 3b may affect determination of weak angles. As its size relative to the tree amplitude T is not well predicted one needs additional experimental data to fix it. An attractive possibility is to enlarge the analysis by including into it the decays $B \to K^0\bar{K}^0$. This requires assuming SU(3) symmetry or, if any, the pattern of its breaking.

Reliable extraction of weak phases requires having rescattering effects under control. The model of refs.[4] was based on a quasi-elastic approximation in which only rescatterings of the type: $K\pi \to K\pi$, $\pi\pi \to \pi\pi$, etc. are considered. All other possible final-state interactions, in particular coupled-channel effects of the type $\pi\pi \to K\bar{K}$ or $K\pi \to K\eta$, were ignored. Since inclusion of the latter processes has been shown to be very important in maintaining SU(3) symmetry [5], one has to analyse their influence on the results of [4]. The analysis performed recently [6,7] was restricted to the coupled-channel effects in the $\pi\pi$, $K\bar{K}$, $\pi^0\eta_8$ and $\eta_8\eta_8$ channels. The latter two have to be included to maintain SU(3) symmetry of the whole scheme.

The rescattering-free amplitudes for B^0 decays may be parametrized as

$$\langle(\pi\pi)_{I=2}|w|B^0\rangle = -\frac{1}{\sqrt{6}}(T+C),$$
$$\langle(K\bar{K})_{I=1}|w|B^0\rangle = -\frac{1}{2}P, \qquad \text{etc.} \tag{6}$$

where the expressions for the decays to the remaining four channels ($(\pi^0\eta_8)_{I=1}$, $(\pi\pi)_{I=0}$, $(K\bar{K})_{I=0}$, $(\eta_8\eta_8)_{I=0}$) are suppressed and only the dominant quark-level amplitudes are exhibited with $|P|, |C| \approx 0.2|T|$ (C corresponds to another possible diagram, not shown in Fig.3).

Due to the coupled-channel effects, the FSI-corrected amplitudes $\langle(\pi\pi)_I|W|B^0\rangle$, $\langle(K\bar{K})_I|W|B^0\rangle$, etc. become linear combinations of quark-level amplitudes in Eq.(6), i.e.:

$$\langle(P_1P_2)_I|W|B^0\rangle = \sum_{(P_1'P_2')_I} \langle(P_1P_2)_I|D_{FSI}|(P_1'P_2')_I\rangle\langle(P_1'P_2')_I|w|B^0\rangle \tag{7}$$

with FSI decribed by

$$\langle (P_1P_2)_I|D_{FSI}|(P_1'P_2')_I\rangle = \sum_{\mathbf{R}} \langle (P_1P_2)_I|\mathbf{R},I\rangle\langle \mathbf{R},I|S_{FSI}|\mathbf{R},I\rangle\langle \mathbf{R},I|(P_1'P_2')_I\rangle \tag{8}$$

In the above equation $\langle (P_1P_2)_I|\mathbf{R},I\rangle$ are the amplitudes for finding the $|(P_1P_2)_I\rangle$ state in the $SU(3)$ eigenstate $|\mathbf{R},I\rangle$, while $\langle \mathbf{R},I|S_{FSI}|\mathbf{R},I\rangle = \rho(\mathbf{R})\exp(i\delta(\mathbf{R}))$ describes SU(3)-symmetric final state interactions in the $|\mathbf{R},I\rangle$ state ($\mathbf{R}$ denotes SU(3) multiplets 1, 8, 27, and I is isospin). It is assumed that the possible renormalization of the *magnitude* of short-distance amplitudes is negligible: $\rho(\mathbf{R}) \approx 1$.

In order to estimate how the quark-level amplitudes T, C, P add up, one has to make assumptions about their relative size. The dominant contribution is provided by the tree amplitude T. We use $T = 3rC$ with $r \approx -3.0$ as appropriate for the QCD-corrected quark model [8]. Assuming furthermore $|P/T|$ in the range of $0.04 - 0.20$ (as estimated in [9]), disregarding the relative phase of P and T, all amplitudes are given in terms of a single parameter T. From Eq.(7) one then obtains FSI phases given in the right-hand side of Table 1.

Table 1 shows that phase difference between the $(\pi\pi)_{I=0,2}$ channels does not depend strongly on the inclusion of coupled-channel effects. This is also the case for the $(K\bar{K})_1$ phase. However, the $(K\bar{K})_0$ phase changes dramatically. Only for $T \ll P$ is the phase change induced by coupled-channel effects small. However, if P is small, the relevant phase change is large. The origin of this effect is as follows: due to coupled-channel effects, the $(\pi\pi)_0$ state created in a quark-level process may be converted into a $(K\bar{K})_0$ final state. Thus, the final $(K\bar{K})_0$ state receives contribution from both the direct $B^0 \to (K\bar{K})_0$ decay (characterized by small amplitude P) with $(K\bar{K})_0$ elastically scattered into the final $(K\bar{K})_0$ state, as well as from the $B^0 \to (\pi\pi)_0$ decay (characterized by large amplitude T) which contributes through coupled-channel effects into the final $(K\bar{K})_0$ state.

TABLE 1. Comparison of calculated phase shifts for B decays

Phase	No coupled channels		Coupled channels, ref.[6]		
	ref.[4]	ref.[6]	$P \gg T, C$	$P = 0.2T$	$P = 0.04T$
$\delta^2_\pi - \delta^0_\pi$	$+11^o \pm 2^o$	18°	14°	18°	19°
δ^1_K		95°	85°	85°	85°
δ^0_K		103°	110°	137°	168°
$\delta^1_K - \delta^0_K$	$-7^o \pm 1^o$	-8°	-25°	-52°	-83°

The net result is interference of the small P amplitude with some admixture coming from the large T amplitude. If the relative size of the two contributions is comparable, one can obtain a wide range of phases for their sum. This lies at the origin of large phases in the $(K\bar{K})_0$ channel of B^0 decays (Table 1).

7 Conclusions: SU(3) breaking

Comparison of predictions of refs.[4] and [6] shows that coupled-channel rescattering effects may have decisive effect on the size of final state phases. As the latter have to be known while extracting weak phases, coupled-channel effects must be included in any SU(3) symmetric analysis.

At present it is not known whether and how SU(3) is broken in B decays. If it is broken, our conclusions would have to be modified. In order to know what these modifications might be, one has to study SU(3) breaking in more detail. This question constitutes the subject of an ongoing work [7].

References

1. J. H. Christenson, J. W. Cronin, V. L. Fitch, and R. Turlay, *Phys. Rev. Lett.* **13**, 138 (1964).
2. M. Kobayashi and T. Maskawa, *Prog. Theor. Phys.* **49**, 652 (1973).
3. L. Wolfenstein, *Phys. Rev.* **D52**, 537 (1995); J. F. Donoghue et al., Phys. Rev. Lett. 77, 2178 (1996).
4. J.-M. Gérard, J. Pestieau, J. Weyers, *Phys. Lett.* **B436**, 363 (1998).
5. P. Żenczykowski, *Acta Phys. Pol.* **B28**, 1605 (1997).
6. P. Żenczykowski, *Phys. Lett.* **B460**, 390 (1999).
7. P. Łach, in preparation.
8. L.-L. Chau, H.-Y. Cheng, *Phys. Rev.* **D36**, 137 (1987).
9. M. Gronau, *Phys. Lett.* **B300**, 163 (1993).

SPINOR ALGEBRAS AND EXTENDED SUPERCONFORMAL ALGEBRAS

M. A. LLEDÓ

Dipartimento di Fisica, Politecnico di Torino, Corso Duca degli Abruzzi 24, I-10129 Torino, Italy, and INFN, Sezione di Torino, Italy.
E-mail: lledo@athena.polito.it

V. S. VARADARAJAN

Department of Mathematics, University of California, Los Angeles, Los Angeles, CA 90095-1555, USA.
E-mail: vsv@math.ucla.edu

We consider supersymmetry algebras in arbitrary spacetime dimension and signature. Minimal and maximal superalgebras are given for single and extended supersymmetry. It is seen that the supersymmetric extensions are uniquely determined by the properties of the spinor representation, which depend on the dimension D mod 8 and the signature $|\rho|$ mod 8 of spacetime.

1 Spacetime Symmetry Algebras and the Super Lie Algebras Containing Them

The idea that one can have a new type of algebraic structure generating symmetries between bosonic and fermionic states goes back to the 1970's when several physicists discovered what are nowadays known as *super Lie algebras.* The super Lie algebra contains the usual spacetime symmetry algebra (Poincaré or conformal algebra) and its representations combine several representations of the spacetime algebra into a single module irreducible under the super algebra, thus giving rise to the concept of a supermultiplet. The central problem is therefore the following: given a spacetime symmetry algebra, to obtain the super Lie algebras that contain the given Lie algebra in its even part.

By spacetime we mean a real vector space with a nondegenerate quadratic form of type (s,t). It is denoted by $V(s,t)$. The dimension of spacetime is denoted by $D=s+t$, and the signature of the quadratic form by $\rho=s-t$. The group of coordinate transformations in V that leaves the metric invariant up to scale transformations is the simple group $\mathrm{SO}(s+1,t+1)$ (for $D=s+t\geq 3$), being the Poincaré group $\mathrm{ISO}(s,t)$ a subgroup of it. The Poincaré group $\mathrm{ISO}(s,t)$ can also be obtained as a contraction of the conformal group in one dimension less $\mathrm{SO}(s,t+1)$.

As an example, we can take a spacetime of type $(s,t) = (10,1)$. We have

$$\mathfrak{so}(10,2) \xrightarrow{\text{contraction}} \mathfrak{iso}(10{,}1) \xrightarrow{\text{subalgebra}} \mathfrak{so}(11,2).$$

The supersymmetric extension of the conformal algebra in $D = 10$ is $\mathfrak{osp}(1|32)$, which gives as a contraction the M-theory superalgebra, that is, the super Poincaré algebra with five and two-brane charges. This can be seen as a subalgebra of $\mathfrak{osp}(1|64)$, the superconformal algebra in $D = 11$.

For a fixed dimension D, theories defined on spacetimes with the same signature $|\rho|$ mod 8 may be related by some duality since the superalgebras are the same. As an example, the theories introduced by Hull [1], M, M* and M', are formulated in 11 dimensional spacetimes with signatures $(s,t) = (10,1), (9,2), (6,5)$ respectively. In all these theories the signature of spacetime is $\rho = s - t = \pm 1$ mod 8.

In this paper we will see how the supersymmetric extensions of spacetime superalgebras are determined by D and $|\rho|$ mod 8. All these results are contained in [2,3].

2 Spin Groups and Spin Modules

For each spacetime $V(s,t)$ there are two spin groups, $\mathrm{Spin}(s,t)$ and $\mathrm{Spin}_{\mathbb{C}}(s,t)$, the real and complex spin groups, the former being a real form of the latter. As it is well known, when D is odd there is only one irreducible spin module of the complex group. It will be denoted by S. When D is even there are two different irreducible spin modules denoted by $S^{\pm}$ (chiral spin modules). For constructing superspacetimes and superfields we need real forms of the complex spin modules. These do not always exist and sometimes we have to combine several copies of them to get a real irreducible module for $\mathrm{Spin}(s,t)$.

The reality issues depend only on $|\rho|$ mod 8. The type of a real irreducible module is the isomorphism class of the commutant of the module. Since this commutant is a division algebra, it has to be one of $\mathbb{R}, \mathbb{C}, \mathbb{H}$. For $\rho \equiv 0,1,7$ the irreducible real spin modules are actually real forms of the complex irreducible spin modules. For $\rho \equiv 3,4,5$ they are quaternionic and on complexification become $2S^{\pm}$ or $2S$ (in the even and odd D cases respectively). For $\rho \equiv 2,6$ they are of complex type and their complexifications are $S^{+} \oplus S^{-}$.

The spin module may admit an invariant form. If this form is symmetric (it is unique up to multiplication by a scalar), then the spin module is said to be orthogonal. If it is antisymmetric the spin module is symplectic and if it has no invariant form it is called linear. The complex spin groups are embedded into the complex orthogonal, symplectic and linear groups respectively. The existence and symmetry properties of this form depend only on D mod 8.

These mod 8 results have been known for a long time both to physicists and mathematicians, but there is now a unified approach based on the super Brauer group [4,5].

3 Non extended supersymmetry

3.1 Super Poincaré Algebras

A non extended ($N = 1$) super Poincaré algebra is a super Lie algebra whose even part is the Poincaré algebra, the odd part consists of one real irreducible spin representation and the anticommutator of two spinors (elements of the spin module) is proportional to the momentum. If we denote a spinor in the form Q_α we have that the anticommutator $\{Q_\alpha, Q_\beta\}$ is symmetric in the indices α, β. The biggest Lie algebra that one can put in the right hand side of the anticommutator is the Lie algebra of symmetric matrices, $\mathfrak{sp}(2n)$ ($2n$ is the real dimension of the spin module). In fact, there is a superalgebra which has $\mathfrak{sp}(2n)$ as bosonic part, it is the orthosymplectic algebra $\mathfrak{osp}(1|2n)$.

In order to see if the required extension exists, we have to determine if there is an abelian subalgebra of $\mathfrak{sp}(2n)$ with the commutation rules of a vector with the generators of the orthogonal algebra. In other words, we have to determine if there is a morphism of $\mathrm{Spin}(s,t)$ modules

$$S \otimes S \longrightarrow V.$$

(It is easy to see that this condition is also sufficient).

In fact one can study the question for existence of maps $S \otimes S \longrightarrow \Lambda^k V$ for all k. As we will see in the next subsection, the case $k = 2$ is important for constructing super conformal algebras, namely super Lie algebras extending $\mathfrak{so}(s+1, t+1)$.

The maps $S \otimes S \longrightarrow \Lambda^k V$ are unique (projectively) if they exist and they are either symmetric (+) or antisymmetric (-). The full details can be worked out without difficulty using the formalism of Deligne [5] and are listed in Table 1.

We denote $\rho = \rho_0 + n8$, $D = D_0 + m8$, with $0 \leq \rho_0, D_0 < 7$, m and n integers.

Real Case. There is a conjugation σ on S commuting with the action of the spin group, so there is a real form of S which is a real $\mathrm{Spin}(s,t)$-module. The anticommutator of two spinors is

$$\{Q_\alpha, Q_\beta\} = \sum_k \gamma_{(\alpha\beta)}^{[\mu_1 \cdots \mu_k]} Z_{[\mu_1 \cdots \mu_k]},$$

D	k even		k odd	
	morphism	symmetry	morphism	symmetry
0	$S^{\pm}\otimes S^{\pm}\to\Lambda^k$	$(-1)^{k(k-1)/2}$	$S^{\pm}\otimes S^{\mp}\to\Lambda^k$	
1	$S\otimes S\to\Lambda^k$	$(-1)^{k(k-1)/2}$	$S\otimes S\to\Lambda^k$	$(-1)^{k(k-1)/2}$
2	$S^{\pm}\otimes S^{\mp}\to\Lambda^k$		$S^{\pm}\otimes S^{\pm}\to\Lambda^k$	$(-1)^{k(k-1)/2}$
3	$S\otimes S\to\Lambda^k$	$-(-1)^{k(k-1)/2}$	$S\otimes S\to\Lambda^k$	$(-1)^{k(k-1)/2}$
4	$S^{\pm}\otimes S^{\pm}\to\Lambda^k$	$-(-1)^{k(k-1)/2}$	$S^{\pm}\otimes S^{\mp}\to\Lambda^k$	
5	$S\otimes S\to\Lambda^k$	$-(-1)^{k(k-1)/2}$	$S\otimes S\to\Lambda^k$	$-(-1)^{k(k-1)/2}$
6	$S^{\pm}\otimes S^{\mp}\to\Lambda^k$		$S^{\pm}\otimes S^{\pm}\to\Lambda^k$	$-(-1)^{k(k-1)/2}$
7	$S\otimes S\to\Lambda^k$	$(-1)^{k(k-1)/2}$	$S\otimes S\to\Lambda^k$	$-(-1)^{k(k-1)/2}$

Table 1. Properties of morphisms.

so the morphism $\gamma^{\mu}_{\alpha\beta}$ must be symmetric in α,β. This happens for

$$\rho_0=0, \qquad D_0=2$$
$$\rho_0=1,7, \qquad D_0=1,3.$$

If we consider also non chiral superalgebras (with $S^{\pm}$ both present), the we find that for $\rho_0=0$ and $D_0=0,4$ there is a morphism $S^{+}\otimes S^{-}\to V$, and the extension is possible

Quaternionic case. There is a pseudoconjugation σ on S commuting with $\mathrm{Spin}(s,t)$. We need two copies $S\oplus S=S\otimes\mathbb{C}^2$ to be able to construct a conjugation $\tilde{\sigma}=\sigma\otimes\sigma_0$ and then to impose a reality condition. The factor $\mathbb{C}^2$ is an internal index space. The biggest group that commutes with σ_0 is $\mathrm{SU}(2)\simeq\mathrm{USp}(2)\simeq\mathrm{SU}^*(2)$.

In the anticommutator the internal index appears,

$$\{Q^i_\alpha,Q^j_\beta\}=\sum_k\gamma^{[\mu_1\cdots\mu_k]}_{[\alpha\beta]}Z^0_{[\mu_1\cdots\mu_k]}\Omega^{ij}+\sum_k\gamma^{[\mu_1\cdots\mu_k]}_{(\alpha\beta)}Z^I_{[\mu_1\cdots\mu_k]}\sigma_I^{ij},$$

where Ω is the symplectic matrix and σ_I are the symmetric Pauli matrices, $I=1,2,3$.

If we impose invariance under SU(2) (R-symmetry group) only the first term (SU(2) singlet) may appear and the morphism $\gamma^{\mu}_{\alpha\beta}$ must be antisymmetric. This happens for

$$\rho_0=3,5, \qquad D_0=5,7$$
$$\rho_0=4 \qquad D_0=6$$
$$\rho_0=4 \qquad D_0=0,4 \quad \text{(non chiral)}.$$

If we want to take the generators in the second term, then the R-symmetry is broken. For example, taking δ^{ij} breaks the symmetry to the subgroup SO*(2). Then we have superalgebras for

$$\begin{aligned} \rho_0 &= 3,5, \qquad & D_0 &= 1,3 \\ \rho_0 &= 4 & D_0 &= 2 \\ \rho_0 &= 4 & D_0 &= 0,4 \quad \text{(non chiral)}. \end{aligned}$$

Both types of R-symmetry will find an interpretation later. We note that non chiral algebras appear with both, SU(2) symmetry and SO*(2) for the same space times. In fact, one can easily see that SO*(2)⊂SU(2) and that they are isomorphic super algebras.

Complex case. There is no conjugation nor pseudoconjugation. We take the direct sum $S^+ \oplus S^-$, and there is a conjugation with the property $\sigma(S^\pm) = S^\mp$. Again we have an R-symmetry group, U(1), with $S^\pm$ having charges ± 1. If we want invariance under U(1) the orthogonal generators must appear in the anticommutator $\{Q^+, Q^-\}$, so the morphism should be of type

$$S^\pm \otimes S^\mp \longrightarrow V.$$

This happens for $\rho_0 = 2,6,\ D_0 = 0,4$.

Otherwise, the morphisms can be of type $S^\pm \otimes S^\pm \to \Lambda^2$, (symmetric) which happens for $\rho_0 = 2,6,\ D_0 = 2$.

We note that for physical spacetimes there is always the possibility of supersymmetric extension.

4 Super Conformal Algebras

A super conformal algebra is a real, simple superalgebra whose even part contains the conformal group. The odd part of the super conformal algebra is a direct sum of N real spinor modules.

A simple superalgebra $\mathfrak{a} = \mathfrak{a}_0 \oplus \mathfrak{a}_1$ has the propriety

$$\{\mathfrak{a}_1, \mathfrak{a}_1\} = \mathfrak{a}_0.$$

For a fixed N, depending on how big is $\mathfrak{a}_0$ we can have from minimal to maximal superconformal algebras.

As in the case of Poincaré, a necessary condition is that

$$\mathfrak{o}(s,t) \subset \mathfrak{sp}(2n). \tag{1}$$

The adjoint of $\mathfrak{o}(s,t)$ is the two fold antisymmetric representation, so one should look for the symmetries of morphisms

$$S \otimes S \longrightarrow \Lambda^2.$$

$\mathfrak{osp}(1|2n)$ is a simple super algebra; it is in fact the maximal super conformal algebra in the cases where condition (1) is satisfied. This happens for

	$\rho_0 = 0$	$D_0 = 4$	
Real case:	$\rho_0 = 1,7$	$D_0 = 3,5$	
	$\rho_0 = 0$	$D_0 = 2,6$	(non chiral)
Quaternionic case:	$\rho_0 = 4$	$D_0 = 8$	
	$\rho_0 = 3,5,$	$D_0 = 1,7$	
	$\rho_0 = 4$	$D_0 = 2,6$	(non chiral)
	$\rho_0 = 4,$	$D_0 = 4$	$(\mathrm{SO}^*(2))$
Complex case:	$\rho_0 = 3,5,$	$D_0 = 3,5$	$(\mathrm{SO}^*(2))$
	$\rho_0 = 2,6$	$D_0 = 2,6$	
	$\rho_0 = 2,6$	$D_0 = 4$	(no $\mathrm{U}(1)$).

As we have mentioned, the complex group is embedded into a larger group, depending whether the spinor representation is orthogonal, symplectic or linear. The spin representation goes into the fundamental representation of the corresponding group (orthogonal, symplectic or linear). The morphisms

$$S \otimes S \longrightarrow \Lambda^0,$$

being the invariant bilinear forms, their existence is governed by D (mod 8).

The interesting question is to see if there is a real form of these groups containing the real $\mathrm{Spin}(s,t)$ group. We have to play with the reality properties of the spinors, which depend on ρ, so we have all the possibilities for D and ρ. The result is that the real form is uniquely determined. It is listed in Table 2.

To make an example, let $\rho_0 = 1,7$ and $D_0 = 1,7$. $\dim(S) = 2^{D-1/2}$ and one can show that

$$\mathrm{Spin}(V) \longrightarrow \mathrm{SO}(2^{(D-3)/2}, 2^{(D-3)/2}) \qquad (\min(s,t) \geq 1).$$

If $D_0 = 3,5$, the spin group goes inside a symplectic group and in fact we have

$$\mathrm{Spin}(V) \longrightarrow \mathrm{Sp}(2^{(D-1)/2}, \mathbb{R}).$$

Some low dimensional examples are

$$\begin{array}{rr} \mathrm{Spin}(4,3) \subset \mathrm{SO}(4,4), & \mathrm{Spin}(8,1) \subset \mathrm{SO}(8,8) \\ \mathrm{Spin}(2,1) \subset \mathrm{SL}(2,\mathbb{R}), & \mathrm{Spin}(3,2) \subset \mathrm{Sp}(4,\mathbb{R}) \\ \mathrm{Spin}(3) \subset \mathrm{SU}(2), & \mathrm{Spin}(5) \subset \mathrm{USP}(4), \\ \mathrm{Spin}(4,1) \subset \mathrm{USp}(2,2) & \mathrm{Spin}(4,3) \subset \mathrm{SO}(4,4), \\ & \mathrm{Spin}(8,1) \subset \mathrm{SO}(8,8). \end{array}$$

The Lie algebra of this group is called the Spin(s,t)-algebra. It happens that for each combination of ρ and D from the ones listed above there is a simple superalgebra whose even part is

$$\mathrm{Spin}(s,t) - \mathrm{algebra} \otimes R - \mathrm{symmetry}.$$

This superalgebra is minimal inside the classical series, and the list of superalgebras is given in Table 2. There is an exceptional superalgebra that is smaller for dimension 5, with bosonic part $\mathfrak{so}(5,2) \oplus \mathfrak{su}(2)$, it is the exceptional super algebra $\mathfrak{f}(4)$. For spacetimes of dimensions 3, 4 and 6 we also have the conformal algebra as a factor in the bosonic part of the superalgebra. We have

$$\begin{aligned} \mathfrak{sp}(4,\mathbb{R}) &\approx \mathfrak{so}(3,2) \\ \mathfrak{su}(2,2) &\approx \mathfrak{so}(4,2) \\ \mathfrak{so}^*(4,\mathbb{R}) &\approx \mathfrak{so}(6,2). \end{aligned}$$

5 Extended supersymmetry

We want to construct now a super Lie algebra whose odd part consists of N copies of the spin module. There is again a maximal super Lie algebra, osp$(1/N2n)$, but the presence of the internal index space $i = 1, \ldots N$ allows extra possibilities in the combinations of ρ and D. For example, in the real case we can have

$$\{Q^i_\alpha, Q^j_\beta\} = \delta^{ij} \sum_k \gamma^{[\mu_1 \cdots \mu_k]}_{(\alpha\beta)} Z_{[\mu_1 \cdots \mu_k]},$$

and an orthogonal R-symmetry group, or

$$\{Q^i_\alpha, Q^j_\beta\} = \Omega^{ij} \sum_k \gamma^{[\mu_1 \cdots \mu_k]}_{(\alpha\beta)} Z_{[\mu_1 \cdots \mu_k]},$$

and symplectic R-symmetry group.

D_0	ρ_0	Spin(V) algebra	Spin(V) superalgebra
1,7	1,7	$\mathfrak{so}(2^{(D-3)/2}, 2^{(D-3)/2})$	
1,7	3,5	$\mathfrak{so}^*(2^{(D-1)/2})$	$\mathfrak{osp}(2^{(D-1)/2})^*\|2)$
3,5	1,7	$\mathfrak{sp}(2^{(D-1)/2}, \mathbb{R})$	$\mathfrak{osp}(1\|2^{(D-1)/2}, \mathbb{R})$
3,5	3,5	$\mathfrak{usp}(2^{(D-3)/2}, 2^{(D-3)/2})$	$\mathfrak{osp}(2^*\|2^{(D-3)/2}, 2^{(D-3)/2})$
0	0	$\mathfrak{so}(2^{(D-4)/2}, 2^{(D-4)/2})$	
0	2,6	$\mathfrak{so}(2^{(D-2)/2}, \mathbb{C})^{\mathbb{R}}$	
0	4	$\mathfrak{so}^*(2^{(D-2)/2})$	$\mathfrak{osp}(2^{(D-2)/2})^*\|2)$
2,6	0	$\mathfrak{sl}(2^{(D-2)/2}, \mathbb{R})$	$\mathfrak{sl}(2^{(D-2)/2}\|1, \mathbb{R})$
2,6	2,6	$\mathfrak{su}(2^{(D-4)/2}, 2^{(D-4)/2})$	$\mathfrak{su}(2^{(D-4)/2}, 2^{(D-4)/2}\|1)$
2,6	4	$\mathfrak{su}^*(2^{(D-2)/2}))$	$\mathfrak{su}(2^{(D-2)/2})^*\|2)$
4	0	$\mathfrak{sp}(2^{(D-2)/2}, \mathbb{R})$	$\mathfrak{osp}(1\|2^{(D-2)/2}, \mathbb{R})$
4	2,6	$\mathfrak{sp}(2^{(D-2)/2}, \mathbb{C})^{\mathbb{R}}$	$\mathfrak{osp}(1\|2^{(D-2)/2}, \mathbb{C})$
4	4	$\mathfrak{usp}(2^{(D-4)/2}, 2^{(D-4)/2})$	$\mathfrak{osp}(2^*\|2^{(D-4)/2}, 2^{(D-4)/2})$

Table 2. Minimal Spin(V) superalgebras.

	D_0	ρ_0	R-symmetry	Spin(s,t) superalgebra
∘	1,7	1,7	$\mathfrak{sp}(2N, \mathbb{R})$	$\mathfrak{osp}(2^{\frac{D-3}{2}}, 2^{\frac{D-3}{2}}\|2N, \mathbb{R})$
⋆	1,7	3,5	$\mathfrak{usp}(2N-2q, 2q)$	$\mathfrak{osp}(2^{\frac{D-1}{2}}{}^*\|2N-2q, 2q)$
⋆	3,5	1,7	$\mathfrak{so}(N-q, q)$	$\mathfrak{osp}(N-q, q\|2^{\frac{D-1}{2}})$
	3,5	3,5	$\mathfrak{so}^*(2N)$	$\mathfrak{osp}(2N^*\|2^{\frac{D-3}{2}}, 2^{\frac{D-3}{2}})$
∘	0	0	$\mathfrak{sp}(2N, \mathbb{R})$	$\mathfrak{osp}(2^{\frac{D-4}{2}}, 2^{\frac{D-4}{2}}\|2N)$
∘	0	2,6	$\mathfrak{sp}(2N, \mathbb{C})_{\mathbb{R}}$	$\mathfrak{osp}(2^{\frac{D-2}{2}}\|2N, \mathbb{C})_{\mathbb{R}}$
⋆	0	4	$\mathfrak{usp}(2N-2q, 2q)$	$\mathfrak{osp}(2^{\frac{D-2}{2}}{}^*\|2N-2q, 2q)$
	2,6	0	$\mathfrak{sl}(N, \mathbb{R})$	$\mathfrak{sl}(2^{\frac{D-2}{2}}\|N, \mathbb{R})$
⋆	2,6	2,6	$\mathfrak{su}(N-q, q)$	$\mathfrak{su}(2^{\frac{D-4}{2}}, 2^{\frac{D-4}{2}}\|N-q, q)$
	2,6	4	$\mathfrak{su}^*(2N, \mathbb{R})$	$\mathfrak{su}(2^{\frac{D-2}{2}}\|2N)^*$
⋆	4	0	$\mathfrak{so}(N-q, q)$	$\mathfrak{osp}(N-q, q\|2^{\frac{D-2}{2}})$
	4	2,6	$\mathfrak{so}(N, \mathbb{C})_{\mathbb{R}}$	$\mathfrak{osp}(N\|2^{\frac{D-2}{2}}, \mathbb{C})_{\mathbb{R}}$
	4	4	$\mathfrak{so}^*(2N)$	$\mathfrak{osp}(2N^*\|2^{\frac{D-4}{2}}, 2^{\frac{D-4}{2}})$

Table 3. Spin(s,t) superalgebras.

These new possibilities fill in the gaps in Table 2. The minimal super Lie algebras have an even part that is again of the form

$$\mathrm{Spin}(s,t) - \text{algebra} \otimes R - \text{symmetry}.$$

A conjugation must exist in the space $S \otimes W$ (R is a module for the R-symmetry group). If there is a conjugation in S, then there is a conjugation in $W$and a pseudoconjugation in S implies a pseudoconjugation in W. If the spacetime group is unitary, then the R-symmetry group is also unitary and if the spacetime group is complex, then the R-symmetry is also complex.

The list is given in Table 3. We mark with a symbol "$\star$" the cases that allow the possibility of a compact R-symmetry group. They correspond to $\mathfrak{so}(s,2)$, that is, the physical conformal groups. We mark with the symbol "$\circ$" the cases that do not arise in the non extended case.

Non chiral superalgebras. In Tables 2,3 we have written a chiral algebra whenever it is possible and a non chiral one for the rest. Nevertheless, one can ask if there is a non chiral algebra in the cases where the chiral one already exists. As an example, consider a spacetime of type $(5,1)$, that is, $D = 6$ and physical signature. The conformal group is SO(6,2), the orthogonal group of a space time of dimension 8 and signature 4. There exists a chiral super algebra, namely $\mathfrak{osp}(8*|2)$. A non chiral superalgebra can be constructed if we go to one dimension more and then we make a dimensional reduction. We can consider the orthogonal groups SO(7,2) or SO(6,3). In both cases, the associated super conformal algebra is $\mathfrak{osp}(16*|2)$. We have the decomposition

$$\begin{array}{ccccc} \mathfrak{so}^*(16) & \xleftarrow{\supset} & \mathfrak{so}(7,2) & \xleftarrow{\supset} & \mathfrak{so}(6,2) \\ \mathbf{16} & \longrightarrow & \mathbf{16} & \longrightarrow & \mathbf{8}_L + \mathbf{8}_R. \end{array}$$

So the super conformal algebra in dimension 7 can be seen as a non chiral superconformal algebra in dimension 6.

Acknowledgments

We want to thank our collaborators, S. Ferrara and R. D'Auria with whom this work was done.

References

1. C. Hull, *JHEP* **9811**, 017 (1998).
2. R. D'Auria, S. Ferrara, M. A. Lledó and V. S. Varadarajan, to appear in the *Jour. Geom. Phys.*; hep-th/0010124.

3. R. D'Auria, S. Ferrara, M. A. Lledó. *Lett. Math. Phys.* **57**, 123 (2001).
4. C. T. C. Wall. *J. Reine Angew. Math.* **213**, 187 (1963/64).
5. P. Deligne, Notes on Spinors, in *Quantum Fields and Strings: A Course for Mathematicians*, American Mathematical Society (1999).

WAVELETS AND TOMOGRAMS IN INFORMATION PROCESSING

MARGARITA A. MAN'KO
P.N. Lebedev Physical Institute, Leninskii Prospect 53, 119991 Moscow, Russia

Some traditional and new types of transforms of analytic signal used in information processing including recently developed method of noncommutative tomography are discussed. The analogy of analytic signal to wave functions and vectors in Hilbert space of states in quantum mechanics is used to present different integral transforms as different obvious relations among the vectors and basis vectors in the linear space. The approach developed permits to construct new relations between wavelet transforms, quasidistibutions and tomograms of analytic signal.

1 Introduction

The aim of this contribution is to review the approach to the problems of signal analysis suggested recently[1,2] where the tomograms and other transforms of an analytic signal were discussed in view of the analogy of analytic signal and wave function of a quantum state. The obvious geometrical properties of the vectors in the Hilbert space of quantum states provide simple mathematical tools to obtain nontrivial relations between different characteristics of the analytic signal contained in wavelets, quasidistributions and tomograms associated with the analytic signal.

The idea of the geometrical approach[2] employes two ingredients: (i) the first one is the signal function which is associated with a vector in the Hilbert space; (ii) the second one is the basis vectors in the Hilbert space which are obtained from an initial vector by means of the action on it by a family of appropriate operators. The projections of the signal vector on the basis vectors completely determine the signal. A choice of the initial vector and the appropriate choice of the family of operators, providing the basis in Hilbert space, determine the signal-vector projections under discussion. As it was shown,[2] the projections can coincide with wavelets, Wigner–Ville-like quasidistributions, tomograms of different types depending on the choice of the initial vector and the family of operators, namely:

(1) In the case of wavelets, the initial vector (fiducial vector) is taken independently on the signal vector. The corresponding projection gives the linear transform of the analytic signal.

(2) In the case of quasidistributions and tomograms, the initial vector is identified with the signal vector. Thus one obtains the quasidistributions and tomograms as bilinear transforms of the analytic signal. The difference

between quasidistributions and tomograms is related to different choices of the family of operators that creates the basis by the action on the initial vector. In the case of tomograms, this family is defined by the Dirac delta-function of the argument which itself is an Hermitian operator. In the case of quasidistributions, the family of operators can be chosen as a set of unitary operators of representations of some Lie groups like affine or symplectic ones.

2 Wavelet-like transforms, quasidistributions and tomograms

The general setting for our construction is as follows.

Signals $f(t)$ are considered to be vectors $\mid f\rangle$ belonging to a Hilbert space. $U(\alpha)$ is a family of operators . In many cases, the family of operators $U(\alpha)$ generates a unitary group. Consider a reference vector $\mid h\rangle$. Out of the set $U(\alpha)\mid h\rangle$, a complete set of vectors can be chosen to serve as a basis. Two of the transforms considered are

$$W_f^{(h)}(\alpha) = \langle U(\alpha)\, h \mid f\rangle; \qquad Q_f(\alpha) = \langle U(\alpha)\, f \mid f\rangle. \tag{1}$$

We will denote the transforms of the $W_f^{(h)}$-type as *wavelet-type* transforms and those of the Q_f-type as *quasidistribution* transforms.

In general, if $U(\alpha)$ are unitary operators, there are self-adjoint operators $B(\alpha)$ such that

$$W_f^{(h)}(\alpha) = \langle h \mid e^{iB(\alpha)} \mid f\rangle; \qquad Q_f^{(B)}(\alpha) = \langle f \mid e^{iB(\alpha)} \mid f\rangle. \tag{2}$$

In this case, since $B(\alpha)$ has a real valued spectrum, another transform may be defined, namely,

$$M_f^{(B)}(X) = \langle f \mid \delta\,(B(\alpha) - X) \mid f\rangle. \tag{3}$$

The delta-operator $\delta\,(B(\alpha) - X)$ is defined by the Fourier transform

$$\delta\,(B(\alpha) - X) = \int_{-\infty}^{\infty} e^{ik(B(\alpha)-X)}\, \frac{dk}{2\pi}\,. \tag{4}$$

Equation (3) defines what we call the tomographic transform of the signal or *tomogram*. In contrast to the quasiprobabilities, the transform $M_f^{(B)}(\alpha)$ is positive and it can be correctly interpreted as a probability distribution.

For a normalized vector $\mid f\rangle$, $\langle f \mid f\rangle = 1$, the tomogram is a normalized function

$$\int M_f^{(B)}\,(X)\, dX = 1. \tag{5}$$

The tomogram is a homogeneous function

$$M_f^{(B/p)}(X) = |p| M_f^{(B)}(pX). \tag{6}$$

The three classes of transforms obviously are mutually related

$$M_f^{(B)}(X) = \frac{1}{2\pi} \int Q_f^{(kB)}(\alpha)\, e^{-ikX}\, dk \tag{7}$$

and

$$Q_f^{(B)}(\alpha) = \int M_f^{(B/p)}(X)\, e^{ipX}\, dX. \tag{8}$$

Wavelet-type transforms, quasidistributions, and tomograms are related by the formulas

$$Q_f^{(B)}(\alpha) = W_f^{(f)}(\alpha), \tag{9}$$

$$W_f^{(h)}(\alpha) = \frac{1}{4} \int e^{iX} \left[M_{f_1}^{(B)}(X) - iM_{f_2}^{(B)}(X) - M_{f_3}^{(B)}(X) + iM_{f_4}^{(B)}(X) \right] dX, \tag{10}$$

where

$$| f_1\rangle =| h\rangle + | f\rangle; \qquad | f_3\rangle =| h\rangle - | f\rangle;$$
$$| f_2\rangle =| h\rangle + i \mid f\rangle; \qquad | f_4\rangle =| h\rangle - i \mid f\rangle.$$

Another important case concerns operators $U(\alpha)$, which can be represented in the form

$$U(\alpha) = e^{ib(\alpha)} P_h e^{-ib(\alpha)}, \tag{11}$$

with P_h being a projector on a reference vector $| h\rangle$. This creates a quasidistribution of the Husimi–Kano type

$$H_f^{(b)}(\alpha) = \langle f \mid U(\alpha) \mid f\rangle.$$

The formulated construction of signal transforms is determined by the choice of the operators $B(\alpha)$ or $b(\alpha)$. If one takes

$$B^{(S)}(\alpha) \equiv B^{(S)}(\mu, \nu) = \mu\hat{t} + \nu\hat{\omega},$$

with $\hat{t} = t$ and $\hat{\omega} = -i\,\partial/\partial t$, the tomogram

$$M_f^{(S)}(X, \mu, \nu) = \langle f \mid \delta\left(\mu\hat{t} + \nu\hat{\omega} - X\right) \mid f\rangle \tag{12}$$

reads[2]

$$M_f^{(S)}(X, \mu, \nu) = \frac{1}{2\,\pi|\nu|} \left| \int \exp\left[\frac{i\mu t^2}{2\,\nu} - \frac{itX}{\nu} \right] f(t)\, dt \right|^2. \tag{13}$$

According to the general scheme, the corresponding quasidistribution is

$$Q_f^{(S)}(\mu,\nu) = \langle f \mid e^{iB^{(S)}(\mu,\nu)} \mid f\rangle = \int M_f^{(S)}(X,\mu,\nu)\, e^{iX}\, dX. \tag{14}$$

This quasidistribution is called the ambiguity function in the signal processing literature.[3]

The tomogram (13) is related to the Wigner–Ville[4,5] quasidistribution $WV(\tau,\omega)$ by the relation[6]

$$M_f^{(S)}(X,\mu,\nu) = \int \exp\Big[-ik(X-\mu\omega-\nu\tau)\Big] WV(\tau,\omega)\, \frac{dk\, d\omega\, d\tau}{(2\pi)^2}\,. \tag{15}$$

The Wigner–Ville quasidistribution is given by the formula

$$WV(\tau,\Omega) = \int f\left(\tau+\frac{u}{2}\right) f^*\left(\tau-\frac{u}{2}\right) e^{-i\Omega u}\, du. \tag{16}$$

The unitary operator $U(\tau,\Omega)$, which determines the Wigner-Ville quasidistribution by

$$WV(\tau,\Omega) = \langle f \mid U^{(WV)}(\tau,\Omega) \mid f\rangle, \tag{17}$$

is

$$U^{(WV)}(\tau,\Omega) = e^{2i\left(\Omega\hat{t}-\tau\hat{\omega}\right)} e^{i\pi\left(\hat{t}^2+\hat{\omega}^2-1\right)/2}, \tag{18}$$

with the generator being

$$B^{(WV)}(\tau,\Omega) = 2\tau\hat{\omega} - 2\Omega\hat{t} + \pi\left(\hat{t}^2+\hat{\omega}^2-1\right)/2. \tag{19}$$

The wavelet transform of a signal $f(t)$ is a linear integral transform decomposing the signal into a set of basis functions

$$W_f^{(A)}(s,\tau) = \int f(t)\, h^*_{s,\tau}(t)\, dt. \tag{20}$$

The wavelets $h_{s,\tau}(t)$ are kernel functions generated from a basic wavelet $h(\tau)$ by means of a translation and a rescaling ($-\infty<\tau<\infty$, $s>0$):

$$h_{s,\tau}(t) = \frac{1}{\sqrt{s}}\, h\left(\frac{t-\tau}{s}\right). \tag{21}$$

Using the operator

$$U^{(A)}(\tau,s) = \exp(i\tau\hat{\omega})\exp(i\log s D), \tag{22}$$

where

$$D = \frac{1}{2}\left(\hat{t}\hat{\omega} + \hat{\omega}\hat{t}\right) = \hat{\omega}\hat{t} + \frac{i}{2} \qquad \text{with} \qquad \hat{\omega} = -i\,\frac{\partial}{\partial t}\,,$$

equation (21) can be represented in the form

$$h_{s,\tau}(t) = U^{(A)\dagger}(\tau, s)h(t). \tag{23}$$

Let us consider the operator (22), with the parameters μ and ν being

$$\nu = \log s \qquad \mu = \frac{\tau \log s}{s-1}\,. \tag{24}$$

We obtain for the unitary operator (22)

$$U^{(A)\dagger}(\tau, s) \equiv U^{(A)}(\mu, \nu) = \exp\left(i\mu\hat{\omega} + i\nu D\right), \tag{25}$$

and the operator $B(\alpha)$ becomes

$$B_1^{(A)} = \mu\hat{\omega} + \nu D. \tag{26}$$

According to the general scheme with the operator (26), the wavelet transform (20) can be rewritten in the form (2)

$$W_f^{(A)}(s, \tau) = \langle h \mid U^{(A)}(\mu, \nu) \mid f\rangle. \tag{27}$$

3 Transforms related to affine group

The diagonal matrix elements determine a quasidistribution for the signal $f(t)$

$$Q_f^{(A)}(s, \tau) = \langle f \mid e^{i(\mu\hat{\omega}+\nu D)} \mid f\rangle, \tag{28}$$

with μ and ν being expressed in terms of shift and scaling parameters (affine group parameters) by equation (24).

The following expression for the quasidistribution in terms of the affine group parameters τ and s can be obtained:

$$Q_f^{(A)}(s, \tau) = \int f^*\left(\frac{t - \tau/2}{\sqrt{s}}\right) f\left(\sqrt{s}\,(t + \tau/2)\right) dt. \tag{29}$$

The wavelet transform (20) may also be written in a similar form

$$W_f^{(A)}(s, \tau) = \int h^*\left(\frac{t - \tau/2}{\sqrt{s}}\right) f\left(\sqrt{s}\,(t + \tau/2)\right) dt, \tag{30}$$

as follows from equation (9).

For the tomograms associated to the affine group, one has

$$M_f^{(A_t)}(S_1,\mu,\nu) = \langle f \mid \delta\left(\mu\hat{t} + \nu\frac{\hat{t}\hat{\omega}+\hat{\omega}\hat{t}}{2} - S_1\right) \mid f\rangle \tag{31}$$

and

$$M_f^{(A_\omega)}(S_2,\mu,\nu) = \langle f \mid \delta\left(\mu\hat{\omega} + \nu\frac{\hat{t}\hat{\omega}+\hat{\omega}\hat{t}}{2} - S_2\right) \mid f\rangle. \tag{32}$$

4 Operator delta-function as a projector density

While constructing tomograms, the nonnegative operator $\delta\left(B(\alpha) - X\right)$ plays an essential role, with $B(\alpha)$ being an Hermitian operator with nondegenerate continuous spectrum. The random variable X takes values on the spectrum of $B(\alpha)$. We may identify $\delta\left(B(\alpha) - X\right)$ with the projector $\mid X\rangle\langle X \mid$

$$\delta\left(B(\alpha) - X\right) = \mid X\rangle\langle X \mid = P_X. \tag{33}$$

From this, it follows

$$M_f^{(B)} = \langle f \mid \delta\left(B(\alpha) - X\right) \mid f\rangle = \langle f \mid X\rangle\langle X \mid f\rangle = |\langle X \mid f\rangle|^2 \tag{34}$$

displaying the positivity of the tomogram.

By a unitary transformation S, $B(\alpha)$ may be transformed to

$$SB(\alpha)S^\dagger = B^{'}(\alpha). \tag{35}$$

Then if $\{\mid Z\rangle\}$ is the set of (generalized) eigenvectors of $B^{'}(\alpha)$, $\left\{S^\dagger \mid Z\rangle\right\}$ is a set of eigenvectors for B. Therefore,

$$M_f^{(B)}(Z) = \langle f \mid \delta\left(B(\alpha) - Z\right) \mid f\rangle = |\langle Z \mid S \mid f\rangle|^2 = \langle f \mid S^\dagger \mid Z\rangle\langle Z \mid S \mid f\rangle. \tag{36}$$

In this case, the operator $U(\alpha)$ in the general scheme described in section 2 would be

$$U(\alpha) = S^\dagger \mid Z\rangle\langle Z \mid S, \tag{37}$$

which is not represented as an exponent of an operator. The form (37), with the presence of a projector operator $P_Z = \mid Z\rangle\langle Z \mid$, also shows the relation of the tomograms to transforms of the Husimi–Kano type.

5 Discrete spectrum

In the case where the operator $B(\alpha)$ has a discrete spectrum, one uses a Kronecker delta-function $\delta_K\left(B(\alpha)-n\right)$ and associates to the operator the Fourier integral on the circle. For example, for the number operator,

$$\delta_K\left(a^\dagger a - n\right) = \frac{1}{2\pi}\int_0^{2\pi} e^{i\varphi\left(a^\dagger a - n\right)}\,d\varphi \qquad n = 0,1,2,3\ldots, \tag{38}$$

and the matrix elements of this operator in the Fock basis $\mid m\rangle$ are

$$\langle m' \mid \delta_K\left(a^\dagger a - n\right) \mid m\rangle = \delta_{m'm}\delta_{mn} \qquad m, m' = 0,1,2,3\ldots, \tag{39}$$

which equal the matrix elements of the projector

$$P_n = \mid n\rangle\langle n \mid . \tag{40}$$

This means that, also in the discrete-spectrum case, the Kronecker delta-function of the operator $B(\alpha)$ is reduced to a projector. The tomogram associated to the Wigner–Ville function by this method is given by the so-called photon-number tomography[7] (see also[8,9]), i.e.,

$$\begin{aligned} w(n,\beta) &= \langle f \mid D^\dagger(\beta) \mid n\rangle\langle n \mid D(\beta) \mid f\rangle \\ &= \frac{1}{\sqrt{\pi}\,2^n n!}\left|\int dt\; e^{-(t^2/2)+i\sqrt{2}\,t\,\mathrm{Im}\,\beta} H_n(t) f\left(t+\sqrt{2}\,\mathrm{Re}\,\beta\right)\right|^2, \end{aligned} \tag{41}$$

where the complex number β is a linear combination of the parameters μ and ν.

One may also construct a tomogram using a Dirac delta-function of the same operator $B(\alpha)$

$$\begin{aligned} M\left(X,\Omega,\tau\right) &= \int dk\,dt'\,dt''\,dt\; f(t'+k\tau) f^*(t''-k\tau) \\ &\quad \times e^{-ik(X+\pi/2)}\frac{i}{2\pi\sin(k\pi/2)}e^{-ik\Omega(t'+t'')} \\ &\quad \times \exp\left[-\frac{i}{2}\cot\frac{k\pi}{2}\left(2t^2+t'^2+t''^2\right)+\frac{it}{\sin(k\pi/2)}(t'+t'')\right]. \end{aligned} \tag{42}$$

The inverse of the transform reads

$$f(t)f^*(t') = \frac{1}{\pi}\int M\Big(X-\frac{\pi}{2},\Omega,\frac{t'-t}{2}\Big)\exp\Big[-i\Omega(t+t')+i\Big(X+\frac{\pi}{2}\Big)\Big]\,d\Omega\,dX. \tag{43}$$

In this case, the Dirac delta-function of $(B(\alpha)-X)$ is not reduced to a projector density.

6 Conclusion

To conclude, the main result in the approach developed is elaborating of a unified view for some linear and nonlinear transforms.[2] The formulation suggested emphasizes the basic unity of these transforms, which are related by the explicit formulas. Nevertheless, for each particular application, one type of transform may be more convenient than the others.

Acknowledgments

The author thanks the Organizers for kind hospitality and the Russian Foundation for Basic Research for partial support under Project No. 01-02-17745.

References

1. V.I. Man'ko and R.V. Mendes, *Phys. Lett. A* **263**, 53 (1999).
2. M.A. Man'ko, V.I. Man'ko and R.V. Mendes, *J. Phys. A* **34**, 8321 (2001).
3. L. Cohen, *Proc. IEEE* **77**, 941 (1989).
4. E. Wigner, *Phys. Rev.*, **40**, 749 (1932).
5. J Ville, *Cables et Transmission A* **2**, 61 (1948).
6. S. Mancini, V.I. Man'ko and P. Tombesi, *Phys. Lett. A* **213**, 1 (1996).
7. S. Mancini, P. Tombesi and V.I. Man'ko, *Europhys. Lett.* **37**, 79 (1997).
8. S. Wallentowitz and W Vogel, *Phys. Rev. A* **53**, 4528 (1996).
9. K. Banaszek and K. Wodkiewicz, *Phys. Rev. Lett.* **76** 4344 (1996).

SPECTRAL TRIPLES FOR THE KRONECKER FOLIATION

R. MATTHES

Fachbereich Physik, Technische Universität Clausthal,
Leibnizstraße 10, 38678 Clausthal-Zellerfeld, Germany
E-mail: ptrm@pt.tu-clausthal.de

O. RICHTER AND G. RUDOLPH

Institut für Theoretische Physik, Universität Leipzig,
Augustusplatz 10/11, 04109 Leipzig, Germany
E-mail: Olaf.Richter@itp.uni-leipzig.de, Gerd.Rudolph@itp.uni-leipzig.de

We describe three spectral triples related to the Kronecker foliation and analyze the related differential calculi Ω_D.

1 Introduction

The notion of spectral triple is central in A. Connes' approach to noncommutative geometry. It encodes information about the topology and the metric and differentiable structure of a noncommutative space. Explicit examples of spectral triples have been considered for commutative algebras of functions over a classical space [1], for matrix algebras [2], for products of both (in particular in the construction of unified field theories [6], [7]) and for the irrational rotation algebra [1].

In [3], Connes and Moscovici have given a general method which associates a spectral triple to the following geometric data: a regular foliation of a compact manifold, orientations and euclidean fibre metrics on the tangent bundle of the leaves and the normal bundle, and a group of diffeomorphisms preserving these structures.

In this note, we present some results in this context for the Kronecker foliation. We choose translation invariant fibre metrics and as diffeomorphisms the group $\mathbb{R}$ acting as translations along the leaves. In this case, we explicitly describe the algebra and the Dirac operator of [3]. We also note that the first order differential calculus related to this spectral triple is almost free, up to some relations for the generators coming from the group action. Next, we show that a modified Dirac operator (linear signature operator) leads also to a spectral triple for the same algebra and Hilbert space. The corresponding differential calculus can be described in all degrees in terms of generators and relations.

Finally, we remark that the methods used to describe this differential calculus can also be used to give a description of the differential calculus related to the well-known spectral triple of the irrational rotation algebra (i.e., the convolution algebra of the reduced holonomy groupoid of the Kronecker foliation, [1], [4]) in terms of generators and relations.

2 The spectral triple related to a foliation

Assume that we are given the following data:

- a compact manifold M with an integrable distribution $V \subset TM$, defining a foliation of M.
- Euclidean fibre metrics on V and the transversal bundle $N := TM/V$, orientations of both bundles.
- a group $\Gamma \subset Diff(M)$ leaving invariant the above structures.

These data obviously define the Hilbert space $\mathcal{H} := L^2(\wedge V_{\mathbb{C}}^* \otimes \wedge N_{\mathbb{C}}^*)$. Moreover, one defines the crossed product $C^\infty(M) \rtimes \Gamma$ as the algebra generated by functions (acting as multiplication operators on $\mathcal{H}$) and by unitaries U_ψ, $\psi \in \Gamma$ (defined by pull-back).

In order to define a Dirac operator, one first defines – making use of the Bott connection – a longitudinal differential d_L. Then, one defines a transversal differential d_H. To this end, one has to choose a transversal complement H of V in TM. This provides an isomorphism $j_H : \wedge V^* \otimes \wedge N^* \longrightarrow \wedge T^*M$, which can be used to transport the exterior differential to a differential on $\wedge V^* \otimes \wedge N^*$. This differential is turned into a derivation d_H of bidegree $(0,1)$ by projection onto a suitable homogeneous component. Now one defines a total signature operator $Q = Q_L(-1)^{\partial_N} + Q_H$, consisting of longitudinal and transversal parts $Q_L = d_L d_L^* - d_L^* d_L$, $Q_H = d_H + d_H^*$. The motivation for using a quadratic longitudinal signature operator is the desire to have no dependence on the choice of H in the principal symbol of Q, which is necessary for a meaningful index theorem [3]. A homotopy argument shows that the longitudinal signature class is not influenced by passing to a quadratic Q_L [3]. The operator Q is a formally selfadjoint operator in $\mathcal{H}$. Finally, the Dirac operator is the selfadjoint operator D uniquely defined by $D|D| = Q$.

Theorem 1 [3] *$(C^\infty(M) \rtimes \Gamma, \mathcal{H}, D)$ is a spectral triple of dimension* $\dim V + 2\dim N$.

In a foliation chart, explicit formulae for the operators d_L, d_H can be given [8]. We note that there is a different method to define a spectral triple for a Riemannian foliation, due to Kordyukov [5], which uses only a (linear) signature operator in transversal direction.

3 A spectral triple for the Kronecker foliation

Let $\mathbb{T}^2$ be the two-torus, and let us choose the following data:

- $M = \mathbb{T}^2$, foliated by straight lines of irrational slope θ. V and N are one-dimensional. Note that there are unique $a, b \in \mathbb{R}$ such that $\theta = \frac{b}{a}$, $a > 0$, $a^2 + b^2 = 1$.
- Translation invariant fibre metrics, and orientations inherited from an orientation of $\mathbb{T}^2$.
- $\Gamma = \mathbb{R}$ acting as translations along the leaves.

The Hilbert space $\mathcal{H}$ is in this case a direct sum of four copies of $L^2(\mathbb{T}^2)$. $C^\infty(\mathbb{T}^2)$ is generated (as a Fréchet $*$-algebra) by the two functions $U_1 = e^{i\vartheta_1}$ and $U_2 = e^{i\vartheta_2}$ (ϑ_j the natural angular coordinates on $\mathbb{T}^2$). Let $\mathcal{O}(\mathbb{T}^2) \subset C^\infty(\mathbb{T}^2)$ be the dense $*$-subalgebra generated by U_1 and U_2.

Proposition 1 *$\mathcal{O}(\mathbb{T}^2) \rtimes \mathbb{R}$ is isomorphic to the $*$-algebra generated by unitary operators U_1, U_2, V_t satisfying*

$$U_1 U_2 = U_2 U_1, \quad V_t V_s = V_{t+s}, \; t, s \in \mathbb{R},$$

$$V_t U_1 = e^{iat} U_1 V_t, \quad V_t U_2 = e^{ibt} U_2 V_t,$$

If we choose as transversal complement H of V the direction orthogonal to the leaves in the natural translation invariant flat metric on $\mathbb{T}^2$, the Dirac operator D defined according to the above general recipe can be diagonalized and has spectrum $\pm\sqrt{(ak+bl)^4 + (bk-al)^2}$, $k, l \in \mathbb{Z}$. The dimension of the resulting spectral triple $(\mathcal{O}(\mathbb{T}^2) \rtimes \mathbb{R}, \mathcal{H}, D)$ is 3.

In order to find the differential calculus Ω_D belonging to this spectral triple, one needs to know relations between the generators U_1, U_2 and V_t and their commutators with D.

Proposition 2

$$[D, V_t] = 0,$$

$$V_t[D, U_1] = e^{iat}[D, U_1]V_t, \quad V_t[D, U_2] = e^{ibt}[D, U_2]V_t.$$

Note that the theory also gives a spectral triple if we choose $\Gamma = \{\mathbf{1}\}$. In that case, the algebra is $C^\infty(\mathbb{T}^2)$. For the differential calculus related to the spectral triple $(\mathcal{O}(\mathbb{T}^2), \mathcal{H}, D)$ we have

Proposition 3 $\Omega^1_D(\mathcal{O}(\mathbb{T}^2)) = \Omega^1(\mathcal{O}(\mathbb{T}^2))$.

Here, Ω^1 on the right is the first order universal differential calculus. For the proof one has to show that there are no other relations between generators and their commutators with D than those coming from the Leibniz rule. The reasoning is based on two lemmas about Hankelian determinants [8], [9], [11]. For the case $\Gamma = \mathbb{R}$, these tools can not be applied. We have

Conjecture 1 $\Omega_D(\mathcal{O}(\mathbb{T}^2) \rtimes \mathbb{R}) = \Omega(\mathcal{O}(\mathbb{T}^2) \rtimes \mathbb{R})/J$, *where the differential ideal J is generated by the relations*

$$dV_t = 0,$$

$$V_t dU_1 = e^{iat} dU_1 V_t, \quad V_t dU_2 = e^{ibt} dU_2 V_t.$$

4 Changing the Dirac operator

Let us replace $Q_L = d_L d_L^* - d_L^* d_L$ by $\tilde{Q}_L = d_L + d_L^*$ (linear signature operator in leaf direction) and use $\tilde{Q} = \tilde{Q}_L + Q_H$ as Dirac operator. Then we have

Proposition 4 $(\mathcal{O}(\mathbb{T}^2) \rtimes \mathbb{R}, \mathcal{H}, \tilde{Q})$ *is a spectral triple of dimension 2.*

To prove this, one first diagonalizes $\tilde{Q}^2$, which gives immediately the eigenvalues $\pm\sqrt{(ak + bl)^2 + (bk - al)^2}$, $k, l \in \mathbb{Z}$, of $\tilde{Q}$ and explicit formulae for $\tilde{Q}$ in the eigenbasis of $\tilde{Q}^2$. The differential calculus $\Omega_{\tilde{Q}}(\mathcal{O}(\mathbb{T}^2) \rtimes \mathbb{R})$ can be described to all orders in terms of generators and relations:

Proposition 5 *(i) $\Omega^1_{\tilde{Q}}(\mathcal{O}(\mathbb{T}^2) \rtimes \mathbb{R})$ is a free left (and right) $\mathcal{O}(\mathbb{T}^2) \rtimes \mathbb{R}$-module with basis $\{dU_1, dU_2\}$. Its bimodule structure is determined by*

$$U_j dU_k = dU_k U_j, \quad \forall j, k \in \{1, 2\},$$

$$V_t du_1 = e^{iat} dU_1 V_t, \quad V_t dU_2 = e^{ibt} dU_2 V_t.$$

Moreover,

$$dV_t = 0.$$

(ii) $\Omega^2_{\tilde{Q}}(\mathcal{O}(\mathbb{T}^2) \rtimes \mathbb{R})$ is a free left (and right) $\mathcal{O}(\mathbb{T}^2) \rtimes \mathbb{R}$-module with basis $\{dU_1 dU_2\}$, with

$$dU_1 dU_2 = -dU_2 dU_1.$$

(iii) $\Omega^k_{\tilde{Q}}(\mathcal{O}(\mathbb{T}^2) \rtimes \mathbb{R}) = 0$ for $k \geq 3$.

Thus, we have a slight modification of a classical calculus on $\mathbb{T}^2$. The above relations are proved easily using explicit formulae for $\tilde{Q}$ and the generators U_1, U_2 and V_t as operators on $\mathcal{H}$. The freeness properties need a bit more work [8]. We conclude that the differential calculus Ω_D in this geometric situation heavily depends on the choice of the Dirac operator D. Nice results like Proposition 5 seem to be exceptions. Below, we describe a result similar to this proposition for the spectral triple of the irrational rotation algebra.

5 A differential calculus for the irrational rotation algebra

Let $\mathcal{O}_\theta$ be the $*$-algebra generated by unitaries U, V subject to the relation $UV = e^{-2\pi i\theta}VU$, θ irrational (irrational rotation algebra [10]). Define $\mathcal{H} := \mathcal{H}_\tau \oplus \mathcal{H}_\tau$, where $\mathcal{H}_\tau$ is the GNS Hilbert space corresponding to the tracial state $\tau(\sum a_{kl}U^kV^l) = a_{00}$ of $\mathcal{O}_\theta$. Let δ_1, δ_2 be the derivations of $\mathcal{O}_\theta$ given by $\delta_1(U) = U$, $\delta_1(V) = 0$, $\delta_2(U) = 0$, $\delta_2(V) = V$, and $\delta_2(U) = 0$. Put $\partial := \frac{1}{\sqrt{2\pi}}(\delta_1 - i\delta_2)$ and $D := \begin{pmatrix} 0 & \partial \\ \partial^* & 0 \end{pmatrix}$. Then it is well known that $(\mathcal{O}_\theta, \mathcal{H}, D)$ is a spectral triple of dimension 2 [1], [4]. Using the same methods as above, one can show

Proposition 6 *i)* $\Omega^1_D(\mathcal{O}_\theta)$ *is a free left (or right)* $\mathcal{O}_\theta$*-module with basis* $\{dU, dV\}$. *The bimodule structure of* $\Omega^1_D(\mathcal{O}_\theta)$ *is given by*

$$UdU = dUU,\ U^*dU = dUU^*,\ UdU^* = dU^*U,\ U^*dU^* = dU^*U^*,$$

$$VdV = dVV,\ V^*dV = dVV^*,\ VdV^* = dV^*V,\ V^*dV^* = dV^*V^*,$$

$$VdU = e^{2\pi i\theta}dUV,\quad UdV = e^{-2\pi i\theta}dVU,$$

$$VdU^* = e^{-2\pi i\theta}dU^*V,\quad U^*dV = e^{2\pi i\theta}dVU^*,$$

$$V^*dU = e^{-2\pi i\theta}dUV^*,\quad UdV^* = e^{2\pi i\theta}dV^*U.$$

ii) $\Omega^2_D(\mathcal{O}_\theta)$ *is a free left (or right)* $\mathcal{O}_\theta$*-module with basis* $\{dUdV\}$, *with*

$$dUdV = -e^{2\pi i\theta}dVdU$$

iii) $\Omega^k_D(\mathcal{A}_\theta) = 0$ *for* $k \geq 3$.

Acknowledgments

R. M. was supported by the Deutsche Forschungsgemeinschaft.

References

1. A. Connes, *Noncommutative geometry*, Academic Press Inc., San Diego (1994).
2. T. Krajewski, Classification of finite spectral triples, *J. Geom. Phys.* **28**, 1 (1998).
3. A. Connes and H. Moscovici, The local index formula in noncommuative geometry, *GAFA* **5**, 174 (1995).
4. J. M. Gracia-Bondía, J. C. Várilly and H. Figueroa, *Elements of noncommutative geometry*, Birkhäuser, Basel (2000).
5. Kordyukov, Y. A.: Noncommutative spectral geometry of Riemannian foliations, dg-ga/9703012
6. J. Lott and A. Connes, The metric aspect of noncommutative geometry, in *New Symmetry Principles in Quantum Field Theory*, eds J. Fröhlich et. al., p. 53, Plenum Press, New York (1992).
7. R. Matthes, G. Rudolph and R. Wulkenhaar, On the structure of a differential algebra used by Connes and Lott, *Rep. Math. Phys.* **38**, 45 (1996).
8. R. Matthes, O. Richter and G. Rudolph, Spectral triples related to the Kronecker foliation, to be published.
9. O. Richter, to be published.
10. M. A. Rieffel, C^*-algebras associated with irrational rotations, *Pac. J. Math.* **93**, 415 (1981).
11. R. Vein and P. Dale, *Determinants and their applications in mathematical physics*, *Appl. Math. Sciences* **134**, Springer (1998).

NON-COMMUTATIVE SPACE-TIME GEOMETRY AND PRINCIPLE OF GAUGE INVARIANCE

R. M. MIR-KASIMOV
Bogoliubov Laboratory of Theoretical Physics
Joint Institute for Nuclear Research
Dubna, Moscow Region, 141980, Russia
E-mail: mirkr@thsun1.jinr.ru

The realization of the two-dimensional Poincare algebra in terms of the non-commutative differential calculus on the commutative algebra of functions is considered. Corresponding algebra of functions is commutative and is defined by the spectrum of unitary irreducible representations of the two-dimensional De Sitter group. Gauge invariance principle consistent with this quantum geometry is considered

In the approach to the quantum space-time considered here[1] the position and time operators are defined as non-commutative differential operators on commutative algebra of functions. By the reasons, explained below such a modification is predefined by the change of the geometry of the momentum space[2–5] and change of the description of the free motion[a].

The version of non-commutative differential calculus considered is an example of the application of the general theory[6–9] singled out by its physical meaning. We shall work in the pseudo-spherical system of coordinates and denote the quantum interval variable by σ and time by n. The common notation for the set (σ, n) - a point of quantum space-time will be ξ.

Let us consider the non-commutative calculus in which the algebra of functions is commutative but differentials don't commute with the coordinates (details see in[1]):

$$[\sigma, n] = 0 \qquad [\sigma, d\sigma] = d\sigma \qquad [n, dn] = dn \tag{1}$$

We consider first the case of the variable σ. Defining the interior derivatives right $\overrightarrow{\partial_\sigma}$ and left $\overleftarrow{\partial_\sigma}$ (see[1]) as

$$df(\sigma) = [f(\sigma), d\sigma] = d\sigma \left(\overrightarrow{\partial_\sigma}\, f(\sigma)\right) = \left(\overleftarrow{\partial_\sigma}\, f(\sigma)\right) d\sigma \tag{2}$$

we develop the differential calculus. The Leibniz rule is fulfilled in the standard form (as compared with usual calculus) for exterior differentiation

$$d(fg) = df\,(g) + (f)\,dg \tag{3}$$

[a]Detailed list of references see in the review article[1]

and in modified form for interior derivatives

$$\begin{aligned} \overrightarrow{\partial_\sigma}\,(fg) &= \left(\overrightarrow{\partial_\sigma}\,f\right) g + f\left(\overrightarrow{\partial_\sigma}\,g\right) + \left(\overrightarrow{\partial_\sigma}\,f\right)\left(\overrightarrow{\partial_\sigma}\,g\right) \\ \overleftarrow{\partial_\sigma}\,(fg) &= \left(\overleftarrow{\partial_\sigma}\,f\right) g + f\left(\overleftarrow{\partial_\sigma}\,g\right) - \left(\overleftarrow{\partial_\sigma}\,f\right)\left(\overleftarrow{\partial_\sigma}\,g\right) \end{aligned} \tag{4}$$

We refer the reader for the detailed theory of non-commutative differential forms on the commutative algebra of functions to [1] and deliver here the only information necessary for introducing the physical operators. Using right and left non-commutative Hodge operators $\overrightarrow{*}$ and $\overleftarrow{*}$ we define the symmetrized interior derivatives in the form

$$\partial_\sigma^- = \frac{1}{2}\left(\overrightarrow{*} + \overleftarrow{*}\right) d\sigma \qquad \partial_\sigma^+ = \frac{1}{2}\left(\overrightarrow{*} - \overleftarrow{*}\right) d\sigma + 1 \tag{5}$$

It can be proved that interior derivatives have the form of finite-difference operations

$$\partial_\sigma^- = \sinh\frac{\partial}{\partial\sigma} \qquad \partial_\sigma^+ = \cosh\frac{\partial}{\partial\sigma} \tag{6}$$

Similar relations exist for the second variable n.

In order to construct desired realization of the Poincare group Lie algebra let us introduce the operators

$$\begin{aligned} p^0 &= -4\,sgn(n)\,\tfrac{\zeta(\sigma,n)}{2\sigma+1}\,\partial_\sigma^- \\ p^1 &= -\tfrac{i}{2\sigma+1}\{(2\sigma+1)\,\partial_\sigma^+\,\partial_n^- + 2n\,\partial_\sigma^-\,\partial_n^+ + \partial_\sigma^-\,\partial_n^-\} \\ p^2 &= \tfrac{1}{2\sigma+1}\{(2\sigma+1)\,\partial_\sigma^+\,\partial_n^+ + 2n\,\partial_\sigma^-\,\partial_n^- + \partial_\sigma^-\,\partial_n^+\} \end{aligned} \tag{7}$$

where

$$\zeta\,(\sigma,n) = \frac{\kappa\,(\sigma,n)}{\kappa\,(\sigma - 1,n)} = \frac{\Gamma\left(\frac{\sigma+n+2}{2}\right)\Gamma\left(\frac{\sigma-n+2}{2}\right)}{\Gamma\left(\frac{\sigma+n+1}{2}\right)\Gamma\left(\frac{\sigma-n+1}{2}\right)} \tag{8}$$

$$\kappa\,(\sigma,n) = \Gamma\left(\frac{\sigma+n+2}{2}\right)\Gamma\left(\frac{\sigma-n+2}{2}\right) \tag{9}$$

Operators (7) possess two important properties
1) They commute with each other

$$\left[p^L, p^K\right] = 0 \tag{10}$$

Using this feature we can consider (7) as the generators of the three dimensional pseudo-Euclidean space or three components of the momentum operator. These operators can be brought simultaneously to diagonal form.
2) They obey the De Sitter condition

$$p_L p^L = p_\mu p^\mu - p_2^2 = -1 \quad L = 0,1,2, \quad \mu = 0,1 \tag{11}$$

It can be proved that the common eigen-functions for operators (7) are the matrix elements of the unitary irreducible representations of the De Sitter group $\langle \xi \mid p \rangle$, for which the surface (11) is a uniform space[b]. They play the role of generalized plane waves. To describe these matrix elements we use the hyper-spherical coordinates

$$\begin{gathered} p^0 = \sinh\zeta \quad p^1 = \cosh\zeta \sin\omega \quad p^2 = \cosh\zeta\cos\omega \\ -\infty < \zeta < \infty \qquad\qquad -\pi < \omega < \pi \end{gathered} \tag{12}$$

In the correspondence limit

$$p^2 \approx 1, \quad p^0 \approx p^1 \approx 0, \qquad \zeta \to 0, \quad \omega \to 0 \tag{13}$$

Below we count the properties of the generalized plane waves

$$\langle \xi \mid p \rangle = \langle \sigma, n \mid \zeta, \omega \rangle = \frac{2^{\sigma+1}\,\kappa(\sigma,n)}{\sqrt{2\pi}\,\cosh\zeta}\, e^{in\omega}\, \mathrm{P}_{n-\frac{1}{2}}^{-(\sigma+\frac{1}{2})}(\pm\tanh\zeta) \tag{14}$$

a) Behavior at the origin

$$\langle \xi \mid 0 \rangle = 1 \tag{15}$$

b) Orthogonality and completeness

$$\begin{gathered} \frac{1}{(2\pi)^2}\int d\Omega_p \langle \xi \mid p\rangle\langle p \mid \xi'\rangle = \mu^{-1}(\sigma)\delta_{\xi^0\xi'^0}\delta_{nn'}\delta(\sigma-\sigma') \\ \frac{1}{(2\pi)^2}\int d\Omega_\xi \langle p \mid \xi\rangle\langle \xi \mid p'\rangle = \delta(p(-)p') = |p_2|\delta(\tilde{p}-\tilde{p}') \\ d\Omega_p = \frac{dp^0 dp^1}{|p_2|} \qquad \mu(\sigma) = -\left(\sigma+\tfrac{1}{2}\right)\tanh\pi\sigma \end{gathered} \tag{16}$$

Symbol $d\Omega_\xi$ denotes over includes the integration over the continuous part of the proper time σ and summation over its discrete part as well as the summation over the discrete time n (*cf.* [5])

[b]Evidently physical constants including the fundamental length l_0 hypothetically equal to the Planck's length must enter the expressions (7), (11) [4]. We use the unit system $\hbar = c = l$. All relations of the theory must go over into the standard ones when l_0 can be considered as a small quantity. This will be called the correspondence limit.

c) The correspondence limit

$$\zeta \simeq p^0, \quad \omega \simeq p^1, \quad p^2 \simeq 1$$

$$n \simeq x^1, \quad \sqrt{\sigma^2 + n^2} \simeq |x^0| \tag{17}$$

$$\langle \xi \mid p \rangle \to e^{ip_\mu x^\mu}$$

Relations (17) suggest our identification of σ and n as quantum analogs of the two-dimensional space-time coordinates (interval and time).

Now we write down the three generators

$$M^{12} = n, \quad M^{10} = 2\, sgn(n)\, \zeta\,(\sigma, n)\, \partial_n^-, \quad M^{20} = 2i\, sgn(n)\, \zeta\,(\sigma, n)\, \partial_n^+ \tag{18}$$

which are as well as (7) the non-commutative differential operators and complete the set (7) up to the Lie algebra of the inhomogeneous 2-dimensional De Sitter group.

Now we consider the gauge transformation

$$\psi'(\xi) = \Omega(\xi)\, \psi(\xi) \qquad \Omega(\xi)^\dagger = \Omega(\xi)^{-1} \tag{19}$$

Unlike the usual theory the gauge transformation intermixes the components of momenta:

$$\Omega^{-1}(\xi)\, P^L \Omega(\xi) = C^L{}_K(\xi)\, P^K \tag{20}$$

For the sake of simplicity we write down the $C^L{}_K(\sigma, n)$-matrix separately for the cases when it depends on only one of the variables ξ or n. For $\Omega = \Omega(\sigma)$ we have

$$C^L{}_K(\sigma) = \left((\partial_\sigma^+ \Omega(\sigma)) - \frac{2i}{2\sigma + 1}\, (\partial_\sigma^- \Omega(\sigma))\, \widehat{\Sigma} \right)^L{}_K \tag{21}$$

where

$$\widehat{\Sigma} = \begin{pmatrix} \frac{i}{2} & M^{10} & M^{20} \\ M^{10} & \frac{i}{2} & M^{12} \\ M^{20} & -M^{12} & \frac{i}{2} \end{pmatrix} \tag{22}$$

For the case $\Omega = \Omega(n) = e^{i\,\lambda(n)}$

$$C(n) = e^{i\,(\partial_n^+ \lambda(n))} \begin{pmatrix} e^{i\,((\partial_n^+ - 1\,)\lambda(n))} & 0 & 0 \\ 0 & \cos(\partial_n^- \lambda(n)) & -\sin(\partial_n^- \lambda(n)) \\ 0 & \sin(\partial_n^- \lambda(n)) & \cos(\partial_n^- \lambda(n)) \end{pmatrix} \tag{23}$$

Also the following the following relations are true

$$\Omega^{-1}(\xi)\, P_L \Omega(\xi) = P_K C_L^{\dagger\,K}(\xi) \qquad C^\dagger = C^{-1} \tag{24}$$

It is easily seen that in a consequence of (20) and (24) the De Sitter condition (11) is invariant in respect to the gauge transformations (19). It can be easily shown[5] that to make the theory gauge invariant we must introduce the complex De Sitter vector $\hat{A}^L$ of electromagnetic field and require that it transforms similarly to (20) and (24):

$$\Omega^{-1}(\xi)\,\hat{A}^L\,\Omega(\xi) = C^L{}_K(\xi)\,\hat{A}^K \qquad \Omega^{-1}(\xi)\,\hat{A}^\dagger_L\,\Omega(\xi) = \hat{A}^\dagger_K C^\dagger_L{}^K(\xi) \tag{25}$$

It follows from(25) that the components of electromagnetic field do not commute with the gauge function and in a consequence do not commute between themselves. Introducing the covariant derivatives

$$\hat{D}^L = -i\left(p^L - \hat{A}^L\right) \tag{26}$$

we obtain the De Sitter invariant equations for the matter fields. The tensor of electromagnetic field is given as

$$\hat{F}^{KL} = \left[\hat{D}^L, \hat{D}^K\right] = \left[\hat{p}^L, \hat{A}^K\right] - \left[\hat{p}^K, \hat{A}^L\right] - \left[\hat{A}^L, \hat{A}^K\right] \tag{27}$$

The action of the electromagnetic field is

$$S = Tr\int \hat{F}^{KL}\hat{F}_{KL}\,d\Omega_\xi \tag{28}$$

and the non-commutative analog of the D'Alembert equation takes the form

$$\left[\hat{D}_K, \hat{F}^{KL}\right] = 0 \tag{29}$$

References

1. R.M.Mir – Kasimov, *Physics of Particles and Nuclei* **31**, 44 (2000).
2. V.G.Kadyshevsky, *Nucl. Phys. B* **141**, 477 (1988).
3. V.G.Kadyshevsky and D.V.Fursaev, *Theor.Math.Phys.* **83**, 197 (1990) .
4. A.Donkov, V.G.Kadyshevsky, M.Mateev and R.M.Mir – Kasimov, *Proceedings of V.A.Steklov Mathematical Institute, Acad.Sci.USSR*, **CXXXVI**, 1975, 85.
5. R.M.Mir – Kasimov, *Phys. Lett. B* **259**, 79 (1991).
6. S.L.Woronowicz, *Commun.Math.Phys.* **122**, 125 (1989).
7. J.Wess and B.Zumino, *Nucl. Phys. B (Proc.Suppl.)* **18**, 302 (1990).
8. A.Connes, *Noncommutative Geometry*, Academic Press, (1994).
9. M.Dubois-Violette, R.Kerner and J.Madore, *Journ.Math.Phys* **31**, 323 (1990).
10. A.Dimakis, F.Müller-Hoissen and T.Striker, *J.Physics A* **26**, 1927 (1993).

AN INTEGRABLE RELATIVISTIC THREE BODY PROBLEM WITH SPIN AND AN APPLICATION

MARCOS MOSHINSKY*

Instituto de Física
Universidad Nacional Autónoma de México
Apartado Postal 20-364, 01000 México D.F., México

1 Introduction

In this paper we will start with a system of three non-interacting relativistic particles with spin, eliminate the center of mass momentum, *i.e.*, introduce Jacobi variables and transform the remaining momenta into creation and annihilation operators so that an interaction is introduced.

We then show how we can determine in a closed analytic form the eigenvalues and eigenfunctions of the problem and compare our results with spectra of non-strange barions with the help of the single parameter at our disposal.

2 The relativistic three body problem with spin

The equation corresponding to a free relativistic three particle system with spin is the sum of three Dirac equations in which we use the units $\hbar = m = c = 1$ and denote the momenta with a prime *i.e.*, $\mathbf{p}'_s, s = 1, 2, 3$, as we reserve the unprimed notation for Jacobi momenta. We thus get[1]

$$\left[(\boldsymbol{\alpha}_1 \cdot \mathbf{p}'_1 + \boldsymbol{\beta}_1) + (\boldsymbol{\alpha}_2 \cdot \mathbf{p}'_2 + \boldsymbol{\beta}_2) + (\boldsymbol{\alpha}_3 \cdot \mathbf{p}'_3 + \boldsymbol{\beta}_3)\right]\boldsymbol{\psi} = \boldsymbol{E\psi}, \quad (1)$$

or

$$\left[(\boldsymbol{\alpha}_1 \cdot \mathbf{p}'_1 + \boldsymbol{\beta}_1) + (\boldsymbol{\alpha}_2 \cdot \mathbf{p}'_2 + \boldsymbol{\beta}_2)\right]\boldsymbol{\psi} = \left[\boldsymbol{E} - (\boldsymbol{\alpha}_3 \cdot \mathbf{p}'_3 + \boldsymbol{\beta}_3)\right]\boldsymbol{\psi}. \quad (2)$$

Defining

$$\hat{\Pi}_s^2 = (p_s'^2 + 1) = (-\nabla_s^2 + 1), s = 1, 2, 3 \quad (3)$$

and squaring both sides we get

*MEMBER OF EL COLEGIO NACIONAL AND THE SISTEMA NACIONAL DE INVESTIGADORES

$$\left[\hat{\mathbf{\Pi}}_1^2 + \hat{\mathbf{\Pi}}_2^2 + 2(\boldsymbol{\alpha}_1 \cdot \mathbf{p}'_1 + \boldsymbol{\beta}_1)(\boldsymbol{\alpha}_2 \cdot \mathbf{p}'_2 + \boldsymbol{\beta}_2)\right]\psi$$

$$= \left[E^2 + \hat{\Pi}_3^2 - 2(\alpha_3 \cdot \mathbf{p}'_3 + \beta_3)E\right]\psi, \qquad (4)$$

or

$$2\left[(\boldsymbol{\alpha}_1 \cdot \mathbf{p}'_1 + \boldsymbol{\beta}_1)(\boldsymbol{\alpha}_2 \cdot \mathbf{p}'_2 + \boldsymbol{\beta}_2) + \boldsymbol{E}(\boldsymbol{\alpha}_3 \cdot \mathbf{p}'_3 + \boldsymbol{\beta}_3)\right]\psi$$

$$= \left[\hat{\Pi}_1^2 + \hat{\Pi}_2^2 - \hat{\Pi}_3^2 - E^2\right]\psi. \qquad (5)$$

We square again and get

$$4\left[\hat{\mathbf{\Pi}}_1^2\hat{\mathbf{\Pi}}_2^2 + \boldsymbol{E}^2\hat{\mathbf{\Pi}}_3^2 + 2\boldsymbol{E}(\boldsymbol{\alpha}_1 \cdot \mathbf{p}_1 + \boldsymbol{\beta}_1)(\boldsymbol{\alpha}_2 \cdot \mathbf{p}'_2 + \boldsymbol{\beta}_2)(\boldsymbol{\alpha}_3 \cdot \mathbf{p}'_3 + \boldsymbol{\beta}_3)\right]\psi$$

$$= (\hat{\Pi}_1^2 + \hat{\Pi}_2^2 - \hat{\Pi}_3^2 - E^2)^2\psi. \qquad (6)$$

Finally passing $4(\hat{\Pi}_1^2\hat{\Pi}_2^2 + E^2\hat{\Pi}_3^2)$ to the right and squaring we obtain

$$64E^2\hat{\Pi}_1^2\hat{\Pi}_2^2\hat{\Pi}_3^2\psi = \left[(\hat{\Pi}_1^2 + \hat{\Pi}_2^2 - \hat{\Pi}_3^2 - E^2)^2 - 4\hat{\Pi}_1^2\hat{\Pi}_2^2 - 4E^2\hat{\Pi}_3^2\right]^2\psi, \qquad (7)$$

and developing this we get[1]

$$\Phi(E^2,\hat{\Pi}_s^2)\psi \equiv \left\{E^8 - 4AE^6 + (4A^2+2B)E^4 - (4C^2+4AB)E^2 + B^2\right\}\psi = 0, \quad (8)$$

where A, B, C are functions of $\hat{\Pi}_s^2, s = 1, 2, 3$ given by

$$\begin{aligned}
A &\equiv \hat{\Pi}_1^2 + \hat{\Pi}_2^2 + \hat{\Pi}_3^2,\\
B &\equiv \hat{\Pi}_1^4 + \hat{\Pi}_2^4 + \hat{\Pi}_3^4 - \hat{\Pi}_1^2\hat{\Pi}_2^2 - \hat{\Pi}_2^2\hat{\Pi}_3^2 - \hat{\Pi}_1^2\hat{\Pi}_3^2 - \hat{\Pi}_2^2\hat{\Pi}_1^2 - \hat{\Pi}_3^2\hat{\Pi}_2^2 - \hat{\Pi}_3^2\hat{\Pi}_1^2 \qquad (9)\\
C^2 &\equiv \tfrac{8}{3}(\hat{\Pi}_1^2\hat{\Pi}_2^2\hat{\Pi}_3^2 + \hat{\Pi}_2^2\hat{\Pi}_1^2\hat{\Pi}_3^2 + \hat{\Pi}_3^2\hat{\Pi}_1^2\hat{\Pi}_2^2 + \hat{\Pi}_1^2\hat{\Pi}_3^2\hat{\Pi}_2^2 + \hat{\Pi}_3^2\hat{\Pi}_2^2\hat{\Pi}_1^2 + \hat{\Pi}_2^2\hat{\Pi}_3^2\hat{\Pi}_1^2).
\end{aligned}$$

The operator $\phi(E^2, \hat{\Pi}_s^2)$ depends through Eq. (3) on the momentum vector $\mathbf{p}'_s$. It is convenient to express the latter in terms of the Jacobi momenta defined by

$$\begin{pmatrix} \mathbf{p}_1 \\ \mathbf{p}_2 \\ \mathbf{p}_3 \end{pmatrix} = \begin{pmatrix} \frac{1}{\sqrt{2}} & -\frac{1}{\sqrt{2}} & 0 \\ \frac{1}{\sqrt{6}} & \frac{1}{\sqrt{6}} & -\sqrt{\frac{2}{3}} \\ \frac{1}{\sqrt{3}} & \frac{1}{\sqrt{3}} & \frac{1}{\sqrt{3}} \end{pmatrix} \begin{pmatrix} \mathbf{p}'_1 \\ \mathbf{p}'_2 \\ \mathbf{p}'_3 \end{pmatrix}. \qquad (10)$$

As the matrix in Eq. (10) is orthogonal we can use its transposed to express $\mathbf{p}'_s$ in terms of $\mathbf{p}_s$. As $\mathbf{p}_3$ is proportional to the total momentum of the system, in the center of mass frame it vanishes, and we can write

$$\mathbf{p}'_1 = \frac{1}{\sqrt{2}}\mathbf{p}_1 + \frac{1}{\sqrt{6}}\mathbf{p}_2, \quad \mathbf{p}'_2 = -\frac{1}{\sqrt{2}}\mathbf{p}_1 + \frac{1}{\sqrt{6}}\mathbf{p}_2, \quad \mathbf{p}'_3 = -\sqrt{\frac{2}{3}}\mathbf{p}_2, \tag{11}$$

As $\phi(E^2, \hat{\Pi}_2^2)$ contains only powers of $\hat{\Pi}_s^2$ defined in (8,9) we can write the latter, using the hermiticity properties of $\mathbf{p}_s$ *i. e.* $\mathbf{p}_s = \mathbf{p}_s^\dagger$, as

$$\begin{aligned}
\hat{\Pi}_1^2 &= \tfrac{1}{2}\mathbf{p}_1^\dagger \cdot \mathbf{p}_1 + \tfrac{1}{6}\mathbf{p}_2^\dagger \cdot \mathbf{p}_2 + \tfrac{1}{2\sqrt{3}}(\mathbf{p}_1^\dagger \cdot \mathbf{p}_2 + \mathbf{p}_2^\dagger \cdot \mathbf{p}_1) + 1, \\
\hat{\Pi}_2^2 &= \tfrac{1}{2}\mathbf{p}_1^\dagger \cdot \mathbf{p}_1 + \tfrac{1}{6}\mathbf{p}_2^\dagger \cdot \mathbf{p}_2 - \tfrac{1}{2\sqrt{3}}(\mathbf{p}_1^\dagger \cdot \mathbf{p}_2 + \mathbf{p}_2^\dagger \cdot \mathbf{p}_1) + 1, \\
\hat{\Pi}_3^2 &= \tfrac{2}{3}\mathbf{p}_2^\dagger \cdot \mathbf{p}_2 + 1.
\end{aligned} \tag{12}$$

We can easily introduce an interaction term in the $\hat{\Pi}_s^2, s = 1, 2, 3$ by defining the creation and annihilation operators

$$\boldsymbol{\eta}_s \equiv \frac{1}{\sqrt{2}}(\omega^{\frac{1}{2}}\mathbf{x}_s - i\omega^{-\frac{1}{2}}\mathbf{p}_s), \qquad \boldsymbol{\xi}_s \equiv \frac{1}{\sqrt{2}}(\omega^{\frac{1}{2}}\mathbf{x}_s + i\omega^{-\frac{1}{2}}\mathbf{p}_s), \quad s = 1, 2, \tag{13}$$

and using in (12) the correspondence

$$\begin{aligned}
\mathbf{p}_s &\to \mathbf{p}_s - i\omega\mathbf{x}_s = -i\omega^{\frac{1}{2}}\sqrt{2}\boldsymbol{\xi}_s, \\
\mathbf{p}_s^\dagger &\to \mathbf{p}_s + i\omega\mathbf{x}_s = i\omega^{\frac{1}{2}}\sqrt{2}\boldsymbol{\eta}_s.
\end{aligned} \tag{14}$$

we get

$$\begin{aligned}
\hat{\Pi}_1^2 &= \omega\big[C_{11} + \tfrac{1}{3}C_{22} + \tfrac{1}{\sqrt{3}}(C_{12} + C_{21})\big] + 1, \\
\hat{\Pi}_2^2 &= \omega\big[C_{11} + \tfrac{1}{3}C_{22} - \tfrac{1}{\sqrt{3}}(C_{12} + C_{21})\big] + 1, \\
\hat{\Pi}_3^2 &= \tfrac{4}{3}\omega C_{22} + 1,
\end{aligned} \tag{15}$$

where the operator $C_{st}, s, t = 1, 2$ are defined by

$$C_{st} = \boldsymbol{\eta}_s \cdot \boldsymbol{\xi}_t. \tag{16}$$

From the fact that

$$[\xi_{it}, \eta_{js}] = \delta_{ij}\delta_{st}\ ; \quad i, j = 1, 2, 3\ ; \quad s = 1, 2, \tag{17}$$

we have the commutation relations

$$[C_{st}, C_{s't'}] = C_{st'}\delta_{s't} - C_{s't}\delta_{st'}, \tag{18}$$

and thus they are generators[2] of a U(2) group. Therefore the operators $\hat{\Pi}_1^2, \hat{\Pi}_2^2, \hat{\Pi}_3^2$, appearing in the equation (15) are linear functions of the generators of this group.

To obtain from Eq. (15) the eigenvalues of the energy for this relativistic oscillator problem we can proceed as follows: First we note that the first order Casimir operator of U(2) group is

$$\hat{N} = C_{11} + C_{22}, \tag{19}$$

and that it has an SU(2) subgroup whose generators are

$$\begin{aligned} \hat{F}_+ &\equiv C_{12}, \\ \hat{F}_o &\equiv \tfrac{1}{2}(C_{11} - C_{22}), \\ \hat{F}_- &\equiv C_{21}, \end{aligned} \tag{20}$$

with a corresponding Casimir operator of the form

$$\hat{F}^2 \equiv \hat{F}_- \hat{F}_+ + \hat{F}_o(\hat{F}_o + 1). \tag{21}$$

By definition $\hat{N}$ and $\hat{F}^2$ commute with all C_{st} and among themselves, so from Eqs.(8,15), they will be integrals of motion of the operator $\Phi(E^2, \hat{\Pi}_s^2)$. Thus the eigenstates of the Eq. (8) can be characterized by the eigenvalues of $\hat{N}, \hat{F}^2$ which we denote respectively by

$$N, \quad \mathcal{F}(\mathcal{F}+1), \tag{22}$$

with $\mathcal{F}$ taking the values $(N/2), (N/2) - 1$, down to 1/2 or 0 depending on whether N is odd or even.

Another operator that commutes with $\hat{N}, \hat{F}^2$ is obviously $\hat{F}_o$ and we shall designate its eigenvalue by

$$\nu = \mathcal{F}, \mathcal{F} - 1, \ldots, -\mathcal{F}, \tag{23}$$

so the eigenstates associated with $\hat{N}, \hat{F}^2, \hat{F}_o$ could be represented by the ket

$$|N\mathcal{F}\nu\rangle, \tag{24}$$

and the solution ψ of Eq. (8) is necessarily a linear combination of these kets *i.e.*

$$\psi = \sum_{\nu=-\mathcal{F}}^{\mathcal{F}} a_\nu |N\mathcal{F}\nu\rangle, \tag{25}$$

as $\hat{N}, \hat{F}^2$ are integrals of motion.

To obtain the eigenvalues of the internal energy E as function of $N, \mathcal{F}$ we need first to consider the matrix elements of the operator Φ of (8) in the basis (24) *i.e.*

$$\langle N\mathcal{F}\nu'|\Phi(E^2,\hat{\Pi}_s^2)|N\mathcal{F}\nu\rangle. \tag{26}$$

These matrix elements can be obtained straightforwardly as from (15, 19, 20) we have

$$\begin{aligned}
\hat{\Pi}_1^2 &= 1+\omega\big[(2/3)(\hat{N}+\hat{F}_o)+(1/\sqrt{3})(\hat{F}_+ + \hat{F}_-)\big],\\
\hat{\Pi}_2^2 &= 1+\omega\big[(2/3)(\hat{N}+\hat{F}_o)-(1/\sqrt{3})(\hat{F}_+ + \hat{F}_-)\big],\\
\hat{\Pi}_3^2 &= 1+\omega\big[(2/3)(\hat{N}-2\hat{F}_o)\big],
\end{aligned} \tag{27}$$

and besides we have the well known angular momentum relations which, in the language used here for the SU(2) group, are

$$\begin{aligned}
\hat{F}_\pm|N\mathcal{F}\nu\rangle &= [(\mathcal{F}\mp\nu)(\mathcal{F}\pm\nu+1)]^{\frac{1}{2}}|N\mathcal{F}\nu\pm 1\rangle,\\
\hat{F}_o|N\mathcal{F}\nu\rangle &= \nu|N\mathcal{F}\nu\rangle.
\end{aligned} \tag{28}$$

To get the internal energy

$$E(N,\mathcal{F},\alpha), \tag{29}$$

with α indicating the rest of the indices, we need to evaluate the determinant of the $(2\mathcal{F}+1)\times(2\mathcal{F}+1)$ matrix whose elements are (26) and equate it to zero. This gives us a numerical equation of degree $4(2\mathcal{F}+1)$ in the variable E^2 and its solution provides us with the values indicated symbolically in (29).

Note that the energy values derived from the determinant of the matrix whose elements are given by Eq. (26) involve both positive and negative parts. To restrict ourselves only to the positive part we musts take the largest eigenvalue of the set. Thus, for example, if we consider $N=0$, this implies $\mathcal{F}=\nu=0$ and

$$<000|\Pi_s^2|000>=1, \quad s=1,2,3. \tag{30}$$

From Eq. (9) we then get for the expectation values of A, B, C with respect to the state $|000>$ that

$$<A>=3, <B>=-3, <C^2>=16 \tag{31}$$

and our equation giving us the energy is

$$E^8-12E^6+30E^4-28E^2+9=0 \tag{32}$$

whose four roots for E^2 are $E^2=1$ repeated three times and $E^2=9$. According to the statement at the beginning of the paragraph we must restrict ourselves tot he largest value of E^2, *i.e.*, 9 as the positive square root is 3, the value we expect for the energy of three particles in our units when there is no motion.

In a similar way we calculate the largest eigenvalues of the energy squared when $N = 1, \mathcal{F} = \frac{1}{2}$ and $N = 2, \mathcal{F} = 1$, and $N = 2, \mathcal{F} = 0$. For $\omega = 1$ these values as function of N are shown in Fig.1, with the allowed angular momenta values indicated above the level together with their degeneracies. On the right hand side of the levels we give the value of $\mathcal{F}$ and, as $\Phi(E, \Pi_s^2)$ is invariant under the permutation group of three particles $S(3)$, we give also the irrep of this group to which they belong.

We have outlined the procedure by which we can get the energy squared in Eq. (8) associated with the quantum numbers N and $\mathcal{F}$. We now proceed to discuss the wave functions.

3 The eigenstates of our problem

The ket $|N\mathcal{F}\nu>$ of definite angular momentum L and projection M can be expressed in terms of the two particle harmonic oscillator state

$$|n_1 l_1, n_2 l_2, LM\rangle = \sum_{m_1, m_2} < \mathbf{x}_1 | n_1 l_1 m_1 >< \mathbf{x}_2 | n_2 l_2 m_2 >< l_1 m_1 l_2 m_2 | LM > \tag{33}$$

as from Eq. (19), (20) and inequalities characterizing L we have

$$N = 2n_1 + l_1 + 2n_2 + l_2, \quad |\nu| = \tfrac{1}{2}|2n_1 + l_1 - 2n_2 - l_2|, \; |l_1 - l_2| \leq L \leq l_1 + l_2 \tag{34}$$

and in Eq. (33) we denote by $< \mathbf{x}|nlm >$ the standard single particle harmonic oscillator state[3] and by $< | >$ a Clebsch Gordan coefficient.

The ket of Eq. (33) is not an eigenstate of F^2 of Eq. (21), but the matrix elements of this operator with respect to these eigenstates are easily evaluated[4] and the diagonalization of the corresponding finite matrix in which $N, |\nu|, L, M$ remain fixed, gives the linear combinations of the kets of Eq. (33) that correspond to a definite $\mathcal{F}$. Furthermore as all the operators used so far are invariant under permutations of the three particle indices of the original coordinates and momenta, we can characterize our states by irreps f of the symmetry group S(3) and Yamanouchi symbols r. Thus we can obtain the states

$$|N\mathcal{F}|\nu|, LM, fr>= \sum_{n_1 l_1 n_2 l_2} |n_1 l_1, n_2 l_2 LM><n_1 l_1 n_2 l_2 L | N\mathcal{F}|\nu|, LM, fr > \tag{35}$$

where the summation extends over all n_1l_1, n_2l_2 for which $N, |\nu|$, and L are fixed.

Fortunately for the examples we want to discuss, that go only up to $N = 2$, the relation between the states in Eq. (35) and those in Eq. (33) are simple and we give them in Table 1 where we suppressed the index M on both sides and also the index $|\nu|$ as it can be obtained from the right hand side of the table using Eq. (34).

States in Eq. (35) $|N\mathcal{F}Lfr\rangle$	States in Eq.(35) in terms of those in Eq.(33)
$|000\{3\}\rangle$	$|00,00,0\rangle$
$|1\frac{1}{2}1\{21\}1\rangle$	$|00,01,1\rangle$
$|1\frac{1}{2}1\{21\}2\rangle$	$|01,00,1\rangle$
$|212\{3\}\rangle$	$\frac{1}{\sqrt{2}}[|00,02,2\rangle + |02,00,0\rangle]$
$|212\{21\}1\rangle$	$(\frac{1}{\sqrt{2}})[|00,02,2\rangle - |02,00,2\rangle]$
$|212\{21\}2\rangle$	$|01,01,2\rangle$
$|210\{3\}\rangle$	$(\frac{1}{\sqrt{2}})[|10,00,0\rangle + |00,10,0\rangle]$
$|210\{21\}1\rangle$	$(\frac{1}{\sqrt{2}})[|00,10,0\rangle - |10,00,0\rangle]$
$|210\{21\}2\rangle$	$-|01,01,0\rangle$
$|201\{111\}\rangle$	$|01,01,1\rangle$

Table 1. We suppresed the index M on both sides of Eq. (35) and also the index $|\nu|$, which can be calculated from the right hand side using its definition Eq. (34). The irreps of S(3) are $f = \{3\}, \{111\}$ that have a single Yamanouchi symbol and thus are suppressed while for $f = \{21\}$ the Yamanouchi symbols $r = (211)$ and $r = (121)$ will be denoted by 1 and 2 respectively.

4 The system of three quarks and the spectra for non-strange barions

It is well known that the wave functions[5] for a system of three quarks have an orbital, spin, flavor and color parts. As all barions are colorless the color part must be antisymmetric under interchange of the quarks. As the latter are fermions the total wave function must also be antisymmetric under interchange of the quarks. Thus the orbital, spin and flavor part taken together must be symmetric under the interchange of the quarks. We are only going to discuss the non-strange barions that are formed from either up(u) or down(d) quarks that we assume of the same mass m, and, in our units, $m = 1$.

The orbital part of our three quark system was discussed in the previous section. The spin part corresponds to three spins of value $\frac{1}{2}$, so if they add up to $\frac{3}{2}$ the spin wave function is symmetric *i. e.* corresponds to the irrep $\{3\}$ of S(3). On the other hand if they add up to spin $\frac{1}{2}$, there are two possibilities, one is that the first two spins add to 1 and the other that they add to 0, and thus we have a two dimensional irrep of S(3) that can only be $f = \{21\}$. Thus the spin part of the ket takes the values

$$|S\sigma fr >= |\tfrac{3}{2}\sigma\{3\} > \text{ or } |\tfrac{1}{2}\sigma\{21\}r > \tag{36}$$

with σ being the projection of the spin.

The ket for the orbital plus spin part of the wave functions becomes then

$$\begin{aligned} &|N\mathcal{F}|\nu|, L, f'; Sf''; JMfr > \\ &= \sum_{m\sigma} |NF|\nu|, Lm, f' > |S\sigma f'' >< Lm, s\sigma|JM > \end{aligned} \tag{37}$$

where f', f'' are the irrep of S(3) corresponding respectively to the orbital and spin part and f is now the total irrep of S(3).

The flavor belongs to the irrep a U(3) group whose states, for non-strange quarks, reduces to that of the isospin, which again can only have values $I = \frac{3}{2}, \frac{1}{2}$ corresponding to the irrep $\{3\}, \{21\}$, of S(3). Thus it can described by just the ket $|f\bar{r} >$ and the total symmetric state is just the coupling of the fr in the ket of Eq. (37) with $|f\bar{r} >$ so as to give the symmetric representation $\{3\}$ of the total wave function.

As the function $\phi(E^2, \Pi_s^2)$ does not depend on the flavor part we just have to deal with its matrix elements with respect to the ket in Eq. (37). The partitions $\{21\}, \{3\}, \{111\}$ characterizing the irrep for spin-orbital part of the eigenfunction Eq. (37) correspond respectively to the what are called the octet, decuplet and singlet states[5].

We start our discussion with the octet states corresponding to $f = \{21\}$. For $N = 0$ the only orbital ket in Table 1 is $|000\{3\} >$ so the spin part must be $|\frac{1}{2}\{21\}r >$ to get the total irrep $\{21\}$ and angular momentum $J = \frac{1}{2}$. Thus for $N = 0$ the only state in Eq. (37) is

$$|0000\{3\}, \tfrac{1}{2}\{21\}\tfrac{1}{2}M > \tag{38}$$

and it corresponds to the lowest level in Fig. 1.

For $N = 1$ the orbital kets in Table 1 are $|1\frac{1}{2}1\{21\}r >$, while the spin part can be either $|\frac{1}{2}\{21\}r >$ or $|\frac{3}{2}\{3\}r >$ if the total resulting irrep of f has

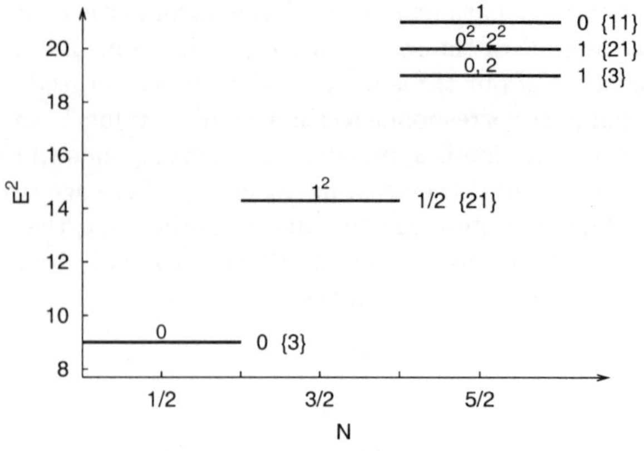

Figure 1. The square of energy E^2, as function of the number of quanta N given when $\omega = 1$. The values of $\mathcal{F}$ are given on the right of each level and the possible orbital angular momenta, with their degeneracies, are given above each level. As the function $\phi(E^2, \Pi_s^2$ is invariant under permutation of the three particles the irreps of $SU(3)$ *i.e.* $\{3\}$, $\{21\}$, $\{111\}$ characterize the levels as discussed in Table 1, and are also indicated in the figure

to be $\{21\}$. Thus the states corresponding tot he spin-orbit part are of the form

$$|1\tfrac{1}{2}\tfrac{1}{2}1\{21\},\ S, JM\{21\} > \tag{39}$$

where we suppressed f'' as it is related to S, and the possible values of (S, J) are

$$(S, J) = (\tfrac{3}{2}, \tfrac{5}{2}), (\tfrac{3}{2}, \tfrac{3}{2}), (\tfrac{3}{2}, \tfrac{1}{2}), (\tfrac{1}{2}, \tfrac{3}{2}), (\tfrac{1}{2}, \tfrac{1}{2}). \tag{40}$$

Thus for $N = 1$, which corresponds to the second level in Fig. 1, there are five states.

For $N = 2$ here are several levels in Fig. 1, but we shall consider only the lowest one for which $f = \{3\}$ and $L = 0, 2$. The spin state has then to be $f'' = \{21\}$ so $S = \frac{1}{2}$ and the spin-orbit states have the form

$$|211L\{3\}, \tfrac{1}{2}, JM >, \tag{41}$$

where we again suppressed the f'', and the possible values of (L, J) are

$$(L, J) = (0, \tfrac{1}{2}), (2, \tfrac{3}{2}), (2\tfrac{5}{2}). \tag{42}$$

Thus for $N = 2, f' = \{3\}$ which corresponds to the third level in Fig. 1 there are three states.

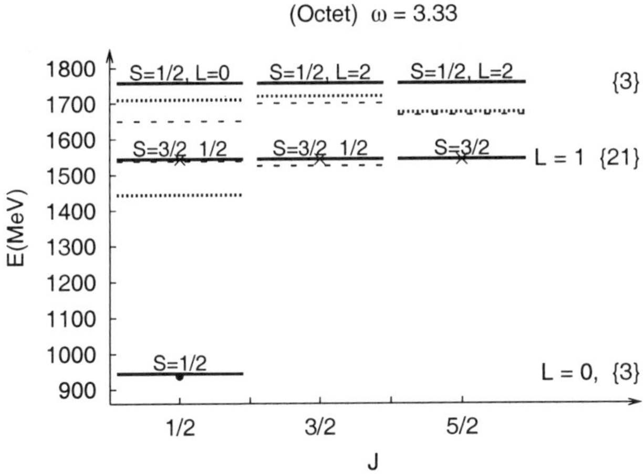

Figure 2. The experimental non-strange octet energy levels (the masses) are indicated by dashed or dotted lines depending on whether in reference[5] they correspond to 1 or 2 quanta. By construction the ground state of the octet coincides with our theoretical result and thus is marked by a full line as do all our other theoretical levels. Our analysis is based on the formalism developed in the paper. The first and the second excited theoretical levels are respectively quintuply and triple denerate but they coincide quite reasonably with experiment. In the abscissa we give the total angular momentum J, while the orbital angular momentum L and S are indicated either above or to the right of the theoretical levels. Also to the right irreps of $SU(3)$ for the orbital part of the state are indicated. In our units $\hbar = m = c = 1$, with m being the mass of the u and d quarks, the frequency ω we use is, an indicated above, $\omega = 3.33$. Only one state, denoted by "?" does not have a theoretical counterpart

The separation between the levels depends on the value of our only parameter ω. In Fig. 2 we draw the experimental energy levels of the octet as function of their total angular momentum J. In another publication[5] these energy levels are also characterized by the number of quanta and thus we indicate by dashed and dotted lines those corresponding respectively to 1 and 2 quanta,. The full lines are the result of our calculation when we choose $\omega = 3.3$. In our units the mass of the quark is taken as 1 so to put our results in MeV we multiply them by (939/3) MeV assuming that the mass of the quark is 1/3 of that of the proton. Above or to the right of the full lines in Fig. 2 we indicate the values of the spin and the orbital angular momenta and also the irrep f' of the orbital part. There is one level indicated by? which is not adjusted by our model.

We now turn our attention to the decuplet case where $f = \{3\}$. For $N = 0$ the only orbital state is $|000\{3\}>$ and the spin part must also correspond to $f'' = \{3\}$, or, equivalently, $S = \frac{3}{2}$ if $f = \{3\}$. Thus we see from Fig. 1 that the lowest level corresponds to the state

$$|0000\{3\}, \tfrac{3}{2}\{3\}, \tfrac{3}{2}M > . \tag{43}$$

For $N = 1$, the orbital state in Table 1 corresponds to $f' = \{21\}$, so the spin state must have $f'' = \{21\}$ for the total irrep of S(3) to be $f = \{3\}$. The spin is then $S = \frac{1}{2}$ and we have the states

$$|1\tfrac{1}{2}\tfrac{1}{2}1\{21\}, \tfrac{1}{2}\{21\}, JM\{3\} > . \tag{44}$$

As J is then $\frac{1}{2}$ or $\frac{3}{2}$ we have two states.

For $N = 2, f' = \{3\}$ we see from Table 1 that the orbital state are $|212\{3\}>$, so the spin must also have $f'' = \{3\}$ or, equivalently, $S = \frac{3}{2}$. Our states then have the form

$$|2112\{3\}, \tfrac{3}{2}\{3\}, JM\{3\} > \tag{45}$$

As J is then $\frac{1}{2}, \frac{3}{2}, \frac{5}{2}, \frac{3}{2}$ we have four states.

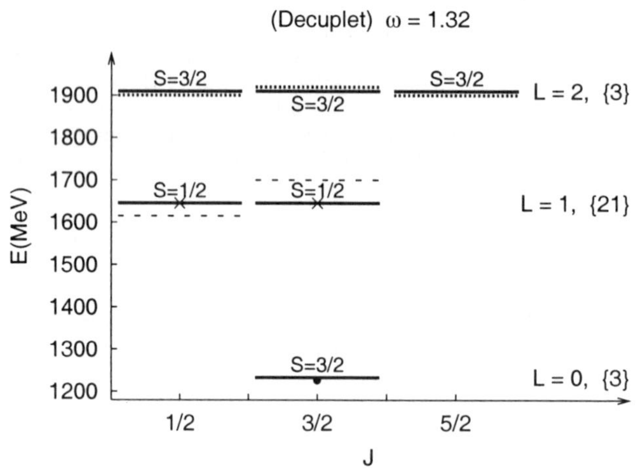

Figure 3. Most of what is mentioned in Fig.2 holds for Fig.3 if we replace the word octet by decuplet. We note that now the first and the second excited theoretical levels are respectively doubly and triple degenerate, and all lowest experimental levels seem to be fitted with $\omega = 1.32$.

In Fig. 3 we draw the experimental energy levels of the decuplet in MeV as function of their total angular momentum J. Again, in another publication[2], these energy levels are also characterized by the number of quanta and this we indicate by dashed and dotted lines corresponding respectively to 1 and 2 quanta. The full lines are the result of our calculation when we choose $\omega = 1.32$ and multiply our levels by (1232/3) MeV so that mass of our quarks instead of 1 becomes 1/3 of the lowest level of the decuplet. We only show the levels up to $J = \frac{5}{2}$ and give above them the spin and to the right the orbital angular momentum and irrep f' of the S(3) group to which they correspond.

5 Conclusion

Starting from the Hamiltonian formed from the sum of three Dirac free particle equations we obtain, by repeatedly squaring, a spin independent operator applied to our wave function. By going to Jacobi momenta and replacing them by creation and annihilation operators we reduce our operator to a function of the generators of an SU(2) group and obtain both its eigenvalues and eigenfunctions.

As an exercise we apply our results to the theoretical discussion of the masses of a three quark model for the octet and decuplet barions and compare them with experiment.

The author would like to acknowledge the support of Consejo Nacional de Ciencia y Tecnología, Project 32521E.

References

1. M. Moshinsky and Yu. F. Smirnov, *The harmonic oscillator in moden physics*, Hardwood Academic Publishers, Amsterdam, The Netherlands (1996), pp. 398-399.
2. M. Moshinsky, *Group theory and the many body problem*, Gordon and Breach, Science Publishers, New York (1968), p. 9.
3. T. A. Brody and M. Moshinsky, *Tables of transformation brackets for harmonic ocillator states*, Gordon and Breach, Science Publishers, New York, Second Edition (1967), p. 46.
4. V. Bargman and M. Moshinsky,*Nucl. Phys.* **18**, 697 (1960), *Nucl. Phys.* **23**,177 (1961).
5. Y. S. Kim and M. E. Noz, *Theory and applications of the Poincaré group*, D. Reichel Publishing Company, Dordreht, Holland (1986), ChapterXI and in particular pp.276-277.

REPRESENTATIONS OF $SO_0(P+1,Q) \times_S I\!R^{P+Q+1}$ FROM ITS DEFORMATIONS

P. MOYLAN
Department of Physics, The Pennsylvania State University, Abington College, Abington, PA 19001, USA
E-mail: pjm11@psu.edu

1 Introduction

We start with a well-known deformation [1] of the Poincaré Lie algebra in $p+q+1$ dimensions $(p+q>0)$, which is defined in terms of the generators $\mathbf{L}_{ij}$ of (pseudo) rotations and the translation generators $\mathbf{P}_i$ by the following (c.f. section 2 for definitions):

$$\mathbf{L}_{ij} \quad \rightarrow \quad \mathbf{L}_{ij} \quad , \tag{1.1a}$$

$$\mathbf{P}_i \quad \rightarrow \quad \mathbf{L}^{\pm}_{p+q+1,i} \quad = \quad \frac{i}{2\,Y}[\mathbf{Q}_2,\ \mathbf{P}_i] \ + \ \mathbf{P}_i \tag{1.1b±}$$

where $\mathbf{Q}_2 = \frac{1}{2}\sum_{i,\ j\ =\ 0}^{p+q} \mathbf{L}_{ij}\ \mathbf{L}^{ji}$ is the second order Casimir operator of $SO_0(p+1,q)$, and Y satisfies $Y^2 = \pm \sum_{i,\ j\ =\ 0}^{p+q} \mathbf{P}_i\ \mathbf{P}^i$. ([,] denotes commutator.) Choice of the plus sign in this equation for Y^2 leads to the Lie algebra of $SO_0(p+2,q)$ and the minus sign gives the commutation relations of $SO_0(p+1,q+1)$. Now eqns. (1.1b±) may be considered as algebraic equations for the translation generators $\mathbf{P}_i$, and we attempt to solve these equations for the $\mathbf{P}_i$. The solution of this problem in the $p=0$, $q=3$ case and with the minus sign for Y^2 is given in ref. (1). It expresses the basis elements of the Poincaré Lie algebra as certain irrational functions of the generators of $SO_0(1,4)$. Here we describe a generalization of this "anti-deformation" for arbitrary p and q $(p+q>0)$ with $p \geq 0$ and $q \geq 0$ (but for $q=0$ we only consider (1.1b−)). The generalization which we describe here holds only for a particular class of representations, which in the four dimensional case considered in ref. (1) corresponds to spinless representations of the Poincaré group.

We briefly mention that we have also obtained results similar to those in ref. (1) for q-deformations in low dimensions e.g. in 2, 3 and 4 dimensions. For example, in two dimensions we start with the Euclidean group, $\mathcal{E}(2)$, with generator $\mathbf{L}_{12}$ (rotation generator) and $\mathbf{P}_i$ $(i = 1, 2)$ (translation generators), and define: $\mathbf{L}_{3i} = \left([2]_{\sqrt{q}}\,(\mathbf{P}_1^2 + \mathbf{P}_2^2)^{1/2}\right)^{-1} [[-i\mathbf{L}_{21}]^2_{\sqrt{q}}\,,\ \mathbf{P}_i] + \mathbf{P}_i$. We obtain the "anti-deformation" by solving these equations for the $\mathbf{P}_i$. As above the solution expresses the translation generators of $\mathcal{E}(2)$ as irrational functions of $U_q(so(2,1))$ generators. [2]

2 $\mathbf{SO_0(p+1,q)}$, $\mathbf{SO_0(p+1,q)} \times_s \mathbb{R}^{p+q+1}$ and their Lie algebras

Let M denote the real vector space $\mathbb{R}^{p+q+1}$ on which the quadratic form

$$Q(\xi) \;=\; \xi_0^2 + \xi_1^2 + \ \ldots\ + \xi_p^2 - \xi_{p+1}^2 - \ \ldots\ - \xi_{p+q}^2\ . \tag{2.1}$$

is defined, and the associated bilinear form of signature (p, q) is given by

$$B(\xi,\, \eta) \;=\; \frac{1}{2}\{Q(\xi + \eta) - Q(\xi) - Q(\eta)\}\ . \tag{2.2}$$

Denote the matrix corresponding to B by β_0: $\beta_0 \;=\; \mathrm{diag}(1,\ 1,\ ..\ 1,\ -1,\ -1,\ ..,\ -1)$, where the right hand side of this equation denotes a diagonal matrix with diagonal entries as shown inside the parentheses, with $p+1$ entries being $+1$ and q entries being -1.

$SO_o(p+1, q)$ is the component connected to identity of

$$SO(p+1, q) \;=\; \{g \in SL(n, \mathbb{R}) \mid g^\dagger\,\beta_0\,g \;=\; \beta_0\}\ . \tag{2.3}$$

(† denotes transpose of a matrix and $n \;=\; p+q+1$.) Its Lie algebra, $so(p+1, q)$, is defined as:

$$so(p+1, q) \;=\; \{X \in sl(n, \mathbb{R}) \mid X^\dagger\,\beta_0 \;+\; \beta_0\,X \;=\; 0\}\ . \tag{2.4}$$

A basis of $\mathcal{G}$ are the generators $\mathbf{L}_{ij}$ $(i,\ j \;=\; 0,\ \ldots,\ p+q,\ i \;<\; j)$. We define $\mathbf{L}_{ij} \;=\; -\mathbf{L}_{ji}$ for $i > j$, and the $\mathbf{L}_{ij}$ satisfy the following commutation relations:

$$[\mathbf{L}_{ij},\ \mathbf{L}_{jk}] \;=\; -\,e_j\,\mathbf{L}_{ik} \tag{2.5}$$

and all other commutators vanish. The e_j are defined as follows:

$$e_0 \;=\; e_1 \;=\; \ldots \;=\; e_p \;=\; 1,\ e_{p+1} \;=\; e_{p+2} \;=\; \ldots \;=\; e_{p+q} \;=\; -1\ . \tag{2.6}$$

Eqn. (2.5) can also be given a well-known form with help of the $g_{ij} \;=\; (\beta_0)_{ij}$:

$$[\mathbf{L}_{ij},\ \mathbf{L}_{kl}] \;=\; g_{ik}\,\mathbf{L}_{jl} + g_{jl}\,\mathbf{L}_{ik} - g_{il}\,\mathbf{L}_{jk} \;-\; g_{jk}\,\mathbf{L}_{il}\ , \tag{2.7}$$

We wish to give explicit matrix expressions of these $\frac{n(n-1)}{2}$ generators in our (β_0) presentation of the fundamental representation of $\mathcal{G}$. Let E_{ij} denote the $n \times n$ matrix whose i, j entry is one, and all other entries are zero. Clearly

$$[E_{ij}, E_{kl}] = \delta_{jk} E_{il} - \delta_{li} E_{kj} \ . \tag{2.8}$$

Then $(q \neq 0)$

$$\mathbf{L}_{ij} = -(E_{i,j} - E_{j,i}) \quad (0 \leq i,\ j \leq p,\ i \neq j) \tag{2.9a}$$

$$\mathbf{L}_{n-i,n-j} = E_{n-i,n-j} - E_{n-j,n-i} \quad (1 \leq i,\ j \leq q,\ i \neq j) \tag{2.9b}$$

$$\mathbf{L}_{i,p+j} = E_{i,p+j} + E_{p+j,i} \quad (0 \leq i \leq p,\ 1 \leq j \leq q). \tag{2.9c}$$

The quadratic Casimir operator of $SO_0(p+1, q)$ is:

$$\mathbf{Q}_2 = \frac{1}{2} \sum_{i,j=0}^{p+q} \mathbf{L}_{ij}\, \mathbf{L}^{ji} = -\frac{1}{2} \sum_{i,j=0}^{p} \mathbf{L}_{ij}\, \mathbf{L}_{ij} -$$

$$-\frac{1}{2} \sum_{i,j=p+1}^{p+q} \mathbf{L}_{ij}\, \mathbf{L}_{ij} + \sum_{i=0}^{p} \sum_{j=1}^{q} \mathbf{L}_{i,p+j}\, \mathbf{L}_{i,p+j} \ . \tag{2.10}$$

The Poincaré group in $p+q+1$ dimensions is the semi-direct product:

$$SO_0(p+1, q) \times_s I\!R^{p+q+1}$$

$$= \{(g, a) |\ g \in SO_0(p+1, q),\ \ a \in I\!R^{p+q+1}\ \} \tag{2.11}$$

with multiplication law given by:

$$(g, a)\ (g', a') = (g\, g',\ a + \phi(g)\ a') \tag{2.12}$$

where $\phi(g)$ is the natural action of $SO_0(p+1, q)$ on $I\!R^{p+q+1}$. Its Lie algebra is

$$\mathcal{P} = so(p+1, q) \oplus_s I\!R^{p+q+1}, \tag{2.13}$$

and a basis of $\mathcal{P}$ is given by:

$$\mathbf{L}_{ij} \ ,\ \mathbf{P}_k \ \ (i,\ j,\ k = 0, \ldots, p+q) \ . \tag{2.14}$$

The commutation relations of these basis elements are: eqns. (2.7) for the $\mathbf{L}_{ij}$ and

$$[\mathbf{L}_{ij}, \mathbf{P}_k] = -g_{jk}\, \mathbf{P}_i + g_{ik}\, \mathbf{P}_j \ , \quad [\mathbf{P}_i, \mathbf{P}_j] = 0 \ . \tag{2.15}$$

A central element of the enveloping algebra $\mathcal{E}(\mathcal{P})$, which will play an important role in what follows, is the following:

$$\mathbf{P}^2 = \sum_{k=0}^{p+q} \mathbf{P}_k\, \mathbf{P}^k = \mathbf{P}_0^2 + \mathbf{P}_1^2 + \ldots + \mathbf{P}_{p+q}^2 - \mathbf{P}_{p+q+1}^2 - \ldots - \mathbf{P}_{p+q}^2 \; . \tag{2.16}$$

3 Representations of $\mathbf{SO_0(p+1,q)\times_s I\!R^{p+q+1}}$ from representations of $\mathbf{SO_0(p+2,q)}$ and $\mathbf{SO_0(p+1,q+1)}$

We would like to solve the equations (to the left of the arrow) in (1.1b±) for the $\mathbf{P}_i$. We have succeeded in doing this by working in a representation $(\pi, \mathcal{H})$ of $SO_0(p+1,q) \times_s I\!R^{p+q+1}$ (strongly continuous) on an Hilbert space $\mathcal{H}$ for which:

$$d\pi(\, \mathbf{P}_0\, \mathbf{\Delta} \;-\; \sum_{i,j=1}^{p+q} \mathbf{P}_j\, \mathbf{L}_{oi}\, \mathbf{L}^{ij}) \;=\; 0 \tag{3.1}$$

where

$$\mathbf{\Delta} \;=\; \frac{1}{2} \sum_{i,j=1}^{p+q} \mathbf{L}_{ij}\, \mathbf{L}^{ji} \; . \tag{3.2}$$

($d\pi$ denotes the representation of $\mathcal{P}$ obtained from π (by differentiation) and also its extension to $\mathcal{E}(\mathcal{P})$, the enveloping algebra of $\mathcal{P}$, and also to any algebraic extension of the Lie field of $\mathcal{P}$(c.f. ref. (1)).) Now the quadratic Casimir operator of $so(p+2,q)$ $[so(p+1,q+1)]$ constructed out of the $\mathbf{L}^+_{p+q+1,i}$ $[\mathbf{L}^-_{p+q+1,i}]$ of (1.1b+) [(1.1b−)] and $\mathbf{L}_{ij}$ is:

$$\mathbf{C}_2^\pm = \mathbf{Q}_2 - \sum_{i=0}^{p+q} \mathbf{L}^\pm{}_{p+q+1,i} \mathbf{L}^{\pm\, p+q+1,i} = \mathbf{Q}_2 \mp \sum_{i=0}^{p+q} \mathbf{L}^\pm{}_{p+q+1,i} \mathbf{L}^\pm{}_{p+q+1}{}^i . \tag{3.3±}$$

(Henceforth the plus sign of any ± refers to the $so(p+2,q)$ case ((1.1b+)) and the minus sign of any ± refers to the $so(p+1,q+1)$ case ((1.1b−)).)

Lemma: For representations $(\pi,\mathcal{H})$ for which (3.1) holds the following is true: $d\pi(\mathbf{C}_2^\pm) \;=\; \pm\, Y^2 \,-\, \left(\frac{p+q}{2}\right)^2 \cdot I \;=\; d\pi(\mathbf{P}^2) \,-\, \left(\frac{p+q}{2}\right)^2 \cdot I$, where now Y is the square root of the operator $\pm d\pi(\mathbf{P}^2)$, which we assume is self-adjoint, and strictly positive, so that it has a unique square root. Furthermore $(n = p+q+1)$:

$$d\pi(\mathbf{D})\; d\pi(\mathbf{P}_0) \;=\; d\pi(\sum_{i=0}^{p+q} \mathbf{A}_0{}^i\; \mathbf{L}^\pm_{n,i}) \tag{3.4±}$$

where

$$\mathbf{D} \;=\; \frac{(n-3)^2}{4}\left\{\mathbf{Q}_2 + \frac{(n-1)(n-3)}{4}\right\} + i\,Y\,(n-3)\,\{\mathbf{Q}_2 + \frac{(n-2)(n-3)}{4}\} - Y^2\left\{\mathbf{Q}_2 - \frac{(n-3)}{2}\right\} + i\,Y^3\,(n-2) - Y^4,$$

and

$$\sum_{i=0}^{p+q}\mathbf{A}_0{}^i\,\mathbf{L}^{\pm}_{n,i} \;=\; \frac{i}{4}\,(n-3)^2\left\{(\frac{n-3}{2})\,\mathbf{L}^{\pm}_{n,0} + \sum_{i=1}^{p+q}\mathbf{L}^{0,i}\mathbf{L}^{\pm}_{n,i}\right\}Y$$

$$-\,2\,(\frac{n-3}{2})\left\{\frac{1}{2}\,(\frac{n-3}{2})\,\mathbf{L}^{\pm}_{n,0} + \sum_{i=1}^{p+q}\mathbf{L}^{0,i}\mathbf{L}^{\pm}_{n,i}\right\}Y^2$$

$$+\,i\left\{(\frac{n-3}{2})\,\mathbf{L}^{\pm}_{n,0} - \sum_{i=1}^{p+q}\mathbf{L}^{0,i}\mathbf{L}^{\pm}_{n,i}\right\}Y^3 - \mathbf{L}^{\pm}_{n,0}\,Y^4\;.$$

Now start with a unitary representation of $SO_0(p+2,q)$ $[SO_0(p+1,q+1)]$ on an Hilbert space $\mathcal{H}$ and we use eqn. (3.4±) to construct a representation of $\mathcal{P}$ out this representation. Denote the (abstract) generators of $SO_0(p+2,q)$ $[SO_0(p+1,q+1)]$ by $\tilde{\mathbf{L}}_{ij}$ $(i=0,1,...,p+q)$ and $\tilde{\mathbf{L}}^{+}_{n,i}$ $[\tilde{\mathbf{L}}^{-}_{n,i}]$ $(n=p+q+1,$ $i=0,1,...,p+q)$.

Theorem: Let $(\pi^{\pm},\mathcal{H})$ be a continuous, unitary representation of $SO_0(p+2,q)$ $[SO_0(p+1,q+1)]$ on an Hilbert space $\mathcal{H}$ and let $(\pi^{\pm},\mathcal{H})$ be such that:

$$\left(\psi,\left\{d\pi^{\pm}(\tilde{\mathbf{C}}_2)+\left(\frac{p+q}{2}\right)^2\cdot I\right\}\psi\right)\gtrless 0\;\forall\,\psi\in\mathcal{D}(d\pi(\tilde{\mathbf{C}}_2))\,, \qquad (3.5\pm)$$

where $\mathcal{D}(\;\;)$ denotes the domain of an operator ($>$ is for (3.5+) and $<$ is for (3.5−)) and

$$d\pi^{\pm}\left(\tilde{\mathbf{L}}^{\pm}_{0n}\,\tilde{\boldsymbol{\Delta}} - \sum_{i,j=1}^{p+q}\tilde{\mathbf{L}}^{\pm}_{nj}\,\tilde{\mathbf{L}}_{oi}\,\tilde{\mathbf{L}}^{ij}\right) \;=\; 0\;.$$

Let $Y:\mathcal{H}\longrightarrow\mathcal{H}$ be a bounded, symmetric linear operator on $\mathcal{H}$, which commutes with all elements of $d\pi^{+}(so(p+2,q))$ $[d\pi^{-}(so(p+1,q+1))]$ and satisfies

$$Y^2 \;=\; \pm\left\{d\pi(\tilde{\mathbf{C}}_2) + \left(\frac{p+q}{2}\right)^2\cdot I\right\}\;. \qquad (3.6\pm)$$

Further, let $\tilde{\mathbf{D}}$ and $\sum_{i=0}^{p+q} \tilde{\mathbf{A}}_0^i \, \tilde{\mathbf{L}}_{n,i}^{\pm}$ be as in eqn. (3.4±), but with $\tilde{\mathbf{L}}_{ij}$ and $\tilde{\mathbf{L}}_{n,i}^{\pm}$ instead of $\mathbf{L}_{ij}$ and $\mathbf{L}_{n,i}^{\pm}$. Then there is formally a skew symmetric representation $d\tilde{\pi}^{\pm}(\mathcal{P})$ of $\mathcal{P} = so(p+1,q) \oplus_s I\!R^{p+q+1}$ on $\mathcal{H}$ with $d\tilde{\pi}^{\pm}(\mathbf{L}_{ij}) = d\pi^{\pm}(\tilde{\mathbf{L}}_{ij})$ $(i = 0, 1, ..., p+q)$, $d\tilde{\pi}^{\pm}(\mathbf{P}_0) = d\pi^{\pm}(\tilde{\mathbf{D}})^{-1} \, d\pi^{\pm}(\sum_{i=0}^{p+q} \tilde{\mathbf{A}}_0^i \, \tilde{\mathbf{L}}_{n,i}^{\pm})$, and $d\tilde{\pi}^{\pm}(\mathbf{P}_i) = [d\tilde{\pi}^{\pm}(\mathbf{L}_{i0}), d\tilde{\pi}^{\pm}(\mathbf{P}_0)] (i = 1, 2, ..., p+q)$, provided $d\pi \left(\left[\tilde{\mathbf{Q}}_2, \sum_{i=0}^{p+q} \tilde{\mathbf{A}}_0^i \, \tilde{\mathbf{L}}_{n,i}^{\pm} \right] \right) \neq 0$ and $d\pi^{\pm}(\tilde{\mathbf{D}})^{-1}$ exists on a suitable dense domain in $\mathcal{H}$.

Finally we show that there are many examples of representations for which the hypotheses of the theorem are satisfied, and, hence, it is possible to obtain non-trivial examples of such representations of $so(p+1,q) \oplus_s I\!R^{p+q+1}$. The $SO_0(p+2,q)$ $(SO_0(p+1,q+1))$ representations are ones associated with functions on real hyperbolic spaces. Real hyperbolic spaces $H^{p,q}$ $(q > 0)$ are: $H^{p,q} = \{\xi \in I\!R^{p+q+1} | \xi_0^2 + \xi_1^2 + \xi_2^2 + ... + \xi_p^2 - \xi_{p+1}^2 - \cdots - \xi_{p+q}^2 \;=\; -1\}$. The isotropy subgroup of the point $\xi^i = 0$ $(i \neq p+q)$, $\xi^{p+q} = 1$ is $H = SO_0(p+1, q-1)$. Since the action of $SO_0(p+1,q)$ is transitive on $H^{p,q}$,we have that $H^{p,q} \cong SO_0(p+1,q)/SO_0(p+1,q-1)$. $H^{p,q}$ are semi-Riemannian spaces of constant curvature. The representation of $G = SO_0(p+1,q)$ is the left regular representation on $H^{p,q}$ i.e. $(\pi(g)f)(\xi) = f(g^{-1}\,\xi)$ for any $g \in G$ and $f \in \mathcal{L}^2(G/H)$. We have on $C^\infty(G/H)$ representations $d\pi(so(p+1,q))$ and $d\pi(\mathcal{E}(so(p+1,q))$ of the Lie algebra and the enveloping algebra, respectively. Let $\mathcal{H}_s = \{f \in C^\infty(G/H) | (d\pi(\mathbf{C}_2)f) = \left(s^2 \;-\; \{\frac{p+q-1}{2}\}^2\right) f\}$ for each $s \in \mathbb{C}$. We then have: [3]

Proposition: $d\pi(\mathbf{C}_2)$ has a self-adjoint closure in $\mathcal{L}^2(G/H)$. For $q > 1$ the spectrum of $d\pi(\mathbf{C}_2) + [(p+q-1)/2]^2 \cdot I$ in $(0, \infty)$ is discrete with eigenvalue $[s + (p+q-1)/2]^2$ for each integer $s > -(p+q-1)/2$. Corresponding to each eigenvalue is an infinite dimensional eigenspace $\mathcal{H}_s$. The completion of $\mathcal{H}_s$ in the $\mathcal{L}^2(G/H)$ norm gives an irreducible, unitary representation of $SO_0(p+1,q)$. The spectrum of $d\pi(\mathbf{C}_2) + [(p+q-1)/2]^2 \cdot I$ in $(-\infty, 0]$ is continuous with eigenvalue $[s]^2$ for each $s \in i\,I\!R$. The completion of $\mathcal{H}_s$ in the $\mathcal{L}^2(G/H)$ norm is the direct sum of two irreducible, unitary representations of $SO_0(p+1,q)$. For $q = 1$ the above is also true except that there is no discrete spectrum.

We leave it to the reader to show that the representations of the proposition fulfill the hypotheses of the theorem. The representations of the discrete spectrum, are the ones we use with (3.5+) (p of Proposition replaced by $p+1$). The representations of the continuous spectrum are the ones we use with (3.5−) (q in Proposition replaced by $q+1$).

References

1. M. Havlicek, P. Moylan, *J. Math. Phys.* **34**, 5320 (1993). (Also *c.f.* P. Bozek, M. Havlicek, O. Navratil, preprint: Universitas Carolina Pragensis, NCITF/1, 1985; *c.f.* A.O. Barut and A. Bohm, *Phys. Rev.* **139**, 1107 (1965) for applications to particle physics.)
2. P. Moylan, *Czechoslovak Jour. Physics* **48**, 1457 (1998).
3. N. Limic, J. Niederle, R. Raczka, *J.Math.Phys.* **8**, 1079 (1967); V. F. Molcanov, *Soviet Math. Dokl.* **9**, 1387 (1968); R.S. Strichartz, *J. Funct. Anal.* **12**, 341 (1973); W. Rossmann, *J. Funct. Anal.* **30**, 448 (1978).

COVARIANCE APPROACH TO QUANTUM THEORY

JAN NAUDTS

Departement Natuurkunde, Universiteit Antwerpen UIA, Universiteitsplein 1, B2610 Antwerpen, Belgium.
E-mail: Jan.Naudts@ua.ac.be

A slight generalization of the C*-algebraic approach to quantum theory is proposed. It is based on the well-known notion of covariance system, also called C*-dynamical system. New definitions of state and of representation of a covariance system are introduced. The concepts are clarified with standard quantum mechanics as an example.

1 Introduction

The C^*-algebraic approach to quantum mechanics and quantum field theory has gone a long way in clarifying the basic structures of quantum theory. See Landsman[4] for a review of the approach, Haag[6] for an in depth exposition. There are some indications that in its present form it is not flexible enough to deal with expected features of a quantum version of general relativity. The present paper tries to broaden the scope of the C^*-algebraic approach by a slight shift in paradigm. The notion of state and representation of a C^*-algebra are generalized to notions of state and representation of a covariance system. As a consequence, also the notion of observable is modified.

2 Covariance systems

A corner stone of the C^*-algebraic approach to quantum mechanics is the notion of C^*-dynamical system, also called covariance system. It is a triple $(\mathcal{A}, G, \alpha)$ consisting of a C^*- algebra $\mathcal{A}$, a locally compact group G, and a strongly continuous action α of G in $\mathcal{A}$ (i.e. α is a group homomorphism of G in aut$(\mathcal{A})$, the group of automorphisms of $\mathcal{A}$; for each $A \in \mathcal{A}$ the element $\alpha_g A$ depends continuously on g). An obvious example, at the origin of the name of dynamical system, is time evolution in standard quantum mechanics. Let $\mathcal{H} = \mathcal{L}_2(\mathbf{R}^d, \mathbf{C})$ be the Hilbert space of square integrable complex functions. Let $\mathcal{A} = \mathcal{B}(\mathcal{H})$ be the C^*-algebra of bounded operators in $\mathcal{H}$. Let $G = \mathbf{R}, +$ be the group of reals. Any bounded self-adjoint operator H in $\mathcal{H}$ defines a strongly continuous action α of $\mathbf{R}, +$ in $\mathcal{A}$ by

$$\alpha_t(A) = e^{i\hbar^{-1}tH} A e^{-i\hbar^{-1}tH}, \qquad A \in \mathcal{A}. \tag{1}$$

The triple $(\mathcal{B}(\mathcal{H}), \mathbf{R}, \alpha)$ is a C^*-dynamical system.

3 Covariant representations and crossed product algebras

Let $(\mathcal{A}, G, \alpha)$ be a covariance system. A *-representation π of the C^*-algebra $\mathcal{A}$ as bounded operators in a Hilbert space $\mathcal{H}$ is said to be *covariant* if there exists a unitary representation $\{U(g), g \in G\}$ of the group G in $\mathcal{H}$ with the property that

$$\pi(\alpha_g A) = U(g)\pi(A)U(g)^* \tag{2}$$

for all $g \in G$ and $A \in \mathcal{A}$.

In the previous example, the trivial representation $\pi(A) = A$ is covariant. Crossed product algebras have been introduced in physics by Doplicher, Kastler and Robinson[2]. A textbook exposition is found in Bratteli-Robinson[3]. The fundamental property of the crossed product algebra $\mathcal{A} \times_\alpha G$ is the one-to-one correspondence between its representations and the covariant representations of $\mathcal{A}$. This correspondence, between a covariant representation π of $\mathcal{A}$, with representation U of G, and the corresponding representation ρ of $\mathcal{A} \times_\alpha G$, is such that

$$\rho(f) = \int_G \mathrm{d}g\, \pi(f(g))U(g) \tag{3}$$

holds for any continuous function f of G in $\mathcal{A}$ with compact support (note that these functions f belong to the crossed product $\mathcal{A} \times_\alpha G$).

4 Standard quantum mechanics formulated as a covariance system

The algebra of continuous complex functions of $\mathbf{R}^d$ vanishing at infinity is denoted $\mathcal{C}_0(\mathbf{R}^d)$. It is a C^*-algebra for the supremum norm. A continuous action α of the group $\mathbf{R}^d, +$ in $\mathcal{C}_0(\mathbf{R}^d)$ is defined by

$$\alpha_a f(q) = f(q - a), \qquad a \in \mathbf{R}^d. \tag{4}$$

The triple $(\mathcal{C}_0(\mathbf{R}^d), \mathbf{R}^d, \alpha)$ is a covariance system. Standard quantum mechanics is described by a covariant representation π of $\mathcal{C}_0(\mathbf{R}^d)$ in the Hilbert space $\mathcal{L}_2(\mathbf{R}^d, \mathbf{C})$ of quadratically integrable complex functions of $\mathbf{R}^d$. In this representation each function f of $\mathcal{C}_0(\mathbf{R}^d)$ is represented by an operator $\pi(f)$ of $\mathcal{B}(\mathcal{H})$, defined as multiplication operator

$$\pi(f)\psi(q) = f(q)\psi(q), \qquad q \in \mathbf{R}^d. \tag{5}$$

The action α is represented by unitary operators $U(a)$, determined by

$$U(a)\psi(q) = \psi(q - a), \qquad a \in \mathbf{R}^d. \tag{6}$$

They form a representation of the group of shifts in $\mathbf{R}^d$.

It is well known that the generators of the unitary representation $U(a)$ of $\mathbf{R}^d, +$ are (proportional to) the momentum operators P_j. Indeed, one has

$$U(a) = \exp(-i\hbar^{-1} \sum_j a_j P_j) \tag{7}$$

On the other hand, the position operators Q_j are affiliated to the algebra $\pi(\mathcal{C}_0(\mathbf{R}^d))''$. Hence, by considering both the representation of functions of position f as operators $\pi(f)$ and the representation of the action α by means of unitary operators $U(a)$, one recovers the essence of standard quantum mechanics, namely position and momentum operators satisfying canonical commutation relations. With the covariant representation π of $\mathcal{A}$ corresponds a representation ρ of the crossed product algebra $\mathcal{C}_0(\mathbf{R}^d) \times_\alpha \mathbf{R}^d$ in $\mathcal{H}$. Position and momentum operators are affiliated with the (weak closure of the) image of this crossed product algebra under the representation ρ.

5 Observables

Quite often, one encounters the statement that the observables of the dynamical system $(\mathcal{A}, G, \alpha)$ are the self-adjoint elements of the C^*-algebra $\mathcal{A}$. In the C^*-algebraic approach to quantum field theory this is usually taken for granted – see e.g. Borchers[8] in these proceedings. In a representation as operators in a Hilbert space, also the self-adjoint unbounded operators affiliated with (the weak closure of) the image of $\mathcal{A}$ are considered to be observable. However, in standard quantum mechanics, the way it is presented here, only functions of position belong to $\mathcal{A}$. According to the above statement, these would be the only observables. It is clear that in conventional quantum mechanics also the momentum operators are observable. Hence, the set of observables is larger than $\mathcal{A}$. It seems therefore to be more correct to call observable all self-adjoint elements of the crossed product algebra $\mathcal{A} \times_\alpha G$, or affiliated with it. In the representation, (the weak closure of) the image of $\mathcal{A} \times_\alpha G$ contains both the operators $\pi(f)$ and the unitary operators $U(a)$. This means for the example above that both position and momentum operators are observable according to the extended notion.

6 Alternative description of standard quantum mechanics

Let us now take $\mathcal{A}$ equal to the complex numbers $\mathbf{C}$ instead of $\mathcal{C}_0(\mathbf{R}^d)$. Let $\mathbf{I}$ denote the trivial action of $\mathbf{R}^d$ in $\mathbf{C}$. Then $(\mathbf{C}, \mathbf{R}^d, \mathbf{I})$ is a rather trivial covariance system. Let V be the unitary representation of $\mathbf{R}^d$ in the Hilbert space $\mathcal{L}_2(\mathbf{R}^d, \mathbf{C})$ defined by

$$V(k)\psi(q) = e^{ikq}\psi(q). \tag{8}$$

It is clear that the trivial representation of $\mathbf{C}$ in $\mathcal{L}_2(\mathbf{R}^d, \mathbf{C})$ is a V-covariant representation. Hence it corresponds with a representation of the crossed product algebra $\mathbf{C} \times_{\mathbf{I}} \mathbf{R}^d$, which itself is isomorphic with $\mathcal{C}_0(\mathbf{R}^d)$. Indeed, the crossed product algebra $\mathcal{C}_0(\mathbf{R}^d) \times_\alpha \mathbf{R}^d$ of bounded observables of standard quantum mechanics, as discussed before, can be written as

$$(\mathbf{C} \times_{\mathbf{I}} \mathbf{R}^d) \times_\alpha \mathbf{R}^d \tag{9}$$

Alternatively, one can consider the covariance system $(\mathbf{C}, \mathbf{R}^d \times \mathbf{R}^d, \mathbf{I})$, where $\mathbf{I}$ is now the trivial action of the group $\mathbf{R}^d \times \mathbf{R}^d$ in $\mathbf{C}$. It is of course not possible to see the standard representation of quantum mechanics as a covariant representation of $\mathbf{C}$. At one hand, the representation of $\mathbf{C}$ is trivial. At the other hand the unitary representation $U(a)V(k)$ of $\mathbf{R}^d \times \mathbf{R}^d$ is projective because of the Weyl commutation relations

$$U(a)V(k) = e^{-iak}V(k)U(a). \tag{10}$$

A generalized notion of representations of a covariance system, discussed below, makes it possible to recover the standard representation of quantum mechanics as a representation of $(\mathbf{C}, \mathbf{R}^d \times \mathbf{R}^d, \mathbf{I})$. There are several reasons why we want the latter possibility. It treats position and momentum observables on the same footing. Instead of intrinsic properties of the particle they become representation-dependent properties. This evolution may be important in the context of general relativity where spacetime is assumed to be generated by particles rather than that particles live in a preexisting flat spacetime. Considering quantum mechanics as a representation of $(\mathbf{C}, \mathbf{R}^d \times \mathbf{R}^d, \mathbf{I})$ is strongly related to Weyl's approach to quantum mechanics (see Petz[5] for an introduction). The complex phase factor e^{-iak} which appears in (10) is a cocycle of the group $\mathbf{R}^d \times \mathbf{R}^d$. It plays an essential rôle both in the construction of the Weyl algebra and in the definition below of a representation of a covariance system.

7 Representations of a covariance system

A representation of the covariance system $(\mathcal{A}, G, \alpha)$ is defined[7] as a triple $(\pi, \mathcal{H}, U)$. It consists of a *-representation π of $\mathcal{A}$ in the Hilbert space $\mathcal{H}$ and a measurable map $g \to U_g$ of the locally compact group G into the unitary operators of $\mathcal{H}$. The following conditions should be satisfied.

- The unitary operators implement the action α, in formulas,

$$\pi(\alpha_g a) = U_g \pi(a) U_g^*, \qquad a \in \mathcal{A}, g \in G. \tag{11}$$

- The map $g \to U_g$ is strongly continuous in the neighborhood of the origin of G.

- There exists a right multiplier ζ of $(\mathcal{A}, G, \alpha)$ with values in the center of (the multiplier algebra of) $\mathcal{A}$ satisfying

$$U_g U_h = U_{gh} \pi(\zeta(g,h)), \qquad g, h \in G. \tag{12}$$

The right multiplier ζ is a measurable map of $G \times G$ into the unitary elements of the center of (the multiplier algebra of) $\mathcal{A}$ satisfying $\zeta(g,e) = \zeta(e,g) = \mathbf{I}$ (e is the neutral element of G) and

$$\alpha_k^{-1} \zeta(g,h) = \zeta(gh,k)^* \zeta(g,hk) \zeta(h,k), \qquad g,h,k \in G. \tag{13}$$

In case $\mathcal{A}$ is the algebra of complex numbers $\mathbf{C}$ then the action α is trivial so that (13) is the cocycle condition satisfied by the phase factor of a projective representation. Hence, in this case any projective representation U of G defines a representation of $(\mathbf{C}, G, \alpha)$.

8 States of a covariance system

Given a representation $(\pi, \mathcal{H}, U)$ of the covariance system $(\mathcal{A}, G, \alpha)$, let ψ be a normalized element of $\mathcal{H}$. It defines correlation functions F by

$$F(a; g; h) = (\pi(a) U_g^* \psi, U_h^* \psi), \qquad a \in \mathcal{A}, g, h \in G. \tag{14}$$

It is obvious to associate the notion of state of a covariance system with these correlation functions. Given F one can reconstruct the representation $(\pi, \mathcal{H}, U)$ up to an isometry[7]. This is a generalization of the well-known GNS-theorem holding for states of a C^*-algebra.

If $\mathcal{A} = \mathbf{C}$ then the first argument of F may be omitted. In particular, a wavefunction ψ in the standard representation of quantum mechanics defines a state of $(\mathbf{C}, \mathbf{R}^d \times \mathbf{R}^d, \mathbf{I})$ by

$$F(a, k; a', k') = (V(k)^* U(a)^* \psi, V(k')^* U(a')^* \psi) \tag{15}$$

Now remember that Moyal's[1] characteristic function $\chi(a,k)$ is given by

$$\chi(a,k) = e^{-iak/2} (V(k) U(a) \psi, \psi). \tag{16}$$

Hence the relation $\chi(a,k) = \exp(-iak/2)F(-a,-k;0,0)$ follows immediately. Conversely, given χ one can calculate F by

$$F(a,k;a',k') = e^{-i(a+a')(k'-k)/2}\chi(a'-a,k'-k) \tag{17}$$

The well-known Wigner function is obtained from χ by Fourier transform

$$\rho(a,k) = (2\pi)^{-d}\int \mathrm{d}a' \int \mathrm{d}k'\, e^{-i(k'a-ka')}\chi(a',k') \tag{18}$$

We conclude that the correlation functions F contain the same information as Wigner's (pseudo-)density function. However, they are useful in a much more general context.

Note by the way that also states of classical mechanics can be expressed by means of a correlation function F. Indeed, let $f(a,k)$ be the probability distribution for finding a particle at position a with speed v given by $\hbar k = mv$ (m is the mass of the particle, $\hbar$ is Planck's constant). Then

$$F(a,k;a',k') = \tilde{f}(a-a',k-k') \tag{19}$$

is a correlation function determining a state of the covariance system $(\mathbf{C},\mathbf{R}^d\times\mathbf{R}^d,\mathbf{I})$ ($\tilde{f}$ is the Fourier transform of f).

References

1. J.E. Moyal, *Quantum Mechanics as a Statistical Theory, Proc. Camb. Phil. Soc.* **45**, 99 (1949).
2. S. Doplicher, D. Kastler, D.W. Robinson, *Covariance Algebras in Field Theory and Statistical Mechanics, Commun. Math. Phys.* **3**, 1 (1966).
3. O. Bratteli, D.W. Robinson, *Operator Algebras and quantum statistical mechanics*, Vol. I, Springer-Verlag, New York (1979).
4. N.P. Landsman, *Quantization and superselection sectors I. Transformation group C^*-algebras, Rev. Math. Phys.* **2**, 45 (1990); *Quantization and superselection sectors II. Dirac monopole and Aharonov-Bohm effect, Rev. Math. Phys.* **2**, 73 (1990).
5. D. Petz, *An Invitation to the Algebra of Canonical Commutation Relations*, Leuven University Press, Leuven (1990).
6. R. Haag, *Local quantum physics*, 2nd ed., Springer (1996).
7. J. Naudts, M. Kuna, *Covariance systems*, math-ph/0009031, *J. Phys. A: Math. Gen.* **34**, 9265 (2001).
8. H.J. Borchers, *Quantum Field Theory as Dynamical System*, proceedings of the conference "Quantum Theory and Symmetries", Krakow, July 2001.

SUPER DUFFIN-KEMMER-PETIAU ALGEBRA

SUSUMU OKUBO
Department of Physics and Astronomy, University of Rochester, Rochester, NY 14627, USA
E-mail: okubo@pas.rochester.edu

We consider a super generalization of the Duffin-Kemmer-Petiau algebra. It is shown that it is related to the Lie superalgebra osp(N, M).

The Duffin-Kemmer-Petiau algebra (hereafter referred to as DKPA) in a N-dimensional carrier space is defined by the relation

$$\beta_\mu \beta_\nu \beta_\lambda + \beta_\lambda \beta_\nu \beta_\mu = \delta_{\mu\nu} \beta_\lambda + \delta_{\lambda\nu} \beta_\mu \tag{1}$$

for $\mu, \nu, \lambda = 1, 2, \ldots, N$. For an interesting historical account of the subject for the physical case of $N = 4$, see the article by Krajcik and Nieto [1]. Fishbach *et al.* [2] have classified the irreducible representations of DKPA in terms of those of the corresponding Lie algebra so$(N+1)$ as follows. Setting

$$\begin{aligned} J_{\mu\nu} &= i\left(\beta_\mu \beta_\nu - \beta_\nu \beta_\mu\right) \quad , \\ J_{0\mu} &= -J_{\mu 0} = \beta_\mu \quad , \\ J_{00} &= 0 \quad , \end{aligned} \tag{2}$$

then J_{AB} now for $A, B = 0, 1, 2, \ldots, N$ will satisfy the so$(N+1)$ Lie algebra if

$$\begin{aligned} J_{AB} &= -J_{BA} \quad , \\ [J_{AB}, J_{CD}] &= i\left(\delta_{BC} J_{AD} - \delta_{AC} J_{BD} - \delta_{BD} J_{AC} + \delta_{AD} J_{BC}\right) \quad . \end{aligned} \tag{3}$$

Since $J_{0\mu}$ $(\mu \neq 0)$ will satisfy

$$\left(J_{0\mu}\right)^3 = J_{0\mu} \quad ,$$

its eigenvalues are restricted only to these values of 1, 0, or -1, which is satisfied only by the totally antisymmetric tensor representations.

The purpose of this note is to analyze the super generalization of DKPA, which we refer to hereafter as SDKPA. Since its full account is already given in [3], we will sketch the result here. However, we can also treat the case of $N = \infty$ in what follows. For that end, let V be a superspace with

$$V = V_0 \oplus V_1 \tag{4}$$

where V_0 and V_1 refer to bosonic and fermionic subspaces respectively, we introduce the grading function σ in V by

$$\sigma(a) = \begin{cases} 0 \text{ if } a \epsilon V_0 \\ 1 \text{ if } a \epsilon V_1 \end{cases} \tag{5}$$

and consider the metric function g_{ab} satisfying conditions

$$\begin{aligned} &\text{(i)} \qquad g_{ba} = (-1)^{\sigma(a)\sigma(b)} g_{ab} \\ &\text{(ii)} \qquad g_{ab} = 0 \quad \text{if} \quad \sigma(a) \neq \sigma(b) \quad . \end{aligned} \tag{6}$$

Then, SDKPA is defined by the relation

$$\beta_a \beta_b \beta_c \pm (-1)^{\Sigma(a,b,c)} \beta_c \beta_a \beta_b = g_{ab} \beta_c \pm g_{bc} \beta_a \tag{7}$$

for $a, b, c \,\epsilon\, V$. Here, the sign function $\Sigma(a,b,c)$ is given by

$$\Sigma(a,b,c) = \sigma(a)\sigma(b) + \sigma(b)\sigma(c) + \sigma(c)\sigma(a) \quad . \tag{8}$$

Note that Eq. (7) is invariant under $a \leftrightarrow c$ in view of Eqs. (6). If Dim $V_0 = N$, and Dim $V_1 = M$ are finite, then the indices $1, 2 \ldots, N$ refer to the bosonic space V_0, which indices $N+1, N+2, \ldots, N+M$ are the fermionic ones. Moreover, the SDKPA relation Eq. (7) is intimately related to the Lie superalgebra osp$(N+1, M)$. The special case of $M = 0$ i.e., $V_1 = 0$ then reproduces the DKPA Lie algebra of so$(N+1)$.

Let Q_a and $\overline{Q}_a$ for $a \,\epsilon\, V$ be generalized annihilation and creation operators satisfying

$$\begin{aligned} &Q_a \overline{Q}_b \pm (-1)^{\sigma(a)\sigma(b)} \overline{Q}_b Q_a = g_{ab} \quad , \\ &Q_a Q_b \pm (-1)^{\sigma(a)\sigma(b)} Q_b Q_a = 0 \quad , \\ &\overline{Q}_a \overline{Q}_b \pm (-1)^{\sigma(a)\sigma(b)} \overline{Q}_b \overline{Q}_a = 0 \quad . \end{aligned} \tag{9}$$

Then, β_a given by

$$\beta_a = \begin{pmatrix} 0 & Q_a \\ \overline{Q}_a & 0 \end{pmatrix} \tag{10}$$

defines SDKPA, i.e., it satisfies Eq. (7) as can be readily verified. Although we can use two sign choices in Eqs. (9), we will consider only the upper sign case in what follows.

Let $|0>$ be the vacuum state so that we have

$$Q_a|0> = 0 \tag{11}$$

and set

$$|a_1 a_2 \ldots a_n> = \overline{Q}_{a_1}\overline{Q}_{a_2}\cdots\overline{Q}_{a_n}|0> \quad . \tag{12}$$

Introducing the completely super antisymmetric tensors of rank n and $n+1$, respectively by

$$\psi^{(n)}_{a_1\ldots a_n} = \begin{pmatrix} |0> \\ |a_1 a_2 \ldots a_n> \end{pmatrix} ,$$

$$\phi^{(n+1)}_{a_1\ldots a_{n+1}} = \begin{pmatrix} |a_1 a_2 \ldots a_{n+1}> \\ 0 \end{pmatrix} , \tag{13}$$

the action of β_a to these tensors is given by

$$\beta_b \psi^{(n)}_{a_1\ldots a_n} = \phi^{(n+1)}_{b a_1 a_2 \ldots a_n}$$

$$\beta_b \phi^{(n+1)}_{a_1\ldots a_{n+1}} = \sum_{j=1}^{n+1} (-1)^{\Sigma_j} g_{b a_j} \psi^{(n)}_{a_1\ldots a_{j-1} a_{j+1}\ldots a_{n+1}} \tag{14}$$

where we have set

$$\Sigma_j = j - 1 + \sigma(b)\left[\sigma(a_1) + \sigma(a_2) + \ldots + \sigma(a_{j-1})\right] \quad . \tag{15}$$

Note that the non-negative integer n specifies representation spaces of β_a's.

The simplest case is when we have Dim $V_1 = 0$ and Dim $V_0 = N < \infty$, corresponding to the original DKPA, then both tensors $\psi^{(n)}_{a_1\ldots a_n}$ and $\phi^{(n+1)}_{a_1\ldots a_{n+1}}$

are now completely antisymmetric and we can restrict the value of n to be limited to

$$0 \leq n \leq \frac{1}{2}\,(N-1) \quad , \tag{16}$$

as we have noted in [3].

The remaining question is whether the representations found here are irreducible or not. For the cases of $M = 0$ and $N = 0$ which correspond to so$(N+1)$ and osp$(1, M)$ algebras, the answer is yes except for the case of $N = 2n + 1$ for the former as we noted in [3]. However, the question remains open for more general cases.

This paper is supported in part by the U.S. Department of Energy Contract no. DE-FG02-91ER40685.

References

1. R.A. Krajcik and M.M. Nieto, *Am J. Phys.* **45**, 818 (1977).
2. E. Fishbach, J.D. Louck, M.M. Nieto, and C.K. Scott, *J. Math. Phys.* **15**, 60 (1974).
3. S. Okubo, *J. Math. Phys.* **42**, 4554 (2001).

DEFORMED JACOBSON GENERATORS OF THE ALGEBRA $\mathrm{U_q[sl(n+1)]}$ AND THEIR FOCK REPRESENTATIONS

T.D. PALEV

Institute for Nuclear Research and Nuclear Energy,
Boul. Tsarigradsko Chaussee 72, 1784 Sofia, Bulgaria
E-mail : tpalev@inrne.bas.bg.

N.I. STOILOVA

Mathematical Physics Group, Department of Physics, Technical University of Clausthal, Leibnizstrasse 10, D-38678 Clausthal-Zellerfeld, Germany
E-mail : ptns@pt.tu-clausthal.de
Permanent address : Institute for Nuclear Research and Nuclear Energy, Boul. Tsarigradsko Chaussee 72, 1784 Sofia, Bulgaria
E-mail : stoilova@inrne.bas.bg

J. VAN DER JEUGT

Department of Applied Mathematics and Computer Science, University of Ghent, Krijgslaan 281-S9, B-9000 Gent, Belgium
E-mail : Joris.VanderJeugt@rug.ac.be

A description of the quantum algebra $U_q[sl(n+1)]$ via a new set of generators, called deformed Jacobson generators, is given. It provides an alternative to the canonical description of $U_q[sl(n+1)]$ in terms of Chevalley generators. The Jacobson generators satisfy trilinear commutation relations and define $U_q[sl(n+1)]$ as a deformed Lie triple system. Fock representations are constructed and the action of the Jacobson generators on the Fock basis is written down. Finally, Dyson and Holstein-Primakoff realizations are given.

1 Deformed Jacobson generators of $U_q[sl(n+1)]$

We describe the quantum algebra $U_q[sl(n+1)]$ in terms of a new set of generators [1], which are different from the known Chevalley generators. The deformation however is the usual Hopf algebra deformation [2]. In the classical case, these new generators describe the Lie algebra $sl(n+1)$ using relations, which were introduced for the first time by Jacobson[3] in the context of Lie triple systems. For this reason we call the new generators for $U_q[sl(n+1)]$ deformed Jacobson generators (JGs).

The definition of deformed JGs can be best presented in the framework of a set of Cartan-Weyl elements e_{ij}, $i, j = 0, \ldots, n$ of $U_q[gl(n+1)]$. These Cartan-Weyl elements can be expressed in terms of the usual Chevalley generators of $U_q[gl(n+1)]$ [4,5]. Recall that a construction of Cartan-Weyl elements is necessary in order to describe a complete basis (PBW-basis) of $U_q[gl(n+1)]$. We refer to e_{ij} as a positive root vector (respectively, negative root vector) if $i<j$ (respectively, $i>j$). Fix a total order for the set of elements e_{ij} : for the positive root vectors

$$e_{ij} < e_{kl}, \text{ if } i<k \text{ or } i=k \text{ and } j<l; \tag{1}$$

for the negative root vectors e_{ij} one takes the same rule (1), and total order is fixed by choosing

$$\text{positive root vectors} < \text{negative root vectors} < e_{ii}. \tag{2}$$

The difference between $U_q[sl(n+1)]$ and $U_q[gl(n+1)]$ is in the elements of the Cartan subalgebra. For $U_q[gl(n+1)]$ the Cartan subalgebra is generated by e_{ii} $(i=0,\ldots,n)$ and for $U_q[sl(n+1)]$ by the elements $H_i = e_{00} - e_{ii}$ $(i = 1,\ldots,n)$. We use also the elements $L_i = q^{H_i}$ and $\bar{L}_i = q^{-H_i}$. Then a complete set of relations between these Cartan-Weyl elements is given by

$$[H_i, H_j] = 0; \quad [H_i, e_{jk}] = (\delta_{0j} - \delta_{0k} - \delta_{ij} + \delta_{ik})e_{jk}; \tag{3}$$

for two positive root vectors $e_{ij} < e_{kl}$:

$$[e_{ij}, e_{kl}]_{q^{\delta_{jl}-\delta_{jk}+\delta_{ik}}} = \delta_{jk} e_{il} + (q-q^{-1})\theta(l>j>k>i) e_{kj} e_{il}; \tag{4}$$

for two negative root vectors $e_{ij} > e_{kl}$:

$$[e_{ij}, e_{kl}]_{q^{-\delta_{jl}+\delta_{jk}-\delta_{ik}}} = \delta_{jk} e_{il} - (q-q^{-1})\theta(i>k>j>l) e_{kj} e_{il}; \tag{5}$$

and finally for a positive root vector e_{ij} and a negative root vector e_{kl} :

$$[e_{ij}, e_{kl}] = \frac{\delta_{il}\delta_{jk}}{q-q^{-1}} \left(L_j \bar{L}_i - \bar{L}_j L_i\right) + \tag{6}$$
$$\Big((q-q^{-1})\theta(j>k>i>l) e_{kj} e_{il} - \delta_{il}\theta(j>k) e_{kj} + \delta_{jk}\theta(i>l) e_{il}\Big) L_i \bar{L}_k +$$
$$L_j \bar{L}_l \Big(-(q-q^{-1})\theta(k>j>l>i) e_{il} e_{kj} - \delta_{il}\theta(k>j) e_{kj} + \delta_{jk}\theta(l>i) e_{il}\Big),$$

where

$$[a,b]_x = ab - xba; \quad \theta(i_1 > i_2 > \ldots > i_r) = \begin{cases} 1, \text{ if } i_1 > i_2 > \ldots > i_r, \\ 0, \text{ otherwise.} \end{cases} \tag{7}$$

Define the Jacobson generators of $U_q[sl(n+1)]$ to be the following Cartan-Weyl vectors :

$$a_i^- = e_{0i}, \qquad a_i^+ = e_{i0}, \qquad H_i, \qquad i = 1,\ldots,n. \tag{8}$$

Then from (6) one obtains :

$$e_{ji} = -\bar{L}_i[a_i^-, a_j^+], \quad (i < j); \qquad e_{ji} = -[a_i^-, a_j^+]L_j, \quad (i > j). \tag{9}$$

In order to be able to order all Cartan-Weyl generators in an arbitrary expression in the right order we need a complete set of relations in terms of the deformed JG's. The result follows :

Theorem 1. A set of Cartan-Weyl elements of $U_q[sl(n+1)]$ is given by H_i, $a_i^\pm$, $[a_i^+, a_j^-]$ $(i \neq j = 1, \ldots, n)$. A complete set of supercommutation relations between these elements is given by :

$$[H_i, H_j] = 0; \qquad [H_i, a_j^\pm] = \mp(1 + \delta_{ij})a_j^\pm; \tag{10}$$

$$[a_i^-, a_i^+] = \frac{L_i - \bar{L}_i}{q - q^{-1}}; \qquad [a_i^\eta, a_j^\eta]_q = 0 \quad (i < j); \tag{11}$$

$$\begin{aligned} [[a_i^\eta, a_j^{-\eta}], a_k^\eta]_{q^{\xi(1+\delta_{ik})}} &= \eta\delta_{jk}L_k^{-\xi\eta}a_i^\eta + \epsilon(j,k,i)(q - \bar{q})[a_k^\eta, a_j^{-\eta}]a_i^\eta \\ &= \eta\delta_{jk}L_k^{-\xi\eta}a_i^\eta + \epsilon(j,k,i)q^\xi(q - \bar{q})a_i^\eta]a_k^\eta, a_j^{-\eta}], \end{aligned} \tag{12}$$

where $(j-i)\xi > 0, \;\; \xi, \; \eta = \pm; \quad \bar{q} = q^{-1}$;

and $\epsilon(j,k,i) = \begin{cases} 1, & \text{if } j > k > i; \\ -1, & \text{if } j < k < i; \\ 0, & \text{otherwise.} \end{cases}$

This description of $U_q[sl(n+1)]$ in terms of deformed JGs is referred to as a deformed Lie triple system.

2 Fock representations

We construct Fock modules using the induced module procedure. $G = U_q[sl(n+1)]$, with Cartan-Weyl elements H_i, $a_i^\pm$ and $[a_i^+, a_j^-]$ $(i \neq j = 1, \ldots, n)$, has a subalgebra $A = U_q[gl(n)]$ with Cartan-Weyl elements H_i and $[a_i^+, a_j^-]$ $(i \neq j = 1, \ldots, n)$. Define a trivial one-dimensional A module as follows :

$$[a_i^-, a_j^+]|0\rangle = 0, \;\; i \neq j = 1, \ldots, n; \;\; H_i|0\rangle = p|0\rangle, \tag{13}$$

where p is any complex number. Let P be the (associative) subalgebra of $G = U_q[sl(n+1)]$ generated by the elements of A and $\{a_i^-; i = 1, \ldots, n\}$. The one-dimensional A module $\mathbf{C}|0\rangle$ can be extended to a one-dimensional P module by requiring $a_i^-|0\rangle = 0, \; i = 1, \ldots, n$. Now the G module $\bar{W}_p$ is defined as $\bar{W}_p = \mathrm{Ind}_P^G \, \mathbf{C}|0\rangle$. Clearly $\bar{W}_p$ is freely generated by the generators a_i^+ $(i = 1, \ldots, n)$ acting on $|0\rangle$. Therefore a basis for $\bar{W}_p$ is given by

$$|p; l_1, l_2, \ldots, l_n\rangle \equiv (a_1^+)^{l_1}(a_2^+)^{l_2} \ldots (a_n^+)^{l_n}|0\rangle, \;\; l_i \in \mathbf{Z}_+, \;\; i = 1, \ldots, n. \tag{14}$$

Theorem 2. The transformation of the basis (14) of $\bar{W}_p$ under the action of the JGs reads ($i = 1, \ldots, n$), denoting $l_1 + l_2 + \ldots + l_n$ by L :

$$H_i|p; l_1, \ldots, l_n\rangle = (p - l_i - L)|p; l_1, \ldots, l_n\rangle, \tag{15}$$

$$a_i^+|p; l_1, \ldots, l_n\rangle = \bar{q}^{l_1+\ldots+l_{i-1}}|p; l_1, \ldots, l_{i-1}, l_i + 1, l_{i+1}, \ldots, l_n\rangle, \tag{16}$$

$$a_i^-|p; l_1, \ldots, l_n\rangle = q^{l_1+\ldots+l_{i-1}}[l_i][p - L + 1] \times|p; l_1, \ldots, l_{i-1}, l_i - 1, l_{i+1}, \ldots, l_n\rangle, \tag{17}$$

where $[x] = (q^x - q^{-x})/(q - q^{-1})$.

The action of the elements H_i and $a_i^{\pm}$ on the basis vectors of $\bar{W}_p$, determined in *Theorem 2* imply that $\bar{W}_p$ has an invariant submodule when p is a nonnegative integer. From now on we shall assume that $p \in \mathbf{Z}_+$. Then we have

Corollary. The $U_q[sl(n+1)]$ module $\bar{W}_p$ has an invariant submodule V_p with basis vectors $|p; l_1, l_2, \ldots, l_n\rangle$, with $L = \sum_{i=1}^n l_i > p$.

The quotient module $W_p = \bar{W}_p/V_p$ carries an irreducible representation for $U_q[sl(n+1)]$ and is referred to as a Fock module. The basis vectors of W_p are given by (the representatives of)

$$|p; l_1, l_2, \ldots, l_n\rangle, \text{ with } L = \sum_{i=1}^n l_i \leq p. \tag{18}$$

Now we consider those Fock modules relevant for physical applications. A representation of $U_q[sl(n+1)]$ is unitary if the representation space is a Hilbert space and the representatives of $a_i^{\pm}$ and H_i satisfy

$$(a_i^+)^\dagger = a_i^-, \qquad (a_i^-)^\dagger = a_i^+, \qquad (H_i)^\dagger = H_i. \tag{19}$$

In each Fock module W_p define a scalar product (,) by postulating :

$$(|0\rangle \ , \ |0\rangle) = \langle 0|0\rangle = 1, \quad (a_i^{\pm}v, w) = (v, a_i^{\mp}w), \quad \forall v, w \in W_p. \tag{20}$$

Then any two vectors $|p; l_1, l_2, \ldots, l_n\rangle$ and $|p; l_1', l_2', \ldots, l_n'\rangle$ with $(l_1, l_2, \ldots, l_n) \neq (l_1', l_2', \ldots, l_n')$ are orthogonal and

$$(|p; l_1, l_2, \ldots, l_n\rangle, |p; l_1, l_2, \ldots, l_n\rangle) = \frac{[p]!}{[p-L]!} \prod_{i=1}^n [l_i]! \ . \tag{21}$$

A careful analysis shows

Theorem 3. The irreducible Fock module W_p ($p \geq 2$) is unitary if and only if q is a phase, i.e. $q = e^{i\phi}$, with $-\frac{\pi}{p} < \phi < \frac{\pi}{p}$.

Under these conditions, define an orthonormal basis of W_p as follows :

$$|p;l_1,l_2,\ldots,l_n) = \sqrt{\frac{[p-L]!}{[p]![l_1]!\ldots[l_n]!}}|p;l_1,l_2,\ldots,l_n\rangle, \quad 0\le L\le p. \tag{22}$$

In the new basis (22) the transformations (15)-(16) read $(i=1,\ldots,n)$:

$$H_i|p;l_1,\ldots,l_n) = (p-l_i-L)|p;l_1,\ldots,l_n), \tag{23}$$

$$a_i^-|p;l_1,\ldots,l_n) = \times q^{l_1+\ldots+l_{i-1}}\sqrt{[l_i][p-L+1]}\ |p;l_1,\ldots l_{i-1},l_i-1,l_{i+1},\ldots,l_n), \tag{24}$$

$$a_i^+|p;l_1,\ldots,l_n) = \times \bar{q}^{l_1+\ldots+l_{i-1}}\sqrt{[l_i+1][p-L]}\ |p;l_1,\ldots l_{i-1},l_i+1,l_{i+1},\ldots,l_n). \tag{25}$$

3 Dyson and Holstein-Primakoff realizations of $U_q[sl(n+1)]$

Denote by $W(n)$ the algebra of all polynomials of n pairs of Bose operators $b_i^\pm$, namely $[b_i^-,b_j^+]=\delta_{ij}$, $[b_i^+,b_j^+]=[b_i^-,b_j^-]=0$. Let $N_i=b_i^+b_i^-$, $N=\sum_{j=1}^n N_j$.

Theorem 4. (Dyson realization). The linear map $\rho : U_q[sl(n+1)] \to W(n)$, defined on the Jacobson generators as

$$\begin{aligned}
\rho(H_i) &= p - b_i^+b_i^- - \sum_{j=1}^n b_j^+b_j^- = p - N_i - N,\\
\rho(a_i^-) &= q^{N_1+\ldots+N_{i-1}}\frac{[N_i+1]}{N_i+1}[p-N]b_i^-,\\
\rho(a_i^+) &= \bar{q}^{N_1+\ldots+N_{i-1}}b_i^+,
\end{aligned} \tag{26}$$

is a homomorphism of $U_q[sl(n+1)]$ into $W(n)$ for any $p\in\mathbf{C}$.

The Dyson realization defines an infinite-dimensional representation of $U_q[sl(n+1)]$ in the Fock space $\mathcal{F}(n)$ with basis :

$$(b_1^+)^{r_1}\ldots(b_n^+)^{r_n}|0\rangle \equiv |r_1,r_2,\ldots,r_n\rangle,\ r_i\in\mathbf{Z}_+,\ i=1,\ldots,n. \tag{27}$$

If p is a positive integer, $p\in\mathbf{N}$, the representation is indecomposable. If $p\notin\mathbf{N}$ the representation is irreducible. In both cases however it is impossible to introduce a scalar product in $\mathcal{F}(n)$ for which the operators (26) satisfy (19). This is a disadvantage of the Dyson realization.

Theorem 5. (Holstein-Primakoff realization). The linear map $\varrho : U_q[sl(n+1)] \to W(n)$, defined on the Jacobson generators as

$$\varrho(H_i) = p - b_i^+b_i^- - \sum_{j=1}^n b_j^+b_j^- = p - N_i - N,$$

$$\varrho(a_i^-) = q^{N_1+\ldots+N_{i-1}}\sqrt{\frac{[N_i+1]}{N_i+1}[p-N]}b_i^-, \tag{28}$$

$$\varrho(a_i^+) = \bar{q}^{N_1+\ldots+N_{i-1}}\sqrt{\frac{[N_i]}{N_i}[p-N+1]}b_i^+,$$

is a homomorphism of $U_q[sl(n+1)]$ into $W(n)$.

If $p \in \mathbf{N}$, then $\mathcal{F}_0(n)$ with basis :

$$\frac{(b_1^+)^{r_1}(b_2^+)^{r_2}\ldots(b_n^+)^{r_n}}{\sqrt{r_1!r_2!\ldots r_n!}}|0\rangle \equiv |r_1,r_2,\ldots,r_n), \quad r_1+\ldots+r_n \leq p \tag{29}$$

is a finite-dimensional irreducible module; the hermiticity condition (19) holds if and only if $q = e^{i\phi}$, $-\frac{\pi}{p} < \phi < \frac{\pi}{p}$. From (9) and (28) one obtains the Holstein-Primakoff realization of the remaining Cartan-Weyl elements of $U_q[sl(n+1)]$:

$$\varrho(e_{ji}) = q^{N_{j+1}+N_{j+2}+\ldots+N_{i-1}}\sqrt{\frac{[N_j][N_i+1]}{N_j(N_i+1)}}b_j^+b_i^-,\ j<i,$$

$$\varrho(e_{ji}) = \bar{q}^{N_{i+1}+N_{i+2}+\ldots+N_{j-1}}\sqrt{\frac{[N_j][N_i+1]}{N_j(N_i+1)}}b_j^+b_i^-,\ j>i. \tag{30}$$

Acknowledgments

N.I. Stoilova is thankful to Prof. H.-D. Doebner for constructive discussions and to the Humboldt Foundation for its support. The work was supported also by the Grant N $\Phi - 910$ of the Bulgarian Foundation for Scientific Research.

References

1. T.D. Palev, P. Parashar, *Lett. Math. Phys.* **43**, 7 (1998).
2. V.G. Drinfeld, in *Quantum Groups*, Proc. Int. Congr. Math. 1, 798 Berkeley (1986); M. Jimbo, *Lett. Math. Phys.* **11**, 247 (1986).
3. N. Jacobson, *Amer. J. Math.* **71**, 149 (1949).
4. M. Rosso, *Commun. Math. Phys.* **124**, 307 (1989).
5. T.D. Palev and V.N. Tolstoy, *Commun. Math. Phys.* **141**, 549 (1991).

COHERENT STATES FROM COMBINATORIAL SEQUENCES

K.A. PENSON

Laboratoire de Physique Théorique des Liquides, Université Pierre et Marie Curie, 75252 PARIS, France
Email: penson@lptl.jussieu.fr

A.I. SOLOMON

Quantum Processes Group, The Open University, Milton Keynes MK7 6AA, United Kingdom
Email: a.i.solomon@open.ac.uk

We construct coherent states using sequences of combinatorial numbers such as various binomial and trinomial numbers, and Bell and Catalan numbers. We show that these states satisfy the condition of the *resolution of unity* in a natural way. In each case the positive weight functions are given as solutions of associated Stieltjes or Hausdorff moment problems, where the moments are the combinatorial numbers.

We describe the construction of coherent states which are generalizations of the standard coherent states defined for complex z by[1]

$$|z\rangle_0 = e^{-\frac{|z|^2}{2}} \sum_{n=0}^{\infty} \frac{z^n}{\sqrt{n!}} |n\rangle, \tag{1}$$

where $|n\rangle$ are eigenfunctions of a Hermitian operator H (usually the system Hamiltonian) with

$$H|n\rangle = \varepsilon_n |n\rangle \quad \langle n|n'\rangle = \delta_{n,n'}, \quad n = 0, 1, \ldots. \tag{2}$$

The generalization consists in defining new states

$$|z\rangle_c = \mathcal{N}_c{}^{-\frac{1}{2}}(|z|^2) \sum_{n=0}^{\infty} \frac{z^n}{\sqrt{c(n)}} |n\rangle, \tag{3}$$

where

$$\mathcal{N}_c(x) = \sum_{n=0}^{\infty} \frac{x^n}{c(n)}, \quad x \le R. \tag{4}$$

In Eq.(3) and Eq.(4), $\mathcal{N}_c(x)$ $(x \equiv |z|^2)$ is the normalization of $|z\rangle$, and $R \le \infty$ is its radius of convergence. We shall assume that in Eq.(3) the positive numbers $c(n)$ for $n = 0, 1, \ldots,$ arise from sequences of combinatorial numbers; that is, integers which count the characteristic properties of an ensemble of n

objects ($c(0)=1$ by convention). In this note we shall refrain from providing the precise combinatorial interpretation of the sequences used, but shall rather limit ourselves to giving their formulae. For more details of the $c(n)$ we refer the interested reader to the classical literature [2,3,4].

In general, the states $|z\rangle_c$ are not orthogonal ($\langle z|z'\rangle_c \neq 0$); the resolution of unity condition is given by the following sum of weighted non-orthogonal projectors

$$\iint d^2z|z\rangle_c W_c(|z|^2)\ {}_c\langle z| = I = \sum_{n=0}^{\infty} |n\rangle\langle n|, \tag{5}$$

where in Eq.(5) the integration is restricted to the part of the complex plane where the normalization converges. Equations (4) and (5) may be succinctly written as

$$\int_0^R x^n \tilde{W}_c(x)dx = c(n), \qquad n = 0, 1, \ldots, \tag{6}$$

with $\tilde{W}_c(x) = \pi W_c(x)/\mathcal{N}_c(x) > 0$. For $R=\infty$, Eq.(6) is the Stieltjes, and for $R<\infty$, the Hausdorff moment problem, with moments $c(n)$.

The motivation for choosing the combinatorial sequences $\{c(n)\}$ as input moments to Eq.(3) is that for many such sequences solutions can be explicitly constructed by considering Eq.(6) as a Mellin transform; see [5,6,7,8] for related instances. For example, in those cases where the $c(n)$ may be expressed solely in terms of gamma functions, the use of properties of Meijer's G function is instrumental in obtaining solutions of Eq.(6). In what follows we shall list a number of sequences $\{c(n)\}$ and simply quote the solutions of Eq.(6), specifying R in each case:

1. $c(n) = (2n)!$; $\tilde{W}_1(x) = \frac{1}{2}e^{-\sqrt{x}}/\sqrt{x}$; $R=\infty$.
2. $c(n) = (2n)!/n!$; $\tilde{W}_2(x) = \frac{1}{2\sqrt{\pi}}e^{-\frac{x}{4}}/\sqrt{x}$; $R=\infty$.
3. $c(n) = \binom{2n}{n}$ (so-called central binomial coefficients) ;
   $\tilde{W}_3(x) = \frac{1}{\pi}[x(4-x)]^{-\frac{1}{2}}$; $R=4$.
4. $c(n) = \binom{2n}{n}/(n+1)$ (so-called Catalan numbers) ;
   $\tilde{W}_4(x) = \frac{1}{\pi}[(4-x)/x]^{\frac{1}{2}}$; $R=4$.
5. $c(n) = (2n)!/(n+1)!$; $\tilde{W}_5(x) = -\frac{1}{2} + \frac{1}{\sqrt{\pi x}}e^{-\frac{x}{4}} + \frac{1}{2}\mathrm{erf}(\frac{\sqrt{x}}{2})$; $R=\infty$; $\mathrm{erf}(y)$ is the error function.

6. $c(n) = (2n)!/(n+1)$; $\tilde{W}_6(x) = \frac{1}{\sqrt{x}}e^{-\sqrt{x}} + \mathrm{Ei}(-\sqrt{x})$; $R = \infty$;
   $\mathrm{Ei}(y)$ is the exponential integral.

7. $c(n) = (3n)!/n!$; $\tilde{W}_7(x) = \frac{1}{3\pi\sqrt{x}}K_{\frac{1}{3}}(2\sqrt{\frac{x}{27}})$; $R = \infty$;
   $K_\nu(y)$ is the modified Bessel function of the second kind.

8. $c(n) = (3n)!/(2n)!$; $\tilde{W}_8(x) = \frac{\sqrt{3}}{27\pi}\exp(-\frac{2x}{27})[K_{\frac{1}{3}}(\frac{2x}{27}) + K_{\frac{2}{3}}(\frac{2x}{27})]$; $R = \infty$.

9. $c(n) = (3n)!/(n!)^3$; (so-called middle trinomial coefficients);
   $\tilde{W}_9(x) = \alpha x^{-\frac{2}{3}}{}_2F_1(\frac{1}{3}, \frac{1}{3}; \frac{2}{3}; \frac{x}{27}) + \beta x^{-\frac{1}{3}}{}_2F_1(\frac{2}{3}, \frac{2}{3}; \frac{4}{3}; \frac{x}{27})$;
   where $\alpha = \frac{1}{3}[\Gamma(\frac{2}{3})]^{-3}$ and $\beta = -\frac{\sqrt{3}}{8\pi^3}[\Gamma(\frac{2}{3})]^3$, $R = 27$.

10. $c(n) = \binom{3n}{n}/(2n+1)$;
    $\tilde{W}_{10}(x) = \frac{\sqrt{3}\,2^{\frac{2}{3}}}{12\pi}\frac{[2^{\frac{1}{3}}(27+3\sqrt{81-12x})^{\frac{2}{3}} - 6x^{\frac{1}{3}}]}{x^{\frac{2}{3}}(27+3\sqrt{81-12x})^{\frac{1}{3}}}$; $R = \frac{27}{4}$.

The above weight functions, which may be seen to be positive by inspection, constitute only a small selection of examples of this type. Moreover, with the $c(n)$ of Examples 1–10, the normalization $\mathcal{N}_c(x)$ of Eq.(4) can always be expressed in terms of known hypergeometric functions. This renders calculations of many expectation values in the state $|z\rangle_c$ entirely analytic.

We now extend our considerations to include the important sequence of Bell numbers, $c(n) = B(n)$, which count the number of partitions of a set of n objects. The Bell numbers are given by the celebrated Dobinski formula[2]

$$B(n) = 1/e\sum_{k=1}^{\infty}\frac{k^n}{k!}, \quad n = 0, 1, \ldots, \tag{7}$$

from which it readily follows that the solution of

$$\int_0^\infty x^n \tilde{W}_B(x)dx = B(n)$$

is

$$\tilde{W}_B(x) = 1/e\sum_{k=1}^{\infty}\delta(x-k)/k!, \tag{8}$$

which is a sum of weighted Dirac delta function spikes situated on the positive integers. Since $x = |z|^2$, the weight is positive on concentric equidistant circles around the origin of the complex plane. The particular form of Eq.(8) permits one to write down the weight function whose moments are $c(n)B(n)$, where

$c(n)$ is any moment from Examples 1–10. For instance, if $c(n)$ are the Catalan numbers, the nth moment of

$$\tilde{W}_{CB}(x) = 1/(2\pi e) \sum_{k=1}^{\infty} \frac{1}{kk!} \sqrt{\frac{4k-x}{x}} H(4 - \frac{x}{k}) \tag{9}$$

equals $\left(\binom{2n}{n} / (n+1)\right) B(n)$; $H(y)$ is the Heaviside function. The powerful convolution property of the Mellin transform[6,7,8] enables one to obtain Eq.(9). It has been shown elsewhere[9,10] that by a different form of parametrization (replacing z by two independent real variables) the states described above may be applied to specific physical systems. A Hamiltonian with energy levels ε_n can be associated with the sequence $\{c(n)\}$ as follows:

$$\varepsilon_0 = 0, \quad \varepsilon_n = c(n)/c(n-1), \ \ n > 0. \tag{10}$$

The wealth of the preceding examples allows for different forms of ε_n, differing from the standard $\varepsilon_n = n$, associated with the standard coherent state characterized by the $|z\rangle_0$ of Eq.(1). These examples are associated with models having discrete non-linear spectra, and may well provide good approximations to some real nonlinear systems.

References

1. J.R. Klauder and B.-S. Skagerstam, *Coherent states; Applications in Physics and Mathematical Physics*, World Scientific, Singapore (1985).
2. L. Comtet, *Advanced Combinatorics*, D. Reidel, Boston (1984).
3. R.P. Stanley, *Enumerative Combinatorics, Vol. 1,2*, Cambridge University Press (1997, 1999).
4. N.J.A. Sloane, *On-line Encyclopedia of Integer Sequences*, published electronically at: http://www.research.att.com/~njas/sequences.
5. J.-M. Sixdeniers, K.A. Penson and A.I. Solomon, em J.Phys.A **32**, 7543 (1999).
6. J.-M. Sixdeniers and K.A. Penson, *J.Phys.A* **33**, 2907 (2000).
7. J.R. Klauder, K.A. Penson and J.-M. Sixdeniers, *Phys.Rev.A* **64**, 013817 (2001).
8. C. Quesne, *Phys.Lett.A* **288**, 241 (2001).
9. J.-P. Gazeau and J.R. Klauder, *J.Phys.A* **32**, 123 (1999).
10. H.Fu, A.I. Solomon, K.A. Penson and J.-P. Gazeau, (to be published).

GENERALIZED COHERENT STATES ASSOCIATED WITH THE C_λ-EXTENDED OSCILLATOR

C. QUESNE

Physique Nucléaire Théorique et Physique Mathématique, Université Libre de Bruxelles, Campus de la Plaine CP229, Boulevard du Triomphe, B-1050 Brussels, Belgium

E-mail: cquesne@ulb.ac.be

Two new types of coherent states associated with the C_λ-extended oscillator, where C_λ is the cyclic group of order λ, are introduced. They satisfy a unity resolution relation in the C_λ-extended oscillator Fock space (or in some subspace thereof) and give rise to Bargmann representations of the latter, wherein the generators of the C_λ-extended oscillator algebra are realized as differential-operator-valued matrices.

1 Introduction

Coherent states (CS) of the harmonic oscillator are known to have properties similar to those of the classical radiation field. In contrast, generalized CS associated with various algebras may have some nonclassical properties.

In the present communication, we shall consider some generalized CS, [1,2] which may be associated with the recently introduced C_λ-extended oscillator. [3] The latter may be considered as a deformed oscillator with a Z_λ-graded Fock space and has proved very useful in the context of supersymmetric quantum mechanics and some of its variants. [3,4]

The main mathematical feature of these new generalized CS is that they satisfy Klauder's minimal set of conditions for generalized CS, [5] including the existence of a unity resolution relation, property that is not shared by generic generalized CS.

2 The C_λ-Extended Oscillator

The C_λ-extended oscillator Hamiltonian is defined by [3]

$$H_0 = \tfrac{1}{2}\left\{a, a^\dagger\right\}, \tag{1}$$

where the creation and annihilation operators $a^\dagger$, a satisfy the commutation relation

$$\left[a, a^\dagger\right] = I + \sum_{\mu=1}^{\lambda-1} \kappa_\mu T^\mu = I + \sum_{\mu=0}^{\lambda-1} \alpha_\mu P_\mu. \tag{2}$$

Here $T = (T^\dagger)^{-1}$ denotes the generator of the cyclic group $C_\lambda = \{T, T^2, \ldots, T^{\lambda-1}, T^\lambda = I\}$ (or, more precisely, the generator of a unitary representation thereof), and $P_\mu = \frac{1}{\lambda}\sum_{\nu=0}^{\lambda-1} \exp(-2\pi i\mu\nu/\lambda)\, T^\nu$, $\mu = 0, 1, \ldots, \lambda-1$, are the projectors on the λ inequivalent unitary irreducible matrix representations of C_λ. The parameters κ_μ are complex and satisfy the conditions $\kappa_\mu^* = \kappa_{\lambda-\mu}$, whereas the α_μ's are real and such that $\sum_{\mu=0}^{\lambda-1}\alpha_\mu = 0$. The commutation rule for T (or P_μ) and $a^\dagger$ is given by

$$a^\dagger T = e^{-2\pi \mathrm{i}/\lambda} T a^\dagger \qquad \text{or} \qquad a^\dagger P_\mu = P_{\mu+1} a^\dagger. \tag{3}$$

The C_λ-extended oscillator algebra also contains a number operator N such that

$$\left[N, a^\dagger\right] = a^\dagger, \qquad [N, T] = 0 \qquad \text{or} \qquad [N, P_\mu] = 0. \tag{4}$$

When T (or P_μ) is realized in terms of N, i.e., $T = \exp(2\pi \mathrm{i} N/\lambda)$, the C_λ-extended oscillator algebra becomes a generalized deformed oscillator algebra (GDOA). In the Fock representation of the latter, there exists a vacuum state $|0\rangle$, such that $a|0\rangle = N|0\rangle = 0$, and the remaining basis states $|n\rangle$, $n = 1, 2, \ldots$, can be obtained from it by successive applications of $a^\dagger$. The Fock space $\mathcal{F}$ is Z_λ-graded: $\mathcal{F} = \sum_{\mu=0}^{\lambda-1}\mathcal{F}_\mu$, where $\mathcal{F}_\mu = \{\,|n\rangle = |k\lambda+\mu\rangle \mid k = 0, 1, 2, \ldots\}$, and P_μ projects on $\mathcal{F}_\mu$.

The C_λ-extended oscillator Hamiltonian H_0, defined in Eq. (1), has a spectrum made of λ infinite sets of equally spaced levels, belonging to $\mathcal{F}_\mu$, $\mu = 0, 1, \ldots, \lambda-1$, respectively. Its spectrum generating algebra (SGA) is generated by the operators $J_+ = (1/\lambda)(a^\dagger)^\lambda$, $J_- = (1/\lambda)a^\lambda$, $J_0 = (1/\lambda)H_0$, and is a C_λ-extended polynomial deformation of su(1,1) for $\lambda > 2$, while it reduces to su(1,1) for $\lambda = 2$. [1]

For $\lambda = 2$, the GDOA corresponding to the C_λ-extended oscillator algebra reduces to the Calogero-Vasiliev algebra, characterized by the commutation relation $[a, a^\dagger] = I + \alpha_0 K$, where $K = T = (-1)^N$. [6] For $\alpha_0 = p-1$, the latter becomes the paraboson algebra of order p. [7]

3 Family of Coherent States $|z;\mu;\alpha\rangle$

The coherent states $|z;\mu;\alpha\rangle$, where $z \in \mathrm{C}$, $\alpha \in \{0, 1, \ldots, [\lambda/2]\}$, $\mu \in \{0, 1, \ldots, \lambda-\alpha-1\}$, are the solutions of the equation [2]

$$\left[a^{\lambda-\alpha} - z\left(a^\dagger\right)^\alpha\right]|z;\mu;\alpha\rangle = 0. \tag{5}$$

As special cases, they include (i) the annihilation-operator CS of the C_λ-extended oscillator SGA, corresponding to $\alpha = 0$ (i.e., $J_-|z;\mu;0\rangle =$

$(z/\lambda)|z;\mu;0\rangle)$, [1] and (ii) the displacement-operator or Perelomov su(1,1) CS, corresponding to $\lambda = 2$, $\mu = 0$, $\alpha = 1$ (i.e., $(a - za^\dagger)|z;0;1\rangle = 0$, where $|z;0;1\rangle \propto \exp(zJ_+)|0\rangle)$. [8]

The states $|z;\mu;\alpha\rangle$ belong to the subspace $\mathcal{F}_\mu$ of $\mathcal{F}$ and can therefore be expanded in the number states $|k\lambda+\mu\rangle$, $k = 0, 1, \ldots,$ as

$$|z;\mu;\alpha\rangle = \left[N_\mu^{(\alpha)}(|z|)\right]^{-1/2} \sum_{k=0}^{\infty} c'_k(z,z^*;\mu;\alpha) z^k |k\lambda+\mu\rangle, \tag{6}$$

where the coefficients $c'_k(z,z^*;\mu;\alpha)$ are given in Ref. [2] and the normalization coefficient can be expressed as

$$N_\mu^{(\alpha)}(|z|) = \\ {}_\alpha F_{\lambda-\alpha-1}\left(\bar\beta_{\mu+1},\ldots,\bar\beta_{\mu+\alpha};\bar\beta_1+1,\ldots,\bar\beta_\mu+1,\bar\beta_{\mu+\alpha+1},\ldots,\bar\beta_{\lambda-1};y\right). \tag{7}$$

In Eq. (7), the variable y is defined by $y \equiv |z|^2/\lambda^{\lambda-2\alpha}$, while the constants $\bar\beta_\mu$ are given in terms of the parameters α_ν of the C_λ-extended oscillator algebra through the relations $\bar\beta_\mu \equiv (\beta_\mu+\mu)/\lambda$, $\beta_\mu \equiv \sum_{\nu=0}^{\mu-1}\alpha_\nu$. From this, we conclude that the states $|z;\mu;\alpha\rangle$ are normalizable on the complex plane for any $\alpha \in \{0,1,\ldots,[(\lambda-1)/2]\}$ and on the unit disc for λ even, $\alpha = \lambda/2$.

In addition, the CS $|z;\mu;\alpha\rangle$ are continous in the label z, i.e., $|z-z'| \to 0 \Rightarrow \left||z;\mu;\alpha\rangle - |z';\mu;\alpha\rangle\right|^2 \to 0$, and for appropriate values of the algebra parameters, they satisfy a unity resolution relation with a positive measure,

$$\int d^2z\, h_\mu^{(\alpha)}(y)|z;\mu;\alpha)(z;\mu;\alpha| = I_\mu, \tag{8}$$

in $\mathcal{F}_\mu$ (with I_μ the unit operator in $\mathcal{F}_\mu$). Alternatively, we may write

$$\sum_{\mu=0}^{\lambda-1} \int d^2z\, h_\mu^{(\alpha)}(y)|z;\mu;\alpha)(z;\mu;\alpha| = I \tag{9}$$

in $\mathcal{F}$. In Eqs. (8) and (9), $|z;\mu;\alpha) \equiv \left[N_\mu^{(\alpha)}(|z|)\right]^{1/2}|z;\mu;\alpha\rangle$ denotes an unnormalized CS and the general form of $h_\mu^{(\alpha)}(y)$ is conjectured to be given in terms of a Meijer G-function as

$$h_\mu^{(\alpha)}(y) = A_\mu^{(\alpha)} G_{\alpha,\lambda-\alpha}^{\lambda-\alpha,0}\left(y \,\middle|\, \begin{array}{c} \bar\beta_{\mu+1}-1,\ldots,\bar\beta_{\mu+\alpha}-1 \\ 0,\bar\beta_1,\ldots,\bar\beta_\mu,\bar\beta_{\mu+\alpha+1}-1,\ldots,\bar\beta_{\lambda-1}-1 \end{array}\right), \tag{10}$$

where $A_\mu^{(\alpha)}$ is some numerical coefficient. Eq. (10) has actually been proved for $\alpha \neq \lambda/2$ and any λ, and for $\alpha = \lambda/2$ and $\lambda = 2, 4, 6$. [2]

Since the CS $|z;\mu;\alpha)$ form a complete (in fact, an overcomplete) set in $\mathcal{F}_\mu$ for some appropriate choice of the algebra parameters if z runs over the complex plane for $\alpha \leq [(\lambda-1)/2]$ or over the unit disc for $\alpha = \lambda/2$ and λ even, we may associate with such a set a realization of $\mathcal{F}_\mu$ as a Hilbert space $\mathcal{B}_\mu^{(\alpha)}$ of entire analytic functions . This gives rise to so-called Bargmann representations [9] of the C_λ-extended oscillator, wherein the operators X defined in $\mathcal{F}_\mu$ become differential operators $\mathcal{X}_\mu^{(\alpha)}$. For the SGA generators J_+, J_-, and J_0, for instance, we get

$$\mathcal{J}_{+\mu}^{(\alpha)} = \lambda^{\alpha-1} z \prod_{\nu=\mu+1}^{\mu+\alpha} \left(z\frac{\partial}{\partial z} + \bar{\beta}_\nu \right), \tag{11}$$

$$\mathcal{J}_{-\mu}^{(\alpha)} = \lambda^{\lambda-\alpha-1} \left[\prod_{\nu=1}^{\mu} \left(z\frac{\partial}{\partial z} + \bar{\beta}_\nu + 1 \right) \right] \left[\prod_{\nu=\mu+\alpha+1}^{\lambda-1} \left(z\frac{\partial}{\partial z} + \bar{\beta}_\nu \right) \right] \frac{\partial}{\partial z}, \tag{12}$$

$$\mathcal{J}_{0\mu}^{(\alpha)} = z\frac{\partial}{\partial z} + \frac{1}{2}(\bar{\beta}_\mu + \bar{\beta}_{\mu+1}). \tag{13}$$

When dealing with the whole Fock space $\mathcal{F}$, we may consider as complete set the collection of CS $|z;\mu;0)$, $\mu = 0, 1, \ldots, \lambda-1$, and realize $\mathcal{F}$ as a Hilbert space $\mathcal{B}^{(0)} = \sum_{\mu=0}^{\lambda-1} \oplus \mathcal{B}_\mu^{(0)}$ of entire analytic functions. It is then convenient to introduce vector CS [10] defined as row vectors, $||z;0)) \equiv \big(|z;0;0), |z;1;0), \ldots, |z;\lambda-1;0)\big)$, the corresponding bras $((z;0||$ being column vectors. In such a case, the operators defined in $\mathcal{F}$ are represented by some operator-valued $\lambda \times \lambda$ matrices.

4 Eigenstates $|z\rangle$ of the Annihilation Operator a

The eigenstates of the C_λ-extended oscillator annihilation operator a, defined by [2]

$$a|z\rangle = z|z\rangle, \tag{14}$$

where $z \in$ C, include as special case the paraboson CS, [11] corresponding to $\lambda = 2$.

They can be constructed as linear combinations of the CS $|\omega;\mu;0\rangle$, $\mu = 0$, $1, \ldots, \lambda-1$, considered in Sec. 3, where one replaces z by $\omega \equiv z^\lambda$,

$$|z\rangle = [\mathcal{N}(|z|)]^{-1/2} \sum_{\mu=0}^{\lambda-1} d'_\mu(z, z^*) \left(\frac{z}{\sqrt{\lambda}} \right)^\mu |\omega;\mu;0\rangle, \tag{15}$$

with $d'_\mu(z,z^*) = \left[N_\mu^{(0)}(|\omega|)/\prod_{\nu=1}^{\mu}\bar{\beta}_\nu\right]^{1/2}$. The normalization coefficient $\mathcal{N}(|z|)$ is given by

$$\mathcal{N}(|z|) = \sum_{\mu=0}^{\lambda-1} {}_0F_{\lambda-1}\left(\bar{\beta}_1+1,\ldots,\bar{\beta}_\mu+1,\bar{\beta}_{\mu+1},\ldots,\bar{\beta}_{\lambda-1};t^\lambda\right)\frac{t^\mu}{\prod_{\nu=1}^{\mu}\bar{\beta}_\nu}, \tag{16}$$

where $t \equiv |z|^2/\lambda$. It is therefore straightforward to see that the CS $|z\rangle$ are normalizable on the complex plane and continuous in their label z.

They satisfy a unity resolution relation of unusual form, [2]

$$\sum_{\mu=0}^{\lambda-1}\int d^2z\, h_\mu(t)|z_\mu)(z_\mu| = \sum_{\mu=0}^{\lambda-1}\int d^2z\, g_\mu(t)|z)\left(ze^{2\pi i\mu/\lambda}\right| = I, \tag{17}$$

generalizing the known one for paraboson CS. [11] Here $|z) = [\mathcal{N}(|z|)]^{1/2}|z\rangle$ denotes an unnormalized CS, $|z_\mu)$ its component in $\mathcal{F}_\mu$, and

$$h_\mu(t) = \lambda^\lambda\left(\prod_{\nu=1}^{\mu}\bar{\beta}_\nu\right)t^{\lambda-\mu-1}h_\mu^{(0)}(t^\lambda), \qquad g_\mu(t) = \frac{1}{\lambda}\sum_{\nu=0}^{\lambda-1}e^{2\pi i\mu\nu/\lambda}h_\nu(t), \tag{18}$$

where $h_\mu^{(0)}(t^\lambda)$ is defined in Eq. (10). It can actually be shown that this nondiagonal CS resolution of unity is entirely equivalent to the diagonal one valid for the set of CS $|z;\mu;0)$, $\mu = 0, 1, \ldots, \lambda-1$, and given in Eq. (9).

Instead of the set of CS $|z;\mu;0)$, $\mu = 0, 1, \ldots, \lambda-1$, we may use $|z_\mu)$, $\mu = 0, 1, \ldots, \lambda-1$, as complet set in $\mathcal{F}$ and introduce a vector notation again, $||z)) \equiv \left(|z_0), |z_1), \ldots, |z_{\lambda-1})\right)$. In the corresponding Bargmann space $\mathcal{B}$, the operators N, $a^\dagger$, a of the C_λ-extended oscillator algebra take a very simple form, namely

$$\mathcal{N} = z\frac{\partial}{\partial z}\mathcal{I}, \tag{19}$$

$$\mathcal{A}^\dagger = \begin{pmatrix} 0 & 0 & \ldots & 0 & z \\ z & 0 & \ldots & 0 & 0 \\ 0 & z & \ldots & 0 & 0 \\ \vdots & \vdots & \ddots & \vdots & \vdots \\ 0 & 0 & \ldots & z & 0 \end{pmatrix}, \qquad \mathcal{A} = \begin{pmatrix} 0 & \frac{\partial}{\partial z}+\frac{\beta_1}{z} & 0 & \ldots & 0 \\ 0 & 0 & \frac{\partial}{\partial z}+\frac{\beta_2}{z} & \ldots & 0 \\ \vdots & \vdots & \vdots & \ddots & \vdots \\ 0 & 0 & 0 & \ldots & \frac{\partial}{\partial z}+\frac{\beta_{\lambda-1}}{z} \\ \frac{\partial}{\partial z} & 0 & 0 & \ldots & 0 \end{pmatrix}, \tag{20}$$

where $\mathcal{I}$ denotes the $\lambda\times\lambda$ unit matrix.

5 Conclusion

In the present contribution, we presented the mathematical properties of some new CS associated with the C_λ-extended oscillator.

Such states have also some interesting physical properties. In Refs. [1, 2], we indeed investigated some of their characteristics relevant to quantum optics, such as their statistical and squeezing properties, for a wide range of parameters and from both viewpoints of real and dressed photons. Their nonclassical features for some parameter ranges were clearly demonstrated.

Acknowledgments

The author is a Research Director of the National Fund for Scientific Research (FNRS), Belgium.

References

1. C. Quesne, *Phys. Lett.* A **272**, 313 (2000); *ibid.* **275**, 313 (2000).
2. C. Quesne, *Ann. Phys. (N.Y.)*, in press.
3. C. Quesne and N. Vansteenkiste, *Phys. Lett.* A **240**, 21 (1998).
4. C. Quesne and N. Vansteenkiste, *Helv. Phys. Acta* **72**, 71 (1999); *Int. J. Theor. Phys.* **39**, 1175 (2000).
5. J.R. Klauder, *J. Math. Phys.* **4**, 1055 (1963); *ibid.* **4**, 1058 (1963).
6. M.A. Vasiliev, *Int. J. Mod. Phys.* A **6**, 1115 (1991).
7. S. Chaturvedi and V. Srinivasan, *Phys. Rev.* A **44**, 8024 (1991).
8. A.M. Perelomov, *Commun. Math. Phys.* **26**, 222 (1972).
9. V. Bargmann, *Commun. Pure Appl. Math.* **14**, 187 (1961).
10. J. Deenen and C. Quesne, *J. Math. Phys.* **25**, 2354 (1984); D.J. Rowe, *ibid.* **25**, 2662 (1984).
11. J.K. Sharma, C.L. Mehta and E.C.G. Sudarshan, *J. Math. Phys.* **19**, 2089 (1978); J.K. Sharma, C.L. Mehta, N. Mukunda and E.C.G. Sudarshan, *ibid.* **22**, 78 (1981).

ASYMMETRIC DOUBLE SINH-GORDON THEORY IN 1+1 DIMENSIONS

A. RADOSZ, K. OSTASIEWICZ, P. MAGNUSZEWSKI AND P. MACHNIKOWSKI

Institute of Physics, Wrocław University of Technology, Wybrzeże Wyspiańskiego 27, 50-370 Wrocław, Poland

Classical thermodynamics of a linear chain with a local asymmetric double sinh-Gordon potential is studied in a continuum limit. Transfer integral method converts the problem of finding the partition function into the pseudo Schrödinger equation with temperature dependent mass and unchanged potential. Due to its hidden symmetry this double well potential belongs to the class of quasi exactly solvable *quantum mechanical* models. In consequence, *classical* partition function and other thermodynamic properties of corresponding (1+1)-dimensional systems are known exactly at several temperatures.

1 Introduction

Structures with a local double well potential have been investigated for years due to their appealing properties. Such systems reveal spontaneously broken symmetry phenomenon and all of the associated effects: in multi-component systems, Goldstone boson and Higgs mechanism; in condensed matter physics, large class of structural phase transitions have been described in terms of the models with double well, on-site potential with characteristic soft mode behaviour. In one-dimensional (1D)/quasi one-dimensional hydrogen bonded chains, protons are moving within local double well potential: thermodynamics and dynamical properties are then well understood in the picture of coexistence of extended, phonons and localised, kinks, excitations. In this paper we will discuss thermodynamics of the systems with a local asymmetric double well potential. Such systems, present in a class of hydrogen bonded chains, had seemed to have rather limited range of applications. Quite recently, however, it has become clear that local, asymmetric double well potential is an inherent feature of complexes commonly referred to as photo-induced phase transitions. Our main purpose is to investigate features characterising thermodynamic behaviour of 1D molecular chains with a local asymmetric double well potential. This potential is chosen as a double sinh-Gordon one: there are some experimental and theoretical predictions [1,2] for that choice instead of other models. Particular property of double sinh-Gordon potential is its quasi-exact solvability. Our approach in this paper is to study the classical, continuous limit and in such a case problem of calculating the partition function is converted into the task of finding of the ground state of pseudo-Schrödinger equation; then due to exact solvability partition function and free energy are given at several temperatures. The paper is organised as follows: in the following section the model is defined and the problem of finding the partition function and free energy function is formulated. Next exact solvability of the local potential is briefly discussed. The results obtained in term of a semi-classical approach, numerical (using "Invention" system [3]) and exact results are presented and discussed in three following Sections; concluding remarks are given in a closing section.

2 Model

Classical partition function of 1D, one-component system described by the Hamiltonian:

$$H = \sum_n Al \left[\frac{1}{2}\dot{\varphi}_n^2 + \frac{1}{2}\frac{c_0^2}{l^2}(\varphi_{n+1} - \varphi_n)^2 + \omega_0^2 V(\varphi_n) \right], \tag{1}$$

where l — lattice constant, c_0 — speed of sound, A — a certain constant, in the continuum limit takes the form (see e.g. [4]):

$$Z_{cl} = \left[\frac{2\pi l}{\beta h c_0} \right]^N \sum_n \exp\left[-\beta N l A \omega_0^2 E_n \right]. \tag{2}$$

where E_n are the eigenvalues of the pseudo–Schrödinger equation,

$$\left[-\frac{1}{2}\frac{1}{m^*}\frac{\partial^2}{\partial \varphi^2} + V(\varphi) \right] \Psi_n(\varphi) = E_n \Psi_n(\varphi), \tag{3}$$

$$m^* = \beta^2 A^2 c_0^2 \omega_0^2. \tag{4}$$

In the thermodynamic limit ground state eigenvalue of that equation determines partition function and free energy while square of the normalised ground state eigenfunction is the probability distribution function (PDF) for the field. In this paper we shall consider asymmetric version of the local DM, potential [5]:

$$V(\varphi) = \frac{1}{8}\left[(B\cosh\varphi - n)^2 + 2BC\sinh\varphi \right], \quad B < n, \tag{5}$$

which is composed of two mirror copies of Morse potential shifted against each other (see below). This potential belongs to the class of quasi exactly solvable therefore exact results for free energy would be given at several temperatures.

3 Morse potential and double-Morse potential

It is well known that Morse potential

$$U(\varphi) = U_0(e^{2(\varphi-\varphi_0)} - 2e^{(\varphi-\varphi_0)}) + \frac{1}{4}b_3^2 \tag{6}$$

is a quasi exactly solvable model and some lowest eigenvalues and eigenstates are known exactly. In fact the Hamiltonian of the particle of mass $m = 1$ with the above potential appears to be composed of operators belonging to $sl(2)$ algebra

$$[S_3, S_+] = S_+, \quad [S_3, S_-] = -S_-, \quad [S_+, S_-] = 2S_3$$

$$\begin{aligned} S_+ &= e^{\varphi}\frac{\partial}{\partial\varphi} + [\frac{1}{2}(b_+e^{\varphi} + b_3) - S]e^{\varphi} \\ S_- &= -e^{-\varphi}\frac{\partial}{\partial\varphi} - [\frac{1}{2}(b_+e^{\varphi} + b_3) + S]e^{-\varphi} \\ S_3 &= \frac{\partial}{\partial\varphi} + \frac{1}{2}(b_+e^{\varphi} + b_3) \end{aligned}$$

and n lowest eigenstates are found in an analytic form. Double Morse potential composed of two Morse potentials:

$$\begin{aligned} U(\varphi) &= U_+(e^{2(\varphi-\varphi_+)} - 2e^{(\varphi-\varphi_+)}) \\ &\quad + U_-(e^{2(\varphi-\varphi_-)} - 2e^{(\varphi-\varphi_-)}) + \frac{1}{4}b_3^2 - \frac{1}{2}b_+b_- \end{aligned}$$

$$U_\pm = \frac{1}{4}(n \mp b_3)^2, \quad \varphi_\pm = \ln\frac{(n \mp b_3)}{b_\pm}$$

$$b_+ = b_- = \frac{1}{2}B, \quad b_3 = C$$

turns out to be quasi exactly solvable. Hamiltonian:

$$H = -\frac{\partial^2}{\partial\varphi^2} + V(\varphi) \tag{7}$$

expressed via the following operator algebra,

$$[S_3, S_+] = S_+, \quad [S_3, S_-] = -S_-, \quad [S_+, S_-] = 2S_3$$

$$S_+ = e^{\varphi}\frac{\partial}{\partial\varphi} + [\frac{1}{2}(b_+e^{\varphi} - b_-e^{-\varphi} + b_3) - S]e^{\varphi}$$
$$S_- = -e^{-\varphi}\frac{\partial}{\partial\varphi} - [\frac{1}{2}(b_+e^{\varphi} - b_-e^{-\varphi} + b_3) + S]e^{-\varphi}$$
$$S_3 = \frac{\partial}{\partial\varphi} + \frac{1}{2}(b_+e^{\varphi} - b_-e^{-\varphi} + b_3)$$

is converted into spin-like hamiltonian:

$$H = -S_3^2 + b_3S_3 + b_+S_+ + b_-S_-$$

Eigenvalues (and eigenfuctions) of the spin-like Hamiltonian, and an original Hamiltonian (7) may be found in an analytic way for $n = 2S + 1$ no greater than 5; for the higher values of n numerical methods may be applied.

4 Exact values of partition function

Pseudo-Schrödinger equation for the transfer operator with temperature dependent mass parameter m^* (4) possess, for integer n, n exact solutions corresponding to temperature $2m^* = 1$. These solutions for fixed n at at fixed temperature $2m^* = 1$ are the same as ground state eigenvalue and eigenfunction of another model (different n and B, C) at certain values of temperature. To show this let us make a simple rescaling: for fixed n Eq.(3) transforms into

$$\left[-\frac{1}{2}\frac{1}{m^*n^2}\frac{\partial^2}{\partial\varphi^2} + \hat{V}(\varphi)\right]\Psi_n(\varphi) = \hat{E}_n\Psi_n(\varphi)$$

$$\hat{V}(\varphi) = \left(\frac{B}{n}\cosh\varphi - 1\right)^2 + 2\frac{B}{n}\frac{C}{n}\sinh\varphi$$

$$\hat{E}_n = E_n/n^2$$

and dividing by factor standing at the second derivative and demanding its inverse root to be integer,

$$2m^*n^2 = p^2$$

one converts fixed n-model into p-model:

$$\left[-\frac{\partial^2}{\partial\varphi^2} + \breve{V}(\varphi)\right]\Psi_n(\varphi)$$

$$\breve{V}(\varphi) = \left(p\frac{B}{n}\cosh\varphi - p\right)^2 + 2p^2\frac{B}{n}\frac{C}{n}\sinh\varphi$$
$$= (B\sqrt{2m^*}\cosh\varphi - p)^2 + 2m^*BC\sinh\varphi$$

$$\hat{E}_n = \frac{p^2}{n^2}E_n = 2m^*E_n$$

Hence the ground state eigenvalue and eigenfunction for n-model, is known for all integer p. One can see that for the arbitrary value of n, not necessarily integer, ground state of the original problem (3) is known at temperatures fulfilling the condition $(2m^*n = p)$. Therefore,

$$T_n = \sqrt{2}k_B A c_0 \omega_0 n$$

PDF and free energy function are given at these points; the highest temperature $T(n)$, at which exact solution is given corresponds to the lowest value $p = 1$.

5 Semiclassical approximation

Equation (3) determining the classical thermodynamic properties of the system may also be solved in the framework of the semiclassical approximation. Standard WKB method has usually been applied in the case of the system with symmetric double well potential. In the low temperature range effective mass is large and the lowest eigenstates, located deep within the well, consist of pairs of levels split due to tunnelling. The ground state eigenvalue, proportional to the free energy of the system, may be then expressed as a sum of two terms, first one corresponding to small fluctuations within the well and the second one, exponentially small referred to as tunnelling term. That result is subsequently interpreted in a phenomenological approach where one assumes that there are two subsystems of excitations in the system: extended, phonons and localised, solitary waves called kinks. In the approximation of weak interaction of these two subsystems, partition function factorises and free energy, consisting of phonon and kink contributions appears to be equal the above described ground state eigenvalue. Phonon subsystem corresponds to small fluctuations within the well, while kink subsystem contribution corresponds to the tunnelling factor of the pseudo-Schrödinger equation (3). Profound argument of such a division is the characteristic behaviour of specific heat: this function exhibits a characteristic hump being related to kinks presence. In the case of the asymmetric double well potential WKB approximation is not a suitable approach; another approximation, origination from the path integral

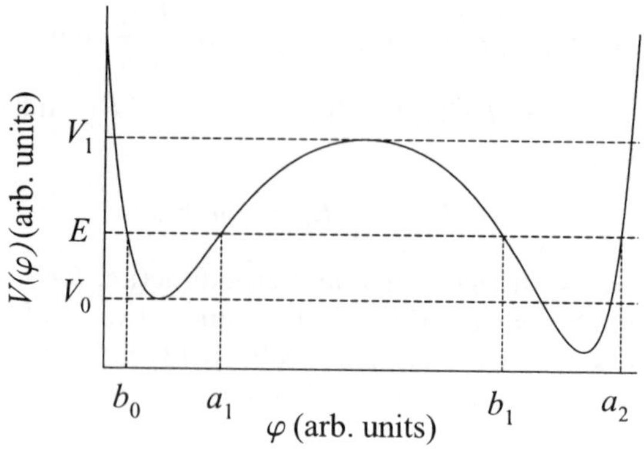

Figure 1. The asymmetric double-Morse potential.

picture was derived - real trajectories in complex time (RTCT). It was shown that this method is free of inconsistencies characteristic of WKB approach (complex eigenvalues) and may be successfully applied to the multiple-well potentials (symmetric and asymmetric). Low energy levels placed below the potential barrier $V(0)$, being the poles of the appropriate Green's function, are determined as the solutions of the equation (see [6]),

$$\cos(W_1)\cos(W_2) - \frac{1}{4}\Gamma^2 \sin(W_1)\sin(W_2) = 0, \tag{8}$$

where (see Fig. 1),

$$W_i(E) = \int_{b_{i-1}}^{a_i} d\varphi \sqrt{2m^*(E - V(\varphi))}, \tag{9}$$

$$\Gamma(E) = \exp\left[-\int_{a_1}^{b_1} d\varphi \sqrt{2m^*(V(\varphi) - E)}\right]. \tag{10}$$

RTCT method extends WKB approach what may be illustrated for the case of large barrier. In the zero order approximation, $b = 0$, the set of low lying energy levels consists of two subsets of levels in the left and in the right well,

$$\cos[W_1(E^L)] = 0$$

$$\cos[W_2(E^R)] = 0$$

Taking into account the barrier penetration and regarding b a small but non-vanishing parameter, one finds energy levels as got mixed of two subsets, $E(n, L)$ and $E(n, R)$: close levels belonging to different wells are getting repelled. In the systems with smaller potential barriers, larger value b, energy levels found from (3) are only weakly connected with zero order solutions.

6 Results and discussion

Some characteristic features of the thermodynamic properties are revealed in the behaviour of free energy, specific heat and probability distribution function as the functions of temperature. They are expressed via ground state of the Eq.(3) as follows:

$$F = -kT \ln Z_{cl} = kTN \ln(\beta\hbar \frac{c_0}{l}) + NlA\omega_0^2 E_0$$

$$C = \frac{\partial}{\partial T}(\frac{\partial}{\partial \beta}(\beta F)) = Nk + NlA\omega_0^2 \frac{\partial}{\partial T}(E_0 + \beta \frac{\partial}{\partial \beta}(E_0))$$

The ground state eigenvalue and eigenfunction of the pseudo-Schrödinger equation (3) has been found numerically and within RTCT method; exact analytical results have also been found at several temperatures. The results for a chosen set of model parameters are presented in Figs. (2,3,4). There are two interesting features worth to be underlined. First, one observe satisfactory coincidence between RTCT and numerical results; one should emphasise that the specific heat, as a second derivative function is a very sensitive for the smallest inadequacies of the numerical treatment [7] - the coincidence between numerical and RTCT results for that case underlines reliability of both methods. Second, the most significant observation relates to hump in the specific heat. In the shape and magnitude it is similar to its counterpart in the system with a symmetric double well potential (also illustrated in Fig.3). On the first sight this similarity appears to be surprising: in the systems with a symmetric double well potential, localised excitations, solitary waves called kinks, turned out to be responsible for that peculiar property of the specific heat. In the model with asymmetric double well potential the situations differs significantly comparing to the former case. Solitary waves, in that case called bell shapes are unstable and it makes no sense to consider them as a subsystem of extended excitations whose free energy would contribute to the overall free energy. This dramatic difference between dynamical properties of the systems with symmetric and asymmetric potentials appears to convert into a minor difference in the thermodynamic context. Ignoring a phenomenological picture which is missing in that latter case, semiclassical

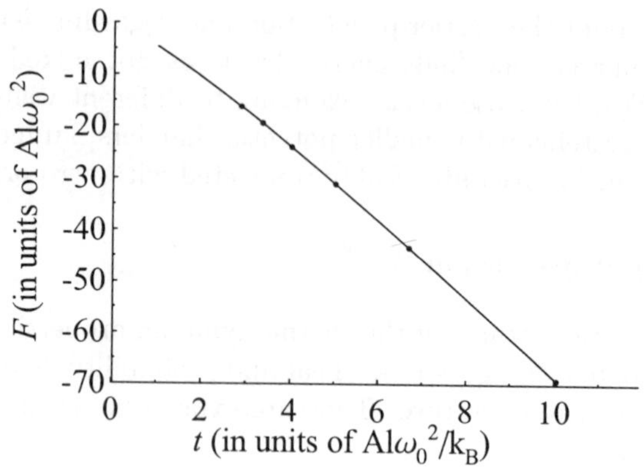

Figure 2. Free energy as a function of temperature t for B=0.1, C=-0.01: RTCT result (solid line) and analytical results (dots)

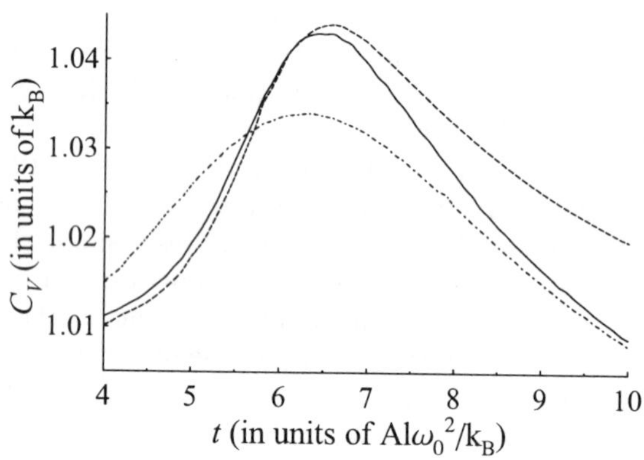

Figure 3. Specific heat for B=0.1, C=-0.01 — exact results (solid line) and RTCT approximation (dashed line); C=0 (symmetric DM) — dashed–dotted line

RTCT analysis provides us with an interesting hint in explanation of such an unexpected behaviour of the specific heat. In order to investigate the origin of the hump in the specific heat let us identify following temperature ranges. At low temperatures, where the effective mass is large enough, ground state eigenvalue, lies below the bottom of the higher well (see Fig. 1),

$$E_0 \approx \frac{1}{2}(\beta A c_0 \omega 0) + V_{min} \leq V_1$$

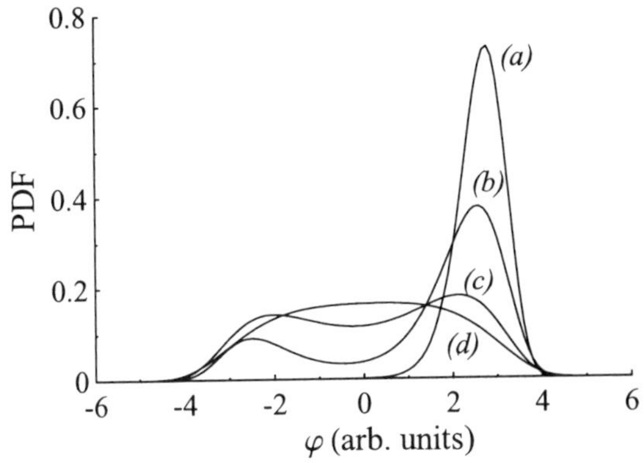

Figure 4. PDF for B=0.1, C=-0.01: t=4.02 (a), 6.7 (b), 9.5 (c), 15.0 (d)

and the corresponding wave function, PDF is expected to be localised in the lower potential well. At high temperatures, the effective mass is small enough and ground state eigenvalue lies well above the potential barrier, and PDF is localised at the centre of the local potential. In the intermediate temperature regime the ground level intersects both wells, being under growing influence of the mechanism of left-right well energy levels mixing. As was stressed in the previous paragraph this should result in the left ground level—right ground level repelling, the effect being growing with temperature. This effect of repelling of the two lowest energy levels, common for both type of systems, with symmetric and asymmetric potentials, makes their thermodynamic properties so much alike, while their dynamic properties are so much different. Confirmation of the above hypothesis may be found in the thermal evolution of the PDF. This function, with a single-peak structure (left, lower well) at low temperatures, gradually leaks to the two-well structure, characteristic for the case of the system with a symmetric potential, eventually reaching the usual form of single, central peak.

7 Conclusion

Asymmetric double sinh-Gordon potential belongs to the class of quasi exactly solvable models.$1D$ lattice model with a local potential of this type, due to various applications seems to be an interesting object for studies, whose thermodynamic properties have not been well–known. In this paper we have described free energy, specific heat and PDF for such a system in 1D continuum limit. By using numerical and semiclassical approaches, we have been able to give reasonable interpretation of the appearance of the hump in the specific heat function. The description is based on the semicalssical treatment of a pseudo–Schrödinger equation otained within transfer matrix method. We have found that interference (repelling) of the two lowest energy levels in the left and right well of a asymmetric double well potential is similar to the split of the energy levels in a symmetric double well potential. In effect the similar thermodynamic behaviour is observed: the specific heat function develops characteristic hump. If the conclusion about the relation between the hump and repelling of the energy levels is true, then another interesting feature, unique for the asymmetric systems would be observed. Let us suppose that the double well potential consists of two wells, the higher one being much wider of the other, lower and narrower one. Then ground levels of left and right potential wells are moving up differently when temperature increases. Left well ground level, corresponding to lower and narrower well, moves up quicker then the other one, corresponding to the higher and wider well. If these two levels intersect each other before reaching the edge of the potential barrier $V(0)$, that should lead to interesting effect, not observed in the systems with symmetric double well potential. That problem will be discussed elsewhere.

References

1. E. S. Kryachko, *Chem. Phys.* **143**, 359 (1990).
2. V. Antonchenko, A. Davydov, A. Zolotaryuk, *Phys. Stat. Sol. (b)*, **115**, 631 (1983).
3. http://dyna.if.pwr.wroc.pl/invention.
4. J. F. Currie, J. A. Krumhansl, A. R. Bishop, S. E. Trullinger, *Phys. Rev. B* **22**, 477 (1980).
5. H. Konwent, P. Machnikowski, A. Radosz, *J. Phys.: Condens. Matter* **8**, 4325 (1996).
6. H. Konwent, P. Machnikowski, P. Magnuszewski, A. Radosz, *J. Phys. A* **31**, 7541 (1998).
7. A. Khare, S. Habib, A. Saxena, *Phys. Rev. Lett.* **79**, 3797 (1997).

QUANTUM FIELD THEORY IN ABSENCE OF THE POSITIVITY CONDITION

S. ROUHANI AND M.V. TAKOOK

Department of Physics, Razi University, Kermanshah, IRAN and Plasma Physics Research Centre, Islamic Azad University, P.O.BOX 14835-157, Tehran, IRAN

E-mail: rouhani@daneshafshan.com,takook@razi.ac.ir, takook@ccr.jussieu.fr

Unlike the standard quantum field theory (QFT) in this review paper the positivity condition is alternatively overlooked. Consequently indefinite metric quantization and the presence of negative norm states as the auxiliary unphysical states has been included. The elimination of the negative norm states in the standard QFT, results in the appearance of the following anomalies: (1) infrared and ultraviolet divergence and (2) breaking of the de Sitter covariance for the minimally coupled scalar field and the breaking of the de Sitter covariance for linear quantum gravity. Inclusion of negative norm states automatically removes the divergence and symmetry breaking. Consequently the QFT is automatically renormalized and the covariance of the QFT in de Sitter space is conserved. The compatibility of this method with the results of some existing calculation is discussed in the last section.

1 Introduction

In the axiomatic QFT, one of the principle is the positivity condition [1], which stems of the concept of probability. Consideration of this principle (positivity condition) for the gauge QFT (GQFT) results in appearance of the breaking of the Lorentz covariance and the appearance of the infrared divergence. It has already been shown that the indefinite metric quantization or presence of the negative norm states (NNS) are necessary for obtaining a causal and covariant GQFT [2]. As a result the positivity condition must be eliminated in GQFT. The general properties of this quantization, *i.e.* the indefinite metric Fock quantisation were studied by Mintchev [3].

The same problems of the GQFT appear in the quantization of the free massless minimally coupled scalar field in de Sitter (dS) space-time (*i.e.* appearance of the infrared divergence [4] and the breaking of the dS covariance [5]). It has been shown that the presence of negative norm states are indispensable for a fully covariant quantization of the minimally coupled scalar field in de Sitter space [6,7]. This type of NNS is different from of the NNS in the GQFT. So we call this construction a "new QFT" (NQFT). This procedure, while leaving the physical content of the theory unchanged, allows us to avoid any infrared and ultraviolet divergence. In other words the vacuum energy remains convergent [7] and consequently the normal ordering procedure is not necessary any more. The old infrared divergence in the two-point function disappears as well [8]. The crucial element in the NQFT is the presence of negative frequency solutions, which is indispensable for preservation of the full covariance of the problem. It has also been shown that if this method is used for the free "massive" scalar field in de Sitter space, an automatic and covariant renormalization of the vacuum energy divergence is obtained[7].

The linear quantum gravity in dS space can be constructed from a projector tensor and a massless minimally coupled scalar field [9,10]. However, in so doing the same problems appear again [11,?]. It is also shown that the infrared divergence or the pathological behavior of the graviton propagator is gauge dependent and therefore should not appear in an effective way as a physical quantity [9,?,?,15,?,?]. An explicit construction of the covariant quantization of the traceless rank-2 "massless" tensor field in dS space (linear covariant quantum gravity) has already been presented [10]. The main ingredient of this construction is the presence of two different types of NNS. The first one is similar to those non physical states, dual to gauge states, which appear in GQFT in Minkowski space. The other is due to the consideration of negative frequency solutions in order to preserve de Sitter covariance. This is similar to the case of free massless minimally coupled scalar field in dS space. This construction allows us to avoid any infrared divergence and enables us to obtain a covariant two point function[14].

There are some similarity between GQFT in Minkowski space and the calculation of massless minimally coupled scalar field in dS space: the appearance of the infrared divergence and the necessity of the inclusion of the NNS for preservation of the covariance of the problem. The difference between the above problems is that in the first case a local symmetry and an interaction field are present where as in the second case (*i.e.* dS space) a global symmetry and a free field are the central elements of the problem. The infrared divergence disappears alone automatically in the second case. In other words there are two different types of NNS, one is due to the local gauge invariance (sign of the metric) and the other is due to the global gauge invariance (negative frequency solution). The important question, which deserves to be considered is: *is it possible to generalize the free NQFT in dS space to the interaction field in dS space or Minkowski space?*

Extending the previously proposed method to the interacting quantum field in Minkowski space-time ($\lambda\phi^4$ theory) a natural renormalization has been achieved [18]. For the free field in Minkowski space-time, a very interesting consequence is that the Schwinger commutator function, the retarded and the advanced Green's functions are one and the same in both formalism. Another remarkable result is that the vacuum energy automatically reduces to zero [18].

In the interaction case the new Feynman rule for the diagram with the n

vertices and I internal propagators is: if $(n+I)$ is an even (odd) number the new diagram is the real (imaginary) part of the old diagram [18]. Thus in the tree order S-matrix elements (with two vertices and one internal propagator), the new diagram is imaginary part of the old one. The old diagram however is pure imaginary. Therefore the application of the above new method of field quantization does not change the tree order S-matrix elements. For the one-loop diagram (in which the divergence appears in the imaginary part of the old Feynman diagram) the theory is automatically renormalized, since the imaginary part of the old diagram is eliminated in the new approach. A very small difference between the physical quantities in the old and new QFT in the one-loop approximation exists. There also exist a very small difference between the physical quantities in the old QFT calculation and the experimental measurement. The basic question, which naturally arises, is: *which one of these results have a greater compatibility with the experimental measurements*? In the forthcoming paper the physical quantities for the QED in the new QFT will be studied and a comparison with the experimental results is examined.

It should be noted that by adopting this NQFT framework, various Green's functions for a scalar field, in a general curved space-time, appear to be convergent in the ultraviolet limit. A natural renormalized one-loop effective action has been also obtained [19] in some further studies. In the one-loop approximation, the Einstein's field equation does not need to be modified in view of divergence elimination. Similar to the cosmological constant in the Einstein's field equations a constant term appears in the effective action of this problem. This means a de Sitter metric is the background in the first approximation. In other words a natural renormalized linear quantum gravity in the one-loop approximation is possible without changing the Einstein's field equation.

In the next section by using this method, calculation of the tree-level scattering amplitudes of the scalar field with one graviton exchange in the de Sitter space, is presented. It is shown that the previous divergence disappears and the theory is automatically renormalized.

2 Tree-level scattering amplitudes

Recalling the previous calculation of the tree-level scattering amplitudes of two scalar particles in the one-graviton-exchange approximation [20]. An expression of the following form must be calculated

$$\mathcal{M} = \int T^{\mu\nu}(x) G_{\mu\nu,\mu'\nu'}(x,x') T^{\mu'\nu'}(x') dV_x dV_{x'}, \tag{1}$$

where $T^{\mu\nu}$ is a conserved, physical energy-momentum tensor of the scalar field and $G_{\mu\nu,\mu'\nu'}$ is the graviton two-point function. By explicit computation it was shown that the only growing pieces for large z in $\mathcal{M}$ comes from the transverse and traceless part of the graviton two-point function

$$G_{TT}^{\mu\nu,\mu'\nu'}(x,x') \sim \ln(1-z)\mathcal{D}^{\mu\nu,\mu'\nu'} \quad \text{for large } z, \tag{2}$$

where $\mathcal{D}$ is a maximally symmetric bi-tensor.

For simplicity, a conformally coupled scalar field was considered

$$\phi(x) = \sum_k a_k f_k(t) e^{i\vec{k}.\vec{x}} + C.C., \tag{3}$$

where $f_k(t) = \sqrt{\frac{2|\vec{k}|}{4\pi^2}}t^2 \frac{\sin t|\vec{k}|}{t|\vec{k}|}$. A particular coordinate system is chosen

$$dS^2 = -\frac{1}{t^2}dt^2 + \frac{1}{t^2}d\vec{x}^2, \tag{4}$$

in which $z = 1 + \frac{1}{4tt'}[(t-t')^2 - (\vec{x}-\vec{x}')^2]$. For the S-channel diagram the only term, which depend on the scattering angle $\vec{k}.\vec{k}'$, gives

$$\mathcal{M}^{\theta\text{-dep.part}} \sim -8(\vec{k}.\vec{k}')^2 \int dV_x dV_{x'} t^2 t'^2 f_k^2(t) f_{k'}^2(t')$$

$$\ln(1-z)\left(1 + \frac{4}{3}\frac{(\vec{x}-\vec{x}')^2}{(\vec{x}-\vec{x}')^2-(t-t')^2} + \frac{8}{15}\left[\frac{(\vec{x}-\vec{x}')^2}{(\vec{x}-\vec{x}')^2-(t-t')^2}\right]^2\right), \tag{5}$$

where $dV_x = \frac{1}{t^4}dtd^3x$. This term has a $\cos^2\theta$ dependence characteristic of a spin-two exchange, but due to the $\ln(1-z)$ element, it appears to be divergent.

Now the new method of field quantization (maintaining the negative norm states as auxiliary states) is implemented. In this method the field operator (3) is defined by [7]

$$\phi(x) = \frac{1}{\sqrt{2}}\sum_k a_k f_k(t) e^{i\vec{k}.\vec{x}} + b_k f_k^*(t) e^{-i\vec{k}.\vec{x}} + C.C., \tag{6}$$

where

$$\begin{aligned} &[a(\vec{k}), a^\dagger(\vec{k}')] = \delta(\vec{k}-\vec{k}'), \quad a^\dagger(\vec{k}) \mid 0 >=\mid 1_{\vec{k}} >=\mid \text{one-particle state} >, \\ &[b(\vec{k}), b^\dagger(\vec{k}')] = -\delta(\vec{k}-\vec{k}'),; \quad b^\dagger(\vec{k}) \mid 0 >=\mid \bar{1}_{\vec{k}} >=\mid \text{unphysical state} > \\ &\qquad b(\vec{k}) \mid \text{physical state} >= 0; \quad a(\vec{k}) \mid \text{unphysical state} >= 0, \\ &\qquad\qquad < \text{unphysical state} \mid \text{physical state} >= 0. \end{aligned} \tag{7}$$

By using the above relations it can be shown that the stress tensor for the physical state is not affected by the negative norm state ([7] section 6). Therefore the only difference appears in the graviton two-point function. In this case the logarithmic term ($\ln(1-z)$) must be replaced by [10,?]

$$\text{const. } \epsilon(X^0 - X'^0)\theta(z-1), \tag{8}$$

where

$$X^0 = \sinh t + \frac{1}{2}e^t|\vec{x}|^2, \quad \epsilon(X^0 - X'^0) = \begin{cases} 1 & X^0 > X'^0 \\ 0 & X^0 = X'^0 \\ -1 & X^0 < X'^0. \end{cases}, \tag{9}$$

and θ is the Heaviside step function. This expression (eq.(8)) can also be obtained by using the fact that the new Feynman propagator is the real part of the previous one (eq.(13a) of [21].) By replacing the equation (8) in the integral (5), we obtained

$$\mathcal{M}^{\theta\text{-dep.part}} \sim \text{const. } (\vec{k}.\vec{k}')^2,$$

which is free of any infrared divergence.

By the very same procedure it can also be shown that the effective action in the one-loop approximation for the large-distance behavior is zero. This result reaffirms the Iliopoulos's calculation [11].

3 Conclusion

The appearance of singularity in physical quantities of QFT, is clearly a manifestation of an anomaly in QFT. For eliminating this anomalies (or singularity in the physical quantities) the new rules are added to the QFT *i.e.* the normal ordering procedure and the renormaization techniques. This anomaly could be a consequence of imposing the positivity condition *i.e.* elimination of negative frequency solution from field equation. In a series of papers we studied the QFT in absence of the positivity condition and their consequences. In the free QFT case without changing the physical content of the theory an automatically renormalized theory was obtained. Following this procedure the covariant principle in dS space was conserved as well.

In the interaction case, an self-renormalized theory was also obtained. There appears to be a very small difference between the QFT and NQFT calculation in the one-loop approximation. Proposed area to be studied in future are the higher order calculations of the new QED such as Lamb Shift, Magnetic anomaly, etc, and comparison of the theoretical and the experimental results.

If the appearance of anomalies are due to removed of negative frequency solution (*i.e.* implementation of positivity condition), a suitable ground is provided for the combination of general relativity of NQFT, free of anomalies.

References

1. R.F. Streater, A.S. Wightman, *PCT, Spin, Statistics, and all that*, Oxford University Press, Inc. (1964).
2. F. Strocchi, *Phys. Rev. D* **17**, 2010 (1978)
3. M. Mintchev, *J. Phys. A: Math. Gen.* **13**, 1841 (1979).
4. B. Ratra, *Phys. Rev. D* **31**, 1931 (1985).
5. B. Allen, *Phys. Rev. D* **32**, 3136 (1985).
6. S. De Bièvre, J. Renaud, *Phys. Rev. D* **57**, 6230 (1998).
7. J.P. Gazeau, J. Renaud, M.V. Takook, *Class. Quantum Grav.* **17**, 1415 (2000), gr-qc/9904023.
8. M.V. Takook, *Mod. Phys. Lett. A* **16**, 1691 (2001), gr-qc/0005020.
9. M.V. Takook, *Théorie quantique des champs pour des systèmes élémentaires "massifs" et de "masse nulle" sur l'espace- temps de de Sitter.*, Thèse de l'université Paris VI (1997).
10. T. Garidi, J.P. Gazeau, J. Renaud, S. Rouhani, M.V Takook, *Linear covariant quantum gravity in de Sitter space*, in preparation.
11. I. Antoniadis, J. Iliopoulos and T.N. Tomaras, *Nucl. Phys. B* **462**, 437 (1996).
12. B. Allen, M. Turyn, *Nucl. Phys. B* **292**, 813 (1987).
13. H.J. de Vega, J. Ramirez and N. Sanchez, *Phys. Rev. D* **60**, 044007 (1999), astro-ph/9812465.
14. M.V. Takook, *Proceedings of the Wigsym6, 16-22 August 1999, Istanbul, Turkey*, gr-qc/0001052.
15. S.W. Hawking, T. Hertog, N. Turok, *Phys. Rev. D* **62**, 063502 (2000), hep-th/0003016.
16. A. Higuchi, S.S. Kouris, *Class. Quantum Grav.* **17**, 3077 (2000), gr-qc/0004079.
17. A. Higuchi, S.S. Kouris, *On the scalar sector of the covariant graviton two-point function in de Sitter spacetime*, gr-qc/0011062.
18. M.V. Takook, *Negative norm states in de Sitter Space and QFT without Renormalization Procedure*, gr-qc/0006019.
19. M.V. Takook, *Natural Renormalized One-loop Effective Action for the scalar field in a curved space-time*, gr-qc/0006052.
20. E.G. Floratos, J. Iliopoulos and T.N. Tomaras, *Phys. Lett. B* **197**, 373 (1987).
21. N.C. Tsamis, R.P. Woodard, *Nuclear Physics B* **474**, 235 (1996).

ON THE GAUGE ORBIT SPACE STRATIFICATION

G. RUDOLPH, M. SCHMIDT
Institute for Theoretical Physics, University of Leipzig
Augustusplatz 10/11, 04109 Leipzig, Germany
E-mail: matthias.schmidt@itp.uni-leipzig.de

I.P. VOLOBUEV
Nuclear Physics Institute, Moscow State University
119899 Moscow, Russia

A method for determining the orbit types, as well as their partial ordering, of the action of the group of gauge transformations on the space of connections for gauge theories with gauge group SU(n) in space-time dimension $d \leq 4$ is presented. The method enables one to determine explicitly the corresponding stratification of the gauge orbit space. As an application, we discuss a criterion that characterizes kinematical nodes for physical states in 2+1-dimensional Chern-Simons-Yang-Mills theory proposed by Asorey *et al.* .

1 Introduction

We consider a compact principal SU(n)-bundle P over the space-time manifold M. Let $\mathcal{A}$ denote the space of connections in P and let $\mathcal{G}$ denote the group of local gauge transformations, acting on $\mathcal{A}$ by

$$A^{(g)} = g^{-1}Ag + g^{-1}\mathrm{d}g\,, \quad A \in \mathcal{A}, g \in \mathcal{G}\,.$$

By passing to appropriate Sobolev classes of connections and gauge transformations, $\mathcal{A}$ becomes an affine Hilbert space, $\mathcal{G}$ becomes a Lie group, and the action becomes smooth. The gauge orbit space is the topological quotient $\mathcal{M} = \mathcal{A}/\mathcal{G}$. It consists of the physical degrees of freedom of the gauge theory. It is well known that $\mathcal{M}$ may contain singularities, due to the several orbit types the action of $\mathcal{G}$ on $\mathcal{M}$ may possess. Let Σ denote the set of orbit types.

It was shown[5] that the subsets of constant orbit type σ, $\mathcal{M}_\sigma$, are Hilbert manifolds and that the decomposition

$$\mathcal{M} = \bigcup\nolimits_{\sigma\in\Sigma} \mathcal{M}_\sigma$$

is countable and satisfies:

- the condition of the frontier: $\mathcal{M}_\sigma \cap \overline{\mathcal{M}_{\sigma'}} \neq \emptyset \quad \Rightarrow \quad \mathcal{M}_\sigma \subseteq \overline{\mathcal{M}_{\sigma'}}$,
- the condition of regularity: $\mathcal{M}_\sigma \cap \overline{\mathcal{M}_{\sigma'}} \neq \emptyset \;\Rightarrow\; \mathcal{M}_\sigma$ closed in $\mathcal{M}_\sigma \cup \mathcal{M}_{\sigma'}$.

Thus, it is the set Σ that encodes the stratification: its elements characterize the strata and its natural partial ordering determines how the strata are linked.

Little is known about physical effects caused by the stratification of the gauge orbit space. Nevertheless, there exist several concrete hypotheses. Two of these shall be cited here.

Asorey *et al.*[2] discussed Yang-Mills-Chern-Simons theory in the canonical formalism. The authors found that the Gauss law forces physical states to vanish on reducible connections with nonvanishing magnetic charge. Thus, these connections form kinematical quantum nodes. Since they do so only in the presence of the Chern-Simons term, i.e., in the deconfined phase, the authors conjecture that these connections may play a major role in the confinement mechanism. Below, the result of Asorey *et al.* will be discussed in the light of our results.

Emmrich and Römer[3] showed that in quantum mechanics on a double cone over some Riemannian manifold, stationary states show a clear tendency to concentrate around the singularity. The authors argue that gauge models should be expected to show a similar behaviour.

Comparing the results of Asorey *et al.* and Emmrich and Römer, one could wonder whether, in general, some of the strata attract physical states, whereas others reject them.

To clarify the physical role of the stratification, a systematic investigation is necessary. As a first step, we determine the orbit types and their partial ordering, thus deriving a method to determine which strata are present, depending on the topology of P, and how they are glued together to build up the orbit space. Besides, this provides a variety of examples to test hypotheses.

The determination of orbit types proceeds in the following two steps:

- Characterizing orbit types by Howe subbundles,
- Characterizing Howe subbundles by characteristic classes.

We do not comment on the derivation of the partial ordering. Instead, the interested reader is referred to our paper[7]. The results presented in this contribution are proved and explained in detail in our paper[6].

2 Characterizing orbit types by Howe subbundles

A Howe subbundle of a principal G-bundle P is a reduction of P to a Howe subgroup of G. A Howe subgroup of G is a subgroup H obeying $C_G^2(H) = H$.
Theorem 1. *Let P be a principal* $\mathrm{SU}(n)$*-bundle over M,* $\dim M \geq 2$*. The assignment* $Q \mapsto \{g \in \mathcal{G} \ : \ g|_Q = \mathrm{const.}\}$ *induces a bijection from the set of*

Howe subbundles of P, modulo isomorphy and $\mathrm{SU}(n)$*-action onto the set of orbit types. The bijection inverts the partial ordering.*

Remark. In general, bundle reductions that characterize orbit types have to obey a stronger condition. Namely, in the terminology of Kondracki and Rogulski[5], they must be evolution bundles of connections. In the case of $\mathrm{SU}(n)$ one can show that this is equivalent to being a Howe subbundle.

3 Characterizing Howe subbundles by characteristic classes

3.1 The Howe subgroups of $\mathrm{SU}(n)$

At first, one has to determine the Howe subgroups of $\mathrm{SU}(n)$ up to conjugacy.

Lemma 1. *The assignment* $\mathfrak{h} \mapsto \mathrm{SU}(n) \cap \mathfrak{h}$ *yields a bijection from the von Neumann-algebras in* $\mathcal{M}_n(\mathbb{C})$ *onto the Howe subgroups of* $\mathrm{SU}(n)$.

According to this, the double commutant theorem and the structure theorem for finite-dimensional von Neumann algebras yield

Theorem 2. *Up to conjugacy, the Howe subgroups are given by sequences of integers* $\mathbf{k} = (k_1, ..., k_r)$, $\mathbf{m} = (m_1, ..., m_r)$ *of length* $r = 1, 2, 3, \ldots$, *obeying*

$$\mathbf{k} \cdot \mathbf{m} = \sum\nolimits_{i=1}^{r} k_i m_i = n\,. \tag{1}$$

Representatives are provided by the subgroups $\mathrm{SU}_{\mathbf{k}}^{\mathbf{m}} = \mathrm{SU}(n) \cap \mathrm{U}_{\mathbf{k}}^{\mathbf{m}}$, *where*

$$\mathrm{U}_{\mathbf{k}}^{\mathbf{m}} = \left\{ \bigoplus\nolimits_{i=1}^{r} B_i \otimes \mathbb{1}_{m_i} \;\middle|\; B_i \in \mathrm{U}(k_i) \right\}.$$

Remark. Obviously, the Howe subgroups $\mathrm{SU}_{\mathbf{k}}^{\mathbf{m}}$ and $\mathrm{SU}_{\mathbf{k}'}^{\mathbf{m}'}$ are conjugate if and only if $\mathbf{k}'$ and $\mathbf{m}'$ can be obtained from $\mathbf{k}$ and $\mathbf{m}$, respectively, by a simultaneous permutation.

3.2 Characteristic classes for principal $\mathrm{SU}_{\mathbf{k}}^{\mathbf{m}}$*-bundles*

Next, one has to construct characteristic classes for principal bundles Q with structure group $\mathrm{SU}_{\mathbf{k}}^{\mathbf{m}}$. One immediately realizes that Q possesses two natural associated bundles:

- $\tilde{Q}$ via extension by means of the embedding $\mathrm{SU}_{\mathbf{k}}^{\mathbf{m}} \hookrightarrow \mathrm{U}_{\mathbf{k}}^{\mathbf{m}}$.

  Since $\mathrm{U}_{\mathbf{k}}^{\mathbf{m}} \cong \mathrm{U}(k_1) \times \cdots \times \mathrm{U}(k_r)$, in space-time dimension $d \leq 4$, $\tilde{Q}$ is classified up to isomorphy by the collection of the (total) Chern classes of its elementary factors, $\alpha = (\alpha_1, ..., \alpha_r)$.

- Q_0 via factorization by the identity component of $\mathrm{SU}_{\mathbf{k}}^{\mathbf{m}}$. Q_0 has the group of connected components of $\mathrm{SU}_{\mathbf{k}}^{\mathbf{m}}$ as its structure group.

Let $d = \text{g.c.d.}\{m_1, ..., m_r\}$. Define $\lambda : \mathrm{U}_{\mathbf{k}}^{\mathbf{m}} \to \mathrm{U}(1)$ by

$$\lambda\left(\bigoplus\nolimits_{i=1}^{r} B_i \otimes \mathbb{1}_{m_i}\right) = \prod_{i=1}^{r} (\det B_i)^{m_i/d} .$$

Then

$$\det(B) = \lambda(B)^d , \quad \forall B \in \mathrm{U}_{\mathbf{k}}^{\mathbf{m}} .$$

In particular, λ restricts to a homomorphism $\tilde{\lambda} : \mathrm{SU}_{\mathbf{k}}^{\mathbf{m}} \to \mathbb{Z}_d$.

Lemma 2. *$\tilde{\lambda}$ projects to an isomorphism of the group of connected components of $\mathrm{SU}_{\mathbf{k}}^{\mathbf{m}}$ onto $\mathbb{Z}_d$.*

Thus, Q_0 is classified up to isomorphy by a characteristic class $\xi \in H^1(M, \mathbb{Z}_d)$.

We view α and ξ as characteristic classes of Q, rather than of $\tilde{Q}$ and Q_0. Since, according to the commutative diagram

$$\begin{array}{ccc} \mathrm{SU}_{\mathbf{k}}^{\mathbf{m}} & \longrightarrow & \mathrm{U}_{\mathbf{k}}^{\mathbf{m}} \\ {\scriptstyle\tilde{\lambda}}\downarrow & & \downarrow{\scriptstyle\lambda} \\ \mathbb{Z}_d & \longrightarrow & \mathrm{U}(1) \end{array}$$

the bundles $\tilde{Q}$ and Q_0 are not independent, neither are α and ξ.

Lemma 3. *The characteristic classes α and ξ obey the relation*

$$\sum\nolimits_{i=1}^{r} \frac{m_i}{d}\ \alpha_i^{(2)} = \beta_d(\xi) . \tag{2}$$

Here $\beta_d : H^1(M, \mathbb{Z}_d) \to H^2(M, \mathbb{Z})$ is the Bockstein homomorphism related to the exact sequence of coefficient groups $0 \to \mathbb{Z} \to \mathbb{Z} \to \mathbb{Z}_d \to 0$.

Theorem 3. *Let $\dim M \leq 4$. Principal $\mathrm{SU}_{\mathbf{k}}^{\mathbf{m}}$-bundles over M are classified up to isomorphy by the collection of characteristic classes (α, ξ) obeying* (2).

Remark. That α and ξ can distinguish between non-isomorphic Q was proved by building the Postnikov tower of the classifying space of $\mathrm{SU}_{\mathbf{k}}^{\mathbf{m}}$ up to the 5th stage.

3.3 The Howe subbundles of P

Finally, one has to express the condition that Q is a bundle reduction of P in terms of the characteristic classes α, ξ.

Theorem 4. *Let $\dim M \leq 4$. The principal $\mathrm{SU}_{\mathbf{k}}^{\mathbf{m}}$-bundle with characteristic classes α, ξ is a reduction of P if and only if*

$$\alpha_1^{m_1} \cdots \alpha_r^{m_r} = c(P) . \tag{3}$$

Thus, the Howe subbundles of P are classified up to isomorphy by systems of invariants $(\mathbf{k}, \mathbf{m}, \alpha, \xi)$, subject to the Equations (1)–(3).

4 The result

By checking how the action of the structure group $\mathrm{SU}(n)$ on Howe subbundles expresses itself on the level of the invariants, one arrives at the final result.

Theorem 5. *The orbit types of the action of $\mathcal{G}$ on $\mathcal{A}$ (hence, the strata of $\mathcal{M}$) are classified by systems of invariants $(\mathbf{k}, \mathbf{m}, \alpha, \xi)$ which are subject to Equations (1)–(3), modulo simultaneous permutations of $\mathbf{k}$, $\mathbf{m}$, and α.*

5 Application: Kinematical quantum nodes in Yang-Mills-Chern-Simons theory

We consider $\mathrm{SU}(n)$-gauge theory on a 2 + 1-dimensional space-time $M \times \mathbb{R}$. The action functional contains the Yang-Mills and Chern-Simons terms,

$$S(\tilde{A}) = \frac{1}{2}\left(\tilde{F}, \tilde{F}\right)_{M\times\mathbb{R}} + \frac{\lambda}{8\pi}\left\{\left(\tilde{A}, *\tilde{F}\right)_{M\times\mathbb{R}} - \frac{4}{3}\left(\tilde{A}, *[\tilde{A}, \tilde{A}]\right)_{M\times\mathbb{R}}\right\}.$$

Here $(\cdot,\cdot)_{M\times\mathbb{R}}$ is the scalar product of $\mathrm{su}(n)$-valued fields on $M \times \mathbb{R}$, defined by an invariant scalar product on $\mathrm{su}(n)$ and a Riemannian metric on M. We work in the Hamiltonian formalism. Physical states are given by functionals $\psi : \mathcal{A} \to \mathbb{C}$, where $\mathcal{A}$ is the space of connections in the principal bundle $M \times \mathrm{SU}(n)$, subject to the Gauss law[1]

$$\left(\mathrm{d}_A^* \frac{\delta}{\delta A} + \frac{\mathrm{i}\lambda}{4\pi} * \mathrm{d}A\right) \psi = 0, \quad \forall\, A \in \mathcal{A}.$$

Asorey *et al.*[2] prove the following.

If A is reducible to a subbundle of P with nontrivial first Chern class then $\psi(A) = 0$ for all physical states ψ. ('A is a kinematical quantum node'.)

We observe that the assumption holds if and only if (the orbit of) A belongs to a stratum labelled by $(\mathbf{k}, \mathbf{m}, \alpha, \xi)$, where some $\alpha_i^{(2)} \neq 0$. Thus,

To be a kinematical node is a property of strata and can be read off directly from the invariants.

The following figure shows the Hasse diagram of the set of orbit types (strata of $\mathcal{M}$) for gauge group $\mathrm{SU}(2)$. Nodal strata are surrounded by a circle. Vertices in the left and the central columns represent strata with stabilizer isomorphic to $\mathrm{SU}(2)$ and $\mathrm{U}(1)$, respectively. They correspond 1 : 1 to elements of $H^1(M, \mathbb{Z}_2)$ and to nonnegative integers, respectively. The vertex on the rhs. corresponds to the generic stratum. The partial ordering increases from left to right.

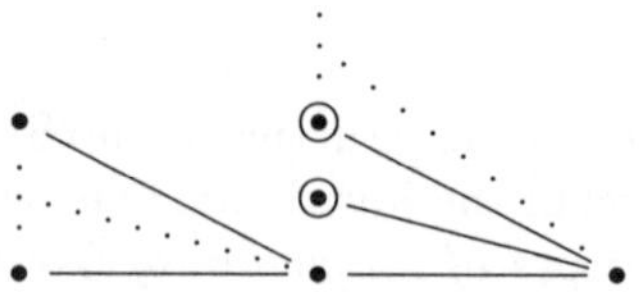

6 Outlook

In order to systematically investigate the physical role of nongeneric strata, one needs to construct a quantization of the gauge theory where the stratification is explicitly visible. For that purpose, it seems promising to us to view the theory as a Hamiltonian system acted upon by the Hamiltonian action of $\mathcal{G}$ and to try to apply the method of singular Marsden-Weinstein reduction[8] to construct the reduced phase space, which is then a stratified symplectic space. Provided one succeeds with this, one can then try to apply recent ideas from geometric quantization of singular Marsden-Weinstein quotients[4] to obtain a nonperturbative quantization of the gauge theory.

References

1. Asorey, M., Mitter, P.K.: *Phys. Lett. B* **153**, 147 (1985).
2. Asorey, M., Falceto, F., López, J.L., Luzón, G.: *Phys. Lett. B* **349**, 125 (1995).
3. Emmrich, C., Römer, H.: *Commun. Math. Phys.* **129**, 69 (1990).
4. Huebschmann, J.:
   1. *J. reine angew. Math.* **408**, 57 (1990);
   2. *Kähler spaces, nilpotent orbits, and singular reduction* math-dg/0104213.
5. Kondracki, W., Rogulski, J.: *On the Stratification of the Orbit Space for the Action of Automorphisms on Connections. Dissertationes Mathematicae* **250**, Warszawa (1986).
6. Rudolph, G., Schmidt, M., Volobuev, I.P.: *Classification of gauge orbit types for* SUn*-gauge theories*, math-ph/0003044.
7. Rudolph, G., Schmidt, M., Volobuev, I.P.: *Partial ordering of gauge orbit types for* SUn*-gauge theories*, math-ph/0009018, accepted for publication in *J. Geom. Phys.*.
8. Sjamaar, R., Lerman, E.: *Ann. of Math. (2)* **134**, 375, (1991).

GROUP-THEORETICAL ASPECTS OF CONTROL OF QUANTUM SYSTEMS

S. G. SCHIRMER AND A. I. SOLOMON

Quantum Processes Group, The Open University, Milton Keynes, MK7 6AA, UK
E-mail: S.G.Schirmer@open.ac.uk, A.I.Solomon@open.ac.uk

Group-theoretical aspects of control of quantum systems are presented and a concise summary of some important results on controllability of quantum systems is provided.

1 Introduction

The subject of control of quantum-mechanical systems has been a fruitful area of investigation lately. The growing interest of researchers in various aspect of control of quantum systems can be attributed at least partly to the fact that the ability to manipulate quantum systems in a controlled fashion will be an essential prerequisite for many exciting new applications from control of chemical reactions to quantum computing devices. One issue of interest, which we shall address in this paper, is the question which target states can be dynamically reached from a given initial state of the system by application of suitable control fields, given a certain type of interaction of the system with those fields.

This paper is organized as follows. In section 2, the concept pure and mixed quantum states is introduced and examples of both are given. Section 3 deals with the evolution of quantum states and the partitioning of states into kinematical equivalence classes arising from the constraint of unitary evolution. In section 4, we discuss the action of the dynamical Lie group on the kinematical equivalence classes and its implications for dynamical reachability of states. In section 5, various degrees of controllability are defined and it is shown how these notions are related to the dynamical Lie group of the system. In section 6, we present some useful results on controllability as well as examples. We conclude with section 7.

2 States of quantum system

The state of a quantum system can be represented by a density operator $\hat{\rho}$ acting on a Hilbert space $\mathcal{H}$. Since a density operator is a positive operator with trace one, it has a pure point spectrum with eigenvalues w_n satisfying

$$\sum_n w_n = 1, \quad 0 \leq w_n \leq 1, \tag{1}$$

and a spectral resolution

$$\hat{\rho} = \sum_{n=1}^{N} w_n |\Psi_n\rangle\langle\Psi_n|, \tag{2}$$

where $|\Psi_n\rangle$ are the eigenstates and $\langle\Psi_n|$ their duals.

Definition 1 *A density operator $\hat{\rho}$ represents a* pure state *if it has rank one. Otherwise, it represents a mixed state.*

From the definition, we see immediately that $\hat{\rho}$ represents a pure state if and only if its eigenvalues are 0 and 1 and 0 occurs with multiplicity $N-1$, where N is the dimension of the Hilbert space of the system. Therefore, the density matrix associated with a pure state is a projector onto a Hilbert space vector $|\Psi\rangle$,

$$\hat{\rho} = |\Psi\rangle\langle\Psi|.$$

The element $|\Psi\rangle$, usually called a wavefunction, completely determines the density operator of the state and is therefore an equivalent representation of the state of the system.

Example 1 (Pure States vs Mixed States) *Consider a simple quantum system whose underlying Hilbert space has dimension four. Choose an arbitrary basis and consider the density matrices*

$$\hat{\rho}_1 = \begin{pmatrix} 1\,0\,0\,0 \\ 0\,0\,0\,0 \\ 0\,0\,0\,0 \\ 0\,0\,0\,0 \end{pmatrix}, \qquad \hat{\rho}_2 = \begin{pmatrix} \frac{1}{2}\,0\,0\,\frac{1}{2} \\ 0\,0\,0\,0 \\ 0\,0\,0\,0 \\ \frac{1}{2}\,0\,0\,\frac{1}{2} \end{pmatrix}.$$

Both density matrices represent pure states. In case of $\hat{\rho}_1$ this is obvious. To see that $\hat{\rho}_2$ is indeed a pure state, note that

$$\hat{\rho}_2 = |\Psi\rangle\langle\Psi| \qquad \textit{for} \qquad |\Psi\rangle = \frac{1}{\sqrt{2}}\begin{pmatrix} 1 \\ 0 \\ 0 \\ 1 \end{pmatrix}.$$

On the other hand, consider the density matrices

$$\hat{\rho}_3 = \begin{pmatrix} \frac{1}{2}\,0\,0\,0 \\ 0\,\frac{1}{2}\,0\,0 \\ 0\,0\,0\,0 \\ 0\,0\,0\,0 \end{pmatrix}, \qquad \hat{\rho}_4 = \begin{pmatrix} \frac{1}{4}\,0\,0\,\frac{1}{4} \\ 0\,\frac{1}{4}\,\frac{1}{4}\,0 \\ 0\,\frac{1}{4}\,\frac{1}{4}\,0 \\ \frac{1}{4}\,0\,0\,\frac{1}{4} \end{pmatrix}.$$

as well as

$$\hat{\rho}_5 = \begin{pmatrix} 0.4 & 0 & 0 & 0 \\ 0 & 0.3 & 0 & 0 \\ 0 & 0 & 0.2 & 0 \\ 0 & 0 & 0 & 0.1 \end{pmatrix}$$

It can easily be verified that all of these density matrices have eigenvalues between 0 *and* 1 *and therefore represent mixed states of the system.*

3 Evolution of quantum states

The evolution of a non-dissipative quantum system is determined by an evolution operator $\hat{U}(t, t_0)$, which satisfies the *Schrodinger equation*

$$i\hbar \frac{d}{dt}\hat{U}(t, t_0) = \hat{H}[\mathbf{f}(t)]\hat{U}(t, t_0), \tag{3}$$

where $\hat{H}[\mathbf{f}(t)]$ is the total Hamiltonian of the system. If the system is subjected to external control fields $f_m(t)$ then $\hat{H}$ depends on a set of time-dependent control functions $\mathbf{f}(t) = (f_1(t), \ldots, f_M(t))$. However, in any case, conservation of energy and probability requires the evolution operator of the system to be unitary at all times, no matter what fields are applied.

Hence, the time-evolution operator $\hat{U}(t, t_0)$ of the system is restricted to the unitary group $U(N)$. This constraint of unitary evolution imposes restrictions on the quantum states that can be reached from a given initial state. Since these restrictions are independent of the control fields applied, we shall refer to them as kinematical constraints, to distinguish them from dynamical constraints arising from constraints on the control fields.

Precisely, given an initial state $\hat{\rho}_0$ and a target state $\hat{\rho}_1$, the target state is kinematically admissible if and only if there exists a unitary operator $\hat{U}$ such that

$$\hat{\rho}_1 = \hat{U}\hat{\rho}_0\hat{U}^\dagger. \tag{4}$$

(4) defines an equivalence relation, which partitions the set of density operators into kinematical equivalence classes.

Definition 2 *Two density matrices $\hat{\rho}_1$, $\hat{\rho}_2$ are* kinematically equivalent *if they are unitarily equivalent, i.e., if $\hat{\rho}_2 = \hat{U}\hat{\rho}_1\hat{U}^\dagger$ for some unitary operator $\hat{U}$.*

It follows immediately that two density matrices are kinematically equivalent if and only if they have the same eigenvalues.

Example 2 (Kinematical Equivalence of States)

- *The pure states $\hat{\rho}_1$ and $\hat{\rho}_2$ in example 1 both belong to the kinematical equivalence class of pure states.*

- *The mixed states $\hat{\rho}_3$ and $\hat{\rho}_4$ are both members of the same kinematical equivalence class since they have the same eigenvalues $\{\frac{1}{2}, \frac{1}{2}, 0, 0\}$.*

- *The state $\hat{\rho}_5$, however, is not kinematically equivalent to any of the states above since its eigenvalues are different.*

4 Dynamical reachability of states

For any initial state, only target states in the same kinematical equivalence class can possibly be dynamically reached by applying external control fields. However, clearly, not every kinematically admissible target state is necessarily dynamically reachable. The question thus arises what subsets of the kinematical equivalence class are dynamically reachable for a given initial state.

The dynamically reachable sets depend on the dynamical Lie group S of the system. If S is the unitary group $U(N)$ then the dynamically reachable set for every initial state comprises the entire kinematical equivalence class it belongs to, since $U(N)$ acts transitively on every equivalence class.

Furthermore, if there exists a unitary operator $\hat{U}$ such that (4) is satisfied then there always exists a unitary operator $\tilde{U}$ with determinant one such that (4) is satisfied for $\tilde{U}$. Hence, $SU(N)$ also acts transitively on every kinematical equivalence class.

However, the only proper subgroups of $SU(N)$ that act transitively on the class of pure states are $Sp(\frac{1}{2}N)$ and $Sp(\frac{1}{2}N) \otimes U(1)$, where $U(1) = e^{i\theta}\hat{I}_N$ is a multiple of the identity matrix $\hat{I}_N$. Furthermore, $Sp(\frac{1}{2}N)$ — and hence $Sp(\frac{1}{2}N) \otimes U(1)$ — does *not* act transitively on many other equivalence classes of mixed states.

Example 3 (Non-transitive action of $Sp(\frac{1}{2}N)$ for mixed states)
Consider a four-level system with $\hat{H} = \hat{H}_0 + f(t)\hat{H}_1$, where

$$\hat{H}_0 = \begin{pmatrix} -\frac{3}{2} & 0 & 0 & 0 \\ 0 & -\frac{1}{2} & 0 & 0 \\ 0 & 0 & +\frac{1}{2} & 0 \\ 0 & 0 & 0 & +\frac{3}{2} \end{pmatrix}, \qquad \hat{H}_1 = \begin{pmatrix} 0 & +1 & 0 & 0 \\ +1 & 0 & +1 & 0 \\ 0 & +1 & 0 & -1 \\ 0 & 0 & -1 & 0 \end{pmatrix}.$$

Note that both $i\hat{H}_0$ *and* $i\hat{H}_1$ *satisfy* $\hat{x}^T\hat{J} = -\hat{J}\hat{x}$ *for*

$$\hat{J} = \begin{pmatrix} 0 & 0 & 0 & +1 \\ 0 & 0 & +1 & 0 \\ 0 & -1 & 0 & 0 \\ -1 & 0 & 0 & 0 \end{pmatrix}.$$

Therefore, the dynamical Lie algebra of the system is a subalgebra of $sp(2)$[a] *and any* $\hat{U}$ *in the dynamical Lie group must preserve* $\hat{J}$ *in the sense that*

$$\hat{U}^T\hat{J}\hat{U} = \hat{J}.$$

Thus, two kinematically equivalent states $\hat{\rho}_0$ *and* $\hat{\rho}_1$ *are dynamically reachable from each other only if there exists a unitary operator* $\hat{U}$ *that preserves* $\hat{J}$ *such that*

$$\hat{\rho}_1 = \hat{U}\hat{\rho}_0\hat{U}^\dagger.$$

Combining the last two equations shows that, in order to be dynamically equivalent, $\hat{\rho}_0$ *and* $\hat{\rho}_1$ *must satisfy*

$$(\hat{J}^\dagger\hat{\rho}_1\hat{J})^* = \hat{U}(\hat{J}^\dagger\hat{\rho}_0\hat{J})^*\hat{U}^\dagger$$

This shows that

$$\hat{\rho}_0 = \begin{pmatrix} \frac{1}{2} & 0 & 0 & 0 \\ 0 & \frac{1}{2} & 0 & 0 \\ 0 & 0 & 0 & 0 \\ 0 & 0 & 0 & 0 \end{pmatrix} \quad \text{and} \quad \hat{\rho}_1 = \begin{pmatrix} \frac{1}{2} & 0 & 0 & 0 \\ 0 & 0 & 0 & 0 \\ 0 & 0 & 0 & 0 \\ 0 & 0 & 0 & \frac{1}{2} \end{pmatrix}$$

are not dynamically reachable from each other since

$$(\hat{J}^\dagger\hat{\rho}_1\hat{J})^* = \hat{\rho}_1 \quad \text{but} \quad (\hat{J}^\dagger\hat{\rho}_0\hat{J})^* \neq \hat{\rho}_0$$

and it is impossible to find a unitary transformation that maps both $\hat{\rho}_0$ *and* $(\hat{J}^\dagger\hat{\rho}_0\hat{J})^*$ *to* $\hat{\rho}_1$.

This example shows that there is no proper subgroup of $SU(N)$ that acts transitively on *all* the kinematical equivalence classes.

5 Degrees of Controllability

Let us now turn our attention to various degrees of controllability.

[a] It can actually be verified that the dynamical Lie algebra is $sp(2)$

Definition 3 (Complete controllability) *A quantum system is completely controllable if any unitary evolution is dynamically realizable, i.e., if given any unitary operator* $\hat{U}$, *there exists* $t_F > 0$ *and an admissible control-trajectory pair* $(\mathbf{f}, \hat{U}(t, t_0))$ *such that* $\hat{U} = \hat{U}(t_F, t_0)$

Definition 4 (Density matrix controllability) *A quantum system is density matrix controllable if given any two density matrices* $\hat{\rho}_0$, $\hat{\rho}_1$ *in the same kinematical equivalence class, there exists* $t_F > 0$ *and an admissible control-trajectory pair* $(\mathbf{f}, \hat{U}(t, t_0))$ *such that* $\hat{\rho}_1 = \hat{U}(t_F, t_0)\hat{\rho}_0\hat{U}(t_F, t_0)^\dagger$.

Definition 5 (Pure-state controllability) *A quantum system is pure-state controllable if for any two pure states (wavefunctions)* $|\Psi_0\rangle$, $|\Psi_1\rangle$, *there exists* $t_F > 0$ *and an admissible control-trajectory pair* $(\mathbf{f}, \hat{U}(t, t_0))$ *such that* $|\Psi_1\rangle = \hat{U}(t_F, t_0)|\Psi_0\rangle$.

From our considerations in the previous section, it is obvious that only a system with dynamical Lie group $U(N)$ is completely controllable. A system with dynamical Lie group $SU(N)$ or $U(N)$ is density matrix controllable since these groups act transitively on every equivalence class of density matrices. Furthermore, a system with dynamical Lie group $Sp(\frac{1}{2}N)$, $Sp(\frac{1}{2}N) \otimes U(1)$, $SU(N)$ or $U(N)$ is pure-state controllable since these groups act transitively on the equivalence class of pure states.

6 Important results on controllability and examples

For a quantum system with Hamiltonian of the form

$$\hat{H}[\mathbf{f}(t)] = \hat{H}_0 + \sum_{m=1}^{M} f_m(t)\hat{H}_m, \tag{5}$$

the dynamical Lie group is determined by the dynamical Lie *algebra* $\mathcal{L}$ generated by the skew-Hermitian operators (matrices) $i\hat{H}_m$ and we can use the observations made above to derive Lie algebraic criteria for various degrees of controllability.

Theorem 1 [1,2]
A quantum system with Hamiltonian (5) is

- *completely controllable if and only if* $\mathcal{L} \simeq u(N)$.
- *density matrix controllable if and only if* $\mathcal{L} \simeq su(N)$ *or* $\mathcal{L} \simeq u(N)$.
- *pure-state controllable if and only if* $\mathcal{L}$ *is isomorphic to either* $u(N)$, $su(N)$, *or if* $N = 2\ell$, $sp(\ell)$, $sp(\ell) \oplus u(1)$.

Complete controllability implies density matrix controllability and the latter implies pure-state controllability. The converse is clearly false. For a system that is not density matrix controllable, there exist initial states for which the set of dynamically accessible target states does not comprise the entire kinematical equivalence class, i.e., at least some of the kinematical equivalence classes are partitioned into disjoint subsets of dynamically reachable states. If the system is not pure-state controllable then there exists such a partition for the equivalence class of pure states. However, lack of controllability does not necessarily imply that the dynamical Lie group of the system does not act transitively on any kinematical equivalence class. For instance, *every* dynamical Lie group clearly acts transitively on the equivalence class of completely random ensembles, which consists of the single element $\hat{\rho}_R = \frac{1}{N}\hat{I}_N$, where $\hat{I}_N$ is the identity matrix of dimension N.

Theorem 2 [3,4] *A quantum system with Hamiltonian* $\hat{H} = \hat{H}_0 + f(t)\hat{H}_1$,

$$\hat{H}_0 = \mathrm{diag}(E_1, \ldots, E_N), \quad \hat{H}_1 = \sum_{n=1}^{N-1} d_n(|n\rangle\langle n+1| + |n+1\rangle\langle n|) \qquad (6)$$

is density matrix controllable if $d_n \neq 0$ *for* $1 \leq n \leq N-1$ *and either*

1. *there exists* p *such that* $\omega_n \neq \omega_p$ *for* $n \neq p$ *where* $\omega_n = E_{n+1} - E_n$*; or*

2. $\omega_n = \omega$ *for all* n *but there exists* p *such that* $v_n \neq v_p$ *for* $n \neq p$*, where* $v_n \equiv 2d_n^2 - d_{n-1}^2 - d_{n+1}^2$.

If $N = 2p$ *then* $d_{p-k}^2 \neq d_{p+k}^2$ *for some* $k \neq 0$ *is required as well.*
If in addition $\mathrm{Tr}(\hat{H}_0) \neq 0$ *then the system is completely controllable.*

This theorem can be used to show that many systems of physical interest are indeed completely controllable.

Example 4 (Completely controllable systems)
The following systems are completely controllable:

- *N-level Morse oscillators with arbitrary (non-zero) dipole moments*
- *N-level harmonic oscillators with non-uniform, non-zero dipole moments*
- *N-level one-electron atoms with arbitrary (non-zero) dipole moments*
- *A confined particle such as an electron in a box (quantum well)*

Pure-state controllability can also be related to connected graphs

Theorem 3 [5,6]
Consider a quantum system with N non-degenerate energy levels E_n and transitions $|n\rangle \to |m\rangle$ with transition frequency ω_{mn} and corresponding dipole moment d_{mn}. Let the energy levels be the vertices of a graph and the transitions between them be the edges. Eliminate the edges corresponding to any degenerate transitions, i.e., transition pairs with $\omega_{mn} = \omega_{m'n'}$. Then the system is pure-state controllable if the resulting graph is connected.

This theorem provides an useful tool for quick identification of pure-state controllable systems.

Example 5 (Pure-state controllable systems)
Each of the following transition diagrams correspond to pure-state controllable systems

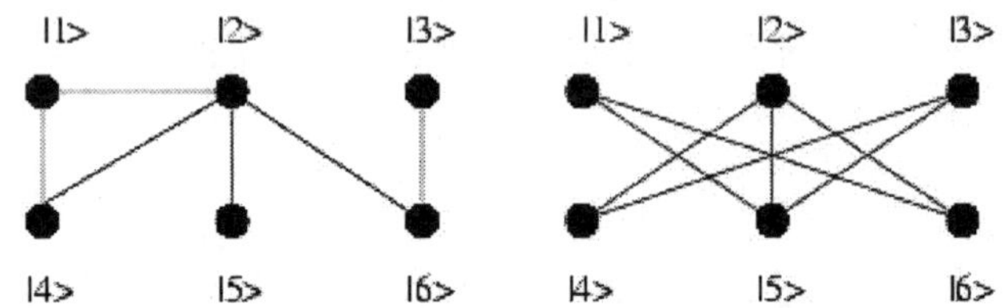

provided that the energy levels and transition frequencies are non-degenerate.

Note, however, that the converse of the theorem is not true, i.e., if the graph is disconnected the system may still be controllable.

Example 6 *Consider a sequentially coupled N-level system with equally spaced energy levels and equal transition probabilities. This system does not satisfy the hypothesis of theorem 3 since there is only one transition frequency. Therefore, removing the edges corresponding to the degenerate transition frequency leaves us with a completely disconnected graph. However, it can be shown that the dynamical Lie algebra of this system is*

- $sp(\frac{1}{2}N)$ *or* $sp(\frac{1}{2}N) \oplus u(1)$ *if* $N = 2\ell$
- $so(N)$ *or* $so(N) \oplus u(1)$ *if* $N = 2\ell + 1$

Therefore, the system is pure-state controllable if N is even.

7 Conclusion

We have shown that the kinematical constraint of unitary evolution partitions the states of a quantum system into kinematical equivalence classes. The dynamically reachable sets are subsets of the kinematical equivalence classes determined by the action of the dynamical Lie group of the system. If the dynamical Lie group acts transitively on all the equivalence classes then the dynamically reachable sets correspond to the kinematical equivalence classes.

Acknowledgments

We sincerely thank J. V. Leahy (University of Oregon), V. Ramakrishna (University of Texas at Dallas), H. Rabitz (Princeton University) and G. Turinici (University of Paris) for helpful discussions and comments. AIS acknowledges the hospitality of the Laboratoire de Physique Théorique des Liquides, University of Paris VI, where he is currently a visiting faculty member.

References

1. F. Albertini and D. D'Alessandro, Notions of controllability for quantum-mechanical systems, quant-ph/0106128 (2001).
2. S. G. Schirmer, J. V. Leahy, and A. I. Solomon, Degrees of controllability for quantum systems and applications to atomic systems, quant-ph/0108114 (2001).
3. H. Fu, S. G. Schirmer, and A. I. Solomon, *J. Phys.* A**34**, 1679 (2001).
4. S. G. Schirmer, H. Fu, and A. I. Solomon, *Phys. Rev. A* **63**, 063410 (2001).
5. G. Turinici, Controllable quantities for bilinear quantum systems, in *39th IEEE CDC Proceedings*, Causal Productions, Adelaide, Australia (2000).
6. G. Turinici and H. Rabitz, *Chem. Phys.* **267**, 1 (2001).

CAN THE BREAKING OF TIME-REVERSAL SYMMETRY INTRODUCE BIFURCATIONS IN QUANTUM MECHANICS?

D. A. SCHUCH
Institut für Theoretische Physik, J. W. Goethe–Universität, Robert–Mayer–Strasse 8-10, D-60054 Frankfurt am Main, Germany
E-mail: Schuch@em.uni-frankfurt.de

Bifurcation is a common phenomenon in nonlinear dynamics and a first step into the domain of chaos. In quantum mechanics, which is supposedly a linear theory, nonlinear equations also occur. So the dynamics of position and momentum uncertainties can be described by a nonlinear differential equation of Riccati-type. Despite the nonlinearity of this equation, no bifurcation occurs, as long as the conservative standard problems that can be solved analytically are considered. However, the situation changes drastically if the influence of a dissipative environment is taken into account in an effective way (via a nonlinear Schrödinger equation). The introduction of an additional parameter - the friction coefficient - leads to bifurcation in the dynamics of the quantum fluctuations. The consequences, e.g. for the wave-packet solutions of the damped free motion, will be discussed.

Considering dynamical systems with nonlinear (NL) evolution equations, the occurrence of bifurcations is well-known and can even be a first step into the domain of chaos. Quantum mechanics, however, is supposedly a linear theory, therefore, no nonlinearities, and thus, no bifurcation should occur. Since we want to study the influence of time-reversal symmetry on quantum systems, we have to consider the time-dependent (TD) Schrödinger equation (SE), indeed, a linear differential equation. In a first step, it shall be shown that already in conventional quantum mechanics NL equations occur in the description of the dynamics of the system, but without bifurcations. In a second step, the effect of the interaction with some dissipative environment will be taken into account that breaks the time-reversal symmetry and makes it possible to obtain bifurcations in the above-mentioned NL quantum evolution equations. The consequences for the dynamics and energetics of the quantum system will be discussed. To make the influence of the environment as transparent as possible, simple model-systems with analytical solutions will be chosen, as well as an effective model for the interaction with the environment that allows for the description of a system by a wave-function that can always be normalized - even if it is in contact with some environment.

Therefore, we first consider the TDSE for the one-dimensional free motion and harmonic oscillator (HO) (extension to higher dimensions is straightforward). In particular, we will start with the equations for the HO and can obtain the results for the free motion in the limit $\omega \to 0$. The TDSE

$$i\hbar\frac{\partial}{\partial t}\Psi_L(x,t) = \left[-\frac{\hbar^2}{2m}\frac{\partial^2}{\partial x^2}+\frac{m}{2}\omega^2 x^2\right]\Psi_L(x,t) \tag{1}$$

possesses exact analytic solutions of Gaussian-wave packet (WP)-type,

$$\Psi_{WP,L}(x,t) = N(t)\,exp\Big\{i\Big[y(t)\,\tilde{x}^2+\frac{<p>}{\hbar}\tilde{x}+K(t)\Big]\Big\} \tag{2}$$

(with $<...>=\int\Psi^*...\Psi dx$ and $\tilde{x}=x-<x>=x-\eta(t)$). The WP-*maximum* at position $<x>$ follows the classical trajectory $\eta(t)$, obeying $\ddot{\eta}+\omega^2\eta=0$. The explicit form of N and $K(t)$ is not relevant to the following discussion. The WP-*width*, $\sqrt{<\tilde{x}^2>}$, (where $<\tilde{x}^2>=<x^2>-<x>^2$) is connected with the imaginary part of $y(t)=y_R(t)+iy_I(t)$ via

$$\frac{2\hbar}{m}y_I = \frac{\hbar}{2m<\tilde{x}^2>} = \frac{1}{\alpha^2(t)}\;. \tag{3}$$

To determine the time-dependence of the WP-width and, thus, the position uncertainty, the complex quadratically nonlinear equation of Riccati-type that is obtained when inserting (2) in (1),

$$\frac{2\hbar}{m}\dot{y}+\left(\frac{2\hbar}{m}y\right)^2+\omega^2 = 0\;, \tag{4}$$

must be solved. Here, we find the desired nonlinear differential equation in the context of conventional quantum mechanics, but no bifurcations occur as will be shown below.

What changes if we take the dissipative environment into account? Since we are not interested in the details of the system-environment interaction - but only in the effect of the environment on the observable system - we choose a model that concentrates on this aspect and, at the same time, also possesses exact analytical solutions in the form of Gaussian WPs. In this model, the breaking of time-reversal symmetry is achieved by introducing a diffusion term into the continuity equation (CE) for the density $\varrho(x,t)=\Psi^*(x,t)\,\Psi(x,t)$. The resulting Fokker-Planck-type equation (FPE)

$$\frac{\partial}{\partial t}\varrho + \frac{\partial}{\partial x}j(x,t) - D\frac{\partial^2}{\partial x^2}\varrho = 0 \tag{5}$$

(with convection current density j, defined as $j=\left(\frac{\hbar}{2mi}\right)\left(\Psi^*\frac{\partial}{\partial x}\Psi-\Psi\frac{\partial}{\partial x}\Psi^*\right)$), can be separated into a pair of complex conjugate SEs by modifying a method of Madelung [1] and Mrowka [2] and imposing the additional condition

$$-D\frac{\frac{\partial^2}{\partial x^2}\varrho}{\varrho} = \gamma\left(\ell n\varrho - < \ell n\varrho >\right) \tag{6}$$

for the diffusion term (for details see [3-5]), thus leading to the NLSE

$$i\hbar\frac{\partial}{\partial t}\Psi_{NL} = \left[-\frac{\hbar^2}{2m}\frac{\partial^2}{\partial x^2} + \frac{m}{2}\omega^2 x^2 + \gamma\frac{\hbar}{2i}\left(\ell n\Psi_{NL} - < \ell n\Psi_{NL} >\right)\right]\Psi_{NL} \; . \tag{7}$$

The logarithmic nonlinearity corresponds to a linear velocity-dependent frictional force with friction coefficient γ, caused by the interaction with a dissipative environment that does not appear explicitly. Despite the imaginary contribution from the logarithmic term, the solutions of the NLSE are normalizable and do not suffer the shortcomings that occur in similar models. For the damped free motion and the damped HO, exact Gaussian WP solutions of the form (2) exist. The maximum at $< x >$ now follows the classical trajectory η according to $\ddot{\eta} + \gamma\dot{\eta} + \omega^2\eta = 0$, i.e., including a linear velocity-dependent frictional force. More interesting for our purpose is, however, the time-dependence of $y(t)$ (and thus the WP-width) that now fulfills the modified Riccati equation (RE)

$$\frac{2\hbar}{m}\dot{y} + \gamma\left(\frac{2\hbar}{m}y\right) + \left(\frac{2\hbar}{m}y\right)^2 + \omega^2 = 0 \; . \tag{8}$$

For certain choices of the occurring parameters (e.g. constant ω and γ), the solution of the imhomogeneous RE can be reduced to the solution of a homogeneous complex Bernoulli equation (BE)

$$\frac{2\hbar}{m}\dot{v} + A\left(\frac{2\hbar}{m}v\right) + \left(\frac{2\hbar}{m}v\right)^2 = 0 \tag{9}$$

using the ansatz $\left(\frac{2\hbar}{m}\right)y(t) = \left(\frac{2\hbar}{m}\right)\hat{y} + \left(\frac{2\hbar}{m}\right)v(t)$, where the time-dependent particular solution of the RE can take the values $\frac{2\hbar}{m}\hat{y}_\pm = -\left(\frac{\gamma}{2}\right) \pm \left(\frac{\gamma^2}{4} - \omega^2\right)^{1/2}$. The coefficient A of the linear term in the BE can also have two different values,

$$A_\pm = 2\left(\frac{\gamma}{2} + \frac{2\hbar}{m}\hat{y}_\pm\right) = \pm 2\left(\frac{\gamma^2}{4} - \omega^2\right)^{1/2} \; , \tag{10}$$

which will be of importance subsequently. Eq.(9) can be solved analytically and the solution given in the form

$$\frac{2\hbar}{m}v\left(t\right) = \frac{\left(\frac{2\hbar}{m}v_0\right)Ae^{-At}}{\left(\frac{2\hbar}{m}v_0\right)\left(1-e^{-At}\right)+A} . \tag{11}$$

It is interesting that Eq.(9) is also obtained in a well-known problem in NL dynamics (only for a real instead of a complex variable, see e.g. [6]), where the quantity corresponding to A in Eq.(9) plays the role of a bifurcation parameter, leading, for values ≥ 0, to a time-evolution that approaches a fixed point whereas, for negative values, it spirals towards a limit cycle. Therefore, the possible values of A and its consequences shall now be analyzed.

Considering the reversible case, i.e. $\gamma = 0$, it is evident that for $\omega \neq 0$ (i.e. the undamped HO) the parameter A is purely imaginary, therefore, solution (11) will show oscillatory behaviour and not approach attractors, as in the classical case.

For $\omega = 0$ (i.e. the free motion), two solutions corresponding to two different states with respective uncertainties and energies should, in principle, be possible. However, since, in this case, the bifurcation parameter takes the value $A_\pm = \pm 0$, they degenerate to only one. The corresponding position fluctuation (uncertainty) and the quantum mechanical energy contribution (ground-state energy) $\tilde{E}$ of the WP can be obtained with the help of Eqs.(11) and (3) and are given by

$$< \tilde{x}^2 >_L = < \tilde{x}^2 >_0 \left\{1 + \left(\beta_0 t\right)^2\right\} \tag{12}$$

$$\tilde{E}_L = \frac{\hbar}{4}\,\beta_0 . \tag{13}$$

Hence, the WP is spreading quadratically in time and its quantum mechanical energy contribution $\tilde{E}$ is constant and related to the initial width by Eq.(13), where $\beta_0 = \left(\hbar/2m < \tilde{x}^2 >_0\right)$.

In the irreversible case, i.e. $\gamma \neq 0$, and for $\omega > \gamma/2$, the same applies in principle as in the reversible case. However, for $\omega = 0$, the positive and negative values of the parameter A are different, $A_\pm = \pm\gamma$, and hence, two solutions exist. In particular, for $A_+ = +\gamma$ follows

$$< \tilde{x}^2 >_+ = < \tilde{x}^2 >_0 \left\{e^{\gamma t} + \left(\frac{\beta_0}{\gamma/2}\right)^2 \; sinh^2\frac{\gamma}{2}t\right\} \tag{14}$$

$$\tilde{E}_+ = \frac{\hbar}{4}\,\beta_0\, e^{-\gamma t} \tag{15}$$

and for $A_- = -\gamma$

$$< \tilde{x}^2 >_- \;=\; < \tilde{x}^2 >_0 \; \left\{ e^{-\gamma t} + \left(\frac{\beta_0}{\gamma/2} \right)^2 \; sinh^2 \frac{\gamma}{2} t \right\} \tag{16}$$

$$\tilde{E}_- \;=\; \frac{\hbar}{4} \; \beta_0 \; \left[1 + \left(\frac{\gamma}{\beta_0} \right)^2 \right] e^{-\gamma t} \; . \tag{17}$$

In both cases, the corresponding WPs are spreading faster in time than in the reversible case - due to the $sinh^2$-term - and, in the case of $< \tilde{x}^2 >_+$, even exponentially; a formal explanation will be given below. There are, however, not only similarities but, also, distinct differences between the two WP solutions. The initial value of the quantum mechanical energy contribution $\tilde{E}_+$, $\tilde{E}_+(0) \;=\; \frac{\hbar}{4}\,\beta_0$, is identical to the (constant) value of $\tilde{E}_L$ in the reversible case, whereas $\tilde{E}_-$ has a larger initial value given by $\tilde{E}_-(0) \;=\; \frac{\hbar}{4}\;\beta_0\;\left[1 + \left(\frac{\gamma}{\beta_0}\right)^2\right] >$ $\tilde{E}_+(0) = \tilde{E}_L$. The difference in energy, $\Delta\tilde{E}_0$, expressed as

$$\Delta\tilde{E}_0 = \tilde{E}_-(0) - \tilde{E}_+(0) = \frac{\hbar}{4}\frac{\gamma^2}{\beta_0} = \frac{m}{2}\gamma^2 < \tilde{x}^2 >_0 \;, \tag{18}$$

is independent of $\hbar$. Comparing the initial value of the diffusion coefficient D in the FPE (5), $D_0 = \gamma < \tilde{x}^2 >_0 /2$, (for details see [4]) with the Einstein relation $D = kT/m\gamma$, gives the impression that $\Delta\tilde{E}_0$ can be interpreted as thermal energy kT of the environment with temperature T.

Concerning the faster spreading of the WPs, it should be mentioned that for the damped free motion the NLSE (7) has not only Gaussian WP solutions but, also, solutions similar to plane waves,

$$\psi_{k,NL} \;=\; \frac{1}{\sqrt{2\pi}} \; exp\left\{ i \left[kx - \int_0^t \frac{\hbar k^2(t)}{2m} dt' \;+\; \gamma \int_0^t k < x > dt' \right] \right\} , \tag{19}$$

but with time-dependent k-values turning, for $\gamma \to 0$, into plane waves $\psi_{k,L} = \frac{1}{\sqrt{2\pi}} exp\left\{ i \left[kx - \frac{\hbar k^2}{2m} \right] \right\}$. Similar to the WP of the free motion that can be expressed as a superposition of plane waves $\psi_{k,L}$ in the form

$$\Psi_{WP}\,(x,t) \;=\; \frac{1}{\sqrt{2\pi}} \; \int dk \; C_L(k) \; \psi_{k,L}(k,t) \tag{20}$$

with *constant* coefficients $C_L(k)$, also the WP solution of our NLSE can be expressed in the same form, only replacing $\psi_{k,L}$ by the $\psi_{k,NL}$ (19), and $C_L(k)$

by coefficients $C_{NL}(k,t)$ that are now explicitly $time - dependent$ and also time-dependent via $k(t) = k_0 \; exp(-\gamma t)$ (for details see [4]). Whereas, in the usual SE, the WP-spreading is caused by a dephasing due to different k-values of its components, in our case, it is not only the k-values that are different for each $\psi_{k,NL}$, but they are also time-dependent as are the coefficients C_{NL}, thus leading to a faster dephasing. Therefore, the WPs of the interacting quantum systems are spreading faster than the WP of the isolated system. This shows some similarity with experimental results by Nimtz [7] where Gaussian-like pulses reach a higher signal velocity in the presence of barriers - i.e. obstacles like collision partners of some environment - than without. Whether there are any connections between our theoretically-predicted, and this, experimental, effect will require further investigation.

In addition, the tunnelling behaviour - according to an $\dot{\alpha}/\alpha$-term in the currents of the CE/FPE - is affected by this deviating spreading beahaviour (see Eq.(3)).

In conclusion, it can be stated that the breaking of time-reversal symmetry by inclusion of an effective interaction with a dissipative environment, expressed by an additional friction coefficient γ, makes it possible for the parameter A in the BE (9) to really act as a bifurcation parameter leading to two systems with different time-evolution of the quantum fluctuations. That these are two physically distinguishable states follows, not only from the different spreading behaviour of the corresponding WPs, but also from the difference in their ground-state energies. In principle, this energy gap should be measurable but a proper experimental set-up has yet to be identified.

References

1. E. Madelung, *Die mathematischen Hilfsmittel des Physikers* (Springer, Berlin, 1950) p.432.
2. B. Mrowka, *Z. Phys.* **130**, 164 (1951).
3. D. Schuch, K.-M. Chung, and H. Hartmann, *J. Math. Phys.* **24**, 1652 (1983).
4. D. Schuch, K.-M. Chung, and H. Hartmann, *J. Math. Phys.* **25**, 3086 (1984).
5. D. Schuch and K.-M. Chung, *Int. J. Quantum Chem.* **29**, 1561 (1986).
6. H.G. Schuster, *Deterministic Chaos* (Physik-Verlag, Weinheim, 1984) p.112.
7. H.Aichmann, A. Haibel, W. Lennartz, G. Nimtz and A. Spanoudaki in *Quantum Theory and Symmetries (Proceedings of the Intern. Symp.)*, ed H.-D. Doebner *et al* (World Scientific, Singapore, 2000) p. 605.

LOOKING FOR A TIME OPERATOR

MARCIN SKULIMOWSKI
Department of Theoretical Physics, University of Łódź,
ul. Pomorska 149/153, 90-236 Łódź, Poland
e-mail: mskulim@merlin.fic.uni.lodz.pl

In spite of widespread opinion it is possible to construct a spectral measure covariant with respect to the time translations. This spectral mesure corresponds to the time operator. However, its physical interpretation is not clear.

1 Introduction

The question about the existence of a time operator is very old. The first one who answered the question was W.Pauli [1]. He has proved that there does not exist any self-adjoint time operator if the energy spectrum is bounded from below. This result is the main obstacle for obtaining a time operator. But this is also a starting point of new concept. To each self-adjoint operator there corresponds a spectral measure. So the problems with finding the time operator are problems with finding an adequate spectral measure. It turns out however that it is possible to consider some generalization of spectral measure, i.e. POV-measure. This idea was explored by many authors [2,3,4,5]. It turns out that the generalization is compatible with the probabilistic interpretation of quantum mechanics. Our paper refers to this idea. We show that it is possible to construct a POV-measure linked with a time for an arbitrary system with a Hamiltonian [6]. Moreover this POV-measure can be „extended" to some spectral measure [7] which therefore can be eventually considered as a spectral measure corresponding to some time operator. However the physical interpretation of this operator is still not well understood.

2 Covariant POV-measure

The question arises: how to construct a POV-measure linked with a time? We will answer the question in this section. First we define a POV-measure on R. Let $B(R)$ be an algebra of Borel sets on R, $L(\mathcal{H})$ the algebra of bounded operators on Hilbert space $\mathcal{H}$, and $L(\mathcal{H})^+$ the cone of positive ones. A *normalized positive operator valued measure on* R (POV-measure) is a map $\tau : B(R) \to L(\mathcal{H})^+$ such that
1.$\tau(I) \geq \tau(\emptyset) = 0$, where $I \subset R$.
2.$\tau(\bigcup I_i) = \sum \tau(I_i)$ where $\{I_i\}$ is a countable set of disjoint Borel sets.
3.$\tau(R) = \hat{1}$.

Assume that $U(\varepsilon) = \exp(-i\varepsilon\hat{H})$ $(\varepsilon \in R)$ is a representation of the one parameter group of time translations.[a] POV-measure τ linked with a time is *covariant*$\pm$, i.e.

$$\langle \phi \mid \exp(-i\varepsilon\hat{H})\tau(I)\exp(i\varepsilon\hat{H}) \mid \phi\rangle = \langle \phi \mid \tau(I \pm \varepsilon) \mid \phi\rangle \,. \tag{1}$$

Covariant$\pm$ POV-measure is called POV-time$\pm$. Using POV-time$\pm$ τ we can write a generalized spectral representation of a time observable

$$\hat{T} = \int t\tau(dt)\,. \tag{2}$$

Since τ is a POV-measure the observable $\hat{T}$ cannot be self-adjoint. The probability that a result of time measurement for a state $\phi \in \mathcal{H}$ is included in the interval $I \subset R$ is given by

$$p(I) = \langle \phi \mid \tau(I) \mid \phi\rangle \,. \tag{3}$$

In the case when $\sum_{i=1}^{n} I_i = R$, $I_i \cap I_j = \emptyset$, for $i \neq j$, the observables

$$\tau(I_1), ..., \tau(I_n)\,, \tag{4}$$

represent a *multi-bin* test. For any multi-bin test one can find a corresponding measuring instrument [5,8].

In [6] we have proposed the following method of constructing a POV-time$\pm$. Consider a Hilbert space and a complete system of commuting observables which contains the Hamiltonian

$$\hat{H}, \hat{O}_1, ..., \hat{O}_n\,. \tag{5}$$

The eigenvalues of the system are $E, \lambda_1, ..., \lambda_n$. Eigenstates of the Hamiltonian connected with our system are denoted by

$$\mid E, \lambda_1, ..., \lambda_n\rangle \equiv \mid E, \lambda\rangle\,. \tag{6}$$

We assume that the spectrum of energy is continuous (bounded or not) and that the following relations are satisfied

$$\langle E, \lambda_1, ..., \lambda_n \mid E', \lambda_1', ..., \lambda_n'\rangle = \delta(E - E')...\delta(\lambda_n - \lambda_n')\,, \tag{7}$$

---

[a] We assume that $\hbar = 1$.

$$\int_L | E, \lambda\rangle\langle E, \lambda | \, d\lambda \, dE = \hat{1}\,, \tag{8}$$

where $L = Sp(\hat{H}) \times Sp(\hat{O}_1) \times ... \times Sp(\hat{O}_n) = Sp(\hat{H}) \times Sp(\hat{O})$ is the Cartesian product of the spectra of the observables $\hat{H}, \hat{O}_1, ..., \hat{O}_n$. States $| S(t), \lambda\rangle$ defined by

$$| S(t), \lambda\rangle := \frac{1}{\sqrt{2\pi}} \exp(-i\hat{H}t) \int_{Sp(\hat{H})} dE \, | E, \lambda\rangle\,, \tag{9}$$

are called *states of time* (see also [9,10]). Let us notice that these states possess the following property

$$\exp(-i\hat{H}\epsilon) \, | S(t), \lambda\rangle = | S(t+\epsilon), \lambda\rangle\,. \tag{10}$$

In the cases when some of operators $\hat{O}_1, ..., \hat{O}_n$ have discrete spectrum, appropriate integrals in formula (8) and (9) have to be replaced by sums.

Example 1

Consider the Hamiltonian of a free particle

$$H = \frac{p^2}{2m}\,. \tag{11}$$

We restrict ourselves to one spatial dimension. We choose the complete system of commuting observables to be

$$\hat{H}, \; sgn(\hat{p})\,. \tag{12}$$

Now according to (9) we define the states of time

$$| S(t), \pm\rangle := \frac{1}{\sqrt{2\pi}} \int_{Sp(\hat{H})} dE \exp(-iEt) \, | E, \pm\rangle\,, \tag{13}$$

where $Sp(\hat{H}) = \{E : 0 \leq E \leq +\infty\}$. In the momentum representation we obtain

$$\langle p \, | \, S(t), \pm\rangle = \frac{1}{\sqrt{2\pi}} \theta(\pm p) \sqrt{| \, p \, |} \exp(-\frac{itp^2}{2})\,. \tag{14}$$

These states can also be obtained by solving the eigenstates problem for the Aharonov-Bohm time operator (see [11,4]).

Example 2

Consider the following Hamiltonian

$$H = \omega J\,. \tag{15}$$

where J is an action variable and ω is an angular frequency. Then one has

$$\hat{J} \mid m\rangle = m \mid m\rangle, \quad \langle m \mid n\rangle = \delta_{mn} \tag{16}$$

and

$$\hat{H} \mid m\rangle = m\omega \mid m\rangle\,. \tag{17}$$

By virtue of (9) we can now define states of time for our system by

$$\mid S(t)\rangle = \frac{\omega}{2\pi} \sum_m e^{-im\omega t} \mid m\rangle\,, \tag{18}$$

where $\frac{\omega}{2\pi}$ is a normalization constant. In the cases when the energy spectrum is continuous this constant is equal $\frac{1}{\sqrt{2\pi}}$ (see (13)). However, now we consider a periodic system with a period $T = \frac{2\pi}{\omega}$. That is why the normalization in time is rather done on the interval $(t, t+T)$ than on the whole R. Using states $\mid S(t)\rangle$ it is possible to construct a "clock time" operator [12], [13].

Using the states of time (9) we define the following function

$$\tau(I) := \int_{I\times Sp(\hat{O})} \mid S(t), \lambda\rangle\langle\lambda, S(t) \mid dt d\lambda\,, \tag{19}$$

where $I \subset R$. It turns out that τ is a POV-time+ (for proof see [6]). Let us notice that if we change the sign of time $t \to -t$ in the evolution operator in states (9), then τ will be *covariant*− (a $POV - time-$). The following theorem ([4,6]) gives us a very interesting property of POV-times±.

Theorem 1 *If τ is a POV-time± and the generator of the time translations $\hat{H}$ is bounded from below then, for each state $\phi \in H$, $\phi \neq 0$ and for each interval $[a,b] \subset R$,*

$$\langle \phi \mid \tau([a,b]) \mid \phi\rangle > 0 \quad \Box\,. \tag{20}$$

This theorem is a generalization of Pauli's Theorem. It follows that a POV-time± τ must not be a spectral measure. However it is possible for a given POV-time± τ to define a spectral measure which is in some sense an extension of τ. We consider this idea in the next section.

3 Covariant spectral measure

Assume that μ is a spectral measure in $\mathcal{H}$, $\mathcal{K}$ is a closed subspace of $\mathcal{H}$ and τ is a POV-measure in $\mathcal{K}$. Let $\hat{P}: \mathcal{H} \to \mathcal{K}$ be a projector and τ be $\hat{P}\mu$ restricted to $\mathcal{K}$. We shall say that τ is the *compression* of μ to $\mathcal{K}$ and that μ is a *dilation* of τ. Assume that $\mathcal{H}_{ph}$ is a Hilbert space of considered quantum system. The following theorem holds [14].

Theorem 2 *Let τ be a nomalized POV-measure in $\mathcal{H}_{ph}$. There exists a Hilbert space $\mathcal{H}_{ext}$ such that $\mathcal{H}_{ph} \subset \mathcal{H}_{ext}$ and a spectral measure μ in $\mathcal{H}_{ext}$ such that τ is the compression of μ to $\mathcal{H}_{ph}$.*

The construction of the spectral measure μ is very interesting. Let $\mathcal{H}_0$ be a space of functions $f: R \to \mathcal{H}_{ph}$ which are of the form

$$f(x) = \sum_{\sigma} \chi_\sigma(x)\phi_\sigma, \quad \sigma \subset R, \; \phi_\sigma \in \mathcal{H}_{ph}\,, \tag{21}$$

where $\chi_\sigma(x)$ is a characteristic function of the Borel set σ and the sum is finite (as was shown in [7] we can assume without any loss of generality that the intersection of any two different sigma sets in (21) is empty). We define an inner product in $\mathcal{H}_0$ as follows

$$(g, f) = \sum_{\sigma,\epsilon} \langle \psi_\epsilon \mid \tau(\epsilon \cap \sigma) \mid \phi_\sigma \rangle\,, \tag{22}$$

where $g(x) = \sum_\epsilon \chi_\epsilon(x)\psi_\epsilon$, $\psi_\epsilon \in \mathcal{H}_{ph}$. Let

$$\mathcal{D} := \{f \in \mathcal{H}_0 : \; (f, f) = 0\}. \tag{23}$$

In the quotient space $\mathcal{H}_0/\mathcal{D}$ we define a positive definite inner product by

$$(f + \mathcal{D}, g + \mathcal{D}) := (f, g), \tag{24}$$

where $f, g \in \mathcal{H}_0$. The completion of $\mathcal{H}_0/\mathcal{D}$ with respect to the inner product (24) is of course a Hilbert space which will be denoted by $\mathcal{H}_{ext}$. We define in $\mathcal{H}_0/\mathcal{D}$ the operator μ as follows

$$\mu(I)(f+\mathcal{D}) := \chi_I f + \mathcal{D}, \quad f \in \mathcal{H}_0\,, \tag{25}$$

where $I \subset R$. It is easy to see that μ can be extended to the space $\mathcal{H}_{ext}$. Notice also that

1. $(\mu(I))^2 = \mu(I)$,
2. $(\mu(I)(f+\mathcal{D}), g+\mathcal{D}) = (f+\mathcal{D}, \mu(I)(g+\mathcal{D}))$,
3. $0 \leq (\mu(I)(f+\mathcal{D}), f+\mathcal{D}) \leq (f+\mathcal{D}, f+\mathcal{D})$.

Thus the operator μ is a spectral measure on $\mathcal{H}_{ext}$. Let us define now $\hat{G} : \mathcal{H}_{ph} \to \mathcal{H}_{ext}$ as follows

$$\hat{G}(\phi) = \chi_R \phi + \mathcal{D}\,. \tag{26}$$

We can see easily that

$$(\hat{G}(\phi), \hat{G}(\psi)) = \langle \phi \mid \tau(R) \mid \psi \rangle = \langle \phi \mid \psi \rangle\,. \tag{27}$$

So $\hat{G}$ is unitary. We have also

$$(\hat{G}(\phi), \mu(I)\hat{G}(\psi)) = (\chi_R\phi + \mathcal{D}, \mu(I)(\chi_R\psi + \mathcal{D})) = (\chi_R\phi, \chi_I\psi) = \langle \phi \mid \tau(I) \mid \psi \rangle\,. \tag{28}$$

Hence τ is unitarily equivalent to the compression of μ to $\hat{G}(\mathcal{H}_{ph}) \subset \mathcal{H}_{ext}$. So we have defined the extension of the POV-measure τ from Hilbert space $\mathcal{H}_{ph}$ to the spectral measure μ on $\mathcal{H}_{ext}$. It easy to see that states from $\mathcal{H}_{ph}$ we may identify with functions

$$f(x) = \chi_R(x)\psi\,. \tag{29}$$

A subspace of functions of the above form is called *a space of physical states* and denoted $\mathcal{H}_0^{ph} \subset \mathcal{H}_0$. It is clear that, for $f, g \in \mathcal{H}_0^{ph}$, $f = \chi_R\psi$ and $g = \chi_R\phi$, we have

$$(f, g) = \langle \psi \mid \tau(R) \mid \phi \rangle = \langle \psi \mid \phi \rangle\,. \tag{30}$$

Now we define on $\mathcal{H}_0/\mathcal{D}$ the following operator

$$\hat{U}(\lambda)(f+\mathcal{D}) := \sum_\sigma \chi_{\sigma+\lambda} \hat{U}_{ph}(\lambda)\psi_\sigma + \mathcal{D}\,, \tag{31}$$

where in the right hand side stays the evolution operator $\hat{U}_{ph}(\lambda) = \exp(-i\hat{H}\lambda)$ in $\mathcal{H}_{ph}$. It is not difficult to see that the operator $\hat{U}$ is unitary [7]. This operator can also be extended to the Hilbert space $\mathcal{H}_{ext}$. Assume that POV-measure τ satisfies covariance+ condition (1). Thus

$$\hat{U}_{ph}(\lambda)\tau(I) = \tau(I+\lambda)\hat{U}_{ph}(\lambda)\,. \tag{32}$$

It turns out that μ has also the covariance+ property. Namely, for $f = \sum_\sigma \chi_\sigma \psi_\sigma \in \mathcal{H}_0$ we have

$$\begin{aligned}
\hat{U}(\lambda)\mu(I)(f+\mathcal{D}) &= \hat{U}(\lambda)\chi_I \sum_\sigma \chi_\sigma\psi_\sigma + \mathcal{D} = \hat{U}(\lambda)\sum_\sigma \chi_I\chi_\sigma\psi_\sigma + \mathcal{D} = \\
&= \hat{U}(\lambda)\sum_\sigma \chi_{I\cap\sigma}\psi_\sigma + \mathcal{D} = \sum_\sigma \chi_{(I\cap\sigma)+\lambda}\hat{U}_{ph}(\lambda)\psi_\sigma + \mathcal{D} = \\
&= \sum_\sigma \chi_{(I+\lambda)\cap(\sigma+\lambda)}\hat{U}_{ph}(\lambda)\psi_\sigma + \mathcal{D} = \\
&= \chi_{I+\lambda}\sum_\sigma \chi_{\sigma+\lambda}\hat{U}_{ph}(\lambda)\psi_\sigma + \mathcal{D} = \\
&= \mu(I+\lambda)\hat{U}(\lambda)(f+\mathcal{D})\,.
\end{aligned} \tag{33}$$

The spectral measure μ is constructed from POV-measure τ. From Theorem 1 it follows that in the case when μ is *covariant*, the generator of the time translations must be unbounded from below. Let us now take $f_0 \in \mathcal{H}_0$ such that

$$f_0 = \chi_J\phi\,, \tag{34}$$

where $I \cap J = \emptyset$. Now we have

$$(f_0 + \mathcal{D}, \mu(I)(f_0+\mathcal{D})) = (f_0, \chi_I f_0) = 0\,. \tag{35}$$

So we conclude that the generator of the evolution $\hat{U}$ is really unbounded from below.

We see that for an arbitrary covariant POV-measure (POV-time) we can construct a spectral measure. This spectral measure is also covariant. It is possible, because the generator of the evolution is unbounded from below. Using this spectral measure we can define a time operator $\hat{T}$ on space $\mathcal{H}_{ext}$. Notice that the operator $\hat{T}$ is not conjugate to the Hamiltonian $\hat{H}$ of the system.

4 Conclusion

We have proposed a method of construction of a time obsevable in the formalism of POV-measures. We have also constructed a covariant spectral measure which is a extension of this POV-measure. However an interpretation of states from $\mathcal{H}_{ext}$ is not clear. We discuss this problem in [7].

This work was supported by the University of Lodz grant.

References

1. W.Pauli, *Die allgemeinen Prinzipien der Wellenmechanik*, Handbuch der Physik, vol. V/1, Springer Verlag, Berlin, 60.
2. M.D.Sirnivas, R.Vijayalakshmi, *Pramana* **16**, 173 (1981.
3. P.Bush, M.Grabowski, P.J.Lahti, *Phys.Lett A* **191**, 357 (1994.
4. R.Giannitrapani, *Int.J.Theor.Phys.* **36**, 1575(1997).
5. M.Toller, *Phys.Rev.A* **59**, 960 (1999).
6. M.Skulimowski, Construction of time covariant POV-measures, accepted for publication in *Physics Letters A*.
7. M.Skulimowski, Spectral measures and time, preprint KFT, Lodz 2002, to be submitted to *Physics Letters A*.
8. E.Lubkin, *J.Math.Phys.* **15**, 663 (1974).
9. A.A.Nowicki, *Bull.Acad.Polon.Sci.,Sér.Sci.Math.Astronom.Phys.,* **10**, 1045 (1977).
10. I.Fujiwara, *Prog.Theor.Phys.* **1**, 18 (1980.
11. Y.Aharonow, D.Bohm, *Phys. Rev.***122**, 1649 (1961).
12. A.Peres, *Am.J.Phys.* **7**, 552 (1980).
13. H.Salecker, E.P.Wigner, *Phys. Rev.* **109**, 571 (1958).
14. R.Beals, *Topics in operator theory*, The University of Chicago Press, Chicago 1971, 80.

QUANTUM AND CLASSICAL MODELS BASED ON $GL(N)$–SYMMETRY

JAN J. SLAWIANOWSKI

Institute of Fundamental Technological Research
Polish Academy of Sciences
Świętokrzyska 21, 00-049 Warsaw
E-mail: jslawian@ippt.gov.pl

Discussed are classical and quantum rigorously solvable one-dimensional lattice models. The approach is based on Hamiltonian systems on the groups $GL(n, \boldsymbol{R})$ and $U(n) \subset GL(n, \boldsymbol{C})$.

The theory of one-dimensional integrable lattices was developed some more than twenty years ago as a very special field within the general theory of completely integrable Hamiltonian systems [1, 2, 3, 5]. At the same time they were considered as an academic model of intermolecular interactions in the one-dimensional world. They were investigated in terms of Lax pairs. But there was also another approach based on co-adjoint orbits of certain matrix groups [4]. In this formulation one proceeds in a way opposite to the traditional analysis based on reducing the number of dynamical parameters. On the contrary, the idea was to inject a small system into greater one. The latter, however, possessing evident constants of motion, usually Casimirs of the mentioned linear groups. The most known examples are lattices due to Calogero–Moser, Sutherland, and the hyperbolic version of Sutherland; they are, respectively, one dimensional n-body problems with Hamiltonians

$$H_{CM} = \sum_{i=1}^{n} \frac{1}{2} {P_i}^2 + \sum_{i \neq j} \frac{1}{(Q^i - Q^j)^2}, \tag{1}$$

$$H_S = \sum_{i=1}^{n} \frac{1}{2} p_i^2 + \sum_{i \neq j} \frac{1}{\sin^2 \frac{q^i - q^j}{2}}, \tag{2}$$

$$H_{Sh} = \sum_{i=1}^{n} \frac{1}{2} p_i^2 + \sum_{i \neq j} \frac{1}{\mathrm{sh}^2 \frac{q^i - q^j}{2}} \tag{3}$$

(1) and (3) are somehow related to $GL(n, \boldsymbol{R})$, whereas (2) is ruled by $U(n)$. The relationship between them seems to have something to do with $GL(n, \boldsymbol{R})$ and $U(n)$ being two different, in a sense opposite real forms of the same

complex group $GL(n,\mathbb{C})$. In (1) and (3) the "material points" move on the straight-line, whereas in (2) – on the circle. Their coordinates are related to eigenvalues of certain matrices. A disadvantage of (1) and (3) as academic models of condensed matter is that their potentials are purely repulsive, and there is no interplay of attraction and repulsion, necessary for the stable configurations. Apparently it is also so in (2), however on the circle the repulsive potential is compatible with the stability.

By the way, one can help against the difficulty for (1), (3) by assuming that q^1, q^n are placed respectively at the negative and positive spatial infinity. These boundary-asymptotic restrictions for solutions imply that for $q^2, \ldots, q^{n-1}$ the infinities become effectively repulsive. This is, however, a relatively artificial solution.

But it turns out that considering purely geodetic Hamiltonian models on $GL(n, \boldsymbol{R})$ and $U(n)$ one can obtain one-dimensional n-body problems which are in principle rigorously solvable and free of the mentioned drawback. Interactions are then encoded in kinetic energy expression, i.e. metric on the group. This resemble the Maupertuis–Jacobi variational principle. Although coordinates of "particles" also have something to do with certain eigenvalues, the model is essentially different from (2), (3) and the status of $GL(n, \boldsymbol{R})$, $U(n)$ is also different from that in [2]. The idea goes back to our old papers about so-called affinely-rigid body [6, 7, 8, 9], although in this form it was not formulated there.

Let us consider purely geodetic Hamiltonian models on $L(n, \boldsymbol{R})$, $GL(n, \boldsymbol{R})$ and $U(n)$. Motions are given by curves $t \to \varphi(t)$ in these manifolds. We introduce also Lie-algebraic objects

$$\Omega := \dot{\varphi}\varphi^{-1}, \qquad \Omega' := \varphi^{-1}\dot{\varphi} = \varphi^{-1}\Omega\varphi. \tag{4}$$

These are general matrices when dealing with $GL(n, \boldsymbol{R})$ and anti-hermitian in $U(n)$. We define kinetic energies, i.e. metric tensors on $L(n, \boldsymbol{R})$, $GL(n, \boldsymbol{R})$, $U(n)$ by the following expressions

$$L(n, \boldsymbol{R}): \quad T = \frac{m}{2}\,\mathrm{Tr}\,(\dot{\varphi}^{\mathrm{T}}\dot{\varphi}), \tag{5}$$

$$GL(n, \boldsymbol{R}): \quad T = \frac{m}{2}\,\mathrm{Tr}\,(\Omega^2) = \frac{\mathrm{m}}{2}\,\mathrm{Tr}\,(\Omega'^2), \tag{6}$$

$$U(n): \quad T = -\frac{m}{2}\,\mathrm{Tr}\,(\Omega^2) = -\frac{\mathrm{m}}{2}\,\mathrm{Tr}\,(\Omega'^2). \tag{7}$$

Obviously, (5), (7) are positively define (assuming $m > 0$), whereas (6) has a hyperbolic structure of signature $\frac{1}{2}n(n-1)$ minuses and $\frac{1}{2}n(n+1)$ pluses.

The defining peculiarity of the metrics (5), (6,) (7) is that they are invariant, respectively, under transformation groups

$$\varphi \mapsto U\varphi V + \alpha, \tag{8}$$

$$\varphi \mapsto A\varphi B, \tag{9}$$

where φ, $\alpha \in L(n, \boldsymbol{R})$, U, $V \in SO(n, \boldsymbol{R})$ for (5), A, $B \in GL(n, \boldsymbol{R})$, $\varphi \in GL(n, \boldsymbol{R})$ for (6) and $\varphi\, A$, $B \in U(n)$ for (7).

Let us use the two-polar decomposition

$$\varphi = LDR^T, \tag{10}$$

where L, $R \in SO(n, \boldsymbol{R})$, D is diagonal and

$$D_{jj} = \exp{(q^j)}, \quad D_{jj} = \exp{(iq^j)} \tag{11}$$

respectively for $GL(n, \boldsymbol{R})$ (or $L(n, \boldsymbol{R})$ and $U(n)$); q^j are obviously real and i denotes the imaginary unit. In this way, the motion on our group manifolds splits into the motion of n "material points" on the straight line or circle and the motion of two n-dimensional fictitious rigid bodies (the L- and R-variables). Besides of the transiational velocities $\dot{q}^i$ we shall use angular velocities of the L-, R-tops in the comoving representation

$$\lambda := L^{-1}\dot{L} = -\lambda^T, \quad \rho := R^{-1}\dot{R} = -\rho^T. \tag{12}$$

Their conjugate canonical spins, i.e. Hamiltonian generators of the regular translations

$$L \mapsto LU, \quad R \mapsto UR, \quad U \in SO(n, \boldsymbol{R}) \tag{13}$$

will be denoted respectively by $j = -j^T$, $k = -k^T$. It is convenient to introduce the auxiliary variables

$$R := -j - k = -R^T, \quad A := j - k = -A^T. \tag{14}$$

After performing the Legendre transformations on the kinetic energies (5), (6), (7) one obtains, respectively, the following Hamiltonians

$$H = \frac{1}{2m}\sum_i P_i^2 + \frac{1}{8m}\sum_{ij}\frac{R_{ij}^2}{(Q^i - Q^j)^2} + \frac{1}{8m}\sum_{ij}\frac{A_{ij}^2}{(Q^i + Q^j)^2}, \tag{15}$$

$$H = \frac{1}{2m}\sum_i p_i^2 + \frac{1}{32m}\sum_{ij}\frac{R_{ij}^2}{\operatorname{sh}^2\frac{q^i - q^j}{2}} - \frac{1}{32m}\sum_{ij}\frac{A_{ij}^2}{\operatorname{ch}^2\frac{q^i - q^j}{2}}, \tag{16}$$

$$H = \frac{1}{2m}\sum_i p_i^2 + \frac{1}{32m}\sum_{ij}\frac{R_{ij}^2}{\sin^2\frac{q^i-q^j}{2}} + \frac{1}{32m}\sum_{ij}\frac{A_{ij}^2}{\cos^2\frac{q^i-q^j}{2}}. \tag{17}$$

Equations of motion may be written in terms of Poisson brackets

$$\frac{df}{dt} = \{\, f,\, H\,\}, \tag{18}$$

where q^i, p_i satisfy the standard positions-momenta rules, q^i, p_i Poisson comute with R_{ij}, A_{ij} and the rules for R, A follow immediately from those for j, k, the latter are simply given by structural constants for $SO(n,\boldsymbol{R})$. Without the extra potential term, (15) is so unstable like (1), however (16) and (17) predict the repulsion-attraction balance and predict the stable situations. In (1), (2), (3) the coupling parameters where constant and just identical. In (15), (16), (17) they become dynamical variables. Let us observe, however, that in a sense (1), (2), (3) and their integrability provide the special case of (15), (16), (17). Namely, there exist special solutions for the latter ones (one can show, this follows from (18)) corresponding to the Ansatz

$$A_{ij} = 0, \quad R_{ii} = 0, \quad \mid R_{ij} \mid = (\text{fixed})R, \quad \text{when } i \neq j. \tag{19}$$

And these solutions are ruled just by (1), (2), (3). The complete integrability of (1), (2), (3) follows from the rigorous solvability of (15), (16), (17). For (16), (17) this solvability is given by the exponential formulas

$$\varphi(t) = \exp{(Et)}\varphi_0 = \varphi_0 \exp{(\varphi_0^{-1}E\varphi_0 t)}, \tag{20}$$

E being arbitrary real for $GL(n,\boldsymbol{R})$ and antihermitian for $U(n)$. For (15) the solutions correspond simply to

$$\varphi(t) = Vt + \varphi_0. \tag{21}$$

The above models may be beautifully and effectively quantized in the Schrödinger sense. Namely, Hamiltonians are given by the Laplace–Beltrami operators corresponding to (5), (6), (7), obviously, multipled by $-\frac{\hbar^2}{2m}$. Hilbert spaces of wave functions are L^2-spaces in the sense of natural measures on the underlying group manifolds

$$L(n,\boldsymbol{R}): \quad d\lambda(\varphi) = d\varphi_{11}\dots d\varphi_{nn}, \tag{22}$$

$$GL^+(n,\boldsymbol{R}): \quad d\chi(\varphi) = (\det\, \varphi)^{-n}\, d\lambda(\varphi), \tag{23}$$

i.e. the Haar measure on $GL^+(n,\boldsymbol{R})$,

$$U(n): \quad d\kappa(\varphi) \tag{24}$$

the Haar measure on $U(n)$. The above measures coincide with the usual Riemannian procedures for (5), (6), (7). It is convenient to write them in the form

$$d\lambda(\varphi) = P_\lambda \, d\mu(L) \, d\mu(R) \, dQ^1 \ldots dQ^n, \tag{25}$$

$$d\chi(\varphi) = P_\chi \, d\mu(L) \, d\mu(R) \, dq^1 \ldots dq^n, \tag{26}$$

$$d\kappa(\varphi) = P_\kappa \, d\mu(L) \, d\mu(R) \, dq^1 \ldots dq^n, \tag{27}$$

where $d\mu$ is the Haar measure element on $SO(n, \boldsymbol{R})$, and

$$P_\lambda = \prod_{i \neq j} \mid Q^{i^2} - Q^{i^2} \mid, \quad P_\chi = \prod_{i \neq j} \mathrm{sh} \mid q^i - q^i \mid, \quad P_\kappa = \prod_{i \neq j} \sin \mid q^i - q^i \mid, \tag{28}$$

According to the Peter–Weyl theorem, the dependence of wave functions may be expanded in the following way

$$\Psi(\varphi) = \sum_{\alpha,\beta \in \Omega} \sum_{m,n=1}^{N(\alpha)} \sum_{m,n=1}^{N(\beta)} D^{\alpha}{}_{mn}(L) \, \Psi_{\binom{nk}{ml}}{}^{\alpha\beta}(q) \, D^{\beta}{}_{kl}(R^T), \tag{29}$$

where Ω denotes the set of irreducible unitary representations (or rather their equivalence classes), $N(\alpha)$ denote their dimensions and the reduced Ψ–amplitudes depend only on lattice point positions. As $D^{\alpha}{}_{mn}$–functions may be considered as in principle known, and due to the exponentiation of Lie algebra representations their status is in principle algebraic, the Schrödinger equation becomes in principle reducible to q–dependent partial amplitudes, i.e. to the Cartan subgroup (the maximal compact subgroup of $GL^+(n, \boldsymbol{R})$ or $U(n)$). In this respect, the quantum problem is simpler that the classical one, where as yet we were unable to achieve such a reduction. If we use the substitution

$$f^{\alpha\beta} = \sqrt{P} \, \Psi^{\alpha\beta}, \tag{30}$$

indices $(nkml)$ for simplicity omitted, P given by (28), we obtain the following effective scalar products for the matrix valued amplitudes f (they are matrices in the sense of (nk)–indices, (ml) kept fixed)

$$\langle f_1 \mid f_2 \rangle = \sum_{\alpha,\beta \in \Omega} \mathrm{Tr} \left(\mathrm{f}_1{}^{\alpha\beta}(\mathrm{q})^{+} \mathrm{f}_2{}^{\alpha\beta}(\mathrm{q}) \right) \mathrm{dq}^1 \ldots \mathrm{dq}^2. \tag{31}$$

The reduced stationary Schrödinger equations for the amplitudes $\Psi^{\alpha\beta}(q)$ have the form

$$\hat{H}^{\alpha\beta} \, \Psi^{\alpha\beta} = E^{\alpha\beta} \, \Psi^{\alpha\beta}, \tag{32}$$

where, for the models (5), (6), (7) $\hat{H}^{\alpha\beta}$ are operators involving second-order differentiation with respect to the q-variables and the effective potentials with couplings controlled by some constants resulting from irreducible representations of the Lie algebra $SO(n, \boldsymbol{R})$! They are respectively given by

$$\hat{H}^{\alpha\beta}_{\lambda}\,\Psi^{\alpha\beta} = -\frac{\hbar}{2m}\,\Delta_{\lambda}\,\Psi^{\alpha\beta} + \frac{1}{8m}\sum_{ij}\frac{(\overset{\leftarrow\beta}{S}{}_{ij} - \overset{\rightarrow\alpha}{S}{}_{ij})^2\,\Psi^{\alpha\beta}}{(Q^i - Q^j)^2} + \frac{1}{8m}\sum_{ij}\frac{(\overset{\leftarrow\beta}{S}{}_{ij} + \overset{\rightarrow\alpha}{S}{}_{ij})^2\Psi^{\alpha\beta}}{(Q^i + Q^j)^2},$$

$$\hat{H}^{\alpha\beta}_{\chi}\,\Psi^{\alpha\beta} = -\frac{\hbar}{2m}\,\Delta_{\chi}\,\Psi^{\alpha\beta} + \frac{1}{32m}\sum_{ij}\frac{(\overset{\leftarrow\beta}{S}{}_{ij} - \overset{\rightarrow\alpha}{S}{}_{ij})^2\,\Psi^{\alpha\beta}}{\mathrm{sh}^2\,\frac{q^i - q^j}{2}} - \frac{1}{32m}\sum_{ij}\frac{(\overset{\leftarrow\beta}{S}{}_{ij} + \overset{\rightarrow\alpha}{S}{}_{ij})^2\Psi^{\alpha\beta}}{\mathrm{ch}^2\,\frac{q^i - q^j}{2}},$$

$$\hat{H}^{\alpha\beta}_{\kappa}\,\Psi^{\alpha\beta} = -\frac{\hbar}{2m}\,\Delta_{\kappa}\,\Psi^{\alpha\beta} + \frac{1}{32m}\sum_{ij}\frac{(\overset{\leftarrow\beta}{S}{}_{ij} - \overset{\rightarrow\alpha}{S}{}_{ij})^2\,\Psi^{\alpha\beta}}{\sin^2\frac{q^i - q^j}{2}} + \frac{1}{32m}\sum_{ij}\frac{(\overset{\leftarrow\beta}{S}{}_{ij} + \overset{\rightarrow\alpha}{S}{}_{ij})^2\Psi^{\alpha\beta}}{\cos^2\frac{q^i - q^j}{2}},$$

where the meaning of symbols is as follows

$$\Delta_{\rho} = \frac{1}{P_{\rho}}\sum_{i}\frac{\partial}{\partial q^i}\,P_{\rho}\,\frac{\partial}{\partial q^i}, \qquad \rho = \chi,\ \kappa,$$

$$\Delta_{\lambda} = \frac{1}{P_{\lambda}}\sum_{i}\frac{\partial}{\partial Q^i}\,P_{\lambda}\,\frac{\partial}{\partial Q^i},$$

$$\overset{\rightarrow\alpha}{S}{}_{ij}\,\Psi^{\alpha\beta} = S^{\alpha}{}_{ij}\,\Psi^{\alpha\beta}, \qquad \overset{\leftarrow\beta}{S}{}_{ij}\,\Psi^{\alpha\beta} = \Psi^{\alpha\beta}\,S^{\beta}{}_{ij},$$

$$S^{\alpha}{}_{ab} = \frac{\hbar}{i}\,M^{\alpha}{}_{ab},$$

and $M^{\alpha}{}_{ab}$ are basic matrices generating in the exp–sense the α-th representation of $SO(n, \boldsymbol{R})$.

The above preliminaries to the quantization of integrable lattices (15), (16), (17) are interesting in that the q-parametrized "particles" are indistinguishable parastatical objects. This is due to the fact that two-polar decomposition (10) is partially non-unique.

Acknowledgements

The author is greatly indebted to the Organizers of the conference QTS2, first of all to professors Horzela and Kapuścik for the invitation. The paper and participation in the Conference were supported by the Polish Committee of Scientific Research (KBN) within the framework of the grant 8 D07A 047 20.

References

1. M.J. Adler, *J. Math. Phys.*, **20**, 60 (1979).
2. F. Calogero, C. Marchioro, *J. Math. Phys.*, **15**, 1425 (1974).
3. V. Guillemin, S. Sternberg, *Sympletic Techniques in Physics*, Cambridge University Press, Cambridge, (1980).
4. D. Kahzdan, B. Constant, S. Sternberg, *Commun. Pure and Applied Math.*, **31**, 481 (1978).
5. J. Moser, *Adv. Math.*, **16**, 1 (1975).
6. J.J. Slawianowski, *Arch. Mech.*, **26**, 569 (1974).
7. J.J. Slawianowski, *Arch. Mech.*, **27**, 93 (1975).
8. J.J. Slawianowski, *Intern. J. Theor. Phys.*, **12**, 271 (1975).
9. J.J. Slawianowski, *ZAMM*, **62**, 229 (1982).

Q-BOSON REALIZATION OF THE POSITIVE DISCRETE SERIES OF REPRESENTATIONS OF THE $U_Q(2,1)$ QUANTUM ALGEBRA

YU.F. SMIRNOV

Instituto de Ciencias Nucleares,
Universidad Nacional Autónoma de México
Circuito Ext., C.U., A.P. 70-543, 04510 México, D.F.

YU.I. KHARITONOV

Petersburg Institute of Nuclear Physics
Russian Academy of Sciences, Gatchina, 188350, Russia.

The q-boson realization of the $U_q(2,1)$ irreducible representations belonging to the positive discrete series is constructed. The complementary relation between $U_q(2,1)$ and $SU_q(2)$ algebras is used to find the connection of Weyl coefficients for $SU_q(2,1)$ with Racah coefficients for $SU_q(2)$.

The quantum algebras and groups, discovered more than 15 years ago, continue to generate a deep interest among both of theoreticians and mathematicians. During first years the investigations were concentrated in main part on compact quantum algebras and their representations. Now the centre of studies displaces in the direction of noncompact quantum algebras.

Following this trend we considered in [1] the structure of unitary irreducible representations (UIR) of the noncompact quantum algebra $U_q(2,1)$, belonging to discrete series (DS). In particular it was shown that the UIRs from positive DS can be labelled by the lowest weight $[f] = [f_1 f_2 f_3]$ whose components satisfy the conditions $f_1 \geq f_2 \geq f_3 + 2$. Two bases of these UIRs were analyzed. One of them (U-basis) corresponds to the reduction

$$U_q(2,1) \supset SU_q^{12}(2) \quad . \tag{1}$$

The second one (T-basis) is adapted to the reduction

$$U_q(2,1) \supset SU_q^{23}(1,1) \quad . \tag{2}$$

The matrix elements of generators $A_{ik}(i,k = 1,2,3)$ were calculated for both bases. Also the transformation brackets (Weyl coefficients) $< U|T >$, connecting two bases, were found. It was conjectured that the $U_q(2,1)$ Weyl coefficients coincide with Racah coefficients for the compact quantum $SU_q(2)$.

Our aim is to prove this conjecture in this paper. To start let us remind the structure of two bases mentioned above.

The basis vectors of the UIR $D^{[f]}$ can be obtained from the lowest vector $|L>$ by repeating action of the generators A_{21}, A_{13}, A_{23} $i<k$. The vector $|L>$ satisfies the relations

$$\begin{aligned} &A_{ii}|L>=f_i|L> \quad , \\ &A_{12}|L>=A_{31}|L>=A_{32}|L>=0 \quad , \\ &<L|L>=1 \quad . \end{aligned} \tag{3}$$

The general vector of the U-basis is given by the expression

$$|[f]m_3UM_U>=\frac{1}{N_{kl}}A_{21}^{U-M_U}P^U A_{23}^k A_{13}^l\,|L> \tag{4}$$

where $0\le k\le f_1-f_2$, $l=0,1,2,\ldots$, $U=\frac{1}{2}(f_1-f_2+k-l)$, $-U\le M_U\le U$. This vector is characterized by the weight $(m)=(m_1m_2m_3)$ with components

$$m_1=f_1+k-U+M_U,\quad m_2=f_2+l+U-M_U,\quad m_3=f_3-k-l\quad .$$

The value U of U-spin indicates the UIR D^U of the $SU_q^{12}(2)$ with generators $U_+=A_{12}$, $U_-=A_{21}$, $U_0=(A_{11}-A_{22})/2$ to which the vector (4) belongs.

The T-basis consists from vectors

$$|[f]m_1TM_T>=\frac{1}{N_{sp}}A_{23}^{M_T-T-1}P^T A_{13}^s\;\;A_{21}^p|L>_+ \tag{5}$$

where $0\le p\le f_1-f_2$, $s=0,1,2,\ldots$, $T=\frac{1}{2}(f_2-f_3+p+s-2)$,
$M_T\ge T+1$, $m_1=f_1-p+s$, $m_2=f_2+p+M_T-T-1$,
$m_3=f_3-s+T-M_T+1$.
The value T of T-spin characterizes the infinite dimensional UIR D^{T+} of the $SU_q^{23}(1,1)$ algebra from the positive DS to which this vector belongs. The algebra $SU_q(1,1)$ is generated by $T_+=A_{23}$, $T_-=A_{32}$, $T_0=\frac{1}{2}(A_{22}-A_{33})$. The explicit expressions for the normalization factors N_{kl}, N_{sp} and for projectors P^U, P^T can be found in [1]. Weyl coefficients $<U|T>$ were calculated in [1] as a scalar products of vectors (4) and (5). However they can be found also using the q-boson realization which was applied by us in [2,3] to obtain the $U_q(3)$ Weyl coefficients. This realization includes six species of q-boson operators $b_i^{\alpha+}$, b_i^{α}, where $i=1,2,3$ and $\alpha=1,2$. These operators satisfy the relations [4,5]:

$$\begin{aligned} (b_i^{\alpha+})^+=b_i^\alpha,\ [b_i^\alpha,b_k^\beta]=[b_i^{\alpha+},b_k^{\beta+}]=0 \quad , \\ \left[b_i^\alpha,b_k^{\beta+}\right]=0 \quad \text{if} \quad i\neq k \quad \text{and/or} \quad \alpha\neq\beta \quad , \\ b_i^\alpha b_i^{\alpha+}=[N_i^\alpha+1] \quad , \quad b_i^{\alpha+}b_i^\alpha=[N_i^\alpha] \end{aligned} \tag{6}$$

The following commutators hold for the number of boson operators N_i^α

$$[N_i^\alpha, b_k^{\beta+}] = \delta_{ik}\delta_{\alpha\beta}\, b_k^{\beta+} \quad ,$$
$$[N_i^\alpha, b_k^{\beta}] = -\delta_{ik}\delta_{\alpha\beta}\, b_k^{\beta} \quad .$$

The UIRs of $U_q(3)$ as well as $U_q(2,1)$ (see below) are realized in a Fock space of vectors

$$|n_i^\alpha >= \prod_{\substack{i=1,2,3\\ \alpha=1,2}} \frac{1}{\sqrt{[n_i^\alpha]!}} (b_i^{\alpha+})^{n_i^\alpha} |0> \quad . \tag{7}$$

Standard q-numbers (or q-operators) and q-factorials are denoted by symbols

$$[N] = \frac{q^N - q^{-N}}{q - q^{-1}} \quad ,$$
$$[N]! = [N][N-1]\ldots[1] \quad .$$

Here q is generic.

The results [2,3] for the $U_q(3)$ algebra can be easily adapted to the case of $U_q(2,1)$ algebra by the substitution

$$b_i^{\alpha+} \to i b_i^\alpha \quad , \quad b_i^\alpha \to i b_i^{\alpha+} \quad . \tag{8}$$

The rest boson operators remain as before. It gives the following expressions for the generators of $U_q(2,1)$ and $SU_q(2)$ algebras:

$$\begin{aligned}
U_q(2,1):\quad & A_{ii} = N_i = N_i^1 + N_i^2 \quad , \quad i = 1,2 \quad , \\
A_{33} &= -N_3^1 - N_3^2 - 2 \quad , \\
A_{12} &= b_1^{1+} b_2^1 q^{\frac{1}{2}(N_1^2 - N_2^2)} + b_1^{2+} b_2^2 q^{-\frac{1}{2}(N_1^1 - N_2^1)} \quad , \\
A_{23} &= i b_2^{1+} b_3^{1+} q^{\frac{1}{2}(N_2^2 + N_3^2 + 1)} + i b_2^{2+} b_3^{2+} q^{-\frac{1}{2}(N_2^1 + N_3^1 + 1)} \quad , \\
A_{21} &= b_2^{1+} b_1^1 q^{\frac{1}{2}(N_1^2 - N_2^2)} + b_2^{2+} b_1^2 q^{-\frac{1}{2}(N_1^1 - N_2^1)} \quad , \\
A_{32} &= i b_2^1 b_3^1 q^{\frac{1}{2}(N_2^2 + N_3^2 + 1)} + i b_2^2 b_3^2 q^{-\frac{1}{2}(N_2^1 + N_3^1 + 1)} \quad , \\
A_{13} &= A_{12} A_{23} - q A_{23} A_{12} \quad , \\
A_{31} &= A_{32} A_{21} - q^{-1} A_{21} A_{32} \quad ; \\
SU_q(2): S_+ &= b_1^{1+} b_1^2 q^{\frac{1}{2}(N_2^1 - N_2^2 - N_3^1 + N_3^2)} \\
&\quad + b_2^{1+} b_2^2 q^{\frac{1}{2}(-N_1^1 + N_1^2 - N_3^1 + N_3^2)} \\
&\quad - b_3^1 b_3^{2+} q^{-\frac{1}{2}(N_1^1 - N_1^2 + N_2^1 - N_2^2)} \quad , \\
S_- &= b_1^{2+} b_1^1 q^{\frac{1}{2}(N_2^1 - N_2^2 - N_3^1 + N_3^2)}
\end{aligned} \tag{9}$$

$$+ b_2^{2+}b_2^1 q^{\frac{1}{2}(-N_1^1+N_1^2-N_3^1+N_3^2)}$$
$$- b_3^2 b_3^{1+} q^{-\frac{1}{2}(N_1^1-N_1^2+N_2^1-N_2^2)} \quad ,$$
$$S_0 = \frac{1}{2}(N_1^1 - N_1^2 + N_2^1 - N_2^2 - N_3^1 + N_3^2) \quad . \tag{10}$$

Direct calculation shows that the generators A_{ik} satisfy the commutation relations characteristic of the $U_q(2,1)$ and that the operators $S_{0,\pm}$ generate the $SU_q(2)$ algebra, which is complementary to $U_q(2,1)$. (The extension of the complementary concept between algebras, introduced by Moshinsky and Quesne [6] and Howe [7], on the quantum algebras was done in [2,3,8,9].

It can be proved that in a space of Fock vectors (2) only the UIRs $D^{[f]} \times D^S$ of the $U_q(2,1) \times SU_q(2)$ algebra can be realized which are characterized by the following values of the lowest weight $[f] = [f_1 f_2 f_3]$ and quasispin S:

$$[f_1 f_2 -2], \quad S = \frac{1}{2}(f_1 - f_2), \quad f_2 = 0,1,2,\ldots, \quad f_1 = f_2, f_2+1, \ldots, \tag{11}$$

$$[f_1 0 f_3], \quad S = \frac{1}{2}(f_1 - f_3 - 2), \quad f_1 = 0,1,2,\ldots, \quad f_3 = -3,-4,\ldots . \tag{12}$$

The representations (11) and (12) are induced by the extremal vectors

$$\psi_1 = (b_1^{1+}b_2^{2+}q^{\frac{1}{2}(N_1^2+N_2^1)} - q^{-1}b_2^{1+}b_1^{2+}q^{-\frac{1}{2}(N_1^1+N_2^2)})^{f_2}(b_1^{1+})^{f_1-f_2}|0> \tag{13}$$

and

$$\psi_2 = (b_1^{1+})^{f_1}(b_3^{2+})^{-f_3-2}|0> \tag{14}$$

respectively. Then extremal vectors are unique except for common factor. They are annihilated by generators A_{12}, A_{31}, A_{32} and S_+.

The $SU_q^{12}(2)$ algebra is complementary to the algebra $\overline{SU}_q^{12}(2)$ determined by generators S_+^{12}, S_-^{12}, S_0^{12} which can be obtained from (5) if one put $N_3^1 >= N_3^2 = 0$ and omit all terms containing $b_3^{\alpha+}$, b_3^α.

The UIR of $\overline{SU}_q^{12}(2)$ are labelled by the quasispin $S_{12} = U$.

The $SU_q^{23}(1,1)$ algebra is complementary to the algebra $\overline{SU}_q^{23}(2)$ with generators S_+^{23}, S_-^{23}, S_0^{23} which can be obtained from (5) if one put $N_1^1 = N_2^1 = 0$ and omit all terms containing $b_1^{\alpha+}$, b_1^α. The $\overline{SU}_q^{23}(2)$ UIR are labelled by the quasispin $S_{23} = T$.

At last it should be noted that the subalgebras $U^i(1)$ with generators A_{ii} are complementary to the $SU_q^i(2)$ subalgebras formed by generators

$$S_+^i = b_i^{1+}b_i^2, \quad S_-^i = b_i^{2+}b_i^1, \quad S_0^i = \frac{1}{2}(N_i^1 - N_i^2), \quad i = 1,2 \ ;$$

$$S_+^3 = -b_3^{2+}b_3^1, \quad S_-^3 = -b_3^{1+}b_3^2, \quad S_0^3 = \frac{1}{2}(N_3^2 - N_3^1) \quad . \tag{15}$$

The UIRs of the $SU^i(2)$ subalgebras are enumerated by quasispins S^i:

$$S^i = \frac{1}{2} n_i = (n_i^1 + n_i^2)/2, \quad i = 1, 2, 3 \ . \tag{16}$$

It follows from the complementarity of algebras mentioned above that the vectors (7), corresponding to the U-basis, can be obtained by the vector coupling of three quasispins S^i in accordance with a coupling scheme $(\mathbf{S}^1 + \mathbf{S}^2) + \mathbf{S}^3$. As a result we obtain

$$\begin{aligned} &|[f](n_1, n_2, -n_3 - 2)U; \quad SM_s >= \\ &= |\frac{n_1}{2}, \frac{n_2}{2}(U); \frac{n_3}{2} : \quad SM_s > \ . \end{aligned} \tag{17}$$

Similarly the vectors of T-basis can be obtained using the coupling scheme $\mathbf{S}^1 + (\mathbf{S}^2 + \mathbf{S}^3)$:

$$\begin{aligned} &|[f](n_1, n_2, -n_3 - 2)T; \quad SM_s >= \\ &= |\frac{n_1}{2}; \frac{n_2}{2} \frac{n_3}{2}(T) : \quad SM_s > \ . \end{aligned} \tag{18}$$

Therefore the Weyl coefficients $< U|T >$, *i.e.* transformation brackets between U- and T- bases coincide with the recoupling (Racah) coefficients for the $SU_q(2)$ algebra, which is complementary to $U_q(2,1)$: $< U|T > \equiv < [f](n_1, n_2, -n_3 - 2)U|[f](n_1, n_2, -n_3 - 2)T >=$

$$= (-1)^k U_q(\frac{n_1}{2} \frac{n_2}{2} S \frac{n_3}{2}; \ UT) \ . \tag{19}$$

The values of the quasispin S corresponding to the UIR $[f]$ are given in (7). It should be remarked that the general UIR of the $U_q(2,1)$ algebra from the positive DS is characterized by three quantum numbers $f_1 \geq f_2 \geq f_3 + 2$ [2]. Meanwhile only a part of these UIRs, enumerated by two numbers, can be realized in Fock space (see Eq. (11), (12)).

In this point it is useful to go from $U_q(2,1)$ to its subalgebra $SU_q(2,1)$ whose UIR $D^{(\lambda\mu)}$ are characterized by two independent integer numbers $\lambda = f_1 - f_2 = 0, 1, 2, \ldots$, $\mu = f_2 - f_3 = 2, 3, 4, \ldots$. It is clear that all UIRs of $SU_q(2,1)$ can be realized in the Fock space of six species of q-bosons. Therefore the formula (12) allows to find Weyl coefficients for the arbitrary UIR of $SU_q(2,1)$. If we express all "angular momenta" in the $6j$-symbol (19) in terms of k, l, p, s (see (4), (5)) and λ, μ we obtain the result which is identical with the analytical formula for Weyl coefficients derived in [2] by direct calculation

$$
\begin{aligned}
&< U|T >_q= \\
&= \left\{ \frac{[2U+1][2T+1][k]![-T+M_T-1]![U+M_U]![T+M_T]!}{[s]![p]![l]![U-M_U]![\lambda+\mu+s-1]!} \right. \\
&\times \left. \frac{[\lambda-k]![\lambda+l+1]![\mu+s-2]![\mu+p-2]!}{[\lambda-p]![\mu+k-2]![\lambda+\mu+l-1]!} \right\}^{1/2} \\
&\times \sum_r \frac{(-1)^r[U-M_U+r]![l+r]![\lambda+\mu+l+r-1]!}{[r]![2U+r+1]![k-r]![l-s+r]![\mu+p+l+r-1]!} \ . \qquad (20)
\end{aligned}
$$

In conclusion the authors express their gratitude to the Russian Foundation of Basic Research for the financial support (grant RFFI-99-01-01163).

References

1. Yu.F. Smirnov, V.N. Tolstoy, in *Group Theory and Special Symmetries in Nuclear Physics*, eds. J.P. Draayer, J. Janeacke, World Scientific, Singapore (1992), p. 375.
2. Yu.F. Smirnov, Yu.I. Kharitonov, *Yad. Fiz.* **64**, N12 (2001).
3. Yu.F. Smirnov, A.A. Malashin, Yu.I. Kharitonov *Yad. Fiz.* **58**, 1105 (1995).
4. L.C. Biedenharn, *J. Phys. A* **22**, L873 (1989).
5. A.J. Macfarlane, *J. Phys. A* **22**, 4581 (1989).
6. M. Moshinsky, C. Quesne, *J. Math. Phys.* **12**, 1772 (1971).
7. R. Howe, *Lecture Appl. Math.* **21**, 179 (1985).
8. C. Quesne, *J. Phys. A* **25**, 5977 (1992).
9. G. Green, *J. Math. Phys.* **40**, 1074 (1999).

ISOTROPIC QUANTIZED MINKOWSKI SPACE AND POINCARÉ GROUP

KORDIAN ANDRZEJ SMOLIŃSKI

Department of Theoretical Physics, University of Lodz, ul. Pomorska 149/153, 90-235 Łódź, Poland

E-mail: xmolin@merlin.fic.uni.lodz.pl

The most general isotropic quantum deformation of Minkowski space is constructed. It depends on two deformation parameters. The four-dimensional differential calculi on the quantum Minkowski space are obtained and classified. Quantum Poincaré group acting covariantly on two-parameter quantum Minkowski space is constructed. The action of this group on the space is braided and the group itself is quantized braided group. The antipode in the group algebra is constructed with help of a pair of quantum metric tensors.

1 Introduction

In a number of recent papers various approaches to quantum deformation of the space-time structures and symmetries were considered. In this paper we construct a quantum deformation of Minkowski space without any quantized algebraic structure in the foundations. We define quantum Minkowski space choosing the set of commutation relations for its generators and then we use this quantum space as a basis for differential calculus, quantum Poincaré group, etc. Because of physical reasons (there is no suggestion that the rotational symmetry of the space is broken), we consider only isotropic deformations of Minkowski space. This approach to quantized relativistic symmetries is essentialy different from other approaches in which the quanutm Minkowski space bases on quantum Poincaré group,[1,2] quantum spinor group[3,4] or quantum Poincaré algebra.[5,6]

The paper is organized as follows. In Sec. 2 we construct quantum Minkowski space. In Sec. 3 we find its differential calculi. In Sec. 4 we build the quantized Poincaré group acting covariantly on the two-parameter quantum Minkowski space and its differential calculus. Finally, in Sec. 5, we make final conclusions.

2 Quantum Minkowski Space

To build quantum Minkowski space we follow the definition of quantum plane. Let $\mathbb{C}[x^\mu]$ be a free associative algebra over $\mathbb{C}$ generated by $\{x^\mu\}_{\mu=0}^{3}$ and let

I be an ideal in $\mathbb{C}[x^\mu]$ generated by elements of the form

$$x^\mu x^\nu - \left(B^{\mu\nu}_{\alpha\beta} x^\alpha x^\beta + A^{\mu\nu}_\alpha x^\alpha + C^{\mu\nu}\right). \tag{1}$$

If the quotient algebra $\mathbb{C}[x^\mu]/I$ is an associative algebra it forms a *quantum space*. The elements (1) define the commutation relations in the algebra. If we introduce introduce to the algebra the *-conjugation (anti-linear and anti-involutive) we get a real structure on the quantum space.

For *quantum Minkowski space* we take the isotropic four-dimensional quantum space with self-conjugated generators. *Isotropy* means that the rotation group $O(3)$ acting on "spatial" generators x^i $(i = 1, 2, 3)$ is an automorphism group of the quantum space.

The most general form of commutation relations fulfilling the above requirements is the following[7]

$$x^i x^0 = q x^0 x^i + i q^{\frac{1}{2}} h x^i, \quad x^i x^j = x^j x^i \tag{2}$$

$(i = 1, 2, 3)$. We have two deformation parameters: q and h; q is a phase and it is dimensionless, h is real and it has dimension of time.

Two subcases of quantum Minkowski space introduced here, have been already considered. The choice $q = 1$ and $h = \kappa^{-1}$ gives the κ-Minkowski space,[5] while the choice $h = 0$ gives Rembieliński's deformation.[8]

3 Quantum Differential Calculi on Quantum Minkowski Space

Now we find differential calculi on quantum Minkowski space. We follow the construction of the differential calculus on quantum spaces[9,10]. We extend the algebra of quantum Minkowski space by the set of generators $\{dx^\mu\}_{\mu=0}^3$ with commutation relations providing associativity and consistency with the differential operator d, which is linear and nilpotent, it obeys Leibniz rule with gradation, i.e. for every two elements f, g of the algebra

$$d(fg) = df g + (-1)^p f dg \tag{3}$$

(here p is the number of dx's in f) and it acts on x^μ such that $d(x^\mu) = dx^\mu$. We require additionaly that the generators dx^μ are self-conjugated.

The condition of isotropy requires that both sides of commutation relation must be the $SO(3)$-tensors of the same type. This reduces the most general form of commutation relations between dx's and x's, and dx's themselves to

$$x^0\, dx^0 = \alpha_1 dx^0\, x^0 + \alpha_2 \sum_k dx^k\, x^k + \alpha_3 dx^0, \tag{4a}$$

Table 1. Quantum differential calculi on quantum Minkowski space (s and p are phases, the other parameters are real).

			$q=1$			$q=1,\ h=0$		
	I	II	III	IV	V	VI	VII	VIII
α_1	qs	p	1	1	1	s	p	1
α_2	0	0	ir	ir	0	0	0	0
α_3	$iq^{\frac{1}{2}}h\frac{qs-1}{q-1}$	$iq^{\frac{1}{2}}h\frac{p-1}{q-1}$	it	it	$-2ih$	$is^{\frac{1}{2}}t$	$ip^{\frac{1}{2}}t$	0
α_4	q	q	1	1	1	1	1	1
α_5	$iq^{\frac{1}{2}}h\frac{qs-1}{q-1}$	0	it	0	0	$is^{\frac{1}{2}}t$	0	0
α_6	s	q^{-1}	1	1	1	s	1	1
α_7	$iq^{\frac{1}{2}}h\frac{s-1}{q-1}$	$-iq^{-\frac{1}{2}}h$	$i(t-h)$	$-ih$	$-ih$	$is^{\frac{1}{2}}t$	0	0
α_8	1	1	1	1	1	1	1	1
α_9	0	0	0	0	0	0	0	0
α_{10}	0	0	0	0	0	0	0	il
α_{11}	0	0	0	0	ik	0	0	ik
α_{12}	s^{-1}	q	1	1	1	s^{-1}	1	1

$$x^i\,dx^0 = \left(\frac{\alpha_4}{\alpha_{12}} - 1\right) dx^i\,x^0 + \alpha_4 dx^0\,x^i + \alpha_5 dx^i, \tag{4b}$$

$$x^0\,dx^i = (\alpha_6\alpha_{12} - 1)dx^0\,x^i + \alpha_6 dx^i\,x^0 + \alpha_7 dx^i, \tag{4c}$$

$$x^i\,dx^j = (\alpha_8 - 1)dx^i\,x^j + \alpha_8 dx^j\,x^i \tag{4d}$$

$$+\left(\alpha_9 \sum_k dx^k\,x^k + \alpha_{10}dx^0\,x^0 + \alpha_{11}dx^0\right)\delta^{ij}, \tag{4e}$$

$$(dx^0)^2 = 0, \quad dx^i\,dx^0 = -\alpha_{12}dx^0\,dx^i, \quad dx^i\,dx^j = -dx^j\,dx^i. \tag{4f}$$

Now we have to find and solve a set of algebraic equations required by the consistency conditions. The solutions are presented in the Table 1. Note, that for the situations that $q=1$ or $h=0$ there exist appropriate limits of calculi with arbitrary q or h.

4 Quantum Poincaré Group

Now we construct the quantum inhomogeneous group acting covariantly on the two-parameter quantum Minkowski space with its differential calculus, i.e. the quantum Poincaré group. We allow braiding[11,12] in action of quantum Poincaré group on quantum Minkowski space and in quantum Poincaré group.

We look for the quantum group generated by the set of self-conjugated generators $\{\Lambda^\mu_\nu, a^\mu\}^3_{\mu,\nu=0}$, which acts on the quantum Minkowski space in the following way

$$\rho(x^\mu) = \Lambda^\mu_\nu \otimes x^\nu + a^\mu \otimes \mathbf{1}, \quad \rho(dx^\mu) = \Lambda^\mu_\nu \otimes dx^\nu. \tag{5}$$

This action is braided, where we can assume that the braiding Ψ is of the form

$$\Psi(x^\alpha \otimes \Lambda^\beta_\gamma) = Z_1{}^{\alpha\beta\nu}_{\gamma\mu\xi}\Lambda^\mu_\nu \otimes x^\xi, \quad \Psi(x^\alpha \otimes a^\beta) = Z_2{}^{\alpha\beta}_{\mu\nu} a^\mu \otimes x^\nu \tag{6}$$

(and the same for dx^μ). The isotropy requires that the matrices Z_1 and Z_2 must provide that both sides of each of equations (6) must be the same $SO(3)$ tensors.

The covariance and consistency of braiding with the algebra structure fixes the braided action and the commutation relations in quantum Poincaré group. For the two-parameter quantum Minkowski space we get that the quantum group acting covarianlty on the two-parameter quantum Minkowski space *exists only if the quantum differential calculus is the common calculus of families I and II* (i.e. $s = q^{-1}$, $p = 1$).

The braid map between quantum Poincaré group and quantum Minkowski space consistent with their algebraic structure is of the form

$$\Psi(x^i \otimes \Lambda^0_k) = q^{-1}\Lambda^0_k \otimes x^i, \quad \Psi(x^i \otimes \Lambda^j_0) = q^{-1}\Lambda^j_0 \otimes x^i, \tag{7a}$$

$$\Psi(x^i \otimes \Lambda^j_k) = q^{-2}\Lambda^j_k \otimes x^i, \quad \Psi(x^i \otimes a^j) = q^{-1}a^j \otimes x^i \tag{7b}$$

(we have listed only the non-trivial relations) and the same relations substituting dx^μ for x^μ.

The non-trivial (i.e. non-commuting) commutations relation between generators of quantum Poincaré group read

$$\Lambda^0_k\Lambda^0_0 = q^{-1}\Lambda^0_0\Lambda^0_k, \quad \Lambda^i_0\Lambda^0_0 = q\Lambda^0_0\Lambda^i_0, \quad \Lambda^i_0\Lambda^0_k = q\Lambda^0_k\Lambda^i_0, \tag{8a}$$

$$a^0\Lambda^0_k = \Lambda^0_k\left(a^0 + iq^{\frac{1}{2}}h\Lambda^0_0\right), \qquad a^0\Lambda^j_0 = q^{-1}\Lambda^j_0\left(a^0 - iq^{\frac{1}{2}}h\mathbf{1}\right) \tag{8b}$$

$$a^0\Lambda^j_k = q^{-1}\Lambda^j_k\left(a^0 + iq^{\frac{1}{2}}h(\Lambda^0_0 - \mathbf{1})\right), \qquad a^i\Lambda^0_0 = q\Lambda^0_0 a^i, \tag{8c}$$

$$a^i\Lambda^j_k = q^{-1}\Lambda^j_k\left(a^i + iq^{\frac{1}{2}}h\Lambda^i_0\right), \qquad a^i\Lambda^0_k = \Lambda^0_k\left(a^i + iq^{\frac{1}{2}}h\Lambda^i_0\right), \tag{8d}$$

$$a^i a^0 = qa^0a^i + iq^{\frac{1}{2}}ha^i. \tag{8e}$$

Comultiplication map is defined in the standard way as

$$\Delta(\Lambda^\mu_\nu) = \Lambda^\mu_\xi \otimes \Lambda^\xi_\nu, \quad \Delta(a^\mu) = \Lambda^\mu_\xi \otimes a^\xi + a^\mu \otimes \mathbf{1} \tag{9}$$

but it must be braided to be an algebra homomorphism and a coalgebra cohomomorphism. The explicit relations for the braid map are the following

$$\Psi(\Lambda^0_k \otimes \Lambda^0_l) = q^{-1}\Lambda^0_l \otimes \Lambda^0_k, \qquad \Psi(\Lambda^0_k \otimes \Lambda^i_0) = q\Lambda^i_0 \otimes \Lambda^0_k, \tag{10a}$$
$$\Psi(\Lambda^i_0 \otimes \Lambda^0_k) = q^{-1}\Lambda^0_k \otimes \Lambda^i_0, \qquad \Psi(\Lambda^i_0 \otimes \Lambda^j_0) = q^{-1}\Lambda^j_0 \otimes \Lambda^i_0, \tag{10b}$$
$$\Psi(\Lambda^i_0 \otimes \Lambda^j_k) = q^{-2}\Lambda^j_k \otimes \Lambda^i_0, \qquad \Psi(\Lambda^i_k \otimes \Lambda^0_l) = q^{-2}\Lambda^0_l \otimes \Lambda^i_k, \tag{10c}$$
$$\Psi(\Lambda^i_k \otimes \Lambda^j_l) = q^{-2}\Lambda^j_l \otimes \Lambda^i_k, \qquad \Psi(\Lambda^0_k \otimes a^i) = qa^i \otimes \Lambda^0_k, \tag{10d}$$

the relations for $\Psi(a^\mu \otimes \Lambda^\nu_\xi)$ and $\Psi(a^\mu \otimes a^\nu)$ have the same form as relations for $\Psi(x^\mu \otimes \Lambda^\nu_\xi)$ and $\Psi(x^\mu \otimes a^\nu)$, the other relations are trivial.

The counit is of the standard form and it reads

$$\epsilon(\Lambda^\mu_\nu) = \delta^\mu_\nu, \quad \epsilon(a^\mu) = 0. \tag{11}$$

We define the antipode in analogy to the Poincaré group:

$$S(\Lambda^\mu_\nu) = \eta_{\nu\beta}\Lambda^\beta_\alpha\hat{\eta}^{\mu\alpha}, \quad S(a^\mu) = -S(\Lambda^\mu_\xi)a^\xi \tag{12}$$

where $\eta = \mathrm{diag}(\gamma, -g, -g, -g)$ and $\hat{\eta} = \mathrm{diag}(\hat{\gamma}, -\hat{g}, -\hat{g}, -\hat{g})$ are the analogs of the metric tensor[7] with covariant and contravariant components, respectively.

The antipode (12) fulfills the defining condition[13]

$$\mu \circ (S \otimes \mathrm{id}) \circ \Delta = \mu \circ (\mathrm{id} \otimes S) \circ \Delta = \iota \circ \epsilon \tag{13}$$

if the entries of quantum metric tensors obey the following commutation relation

$$\Lambda^0_0 g = q^2 g\Lambda^0_0, \qquad \Lambda^0_0\hat{g} = q^{-2}\hat{g}\Lambda^0_0, \qquad \Lambda^0_k\hat{\gamma} = q^2\hat{\gamma}\Lambda^0_k, \tag{14a}$$
$$\Lambda^i_0\hat{\gamma} = q^{-2}\hat{\gamma}\Lambda^i_0, \qquad \Lambda^i_0 g = q^2 g\Lambda^i_0, \qquad \Lambda^i_0\hat{g} = q^{-2}\hat{g}\Lambda^i_0, \tag{14b}$$

$$a^0 g = q^2 g a^0 + i(q^{\frac{1}{2}} + q^{\frac{3}{2}})hg, \qquad a^0\hat{g} = \hat{g}a^0 - i(q^{-\frac{3}{2}} + q^{-\frac{1}{2}})h\hat{g}\Lambda^0_0, \tag{14c}$$
$$a^i\hat{\gamma} = q^{-2}\hat{\gamma}a^i, \qquad a^i g = q^2 g a^i, \tag{14d}$$
$$a^i\hat{g} = \hat{g}a^i - i(q^{-\frac{3}{2}} + q^{-\frac{1}{2}})h\hat{g}\Lambda^i_0 \qquad g\hat{\gamma} = q^4\hat{\gamma}g \tag{14e}$$

and they commute in remaining cases. The elements γ, g and $\hat{\gamma}$, $\hat{g}$ are self-conjugated. Note that these commutation relations do not allow us to choose $\hat{\eta}$ as the inverse of η. Therefore it remains open how to use these tensors to construct the invariant line element in quantum Minkowski space.

5 Conclusions

We have shown that the most general isotropic quantum deformation of Minkowski space is the two-parameter quantum Minkowski space from Sec. 2. Next, we have found quantum differential calculi on quantum Minkowski space. We would like to point out that these differential calculi are isotropic, four-dimensional and covariant under the action of quantum Poincaré group and contrary to the differential calculus on κ-Minkowski space. We build quantum Poincaré group acting covariantly on quantum Minkowski space in the case two-parameter deformation. The action of this group must be braided and the group itself is a quantized braid group. We have found the explicit forms of the braid maps and shown that they can be chosen for space generators and translation generators in a unified way. The definition of the antipode requires the existence of a pair of quantum metric tensors.

Acknowledgments

I would thank to J. Rembieliński and P. Maślanka for their comments. This work is supported by University of Lodz grant.

References

1. P. Podleś and S. L. Woronowicz, *Commun. Math. Phys.* **178**, 61 (1996).
2. P. Podleś in *Particles, Fields, and Gravitation*, ed. J. Rembieliński, AIP, Woodbury (1998).
3. U. Carow-Watamura, M. Schlieker, M. Scholl, and S. Watamura, *Z. Phys. C* **48**, 159 (1990).
4. S. Majid and U. Meyer, *Z. Phys. C* **63**, 357 (1994).
5. S. Zakrzewski, *J. Phys. A: Math. Gen.* **27**, 2075 (1994).
6. P. Kosiński, P. Maślanka, and J. Sobczyk, *Czech. J. Phys.* **46**, 201 (1996).
7. K. A. Smoliński, *Czech. J. Phys.* **44**, 1101 (1994).
8. J. Rembieliński, in *Quantum Groups. Formalism and Applications*, ed. J. Lukierski, Z. Popowicz, and J. Sobczyk, PWN, Warszawa (1995).
9. J. Wess and B. Zumino, *Nuclear Phys. B (Proc. Suppl.)* **18**, 302 (1990).
10. T. Brzeziński, H. Dąbrowski, and J. Rembieliński, *J. Math. Phys.* **33**, 19 (1992).
11. S. Majid, *J. Math. Phys.* **32**, 3246 (1991).
12. L. Hlavatý, *J. Math. Phys.* **35**, 2560 (1994).
13. V. Chari and A. N. Pressley, *A Guide to Quantum Groups*, Cambridge University Press, Cambridge (1994).

CONSTRUCTION OF RAISING AND LOWERING OPERATORS FOR $\mathcal{U}_q(\mathfrak{sl}(2,\mathbb{R}))$

PAVEL ŠŤOVÍČEK

Department of Mathematics
Faculty of Nuclear Science, Czech Technical University
Trojanova 13, 120 00 Prague, Czech Republic
E-mail: stovicek@KMLinux.fjfi.cvut.cz

In the case of $\mathcal{U}_q(\mathfrak{sl}(2,\mathbb{R}))$ at root of unity q-deformed analogues are proposed for the generator of the maximal compact subalgebra, J, and for the raising and lowering operators. We prove an algebraic identity which implies that J has similar properties as in the nondeformed case.

1 Introduction

Despite of an intensive development of the theory of quantum groups during the last decade the results concerning real noncompact forms are rather rare and far away of being complete. Even in the lowest rank cases one encounters serious difficulties (see, for example, references [1,2]). Naturally, most attention has been paid to groups with a physical interpretation [3,4]. Here we concentrate on the quantum real group $\mathcal{U}_q(\mathfrak{sl}(2,\mathbb{R}))$. Even this simple example lacks, to the author's knowledge, a consistent theory on the q-deformed level.

Let us first recall some basic facts concerning the irreducible unitary representations of the classical Lie group $SL(2,\mathbb{R})$ [5,6]. Let H, E, F be the standard basis elements of $\mathfrak{sl}(2,\mathbb{R})$. All of them are skew self-adjoint with respect to the corresponding $*$-involution in $\mathcal{U}(\mathfrak{sl}(2,\mathbb{C}))$. Set

$$J = E - F,\ W^{\pm} = H \pm \mathrm{i}(E + F).$$

For any irreducible representation ϱ of $SL(2,\mathbb{R})$ in a Hilbert space $\mathcal{H}$ the spectrum of $-\mathrm{i}\varrho(J)$ is simple and contained in $\mathbb{Z}$. If $\{v_n\}_{n\in M}$, $M \subset \mathbb{Z}$, is a basis of $\mathcal{H}$ formed by suitably normalised eigenvectors of $-\mathrm{i}\varrho(J)$ then

$$-\mathrm{i}\varrho(J)v_n = n\,v_n,\ \varrho(W^{\pm})v_n = (s+1\pm n)v_{n\pm 2},$$

where $s \in \mathbb{C}$ is a fixed parameter.

Our main goal here is to propose q-deformed analogues for the generator of the compact subalgebra $\mathfrak{so}(2)$, J, and for the raising and lowering operators $W^{\pm}$. This short contribution does not aim at all, however, to develop a complete representation theory for $\mathcal{U}_q(\mathfrak{sl}(2,\mathbb{R}))$; this task is deferred to another publication.

2 $\mathcal{U}_q(\mathfrak{sl}(2,\mathbb{R}))$ at root of unity

We shall consider the case when q is a root of unity. We assume that

$$q = \exp(2\pi \mathrm{i} P/Q)$$

where $Q \in \mathbb{N}$ is odd, $P \in \{1, 2, \ldots, Q-1\}$ and P and Q are relatively prime integers. Let K, K^{-1}, E, F be the standard generators of the complex Hopf algebra $\mathcal{U}_q(\mathfrak{sl}(2,\mathbb{C}))$. The $*$-involutions defining the real form is prescribed by the relations

$$K^* = K,\ E^* = -q^{-1}E,\ F^* = -q\,F.$$

Here we prefer another set of generators generating a "slightly smaller" Hopf algebra $\mathcal{U}$, namely

$$X = -\mathrm{i}\,q^{-1}E\,K^{-1},\ Y = -\mathrm{i}\,q\,F\,K^{-1},\ Z = K^{-2}.$$

Thus the defining relations for $\mathcal{U}$ are

$$\begin{aligned} &ZZ^{-1} = Z^{-1}Z = 1,\ ZX = q^{-2}XZ,\ ZY = q^2\,YZ, \\ &q^{-1}XY - q\,YX = \frac{1}{q-q^{-1}}(Z^2-1), \end{aligned} \tag{1}$$

the comultiplication is defined by

$$\Delta Z = Z \otimes Z,\ \Delta X = 1 \otimes X + X \otimes Z,\ \Delta Y = 1 \otimes Y + Y \otimes Z,$$

the antipode is defined by

$$S(Z) = Z^{-1},\ S(X) = -XZ^{-1},\ S(Y) = -YZ^{-1},$$

and the counit is defined by

$$\varepsilon(Z) = 1,\ \varepsilon(X) = \varepsilon(Y) = 0\,.$$

The $*$-involution defining the real form now reads

$$Z^* = Z,\ X^* = X,\ Y^* = Y\,. \tag{2}$$

The irreducible representations are divided into two families (see the original paper [7] or the monographes [8,9]). The representations from the first family are labelled by a discrete index $r = 0, 1, \ldots, Q-1$, the representation space is $\mathcal{H}_r = \mathrm{span}\{v_0, v_1, \ldots, v_r\}$, and the representation is defined by

$$Z \cdot v_j = \pm q^{r-2j} v_j,\ X \cdot v_j = -q^{r-2j-1}[r-j]_q v_{j+1},\ Y \cdot v_j = [j]_q v_{j-1}. \tag{3}$$

The representations from the second family are labelled by 3 complex parameters λ, a, b, the representation space is always $\mathcal{H} = \{v_0, v_1, \ldots, v_{Q-1}\}$, the representation is defined by

$$\begin{aligned}
Z \cdot v_j &= \lambda\, q^{2j} v_j, \\
X \cdot v_j &= -\mathrm{i}\, q^{j-1} \left(ab - [j]_q \frac{\lambda\, q^{j-1} - \lambda^{-1} q^{-j+1}}{q - q^{-1}} \right) v_{j-1} \quad \text{for } j \neq 0, \\
X \cdot v_0 &= -\mathrm{i}\, q^{-1} a\, v_{Q-1}, \\
Y \cdot v_j &= -\mathrm{i}\, \lambda\, q^{j+1} v_{j+1} \quad \text{for } j \neq Q-1, \\
Y \cdot v_{Q-1} &= -\mathrm{i}\, \lambda\, b\, v_0.
\end{aligned} \tag{4}$$

The two families intersect for the values of parameters $r = Q - 1$ in (3) and $a = b = 0$, $\lambda = \pm q^{-Q+1}$ in (4).

Unfortunately, as pointed out in reference [10], the only representations from this list which are unitarizable with respect to the $*$-involution (2) are one-dimensional. This absolutely unsatisfactory situation of course appeals for a modified definition of $\mathcal{U}_q(\mathfrak{sl}(2, \mathbb{R}))$. A possible way out has been proposed in reference [10]. The set of generators is augmented by introducing an additional generator T which satisfies

$$T^2 = 1,\ \Delta T = T \otimes T,\ S(T) = T,\ \varepsilon(T) = 1.$$

The augmented Hopf algebra will be called $\mathcal{U}^\natural$. The modified $*$-involution takes the form

$$X^* = T\,X\,T,\ Y^* = T\,Y\,T,\ Z^* = T\,Z\,T,\ T^* = T. \tag{5}$$

As shown in papers [10,11], all representations from the first family (3) are unitarizable in this modified sense and may be regarded as q-deformations of representations of $SL(2, \mathbb{R})$ belonging to the discrete series.

3 Raising and lowering operators

Set

$$J = (qX - q^{-1}Y)Z^{-1} = Z^{-1}(q^{-1}X - qY).$$

One easily finds that

$$\Delta J = Z^{-1} \otimes J + J \otimes 1,\ S(J) = -Z\,J,\ \varepsilon(J) = 1\,. \tag{6}$$

For the extended algebra $\mathcal{U}^\natural$ it seems to be reasonable to impose the condition

$$JT = TJ,$$

i.e. $J^* = J$.

Note that conversely one can express

$$X = \frac{1}{q^2 - q^{-2}}(qJZ - q^{-1}ZJ),\ Y = \frac{1}{q^2 - q^{-2}}(q^{-1}JZ - qZJ)\,.$$

It is also straightforward to derive the relations between J and Z. The second and third relations in (1) give

$$Z^2J - (q^2 + q^{-2})ZJZ + JZ^2 = 0 \tag{7}$$

while the last one implies

$$(q^2 + 1 + q^{-2})\,ZJ^2Z - JZJZ - JZ^2J - ZJZJ = (q + q^{-1})^2(Z^2 - 1)\,. \tag{8}$$

All the above manipulations are reversible and this shows

Proposition 1. *The Hopf algebra $\mathcal{U}$ is isomorphic to the Hopf algebra $\mathcal{U}'$ generated by Z and J with the defining relations (7), (8), (6) and the known relations for Z ($Z\,Z^{-1} = Z^{-1}Z = 1$, $\Delta Z = Z \otimes Z$, $S(Z) = Z^{-1}$, $\varepsilon(Z) = 1$).*

The quantum numbers $[x]_q = (q^x - q^{-x})/(q - q^{-1})$ are supposed to be defined for any $x \in \mathbb{C}$ for we set, by definition, $q^x = \exp(2\pi\mathrm{i}(P/Q)x)$. The main result of this contribution is the following identity.

Proposition 2. *For all $x \in \mathbb{C}$ it holds true that*

$$\begin{aligned} &Z\big(J - [x+2]_q\big)\big(J - [x]_q\big)\big(J - [x-2]_q\big)Z \\ &\quad = \Big(\big(J - [x]_q\big)Z\big(J - [x]_q\big)Z - {[2]_q}^2\Big)\big(J - [x]_q\big)\,, \end{aligned} \tag{9}$$

Proof. To prove (9) we expend the both sides in the variable $y = q^x$,

$$\mathrm{LHS} - \mathrm{RHS} = c_{-2}\,y^{-2} + c_{-1}\,y^{-1} + c_0 + c_1\,y + c_2\,y^2,$$

and we find that

$$\begin{aligned} (q - q^{-1})^2 c_{-2} &= (q - q^{-1})^2 c_2 = -JZ^2 + (q^2 + q^{-2})ZJZ - Z^2J, \\ (q - q^{-1})c_{-1} &= -(q - q^{-1})c_1 \\ &= (q^2 + 1 + q^{-2})ZJ^2Z - JZJZ - JZ^2J - Z\,JZJ - {[2]_q}^2(Z^2 - 1), \end{aligned}$$

and

$$\begin{aligned} (q - q^{-1})^2 c_0 = {} & 2\,(JZ^2 - (q^2 + q^{-2})ZJZ + Z^2J) \\ & + (q - q^{-1})^2\big(ZJ^3Z - {[2]_q}^2 Z\,JZ - JZJZJ + {[2]_q}^2 J\big). \end{aligned}$$

Thus the coefficients $c_{\pm 2}$, $c_{\pm 1}$ vanish owing to (7) and (8), the coefficient c_0 vanishes if it is true in the particular case $x = 0$. The following lemma concludes the proof. □

Lemma 3. *It holds true that*

$$Z\big(J^2-(q+q^{-1})^2\big)JZ=\big(JZJZ-(q+q^{-1})^2\big)J\,.$$

Proof. To prove the particular case of (9) with $x=0$ set

$$V=ZJ^3Z-[2]_q{}^2ZJZ+[2]_q{}^2J-JZJZ\,J.$$

Then it holds

$$\begin{aligned}
&ZV+VZ=\\
&\quad\big((q^2+1+q^{-2})\,ZJ^2Z-JZ^2J-ZJZJ-JZJZ+(q+q^{-1})^2(1-Z^2)\big)\\
&\quad\times JZ+ZJ\\
&\quad\times\big((q^2+1+q^{-2})\,ZJ^2-JZ^2J-ZJZJ-JZJZ+(q+q^{-1})^2(1-Z^2)\big)\\
&\quad+(JZ^2-(q^2+q^{-2})ZJZ+Z^2J)J^2Z\\
&\quad+ZJ^2(JZ^2-(q^2+q^{-2})\,ZJZ+Z^2J)
\end{aligned}$$

and thus $Z\,V+V\,Z=0$. This clearly implies

$$Z^kV=(-1)^k\,VZ,\quad\text{for all } k\in\mathbb{Z}.$$

In particular, $Z^QV=-VZ^Q$. But Z^Q belongs to the centre of the algebra $\mathcal{U}$ and therefore $Z^QV=VZ^Q$. Hence $Z^QV=0$ and consequently $V=0$. □

Some straightforward consequences follow from identity (9). Consider an irreducible representation of $\mathcal{U}$ whose dimension equals d. If v is an eigenvector of J with an eigenvalue $[x]_q$ then the vectors

$$\big(J-[x]_q\big)\big(J-[x-2]_q\big)Z\cdot v,\ \ \big(J-[x]_q\big)\big(J-[x+2]_q\big)Z\cdot v, \tag{10}$$

either vanish or are eigenvectors of J corresponding to the eigenvalues $[x+2]_q$ and $[x-2]_q$, respectively. The spectrum of J is of the form

$$[x]_q,[x+2]_q,\ldots,[x+2d-2]_q,$$

and hence the operator J has similar properties (up to the factor i) as its nondeformed counterpart. For example, in the case of the representations from the first family (3) the operator J is known to have the spectrum (10) with $x=Q-d+1$ [10]. Furthermore, the operators $(J-[x]_q)(J-[x-2]_q)Z$ and $(J-[x]_q)(J-[x+2]_q)Z$ play the role of raising and lowering operators, respectively. Finally, the matrix Z, if expressed in the corresponding eigenbasis of J, is tridiagonal in the case of the first family of representations (3) or tridiagonal with the cyclic convention modulo Q in the case of the second family (4).

Acknowledgements

The research was partially supported by Grant GACR 201/01/01308.

References

1. S.L. Woronowicz, *Lett. Math. Phys.* **23**, 251 (1991).
2. L.I. Korogodsky, *Commun. Math. Phys.* **163**, 433 (1994).
3. V.K. Dobrev, P. Moylan, *Phys. Lett. B* **315**, 292 (1993).
4. V.K. Dobrev and R. Floreanini, *J. Phys. A* **27**,4831 (1994).
5. V. Bargmann, *Ann. Math.* **48**, 568 (1947).
6. S. Lang, $SL_2(\mathbf{R})$ Addison-Wesley, Reading (1975).
7. P. Roche, D. Arnaudon, *Lett. Math. Phys.* **17**, 295 (1989).
8. V. Chari, A. Pressley, *A Guide to Quantum Groups*, Cambridge University Press, Cambridge (1994).
9. A. Klimyk, K. Schmüdgen, *Quantum Groups and Their Representations*, Springer-Verlag, Berlin Heidelberg (1997).
10. P. Šťovíček, *Czech. J. Phys.* **50**, 1353 (2000).
11. P. Šťovíček, in *Proc. Internat. Sympos. Quantum Theory and Symmetries, Goslar 1999*, Eds. H.-D. Doebner et al., World Scientific, Singapore (2000), p.470.

EXACTLY SOLVABLE TIME-DEPENDENT POTENTIALS IN QUANTUM MECHANICS

A. A. SUZKO

Joint Institute for Nuclear Research, Dubna, Russia
Radiation Physics and Chemistry Problems Institute, Academy of Sciences of Belarus, Minsk
E-mail: suzko@thsun1.jinr.ru

A method is elaborated of constructing a time-dependent periodic Hamiltonian for which a system of Schrödinger equations admits analytic solutions. Time-independent soluble problems are transformed into time-dependent ones by a set of unitary time-dependent transformations and a proper choice of initial states. A new class of 2×2 periodic time-dependent Hamiltonians with cyclic solutions is constructed in a closed analytic form. As an application of the method, the expectation value of Hamiltonian, total, dynamical, and geometric phases are derived in terms of the obtained solutions.

1 Introduction

The technique of Bargmann-Darboux transformations and the supersymmetry method in quantum mechanics yield a variety of soluble stationary models [1,2,3]. Each of these models with a time-independent exactly soluble Hamiltonian can be generalized to obtain the corresponding family of time-dependent exactly soluble Hamiltonians. Our main findings are as follows. First, we explicitly show how to construct exactly soluble time-dependent generalizations of any exactly soluble time-independent model for a 2×2 Hamiltonian. We do this by using unitary time-dependent transformations which change time-independent problems into time-dependent ones and by choosing initial states so that to derive cyclic or recurrent solutions which require the initial states to be eigenstates of a time-independent Hamiltonian. Second, we show how to generate a large family of time-dependent models from one time-independent model by employing a family of special time-dependent operators of transformation. Finally, as an application of our method, we present explicit expression of an expectation value of Hamiltonian, the nonadiabatic geometric phase, total and dynamical phases which are derived in terms of the obtained cyclic solutions. In particular, we demonstrate how to construct a set of periodic time-dependent Hamiltonians whose expectation values do not depend on time. This procedure can be used for modeling quantum wells and wires.

2 Construction of a time-dependent Hamiltonian

To simplify the description below, we give a scheme of generation of a time-dependent Hamiltonian with corresponding solutions from a time-independent one [4]. Suppose that the state $|\Psi(t)>$ of a dynamical system evolves according to the matrix Schrödinger equation

$$i\frac{\partial|\Psi(r,t)>}{\partial t} = H(r,t)|\Psi(r,t)> \tag{1}$$

with $\hbar = 1$ and T periodic time-dependent Hamiltonian, $H(t) = H(t+T)$,

$$H(r,t) = -\nabla_r^2 + V(r,t), \tag{2}$$

the potential matrix $V(r,t) = \{V_{ij}(r,t)\}$ is Hermitian. Our goal is to give the procedure of obtaining a wide class of time-dependent Hamiltonians $H(t)$ for which exact solutions of (1) can be found. To this end, we use the time-independent Hamiltonian

$$\widetilde{H}(r) = -\nabla_r^2 + V(r) \tag{3}$$

with a real symmetric potential matrix $V(r)$ and $\hbar^2/2m = 1$, and a unitary time-dependent transformation $\mathcal{S}(t)$ [5,4]

$$|\Psi(r,t)>= \mathcal{S}(t)|\Phi(r,t)>, \tag{4}$$

by means of which the known time-independent Hamiltonian (3) is changed into the time-dependent one

$$H(t) = \mathcal{S}(t)\widetilde{H}\mathcal{S}^\dagger(t) + i\dot{\mathcal{S}}(t)\mathcal{S}^\dagger(t). \tag{5}$$

Here $|\Phi(r,t)>$ satisfies the equation of motion (1) with the time-independent Hamiltonian $\widetilde{H}(r)$ and it is of the form

$$|\Phi(r,t)>= \exp(-i\widetilde{H}(r)t)|\Phi(r,0)> . \tag{6}$$

It is evident, the solutions $|\Psi(r,t)>$ and $|\Phi(r,t)>$ can be properly defined by solutions of the time-independent problem

$$\widetilde{H}|\Phi(\widetilde{\mathcal{E}})>= \widetilde{\mathcal{E}}|\Phi(\widetilde{\mathcal{E}})> . \tag{7}$$

If the system of Schrödinger equations (7) with some known time-independent Hamiltonian $\widetilde{H}(r)$ is exactly soluble, the system of equations (1) with the time-dependent Hamiltonian (5) admits exact solutions too. The result also depends on the choice of initial states.

Now consider reconstruction of the 2×2 periodic time-dependent Hamiltonian taken in the form (2). We start with the time-independent Hamiltonian

(3) with the 2×2 real symmetric potential matrix $V(r)$, $V_{12}(r) = V_{21}(r)$. By means of a unitary time-dependent transformation taken in the form

$$\mathcal{S}(t) = \exp(-i\mathbf{s}\cdot\mathbf{h}(t)) = \exp(-i\sum_{i=1}^{3} s_i h_i(t)) \tag{8}$$

the time-independent Hamiltonian (3) with regard to equations (1) and (5) turns into the time-dependent Hamiltonian

$$H(r,t) = -\nabla_r^2 + \exp(-i\mathbf{s}\cdot\mathbf{h}(t))V(r)\exp(i\mathbf{s}\cdot\mathbf{h}(t)) + (\mathbf{s}\cdot\dot{\mathbf{h}}(t)). \tag{9}$$

Here $\mathbf{s} = (1/2)\boldsymbol{\sigma}$ is the spin operator, $\boldsymbol{\sigma} = (\hat{\sigma}_1, \hat{\sigma}_2, \hat{\sigma}_3)$ and $\hat{\sigma}_i$ are the Pauli matrices and a dot means a time-derivative. The solutions of (1) with Hamiltonian (9), according to (4), are represented as

$$|\Psi(r,t)> = \exp(-i\mathbf{s}\cdot\mathbf{h}(t)\exp(-i\widetilde{H}(r)t)|\Phi(r,t=0)>. \tag{10}$$

It is convenient to present the 2×2 Hamiltonian (3) by the sum of diagonal and zero trace matrices as

$$\widetilde{H}(r) = -\nabla_r^2 + \begin{pmatrix} V_{11}(r) & V_{12}(r) \\ V_{21}(r) & V_{22}(r) \end{pmatrix} = \left(-\nabla_r^2 + q(r)\right)\hat{I} + 2(\mathbf{s}\cdot\mathbf{B}(r)), \tag{11}$$

with the evident notations: $q(r) = (V_{11}(r) + V_{22}(r))/2$, $B_1(r) = V_{12}(r)$, $B_2(r) = 0$, $B_3(r) = (V_{11}(r) - V_{22}(r))/2$ and $\hat{I}$ is the identity matrix. Then the time-dependent Hamiltonian (9) can be represented as

$$H(r,t) = \left(-\nabla_r^2 + q(r)\right)\hat{I} + 2\exp(-i\mathbf{s}\cdot\mathbf{h}(t))(\mathbf{s}\cdot\mathbf{B}(r))\exp(i\mathbf{s}\cdot\mathbf{h}(t)) + (\mathbf{s}\cdot\dot{\mathbf{h}}(t)). \tag{12}$$

Obviously, the transformation (8) does not change the first term of (11) and transforms the second term. In terms of the evolution operator $\mathcal{U}(t) = \mathcal{U}(t,0)$, the solution $|\Psi(r,t)>$ is

$$|\Psi(r,t)> = \mathcal{U}(t)|\Psi(r,0)>, \quad \mathcal{U}(0) = 1. \tag{13}$$

In the case when $|\Psi(r,0)> = |\Phi(r,0)>$, it is easy to find from (10) and (13) a very important relationship between the operators $\mathcal{U}(t)$ and $\mathcal{S}(t)$

$$\mathcal{U}(t) = \mathcal{S}(t))\exp -i\widetilde{H}t = \exp(-i\mathbf{s}\cdot\mathbf{h}(t))\exp(-i\widetilde{H}t). \tag{14}$$

The evolution operator in one period is written as

$$\mathcal{U}(T) = \exp(-i\mathbf{s}\cdot\mathbf{h}(T))\exp(-i\widetilde{H}T). \tag{15}$$

In assumption that $\mathcal{S}(T)$ and $\widetilde{H}$ commute $[\mathcal{S}(T)\widetilde{H}] = 0$, we easily obtain that $\mathcal{U}(T)$ and $\widetilde{H}$ commute. Indeed,

$$\mathcal{U}^\dagger(T)\widetilde{H}\mathcal{U}(T) = \mathcal{S}^\dagger(T)\exp(-i\widetilde{H}T)\widetilde{H}\exp(i\widetilde{H}T)\mathcal{S}(T) = \mathcal{S}^\dagger(T)\widetilde{H}\mathcal{S}(T) = \widetilde{H},$$

i.e.,

$$[\mathcal{U}(T)\widetilde{H}] = 0 \tag{16}$$

and therefore $\mathcal{U}(T)$ and $\widetilde{H}$ have common eigenvectors $|\Phi_\nu>$.

Now consider cyclic solutions that after one period T ($T = 2\pi/\omega$) are recovered up to the phase, i.e., initial states $|\Psi_\nu(0)>$ are eigenvectors of $\mathcal{U}(T)$ and because of (16) they are eigenvectors of the effective Hamiltonian $\widetilde{H}$. This leads to

$$|\Psi_\nu(r,T)>= \mathcal{U}(T)|\Psi_\nu(r,0)>= \exp(-i\beta_\nu)|\Psi_\nu(r,0)>, \tag{17}$$

$$|\Psi_\nu(0)>= |\Phi_\nu>, \tag{18}$$

where β_ν is the total phase. ¿From this, the matrix representation of $\mathcal{U}(T)$ can be written as $\mathcal{U}(T) = \sum_\nu \exp(-i\beta_\nu)|\Phi_\nu><\Phi_\nu|$ and $\mathcal{U}(T)$ has a set of linearly independent eigenvectors $|\Phi_\nu>$ with eigenvalues $\exp(-i\beta_\nu)$. With allowance for (10) the recurrent solutions at any time are written as

$$|\Psi_\nu(r,t)>= \exp(-i\mathbf{s}\cdot\mathbf{h}(t))\exp(-i\widetilde{E}_\nu t)|\Phi_\nu(r)> . \tag{19}$$

It is evident now, to determine the cyclic solutions we need time-independent solutions. Thus, whenever $\widetilde{H}(r)$ is an exactly soluble time-independent Hamiltonian, the properly generated time-dependent system of equations has cyclic exact solutions. The evolution of an arbitrary initial state $|\Psi(r,0)>= \sum_\nu \alpha_\nu|\Psi_\nu(r,0)>$ can be represented as a superposition of a basis set of recurrent linearly independent vector solutions, $|\Psi(r,t)>= \sum_\nu \alpha_\nu|\Psi_\nu(r,t)>$.

Now we can proceed to calculations of geometric phase, associated with the evolution of cyclic solutions, dynamical phase and the expectation value of Hamiltonian. To find the dynamical phase δ_ν, we need the expectation value $\epsilon_\nu(t)$ of $H(t)$,

$$\begin{aligned}\delta_\nu = \int_0^T \epsilon_\nu(t)dt &= \int_0^T <\Psi_\nu(r,t)|H(t)|\Psi_\nu(r,t)> dt \\ &= \int_0^T <\Psi_\nu(r,0)|\mathcal{U}^\dagger(t)H(t)\mathcal{U}(t)|\Psi_\nu(r,0)> dt.\end{aligned} \tag{20}$$

By using (14) and (5) we have

$$\mathcal{U}^\dagger(t)H(t)\mathcal{U}(t) = e^{i\widetilde{H}(r)t}\Big(\widetilde{H}(r) - i\mathcal{S}^\dagger(t)\dot{\mathcal{S}}(t)\Big)e^{-i\widetilde{H}(r)t}.$$

Since for cyclic solutions $|\Psi_\nu(r,0)>=|\Phi_\nu(r)>$ and $|\Phi_\nu(r)>$ is an eigenstate vector of $\widetilde{H}(r)$ with an eigenvalue $\widetilde{\mathcal{E}}_\nu$ we express the expectation value $\epsilon_\nu(t)$ and the dynamical phase δ_ν as

$$\epsilon_\nu(t)=\widetilde{\mathcal{E}}_\nu-<\Phi_\nu(r)|i\mathcal{S}^\dagger(t)\dot{\mathcal{S}}(t)|\Phi_\nu(r)>, \tag{21}$$

$$\delta_\nu=\widetilde{\mathcal{E}}_\nu T-\int_0^T<\Phi_\nu(r)|i\mathcal{S}^\dagger(t)\dot{\mathcal{S}}(t)|\Phi_\nu(r)>dt. \tag{22}$$

Hence, the geometric phase φ_ν, given by removing the dynamical phase from the total phase β_ν, is

$$\varphi_\nu=\beta_\nu-\delta_\nu=\beta_\nu-\widetilde{\mathcal{E}}_\nu T+\int_0^T<\Phi_\nu(r)|i\mathcal{S}^\dagger(t)\dot{\mathcal{S}}(t)|\Phi_\nu(r)>dt. \tag{23}$$

It should be noted that for a time-periodic and unitary $\mathcal{S}(t)$, the expression (23) for the geometric phase is simplified

$$\varphi_\nu=\int_0^T<\Phi_\nu(r)|i\mathcal{S}^\dagger(t)\dot{\mathcal{S}}(t)|\Phi_\nu(r)>dt. \tag{24}$$

Really, in this case in contrast with (15) the time evolution operator after one period is $\mathcal{U}(T)=\exp(-i\widetilde{H}(r)T)$. Since $|\Psi_\nu(0,r)>$ is an eigenstate of $\mathcal{U}(T)$ with eigenvalue $\exp(-i\beta_\nu)$ then $|\Psi_\nu(0,r)>$ is also eigenstate of $\widetilde{H}$ with eigenvalue β_ν/T. At the same time, since eigenvalues of $\widetilde{H}$ are $\widetilde{\mathcal{E}}_\nu$, then $\beta_\nu/T=\widetilde{\mathcal{E}}_\nu$ and (24) is valid.

Examples. Consider some particular examples of constructing a time-dependent Hamiltonian (12) with the corresponding solutions by using $SU(2)$ transformation (8) in which the components of $\mathbf{h}(t)$ are linear functions of time, $h_i(t)=\omega_i t$, $\mathcal{S}(t)=\prod_{i=1}^3\mathcal{S}_i(t)=\exp(-i\mathbf{s}\cdot\mathbf{w}t)$.
1 case. Let $\mathcal{S}(t)=\mathcal{S}_3(t)=\exp(-i\hat{\sigma}_3\omega t/2)$ where ω is some constant angular velocity. Then, in accordance with (5) or (12) the time-independent Hamiltonian (11) is transformed into the time-dependent one as

$$H(r,t)=\left(-\nabla_r^2+q(r)\right)\hat{I}+\begin{pmatrix}\frac{V_{11}(r)-V_{22}(r)+\omega}{2} & V_{12}(r)\exp(-i\omega t)\\ V_{21}(r)\exp(i\omega t) & -\frac{V_{11}(r)-V_{22}(r)+\omega}{2}\end{pmatrix}. \tag{25}$$

It is evident that $H(r,t)$ is T-periodic, $H(r,t=0)=H(r,t=T), T=2\pi/\omega$. Its Routhian is $\widetilde{H}(r)$ connected with $H(r,t=0)$ by $H(r,0)=\bar{H}(r)+\frac{\omega}{2}\hat{\sigma}_3$. By using (19) the cyclic solutions can be written as

$$|\Psi_\nu(r,t)>=\begin{pmatrix}\exp(-i(\widetilde{\mathcal{E}}_1^\nu+\omega_3/2)t) & 0\\ 0 & \exp(-i(\widetilde{\mathcal{E}}_2^\nu-\omega_3/2)t)\end{pmatrix}|\Phi_\nu(r)>, \tag{26}$$

where $\widetilde{\mathcal{E}}_\alpha^\nu = (\widetilde{\mathcal{E}}_\nu - \Delta_\alpha)$ and Δ_α can be the energy of inner structure (threshold).
2 case. If $\mathcal{S}(t)$ is chosen as an operator of rotation around y-axis

$$\mathcal{S}_2(t) = \exp(-i\hat{\sigma}_2\omega_2 t/2) = \begin{pmatrix} \cos(\omega_2 t/2) & -\sin(\omega_2 t/2) \\ \sin(\omega_2 t/2) & \cos(\omega_2 t/2) \end{pmatrix}, \tag{27}$$

then in accordance with (12) the time-dependent Hamiltonian $H(t)$ takes another form

$$H(r,t) = (-\nabla_r^2 + q(r))\hat{I} + \tag{28}$$

$$+\begin{pmatrix} \frac{V_{11}-V_{22}}{2}\cos\omega_2 t - V_{12}\sin\omega_2 t & V_{12}\cos\omega_2 t + \frac{V_{11}-V_{22}}{2}\sin\omega_2 t - \frac{i\omega_2}{2} \\ V_{12}\cos\omega_2 t + \frac{V_{11}-V_{22}}{2}\sin\omega_2 t + \frac{i\omega_2}{2} & -(\frac{V_{11}-V_{22}}{2}\cos\omega_2 t - V_{12}\sin\omega_2 t) \end{pmatrix}.$$

In accordance with (19) the recurrent solutions of equation (1) are immediately written as

$$|\Psi_\nu(r,t)>= \begin{pmatrix} \cos(\omega_2 t/2)\exp(-i\widetilde{\mathcal{E}}_1^\nu t) & -\sin(\omega_2 t/2)\exp(-i\widetilde{\mathcal{E}}_2^\nu t) \\ \sin(\omega_2 t/2)\exp(-i\widetilde{\mathcal{E}}_1^\nu t) & \cos(\omega_2 t/2)\exp(-i\widetilde{\mathcal{E}}_2^\nu t) \end{pmatrix}|\Phi_\nu(r)> \tag{29}$$

3 case. Let $\mathcal{S}(t)$ be chosen as an operator of rotation around x-axis

$$\mathcal{S}_1(t) = \exp(-i\hat{\sigma}_1\omega_1 t/2) = \begin{pmatrix} \cos(\omega_1 t/2) & -i\sin(\omega_1 t/2) \\ -i\sin(\omega_1 t/2) & \cos(\omega_1 t/2) \end{pmatrix}. \tag{30}$$

Then, from (12) the time-dependent Hamiltonian $H(t)$ is

$$H(r,t) = \left(-\nabla_r^2 + q(r)\right)\hat{I} \tag{31}$$

$$+\begin{pmatrix} \frac{V_{11}-V_{22}}{2}\cos(\omega_1 t) & V_{12} - \frac{\omega_1}{2} + i\frac{V_{11}-V_{22}}{2}\sin(\omega_1 t) \\ V_{12} - \frac{\omega_1}{2} - i\frac{V_{11}-V_{22}}{2}\sin(\omega_1 t) & -\frac{V_{11}-V_{22}}{2}\cos(\omega_1 t) \end{pmatrix}.$$

The cyclic solutions of (1) with account of (19) are written as

$$|\Psi_\nu(r,t)>= \exp(-i\hat{\sigma}_1\omega_1 t/2)\exp(-i\widetilde{\mathcal{E}}^\nu t)|\Phi_\nu(r)>. \tag{32}$$

Thus, by means of a different choice of transformation operators $\mathcal{S}_i(t), i = 1,2,3$, three families of time-dependent potential matrices with the corresponding cyclic solutions are generated in a closed form from one family of time-independent potential matrices. It is evident that a more general transformation can be taken: i) as a direct product of the $\mathcal{S}_i(t)$, $\mathcal{S}(t) = \prod_{i=1}^3 \mathcal{S}_i(t)$, ii) with a more complicated dependence of $\mathcal{S}$ on time.

Whenever $V(r)$ is an exactly soluble potential matrix for the ordinary time-independent system of Schrödinger equations, a family of exactly soluble time-dependent Hamiltonians of the Schrödinger equations (1) can be generated. Let $V(r)$ be a Bargmann potential matrix for which the system of equations (7) has exact solutions. For example, the matrix elements of the transparent potential and the relevant Jost solutions are [2]

$$V_{\alpha\alpha'}(r) = 2\frac{d}{dr}\sum_{\nu\lambda} \exp(-\kappa_\alpha^\nu r)\gamma_\alpha^\nu P_{\nu\lambda}^{-1}(r)\gamma_{\alpha'}^\lambda \exp(-\kappa_{\alpha'}^\lambda r), \tag{33}$$

$$\begin{aligned} F_{\alpha\alpha'}^{\pm}(k,r) &= \exp(\pm ik_\alpha r)\delta_{\alpha\alpha'} \\ &- \sum_{\nu\lambda} \gamma_\alpha^\nu \exp(-\kappa_\alpha^\nu r)P_{\nu\lambda}^{-1}(r)\int_r^\infty \gamma_{\alpha'}^\lambda \exp(-(\kappa_{\alpha'}^\lambda \pm ik_{\alpha'})r')dr', \end{aligned} \tag{34}$$

where $P_{\nu\lambda} = \delta_{\nu\lambda} + \sum_{\alpha'}^m \frac{\gamma_{\alpha'}^\nu \gamma_{\alpha'}^\lambda}{\kappa_{\alpha'}^\nu + \kappa_{\alpha'}^\lambda} \exp(-(\kappa_{\alpha'}^\nu + \kappa_{\alpha'}^\lambda)r)$. The normalized solution for the vector function of the bound state is immediately expressed through the matrix Jost solutions taken at the energy of the bound state $k_\nu^2 = -\mathcal{E}_\nu$, $k_\alpha^\nu = (\widetilde{\mathcal{E}}_\nu - \Delta_\alpha)^{1/2}$ as

$$\Phi_\alpha(\widetilde{\mathcal{E}}_\nu, r) = \sum_{\alpha'}^m F_{\alpha\alpha'}(k,r)\gamma_{\alpha'}^\nu, \quad \alpha, \alpha' = 1, 2. \tag{35}$$

Here, the indices ν and λ correspond to the bound states characterized by the energies $\widetilde{\mathcal{E}}_\nu$ and amplitudes $|\gamma^\nu >$ determine normalizing matrices $|\gamma_\alpha^\nu >< \gamma_{\alpha'}^\nu|$ and the channel indices are labeled as α, α'. Substitution (33) into (25), (28), (31) and (35) into (26), (29), (32) allows determination of three families of potential matrices and pertinent cyclic solutions in an closed analytical form. It is worth noting that the class of exactly solvable stationary problems is very wide. Algebraic Bargmann-Darboux transformations give multiple examples of problems that can be used to obtain the exact solutions of the Schrödinger equations (1) for a large class of explicit time–dependent Hamiltonians.

Now we show that for all the considered examples with $h_i(t) = \omega_i t$ the expectation values of $H(t)$ for corresponding cyclic solutions are independent on time. Really, the relation (21), with account of expressions for potentials and cyclic solutions (25)–(32), gives the expectation values of $H(t)$ as

$$\epsilon_\nu^i =< \Phi_\nu(r)|\widetilde{H}(r)|\Phi_\nu(r) > +\frac{\omega_i}{2} < \Phi_\nu(r)|\hat{\sigma}_i|\Phi_\nu(r) >= \widetilde{\mathcal{E}}_\nu + \frac{\omega_i}{2}\bar{\hat{\sigma}}_\nu^i. \tag{36}$$

It means that equation (1), for obtained family of time-dependent potential matrices, possesses solutions and properties like time-independent ones. For the dynamical phase δ_ν from (36) we get

$$\delta_\nu^i = \int_0^T \epsilon_\nu^i(t)dt = \widetilde{\mathcal{E}}_\nu T + \pi\bar{\hat{\sigma}}_\nu^i. \tag{37}$$

Removing the dynamical phase (37) from the total phase

$$\beta_\nu = \pi + \widetilde{\mathcal{E}}_\nu T. \tag{38}$$

we get the geometric phase φ^i_ν

$$\varphi^i_\nu = (\beta_\nu - \delta^i_\nu) = \pi(1 - \bar{\bar{\sigma}}^i_\nu). \tag{39}$$

Obviously, the geometric phase is determined by the spin-expectation value $\bar{\bar{\sigma}}^i_\nu = < \Psi_\nu(t,r)|\hat{\sigma}_i|\Psi_\nu(t,r) > = < \Phi_\nu(r)|\hat{\sigma}_i|\Phi_\nu(r) >$. It is none other than spin-alignment along the rotating axis and does not depend on time in this case.

3 Conclusion

We have developed the method of solving a system of coupled time-dependent Schrödinger equations in an explicit form by using unitary time-dependent transformations which change time-independent problems into time-dependent ones. In distinction to earlier studies, the method is based on soluble time-independent equations and a special choice of initial functions. The family of time-dependent soluble Hamiltonians, for which a system of two-coupled Schrödinger equations has exact solutions, is generated. The time-dependent Hamiltonians, for which expectation values don't dependent on time, are given.

Acknowledgments

The paper was supported in part by the BRFFI, grant No.F99R-162.

References

1. K. Chadan and P.C.Sabatier, *Inverse Problems in Quantum Scattering Theory*, Springer-Verlag, Berlin Heidelberg/New York (1989), p.493, 2-nd ed.
2. B.N. Zakhariev and A.A. Suzko, *Direct and Inverse Problems. (Potentials in Quantum Scattering)*, Springer-Verlag, Berlin Heidelberg/New York (1990), p.223, 2-nd ed.
3. G. Junker, *Supersymmetric Method in Quantum and Statistical Physics*, Springer-Verlag, Berlin Heidelberg/New York (1996).
4. E. Velicheva, A. Suzko, *Theor.Math.Physics* **115**, 687 (1998).
5. Wang, Shun-Jin, *Phys.Rev.A* **42**, 5107 (1990); ibid, p. 5103.

REPRESENTATIONS FOR SELECTED TYPES OF DIFFUSION SYSTEMS

R. TWAROCK

Department of Mathematics, City University
Northampton Square, London EC1V 0HB

Diffusion systems are algebraic structures underlying simple exclusion models, that is stochastic processes on linear lattices modelling diffusive dynamics. Representations for two representative types of Diffusion systems are considered with the aim to emphasize the structural differences in the representation theory of different families of Diffusion systems.

1 Introduction

Diffusion algebras have first been introduced in the framework of simple exclusion models[1], and the exhaustiveness of this approach has been proven in a related more mathematically oriented paper[2]. Furthermore, Diffusion algebras have been embedded into the more general concept of Diffusion systems, and a classification of Diffusion systems and thus Diffusion algebras up to linear isomorphism has been provided[3].

Diffusion systems are pairs (A, D) where $D = \{D_i | i = 1, \ldots, N\}$ specifies a finite set of generators and A is a unital, associative quadratic algebra with commutation relations of the form

$$g_{\alpha\beta} D_\alpha D_\beta - g_{\beta\alpha} D_\beta D_\alpha = x_\beta D_\alpha - x_\alpha D_\beta \,, \tag{1}$$

with $g_{\alpha\beta} \in \mathbb{R}^{\geq 0}$, $x_\alpha \in \mathbb{R}$, and which is such that it fulfills the PBW property, that is for a given ordering

$$D_{\alpha_1} > D_{\alpha_2} \qquad \alpha_1 > \alpha_2 \in \{1, \ldots, N\} \tag{2}$$

on the set of generators D there exists a PBW-basis of ordered monomials of the form

$$D_{\alpha_1}^{k_1} D_{\alpha_2}^{k_2} \ldots D_{\alpha_n}^{k_n} \,, \qquad k_j \in \mathbb{N}^0 \tag{3}$$

with $\alpha_1 > \alpha_2 > \ldots > \alpha_n$. The algebra A is called a Diffusion algebra.

The classification of Diffusion systems has led to three different families of Diffusion algebras which are characterized by their structural properties[3], and it is the purpose of this contribution to discuss representations for representative members of two different families in detail to emphasize the structural differences between these families. In particular, we will see that while one type allows for finite dimensional representations, the dimensions of which depend on the number of generators in the algebra, the second type has generically infinite dimensional representations. Thus, focus of this contribution lies on pointing out variety rather than striving for completeness, and the interested reader is referred to the more detailed discussion in another reference[4]. Furthermore, consult the following references for a comprehensive discussion of applications of Diffusion algebras in physics[1], or for a more mathematically oriented discussion of Diffusion systems and Diffusion algebras and their classification and graph structure[3].

2 Selected cases of Diffusion Systems

Diffusion systems split into three families, called SYM, $ASYM$ and $PASYM$[3] according to their interpretation in the framework of simple exclusion models as belonging to symmetric, partially asymmetric, or asymmetric exclusion models. We will focus here on selected algebras from the SYM and the $PASYM$ families.

2.1 *An example of SYM type*

A typical Diffusion system of SYM type has a generating set $D = \{D_+, D_i(i = 1, \ldots, N), D_\alpha(\alpha = 1, \ldots, M)\}$ of $N + M + 1$ generators with algebra relations of the form

$$\begin{aligned} [D_+, D_i] &= g_i D_i \\ D_+ D_\alpha &= \lambda_\alpha D_\alpha \\ D_\alpha D_i &= 0 \\ [D_i, D_j] &= 0 \\ [D_\alpha, D_\beta] &= 0\,, \end{aligned} \tag{4}$$

where g_i $(i = 1, \ldots, N)$, λ_α $(\alpha = 1, \ldots, M) \in \mathbb{R}$.

One can represent these relations by $N + M + 1$–dimensional matrices as follows:

$$D_+ = \begin{pmatrix} a + g_1 & 0 & 0 & 0 & 0 & 0 & c_1 \\ 0 & \ldots & 0 & 0 & 0 & 0 & \ldots \\ 0 & 0 & a + g_N & 0 & 0 & 0 & c_N \\ 0 & 0 & 0 & \lambda_1 & 0 & 0 & c_{N+1} \\ 0 & 0 & 0 & 0 & \ldots & 0 & \ldots \\ 0 & 0 & 0 & 0 & 0 & \lambda_M & c_{N+M} \\ 0 & 0 & 0 & 0 & 0 & 0 & a \end{pmatrix} \tag{5}$$

where a, c_k $(k = 1, \ldots, N+M) \in \mathbb{R}$.

If furthermore $E_{k,l}$ denote the matrices with a single entry 1 in the kth row and lth column and entry 0 elsewhere, we have

$$\begin{aligned} D_i &= \mu_i E_{i,N+M+1}, & i &= 1, \ldots, N \\ D_\alpha &= \nu_\alpha E_{N+\alpha,N+M+1}, & \alpha &= 1, \ldots, M, \end{aligned} \tag{6}$$

with μ_i, $\nu_\alpha \in \mathbb{R}$.

These matrices belong to $gl(N+M+1)$ and can be expressed via its generators. Using the Cartan generators, which include the unit matrix $\mathbb{1}$ and the traceless generators $H_k = E_{kk} - E_{k+1,k+1}$, $k = 1, \ldots, N+M$, and the raising generators $X^+_{j,k} = E_{j,k+1}$ $(j \le k)$, $j, k \in \{1, \ldots, N+M\}$, we obtain the following expression for the generators of the Diffusion system (4):

$$\begin{aligned} D_+ &= D_+^c + \textstyle\sum_{j=1}^{N+M} c_j X^+_{j,N+M} \\ D_+^c &= (a + \Delta_{N+M})\mathbb{1} + \textstyle\sum_{k=1}^{N+M} \Delta_k H_k \\ D_i &= \mu_i X^+_{i,N+M}, \quad i = 1, \ldots, N \\ D_\alpha &= \nu_\alpha X^+_{N+\alpha,N+M}, \quad \alpha = 1, \ldots, M \end{aligned} \tag{7}$$

where

$$\Delta_k = \left(1 - \frac{k}{N+M+1}\right) \sum_{j=1}^{k} d_j - \frac{k}{N+M+1} \sum_{j=k+1}^{N+M} d_j, \tag{8}$$

$k = 1, \ldots, N+M$, and

$$d_j = \begin{cases} g_j, & j = 1, \ldots, N \\ \lambda_j - a, & j = N+1, \ldots, N+M. \end{cases} \tag{9}$$

Note that the dimension of this representation is $N+M+1$. We remark that it is possible to use matrices of a higher dimension to represent the relations in (4), which then leads to more involved expressions for the generators, for example a matrix representing D_+ with more off-diagonal elements being different from 0. This gives rise to more complicated expressions in terms of the generators of $gl(n)$[4].

The fact that finite dimensional representations exist is a common characteristic of Diffusion algebras in the SYM family. This is not necessarily true for other families as the following example shows.

2.2 An example of $PASYM$ type

Consider the following algebras generated by $D = \{D_1, \ldots, D_4\}$ with relations

$$\begin{aligned} D_1D_2 + (\xi - 1)D_2D_1 &= D_1 - D_2 \\ \beta D_1D_3 &= D_3 \\ -\alpha D_3D_2 &= D_3 \\ nD_1D_4 + (\xi - n)D_4D_1 &= -D_4 \\ nD_4D_2 + (\xi - n)D_2D_4 &= D_4 \\ \lambda D_3D_4 - \mu D_4D_3 &= 0\,, \end{aligned} \tag{10}$$

where $\xi \neq 1$, α, β, n, λ, $\mu \in \mathbb{R}$.

The representations of the three generator subalgebras generated by $\{D_1, D_2, D_3\}$ and $\{D_1, D_2, D_4\}$ have been considered before[5,6,7,8]. In particular, it has been shown that the $\{D_1, D_2, D_4\}$–subalgebra can be transformed via a linear transformation[8] into

$$\begin{aligned} AB - qBA &= 1 \\ AN - rNA &= A \\ NB - rBN &= B\,. \end{aligned} \tag{11}$$

This is the q-harmonic oscillator, with the second and third line corresponding to the commutation relations for the deformed number operator N. From this it is clear that representations are generically infinite dimensional for this type of algebras. In particular, Fock representations are given by[8]

$$A|0\rangle = \langle 0|B = 0 \tag{12}$$

with

$$A = \begin{pmatrix} 0 & g_1 & 0 & 0 & \ldots \\ 0 & 0 & g_2 & 0 & \ldots \\ 0 & 0 & 0 & g_3 & \ldots \\ \ldots & \ldots & \ldots & \ldots & \ldots \end{pmatrix}, \qquad B = A^T\,, \tag{13}$$

and $N = \text{diag}(p_1, p_2, \ldots)$, where

$$g_n^2 = [n]_q, \qquad [n]_q = \frac{q^n - 1}{q - 1}, \qquad p_n = [n-1]_r\,. \tag{14}$$

Similarly, infinite dimensional representations are known for the subalgebra generated by $\{D_1, D_2, D_3\}$ [5,6,7]. With

$$\begin{aligned}(x;q)_n &:= (1-x)(1-xq)(1-xq^2)\dots(1-xq^{n-1})\\ (x;q)_0 &:= 1\\ (x;q)_\infty &:= \Pi_{j=0}^{\infty}(1-xq^j)\,.\end{aligned} \tag{15}$$

one has

$$D_1 = -\frac{1}{\xi}\begin{pmatrix} 1 & \sqrt{1-q} & 0 & 0 & \dots \\ 0 & 1 & \sqrt{1-q^2} & 0 & \dots \\ 0 & 0 & 1 & \sqrt{1-q^3} & \dots \\ \dots & \dots & \dots & \dots & \dots \end{pmatrix}, \qquad D_2 = -D_1^T \tag{16}$$

with $q = 1 - \xi$, furthermore $D_3 = |V\rangle\langle W|$ with

$$\langle W| = \kappa\left(1, \frac{a}{\sqrt{(q;q)_1}}, \frac{a^2}{\sqrt{(q;q)_2}}, \dots\right) \tag{17}$$

and

$$|V\rangle = \begin{pmatrix} 1 \\ \frac{b}{\sqrt{(q;q)_1}} \\ \frac{b^2}{\sqrt{(q;q)_2}} \\ \dots \end{pmatrix}, \tag{18}$$

where

$$\begin{aligned} a &= -1 + \tfrac{1-q}{\alpha} \\ b &= -1 + \tfrac{1-q}{\beta}\,. \end{aligned} \tag{19}$$

In order to find representations for (10), these representations have to be combined. Starting from $\{D_1, D_2, D_3\}$, one can show[4] that

$$D_4 = \mathrm{diag}(1, h, h^2, \dots) \tag{20}$$

with $h = \frac{n-\xi}{n}$ follows from the conditions implied by the equations coupling D_4 to D_1 and D_2. Considering furthermore the coupling to D_3, one finds that α and β have to be chosen in an appropriate way in order to satisfy the commutation relation in the last line in (10). With $\alpha = \beta = 1 - q$ and $\lambda = \mu = 1$ this requirement is met.

Although this is just one example for an algebra in the $PASYM$ family in the classification scheme, it is nevertheless displaying the main characteristic of these algebras, which is the fact that they have generically infinite dimensional representations, as opposed to the SYM family, a member of which has been discussed in (4). Finally we remark that representations like the one shown here for a four generator algebra in the $PASYM$ family are candidates for a generalization of existing partially asymmetric exclusion models, which are related to three generator algebras and thus to three different species of particles[5,6,7], to models with a larger number of species.

3 Conclusion

We have met an example for a Diffusion system which allows for finite dimensional representations and an example in which representations are generically infinite dimensional. This shows that Diffusion systems and hence Diffusion algebras comprise algebras of a large variety of structures.

Since the simple exclusion models based on quadratic algebras which have been considered so far are related to algebras with a small number of different species, respectively Diffusion systems with a small number of generators, representations for Diffusion systems with a larger number of generators like the ones presented here may be useful tools for a generalization of these models to simple exclusion models with a larger number of species.

Acknowledgements

I would like to thank A.G. Cox, V.K. Dobrev, P.P. Martin and A. Sudbery for many valuable discussions.

References

1. A.P. Isaev, P.N. Pyatov and V. Rittenberg, 'Diffusion algebras', cond-mat/0103603.
2. A.P. Isaev, P.N. Pyatov and R. Twarock, 'Construction Theorem for Diffusion Algebras', in preparation.
3. R. Twarock, 'Diffusion Algebras and the classification of Diffusion Systems', in preparation.
4. R. Twarock, 'Representations for Diffusion Systems', in preparation.
5. T. Sasamoto, 'One-dimensional partially asymmetric simple exclusion process on a ring with a defect particle', cond-math/9910483.
6. T. Sasamoto, *J. Phys. A* **32**, 7109 (1999).
7. R.A. Blythe, M.R. Evans, F. Caloiori and F.H.L. Essler, *J. Phys. A* **33**, 2313 (2000).
8. F. Arndt, T. Heinzel and V. Rittenberg, *J. Phys. A* **31**, 833 (1998).

QUANTUM MECHANICS AT THE PLANCK SCALE

F. WINTERBERG
Department of Physics, University of Nevada, Reno,
Mail Stop 220, Reno, NV 89557, USA
E-mail:winterbe@physics.unr.edu

1 Introduction

According to Planck [1] the final equations of physics should be expressed solely in terms of h (Planck's constant), G (Newton's constant) and c (the velocity of light), setting a smallest length $r_p = \sqrt{\hbar G/c^3} \approx 10^{-33}$ cm. String theories accomplish this only in part. Even though the radius of a closed string is set equal a Planck length, the diameter of the string itself is zero. Here, I believe, string theory is inconsistent. In a theory fully consistent with Planck's reduction postulate, h, G and c should also enter into the differentiation operators, which would have to be replaced by finite difference operators. Moreover, string (or M) theory cannot be the final theory, since from such a theory one should be able to derive quantum mechanics and the theory of relativity, which cannot be done by string (or M) theory.

To sustain microcausality in a relativistic quantum field theory in 3+1 space-time dimensions the interactions must be pointlike. In string theories the zero-dimensional pointlike interactions are replaced by one-dimensional line interactions, but this is possible only in more than three space-dimensions.

In a theory where the infinitesimal differentiation operators are replaced by finite difference operators, microcausality can still be sustained if Lorentz invariance is an asymptotic low energy dynamic symmetry, as in the pre-Einstein theory of relativity by Lorentz and Poincaré. It still assumed the existence of an aether, with Lorentz invariance resulting from true contractions of rods and slower going clocks in absolute motion against the aether, and with the Minkowski space-time an illusion caused by these deformations. This idea can be extended to the general theory of relativity, where one can likewise say that physical deformations of measuring rods and clocks gives us the illusion of a non-Euclidean space-time.

If we accept the thesis that Einstein's gravitational field equations can be unified with quantum theory solely in more than three space dimensions, this can only mean two things: Either we must accept that nature really has more than three space dimensions, or that general relativity, quantum theory, or both are "wrong," resp. arpproximations for energies small compared to the

Planck energy. Since the idea of higher dimensions in physics is something so fantastic, it should be taken as a last resort and only if everything else fails.

2 The Planck Aether Hypothesis

The replacement of the bosonic string theory in 26 dimensions to describe subnuclear forces (dual resonance model) with QCD in 3+1 dimensions of everyday physics, raises the question if string theory should be replaced by something else as well. The analogies between Yang-Mills theories and vortex dynamics [2] suggest that string theory should perhaps be replaced by some kind of vortex dynamics in three dimensions at the Planck scale [3]. This conjecture is supported by the analogies between GRT and condensed matter physics [?], and by the recent success of the author to derive the finestructure constant at the Planck scale from such a model [7].

In giving up Lorentz invariance as an exact kinematic symmetry, it is possible to formulate alternatives to string theory in three space and one time dimension. The probably most simple model I have come up along these lines of thought, I have called the Planck aether hypothesis, named after Planck who in 1911 had concluded that the vacuum of space is filled with a zero point energy, called by Nernst a new kind of aether. It makes the following assumptions [3,8]:

The ultimate building blocks of matter are Planck mass particles which obey the laws of classical Newtonian mechanics but that there are also negative Planck mass particles.

A positive Planck mass particle exerts a short range repulsive and a negative Planck mass particle a likewise attractive force on either a positive or negative Planck mass particle, with the magnitude and range of the force equal to the Planck force F_p and Planck length r_p.

Space is filled with an equal number of positive and negative Planck mass particles whereby each Planck length volume is in the average occupied by one Planck mass particle.

In reducing all of physics to the Planck units of length, mass and time, we used the freedom to assign the Planck mass two different signs. In addition, we have the freedom in the sign of the Planck force $F_p = c^4/G$.

3 The Origin of Quantum Mechanics

The Planck aether hypothesis not only permits a finitistic formulation, with differential operators replaced by finite difference operators [9], but also explains quantum mechanics to have its cause in the hidden negative Planck masses interacting with positive Planck masses by the Planck force over a Planck length. The Planck aether has a spectrum of quasiparticles, both bosonic and fermionic, with Lorentz invariance as a dynamic symmetry.

The Planck aether resembles a plasma, with the positive and negative Planck masses replacing the positive and negative electric charges. As in a plasma, the energy remains unchanged during the collision of a positive with a negative Planck mass, but not the momentum. Under the assumed sign of the Planck force, Planck mass particles of equal sign are accelerated away from each other, whereas those of unequal sign are accelerated towards each other. While this conserves the energy during the collision between a positive and a negative Planck mass, it leads to a momentum fluctuation equal to $\Delta p = m_p c$, where $m_p = \sqrt{\hbar c/G}$is the Planck mass. With the Planck length $r_p = \sqrt{\hbar G/c^3}$, one thus has $\Delta p \cdot r_p = \hbar$, that is the uncertainty relation for a Planck mass particle during its collision with a Planck mass particle of opposite sign. That this result already explains the origin of quantum mechanics is supported by solving the Boltzmann equation for the Planck aether. As in fluid dynamics one obtains there an Euler equation, but which in addition has a quantum potential. Making the Madelung back-transformation one obtains the Schrödinger equation for a Planck mass particle.

4 Heisenberg-Type Operator Field Equation

Approximating for large wave lengths the finite difference operators with differential operators, the Planck aether can be elegantly described by a nonlinear Heisenberg-type operator field equation:

$$i\hbar\frac{\partial\psi_\pm}{\partial t} = \mp\frac{\hbar^2}{2m_p}\nabla^2\psi_\pm \pm 2\hbar c r_p^2(\psi_\pm^\dagger\psi_\pm - \psi_\mp^\dagger\psi_\mp)\psi_\pm \tag{1}$$

where $\psi_\pm$, $\psi_\pm^\dagger$, obey the commutation relations

$$\begin{aligned}\left[\psi_\pm(r), \psi_\pm^\dagger(r')\right] &= \delta(r - r') \\ \left[\psi_\pm(r), \psi_\pm(r')\right] = \left[\psi_\pm^\dagger(r), \psi_\pm^\dagger(r')\right] &= 0\end{aligned} \tag{2}$$

In its groundstate the Planck aether is a highly degenerate two-component Bose fluid with a completely symmetric wave function for each of its components. With the Hartree-Fock approximation one obtains from (1) two coupled nonlinear Schrödinger equations, and from there with the Madelung transformation the Euler equations with quantum potential for two coupled fluids. These equations have luminal acoustic-type wave solutions and quantized vortex solutions. And they not only unify Maxwell's electromagnetic with Einstein's gravitational wave equations, but also lead to Dirac spinor particles. Charge is there explained by the zero point fluctuations of Planck mass particles bound in quantized vortices.

5 Lorentz Invariance and Quantum Mechanics

Here we focus on the explanation of quantum mechanics, not just for the Planck masses but for all masses, $m < m_p$ and $m > m_p$. That Lorentz invariance turns out to be not independent of quantum mechanics, can be seen by the following argument:

1. The zero point fluctuations of all the Planck mass particles lead to an energy spectrum $F(\omega)$. Now, the only spectrum which does not lead to a frictional force on particles moving through the fluctuating zero point field must have the form

$$F(\omega)d\omega = \text{const} \cdot \omega^3 d\omega \tag{3}$$

This is the phonon spectrum in the groundstate of a superfluid, but it is also the only spectrum which is Lorentz invariant.

2. The kinetic energy term $\left(2m/\hbar^2\right)\nabla^2\psi$ in the Schrödinger equation implies the replacement of the classical momentum $m\underline{v}$ by the operator $(\hbar/i)\,\partial/\partial\underline{r}$, whereby the particle has the zero point energy $E_0 = (1/2)\hbar\omega$. But this leads to the quantum mechanical zero point energy spectrum

$$E(\omega)d\omega = \text{const} \cdot (1/2)\hbar\omega \cdot 4\pi\omega^2 d\omega = \text{const} \cdot \omega^3 d\omega \tag{4}$$

explaining why the frictionless spectrum is not only the only one compatible with Lorentz invariance but also with quantum mechanics.

From the Planck Aether Hypothesis it would follow that quantum mechanics becomes invalid for masses $m \gg m_p$, going over into classical mechanics. For particle masses $m \gg m_p$ the zero point vacuum fluctuations are too weak to impose on the particle a "Zitterbewegung" motion, seen as the cause of quantum mechanics.

In the dynamic interpretation of Lorentz invariance, there is an equilibrium between attractive forces communicated by virtual luminal particles (photons) and the repulsive quantum force. With both the wave equations and the zero point energy Lorentz invariant, Lorentz invariance is then established as dynamic symmetry as in the pre-Einstein theory of relativity by Lorentz and Poincaré.

References

1. M. Planck, *Preuss. Akad.*, XXV (1899).
2. F. Winterberg, *Int. J. Theor. Phys* **34**, 399 (1995).
3. F. Winterberg, *Z. Naturforsch.* **52a**, 183 (1997).
4. W.G. Unruh, *Physical Review Letters* **46**, 1351 (1981).
5. T. Jacobsen, *Physical Review D* **48**, 728 (1993).
6. M. Visser, *Class. Quantum Grav.* **15**, 1767 (1995).
7. F. Winterberg, *Z. Naturforsch.* **56a**, 681 (2001).
8. F. Winterberg, *The Planck Aether Hypothesis*, VRI Karlsbad, Germany (2000).
9. F. Winterberg, *Z. Naturforsch.* **47a**, 545 (1992).

GENERALIZED PT SYMMETRY

M. ZNOJIL
Theory Group, Nuclear Physics Institute AS CR, 250 68 Řež, Czech Republic
E-mail: znojil@ujf.cas.cz

Many non-Hermitian $\mathcal{PT}$−symmetric Hamiltonians support the real spectrum. We offer an explanation, viz., the Hermiticity of their Feshbach "effective" reduction. This point of view is new and enables us to propose an immediate generalization of $\mathcal{PT}$ symmetry. A super-integrable model in two dimensions is chosen for illustration.

1 $\mathcal{PT}$ Symmetry

The Hermiticity of a Hamiltonian $H = H^\dagger$ implies the reality of its energies and, in general, the unitarity of the evolution of any bound state in quantum mechanics, $|\psi_n(t)\rangle = \exp(-i\,E_n\,t)\,|\psi_n(0)\rangle$. A puzzle is encountered when a *non-Hermitian* Hamiltonian happens to produce the *real* energies E_n. Bender et al[1] conjectured that within the so called $\mathcal{PT}$ symmetric quantum mechanics, many models with real spectra could be characterized by a weakened Hermiticity

$$H = \mathcal{PT}\,H\,\mathcal{PT}, \qquad \mathcal{P}^2 = \mathcal{T}^2 = 1. \tag{1}$$

Their requirement has a clear physical interpretation since the operator $\mathcal{P}$ is defined as the parity, $\mathcal{P}\,x = -x$, while $\mathcal{T}$ mimics the time reversal, $\mathcal{T}\,i = -i$.

Each eigenstate $|\psi_n\rangle$ of a $\mathcal{PT}$ symmetric Hamiltonian may be multiplied by $\mathcal{PT}$. The new, linearly dependent ket $\mathcal{PT}\,|\psi_n\rangle$ satisfies the same Schrödinger equation $H\,|\psi_n\rangle = E_n\,|\psi_n\rangle$, $n = 0, 1, \ldots$. This enables us to work with the superpositions $|\psi_n^{[\pm]}\rangle = \pm|\psi_n\rangle + \mathcal{PT}\,|\psi_n\rangle$. In the other words, we are free to require, say, that the $\mathcal{PT}$−parity is even. This means that the real and imaginary parts of the wavefunction have a definite parity $\mathcal{P}$,

$$|\psi_n\rangle = |\sigma_n\rangle + i\,|\tau_n\rangle, \qquad \mathcal{P}\,|\sigma_n\rangle = |\sigma_n\rangle, \qquad \mathcal{P}\,|\tau_n\rangle = -|\tau_n\rangle\ .$$

In the harmonic-oscillator (or any similar) basis $\{|n^{(\alpha)}\rangle\}$ numbered by the integers $n = 0, 1, \ldots$ and by the parity index $\alpha = \pm 1$ we expand

$$|\sigma_n\rangle = \sum_{k=0}^{\infty} |n^{(+)}\rangle\, s_k, \qquad |\tau_n\rangle = \sum_{n=0}^{\infty} |n^{(-)}\rangle\, t_k.$$

Our Schrödinger equation acquires the linear algebraic form

$$\begin{pmatrix} F - E\,I & -A \\ A^\dagger & G - E\,I \end{pmatrix} \cdot \begin{pmatrix} \vec{s} \\ \vec{t} \end{pmatrix} = 0 \, . \tag{2}$$

The matrix elements are easily evaluated and remain real due to the $\mathcal{PT}$ symmetry,

$$F_{mk} = \langle m^{(+)} |\, H \,| k^{(+)} \rangle, \qquad A_{rs} = \langle r^{(+)} |\, H \,| s^{(-)} \rangle, \qquad G_{nj} = \langle n^{(-)} |\, H \,| j^{(-)} \rangle.$$

We eliminate $\vec{t} = -\left(G - E\,I\right)^{-1} A^\dagger \vec{s}$. In the spirit of the Feshbach method[2], we get the reduced Schrödinger equation

$$H_{eff}(\varrho)\vec{s} = E\,\vec{s}, \qquad H_{eff}(\varrho) = F + A\,\left(G - E\,I\right)^{-1} A^\dagger \, . \tag{3}$$

In the new effective Hamiltonian we have to guarantee that

$$\varrho = \varrho_n = E_n(\varrho). \tag{4}$$

The energy dependence of $H_{eff}(E_n)$ is re-interpreted as an external constraint.

All the effective Hamiltonians $H_{eff}(\varrho)$ are Hermitian. This is our most important observation. As its consequence, all the $\varrho-$dependent spectra $\{E_n(\varrho)\}$ are real. As long as the spectrum of the original operator H is determined by the mere selfconsistency condition (4), the complicated question of reality of the spectrum of our $\mathcal{PT}$ symmetric system is significantly simplified.

2 Generalized $\mathcal{PT}$ Symmetry

Schrödinger equation (2) can be read as an interaction between the even- and odd-parity subsystems where the sub-Hamiltonians F and G are coupled and the parity $\mathcal{P}$ is manifestly broken. In the weakly coupled limit the spectrum remains real by construction.

In principle, the partitioning in eq. (2) need not necessarily be related to the parity of the basis. Within the original Feshbach's interpretation, the symbols A may be any complex matrices while F and G have to be Hermitian. In addition, the partition into unequal parts is permissible. This is our present key idea. By analogy, the effective energies remain real for any $(m+n) \times (m+n)-$dimensional Schrödinger operator

$$\begin{pmatrix} Y - E\,I & -C \\ C^\dagger & Z - E\,I \end{pmatrix} \tag{5}$$

where, by assumption, $\dim Y = m$, $\dim Z = n$ and $Y = Y^\dagger$, $Z = Z^\dagger$. In the first step we may notice that the resolvent exhibits the same structure,

$$\begin{pmatrix} Y - EI & -C \\ C^\dagger & Z - EI \end{pmatrix}^{-1} = \begin{pmatrix} Y'(E) & -C'(E) \\ (C')^\dagger (E) & Z'(E) \end{pmatrix} .$$

Using mathematical induction, we may increase the dimension by one and add any new one-dimensional array on the top,

$$\begin{pmatrix} X - EI & \alpha A & \beta B \\ A^\dagger & Y - EI & -C \\ B^\dagger & C^\dagger & Z - EI \end{pmatrix} .$$

Then we construct the effective secular equation as the following one-dimensional relation,

$$X - \left[(\alpha A\, \beta B) \begin{pmatrix} Y'(E) & -C'(E) \\ (C')^\dagger (E) & Z'(E) \end{pmatrix} \begin{pmatrix} A^\dagger \\ B^\dagger \end{pmatrix} \right] = E.$$

This leads to the quick conclusion that the left-hand-side expression remains real if and only if

$$\alpha = -\beta. \tag{6}$$

This is the only condition required. Once we increase the dimension of the Hamiltonian matrix by one, the necessary and sufficient condition (6) adds m *or* n minuses in the upper line. We just choose $\alpha = -1$ or $\alpha = +1$, respectively. In the latter case the two by two partitioning is unchanged and we may replace n by $n + 1$ in eq. (5). In the former case we have to permute the basis first, re-shuffling its topmost item to the bottom. This enables us to return to the two-by-two partitioning and to replace the upper dimension m by $m + 1$.

We may summarize that the Feshbach-based reasons for the reality of the spectrum (as mentioned in section 1 above) remain unchanged after the transition to the generalized two-by-two partitioned structure (5). It degenerates to the Hermiticity at $m = 0$ or $n = 0$ and to the $\mathcal{PT}$ symmetry at $m = n \neq 0$.

3 Examples

An alternative version of our analysis could in principle be based on the multiple partitioning of the basis. This does not seem to lead to any new structure of the corresponding non-Hermitian Hamiltonian with the real effective spectrum.

For illustration, let us pick up the three by three partitioning first. Choosing $\alpha, \beta, \gamma = \pm 1$ we may postulate a parallel between the $\mathcal{PT}$ symmetry and Hermiticity in the triply partitioned equation

$$\left(\begin{array}{ccc} X - E\,I & \alpha\,A & \beta\,B \\ A^\dagger & Y - E\,I & \gamma\,C \\ B^\dagger & C^\dagger & Z - E\,I \end{array} \right) \cdot \left(\begin{array}{c} \vec{r} \\ \vec{s} \\ \vec{t} \end{array} \right) = 0 \,.$$

The elimination of $\vec{t} = -\,(Z - E\,I)^{-1}\left(B^\dagger \vec{r} + C^\dagger \vec{s}\right)$ gives us the two by two effective Schrödinger equation

$$\left[\left(\begin{array}{cc} X - E\,I & \alpha\,A \\ A^\dagger & Y - E\,I \end{array} \right) - \left(\begin{array}{c} \beta\,B \\ \gamma\,C \end{array} \right) (Z - \varrho\,I)^{-1} \left(B^\dagger\ C^\dagger\right) \right] \cdot \left(\begin{array}{c} \vec{r} \\ \vec{s} \end{array} \right) = 0 \qquad (7)$$

where $\varrho \equiv E$. We intend to guarantee that the effective Hamiltonians $H_{eff}(\varrho)$ remain Hermitian. At any $\varrho = \mathit{constant}$ the reality of all the energy roots $E_n(\varrho)$ of the linearized eq. (7) near ϱ will be guaranteed by this Hermiticity. It is true for the diagonal blocks in $H_{eff}(\varrho)$. In order to satisfy also the "off-diagonal" condition,

$$\alpha'\,A' = \alpha\,A - \beta\,B\,C^\dagger, \qquad (A')^\dagger = A^\dagger - \gamma\,C\,B^\dagger$$

we choose $\alpha' = \alpha$ and arrive at the constraint

$$\alpha\beta = \gamma \qquad (8)$$

and menu

$$(\alpha, \beta, \gamma) = (+,+,+), (+,-,-), (-,+,-), (-,-,+). \qquad (9)$$

which lists all the available possibilities. The first option represents the Hermitian choice.

A similar classification of all the eligible four by four partitionings can be chosen as preserving the latter rule (8). Once we add the three new variables $\mu, \nu, \rho = \pm 1$ we have to study the more complicated structure

$$\left(\begin{array}{cccc} X - E\,I & \alpha\,A & \beta\,B & \mu\,U \\ A^\dagger & Y - E\,I & \gamma\,C & \nu\,V \\ B^\dagger & C^\dagger & Z - E\,I & \rho\,W \\ U^\dagger & V^\dagger & W^\dagger & K - E\,I \end{array} \right) \cdot \left(\begin{array}{c} \vec{p} \\ \vec{r} \\ \vec{s} \\ \vec{t} \end{array} \right) = 0 \,.$$

The insertion of $\vec{t} = -\,(K - E\,I)^{-1}\left(U^\dagger \vec{p} + V^\dagger \vec{r} + W^\dagger \vec{s}\right)$ reduces the equation to a three by three partitioned Schrödinger problem

$$\left[\left(\begin{array}{ccc} X - E\,I & \alpha\,A & \beta\,B \\ A^\dagger & Y - E\,I & \gamma\,C \\ B^\dagger & C^\dagger & Z - E\,I \end{array} \right) \right.$$

$$- \begin{pmatrix} \mu U \\ \nu V \\ \rho W \end{pmatrix} (K - \varrho I)^{-1} (U^\dagger \, V^\dagger \, W^\dagger) \Bigg] \cdot \begin{pmatrix} \vec{p} \\ \vec{r} \\ \vec{s} \end{pmatrix} = 0, \qquad \varrho \equiv E\,.$$

In order that the effective Hamiltonians $H_{eff}(\varrho)$ preserve the three by three symmetry $H = H^\dagger$, we satisfy the elementary "diagonal" conditions by fixing $\alpha' = \alpha$, $\beta' = \beta$ and $\gamma' = \gamma$. The three "off-diagonal" constraints

$$\alpha\beta = \gamma, \qquad \alpha\mu = \nu, \qquad \beta\mu = \rho, \qquad \gamma\nu = \rho.$$

lead to the following eight solutions,

n	α_n	β_n	γ_n	μ_n	ν_n	ρ_n
1	+	+	+	+	+	+
2	+	+	+	−	−	−
3	+	−	−	+	+	−
4	+	−	−	−	−	+

n	α_n	β_n	γ_n	μ_n	ν_n	ρ_n
5	−	+	−	+	−	+
6	−	+	−	−	+	−
7	−	−	+	−	+	+
8	−	−	+	+	−	−

(10)

The first line is the Hermitian choice and the fourth line reproduces the result of the two by two partitioning. The non-square two by two partitioning gives the lines number two and seven. The remaining four items offer the genuine three by three structures. Two contain the three minuses and the other two the four ones.

As a nontrivial example of applicability of our four-by-four scheme, let us consider the superintegrable[4] two-dimensional Hamiltonian

$$H = -\partial_x^2 - \partial_y^2 + x^2 + y^2 + g/x^2 + g/y^2$$

which degenerates to the Pöschl-Teller-like problem after the separation of variables in polar coordinates,

$$\left(-\frac{d^2}{d\,\varphi^2} + \frac{g}{\cos^2 2\varphi} \right) \psi(\varphi) = k^2\, \psi(\varphi), \qquad \varphi \in (-\pi, \pi).$$

The symmetry

$$[H, \mathcal{R}] = 0$$

of the Hamiltonian with respect to the shift $\mathcal{R} : \varphi \to \varphi + \pi/2$ resembles the parity once we put $\mathcal{P} = \mathcal{R}^2$.

Let us now consider a breaking of the Hermiticity of the Hamiltonian. By a suitable complex deformation of the coordinate line, $\varphi = \varphi(t) = t + i\,\varepsilon(t)$, $t \in (-\pi, \pi)$ we can avoid the barriers (i.e., poles of the potential which lie at the integer multiples of $\pi/2$) so that a tunneling takes place. Still, the energies need not become complex after such a regularization of the potential, provided that we restrict our attention to the complexifications which preserve the commutativity

$$[H, \mathcal{RT}] = 0. \tag{11}$$

As long as we have $\mathcal{P}^2 = \mathcal{R}^4 = 1$, our model is not $\mathcal{PT}$ symmetric. Having a sample bound-state solution $|\psi\rangle$ and the new symmetry (11) we infer that the state $\mathcal{R}|\psi\rangle$ lies within the same Hilbert space and satisfies Schrödinger differential equation at the identical energy. Superpositions

$$|\psi^{[k,l,m,n]}\rangle = (1 + i\mathcal{R})^k (1 - i\mathcal{R})^l (1 + \mathcal{R})^m (1 - \mathcal{R})^n |\psi\rangle$$

such that

$$\mathcal{R}|\psi^{[0,1,1,1]}\rangle = i\,|\psi^{[0,1,1,1]}\rangle, \qquad \mathcal{R}|\psi^{[1,0,1,1]}\rangle = -i\,|\psi^{[1,0,1,1]}\rangle,$$

$$\mathcal{R}|\psi^{[1,1,0,1]}\rangle = -|\psi^{[1,1,0,1]}\rangle, \qquad \mathcal{R}|\psi^{[1,1,1,0]}\rangle = +|\psi^{[1,1,1,0]}\rangle.$$

can be expanded in a basis $|n^{[k,l,m,n]}\rangle$ with $n = 0, 1, \ldots$ and with the superscript which marks the symmetry. Due to the Schur's lemma, the Hermitian Hamiltonian matrix becomes block-diagonal and is partitioned accordingly.

Paralleling the two-by-two partitioning of $\mathcal{PT}$ symmetric Hamiltonians, we now have a freedom of adding interactions compatible with the four by four partitioning specified by the four different $\mathcal{R}-$parities. Such complexifications should obey any one of the conjugations as listed in eq. (10). In the light of what has been said before, we may expect *a priori* that at least a finite number of energies E_n remains real for the sufficiently weak non-Hermitian interactions.

Acknowledgement

Supported by GA AS (Czech Republic), grant Nr. A 104 8004.

References

1. C.M. Bender, plenary talk at this conference, with further references.
2. H. Feshbach, *Ann. Phys. (N.Y.)* **5**, 357 (1958).
3. P. Dorey, C. Dunning and R. Tateo, *J. Phys. A: Math. Gen.* **34**, L3921 (2001);
   Z. Ahmed, *Phys. Lett. A* **282**, 343 (2001);
   M. Znojil, arXiv: math-ph/0104012.
4. P. Tempesta, A. V. Turbiner and P. Winternitz, Exact solvability of superintegrable systems, hep-th/0011209, *J. Math. Phys.* **42**, No. 9 (2001), to appear.

